1st ProSES Symposium 2019

The Role of Modelling and Simulation Towards IR 4.0

IOP Conference Series: Materials Science and Engineering
Volume 702

Kuantan, Malaysia
4 September 2019

ISBN: 978-1-7138-1032-2
ISSN: 1757-8981

Printed from e-media with permission by:

Curran Associates, Inc.
57 Morehouse Lane
Red Hook, NY 12571

Some format issues inherent in the e-media version may also appear in this print version.

Additional copies of this publication are available from:

Curran Associates, Inc.
57 Morehouse Lane
Red Hook, NY 12571 USA
Phone: 845-758-0400
Fax: 845-758-2633
Email: curran@proceedings.com
Web: www.proceedings.com

TABLE OF CONTENTS

RENEWABLE ENERGY

INTEGRATION AND OPTIMIZATION

CONTROL AND MONITORING

CATALYST AND REACTION ENGINEERING

MATERIAL SCIENCE AND ENGINEERING

SEPARATION AND PURIFICATION TECHNOLOGY

SAFETY AND LOSS PREVENTION

IOP Conf. Series: Materials Science and Engineering **702** (2019) 011001 doi:10.1088/1757-899X/702/1/011001

1st Process Systems Engineering & Safety (ProSES) Symposium 2019 Preface

Ruwaida Abdul Rasid*, Mohamad Rizza Othman, Noorlisa Harun, Aainaa Izyan Nafsun, Mohd Yusri Mohd Yunus, Zainatul Bahiyah Handani, Nor Khonisah Daud

Faculty of Chemical and Process Engineering Technology, College of Engineering Technology, Universiti Malaysia Pahang, Gambang Campus, Lebuhraya Tun Razak, 26300 Kuantan, Pahang, Malaysia

*E-mail: ruwaida@ump.edu.my

The special issue of 1st ProSES Symposium is a collection of invited contributions from the presenters of the inaugural Process Systems Engineering & Safety (ProSES) Symposium, organized by Process Systems Engineering & Safety (ProSES) Research Group, Faculty of Chemical and Process Engineering Technology, formally known as the Faculty of Chemical and Natural Resources Engineering, Universiti Malaysia Pahang, on 4th September 2019 in Kuantan, Pahang, Malaysia. The theme for the 1st ProSES Symposium 2019 is The Role of Modelling and Simulation Towards IR 4.0, since the emergence of IR4.0 technology provides the opportunity for dramatically increasing the reliability and efficiency in production operations of many industrial sectors including chemical-based processing. Hence, it is the aim of this conference to provide the platform for intellectual discourse for the application of computer aided tools for modelling and simulation in the area of process systems engineering and safety particularly in the era of IR4.0, and related topics. 1st ProSES Symposium 2019 attracted more than 60 extended abstracts submissions. Two keynote speakers delivered their particularly interesting topics – Professor Ir Dr Haslenda Hashim from Universiti Teknologi Malaysia with a talk on "Process Modelling and Simulation Toward IR4.0"; and Professor Dr Azmi Mohd Shariff from Universiti Teknologi Petronas on "Process Safety in IR4.0".

This Special Issue has compiled 60 invited full-length contributions in the mentioned area with specific scopes of renewable energy; integration and optimization; control and monitoring; catalyst and reaction engineering; material science and engineering; separation and purification technology; safety and loss prevention and related topics. In the area of renewable energy, Ravendran et al. [1], developed a process model using Aspen Plus to simulate the biogas production through anaerobic digestion process, highlighting the economical potentials and environmental benefits of the process. Through their work, they determined the optimum feed rate of 0.36 l/day and operating pressure of 3 bar with hydrogen flow of 180 l/day, in which the maximum amount of methane was achieved. In another work by Abdul Rahman et al. [2], food waste was pre-treated through the conventional torrefaction process while Ismail et al., [3] studied the application of microwave to enhance the torrefaction of food waste to produce biochar, a source of renewable energy that can potentially replace the conventional fossil-based fuel such as coal. On the other hand, Jamin et al., [4] developed the kinetic model on the anhydrous weight loss of green waste, one of the main compositions of municipal solid waste. The model was developed based on the two-steps reaction in series from Di Blasi and Lanzetta, through thermogravimetric analysis (TGA). Bio-based fuels such as bioethanol [5] and biobutanol [6] from oil palm fruit juice have been studied and proven to be useful in the production renewable energy, both of which applied the central composite design (CCD) through response surface methodology (RSM). In addition, levulinic acid was produced from matured coconut water and modeled successfully in Simulink, to evaluate and predicts its production [7]. A process simulation study on co-gasification of coal and empty fruit bunch [8] has been carried out to evaluate the performance of entrained flow gasifiers in the production of synthesis gas, mixture of carbon monoxide and hydrogen, which can be further utilized in power production or specialty chemicals. They developed the process model in Aspen Hysys and found that carbon conversion reached above 90% and the maximum value of cold gas efficiency (CGE) and higher heating value (HHV) were obtained. Another optimization study on biomass gasification [9] has also been done using CCD through RSM. Saleh and Samad [10] has successfully developed a thermodynamic equilibrium model in Excel to evaluate a wide range of biomass gasification process. The model has been validated and was found to be reliable with the root mean square errors (RMSEs) of lesser than 1 has been obtained.

In the research area related to integration and optimization, Changresegaran et al. [11] performed a simulation and optimization study of industrial spent caustic wastewater treatment by wet oxidation method. The process was successfully simulated, and the optimization results shows that 1:9.2 is the minimum spent caustic feed flow rate to air flow rate ratio and optimum operating temperature of flash separator unit is 140 °C for maximum flow rate of CH_3SSCH_3 in off gas stream. Sidek and Othman [12] assessed the performance of distillation column heat pumping technology for oleochemical fatty acid fractionation. Two heat pumping system configurations were simulated using Aspen Plus. Simulation results indicates that both configurations of heat pump can be feasibly integrated with the distillation column. However, to satisfy the heating and cooling requirement, supply of make-up utility is necessary. Mansor et al. [13] integrate ionic liquid to dissolve palm fruit cellulose to improve the extraction of palm oil. By integrating the ionic liquid, it was found that highest average percentage dissolving of cellulose was 10.67 wt% with condition concentration of sulfate-based ILs is 9000 ppm and the mixing time was two hours. In a nutshell, the higher concentration of ILs and more mixing time will achieve higher percentage of cellulose dissolved. Misrol et al. [14] and Saád et al. [15] studied optimal production of biogas and water integration system of an Eco-Industrial Park (EIP). Both studies show that generating biogas and treating wastewater could be potentially feasible to reduce unnecessary pollutant while reduce the water treatment burden. A superstructure optimization of bioethanol production from corn stover was proposed by Azizan et al. [16] to design an optimal processing route. Generic Algebraic Modelling System (GAMS) was used and the optimization results indicate that the optimal processing route include steam explosion, dilute acid followed by fermentation, beer and dehydration of ethanol on zeolite for each process interval with maximum yield of 2238.4 kg/h. Nafees and Abdul Rasid [17] modelled and simulated natural gas powered solid oxide fuel cell using Aspen HYSYS as well as user defined unit operation in MATLAB. In addition, Mansur and co-workers [18] developed a process model to evaluate the performance of the co-gasification of sub-bituminous coal with densified sawdust pellet using Aspen Plus. The result shows that 50% of the sawdust pellet blending ratio in the co-gasification is the best as it produced maximum syngas subsequently the best performance.

In the research area related to control and monitoring, Sulaiman and Ahmad [19] proposed a partial decoupler approach to minimize loop interation in multi input multi output (MIMO) system. Using case study found in literature, they found that the partial decoupler provides better performance as compared to conventional full decoupler. Hasnen et al. [20] developed an adaptive soft sensor based on dynamic PLS method. The method was applied to an industrial water-tube boiler for continuous online prediction of nitric oxides emission. It was found that the approach can significantly improve the prediction accuracy and reliability by 72.7%. Alias and Othman [21] and Othman [22] performed control study on dividing wall distillation column for fatty acid fractionation. The former focuses on temperature control while the latter on controllability analysis. Adnan and Samad [23] extended the method of classes (MOC) approach for solving population balance equations and optimization problem for to achieve desired CSD with minimum fine crystals. Based on the simulations, the MOC have shown comparable results with published literature, which indicates MOC has been successfully extended to solve the PBE and representing CSD for size dependent growth rate system.

In the research related to recent trends in the area of catalyst and reaction engineering, Razak et al. [24] reported on the effect of Ni loading on SBA-15 synthesized from oil palm ash (OPA) catalysts for methane dry reforming process to produce syngas. The increase of Ni loading from 1 to 3wt.% results in lower surface area and pore volume of catalyst as more Ni particles tend to agglomerate and positioning on the SBA-15 pores. Ahmad et al. [25] synthesized Pt/MCNTs and P-dopedPt/MCNTs catalysts via hydrothermal method for glycerol oxidation reaction (GOR). P-dopedPt/MCNTs catalyst is more stable with high activity compared to Pt/MCNTs alone. In addition, P-dopedPt/MCNTs result high product yield and selectivity of tartronic acid and glyceric acid through selective GOR. Abdullah et al. [26] develop the kinetic model using Langmuir-Hinshelwood approach for the photocatalytic reduction of CO_2 to methanol in liquid phase reaction using cerium oxide-titanium dioxide (CeO_2-TiO_2) catalyst.

In the research area related to material science and engineering, Elganidi et al. [27] synthesize novel terpolymers via polymerization of stearyl methacrylate (SMA), benhely acrylate (BA), and maleic anhydride (MA) monomers to inhibit the wax precipitation in pipelines and reservoir formation. The terpolymer applied in pour point depressants (PPD) as wax control strategy shows high yield (87%) at the concentration (1:1:1) of BA-*co*-SMA-*co*-MA. In another work by Elarbe et al. [28], Stearyl Acrylate-Behenyl Acrylate (SABA) copolymer is synthesize using free-radical solution polymerization method. The highest yield of SABA copolymer was obtained at the mass ratio of monomers of 1:1 (w/w), the reaction temperature of 90 °C, the reaction time of 7 h, and concentration of initiator at 1 wt%. Hybat et al. [29] propose membrane anaerobic system (MAS) to treat Palm Oil Mill Effluent (POME). High COD removal efficiency (94%-97%) in a short period of time and methane recovery around 67.8-70.3% was achieved. A MAS is also successfully applied for sugarcane mill effluent (SCME) with the removal efficiency of COD from 94.2% to 93.9% and produced methane gas (CH_4) 75.4%. [30]. Jusoh et al. [31] investigates intermolecular interaction between monomers m-phenylenediamine (MPD) and trimesoyl chloride (TMC) with polymer Nylon 6 6 (N66) in thin film membrane application. High intermolecular interaction at 3:1 of amount of MPD to TMC shows strong deposition of MPD and TMC on N66. Alashwal et al. [32] extract keratin from chicken feather and characterize it using Fourier transform infrared spectroscopy (FTIR), Scanning electron microscopy (SEM) and X-Ray Diffraction. The extracted keratin shows aspheric shaped, small microspheres with smooth surface. Husain et al. [33] used keratin protein extracted from a chicken feather to synthesize hydrogel using physical crosslinking by the freeze-thawing method for potential biomedical applications. A paper authored by Mohd Sofi et al. [34] highlighted preparation of non-soap based grease from waste transformer oil (WTO) by two types of non-soap based thickeners, i.e. fumed silica and bentonite. The best formulation is bentonite grease formulation with ratio of 80:20 (WTO: bentonite) due to it consistency of NLGI grade 2 which is comparable to most commercial greases. Sukaimi et al. [35] investigated low acid concentration (1.5, 2.5 and 3.5 M) and hydrolysis temperature (40, 60 and 80 °C) for the Microcrystalline cellulose (MCC) isolation from oil palm empty fruit bunch (EFB). The optimum condition for the acid hydrolysis process is 1.5M H_2SO_4 at 60 °C with the middle range of crystallinity (74.7 %) and yields (82.4 %). Ismail et al. [36] investigate three types of tea which are green tea (GT), black tea (BT) and oolong tea (OT) as surfactant in liquid phase exfoliation (LPE) of graphite to get high yield of graphene. Green tea is preferable which result high yield of graphene due to small components of polyphenols that present in a green tea. Alhadadi et al. [37] propose a melt blend of poly (lactic acid) (PLA), polyamide (Nylon 66) and graphene nanoplatelets (GnP) to produce nanocomposites with improved thermal stability. On the other hand, Mokaizh and co-workers [38] recovered alumina from aluminum can waste for biodiesel production as alumina is one of the most important ceramic oxides which have vast technological and industrial applications.

In the research related to recent trends in the area of separation and purification technology, several research works were focused on the screening of influencing factors in membrane development using

Design Expert software for better separation efficiency of membrane. Raj and Sunarti [39] developed the mixed matrix membranes (MMMs) comprised of polyvinyl chloride (PVC) and graphene oxide (GO). Screening test was carried out using Design Expert software by employing 2 Level full factorial design where five factors such as weight ratio PVC, weight ratio GO, pressure, type of solvent and time membrane in water was used to determine the performance of the MMMs in gas separation of CO_2 and CH_4. Study revealed that weight ratio of PVC, weight ratio of GO and type of solvents show the most contribution to the selectivity and highest response during interaction among factors. Study on the factors influencing CO_2 and CH_4 gas separation performance of PVC/Pebax thin film composite (TFC) was also conducted by Ghazali et al. [40] using Design Expert software. The effects of factors such as immersion time, concentration of Pebax and number of coating layer on CO_2 permeability and gas selectivity were explored by using a 2^3 fractional factorial design. The results show that the immersion time and Pebax concentration are the most influential factors that affect permeability while Pebax concentration and number of coating layer are the most influencing factors determine the selectivity. The best conditions to maximize permeability and selectivity were identified; immersion time for 15 min, Pebax concentration at 5wt% and 4 layers of coating resulted in CO_2 permeability and gas selectivity of 19612 Barrer and 7.15, respectively. Furthermore, 2^4 fractional factorial design was employed in the factors screening and interaction between factors in PVDF/Pebax film development on CO_2 gas selectivity in Abdul Wahab et al. [41]. Four factors such as Pebax concentration, bath temperature, evaporation times and quenching times was selected and run simultaneously and the effect on the ideal selectivity was evaluated. Pebax concentration shows the most contribution followed by quenching times, bath temperature, and evaporation times. Among the factors, Pebax concentration, quenching time and bath temperature showed positive effect towards gas selectivity. The interaction of evaporation times and Pebax concentration was the only interaction that contribute to the good gas selectivity. The PVDF/Pebax film composite produced in this study obtained higher barrer of CO_2 permeability of 393 and selectivity of 64.75 compared to the previous study. In addition, Malid et al. [42] applied 2^3 full factorial design for factors screening involved during vapour induced phase separation (VIPS) process for the preparation of polyethersulfone-graphene (PES) graphene membrane supported liquid membrane (SLM) system. The influence of three factors such as temperature of water coagulation bath, exposure time and air humidity on extraction percentage of acetic acid was investigated. The air humidity factor gave the highest contribution of 28.96% among the main factor. The interaction effects of the water bath temperature with exposure time show the most significant contribution of 45.01%. The highest extraction percentage of acetic acid using SLM system was 75.95% which was using the membrane prepare at water bath temperature of 30°C, 10 s exposure time and 70 % air humidity.

In the work by Rahman et al. [43], Classical Scaling (CMDS) method was used to develop a set of membrane profile performance for various types of UV-grafted polyethersulfone (PES) membrane using MATLAB software. Euclidean distance was employed to measure the dissimilarity between all samples. Three parameters were considered in creating the profile namely water permeability, solute permeability, and structural parameter. Based on the finding, the modified membranes were easily clustered based on its grafting mechanism. The study concluded that CMDS profiling tends to favor the pore size as the dominant impact factor in characterizing the membrane performance based on the three specified parameters. Isa et al. [44] investigated the potential of promoted potassium carbonate with glycine for CO_2 removal at elevated pressure of 10 bar for different temperature and solvent concentration. The results show that absorption of CO_2 is less favor at higher temperature and lower concentration of solvent. Meanwhile, Chan et al. [45] report on the vanillin adsorption onto resin H103. The adsorption process was modelled using artificial neural network (ANN) approach. Comparison between simulation and experimental data show that ANN has an excellent performance in predicting the experimental data and suitable to be used in the modeling of vanillin adsorption resin H103. Modeling and simulation of pressurized water scrubbing process for biogas purification to produce bioCH_4 was carried out by Abu Seman and Harun [46] using Aspen Plus software. The simulated results show that both absorber column pressure and liquid to gas (L/G) ratio have significant effect on bioCH_4

IOP Conf. Series: Materials Science and Engineering **702** (2019) 011001 doi:10.1088/1757-899X/702/1/011001

purity and percentage of carbon dioxide removal. The simulation of purification process produce 97.6 mole % bioCH_4 purity and 99.9 mole % CO_2 removal and show good agreement with the previous work.

In the study by Abdurahman and Sundarajan [47], separation of the essential oil from dry patchouli leaves have been studied by using Microwave-assisted hydrodistillation (MAHD) and traditional hydrodistillation (HD) techniques. MAHD showed a greater potential for separation of the essential oil from patchouli leaves as compared to HD method. At solid to solvent ratio of 1:12 g/ml and 45 minutes of extraction time, the results show that MAHD method extracted higher percentage of essential oil from patchouli leaves (5 %) compared to 4 % obtained using HD method. In addition, MAHD method can retain 16 compounds compare to 11 chemical compounds using HD. Meanwhile, study by Ismail et al. [48] focuses on the model development of the counter current extraction process for two rare earth (RE) elements, europium (Eu) and gadolinium (Gd) by using 2-ethylhexyl phosphoric acid mono 2-ethylhexyl ester P204 and ionic liquid-based system of [A336][P204]. The modelling was based on the Counter-current theory and the least number of extraction and scrubbing stages were preferred. The results demonstrated that P204 required a total of thirteen stages (eight stages for extraction and five stages for scrubbing) to extract Eu from Gd with 99.99% purity and 90% recovery, while [A336][P204] required only six stages (four for extraction and two for scrubbing). Therefore, the [A336][P204] offered better separation capacities and more favorable compared to P204. Liyananadirah et al. [49] report the effectiveness of acid/alkali reagents in rare earth (RE) leaching in terms of self-diffusion coefficient (Di) and overall diffusion coefficient (DAB) by using Nernst-Haskell equation. The results of Di have good agreement with experimental data from the previous study where most of calculation and experiment data are less than 5% error. Different alkali and acid reagents such as H_2SO_4, HCl, HNO_3, NaOH and Na_2CO_3 were compared in terms of DAB value. The high quantity of DAB shows the high potential acid/alkaline reagent extract RE elements. NaOH has the highest DAB, hence determined as the best reagent among the others.

Finally, the 1st ProSES Symposium 2019 also has received contributions in the area of safety and loss prevention of which comprises of eleven (11) articles. Firstly, Badli and co-workers [50] investigated the explosion severity of polyethylene/ethylene hybrid mixture in a 20-L spherical vessel. The study concluded that the explosion severity of polyethylene/ethylene hybrid mixture steeply increases as the gas concentration and particle size increase to 15 g/m^3 and 125 μm, beyond which the Pmax and dP/dT steadily decrease. Meanwhile, Al Amshawee and Yunus [51] reviewed anaerobic biotreatment as cost-effective alternatives to deliver low sludge accumulation, efficient biodegradation and mineralisation, microbes' reduction, and solids-free effluents. The paper also comprehensively discusses the impact of various factors on both biological and filtration performances and identifying strengths and limitations. A study that was conducted by Shahid et al., [52] on the other hand, focuses on safety culture maturity progression, whereby a specific roadmap that integrates the present general and workplace models has been proposed. Later, this conceptual model, perhaps, can be the basis for further research in order to provide a comprehensive picture of the standard safety culture improvement. Anuar and co-workers [53] argued that there is lacking in terms of systematic approach to implement the Process Knowledge Management (PKM). Hence, this situation has motivated the group to develop a systematic PKMS database which utilizes Risk Based Process Safety (RBPS) guideline and Plan-Do-Check-Act (PDCA) concept as its general framework. The system was validated by applying real plant data from industries as well as received valuable feedbacks for improvement. The developed PKMS allows the end user to store, review, modify and update the data regarding hazardous chemicals, technology and equipment information that involved with the operation.

Daud et al., [54] on the other hand, investigated the effect of impurities in CO_2 compression systems which typically neglected during operation. The study depicts that decrement of the impurities content from 15 to 0.7%v/v in the CO_2 streams reduced the total compression power in the compression system. The study also concludes that the three-stage compressor combined with subcritical liquefaction and pumping can potentially offer higher efficiency than four-stage compressor with 4 intercoolers for almost pure CO_2 streams. The work of Amri et al., [55] produced two risk assessments that conducted to predict and measure the causes of offshore pipeline failure due to the third party damages. Based on the qualitative analysis, the major threats identified in third party damage were categorized into "anchor impact" and "impact" meanwhile the consequences were human safety, marine life ecosystem and economical loss. In addition, Annuar et al., [56] proposed Behavior-Based Safety (BBS) approach to develop a compliance checklist based on the Building Operations and Works of Engineering Construction (BOWEC) requirements. The finding reveals that more than half of the construction activities that investigated were complying with the specified legal requirements. Meanwhile, Abed and co-workers [57] presented an overview of the emulsions, formation, classification, stability, properties. practical demulsification techniques in the petroleum sector, including chemical, microwave irradiation, biological, thermal, membrane, electrical, and ultrasonic techniques for both oilfield and synthetic emulsions. Their finding suggests that chemical demulsification has been the most widely applied and reported in the literature in this regard. In another instance, Kee et al., [58] explored on the psychological impact by means of noise exposure level among the machinery and non-machinery operators within the construction industry of Malaysia. The results discovered that highly level of personal noise exposure was significantly suffered by the machinery operators in contrast to the non-machinery operators based on the sixty one (61) respondents whom participated. Nurud et al., [59] conducted a numerical simulation to study the efficacy of natural smoke vent to confine fire-induced smoke transportation in the stairways of multi-storeys buildings. The study has demonstrated that when the value of heat release rate (HRR) were kept constant, the different vent's size opening had a different influence on the efficacy of smoke vent and an appropriate opening size was obtained and proposed for further action. Last but not least, Rose and colleagues [60] reported on productivity improvement through the integration of basic lean manufacturing and ergonomics in the food company. The questionnaire result has shown that the most of wasted motion is in weighing and measuring activity, in which, all workers in this process were affected by the musculoskeletal disorder at their neck and shoulder.

Acknowledgments

This Special Issue would not have been possible without the extraordinary support received from the 1st ProSES Symposium 2019 Committee and reviewers.

References

[1] R. R. Ravendran, A. Abdulrazik, and R. Zailan, "Aspen Plus simulation of optimal biogas production in anaerobic digestion process," *IOP Conf. Ser. Mater. Sci. Eng.*, 2019.

[2] R. N. U. Abdul Rahman, R. Abdul Rasid, and M. Ismail, "Effect of temperature and residence time on the characteristics of torrefied food," *IOP Conf. Ser. Mater. Sci. Eng.*, 2019.

[3] M. Ismail, R. Abdul Rasid, R. N. U. Abdul Rahman, N. S. Ahmad, and M. A. W. Mohd Radzuan, "Integrating microwave heating pre-treatment to improve torrefaction of food waste," *IOP Conf. Ser. Mater. Sci. Eng.*, 2019.

[4] N. A. Jamin, N. A. F. A. Samad, and S. Saleh, "Anhydrous weight loss kinetics model development for torrefied green waste," *IOP Conf. Ser. Mater. Sci. Eng.*, 2019.

[5] S. H. M. Zani, F. M. Asri, N. S. Azmi, H. W. Yussof, and M. A. K. M. Zahari, "Optimization of process parameters for bioethanol production from oil palm frond juice by Saccharomyces cerevisiae using response surface methodology as a tool," *IOP Conf. Ser. Mater. Sci. Eng.*, 2019.

[6] F. M. Asri, N. Masngut, and M. A. K. M. Zahari, "Biobutanol production from oil palm frond juice in 2 L stirred tank bioreactor with in situ gas stripping recovery," *IOP Conf. Ser. Mater. Sci. Eng.*, 2019.

[7] F. Ismail, N. A. Rahman, and F. Esa, "Catalytic production of levulinic acid from matured coconut water," *IOP Conf. Ser. Mater. Sci. Eng.*, 2019.

[8] M. R. Deraman, R. A. Rasid, M. R. Othman, and L. N. M. Suli, "Co-gasification of coal and empty fruit bunch in an entrained flow gasifier: A process simulation study," *IOP Conf. Ser. Mater. Sci. Eng.*, 2019.

[9] N. H. A. Halim, S. Saleh, and N. A. F. A. Samad, "Optimization of oil palm empty fruit bunch gasification temperature and steam to biomass ratio using response surface methodology," *IOP Conf. Ser. Mater. Sci. Eng.*, 2019.
[10] S. Saleh and N. A. F. A. Samad, "Syngas software for biomass gasification process," *IOP Conf. Ser. Mater. Sci. Eng.*, 2019.
[11] S. Chandraseagar, A. H. Abdulrazik, S. N. Abdulrahman, and M. A. Abdaziz, "Aspen Plus simulation and optimization of industrial spent caustic wastewater treatment by wet oxidation method," *IOP Conf. Ser. Mater. Sci. Eng.*, 2019.
[12] N. M. Sidek and M. R. Othman, "Assessment of heat pumping technology in oleochemical fatty acid fractionation," *IOP Conf. Ser. Mater. Sci. Eng.*, 2019.
[13] K. N. Mansor, M. A. M. Halim, N. Ali, W. Z. W. A. Rahman, and S. B. Abdullah, "Integrating ILS in dissolving palm fruit cellulose," *IOP Conf. Ser. Mater. Sci. Eng.*, 2019.
[14] M. A. Misrol, S. R. W. Alwi, J. S. Lim, and Z. A. Manan, "Synthesis of optimal biogas production system from multiple sources of wastewater and organic waste," *IOP Conf. Ser. Mater. Sci. Eng.*, 2019.
[15] S. F. Sa'ad, R. Zailan, S. R. W. Alwi, J. S. Lim, and Z. A. Manan, "Towards water integration in Eco-Industrial Park: An overview of water recovery from industries," *IOP Conf. Ser. Mater. Sci. Eng.*, 2019.
[16] S. S. Azizan, N. F. I. Effendy, and Z. B. Handani, "Superstructure optimization of bioethanol production from corn stover," *IOP Conf. Ser. Mater. Sci. Eng.*, 2019.
[17] A. Nafees and R. A. Rasid, "Study of natural gas powered solid oxide fuel cell simulation and modeling," *IOP Conf. Ser. Mater. Sci. Eng.*, 2019.
[18] F. Z. Mansur, C. K. M. Faizal, N. A. F. A. Samad, S. M. Atnaw, and S. A. Sulaiman, "Performance modelling and validation on co-gasification of coal and sawdust pellet in research-scale downdraft reactor," *IOP Conf. Ser. Mater. Sci. Eng.*, 2019.
[19] M S Sulaiman and Z. Ahmad, "Minimizing loop interaction in Multi Input Multi Output (MIMO) system using partial decoupler approach," *IOP Conf. Ser. Mater. Sci. Eng.*, 2019.
[20] S. H. Hasnen, H. Zabiri, K. K. Prakash, and T. T. Mat, "Adaptive PLS inferential soft sensor for continuous online estimation of NOx emission in industrial water-tube boiler," *IOP Conf. Ser. Mater. Sci. Eng.*, 2019.
[21] A. Alias and M. R. Othman, "Temperature control of vacuum dividing wall column – case study on oleochemical fatty acid fractionation," *IOP Conf. Ser. Mater. Sci. Eng.*, 2019.
[22] M. R. Othman, "Fractionation of oleochemical fatty acid using vacuum dividing wall column: A controllability analysis," *IOP Conf. Ser. Mater. Sci. Eng.*, 2019.
[23] S. Z. Adnan and N. A. F. A. Samad, "Application of extended method of classes for solving population balance equations and optimization study for crystallization process involving dissolution phenomena," *IOP Conf. Ser. Mater. Sci. Eng.*, 2019.
[24] H. A. Razak, N. Abdullah, H. D. Setiabudi, C. S. Yee, and N. Ainirazali, "Influenced of Ni Loading on SBA-15 Synthesized from Oil Palm Ash Silica for Syngas Production," *IOP Conf. Ser. Mater. Sci. Eng.*, 2019.
[25] M. S. Ahmad, C. K. Cheng, H. R. Ong, H. Abdullah, and M. R. Khan, "Electro-oxidation of renewable glycerol to value added chemicals over phosphorous-doped Pt/MCNTs nanoparticles," *IOP Conf. Ser. Mater. Sci. Eng.*, 2019.
[26] H. Abdullah, M. M. R. Khan, Z. Yaakob, and N. A. Ismail, "A kinetic model for the photocatalytic reduction of CO2 to methanol pathways," *IOP Conf. Ser. Mater. Sci. Eng.*, 2019.
[27] I. Elganidi, B. S. Elarbe, N. Abdullah, and N. Ridzuan, "Characterization of a novel terpolymer containing maleic anhydride-co-stearyl methacrylate-co-benhely acrylate for wax inhibition application," *IOP Conf. Ser. Mater. Sci. Eng.*, 2019.
[28] B. S. Elarbe, N. Ridzuan, K. Yusoh, I. Elganidi, and N. Abdullah, "Screening of the factors for novel pour point depressant copolymer synthesis to improve the copolymer yield," *IOP Conf. Ser. Mater. Sci. Eng.*, 2019.
[29] M. S. Hybat, N. H. Abdurahman, Z. H. Yasmeen, Z. Jemaat, and N. H. Azhari, "Treatment of

Palm Oil Mill Effluent (POME) using Membrane Anaerobic System (MAS)," *IOP Conf. Ser. Mater. Sci. Eng.*, 2019.

[30] Z. H. Yasmeen, N. H. Abdurahman, M. S. Hybat, and R. M. Yunus, "The efficiency of membrane anaerobic system (MAS) in treating sugarcane mill effluent (SCME)," *IOP Conf. Ser. Mater. Sci. Eng.*, 2019.

[31] W. Z. A. W. Jusoh, S. A. Rahman, A. L. Ahmad, and N. M. Mokhtar, "Interaction study between M-Phenylenediamine, Trimesoyl Chloride and Nylon 66 using Molecular Dynamic Simulation," *IOP Conf. Ser. Mater. Sci. Eng.*, 2019.

[32] B. Y. Alashwal, A. Gupta, and M. S. B. Husain, "Characterization of dehydrated keratin protein extracted from chicken feather," *IOP Conf. Ser. Mater. Sci. Eng.*, 2019.

[33] M. S. B. Husain, A. Gupta, and B. Y. Alashwal, "Development of keratin based hydrogels for biomedical applications," *IOP Conf. Ser. Mater. Sci. Eng.*, 2019.

[34] S. N. A. M. Sofi, M. A. A. Aziz, N. S. A. Japar, N. W. A. Rahman, A. R. Abdulhalim, and M. Y. M. Yunus, "Preparation and characterization of grease formulated from waste transformer oil," *IOP Conf. Ser. Mater. Sci. Eng.*, 2019.

[35] J. Sukaimi *et al.*, "Properties of microcrystalline cellulose prepared from oil palm empty fruit bunch at different sulfuric acid concentration and hydrolysis temperature," *IOP Conf. Ser. Mater. Sci. Eng.*, 2019.

[36] F. S. Ismail, K. Yusoh, A. S. Z. Abidin, A. H. Abdullah, and Z. Ismail, "Three different types of tea as surfactant in liquid exfoliation of graphite: green tea, black tea and oolong tea," *IOP Conf. Ser. Mater. Sci. Eng.*, 2019.

[37] W. Alhadadi *et al.*, "Thermal stability of melt-blended poly (lactic acid) (PLA)/ polyamide 66 (PA66)/ graphene nanoplatelets (GnP)," *IOP Conf. Ser. Mater. Sci. Eng.*, 2019.

[38] A. A. Bin Mokaizh, N. Wirman, and J. H. Shariffuddin, "Synthesis of alumina from aluminium can waste to be applied as catalyst support for biodiesel production," *IOP Conf. Ser. Mater. Sci. Eng.*, 2019.

[39] K. Raj and A. Sunarti, "Preliminary Fractional Factorial Design (FFD) study using incorporation of Graphene Oxide in PVC in mixed matrix membrane to enhance CO_2/CH_4 separation," in *IOP Conference Series: Materials Science and Engineering*, 2019.

[40] A. Ghazali, R. Roshan, S. Abd Rahman, M. Abdul Wahab, N. Hazleen, and N. Mandayar, "2^3 fractional factorial design for polymer based thin film composite (TFC) membrane synthesis for CO_2/CH_4 separation," in *IOP Conference Series: Materials Science and Engineering*, 2019.

[41] M. Abdul Wahab, A. Sunarti, N. Ramli, and A. Ahmad, "2^4 fractional factorial design for PVDF/Pebax film composite synthesis on gas selectivity study," in *IOP Conference Series: Materials Science and Engineering*, 2019.

[42] S. A. A. Malid, S. M. Saufi, and N. Harruddin, "Supported liquid membrane for acetic acid extraction: Screening of membrane support preparation factors," in *IOP Conference Series: Materials Science and Engineering*, 2019.

[43] A. F. H. A. Rahman, M. Mohd Yunus, T. S. L. M. T. A. Nasaruddin, N. Harun, and M. Abu Seman, "Performance profiling of UV-grafted forward osmosis polyethersulfone membrane using multivariate classical scaling technique," in *IOP Conference Series: Materials Science and Engineering*, 2019.

[44] F. Isa, H. Zabiri, and M. Shaarif, "CO_2 removal via promoted potassium carbonate with glycine: solubility measurement at elevated pressure," in *IOP Conference Series: Materials Science and Engineering*, 2019.

[45] W. Chan, R. Samah, N. Zainol, A. Fakharudin, S. Aziz, and L. Phang, "Modeling of vanillin adsorption from aqueous solution using resin H103 by artificial neural network," in *IOP Conference Series: Materials Science and Engineering*, 2019.

[46] N. Abu Seman and N. Harun, "Simulation of pressurized water scrubbing process for biogas purification using Aspen Plus," in *IOP Conference Series: Materials Science and Engineering*, 2019.

[47] N. Abdurahman and S. Sundarajan, "Characterization of bioactive compounds extracted from patchouli: Comparing microwave-assisted hydrodistillation and hydrodistillation methods," in

IOP Conf. Series: Materials Science and Engineering **702** (2019) 011001 doi:10.1088/1757-899X/702/1/011001

IOP Conference Series: Materials Science and Engineering, 2019.

[48] N. Ismail, M. Mohd Yunus, M. Abdul Aziz, and M. Abidin, "Comparison of optimal solvent extraction stages between P204 and [A336][P204] for the separation of europium and gadolinium," in *IOP Conference Series: Materials Science and Engineering*, 2019.

[49] M. Liyananadirah, M. Deraman, W. Ibrahim, and N. Harun, "Diffusion coefficients of trivalent rare earth/actinide ions in acid and alkaline aqueous solutions," in *IOP Conference Series: Materials Science and Engineering*, 2019.

[50] H. Badli, M. N. Arifin, S. Z. Sulaiman, J. I. Susaidevas, and N. A. M. K. Nizar, "Assessment on the severity of low density polyethylene (LDPE)/Ethyelene hybrid explosion," *IOP Conf. Ser. Mater. Sci. Eng.*, 2019.

[51] S. K. Al-Amshawee and M. Y. M. Yunus, "A review on possible approaches of anaerobic biological processes for palm oil mill effluent: Process, quality, advantages and limitations," *IOP Conf. Ser. Mater. Sci. Eng.*, 2019.

[52] A. Shahid, M. S. Zaidi, and R. Azizan, "A road map to generative safety culture: An integrated conceptual model," *IOP Conf. Ser. Mater. Sci. Eng.*, 2019.

[53] N. I. P. Anuar, H. A. Aziz, and R. Ahmad, "Integrated chemical, technology & equipment process knowledge management system based on risk based process safety," *IOP Conf. Ser. Mater. Sci. Eng.*, 2019.

[54] N. K. Daud, S. Martynov, H. Mahgerefteh, and N. Nasuha, "Compression power requirements for Oxy-fuel CO2 streams in CCS," *IOP Conf. Ser. Mater. Sci. Eng.*, 2019.

[55] E. Amri, N. S. Sulaiman, and F. A. Alaw, "Risk analysis of offshore pipeline due to third party damages," *IOP Conf. Ser. Mater. Sci. Eng.*, 2019.

[56] S. N. H. S. M. Annuar, E. H. Sukadarin, J. Zakaria, and H. A. Aziz, "Development of compliance checklist system for landed housing construction activities," *IOP Conf. Ser. Mater. Sci. Eng.*, 2019.

[57] S. M. Abed, N. H. Abdurahman, R. M. Yunus, H. A. Abdulbari, and S. Akbari, "Oil emulsions and the different recent demulsification techniques in the petroleum industry – A review," *IOP Conf. Ser. Mater. Sci. Eng.*, 2019.

[58] H. Y. Kee, N. S. Fauzan, and M. Widia, "Psychological impact of noise exposure among machine and non-machine operators in construction industry," *IOP Conf. Ser. Mater. Sci. Eng.*, 2019.

[59] S. S. Nurud, M. Saari, I. Abdullah, A. NorMariah, and B. Rafee, "Influence of natural smoke vent opening in stairway of multi-storey building," *IOP Conf. Ser. Mater. Sci. Eng.*, 2019.

[60] A. N. M. Rose, M. F. F. A. Rashid, N. M. Z. N. Mohamed, and M. A. S. Jubri, "The implementation of lean manufacturing and ergonomics in Small Medium Enterprise – Case study," *IOP Conf. Ser. Mater. Sci. Eng.*, 2019.

IOP Conf. Series: Materials Science and Engineering **702** (2019) 011002 doi:10.1088/1757-899X/702/1/011002

Scientific Committee

Prof. Dr.-Ing. habil. Jens-Uwe Repke (Technische Universitaet Berlin, GER)
Prof. Dr. Yinlun Huang (Wayne State University, USA)
Prof. Dr.-Ing. Fabian Herz (Anhalt University of Applied Sciences, GER)
Prof. Dr. G.P. Rangaiah (National University of Singapore, SIN)
Prof. Ir. Dr. Haslenda Hashim (Universiti Teknologi Malaysia, MAS)
Prof. Ir. Dr. Sharifah Rafidah bt Wan Alwi (Universiti Teknologi Malaysia, MAS)
Prof. Ir Dr. Dominic C. Y. Foo (University of Nottingham Malaysia, MAS)
Prof. Ir. Dr. Zainal Ahmad (Universiti Sains Malaysia, MAS)
Assoc. Prof. Dr. Syamsul Rizal Abd Shukor (Universiti Sains Malaysia, MAS)
Assoc. Prof. Dr. Norliza Abdul Rahman (Universiti Kebangsaan Malaysia, MAS)
Assoc. Prof. Dr. Rafiziana Md Kasmani (Universiti Teknologi Malaysia, MAS)
Assoc. Prof. Dr. Shuhaimi Mahadzir (Universiti Teknologi Petronas, MAS)
Assoc. Prof. Dr. Mimi Haryani Hassim (Universiti Teknologi Malaysia, MAS)
Dr. Jarinah Mohd Ali (Universiti Kebangsaan Malaysia, MAS)

Host Organizing Committee

Chairman
Dr.-Ing. Mohamad Rizza Othman
CEng MIChemE

Co-chairman
Dr. Noorlisa Harun

Secretary
Dr. Ruwaida Abdul Rasid

Bursary
Dr. Nurul Sa'aadah Sulaiman

Logistic & Technical
Dr. Noor Asma Fazli Abdul Samad
Dr. Syamsul Bahari Abdullah
Ir. Ts. Dr. Siti Zubaidah Sulaiman
Dr. Abdul Halim Abdul Razik

Publication
Miss Zainatul Bahiyah Handani
Dr. Ruwaida Abdul Rasid
Dr. Nor Khonisah Daud

Publicity & Promotion
Dr. Noorlisa Harun
Dr.-Ing. Aainaa Izyan Nafsun
Dr. Ezrin Hani Sukadarin
Dr. Hanida Abdul Aziz

Sponsor
Dr. Mohd Yusri Mohd Yunus
Dr. Nurud Suria Suhaimi
Dr. Mirta Widia
Miss Nur Syafiqah Fauzan

Protocol
Dr. Nor Khonisah Daud
Dr. Wan Norlinda Roshana Mohd Nawi

Email us at **proses.symposium@ump.edu.my** for any enquiry and feedback.
Website: **https://sites.google.com/ump.edu.my/prosessymposium/home**

IOP Conf. Series: Materials Science and Engineering **702** (2019) 011003 doi:10.1088/1757-899X/702/1/011003

1st ProSES Symposium 2019 Sponsors

GOLD Sponsor:

ARAS KUASA

Turus Mentari Sdn Bhd, subsidiary of Aras Kuasa Sdn Bhd

SILVER Sponsors:

MYAH MINES

UMP HOLDINGS Innovative Thinking

1ST PROCESS SYSTEMS ENGINEERING AND SAFETY SYMPOSIUM 2019

IOP Conf. Series: Materials Science and Engineering **702** (2019) 011004 doi:10.1088/1757-899X/702/1/011004

Peer review statement

All papers published in this volume of *IOP Conference Series: Materials Science and Engineering* have been peer reviewed through processes administered by the proceedings Editors. Reviews were conducted by expert referees to the professional and scientific standards expected of a proceedings journal published by IOP Publishing.

IOP Conf. Series: Materials Science and Engineering **702** (2019) 012001 doi:10.1088/1757-899X/702/1/012001

Aspen Plus simulation of optimal biogas production in anaerobic digestion process

R R Ravendran, A Abdulrazik* and R Zailan

Faculty of Chemical and Process Engineering Technology, College of Engineering Technology, Universiti Malaysia Pahang, Gambang Campus, Lebuhraya Tun Razak, 26300 Kuantan, Pahang, Malaysia

*E-mail: abdhalim@ump.edu.my

Abstract. The demanding uses of fossil fuels and their associated environmental footprints are driving researches into renewable energy productions from organic resources and waste. Anaerobic digestion (AD) is an environment-friendly and cost-effective method to produce biogas from biomass. This biogas can be used in power generation, heating systems and in a combined heat and power (CHP) system. Nevertheless, biogas produced from AD contains a big fraction of CO_2 and less methane purity. Aspen Plus simulation model was developed for the AD process to produce biogas, highlighting the economical potentials and environmental benefits. Four steps of AD including hydrolysis, acidogenesis, acetogenesis, and methanogenesis with eight reactions were simulated based on the respective stoichiometries. Optimization has involved the search to identify optimum feed flow rate and operating pressure to produce the maximum amount of pure methane. The obtained results showed that optimum feed rate was 0.36 l/day and operating pressure of 3 bar with hydrogen flow of 180 l/day. By using these optimum conditions, maximum amount of methane with high purity was achieved. Otherwise, through biomass natural decomposition, the methane would escape to the atmosphere as one of those significant greenhouse gases.

1. Introduction

Fast population growths and modern livings in general have made energy demands keep increasing. Challenges are to provide the energy in sustainable supply, cheap and not to degrade the environmental qualities. Consolidated efforts to tackle these challenges can be seen through governmental policies, yearly sustainability report by industries, as well as research and developments by universities and research institutions.

The utilization of biomass as an option has pulled interests due to its abundancy and ubiquity for almost everywhere. Energy from biomass can be in the form of solid, liquid and gas. Particularly, biogas is the extractable energy from biomass in the gas state, mainly consists of methane and carbon dioxide. These gases are produced from the decomposition of biomass, called as AD, in the absence of oxygen. Once produced and purified, methane can be used for heating, power generation, CHP and vehicle fuels [1-2]. Vehicle fuels require bio-methane with purity above 95% [3]. There are four stages of AD process that include hydrolysis, acidogenesis, acetogenesis, and methanogenesis [4]. Process performance analyses for these four stages of AD are complex due to diverse digestion conditions [5], and this has made almost impossible without an accurate process simulation model. The significance of having process simulation model in analysis and improvement for biogas production have been highlighted by [6-8].

In terms of substrates used for the AD process, they can be from numerous wet biomass sources such as food and kitchen waste, poultry manure, and primary sludge [9-10]. For the anaerobic digester design, it can be several types such as fixed dome digester [11], horizontal digester [12], flexible balloon digester [13], and so on. As the main motivation was to develop a valid process model, this paper would explain steps in developing Aspen Plus simulation model and optimization for the AD of cow manure to produce biogas. Emphases were given to find optimal flowrates and operating pressure.

2. Simulation Model Development

The process simulation model was constructed primarily based on the hydrolysis, acidogenesis, acetogenesis and methanogenesis stages [4]. These four steps describe on how the complex substrate such as carbohydrates are broken down into simpler substances and lastly to methane and carbon dioxide. The AD process flowsheet was based on the experimental work by [7]. Aspen Plus simulation flowsheet is shown by figure 1. Primarily, all the compounds involved in the flowsheet were obtained, specifically the cow manure that rich in carbohydrates and simulated in the Aspen Plus. In this study, Non-Random Two-Liquid (NRTL) model was selected for the property method since phases involved liquid and gas.

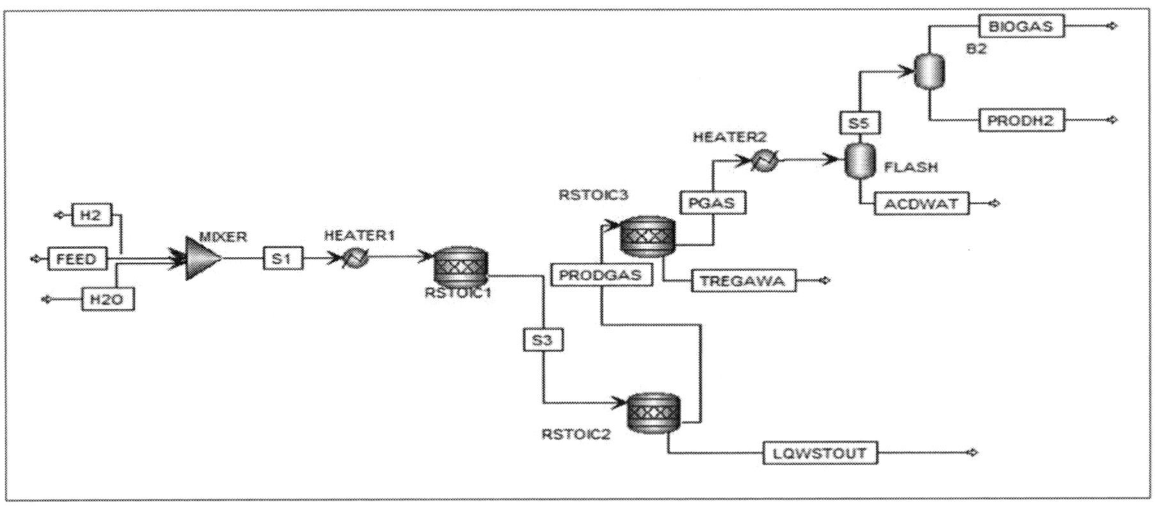

Figure 1. Aspen plus simulation flowsheet of AD process.

All reactor models used stoichiometry reactor (R-Stoic) based on the stoichiometry reactions for the four AD stages. RSTOIC 1 was for hydrolysis, RSTOIC 2 was for both acidogenesis and acetogenesis and RSTOIC 3 was for methanogenesis. The mixer was introduced to mix well the feed with hydrogen and water while the two heaters introduced to increase and maintain temperature of stream in order to achieve efficient conversion in reactors and also for better performance of flash drum. Once after the all reactions complete in three reactors, the product and gases from reactor was flow to flash drum to remove the impurities from biogas. Processing conditions for the units involved in the flowsheet are shown by table 1.

Table 1. Processing conditions in the AD flowsheet.

Model	Conditions		
	Pressure (bar)	Temperature(°C)	Valid phases
Mixer	3	-	Vapor-Liquid
Heater 1	3	37	-
Heater 2	3	53	-
RStoic 1	3	37	Vapor-Liquid
RStoic 2	3	45	Vapor-Liquid
RStoic 3	3	47	Vapor-Liquid
Flash 1	8	53	Vapor-Liquid
Flash 2	10	52	Vapor-Liquid

3. Model Validation

The main criteria for model validations were based on biogas and bio-methane production level from previous works because of the lack of data for the referred flowsheet [7], as well as to have a robust simulation model. Three cases with different substrate and operational data used as to compare and validate. Each case was described as follows:
Case (1): Cow manure has been used as feed which predominantly consists of lignin, cellulose and hemicellulose with input rate of 0.33 l/day with hydraulic retention time of 15 days for hydrolysis reactor or RSTOIC1 of figure 1 [14].
Case (2): Different set up from case 1 with combination of acidogeneic and acetogenic in one step. Cow manure was still being used for hydraulic retention time of 25 days [14].
Case (3): Wastewater generated from industrial and agricultural activities was used to feed the reactors with loading rate of 0.30 l/day and hydraulic retention time of 21 days [15].

For validation purpose, equation (1) was used to calculate percentage of methane production from both experiment and simulation results, and plotted. Given A, the amount of methane production (kg/year), and B, total amount of biogas production (kg/year).

$$Percentage\ of\ methane\ production = \frac{A}{B} x\ 100 \quad (1)$$

The acceptable error was calculated using equation (2) and maximum error tolerance was 10% for validation purpose.

$$Error = [\%\ of\ methane\ production\ (simulation - \%\ methane\ production\ (experiment) \quad (2)$$

4. Determining Anaerobic Digestion Parameters for Optimality

Determination of optimal parameters value was done by comparing plot of variables against biogas production level with one factor at a time, without any constraint. The aim is to find out the best result for pre-determined ranges. The three important factors were identified that have influenced the bio-methane production. These three factors are: i) substrate feed rate, hydrogen introduction and flowrate, and operating pressure [16]. Therefore, for feed rate factor, simulation analysis was carried out by changing substrate feed flow rate that range between 0.03 l/day to 0.51 l/day. To study influence of hydrogen in AD process, simulations were run with the addition of hydrogen. For effect of pressure, the simulations were performed with six different operating pressures; 0.5 bar, 1.0 bar, 1.5 bar, 2.0 bar, 2.5 bar and 3.0 bar. Based on methane production that was influenced by these factors, the graphs were plotted to suggest optimal factors for the process. Then, the relationship between these factors with economic potentials and environmental benefits were further elaborated.

5. Results and Discussions

Validations of the developed simulation model were done for the three cases as mentioned before for the same process flowsheet. The difference percentage or error values were calculated using both equation (1) and (2). The differences between the developed models against case 1, case 2 and case 3 for bio-methane production were 3.64%, 3.7% and 5.49 %, respectively. The tolerance was equal or less than 10%. Since all the differences were less than 10%, the model was validated and could be a basis for further improvement.

5.1 Effect of feed rate

Input feed rate and concentration of the substrates played a vital role in the biogas production rate. Figure 2 shows the simulation results of different feed flowrate from 0.03 to 0.51 l/day at 5% concentration of substrates [17]. It shows the maximum production of methane of 41.57 kg/year was achieved when 0.36 l/day of cow manure was a feed. Beyond that flowrate, methane rate began to decline because the excess organics would act as inhibitor at the given condition. Another reason was the amount of unreacted substrate material without reaction to produce methane increased, as studied experimentally by [18]. The optimal feed rate value will be different for other types of biomass feed as each biomass source poses different ultimate and proximate analyses.

Figure 3 shows composition of methane, carbon dioxide and hydrogen in biogas produced for specific feed rates. Amount of methane rate increases with the increase in feed rate which was from 0.03 to 0.36 l/day at a constant substrates concentration. When the feed rate was increased, the product from the reaction that used feed as a reactant also increase. This enabled the methanogens yielded more methane. By using this model with conversion of 90%, more methane gas was trapped with higher composition which was 0.742 (74.2% purity). This proves that the purity of methane increases by using this model in anaerobic digestion process when compared to natural process.

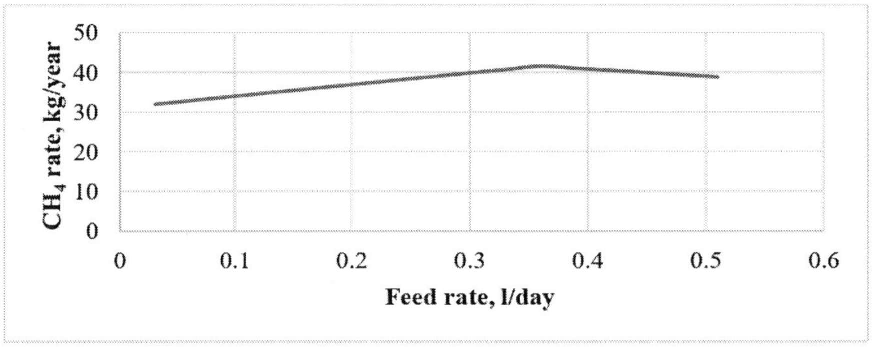

Figure 2. Methane component rate as a function of flow rate for 5% of substrate concentration.

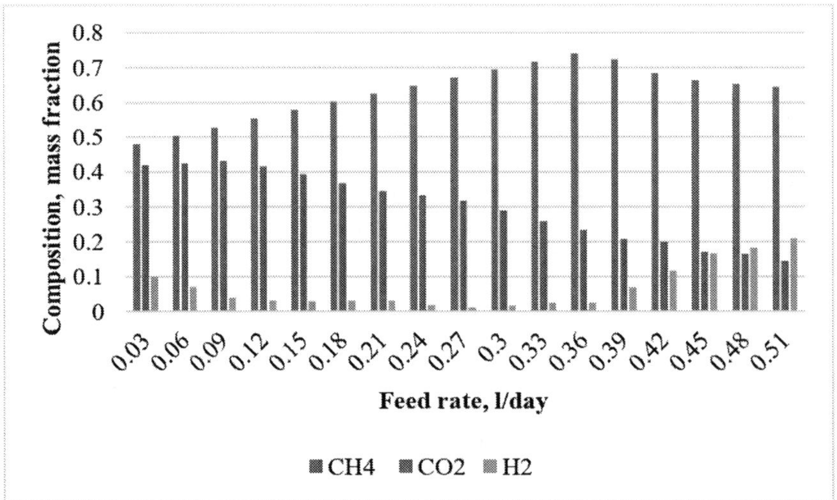

Figure 3. Feed rate against composition of methane, carbon dioxide and hydrogen.

5.2 Effect of hydrogen addition

In studying the effect of hydrogen addition, the previous optimal result i.e. the feed rate at 0.36 l/day was fixed. Addition of a specific amount of hydrogen gas in AD could increase the composition of methane gas in the biogas [16-17]. Introduction of the amount of hydrogen gas however should be equal with amount of hydrogen obtained from the pre-acidification step and also from acidogenic and acetogenic steps. In the developed simulation model, the process run with a supplied hydrogen gas stream and observations were made whether the biogas production and methane gas composition were influenced by hydrogen addition. By theory, the hydrogenotrophic methanogens microorganisms would consume the supplied hydrogen with the carbon dioxide and convert them into pure methane [5].

Once hydrogen gas introduced, the composition of methane in the biogas started to increase until it reached its optimum value as shown by figure 4. After reaching the optimum value, process inhibition occur as the acidity level increases due to greater flow of hydrogen gas into the system. The increasing of acidity level caused decrease in pH value and simultaneously make the hydrogenotrophic methanogens to become inactive. The maximum composition of methane obtained was 0.852 when 180 l/day of hydrogen gas introduced.

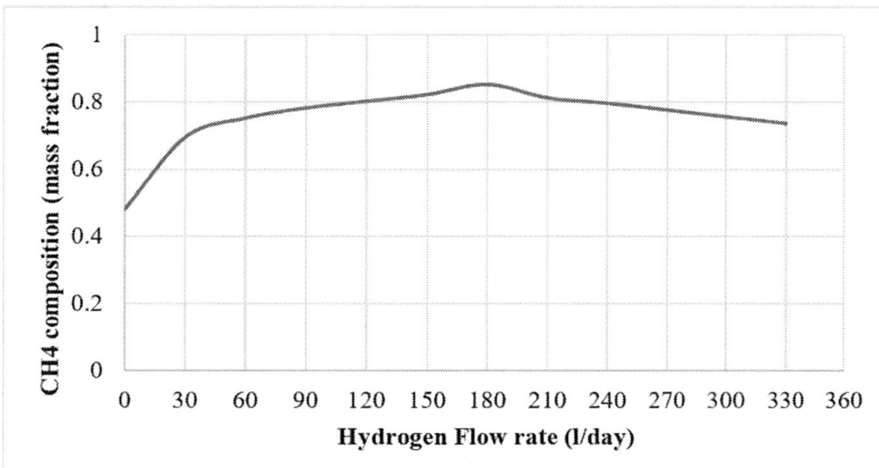

Figure 4. CH_4 composition with H_2 addition.

Figure 5 shows composition of methane, carbon dioxide and hydrogen in biogas produced for specific hydrogen flow rates. It shows more methane was produced and less carbon dioxide released.

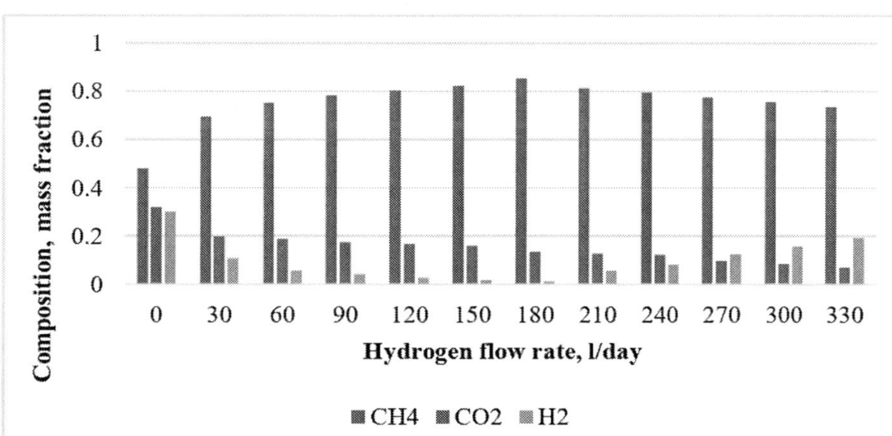

Figure 5. H_2 flowrate against biogas composition.

5.3 Effect of pressure

While the previous results (optimal feed and hydrogen flowrates) were fixed, variations have been made for pressure from 0.5 to 5 bar in the Aspen Plus model. Figure 6 and 7 show the results of varying operating pressures that have affected bio-methane production level and its composition in the biogas, respectively.

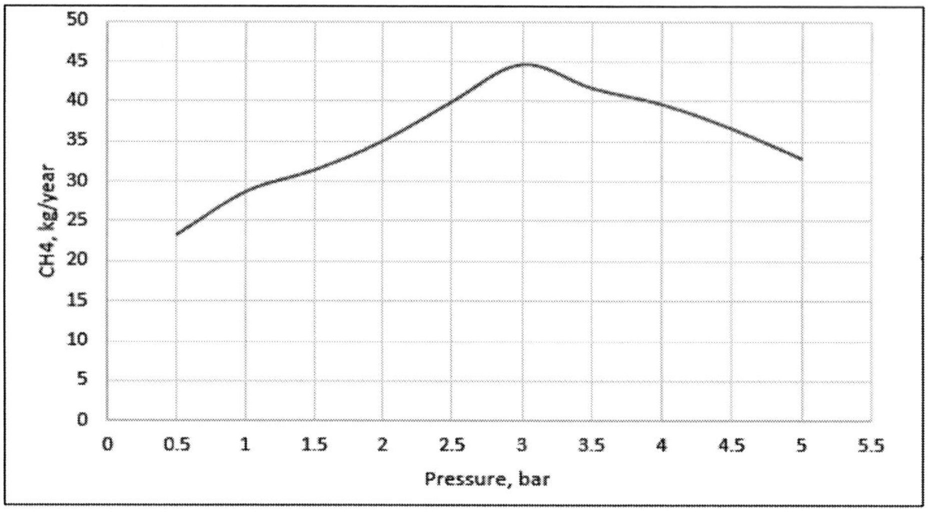

Figure 6. Methane production rate with pressure.

When the operating pressure increased from 0.5 bar to 3 bar, the composition of methane increased from 0.346 to 0.742 and the methane flowrate did from 23.45 kg/year to 44.67 kg/year respectively. From the figure 6, it is observed the optimum pressure is 3 bar with 44.67 kg/hr. Beyond 3 bar of operating pressure, the methane composition and its production rate began to decrease because methanogen could not effectively produce methane and become unstable at high pressure [19]. The effect also can be seen from biogas composition in figure 7.

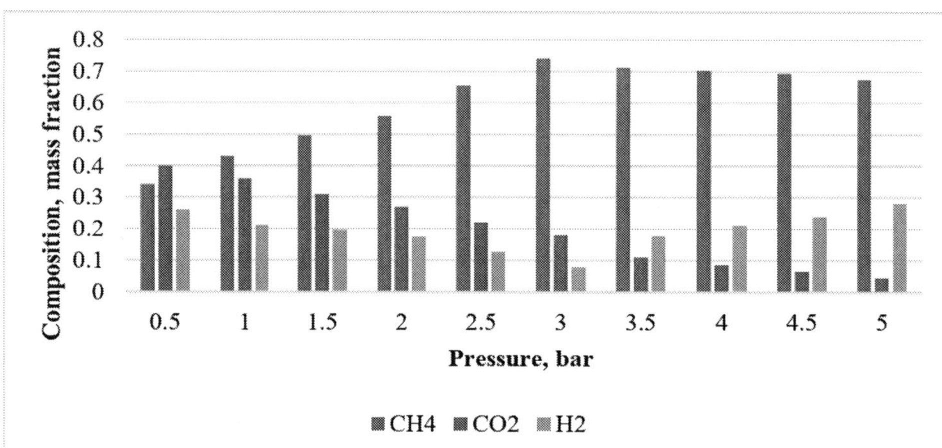

Figure 7. Composition of biogas with pressure variations.

6. Conclusion

An Aspen Plus simulation model has been developed to represent the two-stage AD process that represent the four key steps that convert biomass into biogas. After validation, the model was further simulated to investigate influencing factors to the bio-methane yield. Three factors have identified as suggested and they were feed and hydrogen flowrates, and operating pressure. Simulation results showed that methane composition in the biogas at first increased until reach optimum feed rate and then decreased although the feed rate increases due to the decrease in residence time that the biomass needs to convert to methane. Besides, the simulation results also show that the production of bio-methane was affected by the hydrogen flowrate and operating pressure. The optimal production level of bio-methane (fraction of the biogas) was 0.122 kg/day with 74.2 % purity at the feed flowrate of 0.36l/day, at hydrogen flowrate of 180 l/day, and at pressure of 3 bar. In a nut shell, the produced bio-methane has potentials to substitute natural gas for various applications including heating, power generation and vehicle fuels.

Acknowledgement

This work was funded under the **RDU1703170** research grant and the authors would like to acknowledge Universiti Malaysia Pahang (UMP) for the technical assistances and supports.

References

[1] Rajendran K, Kankanala H R and Lundin M 2014 A novel process simulation model (PSM) for anaerobic digestion using Aspen Plus *Bio-res. Technol.* **168** 7–13
[2] Deublein D and Steinhauser A 2011 *Biogas from Waste and Renewable Resources; An Introduction* (US: John Wiley and Sons)
[3] Papacz W 2011 Biogas as Vehicle Fuel *J. KONES Powertrain and Transport* **18** 403-10
[4] Chen L and Neibling H 2014 *Anaerobic Digestion Basics* (Moscow: University of Idaho Extension) pp 1-6
[5] Meegoda J N, Li B, Patel K and Wang L B 2018 A review of the processes, parameters, and optimization of anaerobic digestion *Int. Environ. Res. Public Health* **15** 2224
[6] Adak A., Mazumder D., and Bandyopadhyay P., 2011 Simulation of a Process design model for anaerobic digestion of municipal solid wastes *Int. J. Environ. and Ecological Eng.* **5** 177-81
[7] Al-Rubaye H, Karambelkar S, Shivashankaraiah M M and Smith J D 2017 Process simulation of two-stage anaerobic digestion for methane production *Biofuels* 1-11
[8] Karambelkar S S 2018 *Effect of Variation of Parameters on Biogass Production using Aspen Plus & Dynamic Simulation using Mimic* Master Thesis (Missouri: Missouri University of Science and Technology)
[9] Iqbal S A, Rahaman S, Rahman M and Yousuf A 2014 Anaerobic digestion of kitchen waste to produce biogas *Proc. Eng.* **90** 657-62
[10] Aguilar M C and Lamidi R O 2017 Biogas from anaerobic co-digestion of food waste and primary sludge for cogeneration of power and heat, *Ener. Proc.* **142** 70-76
[11] Saleh A 2012 *Biogas Potential in Pakistan Biomass Conversion Research Centre* COMSATS Institute of Information Technology
[12] Naik P R 2016 Design, fabrication & performance study of horizontal biogas plant *Int. J. Innovative Res. in Technol.* **3** 311-14
[13] Tumwesige V 2013 The Potential of Flexible Balloon Digester to Improve Livelihoods in Uganda - A Case Study of Tiribogo Master Thesis (Scotland: University of Aberdeen)
[14] Kaparaju P, Ellegaard L and Angelidaki I 2009 Optimisation of biogas production from manure through serial digestion: lab-scale and pilot-scale studies *Bio-res. Technol.* **100** 701-09
[15] Forgacs G. 2012 *Biogas Production from Citrus Wastes and Chicken Feather: Pretreatment and Co-digestion.* PhD Thesis (Sweden: Chalmes University of Technology)
[16] Luo G and Angelidaki I 2012 Integrated biogas upgrading and hydrogen utilization in an anaerobic reactor containing enriched hydrogenotrophic methanogenic culture *Biotechnol. Bioeng.* **109** 2729-736

[17] Ghorbanian R M, Lupitskyy J V, Satyavolu and Berson R E 2014 Impact of supplemental hydrogen on biogas enhancement and substrate removal efficiency in a two-stage expanded granular sludge bed reactor *Environ. Eng. Science* **31** 253–60

[18] Giuliano A, Bolzonella D, Pavan P, Cavinato C. and Cecchi F 2013 Co-digestion of livestock effluents, energy crops and agro-waste: feeding and process optimization in mesophilic and thermophilic conditions," *Bio-res. Technol.* **128** 612–618

[19] Luo G, Johansson S, Boe K, Xie L, Zhou Q and Angelidaki I 2012 Simultaneous hydrogen utilization and in situ biogas upgrading in an anaerobic reactor *Biotechnology and bioengineering*, **109** 1088-94

IOP Conf. Series: Materials Science and Engineering **702** (2019) 012002 doi:10.1088/1757-899X/702/1/012002

Effect of temperature and residence time on the characteristics of torrefied food

R N U Abdul Rahman, R Abdul Rasid* and M Ismail

Faculty of Chemical and Process Engineering Technology, College of Engineering Technology Universiti Malaysia Pahang, 26300 Gambang, Pahang, Malaysia

*E-mail: ruwaida@ump.edu.my

Abstract. Malaysia generated food waste (FW) is approximately 15 000 tons every day and expected to increase every year. Currently, Malaysia is applying the landfilling method to dispose all of the MSW which cause potential threat towards the environment and society. FW may be converted into more valuable products through reutilizing as a renewable energy source through torrefaction. The objective of this study is to investigate the effect of torrefaction temperature and residence time towards the characteristics of torrefied FW. The torrefaction process was carried out in a fixed bed reactor with continuous flow of nitrogen, at various torrefaction temperature (260 to 320 °C) and residence time (15 to 60 minutes). The torrefied FW was evaluated in terms of its mass yield, thermal behaviour and the produced functional groups through the Fourier transform infrared spectroscopy (FTIR). It can be observed that the higher the torrefaction temperature and residence time reduces the mass yield. Meanwhile, the FTIR spectra for raw and torrefied FW at different residence time was executed the functional group of C=C, C=O, C-O and C-H were present in sample. From this study, it is established that there were significant effects of both parameters towards the characteristic of FW.

1. Introduction

Urbanization and population growth have results in increasing amount of solid waste in Malaysia [1]. The daily rate of the solid waste generates was increased gradually from 2000 until 2012, and expected to continue increase up to 41,565 tons in 2025 [2,3]. The constitution of the solid waste in Malaysia are complex which consists of food waste (FW), plastic, bottle, glass, papers, metal, diapers and fabric [4]. The highest amount of waste present in solid form is FW which is about 60%. 4.404 million tons of FW was generated in 2005 and it is estimated to increase to 6.54 million tons in 2020 [5].

Till now, Malaysia still applied landfilling method to dispose all these waste where more than 80% of the wastes are dumped into landfills [2,4]. This increasing trend of solid waste generation has potential threat towards environment, society and the economy. The most serious effect on environment are soil contamination, leachate, air pollution and also green-house gases emission [1]. To overcome all these problems, the FW must be converted into value added products in which FW is reutilized to secondary usage or bio-transformation of waste [6]. However, the characteristics of raw FW is not suitable to directly use as energy source since raw FW has lower higher heating value (HHV), grindability and fuel qualities [7], which will reduce the efficiency of the process. Therefore, torrefaction process that can be applied to improve the properties of raw FW.

Torrefaction is the conversion of biomass into biochar that occur at low temperatures between 200 to 300 °C in inert environment [8]. There are a few studies that have focused on FW as raw material, for example Yuan et al. [7] in Southern China. The FW was torrefied at temperature range of 250 to 450

°C for 30 min. The HHV showed an increasing trend from torrefaction temperature of 250 until 300 °C, after which, the HHV dropped as the temperature continue to increase. This may be the result of the pyrolysis reaction that was just starting. The highest HHV was 19.34 MJ/kg, at 300 °C compared to the HHV of the raw FW at 17.5 MJ/kg. The fixed carbon and ash content in torrefied samples increased with increasing torrefaction temperature compared with raw sample, which was from 14.63 to 28.51% and from 24.94 to 46.24%, respectively. Meanwhile, the volatile content substantially decreased with an increasing torrefaction temperature which was from 60.63 to 25.25%. In terms of the proximate analysis, the carbon (C) content was increased significantly as the torrefaction temperature increased, while the oxygen (O) content was decreased rapidly across torrefaction temperature as the functional group that contained oxygen was easily cracked. The hydrogen (H) content was decreased as a result of hydrocarbon released. Torrefaction also caused the slight changed in nitrogen (N) content and lower amount of sulphur (S). The energy yield also has been discussed in this paper, which the energy yield was decreased gradually until torrefaction temperature increased up to 300 °C. The value was between 70.48 to 87.89%. This happen because of at the temperature range studied, most of the energy was stored in the torrefied sample. The energy yield was below 60% during torrefaction temperature above 300 °C, which means that too much energy was lost and consumed during torrefaction process. The surface structure of the samples was also studied, where by the result of the torrefied sample had smoother surface compared to raw sample.

Poudel et al. [9] studied the torrefaction of FW in South Korea, in terms residence time effect towards the characteristic of torrefied samples. Torrefaction temperatures used was between 250 to 400 °C. The volatile content, mass yield and energy yield were decreased as the torrefaction residence time increased. Meanwhile, the ash content increased with an increasing of torrefaction residence time. In the paper, the authors mentioned that torrefaction residence time had less an effect on torrefaction compare to temperature. This is due to significant decrease in moisture content and volatile matter of the sample. The mass loss increased with longer torrefaction residence time was applied because of decomposition of the less reactive component, while the several reactive components were release during beginning of the torrefaction process.

In another study, Rago et al. [8] also studied about FW which was collected from cafeteria of the University Of Mauritius. The FW was torrefied at temperature between 225 to 300 °C for 2 different residence time which were 1 and 3 hours. The HHV at 300 °C for 1 hour and 274 °C for 3 hours showed a high value of 23.83 MJ/kg and 24.00 MJ/kg respectively. Both conditions showed drastic increased compared to the HHV of raw FW which was 15.08 MJ/kg. The mass loss that has been observed show the more loss of mass occurred at higher residence time at the same temperature. However, the different in mass loss at torrefaction temperature exceeding 250 °C was not drastic which proved the previous study stated that the effect of temperature predominated than residence time. In this work also mentioned that the additional increment in residence time does not affect the properties of torrefied FW. The most mass loss was occurred at torrefaction temperature 300 °C and 3 hours with less than 50% mass loss. This work stated that torrefaction condition 300 °C and 3 hours was energetically inefficient as it will give impact on mass yield, energy yield and the economics of the process. Besides, the energy yield was dropped with increasing in torrefaction temperature and residence time. The lowest energy yield of 56.27% and 48.32% were obtained under condition 300 °C at both durations. Even if the lowest energy yield was attained at 3 hour torrefaction process, the dropped of energy yield was more dramatic on various temperature compared residence time.

The proximate analysis also studied which the moisture content and volatile matter were reduced as the torrefaction temperature and residence time increased. Meanwhile, torrefaction process enhanced the fixed carbon and ash content at increasing torrefaction temperature and residence time. The least volatile matter content and highest fixed carbon content were attained at 300 °C during both torrefaction residence time. The different of volatile matter between 1 hour and 3 hours was significant, 30.26% and 5.02% respectively. The unusual low content of volatile was deemed unsuitable for pretreating the FW. Prominently, the fixed carbon value between 1 hour and 3 hours was insignificant different with 55.86% and 54.02% respectively. The ultimate analysis was also discussed where the element of C was increased and parallel drop in O and H element. The C contain increased from range 41.70% to 55.80% and the highest C contain was obtained at 300 °C and 1 hour. Meanwhile, the lowest O and H content was achieved at torrefaction temperature 300 °C at both torrefaction residence time. Though, the O contained different between 1 hour and 3 hour was significant, 10.93% and 1.67% respectively. Yet, the different of H content was insignificant between both torrefaction residence time

Although there are already several works on the torrefaction of FW, there were limited studies that focused on Malaysia's FW. On the top of that, there were no studies have been conducted to focus on the improvement of FW on based on available functional groups. Therefore, this study seeks to investigate the effect of torrefaction temperature and residence time on the characteristic of FW that collected from cafeterias in University Malaysia Pahang, Malaysia.

2. Methodology

2.1. Material

FW was collected from 4 different cafeterias in University Malaysia Pahang, Malaysia. All these sample had been mixed, then used as sample in this study. Samples were collected at lunch hour between 12 to 1 pm. Samples from each cafeteria were identified. The FW was taken as it is without any specific type of food. The FW was dried in the oven for 24 hours at 105 °C to reduce its moisture content. After that, the dried FW was grinded by using blender. Next, the samples were sieved to make sure that the particles size range from 0.5 mm to 2.0 mm. The sample was stored in air- tight container prior to further use. Table 1 showed the raw

2.2. Experimental Procedure

20 grams of FW was inserted into a tubular reactor and the reactor is placed inside the furnace. The reactor was flushed with N_2 at 2 L/min for 10 minutes to maintain inert atmosphere in the reactor. After that, the temperature was set to 260 °C for the first run. The torrefaction temperature was varied from 260 °C until 320 °C with 20 °C increment. Each run lasted for 15, 30, 45 and 60 min after the sample temperature reached the desired temperature. Subsequently, the reactor was cooled down with N_2 before sample was taken out and weighed. Each run was triplicate to ensure the accuracy of the results. Figure 1 shows the schematic flow diagram of torrrefaction for process.

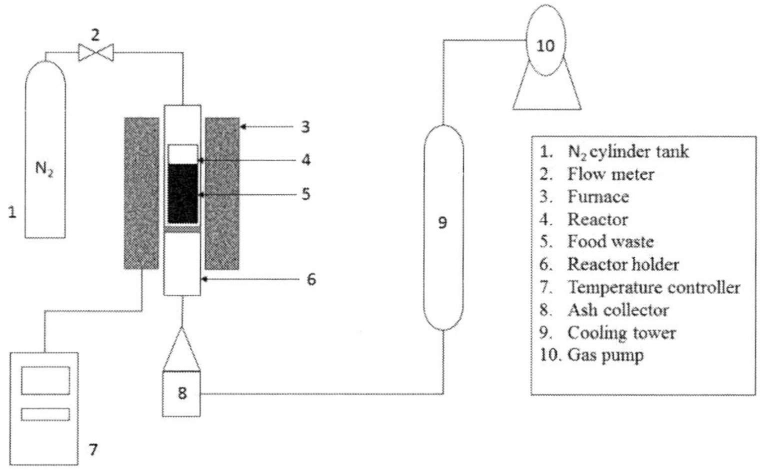

Figure 1. The schematic diagram of torrefaction process.

2.3. Analysis

2.3.1. Mass yield. Mass yield in this study was measured by equation (1) [7].This is used to evaluate the loss of mass during torrefaction.

$$\text{Mass yield } (Y_M) = \frac{\text{mass after torrefaction}}{\text{mass before torrefaction}} \times 100\ \% \quad (1)$$

2.3.2. Fourier transform infrared spectroscopy (FTIR). FTIR analysis is used to determine the functional group based on the carbon chemical functional groups. The analysis was conducted by Nicolet iS5 Thermo Scientific. Samples were placed in the sample slider that filled with KBr. Data acquisition was performed in the range 4000 – 400 cm^{-1}.

2.3.3. Thermogravimetric analysis. Thermogravimetric analysis was conducted with TA TGA Q500-0617. The samples were heated at a heating rate of 10 °C/min and the temperature ranging from 30 °C to 900 °C. TGA analysis were carried out with nitrogen flow of 10 ml/min. This analysis is used to analyse the mass loss of the raw and torrefied FW during the temperature range mentioned.

2.3.4. Scanning electron microscopy (SEM). SEM was used to analyze the surface morphology of raw and torrefied FW. This analysis was conducted by Hitachi TM3030 plus Tabletop Microscpe in Cariff Lab, University Malaysia Pahang. The samples were investigated the morphology surface structure at an accelerating voltage of 15 kV and with magnification of (250x, 500x and 1000x).

3. Result and Discussion

3.1. Physical Observation

The most obvious observation that can be made was the changed of the color of dark-blackish FW as shown in Table 1. The color of the sample changed from light-brownish to dark-blackish across residence time at each temperature. At lower residence time (15 min), all the color of sample across the temperature were dark-brownish in color. At 60 min, all the product turns into blackish color at all temperatures. Meanwhile, the product was into darker-brown sample at 260 and 280 °C at torrefaction residence time of 30 min, while the color at 300 and 320 °C were blackish in color. This observation was also experienced by Mualina et al. [10] where the color of OPF used become darker from the initial brownish color as torrefaction temperature increased. The darker of the sample color indicated the mass loss in biomass. In the paper also stated that the biomass color changed into darker when residence time increased. The color changed was due to destruction of connective, non-structural organic compounds and also variation of C content in the remaining torrefied sample [11].

3.2. Mass Yield

Figure 2 shows the effect of the residence time on mass yield at various temperatures and residence times. It can be observed that mass yield decreased with increasing residence time at each torrefaction temperature. The acceptable range of mass yield for any type of biomass is 24 to 95% of its original weight [8]. As shown in Figure 2, the longer the residence time did not affect the mass yield at lower temperature. There was a gradual decline of mass yield at the lowest torrefaction temperature of 260 °C. Meanwhile, the mass yield was decreased drastically at 280 °C between 30 and 45 min with 9.93% difference. This can be caused by the decomposition of some reactive components [12]. The mass yield continues to decrease between 15 and 30 min at the same torrefaction temperature. At 300 °C, the mass yield dropped drastically after 15 min, with 13.33% difference. However, the mass yield showed insignificant of decline at torrefaction of residence time 45 min with only less than 1% difference. This mean that the loss of reactive components process at 30 and 45 min residence time is the same [13]. Nevertheless, at 60 min, the mass yield drastically decreased again which achieved the lowest mass yield in this study, 66.82%. At torrefaction temperature of 320 °C, the mass yield experienced drastic decrease between 15 and 30 min 300 °C, with 17.06% difference. In this study, the impact of residence time towards mass yield was more significant compared to torrefaction temperature, since at longer residence time, the mass yield was lower.

IOP Conf. Series: Materials Science and Engineering **702** (2019) 012002 doi:10.1088/1757-899X/702/1/012002

Table 1. The color change of raw and torrefied FW at various temperatures and residence times.

Condition	15 min	30 min	45 min	60 min
Raw				
260 °C				
280 °C				
300 °C				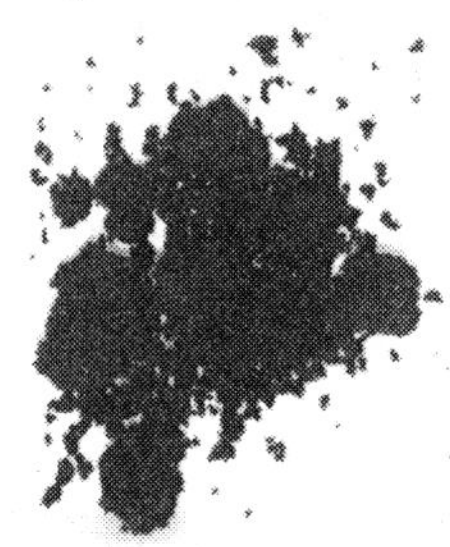
320 °C				

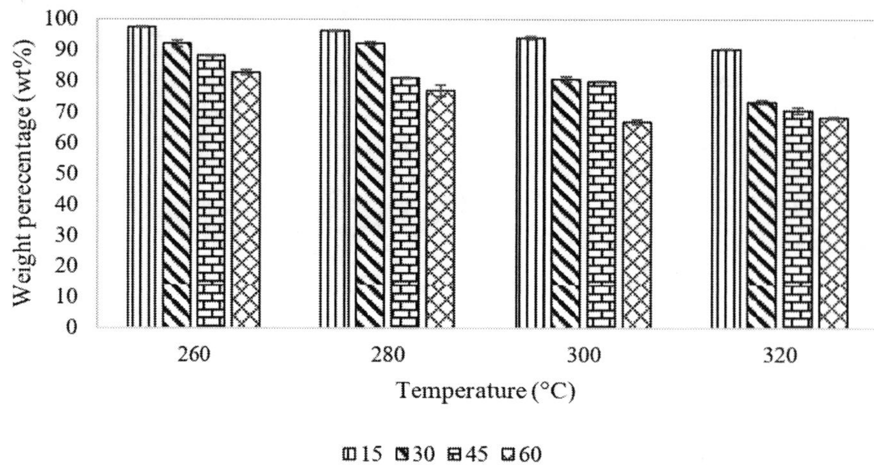

Figure 2. The mass yield across the residence time at specific temperature.

3.3. Functional groups

Figure 3 (a) shows that the FTIR spectra of raw and torrefied sample across the torrefaction temperature for 30 min. The intensity of the peak in the graph was attributed to the release of volatile [9]. All the peak became more intense when the torrefaction temperature increased. The CO_2 peak was slightly appeared which at range 2250 to 2500 cm^{-1} for raw sample until during temperature 260 °C. This was due to more pronounced degradation of biomass structure at mild temperature [9]. However, the CO_2 peak was not existed during severe torrefaction temperature. This is proved that at higher torrefaction temperature, there is no CO_2 components released. Between wavenumber of 1750 to 2000 cm^{-1}, the peak was indicated for C=O and C=C bond stretching [9]. The intensity of the peaks was increased as the torrefaction temperature increased. This may cause by tar formation which the release of phenol, aliphatic compound and carbonyl compounds occurred [16]. Meanwhile, the peaks between wavenumber of 1000 to 1500 cm^{-1} which contained C-O and C-H compound were also increased across the torrefaction temperature. This is because of the release of the phenols and aliphatic compounds and reshuffle of structural in biomass [9]. This can be said that at higher torrefaction temperature, more carbonyl and aromatic compound was released. During torrefaction at higher temperature, emission of carbonyl and aromatic compound was higher which the reason of the decreasing mass and energy yield.

FTIR spectra obtained at torrefaction temperature of 320 °C at various residence times were analysed as shown in Figure 3 (b). At wavelength between 2250 to 2500 cm^{-1}, the peak intensity is increased before 30 min. This occurred due to the released of H_2O during intermolecular dehydration reaction process [14]. The H_2O molecule did not exit after 15 min which all the H_2O molecule was totally dispersed during torrefaction after 30 minutes. At wavelength between 1750 to 2000 cm^{-1}, the peak intensity was increased as the torrefaction residence time increased. This peak is attributed to C=C and C=O bond stretching which lead to the presence of carbonyl and aromatic compounds [13]. According to Kai et al. [14], the stretching of the peak represent the adsorption of aldehyde, ketone and acids group into torrefied product. The peak intensity at wavelength between 1000 to 1500 cm^{-1} was also increased as the torrefaction residence time increased. This peak is associated to the stretching of C-C, C-H and C-O bonds. This stretching occurred due to the removal of aldehyde, ketone and acids groups, yet substituted with the presence of methyl and methylene groups of alkane and alkenes in torrefied FW [15].

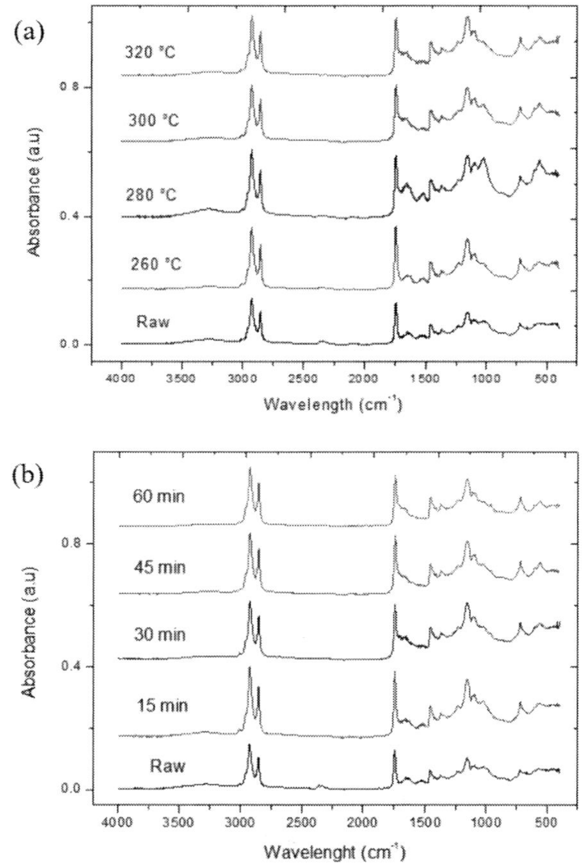

Figure 3. (a) The FTIR spectra of raw with raw and torrefied sample across the torrefaction temperature at 30 min. (b) The FTIR spectra of raw with raw and torrefied sample across the torrefaction residence time at 320 °C.

3.4. Thermal properties

A study to investigate the thermal decomposition of torrefied samples during torrefaction process was carried out. Thermogravimetric experimental curves (TG) for raw and torrefied FW are shown in Figure 4. It showed that the weight loss was decreased as the torrefaction residence time increased at torrefaction temperature 320 °C. During torrefaction temperature 300 °C, the weight loss at 30 and 60 min showed insignificant change. At temperature before 200 °C, the weight loss between raw and torrefied FW was insignificant, except for torrefied FW at 320 °C 30 min. This can be explained that at this temperature, dehydration process occurred which the amount of water content in torrefied FW at 320 °C 30 min was lower than others [16].

Between temperature 200 °C and 300 °C, the weight loss for all torrefaction condition started to have more obvious change. The most weight loss between this temperature was raw FW. Torrefied FW at 320 °C 30 min still had the least weight loss among the other torrefied FW. The weight loss between this temperature was included in the first slope, which means that, the dehydration process was still occurring where the raw FW has the highest amount of water [13]. At the second slope, which is between 300 °C to 500 °C, the weight loss continued to decrease as the torrefaction residence time increased, the decomposition of the volatile components occurred [13]. It can be observed that this occurrence is more rapid compared to the first slope. This is also observed by Qiu et al. [17] which may involve with the release and combustion of the volatiles. In this study, between temperature of 300 °C and 400 °C where the highest less weight loss was determined at 300 °C & 30 min. However, after temperature 400 °C, the least weight loss was experienced by torrefied FW at 320 °C, 60 min. This can be explained that torrefied FW at 320 °C & 60 min has the least amount of volatile components in the sample [18]. On the other hand, 320 °C & 60 min has the least mass yield which can related to the least volatile components presented for all samples. At third stage which is between temperature 500 °C to 900 °C, the weight continued decreased at all sample. During this stage, the torrefied FW at 320 °C & 60 min still has the least weight which caused by limited volatile emission compared to raw and other torrefied FW [18]. This can be concluded that, higher emission of volatile component was occurred during torrefaction process at 320 °C & 60 min which contributed to least weight loss during the thermal decomposition study.

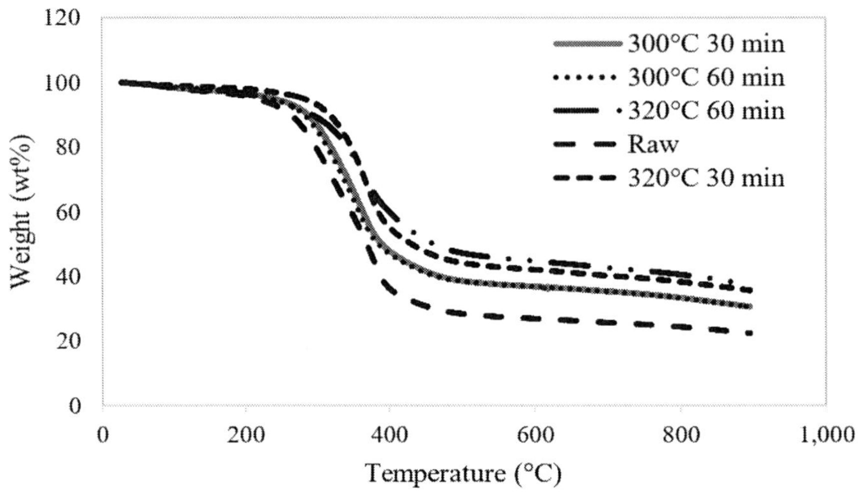

Figure 4. The mass loss across the residence time at specific temperature.

3.5. Surface Morphology

Comparison of SEM images between raw and torrefied FW are shown in Figure 5 at 250x, 500x and 1000x magnification. The surface of the torrefied FW was smothered compare to raw FW. The microparticles that that are present on the surface of raw FW were diminished at 320 °C for 30 min. This may be due to the decomposition of particles and also re-arrangement of the polymer structure [19], causing the sample became more brittle and harder as the torrefaction temperature was increased. Similar observation was made by [7] where the surface of the torrefied sample was found to be smoother. On the other hand, [20] discovered that at mild torrefaction temperature, the sample surface was not significantly changed. However, microparticles were starting to disappear on the torrefied sample structure at 300 °C, which was thought to be due to the combustion on the internal structure of the sample.

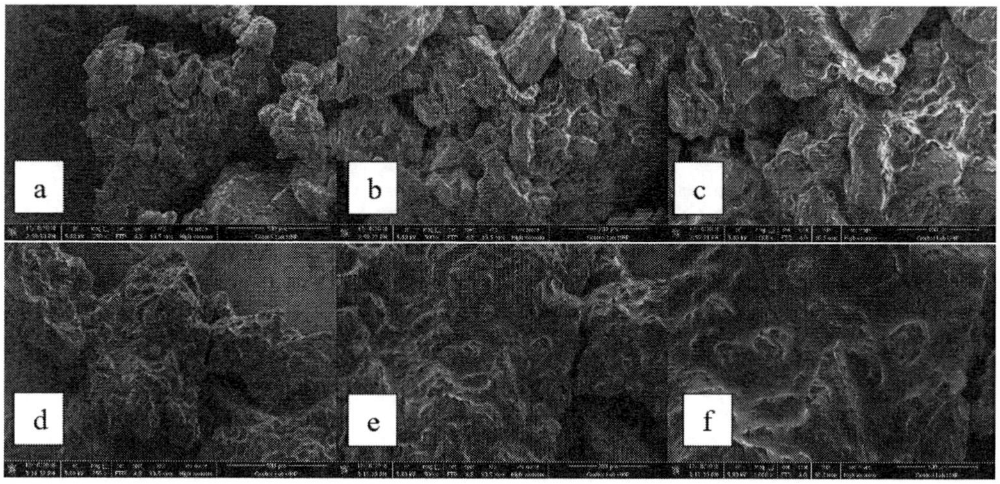

Figure 5. SEM images of (a) raw FW at 250x, (b) raw FW at 500x, (c) raw FW at 1000x, (d) torrefied FW at 320 °C for 30 min at 250x, (e) torrefied FW at 320 °C for 30 min at 500x (f) torrefied FW at 320 °C for 30 min at 1000x.

4. Conclusion

From this study, it has been established that torrefaction of the FW was affected by temperature and residence time which the lowest mass yield was observed at the most severe temperature of 300 °C at the longest residence time of 60 min with value 66.83%. The characteristics of fuel FW can be seen through FTIR analysis which the release of carbonyl and aliphatic compounds are responsible for the fuel formation. Other than that, the thermal treatment helps in improving the FW's characteristics by the weight loss which help rearrange the structure of fuel through volatile released.

Acknowledgement
The authors would like to acknowledge financial support for this work from Universiti Malaysia Pahang (UMP) internal grant (RDU1803110), Faculty of Chemical & Natural Resources Engineering, UMP.

References

[1] Sabariah F, Abd L, Ariffin M and Ho S 2018 Solid waste minimization in Malaysia *PJSRR* **4** 24-41
[2] Tiew K, Basri N E A, Takiyuddin M and Ibrahim B 2015 Assessment of the sustainability level of community waste recycling program in Malaysia *J. Mater. Cycles Waste Manag.* 598–605
[3] Abunama T, Othman F, Ansari M and El-Shafie A 2019 Leachate generation rate modeling using artificial intelligence algorithms aided by input optimization method for an MSW landfill *Environ. Sci. Pollut. Res.* 3368–3381
[4] Tiew K, Ezlin N, Basri A, Deng H, Watanabe K, Zain S and Wang S 2019 Comparative study on recycling behaviours between regular recyclers and non regular recyclers in Malaysia *J. Environ. Manage.* **237** 255–263
[5] Sulaiman N F A B R and Ahmad A 2018 Save the food for a better future : a discussion on food wastage in Malaysia *Int. J. Law Government Commun.* **3** 12-21
[6] Jayawardhana Y, Kumarathilaka P, Mayakaduwa P, Weerasundara L, Bandara L and Vithanage M 2018 Characteristics of municipal solid waste biochar : its potential to be used in environmental remediation *Util. Manag. Bioresour.* 209–220
[7] Yuan H, Yang Q, Wang Y, Gu J, He M and Sun F 2018 Impact of torrefaction on the fuel properties and combustion characteristics of compost of food waste and sawdust *Energy & Fuels.* **3** 3469–3476
[8] Rago Y P, Surroop D and Mohee R 2018 Bioresource technology assessing the potential of biofuel (biochar) production from food wastes through thermal treatment *Bioresour. Technol.* **248**

258–264

[9] Poudel J, Ohm T I and Oh S C 2015 A study on torrefaction of food waste, *Fuel*. **140** 275-281

[10] Maulina S, Silia F, Tafsin M, Khairani Y and Maulina S 2018 Torrefaction of oil palm frond: The effect of process condition to calorific value and proximate analysis, *Mater. Sci. Eng.* **345**

[11] Sadaka S, Sharara M, Ashworth A, Keyser P, Allen F and Wright A 2014 Characterization of Biochar from switchgrass carbonization *Energies* **7** 548-567

[12] Jaafar A A and Ahmad M M 2011 Torrefaction of malaysian palm kernel shell into value-added solid fuels *World Acad. Sci. Eng. Technol. Int. J. Chem. Mol. Eng.* **5** 1106–1109

[13] Pahla G, Ntuli F and Muzenda E 2018 Torrefaction of landfill food waste for possible application in biomass co-firing *Waste Manag.* **71** 512–520

[14] Kai X, Meng Y, Yang T, Li B and Xing W 2019 Effect of torrefaction on rice straw physicochemical characteristics and particulate matter emission behavior during combustion *Bioresour. Technol.* **278** 1–8

[15] Kumar Singh R, Sarkar A and Chakraborty J P 2019 Effect of torrefaction on the physicochemical properties of pigeon pea stalk (Cajanus cajan) and estimation of kinetic parameters *Renew. Energy.* **138** 805–819

[16] Sabil K M, Aziz M A, Lal B and Uemura Y 2013 Effects of torrefaction on the physiochemical properties of oil palm empty fruit bunches, mesocarp fiber and kernel shell *Biomass and Bioenergy*. **56** 351–360

[17] Qiu Y, Jing G, Yazhuo W, Haoran Y, Mingyang H and Fuan S 2018 Impact of torrefaction on the properties of food waste compost *Strategic Study of Chinese Academy of Engineering* **20** 109–116

[18] Zhang Y, Yao A and Song K 2016 Torrefaction of cultivation residue of Auricularia auricula-judae to obtain biochar with enhanced fuel properties *Bioresour. Technol.* **206** 211–216.

[19] Wu K, Tsai C, Chen C and Chen H 2012 The characteristics of torrefied microalgae *Appl. Energy* **100** 52–57

[20] Yang Y, Chyuan H, Chuan T and Chen W 2019 Torrefaction of De-Oiled Jatropha Seed Kernel Biomas For Solid Fuel Production *Energy.* **170** 367–374

Optimization of process parameters for bioethanol production from oil palm frond juice by *Saccharomyces cerevisiae* using response surface methodology as a tool

S H M Zani[1], F M Asri[1], N S Azmi[2], H W Yussof[1] and M A K M Zahari[1,3*]

[1]Faculty of Chemical and Process Engineering Technology, College of Engineering Technology, Universiti Malaysia Pahang, Lebuhraya Tun Razak Gambang, 26300 Kuantan, Pahang, Malaysia
[2]Faculty of Industrial Sciences and Technology, Universiti Malaysia Pahang, Lebuhraya Tun Razak Gambang, 26300 Kuantan, Pahang, Malaysia
[3]Bio Aromatic Research Centre of Excellence, Universiti Malaysia Pahang, Lebuhraya Tun Razak Gambang, 26300 Kuantan, Pahang, Malaysia

*E-mail: ahmadkhushairi@ump.edu.my

Abstract. The aim of this research was to investigate the effect of temperature, medium initial pH and rotation rate on the production of bioethanol from OPF juice without nutrient and nitrogen source supplementation using *Saccharomyces cerevisiae* Kyokai No. 7 (ATCC 26622). A five-level-three-factor central composite design (CCD) was employed in this study and the central point of each process variable was chosen based on the best condition obtained from the one-factor-at-time (OFAT) method. The parameters ranges were set as follows; medium initial pH (5-9), temperature (27.5-37.5°C) and rotation rate (80-120 rpm). Bioethanol and residual sugars concentration were determine using High Performance Liquid Chromatography (HPLC). The optimum conditions for bioethanol production from OPF juice were achieved at medium initial pH (6.62), rotation rate (96.51 rpm) and temperature (33.03°C). Based on the validation experiment, the optimum bioethanol yield was 0.50 ±0.02 g/ g sugars and this value was in close agreement with the model prediction where the difference was only 4.17%. Under the optimal conditions, the bioethanol yield obtained was 47.06% higher compared with non-optimized condition. The promising yield obtained in this study suggests that OPF juice can be used as a renewable and complete fermentation feedstock for bioethanol production.

1. Introduction

In Malaysia, recent developments in palm oil industry have heightened the need for production of value-added product and sustainability environment. Zwart [1] reported that there was a dramatic increase in the production of dry weight oil palm biomass in the oil palm plantation. In 2011, the generation of oil palm biomass such as trunks, fronds, and empty fruit bunches and other biomass fractions were estimated at 80 million tons [1, 31]. These resources however, present an attractive raw material for the newly emerging and potentially profitable renewable energy and bio-based chemicals market [2]. Recently, researchers have shown an increased interest in the used of oil palm trunks sap [3-5] and lignocellulosic biomass [6-8, 31-32] as an alternative to edible food sources such as corn, sugarcane and sugar beets [8-9] for the production of bioethanol. This was due to the fact that bioethanol production from corn, sugarcane and sugars beets has posed a threat to the food supply, and the cost of these raw materials accounts for up to 40% to 70% of the production cost [8]. In addition, tough competition between edible food sources for human consumption and production of bioethanol raise attention for researcher to find an alternative feedstock [2].

Oil palm frond (OPF) is a by-product of the cultivation of oil palm trees is a promising raw material as OPF is the most generated biomass from the oil palm plantation [31]. Previously, it was reported that OPF juice contained renewable sugars such as glucose, sucrose and fructose, which have great potentials as renewable carbon source for the production of poly(3-hydroxybutyrate), P(3HB) [10]. Further evaluation on the potential of OPF juice as fermentation feedstock was tested for bioethanol production by *Saccharomyces cerevisiae* (Baker's yeast) [2]. It was reported that the bioethanol yield obtained from OPF juice without nitrogen source supplementation was slightly lower compared with the fermentation supplemented with nitrogen source under non-optimized condition [2]. Even though, higher bioethanol yield could be obtained by nitrogen source supplementation, it was not recommended at industrial scale production since the addition of nitrogen source and other nutrient in fermentation medium could contribute to high production cost. In order to make the production of bioethanol feasible for industrial application, it is crucial to have high bioethanol production yield without any nutrient and nitrogen source supplementation. One of the alternatives is by optimizing the physical parameters including medium initial pH, temperature and rotation rate which could affect the bioethanol production. It was reported that the optimization study of the fermentation parameters; temperature, rotation rate and pH will contribute to high fermentation efficiency [6, 11-13, 22, 30]. Raikar [22] proposed, pH value has significant influence on alcoholic fermentation. pH of the broth influence bioethanol production in terms of bacterial contamination, yeast growth, fermentation rate and by-product formation [30]. In terms of temperature, lowering the temperature resulted in decrease of ethanol yield meanwhile increase of extremely high temperature resulted in decrease of ethanol yield and enzymes to be easily denatured [6, 30]. Rodmui et al. [11] proposed, agitation is important in fermentation process to enhance cell mass and ethanol productivity. In summary, it had been reported that most of fermentation process employed *S. cerevisiae* for bioethanol production were carried out at temperature between 30°C to 37°C, whereas pH in the range of 4.25 to 6.0 and fermentation process were performed with agitation speed in the range of 50 to 200 rpm [6, 11-13, 22, 30].

In this study, bioethanol production from OPF juice by *Saccharomyces cerevisiae* Kyokai No. 7 (ATCC 26622) without nutrient and nitrogen source supplementation was optimized *via* response surface methodology (RSM) with central composite design (CCD). Prior to that, one-factor-at-time (OFAT) method was employed to screen the best condition of parameters affecting the production of bioethanol. The effects of physical parameters including medium initial pH, temperature and rotation rate were evaluated to improve the bioethanol fermentation performance. This was followed by optimization of fermentation process for bioethanol production by using response surface methodology (RSM) to explore the response pattern. A five-level-three-factor central composite design (CCD) was employed in this study and the central point of each process variable was chosen based on the best condition obtained from the OFAT method.

The OFAT method for parameter design was implemented to obtain the best possible condition. In OFAT method, information on one-factor in each experimental trial was gathered and repeated for all respected factors [14]. Optimization process is defined as a way to enhance the performance of any system or process by obtaining the maximum benefit and investigating suitable parameters which give best response [15]. OFAT is a conventional method in which influences one factor at a time on an experimental response is studied. Only one parameter is changed, others are kept fixed [15]. Wahid and Nadir [16] proposed that OFAT do not require an advanced statistical knowledge in terms of execution and data analysis. An exact optimum can be studied using RSM which illustrates quadratic surfaces such as maximum, minimum, ridge and saddle [17]. This method of optimization is recommended due to its desirability function which gives the optimum performance levels for one or more responses.

2. Materials and Methods

2.1 Preparation of oil palm frond juice
In this study, fresh OPF (petiole part without leaves) were collected from the oil palm plantation at Felda Lepar Hilir 1, Gambang, Pahang, Malaysia during the dry season in the month of May. The OPF juice was collected by crushing the fresh OPF petioles using a conventional sugarcane press machine following a method described earlier by Zahari et al. [10]. It was centrifuged at 15,000 × *g* for 15 min and the supernatant was filtered using a mixed cellulose ester membrane filter with the pore size between 3 and 5 µm (Cole Parmer, Illinois, USA) and stored at -20°C for storage purposes before being used for fermentation [10].

2.2 Chemicals and yeast strains
The sulfuric acid (H_2SO_4), sodium hydroxide (NaOH) and yeast extract were obtained from R&M Marketing Essex (UK). Bacteriological peptone was obtained from Oxoid Ltd. Basingstoke (Hampshire, England). Meanwhile, standard sugars for HPLC analysis such as glucose, sucrose and fructose were obtained from Fisher Scientific (Leicestershire, UK). An industrial strain of *Saccharomyces cerevisiae* (*S. cerevisiae*) Kyokai No. 7 (ATCC 26622) was obtained from American Type of Culture Collection (ATCC).

2.3 YPD (Yeast extract peptone dextrose)
Yeast (*S. cerevisiae* Kyokai No. 7 (ATCC 26622)) was grown and maintained on YPD agar following method described earlier by Kosugi et al. [3] and Ho and Powel [13] with some modification composed of 20 g/L technical agar, 20 g/L dextrose anhydrous, 20 g/L bacteriological peptone and 10 g/L yeast extract. The prepared media were then autoclaved at 121°C for 15 minutes. The YPD agar was poured into sterile Petri dishes and left for solidification. YPD agar also can be prepared by using 28 g/L of nutrient agar. Strains were maintained and sub-cultured into fresh Petri dishes every 5 weeks.

2.4 Inoculum preparation
A 3-loop full of microorganism (*S. cerevisiae* Kyokai No. 7 (ATCC 26622)) from the plate was transferred into the growth medium for a culture contained: 5 g/L glucose, 10 g/L peptone, 5 g/L yeast extract and 200 mL distilled water. The culture was grown in a 250 mL Erlenmeyer flask containing 100 mL growth medium on a shaker incubated at 30°C and 200 rpm for 24 h to reach the exponential phase following the method described earlier by Chin et al. [6] and Bakri et al. [18] The cell concentration was standardized to 0.2-0.4 g/L (OD = 1.5-2.0) determined using a calibrated UV-vis spectrophotometer U-1800 (Hitachi, Japan) at 600 nm. All of the procedures were carried out aseptically and analysis was run in duplicate.

2.5 Screening of process variable using one-factor-at-time (OFAT) method
Preliminary experiment was carried out in the first place to study the effect of sterilization on bioethanol production by employing two set of experiments using autoclaved and non-autoclaved OPF juice as a fermentation substrate and the result obtained from this study was used for subsequent experiment. Both of the experiments were conducted in a rotary shaker (150 rpm) under anaerobic condition at 30°C for 48 h without pH adjustment. Samples were withdrawn every 6 or 12 h from the broth for bioethanol and residual sugars determination.

For the first part of this study, the effect of different parameters on bioethanol production from OPF juice was screened by using one-factor-at-time (OFAT) method. The OPF juice was filtered using 9.0 µm mixed cellulose ester membrane filter to remove the unwanted particles as described by Norhazimah [19]. OPF juice (100% v/v) which comprises of glucose, sucrose and fructose was used as the carbon source and nutrients throughout the study period. Pre-cultured yeast cells (10% v/v) were inoculated into a 250 mL Erlenmeyer flask containing 100 mL of autoclaved OPF juice following the previous method described by Zahari et al. [2] and Kosugi et al. [3] without any nutrient or nitrogen source supplementation. In order to study the effect of medium initial pH on bioethanol production, the initial pH value of OPF juice was adjusted to pH 5.0-9.0 using 2 M NaOH prior to autoclave. Another set of experiment was conducted to study the effect of rotation rate on bioethanol production by investigating several rotation rates at 0, 50, 100, 150 and 200 rpm. For the effect of temperature, various temperatures in the range of 27.5-37.5°C were investigated. Fermentation was run for 24 h under anaerobic condition, and all experiments were conducted in duplicates. Samples were harvested at the end of the fermentation period for bioethanol and residual sugars determination.

2.6 Optimization of process variable using Response Surface Methodology (RSM)
Meanwhile, for the second part of this study, the effect of three independent variables (pH, temperature and rotation rate) on the bioethanol yield was optimized using a factorial Central Composite Design (CCD) of Response Surface Methodology (RSM). A five-level-three-factor CCD was employed in this study and the three most significant process variables namely the medium initial pH, temperature and rotation rate. The CCD and RSM were performed using commercial software, Design Expert Version 7.1.6 (Statease Inc., Minneapolis, Minn., U.S.A.). A total design of 20 runs were set based on computer generated process variable including 6 replicate central points and $\alpha = 2$. The central point of each parameter studied in optimization experiment was selected based on the results obtained from OFAT experiment and the parameters ranges were set as follows; medium initial pH (5-9), temperature (27.5-37.5°C) and rotation rate (80-120 rpm) as shown in table 1.

Table 1. Independent variables and their coded and actual levels used in the RSM studies for optimizing the fermentation conditions of bioethanol production using OPF juice.

Independent variable	Unit	Symbol	Coded level				
			-α	-1	0	1	+α
Temperature	°C	A	22.5	27.5	32.5	37.5	42.5
Initial medium pH		B	3	5	7	9	11
Rotation rate	rpm	C	60	80	100	120	140

For the optimization study, fermentation was run for 24 h under anaerobic condition and all experiments were conducted in duplicates. Samples were withdrawn at the end of the fermentation period and the cells were separated from the fermentation broth by centrifugation at 5,000 rpm for 10 min. The cells were discarded, whereby the supernatant were filtered in vial by using syringe filter 0.2 μm prior for analysis using HPLC.

2.7 Statistical analysis
Response surface methodology (RSM) was statistically analyzed by Design-Expert, Version Version 7.1.6 software (Statease Inc., Minneapolis, Minn., U.S.A.). The coefficients can be obtained through multiple regression analysis. Estimation of coefficients with levels higher than 95% ($p<0.05$) were included in the CCD models. The bioethanol yield can be expressed as a function of independent variables by a second order polynomial equation:

$$Y = \beta_0 + \sum \beta_0 x_j + \sum \beta_{jj} x_j^2 + \sum \beta_{jk} x_j x_k \quad (1)$$

where, Y is the response (bioethanol yield), β_0 β_j, β_{jj} and β_{jk} represent the regression coefficient for intercept, linear, quadratic and interaction terms, respectively. The responses obtained were statistically evaluated by using analysis of variance (ANOVA) and the model was built based on the variables with confidence levels more than 95%.

2.8 Analytical methods
The biomass concentration was determined by using a UV-Vis spectrophotometer (OD = 600 nm) (Hitachi, Japan). Samples for quantitative analysis were centrifuged at 5,000 rpm for 20 minutes to obtain the supernatants. The supernatants were filtered through 0.22 μm membrane filter for the determination of residual sugars and bioethanol concentration. Sugars and bioethanol were quantified

through high performance liquid chromatography (HPLC) (Agilent 1200 series, U.S.A). A Rezex ROA organic acid H^+ (300×7.8 mm) column and RI detector was used for the separation. The chromatography grade 0.005 N H_2SO_4 was used as mobile phase and the flow rate was set at 0.6 mL/min. The column temperature was set at 60°C and RI detector temperature at 40°C. The injection volumes of 10 µl were applied. The components were identified by comparing their retention times with those of authentic standards under analytical conditions and quantified by external standard method [10].

3. Results and Discussion

3.1 Sugars composition of oil palm (OPF) juice

The results of HPLC analysis on sugars concentration in OPF juice is shown in table 2.

Table 2. Initial sugars concentration contained in the oil palm frond (OPF) juice.

Types of sugar	Sugars concentration (g/L)	
	This work	Zahari et al. (2012)
Glucose	44.16	53.95
Sucrose	11.25	20.46
Fructose	1.46	1.68
Total	56.87	76.09

The total sugars concentration in OPF juice used in this study was 56.87 g/L. Glucose was found to be dominant sugar (44.16 g/L) followed by sucrose (11.25 g/L) and fructose (1.46 g/L). The result obtained in this study was almost similar with the findings reported by Zahari et al. [10]. They have reported that glucose was the dominant sugar in OPF juice followed by sucrose and fructose. However, Zahari et al. [10] showed higher total sugars concentration (76.09 g/L) compared to this study where sugars composition was reported as; glucose (53.95 g/L), sucrose (20.46 g/L) and fructose (1.68 g/L). This might be due to the different location of vegetation of the oil palm tree used in this study, which could affect the sugars concentration in the OPF. The OPF in this work was obtained from the oil palm plantation in Gambang, Pahang, meanwhile Zahari et al. [10] obtained their sample from Serdang, Selangor. Another possible explanation which could contribute to the low sugars content in OPF juice in this study could be due to the time of harvest of fresh OPF during the dry season in the month of May. According to Yusof Basiron, chief executive officer of the Malaysian Palm Oil Council (MPOC), moderate amount of rain may provide a good condition to induce the growth of oil palm tree [20] and subsequently influence the sugars content in the OPF as well.

3.2 Preliminary experiment

3.2.1 Effect of sterilization

OPF juice is readily fermentable to produce bioethanol because it contains mixture of sugars and rich in minerals and nutrients which are essential for bacterial growth during fermentation. Bioethanol concentrations and total sugars consumed were analyzed using high performance liquid chromatography (HPLC), whereby the bioethanol yield (g g^{-1}) was calculated based on experimental bioethanol produced and expressed as g bioethanol per total g of sugar utilized (equation (2)):

$$\text{Bioethanol yield}\left(\frac{\text{g bioethanol}}{\text{g sugars}}\right)=\frac{\text{Bioethanol concentration}\left(\frac{\text{g}}{\text{l}}\right)}{\text{Total sugars consumed}\left(\frac{\text{g}}{\text{l}}\right)} \quad (2)$$

To investigate the effect of sterilization on bioethanol production, two sets of experiment was conducted, where autoclaved (sterile) and non-autoclaved (non-sterile) OPF juice were used as a substrate for fermentation. In order to achieve this, yeast, *S. cerevisiae* Kyokai No. 7 (ATCC 26622) was cultured into autoclaved and non-autoclaved OPF juice and the result was depicted in figure 1.

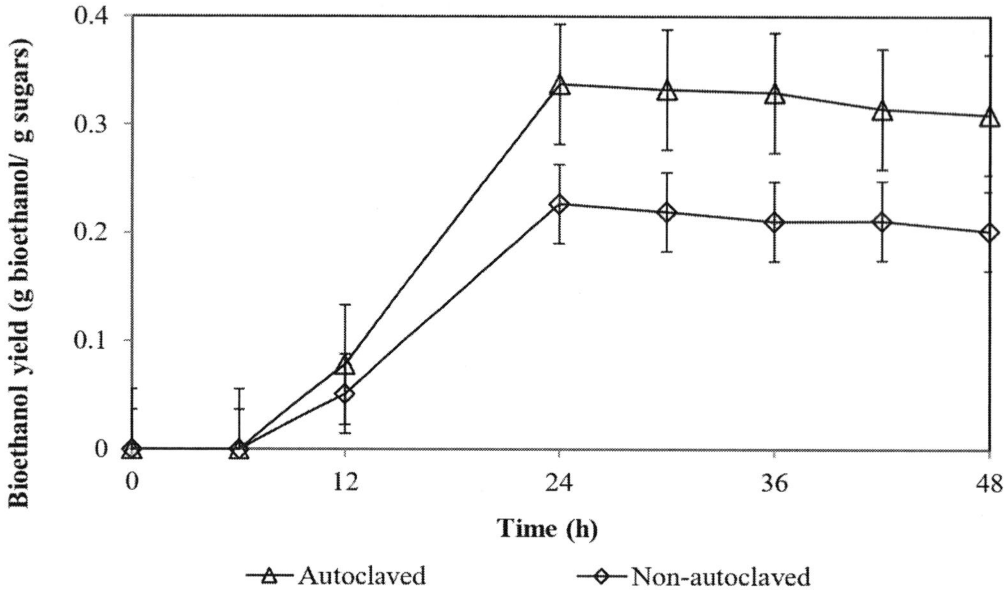

Figure 1. Comparison of bioethanol yield by *S. cerevisiae* Kyokai No. 7 (ATCC 26622) supplemented with autoclaved and non-autoclaved OPF juice. (Experiments were conducted in a rotary shaker (150 rpm) under anaerobic condition at 30°C for 48 h without pH adjustment).

As shown in figure 1, it was noted that highest yield of bioethanol at 0.34 g bioethanol/ g sugars was obtained after 24 h of fermentation period when sterilized (autoclaved) OPF juice was used as fermentation substrate. For the non-sterilized (non-autoclaved) OPF juice, only 0.23 g bioethanol/ g sugars were obtained within the same fermentation period. It was observed that the total bioethanol yield obtained was slightly higher using sterilized OPF juice as a fermentation feedstock. It is worth to mention that, by autoclaving the OPF juice, any unwanted microorganisms which can cause contamination are inhibited or killed. This result is in agreement with other findings whereby heat sterilization may affect bioethanol production from oil palm trunk (OPT) sap by using the similar yeast strain [19]. In the report, it was mentioned that the maximum bioethanol concentration in heat sterilized sap was 29.96% higher than the fermentation in cold sterilized sap, and this was two times higher than the fermentation in non-sterile sap. In addition to that, higher bioethanol yield was obtained in fermentation using sterilized OPF juice might be due to some complex oligosaccharides in OPF juice was also hydrolysed to monosaccharides during autoclaving, resulted in more reducing sugars [33-35]. The reducing sugars produced was further utilized by *S. cerevisiae* Kyokai No. 7 (ATCC 26622) to produce bioethanol hence increased the yield.

3.2.2 Sugars consumption and bioethanol production profile.

Fermentation profile for bioethanol production from sterilized (autoclaved) OPF juice using *S. cerevisiae* Kyokai No. 7 (ATCC 26622) is shown in Figure 2. The bioethanol yield increased proportionally with reducing sugar content from fermentation broth. The highest bioethanol yield of 0.34 g bioethanol/ g sugars was obtained after 24 h of fermentation period. Prolonged time of incubation up to 48 h did not contribute to the increase in production of bioethanol. Whereby, it was slightly decreased to only 0.31 g/g sugars of bioethanol yield, which accounts for approximately 8.82% decrease. In Figure 1, there is a clear trend of decreasing in the production of bioethanol after 24 h of incubation might cause due to the decreased of sugars concentration level in the fermentation broth. During the fermentation, equal molarity of CO_2 and bioethanol was produced hence reduction of sugar consumption leads to weight lost in CO_2. This explained the reduction in bioethanol production in a longer period of time [21]. From this explanation, for the subsequent experiment, the fermentation was conducted for 24 h of incubation period to evaluate the effects of several physical parameters on bioethanol production from OPF juice by *S. cerevisiae* Kyokai No. 7 (ATCC 26622).

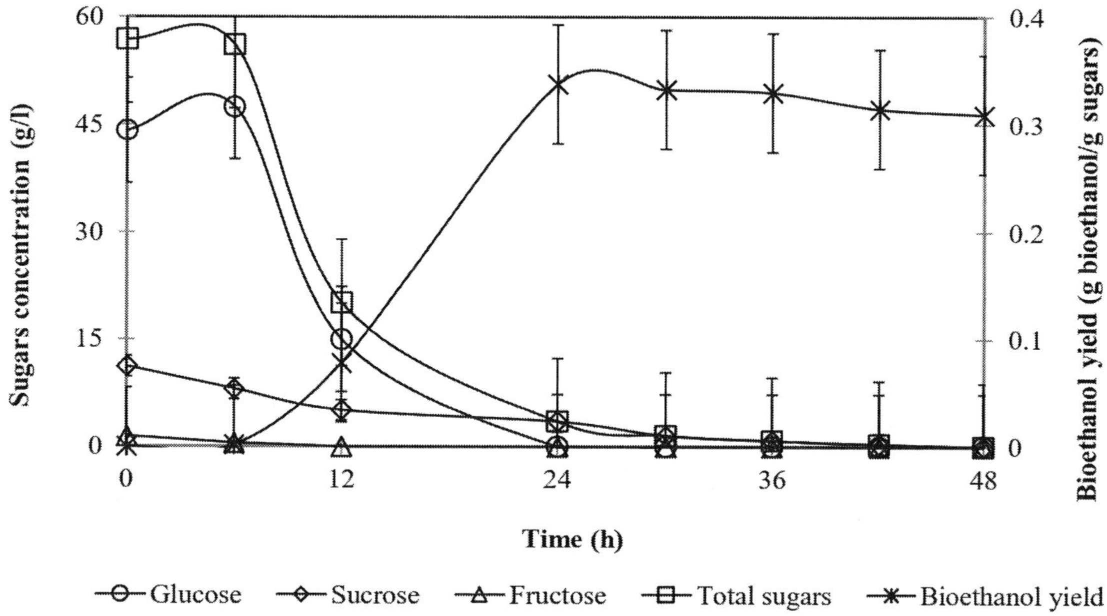

Figure 2. Sugars consumption and bioethanol production profile by *S. cerevisiae* Kyokai No. 7 (ATCC 26622) supplemented with OPF juice.

Figure 2 also demonstrated the profile of sugars consumption by *S. cerevisiae* Kyokai No. 7 (ATCC 26622) in the fermentation broth throughout the incubation period. Overall, sugars in OPF juice was completely consumed by the yeast at the end of fermentation period including sucrose. During the first 6 h until 12 h, the sugars concentration decrease rapidly as bioethanol was produced. Our finding revealed that the concentration of fructose, sucrose and total sugar started to decrease during the first 6 h and then rapidly decreased afterward. This finding highlights the increase in glucose concentration can be attributed to the breakdown of sucrose to its monomer by the presence of invertase during the fermentation as sucrose is a disaccharide composed of glucose and fructose. The present finding also supported by previous studies which concluded that *S. cerevisiae* has the ability to produce invertase enzyme namely perisplamic invertase in which used for conversion of sucrose to glucose and fructose at the yeast cell during the fermentation process [2, 5].

3.3 Screening of parameters affecting bioethanol production from OPF juice

3.3.1 Effect of medium initial pH

The medium initial pH is a key factor which has significant influence on fermentation [22] [23]. All organism and cellular processes are affected by pH; this is mainly due to the concentration of H+ ions in the liquid environment. The cells grow and perform fermentation best within a certain pH range.6 In this study, the effect of medium initial pH on bioethanol production from OPF juice was conducted by adjusting the initial pH value of OPF juice prior to autoclaving between 5.0 and 9.0 with an increment of 1.0. As shown in table 3, bioethanol yield was found to be the highest when the medium initial pH was adjusted at pH 7.0 compared to others after 24 h of fermentation period. The highest bioethanol yield obtained in this experiment was 0.39 g/ g sugars. Thus, pH may be an important factor to achieve maximum bioethanol yield. Optimum pH is essential for bioethanol yield to avoid maximum acidic or basic condition of medium hence retard the metabolic of yeast and cell growth [24].

Table 3. Bioethanol yield at different initial medium pH with rotation rate and temperature were set at 150 rpm and 30°C, respectively.

Initial medium pH	Bioethanol concentration (g/L)[a]	Bioethanol yield (g/ g sugars)
5	19.79	0.35
6	20.14	0.35
7	22.10	0.39
8	18.40	0.32
9	16.77	0.29

[a] Determination by HPLC from filtered supernatant after 24 h of incubation period.

3.3.2 Effect of rotation rate

Rotation of an incubator shaker is necessary for constant mixing of the medium components to provide uniform oxygen transfer rates. Rotation also played significant role in improving bioethanol concentration and yield [12]. The effect of rotation rate is fundamental to obtain successful fermentation by providing adequate mixing, mass transfer and heat transfer [11]. Besides assisting mass transfer between two different phases of the medium, it also enables uniform suspension of microbial cells in homogenous nutrient medium. table 4 shows bioethanol production by *S. cerevisiae* Kyokai No. 7 (ATCC 26622) using OPF juice as the fermentation feedstock at different rotation rate of 0, 50, 100, 150 and 200 rpm.

Table 4. Bioethanol yield at different rotation rate with initial medium pH and temperature were set at 7.0 and 30°C, respectively.

Rotation rate (rpm)	Bioethanol concentration (g/L)[a]	Bioethanol yield (g/ g sugars)
0	19.00	0.33
50	19.47	0.34
100	23.00	0.40
150	22.10	0.39
200	19.86	0.35

[a] Determination by HPLC from filtered supernatant after 24 h of incubation period.

After 24 h of incubation period, rotation rate at 100 rpm gave the best bioethanol yield of 0.40 g/ g sugars compared to the other rotation rate. Thus, maximum productivity in microbial fermentation was achieved at optimum rotation rate. Mittal [25] proposed that rotation creates shear forces by causing morphological changes and disruption of cell structure. High rotation rate is not suitable for successful fermentation as it could contribute to the effect of hydrodynamic stress which can cause leakage of intracellular compounds [18]. However, rotation is needed to improve cell mass and bioethanol activity. Therefore, low rotation rate may contribute to low bioethanol production due to less nutrient consumption by yeast cells in static condition [12]. Optimum rotation rate will enable symmetrical fermentation system hence accelerating nutrient consumption by yeast.

3.3.3 Effect of temperature

In this study, the influence of different temperature on the bioethanol fermentation by *S. cerevisiae* Kyokai No. 7 using OPF juice was studied with regard to bioethanol production. Temperature is one of the most significant parameters that contribute to yeast growth and fermentation performance. *S. cerevisiae* Kyokai No. 7 is a type of yeast which is mesophilic in nature, thus able to withstand temperature up to 48°C. However, Ho and Powel [13] suggested the preferable temperature for *Saccharomyces* yeast is between 25 to 35°C and at a temperature up to 43°C, yeast cells began to lose their capability to be superior ethanologenic yeast strains. Table 5 shows the bioethanol yield at various temperatures from 27.5 to 47.5°C.

Table 5. Bioethanol yield at different temperature with initial medium pH and rotation rate were set at 7.0 and 150 rpm, respectively.

Temperature (°C)	Bioethanol concentration (g/L)[a]	Bioethanol yield (g/ g sugars)
27.5	18.15	0.32
30.0	22.10	0.39
32.5	23.11	0.41
35.0	18.85	0.33
37.5	16.95	0.30

[a] Determination by HPLC from filtered supernatant after 24 h of incubation period.

Highest bioethanol yield was obtained at temperature of 32.5°C (0.41 g bioethanol/ g sugars) and thus regarded as an optimum temperature for production of bioethanol using OPF juice by *S. cerevisiae* Kyokai No. 7. Bioethanol yield however gradually decreased at 37.5°C and this event was previously reported by Fakruddin et al. [26] in which the production of bioethanol by strains *Saccharomyces unisporous* (P), *S. cerevisiae* (C) and (T) gradually decreased.

3.4 Optimization of bioethanol production employing Response Surface Methodology (RSM)

Based on the OFAT experiment, it was observed that the best condition for bioethanol production from OPF juice by *S. cerevisiae* Kyokai No. 7 was obtained at the following parameter's conditions; medium initial pH (7.0), temperature (32.5°C) and rotation rate (100 rpm). These conditions were then selected as the central point for optimization study using CCD. Studies were carried out to establish the range of parameters such as medium initial pH, temperature and rotation rate to be optimized. A design matrix corresponding to the yield of bioethanol was subjected to regression analysis to study the effect of these parameters. The RSM experimental design matrix with three factors at five levels and the experimental results are presented in table 6.

Table 6. The experimental results for bioethanol yield for the central composite design.

Standard Run	Factor A Temperature (°C)	Factor B Initial medium pH	Factor C Rotation rate (rpm)	Response Bioethanol yield (g bioethanol/ g sugars)
1	27.5	5	80	0.35
2	37.5	5	80	0.37
3	27.5	9	80	0.30
4	37.5	9	80	0.30
5	27.5	5	120	0.27
6	37.5	5	120	0.35
7	27.5	9	120	0.27
8	37.5	9	120	0.30
9	22.5	7	100	0.22
10	42.5	7	100	0.26
11	32.5	3	100	0.31
12	32.5	11	100	0.26
13	32.5	7	60	0.35
14	32.5	7	140	0.31
15	32.5	7	100	0.49
16	32.5	7	100	0.46
17	32.5	7	100	0.49
18	32.5	7	100	0.48
19	32.5	7	100	0.46
20	32.5	7	100	0.49

Bioethanol yield was used as a response and was arranged into design expert experiment based on standard run. As shown in table 6, the average value of all bioethanol yields was around 0.22–0.49 g bioethanol/ g sugars.

3.4.1 ANOVA and model development

ANOVA is known as analysis of variance which offers an excellent technique to determine the process variables that gives significant impact on process quality and their possible interaction. ANOVA which includes *F*-value, p-value, R^2 and lack of fit was applied to determine suggested model that fit with experimental data. R^2 is known as coefficients of determination to ensure the quality of fit for the model. The p-values of less than 0.05 were indicated as statistically significant. The significant terms showed whether the parameters studied affects fermentation process. The model and individual coefficient will be more significant if the results show a larger magnitude of *F*-value and a smaller p-value. The relationship between independent variables and response can be performed from analysis of quadratic model as shown in table 7.

Source	Sum of Squares	Degree of Freedom	Mean Square	*F*-Value	p-value, Prob>F
Process order: Quadratic					
Model	0.15	9	0.017	48.74	< 0.0001
A-Temp	0.0027	1	0.0027	7.90	0.018
B-pH	0.0045	1	0.0045	13.06	0.0047
C-Speed	0.0027	1	0.0027	7.90	0.018
AB	0.00061	1	0.00061	1.76	0.21
AC	0.0010	1	0.0010	2.90	0.11
BC	0.00061	1	0.00061	1.76	0.21
A^2	0.094	1	0.094	271.43	< 0.0001
B^2	0.063	1	0.063	181.03	< 0.0001
C^2	0.037	1	0.037	108.87	< 0.0001
Residual	0.0034	10	0.00034		
Lack of Fit	0.0024	5	0.00048	2.22	0.20
Pure Error	0.0010	5	0.00021		
Cor. Total	0.15	19			
Standard Deviation		0.019	R^2		0.97
Mean		0.35	Adjusted R^2		0.95
C.V. %		5.27	Predicted R^2		0.86
PRESS		0.021	Adequate Precision		20.57

Table 7. Analysis of variance (ANOVA) for quadratic model

*Lack of Fit is not significant relative to the pure error

The mathematical model derived from the experimental results for bioethanol yield (Y) was shown in Equation (3):

$$Y = +0.47 + 0.013 \times A - 0.017 \times B - 0.013 \times C - 0.0088 \times A \times B + 0.011 \times A \times C + 0.0088 \times B \times C - 0.061 \times A^2 - 0.05 \times B^2 \quad (3)$$

where Y is bioethanol yield, A is temperature, B is medium initial pH and C is rotation rate. The quadratic model was selected to provide the best fit with the experimental results.

The model presented in table 7 exhibits a high determination coefficient ($R^2 = 0.97$), explaining 97.77% of the variability in the response, as well as a high value of the adjusted determination coefficient (adjusted $R^2 = 0.95$), suggesting a high significance of the model. A very low probability ($p < 0.0001$) obtained from the regression analysis of variance (ANOVA) demonstrated that the model was significant. In this study, all the linear model terms including temperature (A), medium initial pH (B) and rotation rate (C) have significant effect, as the p-values calculated for this factor was less than 0.05. Therefore, changes in this parameter could significantly impact the bioethanol production from OPF juice fermentation. The most significant effect is the linear term of medium initial pH (B), followed by rotation rate (C) and temperature (A). All the two level interactions including temperature and rotation rate (AC), temperature and medium initial pH (AB) as well as medium initial pH and rotation rate (BC) were indicated as significant. In a similar manner, all the second order effects showed the significant results including A^2, B^2 and C^2. Generally, the lack of fit p-value of 0.20 implied that the lack of fit is not significant relative to the pure error. The non-significant lack of fit is positive because it demonstrates a good fit of the model to the data. A good fit means that the generated models adequately explained the variation of data.

3.4.2 Response surface plot

Response surface plots based on equation (3), with the relationships between the response and variables, are presented in Figure 3 (a–c). The plots were constructed by plotting the response (bioethanol yield) on the Z-axis against any two dependent variables while maintaining the other variables at their optimal values.

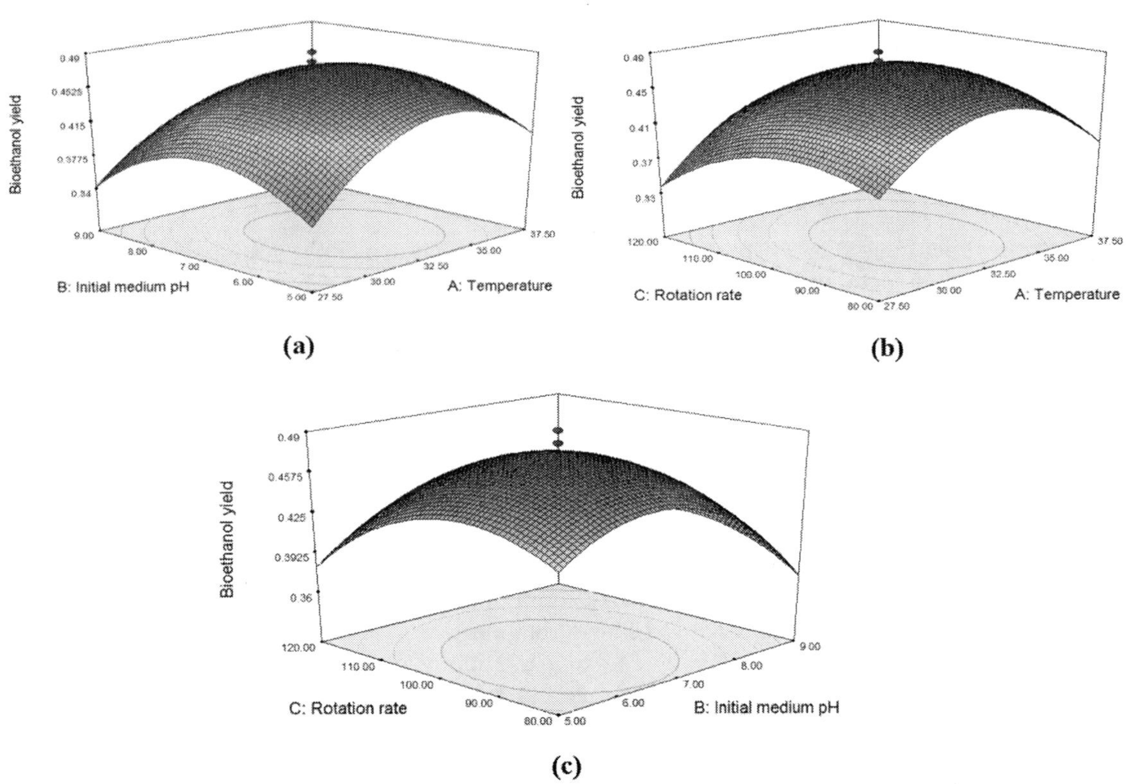

Figure 3. Response surface plots depicting the interaction between variables in the production of bioethanol from OPF juice; (a) temperature (°C) and initial medium pH, (b) temperature (°C) and rotation rate (rpm), (c) initial medium pH and rotation rate (rpm).

Figure 3(a) depicts the interaction between temperature and the medium initial pH (AB) while holding factor C (rotation rate) at 100 rpm. As shown in Figure 3(a) it was observed that bioethanol yield increased when the temperature was changed from 27.5 to 37.5°C as medium initial pH increased from 5.0 to 9.0. In this study, the bioethanol yield decreased by lowering the temperature to 27.5°C with similar effects by increasing the temperature to 37.5°C. Similar event was reported by Chin et al. [6], whereby temperature affected the enzyme activity which explained the facilitation of chemical reactions within the yeast. In contrast, high bioethanol yield was observed at moderate temperature and medium initial pH ranges. Based on the optimum result suggested by the Design-Expert, Version 7.1.6 software, bioethanol yield was relatively high at temperature of 33.03°C and medium initial pH of 6.62. Adnan et al. [27] reported that bioethanol production from glycerol by *Escherichia coli* SS1 was greatly influenced by pH and an optimum pH value of 7.61 was identified. Further increases in the pH resulted in lower bioethanol production. The initial pH is an important factor that influences the NADH to NAD^+ ratio, which greatly affects the metabolic flux under anaerobic conditions [27]. Hence, to obtain optimal bioethanol production, it is necessary to control the medium initial pH under optimum conditions. In general, yeast is able to grow and efficiently ferment substrates into bioethanol at pH values of 3.5–6.0 and temperatures of 28 – 35°C.28 The optimum medium initial pH (6.62) and temperature (33.03°C) obtained in this work is within the range of those reported in the literature especially for Saccharomyces cerevisiae.

The interaction between temperature and rotation rate (AC) while holding medium initial pH (B) at 7.0 towards bioethanol yield in terms of 3D is shown in Figure 3 (b). As shown in Figure 3 (b), the bioethanol yield decreased at higher rotation rate (120 rpm) as the temperature was increased from 27.5 to 37.5°C. In contrast with lower rotation rate (80 rpm), it shows an increasing trend when temperature was increased. The lower bioethanol yield (0.31 g/ g sugars) showed at higher rotation rate (140 rpm) compared to central point rotation rate value (100 rpm) which has much higher bioethanol yield (0.49 g/ g sugars). Rotation rate is known to have an important role in ensuring uniform adequate mixing, mass transfer and heat transfer within the fermenter in medium components [5]. The effects of rotation rate are required for successful fermentation process to improve product yields. The advantages of rotation toward performance and growth of microorganism cells could improve the mass transfer on substrates, products or by-products and oxygen. Better mixing process has the capability to maintain adequate supply of sugars and nutrients to the cells as well as to maintain the concentration gradient between interior and exterior cells in fermentation broth [5].

Meanwhile, Figure 3 (c) depicts the interaction between the medium initial pH and rotation rate (BC) while holding factor A (temperature) at 32.5°C. The interaction between medium initial pH and rotation rate (BC) demonstrated that bioethanol yield decreased when rotation rate changed from 80 to 120 rpm as medium initial pH increased from 5.0 to 9.0. At higher rotation rate (120 rpm), the response yield illustrated a linear decrease with increasing pH value. In contrast, lower rotation rate (80 rpm) showed a pattern of increasing slope. The medium with pH variations may lead to the changes in enzyme activity as well as changes in reaction rate. pH plays a significant role in bioethanol fermentation by *Saccharomyces cerevisiae* as pH affects the growth of yeast, by-product formation and fermentation rate due to the concentration of H^+ ions in the liquid environment [5-6, 29].

In summary, high temperature still showed the production of bioethanol, however, slightly decreased with time of incubation. Changes in the medium initial pH might lead to the changes in the fermentation pathway. Hence, based on the result obtained; medium initial pH showed the highest significant effect towards the production of bioethanol. Minimal rotation rate was required to produce maximum bioethanol yield to ensure uniform mixing and consumption of nutrition. Therefore, it is necessary to enhance optimal temperature, medium initial pH and rotation rate to accelerate cell activities, thus achieve high bioethanol yield.

3.4.3 Confirmation of model prediction

The reproducibility of the model was tested by performing the fermentation under the optimal conditions obtained from the CCD. This validation was also used to verify the accuracy of the model. The bioethanol production model suggested that optimum bioethanol yield could be achieved at medium initial pH of 6.62, temperature of 33.03°C and rotation rate of 96.51 rpm. The predicted bioethanol yield under these optimum conditions was 0.48 g bioethanol/ g sugars. Three replicates of the batch fermentation using OPF juice without nutrient supplementation under the optimized conditions were conducted in shake flask to confirm the model validity. Maximum bioethanol yield of 0.50 ±0.02 g/ g sugars was obtained from the confirmation test. These experimental findings were in close agreement with the model prediction, with a difference of only 4.17%. Hence, it is confirmed that the model developed from the response surface methodology could reliably predict bioethanol yields. According to Adnan et al. [27], differences between experimental and predicted values of less than 10% confirm the validity of a model. The yield obtained in this study was 47.06% higher compared with the bioethanol produce under non-optimized condition (0.33 g bioethanol/ g sugars). These findings suggest that in general, medium initial pH, temperature and agitation speed may indeed play an important role in the bioethanol production by *S. cerevisiae* Kyokai No. 7 (ATCC 2662) utilizing OPF juice as a complete medium. In addition to that, even without the supplementation of nitrogen source into the fermentation medium, the bioethanol yield obtained in this study was almost comparable to those reported by Zahari et al. [2]. They have reported that 0.49 g/g sugars of bioethanol yield were obtained from OPF juice supplemented with 4 g/L of peptone and yeast extract (nitrogen source). The evidence from this study suggests that OPF juice can be used directly as the fermentation medium for bioethanol production at industrial scale.

4. Conclusion

The present work demonstrated that OPF juice can be a potential complete fermentation feedstock for the production of bioethanol. The optimal fermentation conditions for bioethanol production were determined using the methods of OFAT and response surface analysis. Experimental results indicated that the temperature exert significant effects on bioethanol yield. The maximum bioethanol yield of 0.50 g/ g sugars was obtained under the following optimum condition; medium initial pH (6.62), rotation rate (96.51 rpm) and temperature (33.03°C). Compared to the predicted maximum bioethanol yield of 0.48 g/ g sugars, only a small error exists between the predicted value and the actual experimental value. In conclusion, this study demonstrated that high bioethanol yield could be obtained by culturing *S. cerevisiae* Kyokai No. 7 at the optimized condition using OPF juice as the sole renewable fermentation feedstock. Under the optimal conditions, the bioethanol yield obtained was 47.06% higher compared to non-optimized condition.

Acknowledgements

The authors would like to acknowledge Universiti Malaysia Pahang (UMP) and the Ministry of Higher Education (MOHE) for the scholarship and financial support by providing the research grant, FRGS (RDU140139) and RAGS (RDU121406). The authors also would like to acknowledge the Institute of Postgraduate Studies of UMP for motivating the author to publish this work and PhD student, Kamaliah Abdul Samad for supplying the raw material.

References

[1] Zwart D 2013 *Journal of Oil Palm and Environment* **4**(5) 41-62

[2] Zahari M A K M, Abdullah S S S, Roslan A M, Ariffin H, Shirai Y and Hassan M A 2014 *Journal of Cleaner Production* **65** 252-60

[3] Kosugi A, Tanaka R, Magara K, Murata Y, Arai T, Sulaiman O, Hashim R, Hamid Z A A, Yahya M K A, Yusof M N M, Ibrahim W A and Mori Y 2010 *Journal of Bioscience and Bioengineering* **110**(3) 322-25

[4] Norhazimah A H and Faizal C K M 2013 *Research Journal of Chemistry & Environment* **17**(10) 90–93

[5] Shahirah M N N 2014 *Master Thesis* (Universiti Malaysia Pahang Malaysia)

[6] Chin K L, H'ng P S, Wong L J, Tey B T and Paridah M T 2010 *Bioresource Technology Journal* **10** 3287-3291

[7] Mabee W E, McFarlane P N and Saddler J N 2011 *Biomass and Bioenergy* **35** 4519–4529

[8] Pothiraj C, Arumugam R and Gobinath M 2014 *Bioresources and Bioprocessing* **1** 27

[9] Rudolf A, Karhumaa K and Hagerdal B H 2009. *In: T. Satyanarayana, G. Kunze (eds.), Yeast Biotechnology, Diversity and Applications* (New York: Springer Science. Publishing Co.) 489-513
[10] Zahari M A K M, Zakaria M R, Ariffin H, Mokhtar M N, Salihon J, Shirai Y and Hassan M A 2012 *Bioresource Technology Journal* **110** 566-71
[11] Rodmui A, Kongkiattikajorn J and Dandusitapun J 2008. *Kasetsart Journal, Nat Sci.* **42** 285-93
[12] Yan L, Tiansheng Q, Naikun S, Mingzhe G, Yanling J and Hai Z 2009 *Chinese Journal of Applied and Environmental Biology* **15**(4) 563-67
[13] Ho D H N and Powel C 2014 *International Journal of Renewable Energy and Environmental Engineering* **2**(1) 1-6
[14] Frey D, Engelhardt F and Greitzer E M 2003 *Research Engineering Design* **114** 65-74
[15] Bezerra M A, Santelli R A, Oliveira E P, Villar L S and Escaleira L A 2008 *Talanta.* **76**(5) 965-77
[16] Wahid Z A and Nadir N 2013. *World Applied Sciences Journal 21 (Mathematical Applications in Engineering)* **21** 56-61
[17] Sakkas V A, Islam M A, Stalikas C and Albanis T A 2010 *Journal of Hazardous Material* **175** 33-44
[18] Bakri Y, Mekaeel A and Koreih A 2011 *Brazilian Archives of Biology and Technology* **54**(4) 659-64
[19] Norhazimah A H 2012 *Master Thesis* (Universiti Malaysia Pahang, Malaysia)
[20] Basiron Y 2011 *Bloomberg Businessweek*, http://www.bloomberg.com/news/articles/2011-02-24/palm-oil-output-in-malaysia-to-gain-as-weather-boosts-yields-group-says. Accessed July 10, 2019
[21] Krishnamurthy R, Animasaun D A, Ingalhalli R S and Ramani N D 2014 *International Journal of Pure & Applied Bioscience* **2**(22) 174–80
[22] Raikar R V 2012 *International Journal of Environmental Sciences* **3**(2) 776-783
[23] Lin Y, Zhang W, Li C, Sakakibara K, Tanaka S and Kong H 2012 *Biomass and Bioenergy* **47** 395-401
[24] Willaert R and Nedovic V A 2006 *Journal of Chemical Technology and Biotechnology* **81** 1353-67
[25] Mittal G S 1992 *Lancaster: Technomic*
[26] Fakruddin M, Islam M A, Ahmed M M and Chowdhury N 2013 *International Journal of Renewable and Sustainable Energy* **2**(4) 133-139
[27] Adnan N A A, Suhaimi S N, Aziz S A, Hassan M A and Phang LY 2014 *Journal of Renewable Energy* **66** 625-633
[28] Alam M Z, Kabbashi N A and Hussin S N I S 2009 *Journal of Industrial Microbiology & Biotechnology* **36** 801–808
[29] Pramanik K 2003 *Journal of the Chinese Institute of Chemical Engineers* **34** 4) 487–42
[30] Azhar S H M, Abdulla R, Jambo S A, Marbawi H, Gansau J A, Faik A A M and Rodrigues K F 2017 *Biochemistry and Biophysics Reports* **10** 52-61
[31] Ahmad F, Zhang Z, William O S D and O'Hara I M 2019 *Renewable and Sustainable Energy Reviews* **109** 386–411
[32] Hossain N, Zaini J and Mahlia T M I 2017 *International Journal of Technology* **1** 5–18
[33] de Moura F A, Macagnan F T and da Silva L P 2014 *International Journal of Food Science and Technology* **50**(2) 275–81
[34] Bougrier C, Delgenès J P and Carrère H 2008 *Chemical Engineering Journal* **139** 236–244
[35] Hafid H S, Rahman N A, Md Shah U K, Baharudin A S 2015 *Journal of Environmental Management* **156** 290–98

IOP Conf. Series: Materials Science and Engineering **702** (2019) 012004 doi:10.1088/1757-899X/702/1/012004

Biobutanol production from oil palm frond juice in 2 L stirred tank bioreactor with in situ gas stripping recovery

F M Asri, N Masngut and M A K M Zahari*

Faculty of Chemical and Process Engineering Technology, College of Engineering Technology, Universiti Malaysia Pahang, Lebuhraya Tun Razak, 26300 Gambang, Kuantan, Pahang, Malaysia.

*E-mail: ahmadkhushairi@ump.edu.my

Abstract. Biobutanol has generated renewed interest of being a sustainable alternate fuel. As butanol can fully miscible with diesel fuel even at low temperatures, better fuel extender than ethanol, low vapour pressure, low miscibility with water and has higher price, making it most potential solvent compared to acetone and ethanol. Previous study shows that biobutanol can be biosynthesized from oil palm frond (OPF) juice by *Clostridium acetobutylicum* ATCC 824 in the lab scale experiment. Thus, this paper focused on the synthesis and characterization of the OPF juice as fermentation substrate for biobutanol production in 2 L batch fermentation using 50 g/L OPF juice. 14.0 g/L biobutanol was produced in batch fermentation. The research was further extended with fed-batch experiment and in situ gas stripping. 71.5% acetone, 36.5% biobutanol and 18.4% ethanol were stripped out from the fermentation broth during first gas stripping interval (42-48 h). After 120 h, total of 395 g/L biobutanol was stripped out. Thus, it was proven that OPF juice can be used as an alternative carbon source for biobutanol production while fed-batch fermentation with in situ gas stripping method improved the biobutanol production from OPF juice.

1. Introduction

The United States Department of Energy (DOE) and United States Department of Agriculture (USDA) define sustainable biofuels as production of biofuels which are economically competitive, preserve the natural reserves, reduce greenhouse emissions and secure social well-being. According to International Energy Agency (IEA) report, biofuels be it liquid or gaseous fuels, helped in reducing CO_2 emissions in the transportation sector. Currently, biofuels contribute only 2% of energy share in transportation, however, due to rapid growth in the biofuels industry, it is predicted that biofuels would supply 27% of the total transport fuel need by 2050 [1].

Butanol is expected to play a major role in the next generation of biofuels [2]. Biobutanol which is butanol derived from fermentation, has been claimed to be "superior biofuel", as it can be blended into standard gasoline similarly to ethanol but with several advantages. These include higher energy content and lower vapour pressure, which make storage and transportation easier. It is also immiscible with water, has a better blending ability with gasoline and diesel fuel, and can be used in conventional internal combustion engines without modification [2]. With scientific advances in microbiology and dramatic improvements in process technology nowadays, the advanced ABE process is able to produce higher purity n-butanol with a carbon footprint that is over 45% better than petroleum-based butanol.

Despite these promising results, there are some drawbacks that affect its widespread use such as the high cost of substrates, low productivity, low concentrations (less than 20 g/L) due to solvent toxicity, and the high cost of product recovery [3]. High cost of the substrate, low productivity and product yield due to slow fermentation, and uneconomical product recovery are the three areas of greatest factors that affect the economic decline of butanol fermentation [2, 4, 5]. In order to reintroduce an environmentally safe biological process, a cost-effective non-food feedstock with a successful genetically engineered *Clostridia* strain and an effective recovery method are needed.

Previous study shows that biobutanol can be biosynthesized from oil palm frond (OPF) juice by *Clostridium acetobutylicum* ATCC 824 in the lab scale experiment [6]. Under optimized conditions, the highest biobutanol yield was 0.3054 g/g after 144 hours of incubation period obtained in 100 mL serum bottle. However, the potential of OPF juice for biobutanol production at 2 L bioreactor scale has yet to be studied. One of the major bottlenecks faced many researchers to produce biobutanol in large scale is the problem of scaling up the process for industrial scale production. For instance, most of the fermentation process performance could not sustain at larger scale compared to the lab scale. In addition to that, there was no research conducted on the biobutanol recovery from the previous work using oil palm frond juice.

One of the efforts to overcome the issue of butanol toxicity is by coupling the fermentation with in situ butanol recovery. Among the available separation technologies, gas stripping appears to be more economical and superior for separating butanol from ABE fermentation broth. However, it is not applicable to the highly concentrated-substrate batch fermentation over 162 g/L [7] due to the substrate inhibition. In this case, fed-batch fermentation appears to solve the substrate toxicity problem. It is an industrial technique where low substrate concentration is used at initial and concentrated substrate was added after the substrate is consumed to keep the substrate inhibitory under levels. The limitation to fed-batch fermentation is, it can cause butanol toxicity. Therefore, fed-batch fermentation with in situ product recovery methods should be carried out simultaneously to avoid butanol toxicity.

2. Methodology

2.1. Chemical and biological materials
All chemicals are of analytical grade. *Clostridium acetobutylicum* ATCC 824 was procured from American Type Culture Collection (ATCC) in freeze-dried form.

2.2. OPF juice preparation
Fresh OPF (without leaves) was shredded and pressed using conventional sugarcane press machine repeatedly to extract the juice. The obtained juices were subsequently filtered to remove fibrous debris. It was then centrifuged at 15,000 × g for 15 min at 4 °C (Thermo Fisher Scientific, NC, USA) and the supernatant was filtered using a mixed cellulose ester membrane filter with the pore size between 3 and 5 µm (Cole Parmer, Illinois, USA). The supernatant was analyzed using High Performance Liquid Chromatography (HPLC) to determine the sugar content and its elemental constituents using CNHS Analyzer (vario MACRO cube, Elementar, Germany) following ASTM D-5291 method [8]. Nutrients and heavy metals in the OPF and OPF juice were determined using in-house method based on APHA 3010 and the content was then determined using Inductively Coupled Plasma (ICP) (Perkin Elmer, 7300 DV, USA).

2.3. Reinforced Clostridial Medium (RCM) Preparation
RCM was used to revive the bacteria and for inoculum development. 38 g of RCM powder was liquefied in 1 L of deionized water. Solutions were distributed each into 100 mL serum bottles and sparged with oxygen-free nitrogen gas. Serum bottles were fitted with butyl rubber stopper and sealed with aluminium cap before sterilized at 121 °C for 15 min.

2.4. P2 medium
P2 medium containing buffer (KH_2PO_4, K_2HPO_4, and $C_2H_7NO_2$), minerals ($MgSO_4$-$7H_2O$, $MnSO_4$-H_2O, $FeSO_4$-$7H_2O$, and NaCl) and vitamin (*para*-amino-benzoic acid, thiamine, biotin) was prepared according to table 1 [9].

Table 1. Composition of P2 medium [9].

Nutrient contents	Composition (g/L)
Yeast extract	1
Potassium dihydrogen phosphate, KH_2PO_4	0.5
Dipotassium hydrogen phosphate, K_2HPO_4	0.5
Ammonium acetate, $C_2H_7NO_2$	2.2
Magnesium sulfate heptahydrate, $MgSO_4$-$7H_2O$	0.2
Manganese sulfate heptahydrate, $MnSO_4$-H_2O	0.01
Iron (II) sulfate heptahydrate, $FeSO_4$-$7H_2O$	0.01
Sodium chloride, NaCl	0.01
Para-amino-benzoic acid	1 mg
Thiamine	1 mg
Biotin	0.01

2.5. Inoculum Preparation
The inoculum was prepared from a *Clostridium acetobutylicum* spore suspension. The spore suspension was heat-shocked by immersing in 90 °C water bath for 90 s before used. After that, inoculum was initiated by inoculating 10% (v/v, volume of spore suspension per volume of media) into sterile deoxygenated RCM aseptically. The mixture was allowed to grow in static conditions for 16-18 h at 37 °C. The inoculum was deemed to be readily used in the fermentation when the optical density at 600 nm gave a reading between 1.5 and 2.0 and its pH was in the range of 5.2-5.5 [10].

2.6. Research design
A five-level-three-factor central composite design (CCD) was employed in this study. The variable and the selected levels for the fermentation process were: inoculum size (10-30%); temperature (27-47 °C); agitation speed (0-200 rpm). The research design was performed using software Design Expert Version 7.1.6. (Stat-Ease Inc., Statistic made easy, Minneapolis, MN, USA).

2.7. Batch fermentation in 1 L Schott Bottle
The anaerobic batch fermentation was conducted in 1 L Schott (Duran) bottle using 50 g/L of total sugars in OPF juice with modified upper cap consisted of both inlet and outlet ports; one for sample withdrawal and nitrogen sparging, the other one for by-product gas out. The initial pH of OPF juice was adjusted to pH 6.0 with an addition of 2M NaOH before autoclaved. Upon cooling the medium, the bottle was sparged with nitrogen to remove the oxygen. 1 g/L of autoclaved yeast extract and filter-sterilized P2 solutions were aseptically added into the sterilized medium. Subsequently, the batch culture was initiated by inoculation of 16-h grown *Clostridium acetobutylicum* ATCC 824 into the medium. The fermentation bottles were kept in an incubator shaker (Scigenics Biotech, Orbitek) under specific temperature control and different agitation rate as per CCD design table generated by the software. Total 20 set of experiments were done under uncontrolled pH and each experiment was conducted in duplicate and measurements were average values.

2.8. Fed-batch fermentation in 2 L stirred tank bioreactor
Fed-batch fermentation with in situ gas stripping product recovery was set up as in figure 1. Fed-batch fermentation was carried out at 1.5 L working volume. The initial pH of the fermentation broth was adjusted to pH 6.0. The batch culture was initiated by inoculation of medium with 20% (v/v, volume of fresh inoculum per volume of media). Nitrogen gas was flushed through the broth to initiate anaerobic condition. When sugar concentration reached below 20 g/L in the fermentation broth, a fresh OPF juice, additional yeast extract and P2 medium was fed for extended fed-batch fermentation. Throughout the process, the fermentation was maintained at 37 °C and pH was uncontrolled.

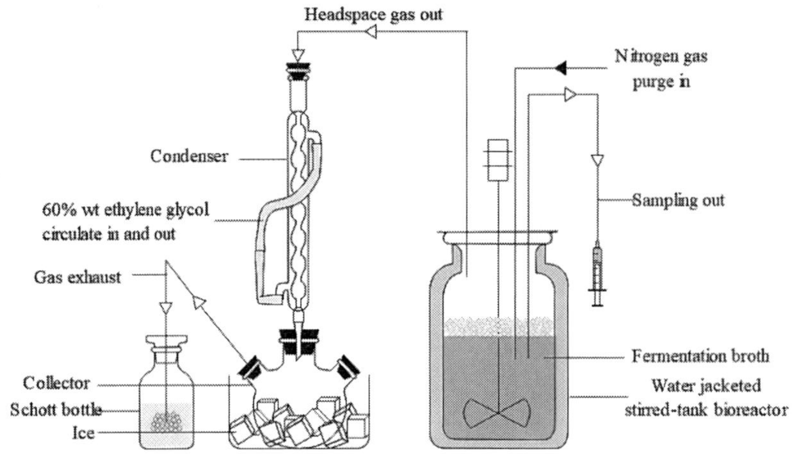

Figure 1. Fed-batch fermentation with in situ gas stripping biobutanol set up.

Gas stripping was first initiated at the intervals of 42-58 h of fermentation followed with second, third and fourth intervals at 66-72 h, 90-96 h and 114-120 h. The nitrogen gas was purged to circulate the headspace fermentation gas through the 36 cm height x 4 cm width condenser. 60%.wt ethylene glycol was circulated at -10 °C in the condenser. The bottom of the condenser was attached to an ace three-neck round-bottom flask that act as a collector of the condensate. The flask was kept at 0 °C in an iced box. One side arm of the flask was closed while the other side arm was fitted with silicon tubing to act as a gas exhaust. The tubing was left in the Schott bottle containing distilled water to check any product loss. The condensate was collected at the end of each gas stripping cycle. The product yield, titer and productivity from the fermentation was estimated based on total sugar consumption and products present in the fermentation broth and condensate collected from gas stripping.

2.9. Analysis procedures
All analysis was made on the supernatants of the samples. The optical density of sample was measured at 600 nm using UV/visible spectrophotometer (Hitachi).

2.9.1. High Performance Liquid Chromatography (HPLC). The sugars present in the OPF juice and fermentation broth were analyzed using an organic acid analysis column (Bio-Rad Monitoring Fermentation) and Agilent Technologies 1200 series high performance liquid chromatography (HPLC) equipped with a refractive index detector (RID) at 60 °C. The mobile phase used was 0.0005N H_2SO_4 at 0.8 mL/min.

2.9.2. Gas Chromatography with Flame Ionization Detector (GC-FID). The separation and concentration of acids and solvents were determined by GC-FID Agilent Technologies 6890N series an Agilent 7693 series autosampler. An internal standard method was used in which 10 g/L methanol was mixed at 1 to 1 ratio with the sample's supernatant liquid [9]. Sample was then auto injected into a capillary column HP INNOWax (30 m × 0.25 mm id × 0.25 μm). Helium was fed constantly at 2.5 mL/min as a carrier gas.

3. Results and Discussion

3.1. Extraction and Characterization of OPF Juice
From 30 kg fresh OPF pressed using sugarcane press machine, 50 wt. % of OPF juice was obtained. pH of untreated OPF juice was around 4.60 to 4.70. Sugars composition in OPF juice was presented in Table 2. Glucose concentration was around 48.67 ±2.25 g/L while sucrose and fructose were 18.12 ± 0.85 and 10.91 ± 0.15 g/L, respectively. Thus, average total amount of sugars found in OPF juice was 76.21 ±

2.25 g/L. The highest butanol yield from butyrate occurs in a medium rich in glucose, thus implied that OPF juice has good impression to generate high solvent yield [11].

Table 2. Comparison of the sugar content analysis in OPF juice.

Glucose (g/L)	Sucrose (g/L)	Fructose (g/L)	Total sugar (g/L)	References
48.67	18.12	9.42	76.21	This study
53.95	20.46	1.68	76.09	[12]
40.50	NA	NA	40.50	[13]

Table 3 showed the elemental constituents (carbon, nitrogen, hydrogen, sulphur), nutrient and heavy metals content in OPF and OPF juice. Based on the results obtained by CNHS analyzer and ICP, OPF and OPF juice contained high carbon which is 46.2% and 41.52%, respectively and suitable C/N ratio >15. Several compounds found in OPF were similar to P2 medium components, for examples, potassium (K), phosphate (P), calcium (Ca), magnesium (Mg), manganese (Mn), iron (Fe), zinc (Zn), aluminium (Al) and sodium (Na). This indicated that OPF juice is suitable to be used as a renewable carbon source for production of various bioproducts.

Table 3. Nutrients and metallic elements compositions.

Composition	OPF juice	OPF juice[a]	OPF juice[b]
C (%)	41.52	39	0.023
N (%)	1.47	0.8	0.0097
C/N	28.24	50	–
S (%)	0.66	0.4	ND
P (%)	NM	0.02	0.119
K (%)	0.35	2.3	19.20
Ca (%)	0.043	2.9	12.17
Mg (%)	0.025	0.5	2.90
B (ppm)	15.91	2	3.80
Mn (ppm)	7.27	2	22.60
Cu (ppm)	0.45	2	0.30
Fe (ppm)	12.27	66	73.25
Zn (ppm)	ND	9	0.50
Al (ppm)	10.28	NM	NM
Mg (ppm)	246.6	NM	NM
Na (ppm)	ND	NM	NM

*Data was the mean of duplicate samples
*NM = Not measured
*ND = Not detected
[a] [12]
[b] [14]

3.2. Optimization of biobutanol production in 1 L batch fermentation

Inoculum size (A), temperature (B) and agitation speed (C) were studied in this optimization process with biobutanol concentration (Y) as the response. Five levels of numeric factor variation were used. The five levels consisted of plus and minus alpha (axial point), plus and minus 1 (factorial design) and the centre point. By using Design Expert 7.1.6 software, 20 sets of experiments were obtained and quadratic model was proposed. Full central composite design for optimization process with three of its variables, and both responses; actual response obtained from experimental works are presented in table 4.

Table 4. CCD results for batch ABE fermentation in 1 L Schott bottle.

Standard Order	Factors						Response
	Inoculum size, A		Temperature, B		Agitation speed, C		Butanol yield, Y (g/g)
	Coded value	Actual value (%)	Coded value	Actual value (°C)	Coded value	Actual value (rpm)	Actual value
1	-1	15	-1	32	-1	50	0.189
2	+1	25	-1	32	-1	50	0.180
3	-1	15	+1	42	-1	50	0.169
4	+1	25	+1	42	-1	50	0.177
5	-1	15	-1	32	+1	150	0.141
6	+1	25	-1	32	+1	150	0.159
7	-1	15	+1	42	+1	150	0.075
8	+1	25	+1	42	+1	150	0.084
9	-2	10	0	37	0	100	0.150
10	+2	30	0	37	0	100	0.158
11	0	20	-2	27	0	100	0.093
12	0	20	+2	47	0	100	0.065
13	0	20	0	37	-2	0	0.169
14	0	20	0	37	+2	200	0.050
15	0	20	0	37	0	100	0.171
16	0	20	0	37	0	100	0.169
17	0	20	0	37	0	100	0.158
18	0	20	0	37	0	100	0.160
19	0	20	0	37	0	100	0.150
20	0	20	0	37	0	100	0.190

It was apparent that an increased in inoculum size from 10% (v/v) to 30% (v/v) resulted in an increased of biobutanol production. Similar trend was found which showed that an increase in inoculum size from 5% to 15% (v/v) elevated biobutanol production [15]. This previous study reported an optimum inoculum size of 16.2% (v/v) using oil palm decanter cake hydrolysate as the substrate. Increasing the inoculum size shorten the lag phase of the clostridial cell growth. This improves clostridial growth by triggering the solventogenic phase faster and increased biobutanol production [16]. However, further increased in inoculum size had no considerable effects on lag phase of microbial growth and cell activity which consequently resulted in the production of biobutanol in concentration close to best inoculum size [17]. The optimum inoculum size was varied depending upon the type of microorganism and substrate used. In this study, the best inoculum size for the production of biobutanol from OPF juice using *Clostridium acetobutylicum* ATCC 824 was 20% (v/v).

As the agitation speed increased from 0 to 200 rpm at an interval of 50 rpm, the biobutanol yield decreased. The maximum biobutanol yield (0.19 g/g-sugar) was observed at agitation speed of 50 rpm. The present finding was equal to the previous study where an increase in agitation speed subsequently reduced the ability of the bacteria to breakdown the glucose or carbohydrate uptake and thus affecting the production of biobutanol [18]. In this study, the best condition for biobutanol production is at 50 rpm agitation speed with highest biobutanol reached up to 0.19 g/L.

The optimum temperature for fermentation of *Clostridium acetobutylicum* strains strongly depends not only on the type of strain, but strongly influenced by various factors such as, medium, substrate, pH and others. The reported optimal growth temperature for most of Clostridia was between 30 and 37 °C. *Clostridium acetobutylicum* have lost the ability to continue the acidogenesis and solventogenesis stage at high temperature [18]. Clostridia grow optimally between 20 and 45 °C. Beyond this range resulted in low efficiency on producing acids as well as solvents. It was observed that the biobutanol yield was

IOP Conf. Series: Materials Science and Engineering **702** (2019) 012004 doi:10.1088/1757-899X/702/1/012004

low outside the temperature range (27 and 47 °C) with yield of 0.06 g/g-sugar and 0.09 g/g-sugar respectively than the optimum temperature of 37 °C which is 0.19 g/g-sugar yield.

3.3. Analysis of variance (ANOVA)

From the ANOVA data in table 5, the Model F-value of 10.37 and Prob>F value less than 0.0500 indicated that the model terms were significant. There was only a 0.05% chance the "Model F-Value" this large occurred due to the noise. In this case, B, C, BC, B^2 and C^2 were significant model terms. While, the model term of A, AB, AC and A^2 were not significant, as their p-value more than 0.05. Values greater than 0.1000 indicated the model terms were not significant. The "Lack of Fit p>F " of 0.1643 implied the Lack of Fit is not significant relative to the pure error. There is a 16.43% chance that a "Lack of Fit p>F" this large could occurred due to the noise.

R^2 obtained was 0.9032 (90.32%). Thus, there was only 0.10% of variability response in this study. The presented R^2 value, therefore, implied a very good fit between experimental and predicted values. Adj R-Squared was 0.8161. For a second-order model to be a good predictive model, the adjusted R-square ≥ 0.8 (Myers et al., 2009). The Pred R-Squared of 0.4066 is not as close to the Adj R-Squared as one might normally expect. Adequate precision ratio of 9.343 for this model indicated an adequate signal, which supported the fitness of the model.

Table 5. Analysis of variance (ANOVA).

Source	Sum of squares	Degree of freedom	Mean square	F-value	Prob>F	
Model	0.033	9	3.652E-003	10.37	0.0005	significant
A	1.056E-004	1	1.056E-004	0.30	0.5960	
B	3.011E-003	1	3.011E-003	8.55	0.0152	
C	0.015	1	0.015	43.65	<0.0001	
AB	7.411E-006	1	7.411E-006	0.021	0.8875	
AC	1.044E-004	1	1.044E-004	0.30	0.5980	
BC	1.773E-003	1	1.773E-003	5.04	0.0487	
A^2	4.168E-005	1	4.168E-005	0.12	0.7379	
B^2	0.010	1	0.010	28.85	0.0003	
C^2	3.960E-003	1	3.960E-003	11.25	0.0073	
Residual	3.521E-003	10	3.521E-004			
Lack of Fit	2.528E-003	5	5.055E-004	2.54	0.1643	Not significant
Pure Error	9.935E-004	5	1.987E-004			
Cor Total	0.036	19				
Std. Dev.	0.019		R-Squared		0.9032	
Mean	0.14		Adj R-Squared		0.8161	
C.V.%	13.15		Pred R-Squared		0.4066	
PRESS	0.022		Adeq Precision		9.343	

*A = inoculum size
*B = Temperature
*C = Agitation speed

3.4. Fed-batch fermentation kinetics with in situ gas stripping

Fed-batch fermentation was applied to processes where a high substrate concentration is toxic to culture. Since butanol is toxic to *Clostridium* cells, the fed-batch fermentation technique cannot be applied unless one of the feasible product recovery technique was applied for simultaneous separation of product. Thus, in order to overcome both substrate and solvent inhibitions, fed-batch fermentation was carried out with in situ gas stripping. A batch fermentation in 1 L working volume was first conducted for 42 h before gas stripping started, followed with first fed time. The additional fresh OPF juice, yeast extract and P2 medium was fed again after the second gas stripping finished by 72 h. Next after that was just gas stripping parts with no more feeding.

Over this period, a total of 38.1 g/L substrate was consumed in fed-batch fermentation. Normally, only 60% of sugars was used for ABE production [19] and the rest was used for cell growth. The fermentation produced 6.69 g/L of butanol, 3.48 g/L of acetone, and 0.87 g/L of ethanol with total ABE of 11.04 g/L. Gas stripping not only eliminate butanol toxicity but altered the fermentation kinetic as well. Typical ABE fermentations produce butanol and acetone at a 2:1 ratio. However, the butanol/acetone ratio was significantly higher than 2 in this study at the end of 240 h fermentation. Good acid reassimilation has been one of the reasons to this increment in evident by the stable concentration levels of both acetic and butyric acids in the fermentation broth throughout the fed-batch fermentation. In this study, the overall biobutanol and ABE productivities in fed-batch fermentation with in situ gas stripping were significantly higher compared to batch fermentation without gas stripping. Biobutanol and ABE productivities by fed-batch fermentation increased 11% and 17%, respectively. Clearly, gas stripping not only prevented the biobutanol toxicity but increased the fermentation rate and the biobutanol yield as well. Table 6 shows the kinetic data for batch fermentation without gas stripping and fed-batch fermentation with gas stripping in 2 L stirred tank bioreactor.

Table 6. Kinetic data for batch fermentation without gas stripping and fed-batch fermentation with gas stripping in 2 L stirred tank bioreactor.

	Batch without gas stripping	Fed-batch with gas stripping
Substrate (g/L)	50	50
Fermentation time (h)	144	42
Substrate consumed (g/L)	49.0	38.1
Substrate consumption rate (g/L/h)	0.34	0.91
Acetic acid produced (g/L)	2.14	3.37
Butyric acid produced (g/L)	0.72	1.42
Total acid produced (g/L)	2.86	4.40
Total acid yield (g/g)	0.06	0.13
Acetone production (g/L)	3.86	3.48
Butanol production (g/L)	7.83	6.69
Ethanol production (g/L)	1.18	0.87
Total ABE production (g/L)	12.9	11.0
Acetone yield (g/g)	0.08	0.09
Butanol yield (g/g)	0.16	0.18
Ethanol yield (g/g)	0.02	0.02
Total ABE yield (g/g)	0.26	0.29
Biobutanol productivity (g/L/h)	0.05	0.16
ABE productivity (g/L/h)	0.09	0.26

3.5. Gas stripping effects on ABE fermentation

All previous studies on integrated ABE fermentation with gas stripping were conducted at relatively low butanol concentrations of 5 g/L or less to minimize butanol toxicity [20]. Based on butanol production profile from batch fermentation, the butanol production of 5 g/L started at 42 h. Thus, gas stripping was started at 42 h as well to intermittently remove ABE from the fermentation when the butanol concentration was 5 g/L and higher. The stripping process in the whole system took out at four time intervals; 42-48, 66-72, 90-96, and 114-120 h. Acids and solvents production profile for fed-batch fermentation in 2 L stirred tank reactor with in situ gas stripping is presented in figure 2. The shaded area in the graph was the time when gas stripping being applied (First gas stripping: GS 1, second gas stripping: GS 2, third gas stripping: GS 3, fourth gas stripping: GS 4).

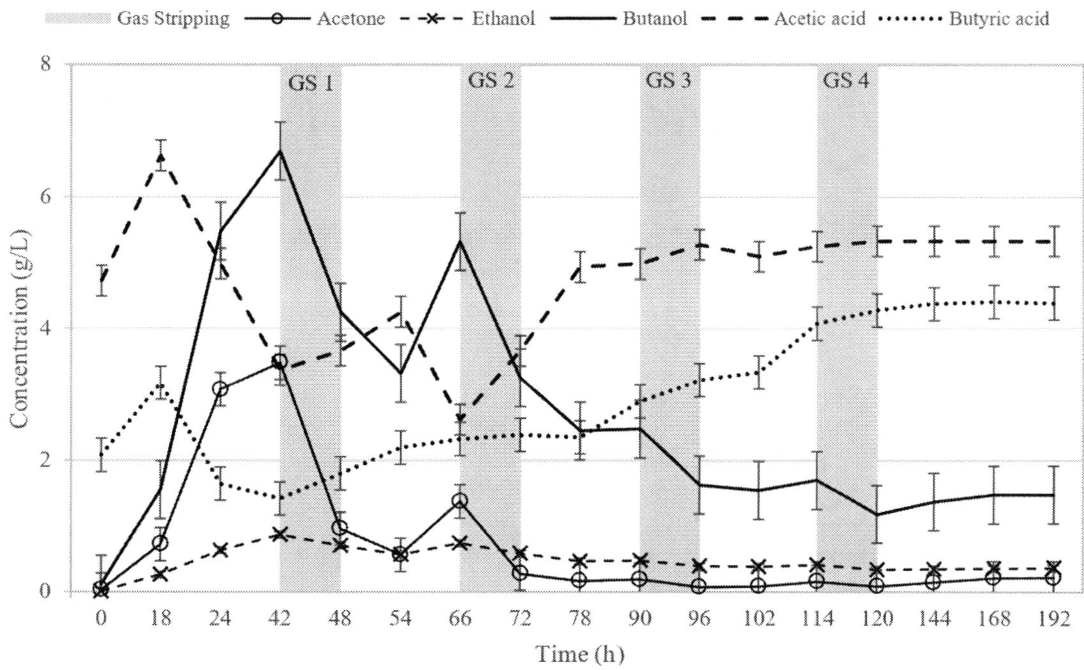

Figure 2. Acids and solvents produced from fed-batch ABE fermentation in 2 L stirred tank bioreactor with in situ gas stripping method.

Based on the graph in figure 2, after which the gas stripping has been initiated, ABE concentrations in the reactor decreased dramatically. This indicates that ABE were successfully stripped out of the reactor. The butanol concentration in the fermentation broth fluctuating between 3 and 5 g/L in the first 72 h, indicating a dynamic equilibrium between solvent production and removal. During first gas stripping, about 10.53 g/L acetone, 135.15 g/L butanol and 10.03 g/L ethanol have been stripped. The stripped butanol and ethanol concentrations was higher in second gas stripping with 136.39 g/L and 14.11 g/L of butanol and ethanol were stripped out, respectively.

There is no further ABE production after 72 h. When the third gas stripping was done, only butanol and ethanol left to be stripped. Total ABE concentration that were successfully stripped and percent recovery is presented in table 5 and Table 6 respectively. The product stripped is displayed in figure 3.

Table 7. ABE concentration that have been stripped from gas stripping.

Gas stripping intervals (h)	Acetone (g/L)	Butanol (g/L)	Ethanol (g/L)
42-48	10.53	135.15	10.03
66-72	8.64	136.39	14.11
90-96	1.25	80.26	8.03
114-120	1.02	42.32	5.75

Table 8. Percent recovery of ABE.

Gas stripping intervals (h)	Acetone (%)	Butanol (%)	Ethanol (%)
42-48	71.5	36.5	18.4
66-72	79.7	38.7	21.6
90-96	57.9	34.0	18.8
114-120	50.0	30.6	17.1

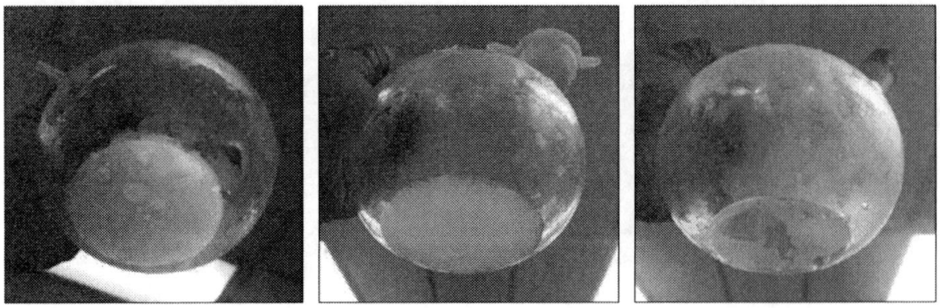

(a) (b) (c)

Figure 3. (a) First strip (b) second strip (c) third strip.

From table 7, the results indicate that gas stripping removed only volatile compounds, which are mainly ABE and did not remove acetic and butyric acid at 37 °C. It was found that the gas stripping process is more selective in separating butanol from water since butanol is the highest concentration in the condensate collected, although butanol has a higher boiling temperature and lower vapor pressure than acetone, ethanol, and water at 37 °C.

The second gas stripping produced higher ABE production and substrate consumption rates, likely due to increased viable cell density in the fed-batch fermentation. However, solvent production decreased after the third gas stripping, indicating decreased in cell viability. It was probably because the dead cells and non-active cells were present with active solvent producing cells in the fermenter [9]. In this experiment, butanol production ceased after extended fermentation hours starting from 72 h and onwards. This finding was matched with previous study where they claimed that there is an inhibitory effect on cell growth and acid production due to the removal of solvents [10]. At 72 h, butanol concentration reached 5.32 g/L. At relatively low butanol concentrations of 5 g/L or less, sporulation occurred and caused culture degeneration [20]. Therefore, to ensure better yield of butanol removal by gas stripping, the butanol production should be maintained above 8 g/L because the gas stripping is more effective in separating butanol from the aqueous solution at higher concentrations. Butanol goes through phase break when the concentration is higher than 8%. Butanol in the condensate boost to 150 g/L when the butanol concentration in the solution was nearly 8 g/L [21].

The product separation results in this study were compared with other methods from previous researches in order to evaluate the performance of gas stripping in recovering the product from the fermentation. Table 9 shows the comparison of several fermentation strategies from other research studies with different modes of fermentation and recovery method used in product recovery. Gas stripping was found to be the most practical process for biobutanol recovery.

Table 9. ABE concentration obtained from gas stripping.

Substrate concentration (g/L)	Fermentation strategy	Concentration (g/L)			References
		Acetone	Butanol	Ethanol	
150 Glucose	Continuous, integrated with pervaporation	5.1	152.0	1.1	[22]
100 Glycerol	Batch, in situ membrane distillation	N.A	29.8	N.A	[23]
86.2 Glucose	Fed-batch, ex situ adsorption	6.7	18.2	0.9	[24]
600 Glucose	Fed-batch, in situ gas Stripping	58	703.4	14	[25]
50 OPF juice	Fed-batch, in situ gas stripping	10.53	135.15	10.03	(This study)

IOP Conf. Series: Materials Science and Engineering **702** (2019) 012004 doi:10.1088/1757-899X/702/1/012004

4. Conclusions

OPF juice can be used as an alternative carbon source for biobutanol production at bioreactor scale. The findings have suggested the optimum operating parameters which leading to high yield and productivity of biobutanol, at 20% inoculum size, 37 °C fermentation temperature and 50 rpm agitation speed. It is evident that the overall biobutanol and ABE productivities in fed-batch fermentation with in situ gas stripping were significantly higher compared to batch fermentation without gas stripping. Biobutanol and ABE productivities by fed-batch fermentation increased 11% and 17%, respectively. Clearly, gas stripping not only prevented the biobutanol toxicity but increased the fermentation rate and the biobutanol yield as well.

References

[1] N.d. Retrieved from https://www.iea.org/newsroom/news/2011/april/biofuels-can-provide-up-to-27-of-world-transportation-fuel-by-2050-iea-report-.html.
[2] Dürre P 2008 *Ann. N. Y. Acad. Sci.* **1125** 1 353-62
[3] Gapes J R 2000 *J. Mol. Microbiol. Biotechnol.* **2** 1 27-32
[4] Green E 2011 *Curr. Opin. Biotechnol.* **22** 3 337-43
[5] Jin C, Yao M, Liu H, Lee C and Ji J 2011 *Renew. Sust. Energ. Rev.* **15** 8 4080-106
[6] Nasrah M N S, Zahari M A K M and Masngut N 2016 *Proc. of The National Conf. for Postgraduate Research (NCON-PGR 2016)* (Universiti Malaysia Pahang Pekan) 24-5 September 2016 p 34-8
[7] Ezeji T, Qureshi N and Blaschek H 2007 *Curr. Opin. Biotech.* **18** 3 220-27
[8] Omar R, Idris A, Yunus R, Khalid K and Aida Isma M I 2011 *Fuel* **90** 1536–1544
[9] Qureshi N and Blaschek H 1999 *Biotechnol. Prog.* **15** 4 594-602
[10] Masngut N 2013 *A Study of Butanol Production in a Batch Oscillatory Baffled Bioreactor (Doctor of Philosophy)* (Newcastle University)
[11] Tashiro Y, Shinto H, Hayashi M, Baba S, Kobayashi G and Sonomoto K 2007 *J. Biosci. Bioeng.* **104** 3 238-40
[12] Zahari M A K M, Zakaria M R, Ariffin H, Salihon J, Shirai Y and Hassan M A 2012 *Bioresource Technol.* **110** 556-71
[13] Aliyu A S, Aziz A A, Yahya A and Lattiff Z A 2015 *J. Teknol.* **74** 10
[14] Abdullah S S S, Shirai Y, Shirai Y, Bahrin E K and Hassan M A 2015 *Ind. Crop. Prod.* **63** 357-61
[15] Razak M A, Ibrahim M, Yee P, Hassan M and Abd-Aziz S 2013 *Bioresources* **8** 2
[16] Shukor H, Al-Shorgani N, Abdeshahian P, Hamid A, Anuar N, Rahman N and Kalil M 2014 *Bioresource Technol.* **170** 565-73
[17] Ranjan A, Mayank R and Moholkar V 201 *Biomass Convers. Bior.* **3** 2 143-55
[18] Khamaiseh E, Hamid A, Yusoff W and Kalil M 2013 *Pak. J. Biol. Sci.* **16** 20 1145-51
[19] Zhang W, Liu Z, Liu Z and Li F 2012 *Lett. Appl. Microbiol.* **55** 3 240-46
[20] Kashket E R and Cao Z Y 1993 *Appl. Environ. Microbiol* **59** 4198-202
[21] Ezeji T, Qureshi N and Blaschek H 2004 *Appl. Microbiol. Biot.* **63** 6 653-58
[22] Rohani A S, Mehrani P and Thibault J 2015 *Can. J. Chem. Eng.* **93** 6 986-97
[23] Lin X, Fan J, Wen Q, Li R, Jin X and Wu J 2013 *J. Chromatogr. Sci.* **52** 3 264-70
[24] Lee S, Eom M, Choi J, Kim S, Kim J, Shin Y and Kim K H 2016 *Bioproc. Biosyst. Eng.* **39** 695-702
[25] Chen Y, Ren H, Liu D, Zhao T, Shi X, Cheng H and Ying H 2014 *Bioresource Technol.* **164** 276-84

IOP Conf. Series: Materials Science and Engineering **702** (2019) 012005 doi:10.1088/1757-899X/702/1/012005

Co-gasification of coal and empty fruit bunch in an entrained flow gasifier: A process simulation study

M R Deraman, R Abdul Rasid*, M R Othman and L N M Suli

Faculty of Chemical and Process Engineering Technology, College of Engineering Technology, Universiti Malaysia Pahang, Lebuhraya Tun Razak, 26300 Gambang, Kuantan, Pahang, Malaysia.

*E-mail: ruwaida@ump.edu.my

Abstract. Co-gasification of coal and biomass is a proven method to improve gasification performance and a platform towards being independent of fossil fuel in power generation. Thus, this research was conducted to assess the feasibility of coal and EFB as fuels in co-gasification using Aspen HYSYS. A sequential model of an EFG was developed to predict the syngas composition and the optimum operating condition of the gasifier. The process was modelled with a set of five reactors to simulate various reaction zones of EFG in accordance with its hydrodynamics. The model considers devolatilization, char and volatile combustion, char gasification and water-gas shift reactions. The model prediction has exhibited excellent agreement with the experimental results. Three parameters of BR, Top and S/F were considered to account for their impacts on syngas composition in the process. The CE, CGE, PE and HHV were adopted as the indicators of process performance. The optimal values of BR, Top and S/F were 50%, 950°C and 0.75, respectively. While the value of CE reached above 90% and the maximum value of CGE, PE and HHV was obtained. This finding should be helpful in designing, operating, optimizing and controlling any co-gasification process especially in the entrained flow system.

1. Introduction

The use of coal as energy source for electricity has been practiced in centuries [1]. As recorded in 2017 alone, the prevalence of coal utilization is greater than renewable sources in total, owing to the fact that it produces higher magnitude of combustion energy [2]. Unfortunately, not only that burning coal as fuel will contribute to greenhouse gas emission and global warming, a far more important concern is that coal supply is depleting rapidly [3]. The International Energy Agency (IEA) came out with an estimated of 860 billion tons of coal available for a 100-years supply of energy before its resources finally drained [4]. In light of the issue, key players worldwide have been researching and starting a massive shift towards alternative long-lasting energy sources for heat and power generation [5].

Ever since biomass waste becomes a hassle to countries that depends on agriculture as the main economical source, institutions begin to engineer a proper beneficial way to utilize these left-overs as a new source of energy. For agricultural waste such as the palm oil empty fruit bunch (EFB), gasification is appropriate to convert the carbonaceous materials into fuel gas with sufficient thermochemical properties [6]. Gasification enables production of product gases that mainly consist of CO and H_2 from carbon solid fuels in a very efficient way that is both low-cost and safer for the environment [7]. Being heterogeneous, burning biomass directly could increase ash content and cause fouling [8]. Concerning the syngas quality, co-gasification of biomass and coal could improve product quality, ensure higher efficiency conversion, optimize logistics, and also economically feasible [9]. Using coal to supplement EFB gasification enhances the reactivity of the biowaste material and produce cleaner syngas [10]. Coal-biomass blending is also attractive because the extra hydrogen in the biomass increase H_2 production, while the additional oxygen content lowers activation energy that improves the rate of reaction [11]. Coal deficiency could be delayed, and the seasonal shortage of biomass supply could also be prevented [12]. Most importantly, co-gasification makes utilization of EFB possible, and solves the issue of excessive landfills in agricultural countries, as well as adjourning fossil fuel depletion [13]. Generally, there are three types of gasification reactors that are commercially available including fixed bed gasifiers, fluidized bed gasifier and entrain flow gasifier (EFG) [14-15]. The pros and cons of each type of gasifiers are illustrated in detail in figure 1. Meanwhile, co-gasification is effectively performed using a gasifier which enables conversion of coal and biomass into syngas to be carried out at high temperature [16]. The EFG is often considered favorable for co-gasification of coal and biomass at 1500°C [17]. The EFG enables high heating rate, as well as extreme operating conditions [18]. In terms of large-scale operations, EFG could ensure generation of high quality syngas and efficiency [19]. Interaction between coal and the biomass is a concern, either it may or may not create synergism which will benefit the process or, otherwise [20]. Therefore, the compatibility of blended fuels with respect to their gasification performance must be properly evaluated.

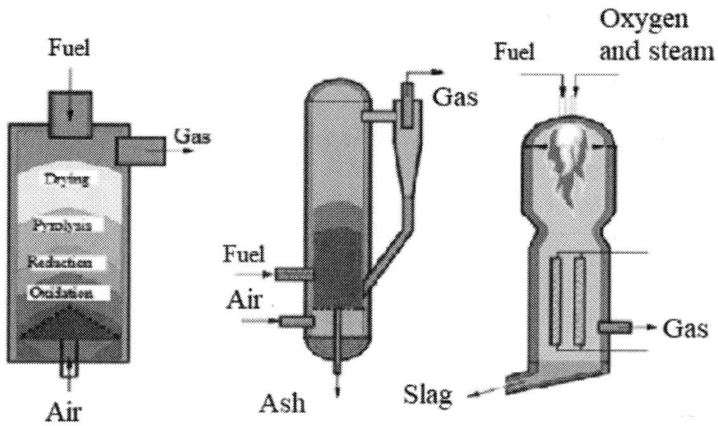

Properties	Fixed Bed Gasifier	Fluidized Bed Gasifier	Entrained Flow Gasifier
Fuel flexibility	No	Yes	Yes
Uniform temperature	No	Yes	Yes
Solid-gas contact	Good	Good	Good
Tar content	High	Low	Low
Resident time	Hours	Minutes	Seconds
Ability to scale-up	Limited	Yes	Yes

Figure 1. Pros and cons of different types of gasification technology (adopted from [14, 21]).

Process simulation and modelling is the most sensible way of testing and validating the compatibility of fuels and to study the thermodynamics of reactions without exhausting time and cost for trial and error in the real system [22]. The accuracy and precision of simulation is very reliable, of which engineers use to predict the direction of a process while maintaining the current operation in the plant and avoid unscheduled disturbances [23-24]. The key element of simulation is the mathematical model that is the representation of the actual process and is used to analyze effects of various operating parameters on process behavior based on the conservation laws of mass, energy and momentum [22, 25]. The complexity of models can range from complex three-dimensional models that take fluid dynamics (CFD

IOP Conf. Series: Materials Science and Engineering **702** (2019) 012005 doi:10.1088/1757-899X/702/1/012005

models) and chemical reactions into consideration to simpler models where mass and energy balances are considered over the entire gasifier to predict the syngas composition [9].

Simulation of co-gasification that comprises multiple reactions in an entrained flow gasifier usually uses Gibbs free energy minimization method [26-27]. This approach is independent of reactor dimension and suitable to study the influence of operating parameters on syngas composition as it reflects the equilibrium model which is suitable for EFG reactors. Although, the model may deviate from the actual data at lower process temperatures, where CO and H_2 are typically overestimated while CO_2, CH_4, char and tar are underestimated [28]. Previously, Kong and her co-workers utilized Aspen Plus to develop a three-stage equilibrium model for coal gasification in the Texaco entrained flow gasifier by taking into account the pyrolysis and combustion reaction, solid-gas reaction, and gas-gas reaction [26]. Meanwhile, Adeyemi and Janajreh [23] approves the competency of Aspen Plus for simulation of coal and biomass co-gasification, including drying, de-volatilization, volatiles combustion and char gasification in the mathematical model [23]. The model considered the fluid dynamic of an entrained flow reactor which uses the kinetic approach in simulating the processes where different experimental data is required. Aspen HYSYS was used by Bassyouni et al. [24] to simulate biomass gasification with the equilibrium approach [24]. The limitation of it model is that the drying process was neglected by considering the biomass as dry-ash-free (daf)'s fuel, thereby only applicable for a downdraft gasification only. Considering all of the three approaches taken by these authors, it is necessary to developed an equilibrium model which simulates co-gasification of coal and biomass in an entrained flow gasifier that includes all process reaction stages and the hydrodynamic of the reactor.

In this work, Aspen HYSYS was used to model and simulate co-gasification of Illinois #6 coal and EFB in an entrained flow gasifier. The co-gasification process was modeled with a set of five reactors to simulate various reaction zones (including de-volatilization, combustion and gasification) of the entrained flow gasifier in accordance with its hydrodynamics. The effects of fuel blending (B/F), operating temperature (T_{op}) and steam-to-fuel ratio (S/F) on syngas composition was studied. The process performance was analyzed using the carbon gasification efficiency (CE), cold gas efficiency (CGE), syngas product efficiency (PE) and the heating value of syngas product (HHV) values.

2. Methodology

2.1. Model assumptions

A multi-zonal steady state model has been developed to simulate co-gasification of coal and biomass using Aspen HYSYS. The unit operations were arranged in accordance with hydrodynamics of entrained flow gasifier. The assumptions made in the present model are summarized in the following [23, 24, 29]:

1) Steady state isothermal process.
2) Feedstock is assumed to contain only carbon, hydrogen and oxygen.
3) Dry-ash-free feedstock are used, so the drying process is assumed to be instantaneous.
4) Uniformly sized particle with sphericity equal to one and did not affect the reaction.
5) Model eliminates NH_3 and H_2S formation and ignores tar formation to simplify the process.

The following figure 2 shows the schematic diagram of Texaco entrained flow coal-gasifier adopted from literature [30].

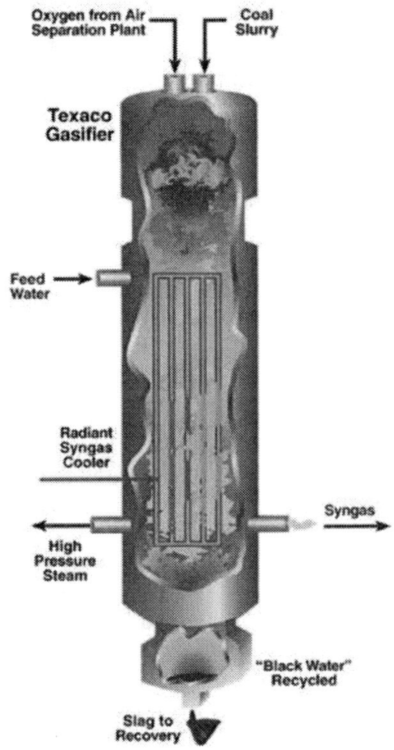

Figure 2. Schematic of Texaco entrained flow coal-gasifier (adopted from [30]).

2.2. *Simulation basis manager*

Simulation basis manager (SBM) is an interface provided in Aspen HYSYS to assists users in defining and choosing pure components, appointing a property package to execute flash and physical properties calculations, and specifying reactions which can be fixed into any unit operations during simulation process [24]. Aspen HYSYS does not have Illinois #6 coal and EFB as library component, thus they were modelled as hypothetical components using the feedstock's analysis in table 1. Peng-Robinson equation of state (EOS) was preferred as property package to estimate the physical properties of components, and has been reported to well estimate the physical properties in an IGCC power plant simulation [24, 29, 31, 32].

2.3. *Model descriptions*

The co-gasification of coal and biomass in an entrained flow gasifier was simulated in three zones that include devolatilization process, combustion process (char and volatiles), and gasification process (char). The feedstock used were Illinois #6 coal and EFB, and their proximate and ultimate analysis are shown in table 1. The gasifying agents were oxygen and steam. The operational condition of reactor is as follows: 0.5 of biomass ratio; 0.3 of equivalent ratio; 0.75 steam/fuel ratio; 950°C; and 24 atm. The simulation was carried out using Aspen HYSYS, and the process flow diagram of the simulation, as well as a detailed description of main unit operation models is shown in figure 3.

Table 1. Ultimate and proximate analysis of empty fruit bunch and American coal, Illinois #6.

Feedstock Analysis	Empty Fruit Bunch	Illinois #6 Coal
Proximate Analysis (wt.%)		
Moisture (wet basis)	55.600	-
Moisture (dry basis)	5.180	-
Volatile matter	82.580	24.460
Fixed carbon	8.970	60.010
Ash	3.450	15.500
Ultimate Analysis (wt.% dry basis)		
C	46.620	74.100
H	6.450	6.210
N	1.210	1.100
O	45.660	1.320
S	0.035	1.770
Molecular Formula	$CH_{1.649}N_{0.022}O_{0.735}S_{0.000}$	$CH_{0.999}N_{0.013}O_{0.013}S_{0.009}$

The model was based on the following global steam-oxygen co-gasification reaction [33]:

$$C_xH_yO_z + aH_2O + bO_2 \rightarrow \alpha_1 H_2 + \alpha_2 CO + \alpha_3 H_2O + \alpha_4 CO_2 + \alpha_5 CH_4 \tag{1}$$

where *x*, *y* and *z* are the molar fraction of carbon, hydrogen and oxygen in the dry-ash-free (daf)'s fuel, which were determined from the ultimate analysis data (table 1), *a* and *b* are the molar flow rate of steam and oxygen fed to the reactor per mole of processed dry fuel. On the right-hand side of equation 1, α_i coefficients are the unknown moles of gaseous species leaving the gasifier (i.e., i = H_2, CO, CO_2, CH_4 and H_2O) for mole of processed coal/biomass.

2.3.1. Devolatilization. As moisture has been removed from the fuel mass, subsequent heating may create volatiles from a process known as devolatilization [23, 34]. Devolatilization is a rather rapid process [35] when initiated at 400°C to 600°C in a reactor without the presence of oxygen or air [18]. In general, the dried fuel is converted into a denser solid known as char, along with CO, CO_2, H_2, H_2O and CH_4 [36]. Thus, for entrained flow gasifier that usually uses temperatures higher than 1000°C, devolatilization is expected to be spontaneous [18, 23] and was modelled with Conversion reactor in Aspen HYSYS. The devolatilization process of Illinois #6 and EFB is represented by the following equation [18].

$$C_xH_yO_z \rightarrow \alpha_1 Char + \alpha_2 CH_4 + \alpha_3 H_2 + \alpha_4 CO + \alpha_5 CO_2 \tag{2}$$

where *x*, *y* and *z* are the molar fraction of carbon, hydrogen and oxygen, while α_n is the number of moles of the species involved in the reaction.

2.3.2. Combustion reactions. The char and volatiles then reacts with oxygen that forms H_2O, CO_2, and CO as the product [23]. However, only CO, H_2 and CH_4 are involved in the volatiles combustion [23], [26]. The char and volatiles undergo a separate combustion reaction, but oxygen is fully utilized to generate sufficient heat from the exothermic reaction in each of the combustion chambers [26]. These char and volatiles combustion reactions were modeled in Aspen HYSYS using Equilibrium and Gibbs reactors, labeled as Char Combustion and Volatiles Combustion, respectively.

Volatiles combustion reactions:

$$H_2 + 0.5O_2 \rightarrow H_2O\ (\Delta H = -242\ kJ/kmol) \tag{3}$$

$$CO + 0.5O_2 \rightarrow CO_2\ (\Delta H = -283\ kJ/kmol) \tag{4}$$

$$CH_4 + O_2 \rightarrow H_2O + CO_2\ (\Delta H = -242\ kJ/kmol) \tag{5}$$

Char combustion reactions:

$$C + 0.5O_2 \rightarrow CO \; (\Delta H = -111 \; kJ/kmol) \tag{6}$$

$$C + O_2 \rightarrow CO_2 \; (\Delta H = -394 \; kJ/kmol) \tag{7}$$

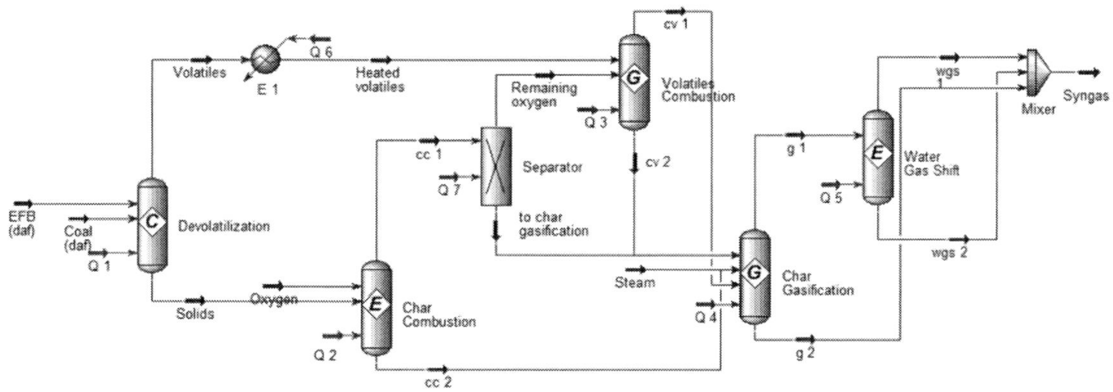

No.	Reaction Zones	HYSYS Reactor/Block ID	Model Description
1	Devolatilization	Conversion/Devolatilization Process	Models the thermochemical conversion of solid fuels into volatiles and char.
2	Combustion	Gibbs/Volatiles Combustion	Models the combustion of volatiles using Gibbs free energy minimization approach.
		Equilibrium/Char Combustion	Models the combustion reaction of char using equilibrium data.
3	Gasification	Gibbs/Char Gasification	Models the main gasification reactions using Gibb free energy minimization approach.
		Equilibrium/Water Gas Shift Reactor	Models water-gas shift and steam-methane reactions using equilibrium data.

Figure 3. Aspen HYSYS flowsheet used for modelling and simulation of co-gasification of coal and biomass in an entrained flow gasifier and its details description of main unit operation models.

2.3.3. Gasification reactions. Carbon (in the form of graphite) from coal/EFB involves in multiple solid-gas reactions that include the carbon-steam, Boudouard, and carbon hydrogenation reactions [26]. Despite the complexity of mixed reactions occurring simultaneously during gasification, the system may reach the state of equilibrium [26, 37]. The char gasification is mainly consisting of reactions with H_2O, CO_2, and H_2 to produce syngas that consists mainly of CO and H_2 [38].
Char gasification reactions:

$$C + H_2O \rightarrow CO + H_2 \; (\Delta H = +131 \; kJ/kmol) \tag{8}$$

$$C + CO_2 \rightarrow 2CO \; (\Delta H = +172 \; kJ/kmol) \tag{9}$$

$$C + 2H_2 \rightarrow CH_4 \; (\Delta H = -75 \; kJ/kmol) \tag{10}$$

Water-gas shift and steam-methane reforming:

$$CO + H_2O \leftrightarrow CO_2 + H_2 \; (\Delta H = -41 \; kJ/kmol) \tag{11}$$

$$CO + 3H_2 \leftrightarrow CH_4 + H_2O \; (\Delta H = +206 \; kJ/kmol) \tag{12}$$

This stage of the co-gasification process was modeled in Aspen HYSYS using coupled Gibbs and Equilibrium reactors. A Gibbs reactor simulates multi-phase reactions including Boudouard, water-gas and methanation reactions using Gibbs free energy minimization method in equilibrium. Meanwhile, Equilibrium reactor simulates water-gas shift and steam-methane reforming reactions, which completes the gasification process.

2.4. Sensitivity analysis and performance measurement
The optimal value of biomass ratio (BR), operating temperature (T_{op}) and steam-to-fuel ratio (S/F) were determined from the sensitivity analysis of results. The biomass ratio defines the ratio of EFB feed rate to the total feed rate (coal + EFB) on carbon basis [39]–[41].

$$BR = \frac{EFB\ feed\ rate\ (mol_{carbon}/hr)}{Total\ feed\ rate\ (mol_{carbon}/hr)} \tag{13}$$

Steam-to-fuel ratio, denoted by S/F, defines the ratio of the mass flow rate of steam to the total feed rate.

$$S/F = \frac{The\ mass\ flow\ rate\ of\ steam\ (kg/hr)}{The\ mass\ flow\ rate\ of\ fuel\ (kg/hr)} \tag{14}$$

Next, performance indicators that include CGE, CE, PE and HHV values of syngas produced by 1 kg feedstock were assessed to evaluate the process characteristic. The CE refers to the percentage of the moles of carbon in syngas to the moles of carbon in solid fuel [40].

$$CE\ (\%) = \frac{moles\ of\ carbon\ in\ syngas}{moles\ of\ carbon\ in\ fuel}\ X\ 100 \tag{15}$$

The CGE refers to the percentage of the HHV value of syngas to the HHV value of feedstock [42-44].

$$CGE\ (\%) = \frac{HHV_{syngas}}{HHV_{feedstock}}\ X\ 100 \tag{16}$$

The HHV of syngas (MJ/Mm^3) and HHV of feedstock (MJ/kg) could be calculated by using the following correlation proposed by [45, 46].

$$HHV\ of\ syngas\ (MJ/Mm^3) = 12.75H_2 + 12.63CO - 39.82CH_4\ /100 \tag{17}$$

$$HHV\ of\ feedstock\ (MJ/kg) = 0.3419x_C\ + 1.1783x_H - 0.1034x_O \tag{18}$$

The PE is the percentage of the moles of CO and H_2 in the syngas to the moles of C and H_2 in the feedstock [47].

$$PE\ (\%) = \frac{n_{CO} + n_{H_2}}{n_{i,C} + n_{i,H_2}}\ X\ 100 \tag{19}$$

where n_{CO} and n_{H_2} are the moles of CO and H_2 in the syngas, respectively. While $n_{i,C}$ and n_{i,H_2} are the mole of C and H_2 in the feedstock.

The percentage of combustible gas in the syngas (n) is defined as the percentage of the moles of CO, H_2 and CH_4 in the syngas to the mole of syngas [47].

$$\text{n}\ (\%) = \frac{n_{CO} + n_{H_2} + n_{CH_4}}{n_{total}}\ X\ 100 \tag{20}$$

where n_{CO} , n_{H_2} and n_{CH_4} are the moles of CO, H_2 and CH_4 in the syngas, respectively. The n_{total} is the mole of syngas.

3. Results and discussion

3.1. Model validation

To evaluate the co-gasification performance, the model was initially tested for coal gasification using Illinois #6 as presented in table 1. Then, the simulation result was validated with practical data from operating a Texaco entrained flow gasifier obtained in the literature, as shown in table 2. Of the total of 9 recorded experimental runs with varying conditions, the simulation results was compared with one experimental set up of 1227°C and 24 atm, with 316.44 kg/hr coal feed; 0.787 O_2/coal ratio; and 0.268 steam/coal ratio [18, 26]. The predicted results were found to satisfactorily compliment the actual data with minor errors. Referring to the percentage of H_2 and CO (which are crucial components that contribute to the calorific value of fuel), when combined, the predicted value is a total of 97.36% against 98.88% of the experiment with a minimum error of 1.74%. The predicted calorific value of 12.34 MJ/m^3 compares reasonably well with the experimental value of 12.54 MJ/m^3 with 1.59% of error.

Table 2. The comparison of syngas composition between simulation results and experimental data of Texaco entrained flow coal gasification.

Syngas composition	Experimental data	Simulation result
Hydrogen (mol.%)	39.76	34.66
Carbon monoxide (mol.%)	59.12	62.70
Carbon dioxide (mol.%)	1.12	2.64
Calorific value (MJ/m^3)	12.54	12.34

3.2. Effect of biomass blend on syngas composition

It is important to understand to what extent the composition of syngas is affected by the amount of EFB used in the biomass/coal blend. Not only that, the results would clarify the intensity of synergy between EFB and Illinois #6 as fuels for co-gasification. The simulation was set to 100 kmol/hr of feedstock at equivalent ratio (ER) of 0.3, S/F of 0.75, operating temperature (T_{op}) of 950°C and operating pressure (P_{op}) of 24 atm. The major chemical reactions occurring in the gasifier are as shown in equations 1 - 11. The results, as illustrated in figure 4 shows that syngas composition was affected in a non-linear manner (especially for H_2) with BR increase. This gives the impression that the properties of EFB compliments that of Illinois #6 which produces a positive synergism as a blend of fuel [41]. Increasing the proportion of EFB in the fuel blend provides higher carbon availability to react with gasifying agents that alleviates product gas yield [41, 48]. However, the production of H_2 started to deplete as BR exceeds 0.5. It appears that beyond 0.5 BR, Boudouard and water-shift reactions are favored, thereby, explains the CO and CH_4 increase and the reduction of H_2/CO ratio in the gas yield. The increase in H_2O could be attributed to higher internal water bond in the EFB [49]. Usually, this can only be achieved from a high temperature reaction at more than 1000°C. However, the simulation analysis suggests that these circumstances are achievable at lower temperature by utilizing co-gasification of coal and EFB. High H_2O in the product gas would not be a matter of concern because water is easily separated and the quality of syngas would not be affected [24]. Other than that, increasing BR did not affect the composition of CO_2 as well as the calorific value of dried syngas. Generally, these results agree with the findings from several literatures that the optimal composition of biomass is 50% of the overall fuel [39, 41].

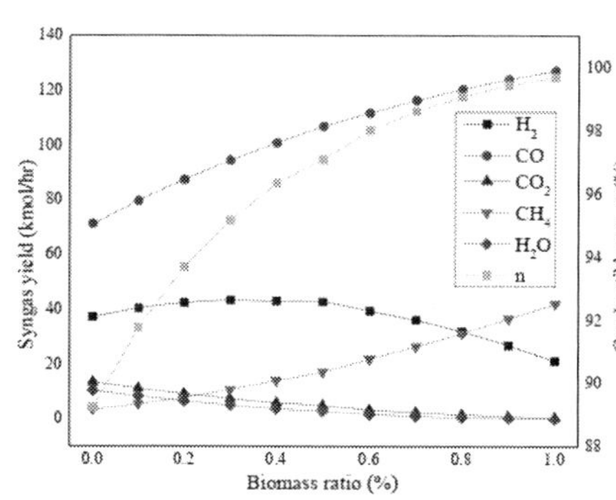

Figure 4. Effect of different biomass ratio on the total product gas yield in co-gasification process.

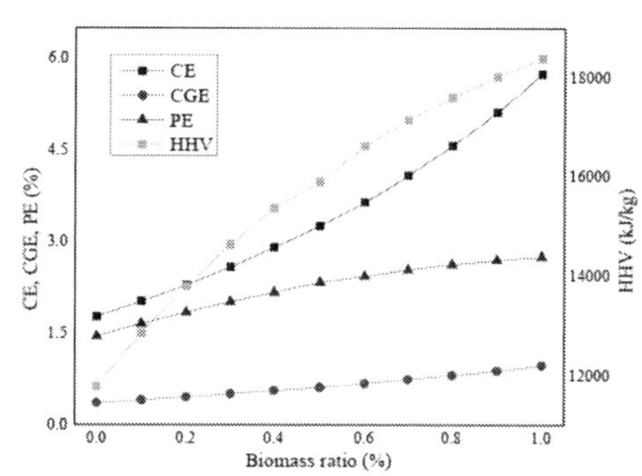

Figure 5. The performance evaluation indexes of the co-gasification of Illinois #6 coal and EFB at different biomass ratio (B/F).

In terms of the performance of the co-gasification of coal and EFB, figure 5 shows remarkable increase of CE, CGE, PE and HHV with the increase of BR (CE: 160% to 580%; CGE: 23.5% to 68.9%; PE: 150% to 170%; and HHV: 12000 kJ/kg to 18250 kJ/kg). Based on figures 2 and 3, it can be concluded that the addition of EFB into coal gasification will improve the performance of gasification. However, considering the BR effect on syngas composition, the optimum value of BR is 0.5. Thus, this finding proves that co-gasification process of coal and biomass is more efficient than gasification of coal alone

3.3. Effect of operating temperature

Gasification of solid fuel to produce syngas involves a few reactions that occur on the same time which makes it rather complicated. Despite the challenge, manipulation of operating temperature could promote a certain type of reaction and decreases the intensity of others. Using the same operating condition (ER, 0.3; S/F, 1.5; P, 24 atm), the temperature was varied from 500°C to 1500°C to study the effect of heat on co-gasification performance and characteristics. figure 6 exhibits the pattern of yield gas that was generated in response to temperature rise. Initially, a notable increase of H_2 denotes that a high temperature is advantageous for high H_2 yield [47]. It is also expected that H_2 increase was also contributed by the primary water gas reaction (PWG) and the methane reforming [47]. Looking at the pattern for CO_2, the gas was high at the lowest temperature setting because PWG and water-gas reaction (WGS) reactions were favored [47]. The amount of CO_2 continued to drop as the temperature was increased, by which BD, reverse WGS and MRF reactions dominated the extreme temperature where CO_2 and H_2 were consumed to produce CO [47, 50]. On the same time, it seems that MRF and the reverse methane formation (rMF) coincided which decreased the production of CH_4 [47]. Thus, it is clear that the percentage of combustible gas in the syngas has a positive effect with increased temperature, and it was above 80% when the temperature was higher than 950°C.

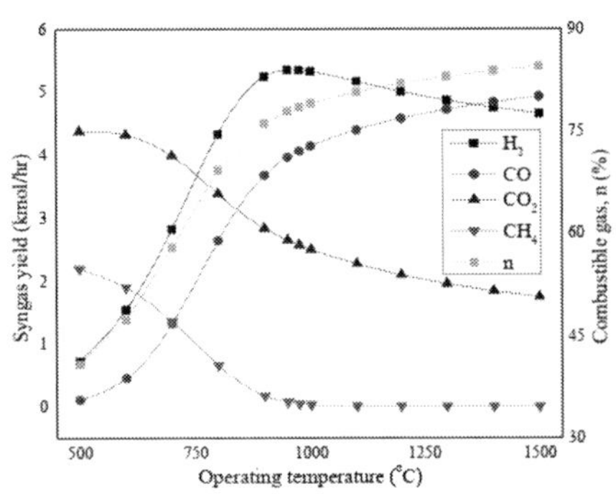

Figure 6. Effect of different operating temperature on syngas composition in co-gasification process.

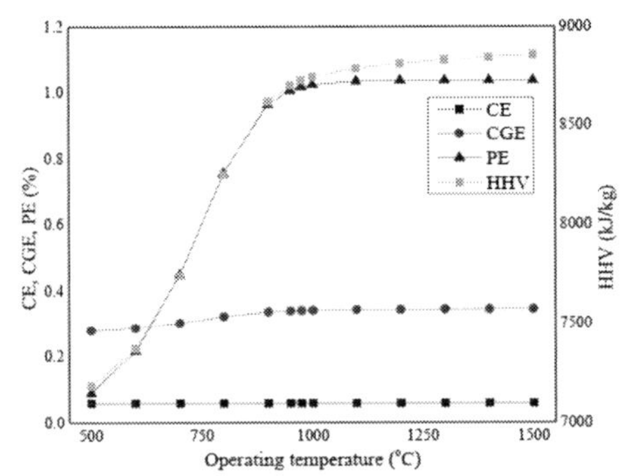

Figure 7. The performance evaluation indexes of the co-gasification at different operating temperature.

Figure 7 exhibits the results of the performance indexes in response to temperature increase. First, CE was found to be constant, which suggest that carbon conversion from coal/EFB was independent of temperature change. CGE increased slightly before it went into constant value of 36% as the temperature reached 950°C. PE increased significantly, also until 950°C when it started to remain constant at 110% as the temperature was raised higher. It is believed that the reason CGE and PE were not affected by temperature beyond 950°C was because of high availability of H_2 and CO [47]. HHV increased in a similar manner as PE where it reached the highest value of 8700 kJ/kg at 950°C. Instead of becoming constant like PE, HHV increased as much as 200 kJ/kg from 950°C to 1500°C which might be related to the value of combustible gas, n in figure 4. It is thus, demonstrated that the most optimum temperature for co-gasification of coal and EFB is 950°C.

3.4. Effect of changing steam/fuel ratio (S/F) on syngas composition

The relevance of S/F is that it could influence the direction of reactions in co-gasification by means of using steam to increase the amount of H_2 and CO [47]. As steam was readily available as the output from tar cracking, utilizing the steam would favor PWG, MRF and WGS reactions. Thus, it is important to study the effect of steam-to-fuel ratio (S/F) on syngas composition. The simulation was set to ER of 0.3, 24 atm, and 950°C, and S/F was varied from 0.0 to 4.0. Supposedly, setting the temperature to 950°C yields more H_2 and CO, but the simulation results in figure 8 exhibit a clear increase of H_2 and

CO_2 instead. It is obvious that H_2 spiked quite rapidly from S/F=0.0 to S/F=0.5 which denotes the significance of steam injection. Seems that the reaction shifts to the right where PWG and WGS took place and generated more H_2 and CO. However, much of the CO were also utilized in WGS which proliferates the amount of H_2 and CO_2. figure 6 also shows that H_2 depletes after S/F=0.5 and this is consistent with the literature [24, 47]. On the other hand, CH_4 exhibits a decreasing trend, almost reach zero when S/F exceeds 0.5. As discussed previously on the subject of figure 6, operating the procedure at 950°C will inhibited the production of CH_4 [47, 51, 52]. Meanwhile, the percentage of combustible gas, n slightly decreased with the increasing of S/F. This is due to the drastic drop in CO component when S/F was increased.

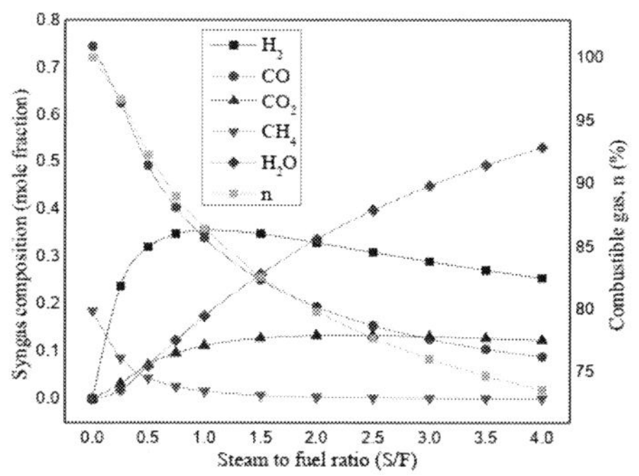

Figure 8. Effect of different steam to fuel ratio (S/F) on syngas composition in co-gasification process.

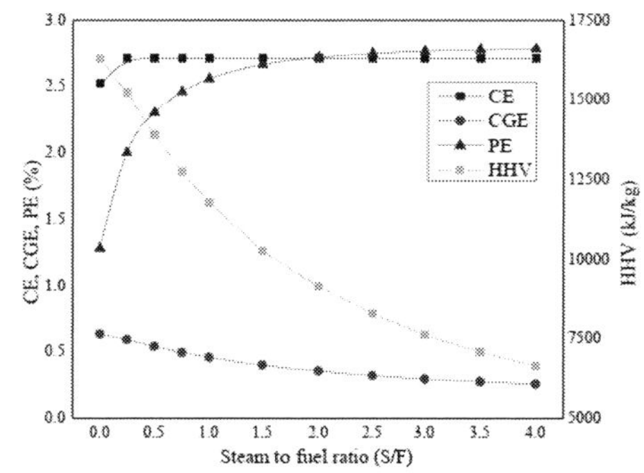

Figure 9. The performance evaluation indexes of the co-gasification process at different steam-to-fuel ratio..

Variation of steam injected into the fuel appears to affect CE and PE in a positive way, but CGE and HHV were affected negatively, as shown in figure 9. The positive trend of CE shows that steam improves carbon conversion, but the increment is minimal. Meanwhile, PE increased steadily until S/F=2.0, then remained constant afterwards. This can be associated to the content of CO and H_2 in the syngas, which means, it depends on the degree of which steam affects carbon conversion of coal/EFB fuel. In accordance to the findings in figure 8, steam injection have increased the interaction of CO and H_2 that caused CO to decrease and H_2 to remain constant when carbon was fully converted, thereby explaining the trend of CE and PE in figure 9 [47]. On the other hand, higher ratio of steam has caused a very apparent reduction of CGE value, which was recorded to decrease from 63.54% to 25.89%. Likewise, HHV also decreased significantly from 17000 kJ/kg to 7000 kJ/kg. The reason may have been caused by the vaporization of extra steam that required additional energy. Generally, these results corroborate the findings of other related studies [23, 24] exhibits the results of the performance indexes in response to temperature increase. First, CE was found to be constant, which suggest that carbon conversion from coal/EFB was independent of temperature change. CGE increased slightly before it went into constant value of 36% as the temperature reached 950 °C. PE increased significantly, also until 950 °C when it started to remain constant at 110% as the temperature was raised higher. It is believed that the reason CGE and PE were not affected by temperature beyond 950 °C was because of high availability of H_2 and CO [47]. HHV increased in a similar manner as PE where it reached the highest value of 8700 kJ/kg at 950 °C. Instead of becoming constant like PE, HHV increased as much as 200 kJ/kg from 950 °C to 1500 °C which might be related to the value of combustible gas, n in figure 6. It is thus, demonstrated that the most optimum temperature for co-gasification of coal and EFB is 950 °C.

Based on the results shown in figures 8 and 9, the highest value of CO and H_2 composition in syngas yield was found at S/F=0.75 which also corresponds to reasonable value of CE, PV, CGE, and HHV. Thus, the optimal S/F value for co-gasification of coal and EFB is 0.75. It is interesting to find that Seo and his co-worker also recorded S/F=0.7 for co-gasification of coal/biomass blend [41]. The value differs slightly due to different operating condition and reactor set up.

4. Conclusions

This study was intended to simulate the synergy of Illinois #6 coal and EFB in co-gasification by manipulating the biomass ratio (BR), operating temperature (T_{op}), and steam-to-fuel ratio (S/F) and evaluate the composition of syngas produced from the reactions. Taken together all strands of findings from this study, it was proven that coal and EFB would be beneficial for mass production of combustible gas. The maximum H_2 yield was achieved with BR of 50%. From the aspect of CE, PE and CGE, the optimal temperature to simulate co-gasification was 950°C. The best ratio of steam to fuel (S/F) is supposedly S/F=0.75.

Acknowledgements
This research was supported by the government of Malaysia under MyBrain15 program. The authors also acknowledge the Faculty of Chemical and Process Engineering Technology, Universiti Malaysia Pahang.

References

[1] Song Y and Wang N 2019 *J. Energy* **168** 1073
[2] Gil M V and Rubiera F 2019 *New Trends in Coal Conversion: Combustion, Gasification, Emissions, and Coking* 117
[3] Singh G K, Rajeev P, Paul D and Gupta T 2018 *Springer* 53
[4] BP 2018 *British Petroluem Statistical Review of World Energy -Electricity generation* Report in June 30 **67**
[5] Islas J, Manzini F, Masera O and Vargas V 2018 *Solid Biomass to Heat and Power*
[6] Vega, F., Alonso-Farinas, B., Baena-Moreno, F.M., Rodríguez, J.A. and Navarrete, B 2019 *New Trend in Coal Conversion: Combustion, Gasification, Emissions and Coking* 141
[7] Beheshti S M, Ghassemi H and Shasavan M R 2015 *Energy Convers. Manag.* **94** 345
[8] de Jong W, Nola G D, Venneker B C H, Spliethoff H and Wójtowicz M A 2007 *Fuel* **86 (15)** 2367
[9] Mikulandrić R, Loncar D, Bohning D and Bohme R 2013 *Biomass gasification process modelling approaches* Proc. of the 13th European Int. Conf. on Polygeneration Strategies Vienna, Austria 1-13
[10] Haykiri-Acma H and Yaman S 2007 *Fuel* **86 (3)** 373
[11] Haykiri-Acma H and Yaman S 2008 *Waste Manag.* **28 (11)** 2077
[12] Yang Z, Wu Y, Zhang Z, Li H, Li X, Egorov R I, Strizhak P A and Gao X 2019 *Renew. Sust. Energ. Rev.* **103** 384
[13] Goldfarb J L and Liu C 2013 *Bioresour. Technol.* **149** 208
[14] Farzad S, Mandegari M A and Görgens J F 2016 *Biofuel Res. J. 3*(**4**), 483-495
[15] Ramos A, Monteiro E, Silva V and Rouboa A 2018 *Renew. Sustain. Energy Rev.* **81** 380
[16] Biagini E, Bardi A, Pannocchia G and Tognotti L 2009 *Ind. Eng. Chem. Res.* **48 (19)** 9028-9033
[17] Valero A and Usón S 2006 *Energy*, **31(10-11)**, 1643-1655
[18] Ali D A, Gadalla M A, Abdelaziz O Y, Hulteberg C P and Ashour F H 2017 *J. Nat. Gas Sci. Eng.* **37** 126
[19] Vascellari M, Roberts D G, Shwe S, Harris D J and Hasse C 2015 *Fuel* **152** 58-73
[20] Moon C, Sung Y, Ahn S, Kim T, Choi G and Kim D 2013 *Exp. Therm. Fluid Sci.* **47** 232
[21] Molino A, Chianese S and Musmarra D 2016 *J. Energy Chem.* 25 (1) 10
[22] Gòmez-Barea A and Leckner B 2010 *Prog. Energ. Combust. Science* **36(4)**, 444-509
[23] Adeyemi I and Janajreh I 2014 *Renew. Energy* **82** 77-84.
[24] Bassyouni M, Hasan S W U, Abdel-Aziz M H, Abdel-hamid S M S, Naveed S, Hussain A and Ani F N 2014 *Energy Convers. Manag.* **88** 693
[25] Mikulandrić R, Lončar D, Böhning D, Böhme R and Beckmann M 2014 *Energ. convers. manage.* **87** 1210-1223

[26] Kong X, Zhong W, Du W and Qian F 2013 *Chinese J. Chem. Eng.* **21 (1)** 79
[27] Kruger M 2014 *Int. J. Thermodyn.* 17(2), 87-95
[28] Puig-Arnavat M, Bruno J C and Coronas A 2010 *Renew. Sustain. Energy Rev.* **14 (9)** 2841
[29] Acevedo J C, Posso F R, Durán J M and Arenas E 2018 *J. Phys. Conf. Ser.* **1126 (1)**
[30] Li S 2010 *Char-SLag Transition During Pulverized Coal Gasification* Thesis University of Utah
[31] Mansaray K G and Ghaly A E 2002 *Energ. Sources* **21 (5)** 453
[32] Pérez-Fortes M, Bojaeski B and Nadal F S 2008 *Comput. Aided Chem. Eng.* **25** 793
[33] Brachi P, Chirone R, Miccio F, Miccio M and Ruoppolo G 2018 *Fuel* **220** 744
[34] Basu P 2018 *Biomass gasification, pyrolysis and torrefaction: practical design and theory* Academic press United Kingdom **3** 479
[35] Hla S S, Arris D J and Roberts D G 2008 *Gasification conversion model : PEFR* CCSD Pullenvale
[36] Svensson H, Tunå P, Hulteberg C and Brandin J 2013 *Fuel* **106** 271
[37] Nguyen T D B, Lim Y I, Song B H, Kim S M, Joo Y J and Ahn D H 2010 *Fuel* **89 (12)** 3901
[38] Sharma M, Attanoor S and Dasappa S 2015 *Fuel Process. Technol.* **138** 435
[39] Kumabe K, Hanaoka T, Fujimoto S, Minowa T and Sakanishi K 2007 *Fuel* **86** 684
[40] Ma X, Zhao X, Gu J and Shi J 2019 *Renew. Sustain. Energy Rev.* **132** 509
[41] Seo M W, Goo J H, Kim S D, Lee S H and Choi Y C 2010 *Energ. Fuel.* **24 (5)** 3108
[42] Ravikiran A, Renganathan T, Pushpavanam S, Voolapalli R K and Cho Y S 2012 *Ind. Eng. Chem. Res.* **51 (4)** 1601
[43] Shabbar S and Janajreh I 2013 *Energy Convers. Manag.* **65** 755
[44] Wang Z, Zhou J, Wang Q, Fan J and Cen K 2006 *Int. J. Hydrogen Energy* **31 (7)** 945
[45] Lide D R 2004 *CRC Handbook of Chemistry and Physics* CRC press **85**
[46] Channiwala S A and Parikh P P 2002 *Fuel* **81** 1051
[47] Duan W, Yu Q, Wang K, Qin Q, Hou L, Yao X and Wu T 2015 *Energy Convers. Manag.* 100 30
[48] Quan C and Gao N 2016 *Biomed Res. Int.* **2016** 1
[49] Lee H V, Hamid S B A and Zain S K 2014 *Sci. World J.* **2014** 1-20
[50] Renganathan T, Yadav M B, Pushpavanam S, Voolapalli R K and Cho Y S 2012 *Chem. Eng. Sci.* **83** 159
[51] Wang X, Wang N, Zhao J and Wang L 2010 *Int. J. Hydrogen Energy* **35 (23)** 12800
[52] Tursun Y, Liu J, Xu S, Wei L and Zou W 2013 *Fuel* **112** 641

Optimization of oil palm empty fruit bunch gasification temperature and steam to biomass ratio using response surface methodology

N H A Halim, S Saleh and N A F A Samad*

Faculty of Chemical & Natural Resources Engineering, Universiti Malaysia Pahang, 26300 Gambang, Pahang, Malaysia

*E-mail: asmafazli@ump.edu.my

Abstract. An experimental work of empty fruit bunch gasification was conducted by using the bubbling fluidized bed to study the effect of the gasification temperature and steam biomass ratio (SBR) on the synthesis gas yield, Lower Heating Value (LHV) and Cold Gas Efficiency (CGE). Response Surface Methodology (RSM) was used to design the gasification experiment from the temperature range of 800 – 1000°C and SBR range of 0.5 – 1.5. Thirteen number of runs were generated based on Central Composite Design (CCD) with five replicated center points. Three regression models for predicting synthesis gas yield, LHV and CGE were developed and Analysis of Variance (ANOVA) was performed in this study. From ANOVA, the most influencing factor was gasification temperature which obtained higher F-value compared to SBR. The numerical optimization was also conducted in order to obtain the optimum condition to maximize the synthesis gas yield, LHV and CGE. From numerical optimization, gasification temperature of 800 °C and SBR of 1.14 were determined as the optimum condition which contributes to the maximum synthesis gas yield, LHV and CGE which are 1.25 Nm^3/kg, 10.49 MJ/Nm^3 and 90.72% respectively. The percentage error between the predicted and actual value of response variables was calculated and the error obtained lesser than 1%. Thus, it confirmed that the models obtained can be used to optimize the gasification of the empty fruit bunch.

1. Introduction

Recent issues on the limitation of fossil fuels and high level of air pollution made renewable energy technologies become more importance [1]. Currently, the depletion of fossil fuel has become global concerns where the energy sources are mainly depending on fossil fuels. Besides, the development of the industries and the growth of population especially in Malaysia are also degrading the environmental quality due to the excessive utilization of fossil fuels [2-3]. Thus, it is necessary to focus on the alternative of renewable energy from green and clean sources in order to obtain a better quality of the environment [3]. Thus, biomass has been introduced to the world as promising clean and green energy sources to replace the use of fossil fuels. The utilization of biomass does not increase carbon dioxide in the atmosphere as it absorbs the same amount of carbon as it releases during the photosynthesis process [3]. In Malaysia, oil palm wastes are the primary biomass resources which contain of empty fruit bunch (EFB), palm kernel shell (PKS) and oil palm fronds (OPF). Among oil palm wastes, EFB is found to have the highest amount produced which are 17.08 million tons per year [4]. Currently, EFB was utilized for the production of fertilizer for agricultural purposes. However, the fertilizer production only covered small amounts of EFB where the unused one are being left rotten. Meanwhile, palm oil mills used EFB for combustion process in their boilers in order to generate electricity which is used to power the milling process [5].

Thermochemical conversion of biomass such as combustion, pyrolysis, liquefaction and gasification into biofuels and useful chemicals would solve the waste problem and energy crisis. Biomass gasification is the process that converts feedstock to fuel gas in the presence of the gasifying agent such as air and steam. However, steam is found to be more efficient as it produces reaction products with higher heating value compared to air [6]. There are several gasification technologies such as fixed bed, fluidized bed and entrained flow gasifiers. However, fluidized bed gasifier promotes heat transfer to the biomass particle which contributes to enhance the reaction rates and conversion efficiency [7]. There are two types of fluidized bed gasifier which are bubbling and circulating fluidized bed. Compared to circulating fluidized bed, bubbling fluidized bed has a minimum fluidizing velocity to promote the mixing of the hot bed material [8]. For biomass gasification, the bubbling fluidized bed gasifier has a simple process thus explains why it was chosen in this work. Biomass gasification is a noteworthy process because it produces high quality of synthesis gas mainly hydrogen, carbon monoxide, carbon dioxide and methane which currently use as fuel for internal combustion engines and fuel cells [6]. There are several parameters that influence the gasification performance as well as synthesis gas production such as gasification temperature and steam to biomass ratio (SBR). Hence, the optimization of the biomass gasification must be conducted in order to obtain the best gasification temperature and SBR.

Response Surface Method (RSM) is the method of experimental design which generate regression equation model by using analysis of variance (ANOVA) of responses and process variables. RSM with Central Composite Design (CCD) can investigate the influence of the interaction of variables on the responses by using Design Expert Software. In previous literature, Razi et al. [1] used RSM to investigate the effect of gasification temperature, biomass blending ratio and steam flow rate on response variables which were gas, tar and char yield from the co-gasification process of pretreated PKS and pretreated Malaysian low rank coal. Besides, Shahbaz et al. also used RSM with CCD to study the influence of temperature, feedstock particle size, CaO/biomass and coal bottom ash on methane production from PKS gasification [9]. However, the optimization of the gasification performance in terms of synthesis gas yield, lower heating value and cold gas efficiency is not carried out by other researchers yet. The higher synthesis gas yield, LHV and CGE obtained from the gasification will indicate the higher efficiency of the gasification performance. Therefore, the objective of this work is to determine optimum operating condition (gasification temperature and steam to biomass ratio) of the gasification of the empty fruit bunch (EFB) in order to maximize the synthesis gas yield, lower heating value (LHV) and cold gas efficiency (CGE). The raw EFB will be used in a bubbling fluidized bed reactor and the experiment will be designed by using Central Composite Design (CCD) and the number of experimental runs will be obtained. Furthermore, the optimization of the process parameters was carried out using Response Surface Method (RSM) in Design Expert software.

2. Materials and Methods

2.1. Materials

Raw EFB was used in this work, which was obtained from Lepar Hilir Palm Oil Mill, Kuantan, Pahang. The sample was dried in the oven for about 4 hours at 105°C to reduce the moisture content from 23.44% to 4.63%. After drying process, the dried EFB was grinded and sieved to the particle size of 0.5 – 1.0 mm and stored in an air-tight container to maintain its moisture content. The higher heating value (HHV), proximate and ultimate analysis of EFB were obtained from the previous work on raw EFB as shown in the Table 1 and Table 2 [10]. Ultimate analysis is the composition of the element in the EFB while proximate analysis is the percentage of the moisture content, fixed carbon, volatile matter and ash of the EFB [11].

Table 1. The proximate analysis of the empty fruit bunch (EFB) [10]

Proximate Analysis (wt%)	Weight percentage (wt%)
Moisture Content	4.63
Volatile Matter	48.44
Fixed Carbon	39.23
Ash	7.70

Table 2. The ultimate analysis of the empty fruit bunch (EFB) [10]

Ultimate Analysis (wt%)	Weight percentage (wt%)
C	54.63
H	5.63
N	6.37
S	0.21
O	36.04
HHV (MJ/kg)	19.60

2.2. Experimental gasification of empty fruit bunch
The gasification of EFB was carried out using a bubbling fluidized bed gasifier which was made from stainless steel. The gasifier with a bed diameter of 60 mm and height of 850 mm is used and silica sand is employed as the bed material since it has high specific heat capacity and could operate at higher temperature [7]. The EFB feedstock with flow rate of 0.25 – 0.4 kg/h was fed into the reactor by using a screw feeder conveyor. The gasifying agent used in this work was steam which was flushed from the bottom of the reactor. Then, the gasifier was heated by using an electric furnace to the desired temperature. The volatile product exited from the top of reactor and proceeded to the separation process at the cyclone separator. The purpose of the separation process is to separate the volatile product from ash and tar. After the separation process, cleaning and drying process were made using a dry ice trap and stored in a gas sampling bag.

2.3. Design of experiment using Response Surface Methodology (RSM)
Response Surface Methodology (RSM) was chosen for experimental design using Design Expert 7 software. The factors of this study (gasification temperature and steam to biomass ratio) were optimized by using Central Composite Design (CCD) in RSM. The gasification temperature in the range from 800 – 1000°C and SBR in the range from 0.5 – 1.5 were selected as the range of the factors based from the literature [12]. The experiment was designed by the software with 13 number of runs for each response variable including 5 center points as shown in Table 2. The number of runs was calculated by [1]

$$N = 2^n + 2n + n_c \tag{1}$$

Where N is the number of runs, n is the number of factors used which are gasification temperature and SBR, and n_c is the number of central points. The responses that were measured are synthesis gas yield, lower heating value (LHV) and cold gas efficiency (CGE). The ANOVA analysis were used to evaluate the analysis of the interaction and individual effect of factors on each response. The regression model and response models in 3D plot surface were also evaluated. The EFB gasification using bubbling fluidized bed was conducted as designed experiments and synthesis gas produced was collected. From the gasification result, the synthesis gas yield was calculated

$$Syn_{yield} = \frac{V_g}{M_f} \tag{2}$$

Where V_g is the volume of the synthesis gas obtained from EFB gasification and M_f is the mass of the EFB used. The energy content in biomass without heat of condition of water is expressed as lower heating value (LHV) and is defined as [4]

$$LHV_{yield} = (30x_{CO} + 25.7x_{H_2} + 85.4x_{CH_4}) \times 4.2 \tag{3}$$

Where x is the mole fraction of the produced gas species. Cold gas efficiency (CGE) is also required to evaluate the gasification performance and is calculated by [4]

$$CGE(\%) = \frac{LHV_{gas} \times Syn_{yield}}{HHV_{feedstock}} \times 100 \tag{4}$$

Where HHV is the higher heating value of the EFB which was shown in Table 1. Table 3 shows the values of synthesis gas yields, LHV and CGE calculated based on 13 different operating conditions for EFB gasification.

Table 3. Experimental design for EFB gasification at different temperature and SBR

No of Run	Gasification Temperature (°C)	SBR	Synthesis gas yield (Nm^3/kg)	LHV (MJ/Nm^3)	CGE (%)
1	800	1.00	1.10	9.97	85.67
2	1000	1.00	1.30	10.31	91.35
3	800	1.00	1.08	10.02	84.36
4	1000	0.50	1.16	9.54	82.76
5	800	1.00	1.11	9.91	85.06
6	600	0.50	0.85	7.84	67.38
7	800	1.50	1.20	10.08	82.78
8	800	1.00	1.09	9.90	83.84
9	1000	1.50	1.23	10.23	89.24
10	600	1.50	0.93	9.28	78.54
11	600	1.00	0.90	8.85	73.21
12	800	1.00	1.07	10.02	85.03
13	800	0.50	0.98	8.63	76.28

3. Result and Discussion

3.1. Statistical analysis

The ANOVA was performed for the evaluation of the response variables with the possible interactions. The validation of the developed model is indicated by P-value which is the probability value from the ANOVA while F-value is Fischer's F-test value which indicates the influences of process parameter on the response. As shown in Tables 3-5, all model found to be significant with high value of F-value and low P-value. For synthesis gas yield, the gasification temperature is more influence with higher F-value (89.50) compared to SBR (12.01). The gasification temperature also is the most influence factor for response variables LHV and CGE with higher F-value compared to the SBR. Based on the result obtained, the synthesis gas yield increased to a maximum value at temperature 1000°C due to the further cracking of liquid and enhanced char reaction with steam that reduce the amount of liquid and char. However, the excessive of steam will reduce the temperature as it will absorb heat thus reducing the synthesis gas yield [1][9]. Furthermore, the results also showed the lack of fit for the three developed models are significantly larger than the pure error. Thus, the lack of fit is signified for all of the terms in the model which are fitted to the response variables.

Table 4. Result of ANOVA for synthesis gas yield

Source	Sum of squares	df	Mean square	F-value	P-value	
Model	0.20	5	0.040	20.08	0.0005	Significant
A-Temperature	0.17	1	0.17	89.50	<0.0001	
B-SBR	0.023	1	0.023	12.01	0.0105	
AB	2.500E-005	1	2.500E-005	0.013	0.9119	
A^2	8.710E-003	1	8.710E-004	0.46	0.5201	
B^2	2.128E-003	1	2.128E-003	1.12	0.3250	
Residual	0.013	7	1.900E-003			
Lack of fit	0.012	3	4.099E-003	16.40	0.0103	Significant
Pure Error	1.000E-003	4	2.500E-004			
Cor Total	0.21	12				
R-squared	0.9369					

Table 5. Result of ANOVA for LHV

Source	Sum of squares	df	Mean square	F-value	P-value	
Model	6.40	5	1.28	79.67	<0.0001	Significant
A-Temperature	2.82	1	2.82	17529	<0.0001	
B-SBR	2.14	1	2.14	133.00	<0.0001	
AB	0.14	1	0.14	8.76	0.0211	
A^2	0.19	1	0.19	11.86	0.0108	
B^2	0.66	1	0.66	40.88	0.0004	
Residual	0.11	7	0.016			
Lack of fit	0.099	3	0.033	9.92	0.0253	Significant
Pure Error	0.013	4	3.330E-003			
Cor Total	6.51	12				
R-squared	0.9827					

Table 6. Result of ANOVA for CGE

Source	Sum of squares	df	Mean square	F-value	P-value	
Model	501.04	5	100.21	42.16	<0.0001	Significant
A-Temperature	325.90	1	325.90	137.12	<0.0001	
B-SBR	97.12	1	97.12	40.87	0.0004	
AB	5.48	1	5.48	2.30	0.1728	
A^2	4.84	1	4.84	2.04	0.1967	
B^2	45.83	1	45.83	19.28	0.0032	
Residual	16.64	7	2.38			
Lack of fit	14.64	3	4.88	9.80	0.0258	Significant
Pure Error	1.99	4	0.50			
Cor Total	517.67	12				
R-squared	0.9679					

The synthesis gas yield, lower heating value (LHV) and cold gas efficiency (CGE) are important for the biomass gasification as the key response variables [13]. Hence, this study investigated the influence of gasification temperature (X_1), steam to biomass ratio (X_2) and the interaction of both process variables to the synthesis gas yield (Y_1), LHV (Y_2) and CGE (Y_3). Other than F-values and P-values, the precision of the model towards the actual data was further evaluated by using regression coefficient (R^2). The value of R^2 obtained were 0.9369, 0.9827 and 0.9679 for synthesis gas yield, LHV and CGE respectively which close to 1. This indicates the reliability of the regression models since the validity of regression model is evaluated based on R^2>0.75 [14]. Furthermore, the residual shows in Table 4 – 6 indicate the difference of predicted and actual values of the response variables. The developed regression models are shown below:

$$Y_1 = -0.11391 + 1.57701E\text{-}03X_1 + 0.36540X_2 - 2.5E\text{-}07X_1X_2 - 4.43966E\text{-}07X_1^2 - 0.11103X_2^2 \quad (5)$$

$$Y_2 = -1.65575 + 0.015803X_1 + 6.59402X_2 - 1.875E\text{-}03X_1X_2 - 6.56466E\text{-}06X_1^2 - 1.95034X_2^2 \quad (6)$$

$$Y_3 = 0.096782 + 0.10149X_1 + 49.99425X_2 - 0.0117X_1X_2 - 3.30862E\text{-}05X_1^2 - 16.29379X_2^2 \quad (7)$$

The residuals obtained for all three response variables were acceptable and indicated the significant of all the regression models.

To visualize the interaction between interaction between two process variables on the response variable, 3D surface plots are studied as shown in Figure 1. The graphical results show that synthesis gas yield is increased with increasing of gasification temperature and SBR. Among those two variables, gasification temperature is the most significant process parameter which influence the synthesis gas yield. At higher temperature, endothermic reactions such as steam methane reforming contributes to more synthesis gas production. Meanwhile, from Figure 1(b), the higher LHV obtained at higher gasification temperature and SBR. At higher temperature, the composition of hydrogen, carbon monoxide and methane gases produced were higher which contributes to the increasing of LHV as shown in Equation (2). The LHV also affected by the SBR which denoted a dominance of steam methane reforming reaction. For CGE, it can be observed that CGE is enhanced by a rise in the gasification temperature as well. However, the increasing of SBR which also increase the oxygen content in the gasifying atmosphere led to slower increment of gasification temperature as oxidation reactions are favored to detriment of the gasification reactions [15]. Figure 1(c) clearly shows that the interaction of the gasification temperature and SBR affected the value of CGE.

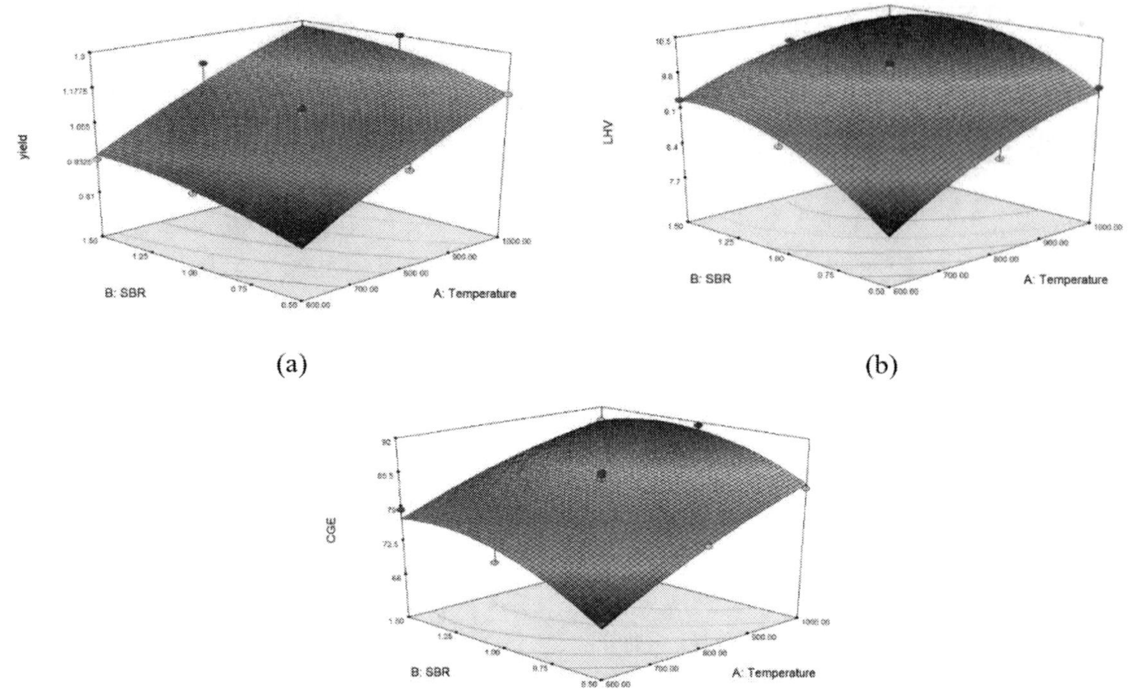

(a) (b)

(c)

Figure 1. The response models in 3D surface plot for responsive variables (a) synthesis gas yield, (b) lower heating value (LHV) and (c) cold gas efficiency (CGE)

3.2. Process optimization and validation

Table 7. The predicted and actual values of synthesis gas yield, LHV and CGE

	Predicted	No of Runs			Average	Error (%)
		1	2	3		
Synthesis gas yield (Nm^3/kg)	1.26	1.27	1.25	1.24	1.25	0.8
LHV (MJ/Nm^3)	10.43	10.27	10.69	10.52	10.49	0.57
CGE (%)	90.99	90.48	90.64	91.05	90.72	0.30

The optimum operating condition for the gasification process was obtained from the numerical optimization using Design Expert software. From optimization, the optimum operating condition was specified at gasification temperature of 800°C and SBR of 1.14 for all responses. The desirability value of 1.000 is obtained from numerical optimization in Design Expert software indicates that the design is suitable for use. Actual experiments were done based on the optimum operating conditions obtained in order to verify the optimum operating condition. Gasification experiments were performed 3 times at gasification temperature of 800°C and SBR of 1.14 and the average value is presented for synthesis gas yield, LHV and CGE. The percentage errors between actual and predicted value of all experiment are shown in Table 7. It can be observed that the percentage errors were in acceptable range which was lesser than 1%. This confirms that the optimum operating condition generated from Design Expert software is indeed reliable in producing maximum synthesis gas yield, LHV and CGE for EFB gasification process.

4. Conclusion

In this study, the biomass gasification of EFB was carried out using bubbling fluidized bed. The effect of gasification temperature from 800 - 1000°C and steam to biomass ratio (SBR) from 0.5 – 1.5 were studied using Response Surface Methodology (RSM). The gasification temperature and SBR are selected as factors and synthesis gas yield, lower heating value (LHV) and cold gas efficiency (CGE) are used as response variables in this work. The experimental work was designed in Design Expert software by using Central Composite Design (CCD). Based on numerical optimization study, the optimum operating condition for EFB gasification was found at gasification temperature of 800 °C and SBR 1.14. The EFB gasification was conducted again based on generated optimum operating conditions in order to verify the synthesis gas yield, LHV and CGE. The average results obtained from the experimental work are 1.25 Nm^3/kg, 10.49 MJ/Nm^3 and 90.72% for synthesis gas yield, LHV and CGE respectively. The percentage errors between predicted and actual values of synthesis gas is 0.8%, LHV is 0.7% and CGE is 0.3% which proved that the developed model is valid and optimum operating condition obtained is significant.

References

[1] Ahmad R, Ishak M A M, Kasim N N and Ismail K 2018 Optimization of co-gasification process parameters of pretreated palm kernel shell and pretreated Malaysian low rank coal using response surface methodology *AIP Conf. Proc.* **2013**

[2] Karmakar M K and Datta A B 2011 Generation of hydrogen rich gas through fluidized bed gasification of biomass *Bioresour. Technol.* **102** 1907–13

[3] Karaci A, Caglar A, Aydinli B and Pekol S 2016 ScienceDirect The pyrolysis process verification of hydrogen rich gas (H e rG) production by artificial neural network (ANN) *Int. J. Hydrogen Energy* **41** 4570–8

[4] Muslim M B, Saleh S and Abdul Samad N A F 2017 Torrefied Biomass Gasification: A Simulation Study by Using Empty Fruit Bunch *MATEC Web Conf.* **131** 03008

[5] Sukiran M A, Abnisa F, Wan Daud W M A, Abu Bakar N and Loh S K 2017 A review of torrefaction of oil palm solid wastes for biofuel production *Energy Convers. Manag.* **149** 101–20

[6] Li Y H and Chen H H 2018 Analysis of syngas production rate in empty fruit bunch steam gasification with varying control factors *Int. J. Hydrogen Energy* **43** 667–75

[7] Kumar A, Jones D D and Hanna M A 2009 Thermochemical biomass gasification: A review of the current status of the technology *Energies* **2** 556–81

[8] Muslim M B, Saleh S and Samad N A F A 2017 Effects of purification on the hydrogen production in biomass gasification process *Chem. Eng. Trans.* **56** 1495–500

[9] Shahbaz M, Yusup S, Pratama A, Inayat A, Patrick D O and Ammar M 2016 Parametric study and optimization of methane production in biomass gasification in the presence of coal bottom ash *Procedia Eng.* **148** 409–16

[10] Harun N H H M, Wahid F R A A, Saleh S and Samad N A F A 2017 Effect of torrefaction on palm oil waste chemical properties and kinetic parameter estimation *Chem. Eng. Trans.* **56** 1195–200

[11] Abdul Wahid F R A, Saleh S and Abdul Samad N A F 2017 Estimation of Higher Heating Value of Torrefied Palm Oil Wastes from Proximate Analysis *Energy Procedia* **138** 307–12

[12] Lahijani P and Zainal Z A 2011 Gasification of palm empty fruit bunch in a bubbling fluidized bed: A performance and agglomeration study *Bioresour. Technol.* **102** 2068–76

[13] Raheem A, Wan Azlina W A K G, Taufiq Yap Y H, Danquah M K and Harun R 2015 Optimization of the microalgae Chlorella vulgaris for syngas production using central composite design *RSC Adv.* **5** 71805–15

[14] Yusup S, Khan Z, Ahmad M M and Rashidi N A 2014 Optimization of hydrogen production in

in-situ catalytic adsorption (ICA) steam gasification based on response surface methodology *Biomass and Bioenergy* **60** 98–107

[15] Fermoso J, Gil M V., Arias B, Plaza M G, Pevida C, Pis J J and Rubiera F 2010 Application of response surface methodology to assess the combined effect of operating variables on high-pressure coal gasification for H2-rich gas production *Int. J. Hydrogen Energy* **35** 1191–204

IOP Conf. Series: Materials Science and Engineering **702** (2019) 012007 doi:10.1088/1757-899X/702/1/012007

Syngas software for biomass gasification process

S Saleh and N A F A Samad*

Faculty of Chemical and Process Engineering Technology, College of Engineering Technology, Universiti Malaysia Pahang, 26300 Gambang, Pahang, Malaysia

*E-mail: asmafazli@ump.edu.my

Abstract. In this work, the thermodynamic equilibrium model has been developed in Excel software for evaluating gasification process. The software is called as Syngas software is applicable to study the effects of gasification temperature either using air or steam as gasifying agents for a wide range of biomass. The application of Syngas software has been highlighted through gasification of EFB using air as gasifying agents. The effects of gasification temperature between 650°C and 900°C has been conducted. The highest amount of hydrogen gas and carbon monoxide gas produced are 29.4 mol% and 44.1 mol% respectively at gasification temperature of 900°C. Meanwhile around 19.5 mol% of carbon dioxide gas and 5.8 mol% of methane gas have been generated at similar gasification temperature. The thermodynamic equilibrium model developed in the Syngas software has been validated against experimental data where root mean square errors (RMSEs) lesser than 1 is obtained indicating a reliable of the developed Syngas software.

1. Introduction

Recently waste to energy (WtE) is becoming one of the innovative technologies for producing sustainable energy with minimum damage to the ecosystems. Through WtE, the waste produced from household, industries and various other agricultural related activities is transformed into a fuel source or in the form of electricity and/or heat [1-2]. One of the attractive WtE conversion technology is biomass gasification. Biomass gasification process is a widely applied technology for converting solid biomass into synthesis gas. The main components of synthesis gas are hydrogen, carbon monoxide, carbon dioxide and methane [1]. The generated synthesis gas is useful for production of heat, power and gaseous fuel. The synthesis gas can also be used to produce methanol, dimethyl ether (DME) and methyl tert-butyl ether (MTBE) for liquid fuel in transportation [1,3].

Thus extensive works either based on experimental or simulation have been performed by numerous researchers in investigating and studying the gasification using various types of biomass. In terms of experimental works, the effects of gasification temperature has been performed on Eucalyptus wood chips [4]. Effect of temperature on gasification using different types of biomass (switchgrass, hardwood, softwood, fiber, cardboard and chicken manure) has been investigated [5]. The influences of temperature and operation time for oil palm frond (OPF) gasification process have been investigated where optimum synthesis gas can be produced at temperature range between 700 and 900°C [6]. In addition to experimental works, the gasification performance can also be studied using simulation models. Through simulation, the effects of various operating conditions on the gasification performance can be evaluated without relying on experimental setup. The model for gasification process can be developed based on kinetic rate models and a thermodynamic equilibrium model. Usually the thermodynamic equilibrium model is preferable due to its simplicity, is easy to develop and the maximum achievable yield under equilibrium conditions can be predicted [7]. Accordingly, a thermodynamic model of a fixed bed gasifier in Cycle-Tempo software has been developed for studying the gasification process [8] and an equilibrium model based on Gibbs free energy minimization has been used in Aspen Plus software to predict the performance of oil palm frond and empty fruit bunch gasification processes [9-10]. In addition the equilibrium model has been developed in order to study thermochemical gasification for agricultural residues [11-12].

Although many experimental and simulation works have been done to study gasification process, but there are some limitations on those approaches. Firstly, the suitability of biomass to be used as a feedstock for gasification. This is depending on its properties such as ultimate analysis and proximate analysis. For example, high moisture content and low fixed carbon of biomass tends to produce low synthesis gas yield [1,9]. Secondly, the optimal operating condition for producing maximum hydrogen gas is different for each biomass. This ultimately consumes a lot of experimental works for finding an optimal condition for biomass to be studied as well as involving trial and error for finding the best biomass to be used as feedstock for gasification process. Based on our literature review, there are most of the model available in literature are applicable for problem specific case study and there is no specific biomass database for studying a wide range of biomass as feedstock for gasification process. Therefore, the objective of this work is to develop a software called as Syngas Software for studying biomass gasification process which is based on thermodynamic equilibrium model. Through this software, it is possible to investigate the synthesis gas production using air or steam. In addition, a database contains around 250 biomass has been developed and integrated with the software in order to applicable for wide range of biomass selection. This in turn helps for evaluating the suitability of biomass as feedstock before it can be implemented in the experimental work.

2. Thermodynamic Equilibrium Model-based Framework for Biomass Gasification Process

In our previous work, the framework for thermodynamic equilibrium model-based has been developed as shown in figure 1 [13].

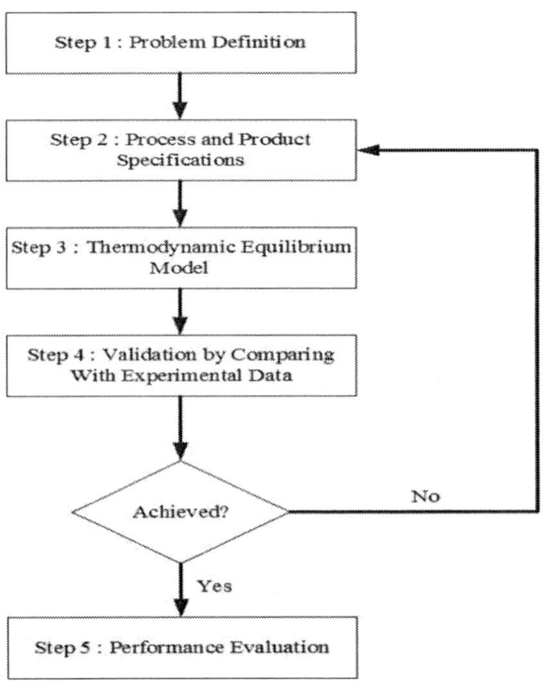

Figure 1. Biomass gasification framework using thermodynamic equilibrium method adopted from [13]

The starting point of this framework is problem definition in terms of objective specification. The example of objectives could be to evaluate gasification performance, to investigate the effects of temperature on synthesis gas production or to compare gasification performance using either air or steam. Based on specified objective, the process and product specifications are performed in step 2. For process specification, the option to select gasifying agents either air or steam is included. In addition the biomass that needs to be investigated will be selected. Here the user is able to select biomass from the developed database as shown in table 1. In the case of biomass data is not available in our database, the ultimate and proximate analysis need to be provided since these properties are essential for thermodynamic equilibrium model. In terms of product specification, the user able to select the components of synthesis gas such as hydrogen gas or all synthesis gas components.

Table 1. Example of biomass properties in database

No.	Type of Biomass	C	H	O	N	S	References
1	Bamboo	40.64	5.64	44.09	0.52	0.03	[1]
2	Empty Fruit Bunch	43.53	7.20	47.09	1.73	0.46	[9]
3	Oil Palm Frond	43.94	6.94	44.88	3.52	0.72	[14]
4	Palm Kernel Shell	49.91	6.94	38.20	3.52	0.72	[15]
5	Pomace	51.31	6.40	35.01	2.00	0.26	[8]
6	Rice Husk	38.43	2.97	36.36	0.49	0.07	[12]
7	Wood Chips	45.69	4.65	50.32	0	0	[4]
8	Wood Pellet	43.52	7.00	43.53	0.23	0.08	[5]

Step 3 involves the calculation of synthesis gas using thermodynamic equilibrium model. Table 2 shows the thermodynamic equilibrium model based on the selected gasifying agents. The steps needed for applying thermodynamic equilibrium model is shown below:

a) Calculate value of w using equation (1)

$$w = \frac{24MC}{18(1-MC)} \tag{1}$$

b) Determine values of ΔA, ΔB, ΔC and ΔD using information of heat capacities.
c) Calculate values of ΔG° and ΔH° at reference temperature.
d) Obtain coefficients of J and I simultaneously using information from a) and b).
e) Calculate equilibrium constants k_1 and k_2.
f) Based on e), the stoichiometries are obtained by using Newton Raphson method.
g) Based on values of stoichiometries, the synthesis gas produced can be calculated using mole balance.

The information for heat capacities, heat of formation and Gibbs of formation at reference temperature can be obtained from [16]. The validation is then performed in step 4 in order to compare the values of synthesis gas obtained from thermodynamic equilibrium model with literature data. The error between the synthesis gas from the experimental work and the thermodynamic equilibrium model is calculated using root mean square error (RMSE) as shown in equation (2).

$$\text{RMSE} = \sqrt{\frac{\sum(X_{\exp} - X_{pred})^2}{N}} \tag{2}$$

Where $X_{\exp}$ and X_{pred} are the experimental data and predicted data of synthesis gas respectively and N is the number of data used. If the RMSE obtained is lesser than 1 then the model is classified as reliable indicating the predicted synthesis gas is in good agreement with literature data [12]. Otherwise the user needs to specify process and product again in step 2 in order to reduce the RMSE. Lastly the performance of gasifier is evaluated based on two factors which are synthesis gas produced at different temperature and different moisture content. Based on these two factors, the best operating conditions can be obtained for improvement and validation of gasification process.

Table 2. General gasification process using thermodynamic equilibrium approach [13]

Gasifying Agent: Air	Gasifying Agent: Steam
$CH_aO_b + wH_2O + mO_2 + 3.76mN_2 = x_1H_2 + x_2CO$ $+x_3CO_2 + x_4H_2O + x_5CH_4 + 3.76mN_2$	$CH_aO_b + wH_2O + mH_2O = x_1H_2 + x_2CO + x_3CO_2$ $+x_4H_2O + x_5CH_4$
Carbon balance: $x_2 + x_3 + x_5 - 1 = 0$	Carbon balance: $x_2 + x_3 + x_5 - 1 = 0$
Hydrogen balance: $2x_1 + 2x_4 + 4x_5 - a - 2w = 0$	Hydrogen balance: $2x_1 + 2x_4 + 4x_5 - a - 2w - 2m = 0$
Oxygen balance: $x_2 + 2x_3 + x_4 - w - 2m - b = 0$	Oxygen balance: $x_2 + 2x_3 + x_4 - w - m - b = 0$

Equilibrium constant:

$$k_1 = \frac{x_5}{(x_1)^2}$$

Equilibrium constant:

$$k_2 = \frac{x_1 x_3}{x_2 x_4}$$

$$\ln k = -\frac{J}{RT} + \Delta A \ln T + \frac{\Delta B.T}{2} + \frac{\Delta C.T^2}{6} + \frac{\Delta D}{2T^2} + I$$

$$\frac{\Delta H^\circ}{R} = \frac{J}{R} + \Delta A.T + \frac{\Delta B.T^2}{2} + \frac{\Delta C.T^3}{3} - \frac{\Delta D}{T}$$

$$\Delta G^\circ = J - RT\left[\Delta A.\ln T + \frac{\Delta B.T}{2} + \frac{\Delta C.T^2}{6} - \frac{\Delta D}{2T^2} + I\right]$$

Data on heat capacity:

$$\Delta A = A_{CH_4} - A_C - 2A_{H_2}$$

$$\Delta B = B_{CH_4} - B_C - 2B_{H_2}$$

$$\Delta C = C_{CH_4} - C_C - 2C_{H_2}$$

$$\Delta D = D_{CH_4} - D_C - 2D_{H_2}$$

3. Application of Syngas Software

In this work, the gasification process based on thermodynamic equilibrium has been constructed in Excel software which is called as Syngas software. The selection of Excel software is due to its simplicity and accessible. All of the mathematical equations as shown in table 2 has been setup in Excel sheet for ease of the solution. In this section the applicability of the Syngas software is highlighted through the gasification of empty fruit bunch (EFB).

3.1. Problem Definition (Step 1)

The objective of this work is to validate the performance of gasification process as shown in figure 2.

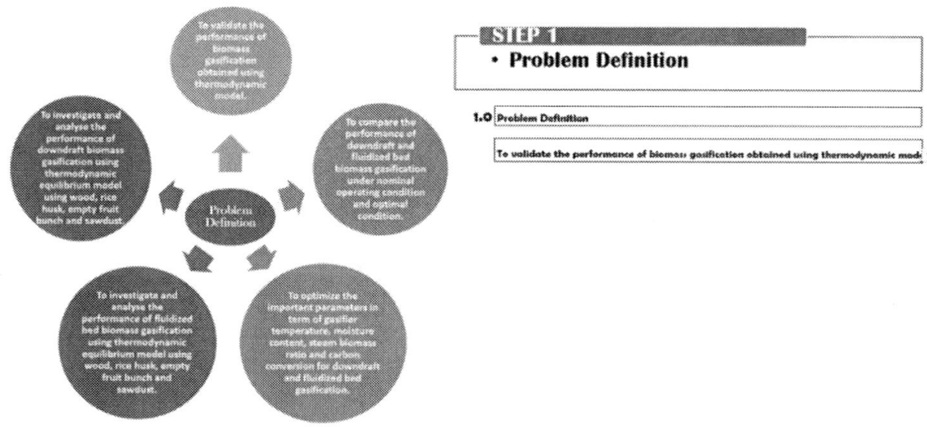

Figure 2. Problem definition (Step 1).

3.2. *Process and Product Specifications (Step 2)*

In terms of process specification, the gasifying agent selected is air and type of biomass to be studied is empty fruit bunch (EFB) as shown in figure 3. Meanwhile the main product for this work is the amounts of synthesis gas produced. The main variables to be studied for operating condition is gasification temperature.

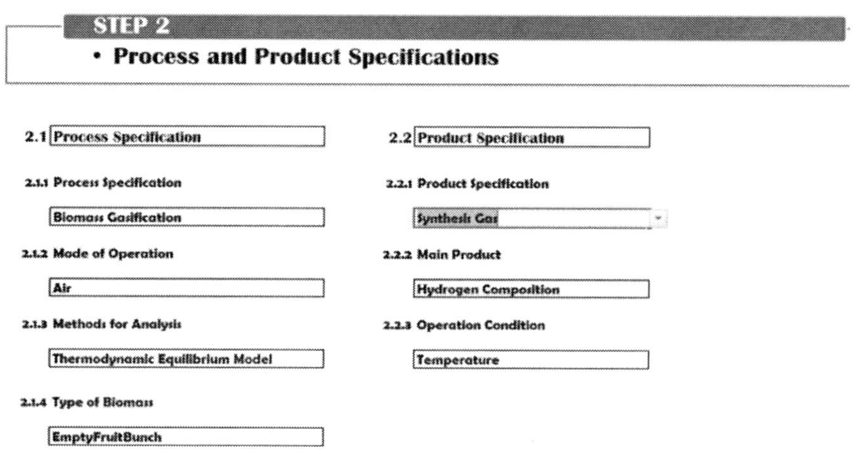

Figure 3. Process and product specifications (Step 2).

3.3. *Thermodynamic Equilibrium Model (Step 3)*

Based on selection in step 2, the chemical formula for EFB is then extracted from database as shown in figure 4. Notes that the elements of S and N are not considered in this work since both amounts are relatively small and thus neglected. Here the user needs to specify the gasification temperature in the range from 600-1000°C. In addition the moisture content used for this case study is 0.2. Based on this specifications, the model is then performed as shown in figure 5 where the values of ΔA, ΔB, ΔC and ΔD are calculated based on heat capacities for each chemical components. Then Gibbs of formation (ΔG°) and heat of formation (ΔH°) at 25°C are calculated. These values are then used for determining variables J and I which subsequently used for calculating equilibrium constants k_1 and k_2 as shown in figures 5 and 6 respectively. Based values of equilibrium constants, the coefficients of x_1 (hydrogen), x_2 (carbon monoxide) and x_3 (carbon dioxide) are obtained by using Newton Raphson method as shown in figure 7. Meanwhile the remaining coefficients of x_4 (water) and x_5 (methane) are then obtained by solving the carbon, hydrogen and oxygen balances as shown in figure 8. Lastly the composition of synthesis gas produced can be calculated using mole balance.

IOP Conf. Series: Materials Science and Engineering **702** (2019) 012007 doi:10.1088/1757-899X/702/1/012007

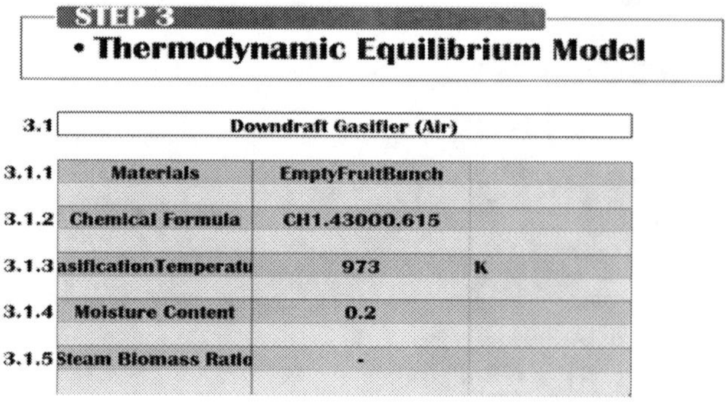

Figure 4. Thermodynamic equilibrium model (Step 3).

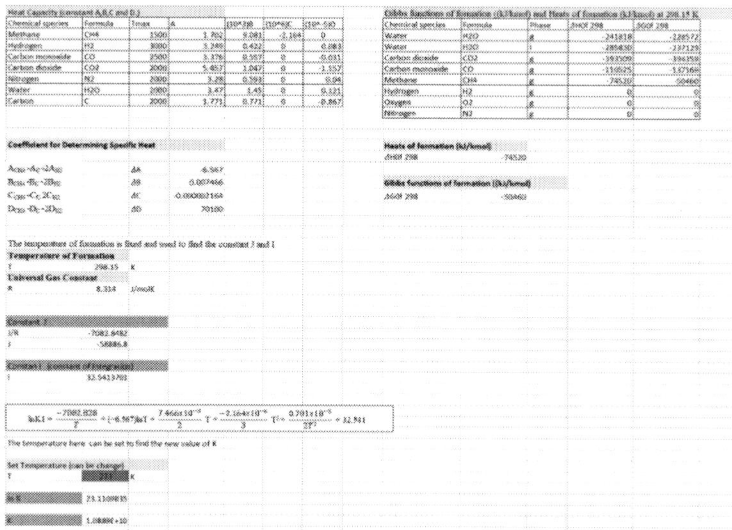

Figure 5. Calculation of equilibrium constant, k_1

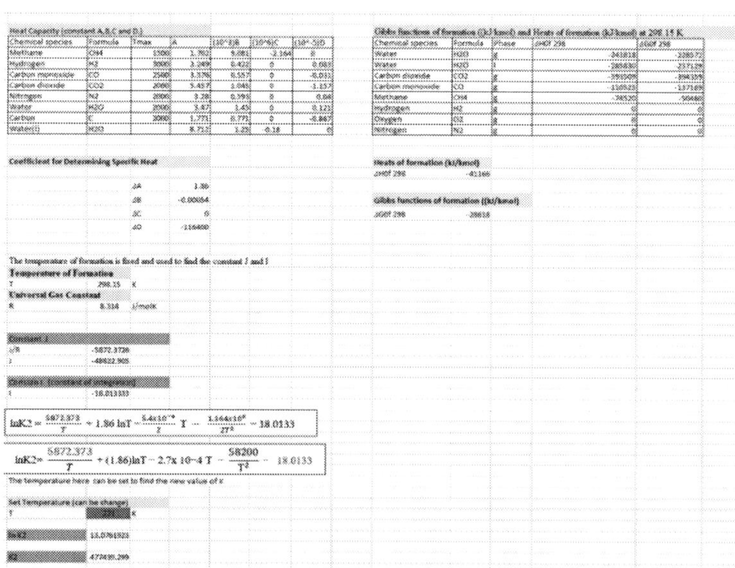

Figure 6. Calculation of equilibrium constant, k_2

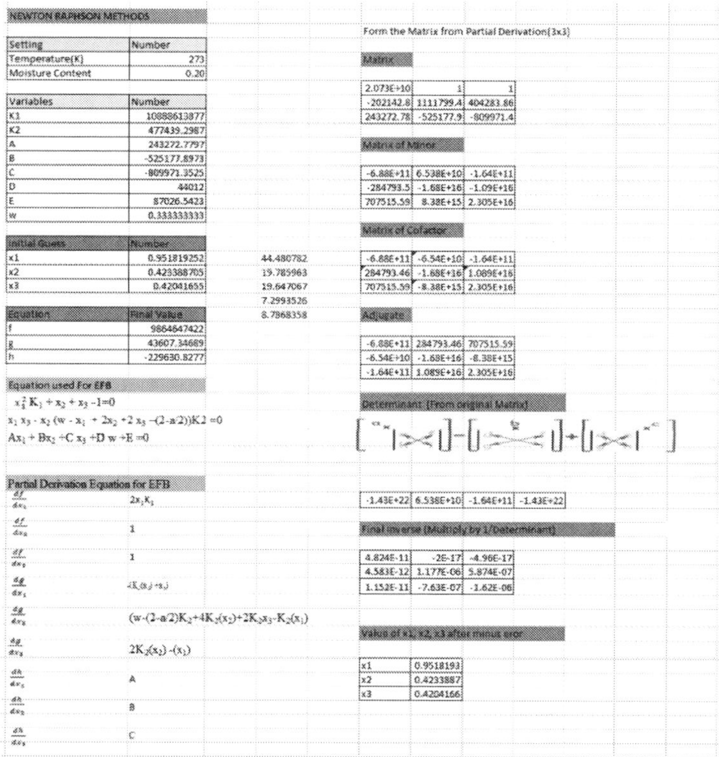

Figure 7. Calculation of coefficients using Newton-Raphson method

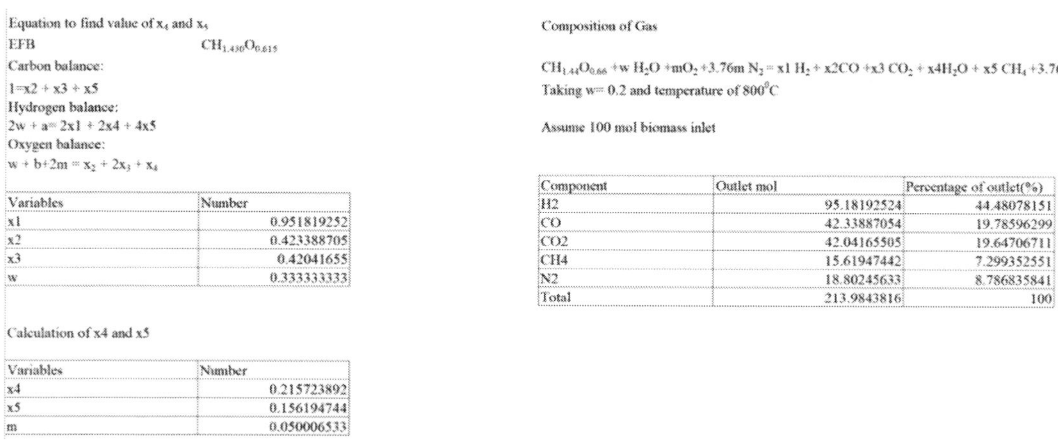

Figure 8. Gas composition calculation (Step 3).

3.4. *Model Validation (Step 4)*

The synthesis gas compositions is compared with EFB gasification experimental data at temperature of 700°C where RMSE of 0.6173 is obtained. In order to further validate the developed thermodynamic equilibrium model, RMSE at different gasification temperature from 650 to 900°C are calculated where RMSEs lesser than 1 have been obtained at all investigated temperature as shown in figure 9. This indicates the thermodynamic equilibrium model is indeed reliable for representing gasification process.

STEP 4

• Model Validation

Synthesis gas composition (%mol dry basis)	Raw EFB		
	Present model	Experimental data	RMSE
H_2	16.31	15.67	0.6173
CO	33.02	32.64	
CO_2	42.16	43.08	
CH_4	8.60	8.62	

Feedstocks	Gasification Temperature (°C)	RMSE
Raw EFB	650	0.5812
	700	0.6173
	750	0.7348
	800	0.8537
	850	0.8892
	900	0.8609

Figure 9. Model validation (Step 4).

3.5. Performance Evaluation (Step 5)

Lastly the gasification performance is evaluated based on synthesis gas compositions at different gasification temperature as shown in figure 10. Overall, a good agreement has been obtained between predicted values and the experimental data of synthesis gas for EFB indicating that the thermodynamic equilibrium model used in this work is able to predict the synthesis gas compositions accurately. The amount of hydrogen gas produced has increased from 14.65 mol% to 29.4 mol% when the gasification temperature is increased from 650°C to 900°C as shown in figure 10. Meanwhile the carbon monoxide composition has also increased from 30.56 mol% to 44.1 mol%. However, the amount of carbon dioxide and methane show an opposite trends as shown in figure 10. At gasification temperature of 650°C around 44.44 mol% of carbon dioxide composition have been produced and the amounts have decreased to 19.5 mol% when the gasification temperature is increased to 900°C. In the same condition the methane gas composition has initially produced approximately 9.72 mol% but has decreased to 5.8 mol%. The increment of carbon monoxide and hydrogen gas compositions can be explained through Boudouard reaction, water gas reaction and water gas-shift reaction as shown in Equations (3) – (5) respectively.

$$C + CO_2 = 2CO \quad (3)$$

$$C + H_2O = CO + H_2 \quad (4)$$

$$CO + H_2O = CO_2 + H_2 \quad (5)$$

As gasification temperature is increased, Boudouard and water gas reactions become more dominant and thus produce more carbon monoxide and hydrogen gas compositions. At low temperatures, carbon monoxide reacts with steam from moisture content of feedstock to produce both hydrogen and carbon dioxide gases. However, at higher temperatures, the water gas-shift reactions is shifted from right to left where carbon monoxide and steam are produced due to the consumption of hydrogen and carbon dioxide gases. Consequently, the amount of carbon dioxide and methane gases are decreased in order to produce carbon monoxide and hydrogen gases as the gasification temperature is increased.

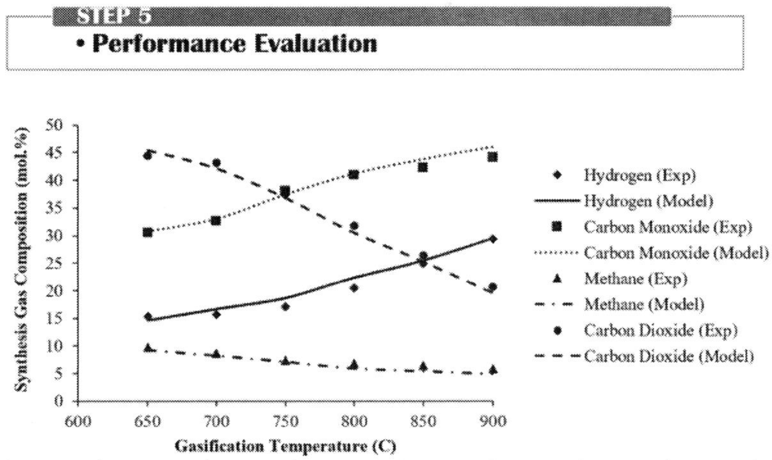

Figure 10. Performance Evaluation (Step 5).

4. Conclusions and Future Perspectives

The gasification model based on thermodynamic equilibrium approach has been integrated in Excel software for studying the gasification process. The database contain biomass properties has been integrated within the software for covering a wide range of biomass as feedstock for gasification process. The applicability of this software has been highlighted in gasification of EFB using air as gasifying agents. A good agreement has been obtained between predicted and experimental data where RMSEs lesser than 1 have been obtained at different gasification temperature. Based on the effects of gasification temperature, it was found that temperature of 900°C produces a highest hydrogen gas production at 29.4%. Although the software can be used to evaluate the effects of gasification temperature for a wide range of biomass but the gasifier's performance is not considered in this work. The performances of gasifier in terms of cold gas efficiency (CGE), lower heating value (LHV) and carbon conversion (CC) are essential in determining the suitability of the biomass as the feedstock for gasification process. In addition, the biomass database developed in this work consists only the raw biomass where some of the biomass have high moisture content and low carbon content. This ultimately reduces the production of synthesis gas. Due to the low carbon content and high moisture content, a further pre-treatment is recommended to improve the properties of biomass before it can be used for the gasification process. One of the pre-treatment methods that can be used is by using the torrefaction process where the moisture content will be significantly reduced and the carbon content as well as the calorific value will be increased [1, 15]. The addition of torrefied biomass in this developed software could provide a good alternative for biomass utilization as energy generation. This could represent the future direction of this work.

References

[1] Kuo P, Wu W and Chen W 2014 Gasification performances of raw and torrefied biomass in a downdraft fixed bed gasifier using thermodynamic analysis *Fuel* **117** 1231-1241

[2] Wahid F R A A, Muslim M B, Saleh S and Samad N A F A 2016 Integrated gasification and fuel cell framework: Biomass gasification case study *ARPN Journal of Engineering and Applied Sciences* **11(4)** 2673-2680

[3] Chen W H, Lin B J, Lee H M and Huang M H 2012 One step synthesis of dimethyl ether from the gas mixture containing CO2 with high space velocity *Appl Energ* **98** 92-101

[4] Ngamchompoo W and Triratanasirichai K 2017 Experimental investigation of high temperature air and steam biomass gasification in a fixed bed downdraft gasifier *Energ Source Part A* **39(8)** 733-740

[5] Madadian E, Orsat V and Lefsrud M 2017 Comparative study of temperature impact on air gasification of various types of biomass in a research scale downdraft reactor *Energ Fuel* **31**

4045-4053

[6] Moni M N Z and Sulaiman S A 2012 Downdraft gasification of oil palm frond: Effects of temperature and operation time *Journal of Applied Sciences* **12(24)** 2574-2579

[7] Patra T K and Sheth P N 2015 Biomass gasification models for downdraft gasifier: A state-of-the-art review *Renew Sust Energ Rev* **50** 583-593.

[8] Fortunato B, Brunetti G, Camporeale S M, Torresi M and Fornarelli F 2017 Thermodynamic model of a downdraft gasifier *Energy Convers. Manag.* **140** 281-294

[9] Muslim M B, Saleh S and Samad N A F A 2017 Torrefied biomass gasification: A simulation study by using empty fruit bunch *MATEC Web of Conferences* **131** 03008

[10] Muslim M B, Saleh S and Samad N A F A 2017 Effects of purification on the hydrogen production in biomass gasification process *Chemical Engineering Transactions* **56** 1495-1500

[11] Azzone E, Morini M and Pinelli M 2012 Development of an equilibrium model for the simulation of thermochemcal gasification and application to agricultural residues *Renew. Energy* **46** 248-254

[12] Karmakar M K and Datta A B 2011 Generation of hydrogen rich gas through fluidized bed gasification of biomass *Bioresour Technol* **102(2)** 1907-1913

[13] Hian G G, Saleh S and Samad N A F A 2016 A generic thermodynamic equilibrium model-based framework for biomass gasification processes *ARPN Journal of Engineering and Applied Sciences* **11(4)** 2222-2229

[14] Wahid F R A A, Harun N H H M, Rashid S R M, Samad N A F A and Saleh S 2017 Physicochemical property changes and volatile analysis for torrefaction of oil palm frond *Chemical Engineering Transactions* **56** 199-204

[15] Harun N H H M, Wahid F R A A, Saleh S and Samad N A F A 2017 Effect of torrefaction on palm oil waste chemical properties and kinetic parameter estimation *Chemical Engineering Transactions* **56** 1195-1200

[16] Robert H P and Don W G 1984 Perry's Chemical Engineers' Handbook 6th edition New York: McGraw Hill

IOP Conf. Series: Materials Science and Engineering **702** (2019) 012008 doi:10.1088/1757-899X/702/1/012008

Anhydrous weight loss kinetics model development for torrefied green waste

N A Jamin, N A F A Samad* **and S Saleh**

Faculty of Chemical and Process Engineering Technology, College of Engineering Technology, Universiti Malaysia Pahang, Lebuhraya Tun Razak, 26300 Kuantan, Pahang, Malaysia

*E-mail: asmafazli@ump.edu.my

Abstract. One of the compositions of municipal solid waste (MSW) is green waste (GW) that collected from landscaping, garden, yard and trimming waste. GW has potential in becoming a biomass feedstock, but poses some drawbacks such as high moisture content, low heating value, high O/C and H/C ratios. Implementation of torrefaction as pre-treatment will improve the GW properties. During torrefaction, biomass is decomposed and leads to anhydrous weight loss (AWL). The estimation model for AWL is significant to study thermal degradation of GW. The aim of this work is to study two steps reaction in series for AWL prediction. GW were torrefied under inert condition at 240-300°C, 10°C/min heating rate and 30 minutes holding time using thermogravimetric analysis (TGA). Two steps reaction series model named Di Blasi and Lanzetta with extended non-isothermal phase is used in developing the AWL model. From initial guess, the parameters of activation energy and kinetic constant are adjusted to fit the calculated AWL to experimental AWL data by applying nonlinear optimization 'lsqcurvefit' routine in Matlab. The estimated kinetic parameters been used for AWL model and later being compared to experimental data from TGA. Good agreement obtained between experimental and model data indicating good kinetic parameters estimation

1. Introduction

Department of Statistics states that Malaysia is having a massive population increment by 1.5% annual growth rate from 2015 to 31.7 million in 2016. As developing countries that currently facing fast accelerating population increment and experiences a rapid industrial growth as well as urbanization, Malaysia also received an adverse effect towards an environment due to increasing amount of municipal solid waste (MSW) generation [1-2]. In present time, Malaysia is generating over 23 000 tonnes of solid waste daily and the amount of MSW generation was expected to reach over 30 000 tonnes per day in 2020 [3]. One of the main organic waste compositions in MSW is green waste (GW) that includes garden, trimming and yard waste which represents as lignocellulosic component. Currently, landfilling has become one of the common methods for GW disposal, however open burning at dump site is the most preferable technique due to its low management cost even though it poses a lot of severe impact toward an environment due to inadequate waste treatment facilities and handling [4]. Prior to this problem, many researches have been done regarding on waste management focusing on Waste-to-Energy (WtE) technologies as GW has a potential as fuel feedstock that can be utilized for energy generation [5,6]. Also known as green technologies, WtE is a process of energy recovery through various process including torrefaction, anaerobic digestion, combustion, gasification, pyrolysis, which transform the waste materials into fuel, heat or electricity [7]. Despites of its potential in energy recovery, GW is having some drawbacks such as relatively high moisture content and high O/C ratio, low calorific value and bulk density as well as high tar production during the process that possibly will reduce the efficiency of WtE process, and it might lead to storage and transportation difficulties [8]. Thus, GW must be subjected to pre-treatment in order to improve its properties as a fuel feedstock before undergoing any WtE process [8,9]. One of the pre-treatment processes that is suitable and has an ability to improve the GW properties is torrefaction [8,10]. Torrefaction involves the combination of two processes which is combustion and partial pyrolysis.

Torrefaction is a thermochemical treatment process that operate at temperature between 200°C to 300°C with residence time between 15 to 60 minutes under inert condition was found to be beneficial for biomass energy properties improvement [11]. Recent research has verified torrefaction process gives a significant impact in improving the properties of biomass which include moisture content reduction, increment of heating value and energy density, decrease the atomic O/C and H/C ratios and improve the grindability of biomass [11-12]. During torrefaction, mass loss is expected which was a result from the decomposition of biomass, volatile release and moisture loss [13-14]. The thermal degradation during torrefaction leads to an anhydrous weight loss (AWL) of a biomass that represent the physical-chemical transformation of the sample during the process [15]. AWL is described as the loss of mass fraction during torrefaction. The mass fraction loss here explains the amount of biomass that has been decomposed by the release of moisture, gas and volatiles throughout the torrefaction process. Hence, it is important to have a model to estimate the AWL in order to study the optimum condition for torrefaction process in achieving the desired properties of torrefied GW.

The most common technique to investigate the thermal degradation and mass loss of biomass during torrefaction process is by using thermogravimetry analysis (TGA). The kinetic evaluation of TGA data can be used in development of simple model which known as one equation global model that may able to describe the reaction accurately at low reaction temperature with only considering the isothermal stage [16-17]. In one step global model, it is only considered direct decomposition of biomass into char and neglecting the intermediate volatile. Next, kinetic model was introduced by Di Blasi and Lanzetta regarding on intrinsic kinetics of isothermal degradation at 200 to 340°C temperature range in inert atmosphere. Di Blasi and Lanzetta developed two-step models by taking volatile and solid compounds into account which include the intermediate volatiles. This model explains that the first step occurs due to degradation of fraction in hemicellulose while the second step involving the degradation of cellulose, lignin and part of hemicellulose [18]. Meanwhile, Repellin et al. have studied the torrefaction of beech and spruce chips of a continuous pilot-scale reactor by applying three different model including simple model (one step global model), Di Blasi and Lanzetta model and Rousset model [15]. As a result, these three models were successfully applied to beech and spruce chips and good correlation between calculated and experimental AWL was obtained [15].

Consequently, the decomposition of a biomass leads to AWL, hence the development of model to estimate the AWL is necessary to study the degradation of biomass during the torrefaction process. Therefore, the objective of this work is to study two steps reaction in series models that can be used to predict the AWL. Thus, the torrefaction of GW using thermogravimetric analysis (TGA) has been done at different temperature (240°C, 270°C and 300°C) in order to obtain the residual mass. In this work Di Blasi and Lanzetta model has been selected in developing the AWL model for green waste (GW) as the model describe more detailed for biomass decomposition. However, the original model of Di Blasi and Lanzetta only consider the isothermal phase and in this work, it has been extended by including the non-isothermal (heating) phase. The model needs to be extended by include non-isothermal phase due to the extractives and reactive hemicellulose fraction start to degrade this phase which also involve countable mass loss of sample. The kinetic parameters have been estimated from residual mass data obtained from TGA for both isothermal and non-isothermal phases. The AWL of GW from predicted model will be compared to the experimental data from TGA for validation purposes.

2. Materials and Methods

2.1. Materials and Thermogravimetric Analysis (TGA)
Yard and trimming waste which are considered as green waste (GW) were collected from orchard and residence nearby Gambang, Pahang, Malaysia. The GW samples being subjected to pre-drying process at temperature of 105°C overnight. Then, dried samples were grinded and sieved to 0.5-1.0 mm particle size. To evaluate the biomass mass loss during torrefaction, TGA/DSC 1 Mettler Toledo analyser was used based on 30 ml/min nitrogen flow for inert condition. GW samples weight was varied from 5 mg to 10 mg and tested at three different torrefaction temperatures of 240 °C, 270 °C and 300 °C with heating rate 10 °C/min. Samples was heated up to desired temperature (240°C, 270°C and 300°C) and were held for 30 minutes. TGA was run three times for each sample. AWL of each samples was calculated based on the data obtained from TGA experiment and was used to compared with modelled data from simulation work.

2.2. Modelling of AWL
Modelling of torrefaction process can be done through AWL model development. Di Blasi and Lanzetta has developed two stages degradation of hemicellulose (xylans) for weight loss prediction. At the beginning, this two stages model predicts one stage of degradation and later continue over the whole process. A two stages in series model as shown in equation (1) is used to study the kinetics parameter estimation. As shown in equation (1), the biomass first degrades into solid intermediate compound (B) through rate constant and volatiles (k_B and k_{V1}).

Table 1. Di Blasi and Lanzetta kinetic model [18].

Two reaction stages	Reaction mechanism
Biomass (A) → Intermediate compound (B) Biomass (A) → Volatiles 1 (V1) Intermediate compound → Char (C) Intermediate compound → Volatiles 2 (V2)	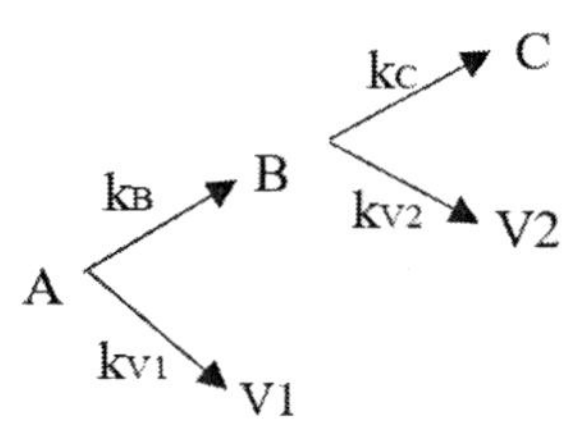 (1)

The torrefaction process sequential step is the conversion of intermediates into char (C) and another volatiles (V2) through k_C and k_{V2}. Total AWL is denoted as the total of A, B and C as shown in equation (2). The torrefaction process was assumed to be in first order reaction as the differential equations as shown in equations (4) – (6) for the decomposition of biomass under isothermal phase reaction. The rate constant is denoted as k (s^{-1}) which determined from equation (3) which known as Arrhenius equation.

$$AWL = [A] + [B] + [C] \tag{2}$$

$$lnk = \ln(A) - \frac{Ea}{RT} \tag{3}$$

$$\frac{d[A]}{dt} = -(k_B + k_{V1})[A] \tag{4}$$

$$\frac{d[B]}{dt} = k_B[A] - (k_C - k_{V2})[B] \tag{5}$$

$$\frac{d[C]}{dt} = k_C[B] \tag{6}$$

Meanwhile, M/M_0 can be determined through TGA experiment by using equation (7). m_0 is initial mass of the sample, m_{ash} is mass of ash in the sample am m_{TGA} is the mass measured by TGA in a function of time.

$$\left(\frac{M}{M_0}\right)_{exp} = \frac{m_{TGA} - m_{ash}}{m_0 - m_{ash}} \tag{7}$$

Many kinetic studies have been done in previous study only considering isothermal stage, however there is weight loss apparently at heating stage which is significant that contribute towards the whole process of degradation especially when using high heating rate [19-20]. The existing isothermal equation do not consider the heating rate of torrefaction process. In this work, the torrefaction using TGA is conducted at three different temperature and the time taken in achieving desired temperature from heating phase to isothermal phase is different which is 4, 7 and 10 minutes for 240°C, 270°C and 300°C respectively. Equations (8) – (10) show the kinetic reaction equation for heating phase which has been included together with the original Di Blasi and Lanzetta model. Based on the modified equations, the main different for non-isothermal model equations and isothermal phase model equations is the heating rate β (s^{-1}) is included in the heating phase model equations.

$$\frac{d[A]}{dt} = \left(\frac{dt}{dT}\right) \times \{-(k_B + k_{V1})[A]\} = \frac{1}{\beta}\{-(k_B + k_{V1})[A]\} \tag{8}$$

$$\frac{d[B]}{dt} = \frac{1}{\beta} \times \{k_B[A] - (k_C + k_{V2})[B]\} \tag{9}$$

$$\frac{d[C]}{dt} = \frac{1}{\beta} \times \{k_C[B]\} \tag{10}$$

The unknown parameters that are needed in order to solve the AWL model is based on equation 1, which is the rate constant (k) for volatile 1 (k_{V1}), intermediate compound B (k_B), volatile 2 (k_{V2}) and char (k_C). Figure 1 shows the overall workflow in estimating the kinetic parameter for GW torrefaction using 'lsqcurvefit' solver in MATLAB in order to minimize the root mean square error between experimental and calculated data. It begins with the first iteration by assuming only biomass A with no intermediate compound B and char C. The activation energy (*Ea*) and pre-exponential factor (*A*) are calculated by Arrhenius equation and subsequently the calculated values were used in order to determine the mass fractions of A, B and C for isothermal and non-isothermal phases. The isothermal phase began from 4th minutes onwards where the torrefaction temperature is maintained at 240°C for 30 minutes until the end of the reaction. Both isothermal phase reaction from equations (4) – (6) and the non-isothermal phase from equations (8) – (10) were integrated for kinetic parameter derivation by using the new calculated A, B and C. From the second iteration onwards the new calculated (*Ea*) and (*A*) is used to provide the initial guess for the optimization. The procedure was repeated until the values of [A], [B] and [C] are converged.

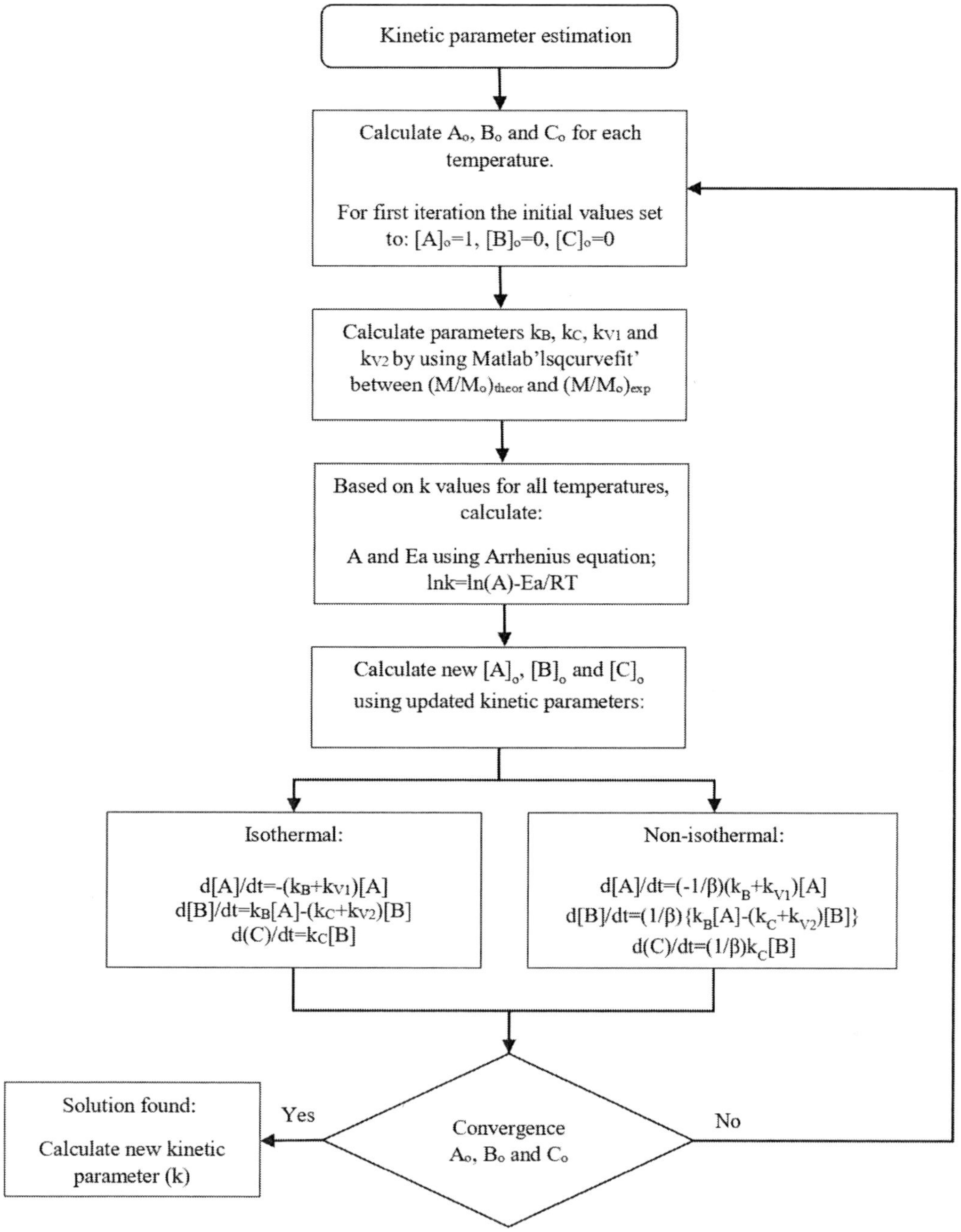

Figure 1. Overall workflow for kinetic parameters estimation of GW torrefaction in Matlab.

3. Results and Discussion

3.1. Estimation of Kinetic Parameters

For estimation of kinetic parameters, the data from TGA test at temperatures of 240°C, 270°C and 300°C were used. The data taken is starting from temperature of 200°C since the thermal degradation of biomass roughly started at 200°C and biomass shows an immediate loss of mass. This indicates that some of the constituent in a biomass start to break down and decomposed. Table 2 shows the kinetic parameters from parameter estimation implemented in MATLAB (R2014a) software. The iteration for kinetic parameters estimation was done using the nonlinear optimization which employing 'lsqcurvefit' routine in MATLAB. For model validation, the model was verified by comparing both experimental data and predicted data of AWL for the whole reaction at different torrefaction temperature.

Table 2. Kinetic parameters for reaction model for green waste at 240°C, 270°C and 300°C

Parameters	A (s^{-1})	Ea ($Jmol^{-1}$)
k_B	1.22 x 10^{-2}	5.27 x 10^{3}
k_C	3.11 x 10^{-8}	-3.94 x 10^{4}
k_{V1}	5.77 x 10^{-1}	8.74 x 10^{5}
k_{V2}	7.56 x 10^{-2}	3.06 x 10^{4}

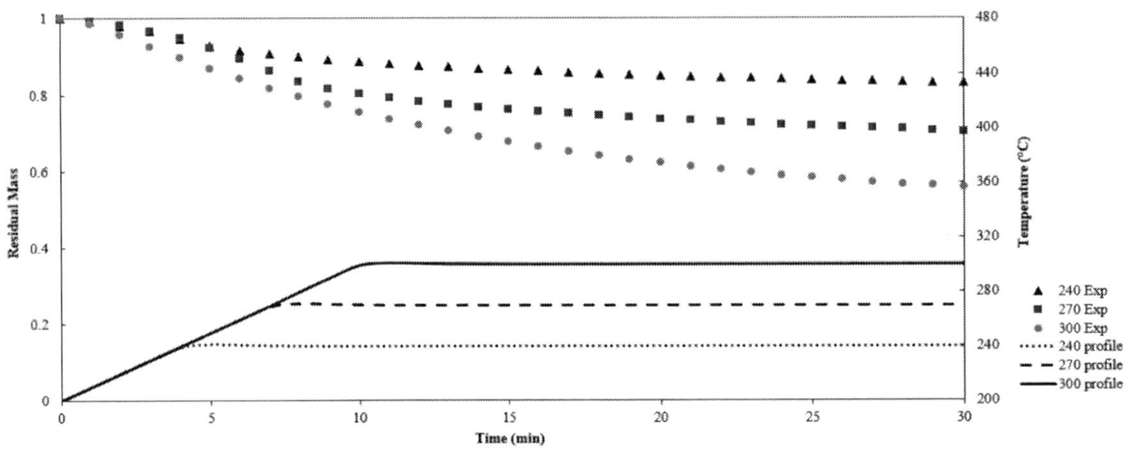

Figure 2. Residual mass of torrefied green waste (GW) at different temperature based on TGA with heating rate 10°C/min.

Based on Figure 2, torrefaction process can be divided into two stages which is heating stage where biomass is heated up at certain temperature and continue with isothermal stage where the process was maintained at desired torrefaction temperature until the end of operational time. Thus, it is significant to develop a model in order to evaluate the AWL of GW torrefaction for both non-isothermal and isothermal phases. Shang et al. suggested to consider the weight loss during non-isothermal (heating) phase in deriving the kinetics parameters [16]. Figure 2 shows the reaction stage involve in torrefaction experiment and modelled from GW at different temperature. Based on temperature of 240°C, the heating phase occurred at the first 4 minutes and the process was held for 30 minutes at isothermal stage.

According to figure 2, the residual mass reduction of biomass is referring to AWL profiles. The AWL curve can be divided into two stages. The first stage is related to rapid increment in AWL while the second stage corresponds to slow increment of AWL. Rapid increment of AWL can be identified at first 4 minutes of reaction which indicates most of a weight loss occurred at this stage. The weight loss and decomposition as shown in figure 2 also explain the degradation rate of green waste. As shown in figure 2, AWL shows slightly increment over 4 minutes of reaction at 240°C and remain slightly constant after 25 minutes. Meanwhile, figure 3 shows the residual mass for all torrefaction temperature of 240°C, 270°C and 300°C for GW. The degradation rate of GW is higher at high torrefaction temperature. 44.25% of weight loss is detected after 30 minutes for temperature of 300°C compared to only 16.82% weight loss for temperature of 240°C. This is due to the degradation of hemicellulose and cellulose contents that contribute to evaporation of water and volatile release [18].

For model verification, figure 3 (a), (b) and (c) show the comparison of predicted residual mass and TGA experimental for both isothermal and non-isothermal phases starting at 200°C to desired torrefaction temperature. Based on figure 3 (a), (b) and (c), extended Di Blasi and Lanzetta model is able to describe the reaction of GW precisely and both experimental results and modelled data shows a good agreement at torrefaction temperature of 240°C, 270°C and 300°C. Meanwhile, figure 3 (d), (e) and (f) explains the mass yields for reaction of A (GW), B (Intermediate compound) and C (Char) for GW at different torrefaction temperature. It can be seen that the conversion of A into intermediate compound B is starting from the beginning and fully decomposed into B within 5 minutes. The intermediate compound (B) initially increase due to the decomposition of A but shows a dramatic decrease until the end of time which indicates rapid conversion of intermediate compound (B) into char (C). The mass yield of (C) is increased steadily as most of the component (B) has been decomposed and converted into char. The mass yield of char (C) produced from intermediate compound (B) at high temperature is much lower compared to low torrefaction temperature. Only 0.46 of mass yield for (C) produce at 300°C compared to 0.83 of mass yield for C at temperature of 240°C. This is due to the more moisture and volatile release at high temperature which also related to the degradation of GW. Figure 4 shows that a good model fit has been obtained based on the comparison of residual mass value obtained by experimental and by modelling for GW. The predicted R^2 value from the analysis of variance (ANOVA) test according to this model is 0.9998. Thus, it can be said that the kinetic parameters for GW is best fit to represent the residual mass.

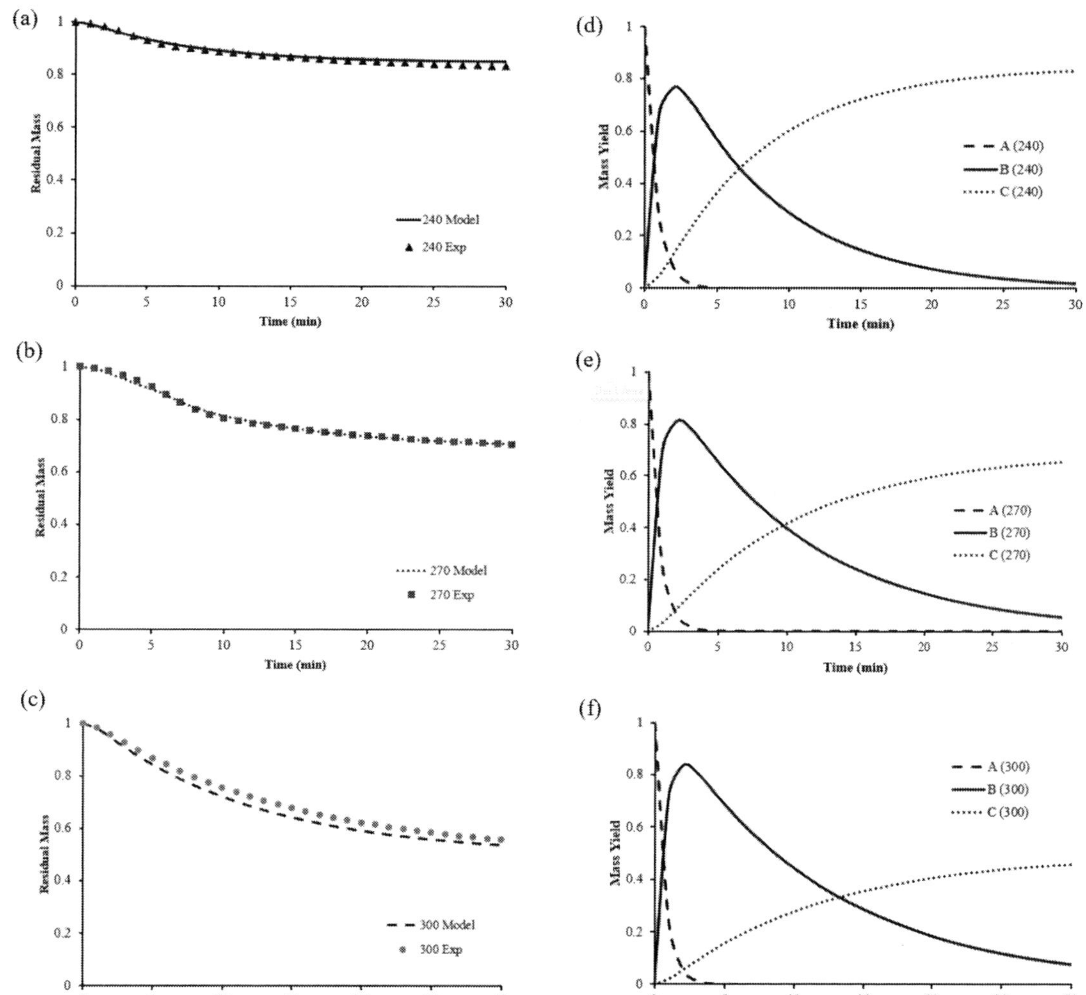

Figure 3. AWL modelling and experimental of GW (a) 240°C; (b) 270°C; (c) 300°C and, mass yield of constituent A, B and C at (d) 240°C; (e) 270°C; (f)300°C of GW torrefaction with heating rate 10°C/min.

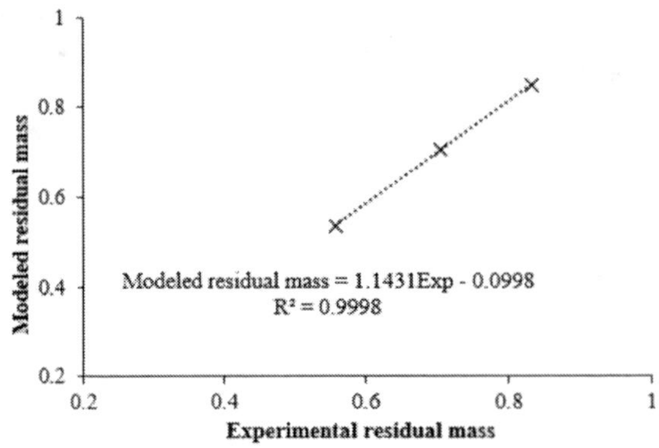

Figure. 4. Correlation between experimental and modelled residual mass of GW.

4. Conclusion

In this study, torrefaction of green waste (GW) were carried out using TGA. Green waste sample has been heated up to desired torrefaction temperatures (240°C, 270°C and 300°C) with heating rate of 10°C/min and 30 minutes holding time. Mass loss data by torrefaction time was determined from TGA and AWL of GW sample was evaluated. Two steps reaction model in series which is known as Di Blasi and Lanzetta model has been used to study the AWL of GW. In this study the original model which only considering the isothermal phase has been extended by taking into account the non-isothermal phase. The kinetic parameter estimations are able to obtain a good residual mass prediction of torrefaction process at all temperature. Based on residual mass results, the rate of degradation at temperature of 300°C is higher compared to the lower torrefaction temperature 240°C. This is due to constituent in GW such as hemicellulose and cellulose which contribute to mass loss in terms of moisture and volatiles during torrefaction process. Overall, the developed model can be concluded as a reliable model in predicting AWL of GW. Good correlation has been obtained between experimental and calculated residual mass of GW with R^2= 0.9998.

Acknowledgement

This work was financially supported by Doctorate Research Scheme (DRS) and Research Grant of RDU1703241 entitled "Influence of torrefaction on the synthesis gas production from gasification of municipal solid waste (MSW)" by Universiti Malaysia Pahang.

References

[1] Saeed M O, Hassan M N and Mejeebu M A 2009 *Waste Manage.* **29** 2209–13
[2] Tarmudi Z, Abdullah M L and Tap A O M 2012 *Jurnal Teknol.* **57** 41–56
[3] Anwar J, Habib A, Haslenda H, Saeed I A and Ramli M 2014 *Mod. Appl. Sci.* **8** 37–49
[4] Agamuthu P and Fauziah S H 2011 *Waste Manage. Res.* **29** 13–19
[5] Saad M, Meghdad B, Azadeh S and Zainal S 2014 *Renew. Sust. Energ. Rev.* **40** 497–504
[6] Yatim P, Mamat M N, Mohamad-Zailani S Hand Ramlee S 2016 *Clean Technol. Envir. Policy* **18** 1685–95
[7] Rawat J, Kaalva S, Rathhore V, Gokak D. T and Bhargava S 2016 *Procedia Environ Sci* **35** 483–90
[8] Chen W H, Peng J and Bi X T 2015 *Renew. Sust. Energ Rev.* **44** 847866
[9] van der Stelt M J C, Gerhauser H, Keil J H A and Ptasinski K J 2011 *Biomass Bioenergy* **35** 3748–62
[10] Jamin N A, Saleh S and Samad N A F A 2019 *Energy Proced.* **158** 836–841
[11] Kingsley L I, Oon-Doo B and Lope G T 2017 *Biomass Bioenergy* **106** 8–20
[12] Raut M K, Basu P and Acharya B 2016 *Int J Renew Energy Biofuels* 1–13
[13] Harun N H H M, Samad N A F A and Saleh S 2017 *Energy Proced.* **105** 744–749
[14] Samad N A F A, Jamin N A and Saleh S 2017 *Energy Proced.* **138** 313–318

[15] Repellin V, Govin A, Rollan M and Guyonnet R 2010 *Biomass Bioenergy* **34** 602–609
[16] Shang L, Ahrenfeldt J, Holm J K, Bach L S, Stelte W and Henriksen U B 2014 *J. Anal. Appl. Pyrol.* **108** 109–116
[17] Rashid S R M, Harun N H H M, Saleh S and Samad N A F A 2017 *Energy Proced* .**138** 319–324
[18] Di Blasi C and Lanzetta M 1997 *J. Anal. Appl. Pyrol.* **40** 287–303
[19] Gonzalez-Pena M M, Curling S F and Hale M D 2009 *Polym. Degrad. Stabil.* **94** 2184–2193

IOP Conf. Series: Materials Science and Engineering **702** (2019) 012009 doi:10.1088/1757-899X/702/1/012009

Integrating microwave heating pre-treatment to improve torrefaction of food waste

M Ismail, R Abdul Rasid, R N U Abdul Rahman, N S Ahmad and M A W Mohd Radzuan

Faculty of Chemical and Process Engineering Technology, Universiti Malaysia Pahang, 26300 Gambang, Pahang, Malaysia.

*E-mail: mazni@ump.edu.my

Abstract. The food waste (FW) disposal is one of the main issues of concern, as it affects environmental and sustainable development in Malaysia. FW should be utilized to produce more valuable products such as energy. However, FW is characterized by its low high heating value (HHV) and high moisture content and which lower the energy efficiency during energy production process. To overcome this problem, torrefaction assisted by microwave heating pre-treatment is used to upgrade FW properties to produce solid biofuel which is called as biochar. The objective of this study is to investigate the feasibility of microwave (MW) heating power at 240 W for 5 min to enhance the quality of torrefied FW. The pre-treated FW is then torrefied at three different temperatures (280, 300 and 320 °C) for 30, 45 and 60 min. The torrefied FW was then characterized in terms of ultimate and proximate analysis, HHV, mass and energy yield as well as its functional groups. The optimum condition of torrefaction in this study was 320°C and 30 min with HHV of 23.82 MJ/kg, 95.54% of energy yield and 84.28% of mass yield. This suggests that torrefied FW with MW heating pre-treatment has the potential to be used as an energy source.

1. Introduction

In 2019, 1.3 billion tons of FW generated every year globally. In Malaysia, it is estimated that 6.7 million tons of food waste (FW) will be generated a year in 2020 [1]. This tremendous amount of the FW is mainly being disposed of at landfill as it is relatively inexpensive and simple to apply. However, once the maximum capacity of the landfill site is reached, identifying suitable locations for new landfills is a serious issue. Moreover, disposing the FW at the landfill has creates various environmental problems such as groundwater contamination, greenhouse gas emissions which contribute to the climate change, unpleasant odour and leachate formation [2].

Conversion of FW into value-added products such as biofuel energy is becoming an interesting option to reduce the accumulating wastes at landfill as well as reducing the dependency in fossil-based fuels. The use of FW biomass as a new renewable energy sources is necessary to reach the energy demand in the world.

The utilization of FW as a renewable energy resource however is quite challenging for several reasons that are related to its high moisture content, lower higher heating value (HHV) and lower energy density [3-4]. These challenges may be overcome through torrefaction. Torrefaction is a thermal pretreatment that converts biomass into biochar at low operating temperature of 200-240 °C in an inert

IOP Conf. Series: Materials Science and Engineering **702** (2019) 012009 doi:10.1088/1757-899X/702/1/012009

condition [4]. It has been found to be effective in improving the energy density which provides a much better fuel quality for gasification and combustion applications [5].

During the torrefaction process, water content and volatile matter in the biomass are released and the biopolymers include cellulose, hemicellulose and lignin are partly decomposed, giving off various types of volatiles [6]. The final product is the remaining solid, dry and blackened material that is referred to as torrefied biomass or biochar.

Torrefaction has been conducted on variation of biomass including lignocellulosic wastes such as oil palm solid waste [7, 8], forest residues [5], coffee residues [9], all of which shows promising results. A limited amount of work done on food waste torrefaction as compared to the lignocellulosic wastes [6]. At the moment, Abdul Samad et al. [10] and Abdul Rasid et al. [11] are among those who studied torrefaction of FW in Malaysia.

The torrefaction of FW can be further improved by integrating microwave (MW) heating pre-treatment prior to the torrefaction process. MW heating has the potential to alter the structure by creating more pores and bigger surface which increase the efficiency of the torrefaction process.

The purpose of this study is to investigate the improvement of torrefied food waste properties by integrating MW heating as a pre-treatment. The MW heating pre-treatment will be carried at 160 and 240 W for 5 min. The pre-treated FW is then torrefied at three different temperatures (280, 300 and 320 °C) for 30, 45 and 60 min.

2. Methodology

2.1. Material

FW utilized in this study was collected from the Universiti Malaysia Pahang cafeteria which contained of bones (74.03%), rice (13.81%), vegetables (10.20%), and others (1.96%). The samples were dried overnight in the oven at 105 °C. The dried FW was ground and sieved to get uniform particles size ranging from 0.5 to 1 mm. The properties of raw FW samples used in this work are tabulated in table 1.

Table 1. Raw FW Properties

Raw FW Properties	
Moisture content (%, wet)	61.43
HHV (MJ/kg, dry)	21.01
Elemental (wt.%, dry):	
C	44.41
H	7.15
N	5.20
O	39.46
S	3.78

2.2. Experimental Procedure

2.2.1. Microwave heating pre-treatment. Microwave oven (SHARP, model: R207EK) is used for the microwave heating pre-treatment. A 20 g of dried FW is placed in the microwave oven and heated at 160 and 240 W for 5 min. After microwave heating, the FW is proceeded with torrefaction process.

2.2.2. Torrefaction process.

Figure 1 shows the schematic diagram of torrefaction set-up. The system is consisting of nitrogen tank, flow controller, reactor, temperature controller, furnace and liquid collection flask. A tubular reactor with diameter of 3.7 cm and length of 10 cm was used in for the torrefaction process. 20 g of the treated food waste is loaded in the reactor. The FW torrefaction process was carried out at 280°C, 300°C and

320 °C for 30, 45 and 60 min. The reactor was purged by nitrogen gas (N_2) with flowrate of 2 L/min for 10 min before start the torrefaction process. This purposed of this step is to create an inert environment. The nitrogen gas was flowed continuously with the same flow rate during the torrefaction. The solid product was collected once the reactor was cooled down. Experiments was repeated thrice under the same conditions to ensure reproducibility of the results. The samples were stored in desiccator in order to avoid moisture reabsorption.

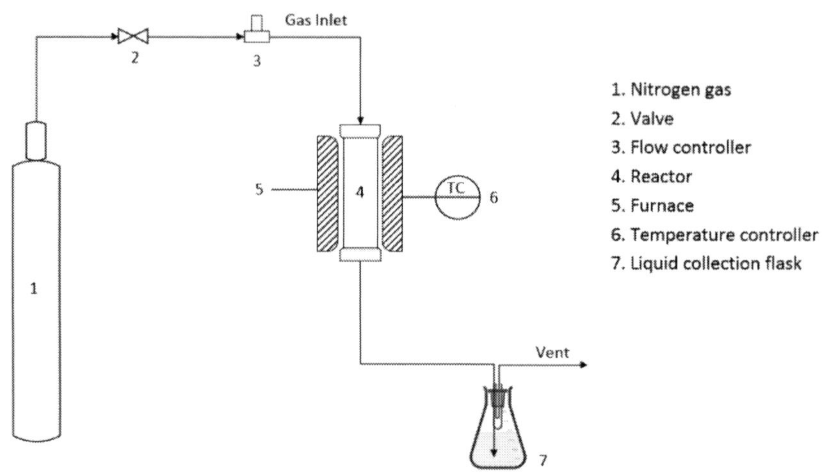

Figure 1. The diagram of torrefaction system.

2.3. Product analysis and characterization.

The proximate and ultimate analysis, high heating value (HHV), mass and energy yield, functional group and morphology analysis of the FW has been performed. The proximate analysis was conducted using TGA (Q500-0617). The samples were heated for 10 °C/min from 30 to 900 °C under the flow of 10 ml/min N_2. Fourier-transform infrared spectroscopy (FTIR) and scanning electron microscopy (SEM) is used for functional group and morphology analysis, respectively. For the HHV, the analysis is conducted using bomb calorimeter (model IKA Calorimeter C2000 Basic). The ultimate analysis is conducted using the CHNS analyzer to determine the C, H, N, O and S contents of the samples. ATR machine (Nicoler iS5 Thermo Scientific ID7) with OMNIC software is used for functional group analysis. The mass and energy yield are calculated using equation (1) and equation (2), respectively.

$$\text{Mass Yield (\%)} = \frac{M_t}{M_o} \times 100\% \quad (1)$$

Where,
M_o = the mass of sample before torrefaction (g)
M_t = the mass of sample after torrefaction (g)

$$\text{Energy Yield (\%)} = \frac{M_t}{M_o}\frac{HHV_t}{HHV_o} \times 100\% \quad (2)$$

Where,
HHV_o = the lower high heating value (MJ/kg)
HHV_t = the high high heating value (MJ/kg)

IOP Conf. Series: Materials Science and Engineering **702** (2019) 012009 doi:10.1088/1757-899X/702/1/012009

3. Result and Discussion

3.1. Preliminary Study

Preliminary study on MW heating pre-treatment is conducted to find out the effect of MW on the structure of FW. Two different MW power were selected for the preliminary study which are 160 and 240 W for 5 min. The Higher Heating Value (HHV) of the torrefied FW (at 320 °C for 30 min) is 23.04 and 23.82 MJ/kg using MW power of 160 and 240 W respectively. This finding shows that, MW heating pre-treatment with 240 W is able to produce higher energy of biochar compared to 160 W.

The scanning electron microscope (SEM) images of the raw FW, with and without exposing to the MW heating are illustrated in Figure 2 and Figure 3. It can be seen that the surface of the raw sample without MW heating is more compact, dense and non-porous. In contrast, MW heating process was altered the structure by creating more pores which produced bigger surface area. MW heating has been reported to rupture cell walls caused by temperature and pressure increases due to the thermal effects of the MW [12].

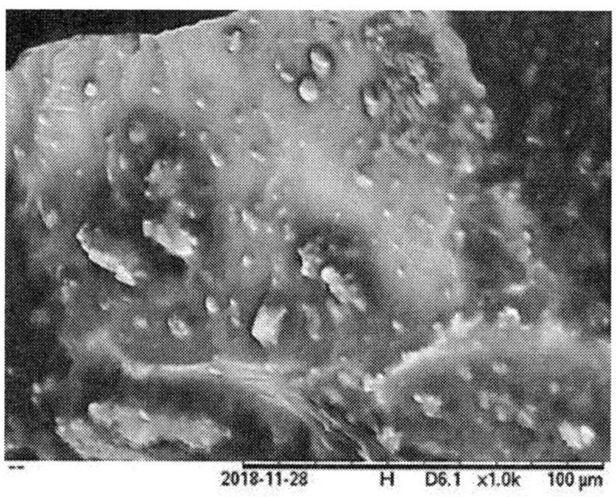

Figure 2. SEM image of raw sample without MW heating (Magnification x1000) .

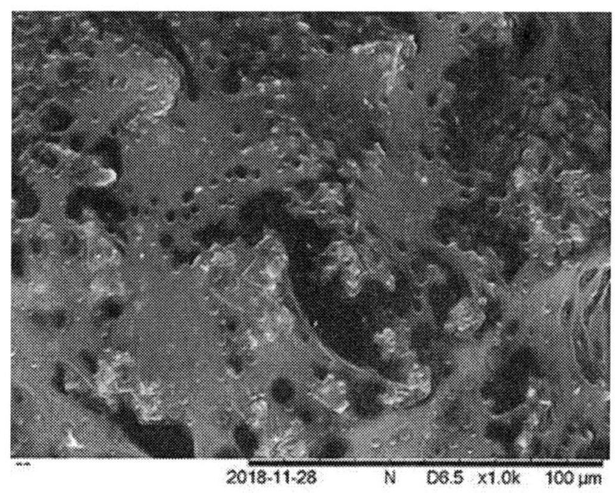

Figure 3. SEM image of raw sample with MW heating (Magnification x1000) .

3.2. Torrefaction Study

The torrefaction of FW is conducted after the optimum condition of MW heating pre-treatment has been selected in the preliminary study (240 W and 5 min). The torrefaction were performed at three different temperatures (280, 300 and 320 °C) for 30, 45 and 60 min. The analysis of the torrefied FW includes ultimate and proximate analysis, HHV, mass and energy yield.

3.2.1. Ultimate analysis. The ultimate analysis is conducted to determine the C, H, O, N and S compositions in the torrefied FW and the results is tabulated in table 2.

Table 2. Ultimate analysis of torrefied FW

Torrefaction conditions		Ultimate analysis (dry wt%)				
Time (min)	Temperature (°C)	C	H	O	N	S
30	280	46.85	6.72	39.81	6.33	0.29
	300	48.37	6.62	38.54	6.19	0.28
	320	53.02	6.49	33.72	6.51	0.26
45	280	50.38	8.82	32.29	6.45	2.06
	300	52.39	8.79	28.89	6.07	3.86
	320	55.50	7.10	28.06	5.83	3.51
60	280	51.85	8.55	30.45	6.36	2.79
	300	52.78	7.53	29.46	6.46	3.77
	320	54.93	7.34	28.38	5.80	3.55

An increase in torrefaction temperature and time resulted an increase in C and decrease in H and O of the biochar. The C content increased up to 55.50 from 44.41 % (in raw FW) while O is reduced from 39.46 to 28.06% after being torrefied. The increase of the C content is contributed by the loss of moisture and other volatile component which contained O and H such as O_2, CO, and CO_2 as explained by Poudel et al. [3], Rago et al. [4] and Abdul Samad et al. [10]. High content of C and low content of O is required in order to produce biochar with high HHV. The N elements only has changed slightly after torrefaction. The lowest element in the torrefied FW is S.

3.2.2. Proximate analysis. The proximate analysis of FW is conducted after being torrefied at 280, 300 and 320 °C for 45 min. The moisture content (MC), volatile matter (VM), ash content (AC) and fixed carbon (FC) is shown in table 3.

Table 3. Proximate analysis of torrefied FW.

Temperature (°C)	Composition (dry wt%)			
	MC	VM	AC	FC
280	2.51	56.27	29.11	12.11
300	2.23	50.17	33.23	14.37
320	1.87	43.87	34.44	19.82

It can be observed that torrefied FW contains more AC and FC but less VM and MC as the temperature increases. A similar trend was reported by and Rago et al. [4] and Samad et al. [10]. The VM of the biochar is reduced from 56.27 to 43.87 % when the temperature increase from 280 to 320 °C. This is due to the devolatilization during the torrefaction. The devolatilization is more intense as the temperature increased. This is also the reason why the FC of the biochar is increases from 12.11 to 19.82 % when the temperature increased.

Torrefaction technique could be used to convert raw FW with high moisture content (61.43 %) to a very low moist biochar (1.87 %). This value is lower than the moisture content of a coal [4]. The reduction of moisture content and the increase of FC as temperature increases will enhance the calorific value of the torrefied FW.

3.2.3. Higher Heating Value (HHV). HHV represent the energy contain in a biochar. The HHV of torrefied FW at different conditions are shown in Figure 4.

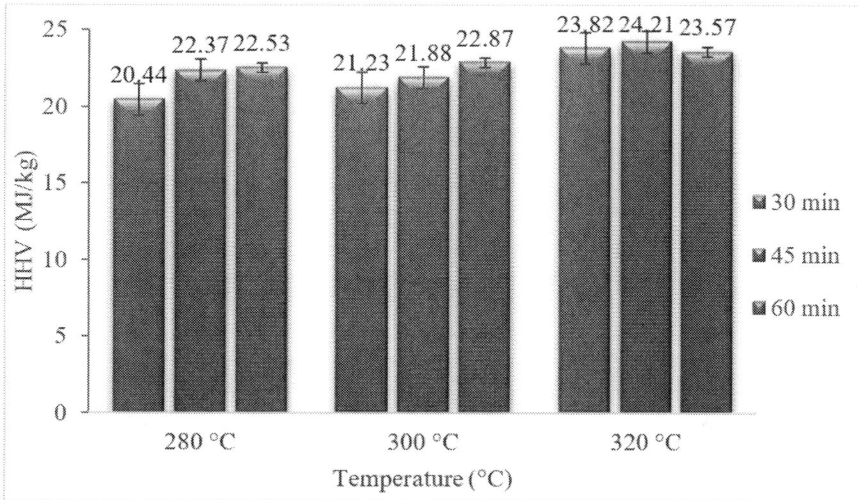

Figure 4. High heating value (HHV) of torrefied FW.

HHV is increases with torrefaction temperature and time accept at 320 °C, where the HHV is start to drop when the time increases from 45 to 60 min. The increased of HHV is due to the loss of moisture and the emission of volatile gaseous which is lead to increase the C content as discussed in the previous section. The increase the C content enhances the value of HHV. The highest HHV, 24.21 MJ/kg was obtained at 320 °C and 45 min.

3.2.4. FTIR Spectra. The functional group of the raw and torrefied FW was analyzed using FTIR as shown in Figure 5.

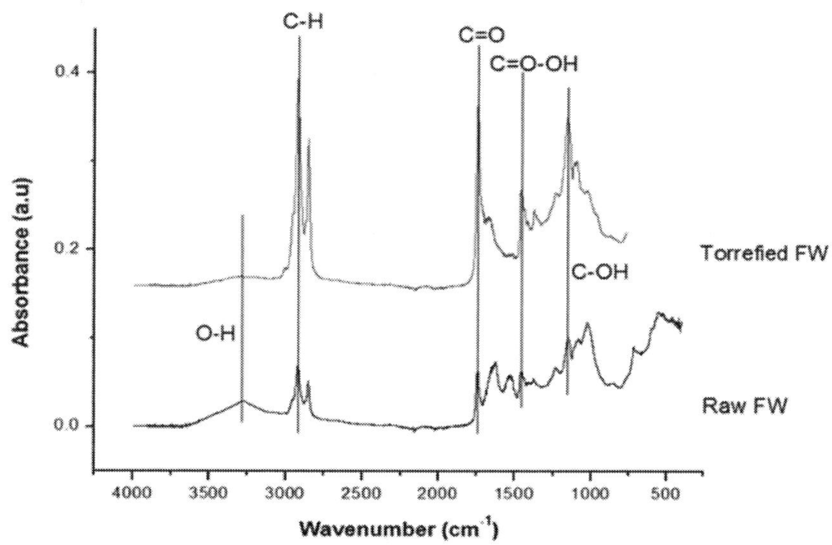

Figure 5. FTIR spectra of raw and torrefied FW.

All the carbon bond peaks are increased after torrefaction accept the O-H group. The intensity of O-H peak is reduced after torrefaction due to dehydration reactions or moisture lost. As the carbon bond contained higher energy compared to O-H bond, the high intensity of carbon bond in the torrefied FW improved the HHV of the FW [6].

3.2.5. Mass Yield. The mass yield of torrefied FW at various operating conditions are demonstrated in Figure 6. It is calculated based on equation (1)

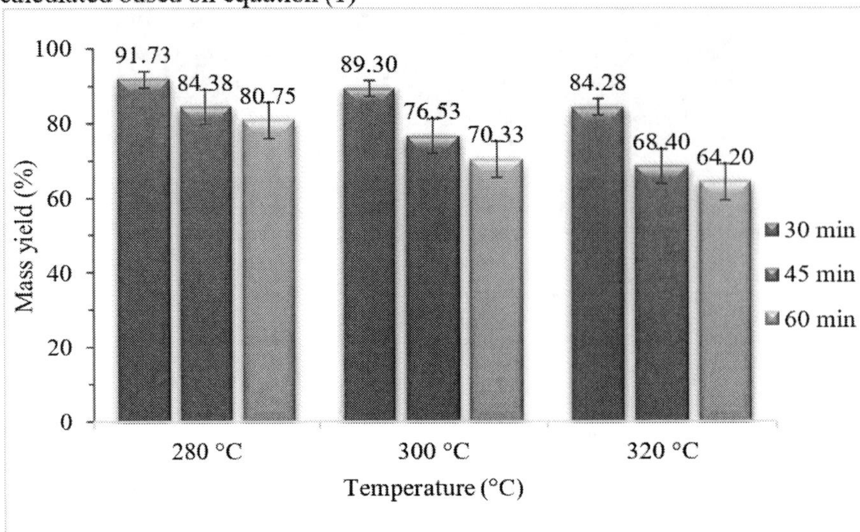

Figure 6. Mass yield of torrefied FW

As can be seen, the mass yield decreased progressively with temperature and time. The mass yield of torrefied FW is drastically decreased at higher temperature and the reduction become higher as the time increased from 30 to 60 min. For example, at 280 °C, the mass yield drop about 7.38% (from 91.73 to 80.75%) as the time increases from 30 to 45 min. However, the mass yield dropped to 12.77% and 15.88% at 300 and 320 °C respectively with the same duration. The mass yield of torrefied FW is the lowest (64.2%) at the most severe condition (320 °C and 60 min). At 280 °C, the reduction of mass yield contributed by moisture lost and some volatiles (O_2, CO and CO_2) emission from the degradation of hemicellulose in the FW. At temperature more than 300 °C, heavier volatiles (methane, acetic acid, formic acid and aromatics) are released from the thermal degradation of cellulose, protein and hydrocarbon which cause more reduction of mass yield [6].

3.2.6. Energy Yield. Energy yield is a parameter which measure the balance between HHV enhancement and mass loss. It was determined using equation (2) and presented in figure 7. The figure shows that, an increase in temperature and time lowers the energy yield.

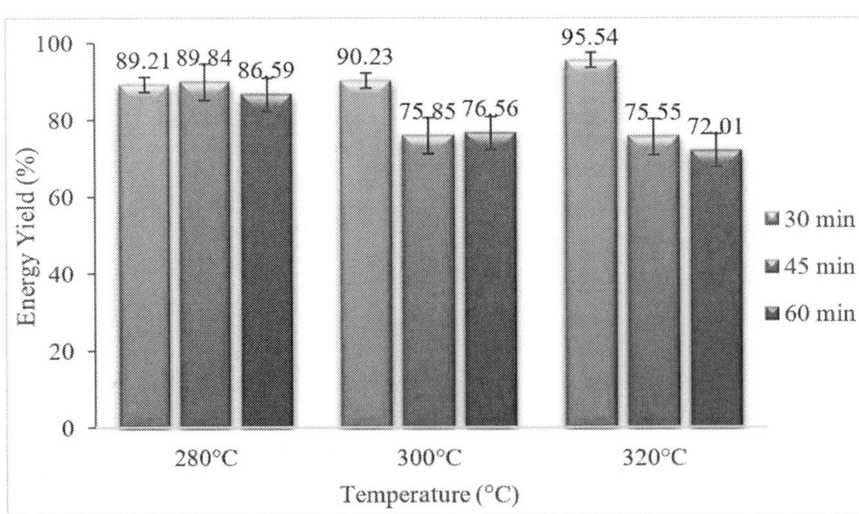

Figure 7. Energy yield for torrefaction process.

Based on the energy yield, temperature of 320 °C and torrefaction time of 30 min is the optimum condition for the torrefaction as it offers the highest energy yield (95.54%).

4. Conclusion

Microwave heating pretreatment increases the surface area of FW by creating pores which increases the thermal degradation during torrefaction. Torrefaction at 320 °C and 45 min produce biochar with highest HHV 24.21 MJ/kg. Even though, the torrefaction principally aims to produce highest HHV, however severe mass loss should be avoided. Therefore, the optimum condition of torrefaction in this study was 320°C and 30 min with HHV of 23.82 MJ/kg, 95.54% of energy yield and 84.28% of mass yield. It can be concluded that torrefied FW pretreated with MW heating has potential for use as an alternative fuel.

Acknowledgement
The authors would like to acknowledge financial support for this work from Universiti Malaysia Pahang (UMP) internal grant (RDU1703171), Faculty of Chemical & Natural Resources Engineering, UMP.

References

[1] Tsang Y F, Kumar V, Samadar P, Yang Yi, Lee J, Ok Y S, Song H, Kim K, Kwon E E and Jeon Y J 2019 Production of bioplastic through food waste valorization *Environment International* **127** 625–44

[2] Wan Ab. Karim Ghani W A, Rusli I F, Awang Biak D R and Idris A 2013. An application of the theory of planned behaviour to study the influencing factors of participation in source separation of food waste *Waste Management* **33** 1276–81

[3] Poudel J, Ohm T and Oh S C 2015 A study on torrefaction of food waste *Fuel* **140** 275–281

[4] Rago Y P, Surroop D and Mohee R 2018. Assessing the potential of biofuel (biochar) production from food wastes through thermal treatment *Bioresource Technology* **248** 258–264

[5] Bach Q, Skreiberg O and Lee C 2017 Process modeling and optimization for torrefaction of forest residues *Energy* 348–354

[6] Pahla G, Ntuli F and Muzenda E 2018 Torrefaction of landfill food waste for possible application in biomass co-firing *Waste Management* **71** 512–520

[7] Abdul Wahid F R A, Saleh S and Abdul Samad N A F 2017 Estimation of higher heating value of torrefied palm oil wastes from proximate analysis *Energy Procedia* **138** 307–312

[8] Sukiran M A, Abnisa F, Wan Daud W M A, Abu Bakar N and Loh S K 2017 A review of torrefaction of oil palm solid wastes for biofuel production *Energy Conversion and Management* **149** 101–120

[9] Buratti C, Barbanera M, Lascaro E and Cotana F 2018. Optimization of torrefaction conditions of coffee industry residues using desirability function approach *Waste Management* 73 523–534

[10] Abdul Samad, N A F, Jamin N A and Saleh S 2017.Torrefaction of municipal solid waste in Malaysia *Energy Procedia* **138** 313–318

[11] Abdul Rasid R, Chin T M, Ismail M and Abdul Rahman R N U 2019. Effect of torrefaction temperature, residence time and particle size on the properties of torrefied food waste *Indones. J. Chem.* **19**(3) 753–760

[12] Liew R K, Nam W L, Chong M Y, Phang, X Y, Su M. H, Yek P N Y, Ma N L, Cheng C K, Chong, C T and Lam S S 2018 Oil palm waste: An abundant and promising feedstock for microwave pyrolysis conversion into good quality biochar with potential multi-applications *Process Safety Environmental Protection* **115** 57–69

IOP Conf. Series: Materials Science and Engineering **702** (2019) 012010 doi:10.1088/1757-899X/702/1/012010

Catalytic production of levulinic acid from matured coconut water

F Ismail, N A Rahman* and F Esa

Chemical Engineering Programme, Faculty of Engineering and Built Environment, Universiti Kebangsaan Malaysia, 43600 UKM Bangi, Selangor Darul Ehsan, Malaysia

*E-mail: norlizajkkp@ukm.edu.my

Abstract. Depletion of fossil as fuel resources due to increasing energy demands has led to a challenging transition to renewable abundance bio-based feedstock. Levulinic acid is a versatile platform chemical that can be utilized to produce fuels, solvents, polymers, pharmaceutical and agrochemical products. It can be synthesized from natural resources such as lignosellulosic biomass, sugars and environmental waste. Matured coconut water is an agro waste that produced mainly from coconut milk industry which contains variety of sugars and minerals. This study was carried out on synthesizing levulinic acid from matured coconut water by focusing on the effect of catalytic acid to coconut water ratio, types of acid and reaction time. The matured coconut water was mixed with sulfuric acid at three different concentrations (0.5 M, 0.75 M, 1.0 M) in 9:1, 8:2, and 7:3 ratios respectively. Mixtures were heated at 100°C for 30 minutes. The ratio decided was carried out into determine the best types of acid (sulfuric acid, hydrochloric acid, phosphoric acid) followed by the reaction time (15, 30, 45, 60 minutes). Levulinic acid production increase as concentration of acid increase. However, low acid percentage in coconut water medium was appropriate enough to give high yield of product (0.135%). Both hydrochloric and sulfuric acid catalyzed more product at high concentration.

1. Introduction

Increasing demand on fossil fuels particularly in petrochemical industry requires new replacement for this resource [1]. Biomass is the only carbon sources that can be used in production of liquid hydrocarbons and chemicals [2]. Levulinic acids has been identified as potential platform chemical in fuel production by the US Department of Energy [5]. It is derived from degradation of sugars or cellulose. Naturally, lignocellulosic biomass consists of three polymers that are cellulose, hemicellulose and lignin that exist in a matrix form to prevent depolymerization and transformation [3]. Acids or enzymes are usually used to degrade the polymer of biomass into formation of simple sugar such as glucose, galactose and xylose.

Waste product from coconut industry such as matured coconut water contains variety of sugars including glucose, fructose and sucrose as it also rich with minerals (table 1) [4]. Under harsh processing conditions, sugar can be converted into useful levulinic acid [3].

Table 1. Composition of coconut water in different stage of maturity [4]

Physicochemical properties	Stage of maturity coconut (months)		
	5-6	8-9	>12
Volume of water (mL)	684	518	332
Total dissolved solid (*Brix)	5.6	6.15	4.85
Titratable acidity (%)	0.089	0.076	0.061
pH	4.78	5.34	5.71
Turbidity	0.031	0.337	4.051
SUGARS CONTENT			
Fructose (mg/mL)	39.04	32.52	21.48
Glucose (mg/mL)	35.43	29.96	19.06
Sucrose (mg/mL)	0.85	6.36	14.37
MINERALS			
Potassium (mg/100mL)	220.94	274.32	351.10
Sodium (mg/100mL)	7.62	5.60	36.51
Magnesium (mg/100mL)	22.03	20.87	31.65
Calcium (mg/100mL)	8.75	15.19	23.98
Iron (mg/100mL)	0.294	0.308	0.322
Protein (mg/mL)	0.041	0.042	0.217
Total phenolic content (mg/mL)	54.00	24.59	25.7

Levulinic acid or gamma-ketovaleric acid is a short-chain fatty acid, having a group of ketone carbonyl (CO) and a carboxylic acid (COOH). It is a versatile chemical with a variety of potential usage such as raw material for synthesis of pharmaceutical products, plastic production, polymer resins, textile dyes, animal feed additives, fuel additives, antifreeze and antimicrobial foods [6]. In recent years, the potential of levulinic acid had attracted attention of many researchers [7]. Girisuta has provided a comprehensive study on the various carbohydrate feedstocks used to produce levulinic acid [8]. The production of levulinic acid from carbohydrates involves dehydration of hexose sugars (such as glucose, fructose, sucrose) for hydroxylation of HMF to levulinic acid and formic (Figure 1).

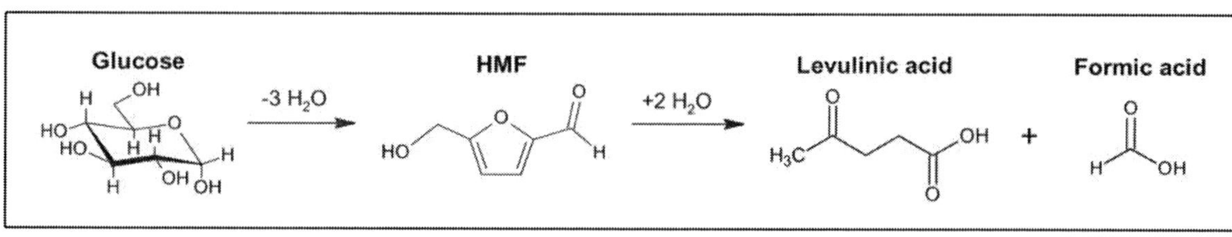

Figure 1. Acid-catalyzed decomposition of glucose to levulinic acid [8]

The most common catalyst used for producing levulinic acid is mineral acid such as sulfuric acid, hydrochloric acid and phosphoric acid [9]. Even though many heterogeneous catalysts have been used in previous study, problems regarding with separation of catalysts from mixtures is still become a concern [10]. In this study, the potential of matured coconut water as raw material for levulinic acid production was investigated with the help of mineral acids as catalyst.

2. Methodology

2.1 Determination of matured coconut water ratio into acid

The sulfuric acid was diluted into three different concentrations that were 0.5 M, 0.75 M and 1.0 M. For each concentration, the acid was added into matured coconut water by 9:1, 8:2 or 7:3 ratios to make 100 ml of mixture. The solution was placed in 500 ml round bottom flask and heated at 100 °C for 30 minutes to measure the amount of levulinic acid produced. The yield was analyzed by High Performance Liquid Chromatography (HPLC).

2.2 Determination of potential acid catalyst

Hydrochloric acid, sulfuric acid and phosphoric acid were diluted into 0.5 M, 0.75 M and 1.0 M concentration. The catalyst was added into matured coconut water to make 100 ml solution according to the best ratio decided from method above. The mixture was placed in 500 ml round bottom flask and heated at 100 °C for 30 minutes to measure the amount of levulinic acid produced. The yield was analyzed by High Performance Liquid Chromatography (HPLC). The best acid catalyst was decided.

2.3 Effects of reaction time on levulinic acid production

The selected acid in three concentrations (0.5, 0.75, 1.0 M) with appropriate ratio was heated in different duration that were 15, 30, 45 and 60 minutes respectively.

2.4 Levulinic acid and formic acid analysis

All samples were filtered to remove the particulate residues and injected into a high performance liquid chromatography (HPLC) system. The analytical column used was an Phenomenex C18 250 mm column. The operating temperature was maintained at 40 °C using 95% of 20 mM sulfuric acid and 5% acetonitrile mobile phase while the volumetric flow rate was maintained at 0.6 ml/min.

2.5 Modelling technique for simulation of levulinic acid production

The kinetic parameters for the proposed kinetic model were estimated using a maximum-likelihood approach, which is based on minimization of errors between the experimental data and the kinetic model. Minimization of errors was initiated by providing initial guesses for each kinetic parameter. The best estimates were obtained using the MATLAB toolbox fminsearch, which is based on the Nelder–Mead optimization method.

The yield of LA on a molar base (YLA) is defined as the ratio of the LA concentration in the reaction product (C_{LA}) and the initial concentration of the C6-sugars in the matured coconut water ($C_{C6,0}$):

$$Y_{LA}\,(\text{mol }\%) = \frac{C_{LA}}{C_{C6,0}} \qquad (1)$$

$C_{C6,0}$ is the sum of the available C6-sugar monomers (glucose and galactose) in the coconut water. It is also possible to define the yield of LA on a weight base (Y_{LA},wt), which is defined as the mass ratio of LA and the available C6-sugars in the coconut water.

$$Y_{LA,wt}\,(\text{wt}\%) = \frac{C_{LA} \times M_{LA}}{C_{C6,0} \times M_{C6,sugar}} \qquad (2)$$

the terms M_{LA} and $M_{C6\text{-}sugars}$ represent the molecular weight of LA (116 g mol^{-1}) and C6-sugars (180 g mol^{-1}), respectively. Chemical balance (1) to (4) showed the chemical reaction of glucose to levulinic acid. The equation (5) to (16) is the mathematical kinetic model of levulinic acid production from glucose that was for simulation in MATLAB R2013b (Simulink).

IOP Conf. Series: Materials Science and Engineering **702** (2019) 012010 doi:10.1088/1757-899X/702/1/012010

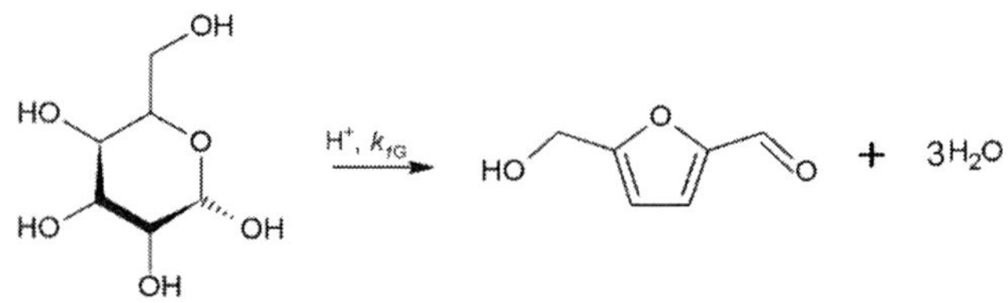

(3)

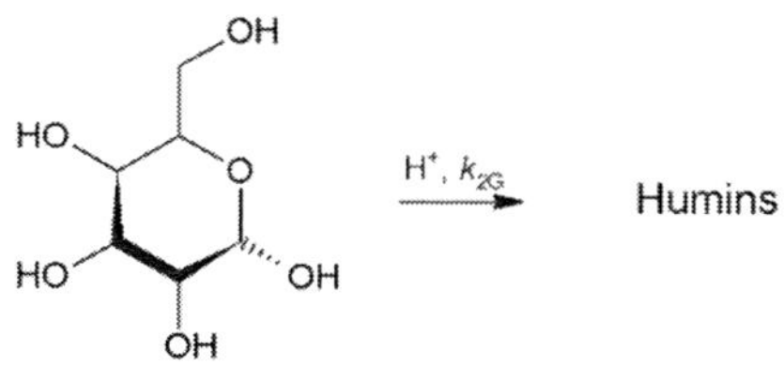

(4)

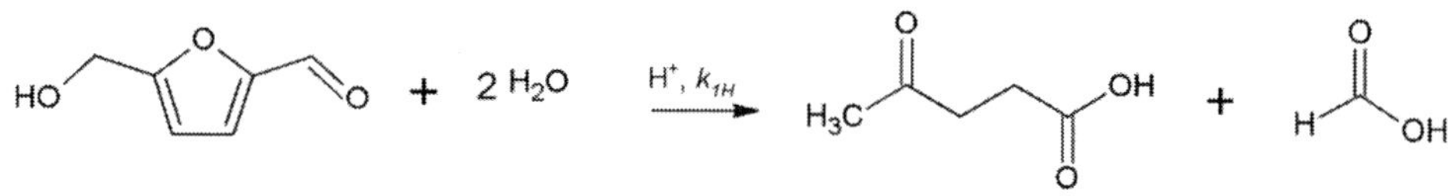

(5)

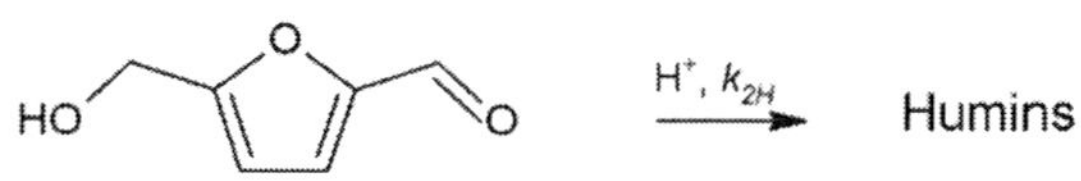

(6)

$$R_1 = k_{1G}(C_{GLC})^{aG}(C_{H^+})^{\alpha G} \tag{7}$$

$$R_2 = k_{2G}(C_{GLC})^{bG}(C_{H^+})^{\beta G} \tag{8}$$

$$R_3 = k_{1H}(C_{HMF})^{aH}(C_{H^+})^{\alpha H} \tag{9}$$

$$R_4 = k_{2H}(C_{HMF})^{bH}(C_{H^+})^{\beta H} \tag{10}$$

$$K_{1G} = k_{1RG}\ exp^{[-(E_{1G}/R)(\frac{1}{T}-\frac{1}{T_R})]} \tag{11}$$

$$K_{2G} = k_{2RG}\ exp^{[-(E_{2G}/R)(\frac{1}{T}-\frac{1}{T_R})]} \tag{12}$$

$$K_{1H} = k_{1RH}\ exp^{[-(E_{1H}/R)(\frac{1}{T}-\frac{1}{T_R})]} \tag{13}$$

$$K_{1H} = k_{2RH}\ exp^{[-(E_{2H}/R)(\frac{1}{T}-\frac{1}{T_R})]} \tag{14}$$

$$C_{H^+} = C_{H_2SO_4} + \frac{1}{2}(-K_{a,HSO_4^-} + \sqrt{(K_{a,HSO_4^-}^2 + 4C_{H_2SO_4}K_{a,HSO_4^-}}) \tag{15}$$

$$\frac{dC_{GLC}}{dt} = -(R_1 + R_2) \tag{16}$$

$$\frac{dC_{HMF}}{dt} = R_1 - (R_3 + R_4) \tag{17}$$

$$\frac{dC_{LA}}{dt} = R_3 \tag{18}$$

IOP Conf. Series: Materials Science and Engineering **702** (2019) 012010 doi:10.1088/1757-899X/702/1/012010

3. Results and Discussions

Levulinic acid was produced from sugar through acid hydrolysis along with heating. In this process, glucose was decomposed into hydroxy methyl furfural (HMF) followed by conversion to levulinic acid and formic acid. Sulfuric acid is a well-known catalyst in process but its efficiency may vary depends on the raw material [11] and reaction conditions such as temperature and acid concentration. Thus, a preliminary study was conducted prior to simulation modelling on levulinic acid production by manipulating reaction time and catalyst concentration. The concentration profile of acid sulfuric-catalysed hydrolysis on matured coconut water are given in table 2 where the reactions were carried for 30 minutes at 100°C. For simple sugars such as glucose and fructose, temperature ranging from 90-120°C was sufficient for their conversion to levulinic acid [8].

At the end of reaction, 10 % of 1 M sulfuric acid mixture give the most production of levulinic acid. The yield however decrease when lower concentration of acid were used and same pattern was observed as portion of coconut water in mixture decrease. Eventhough 10 % of 1 M sulfuric acid has the highest product, the yield percentage was relatively low (less than mol 1%). The possible explaination for this finding is duration for monomeric C6-sugars' reduction was not long enough for reaction. This is consistent with Rackemann [3] which suggest that an improvement trend in the yield of levulinic acid will happen by increasing catalyst concentration and response time. In further experiment, ratio of 9: 1 old coconut water to sulfuric acid was used.

Table 2. The yield of levulinic acid in different ratio of coconut water to sulfuric acid mixture

Ratio	Concentration of acid catalyst (M)	Yield of levulinic acid (mol%)
9:1	0.50	0.096
8:2	0.50	0.094
7:3	0.50	0.054
9:1	0.75	0.102
8:2	0.75	0.094
7:3	0.75	0.093
9:1	1.00	0.134
8:2	1.00	0.098
7:3	1.00	0.094

To determine the other potential catalyst and enhance the levulinic acid production, a comparison was made between three types of strong acids that were hydrochloric acid (HCl), sulfuric acid (H_2SO_4) and phosphoric acid (H_3PO_4). As these acids were diluted into different concentrations, the acid concentration of 1.0 M gave the highest yield respectively (table 3).

HCl and H_2SO_4 catalyst gave higher yield than H_3PO_4 due to its acidity strength. These acids have different dissociation constant (pKa) (HCl: -7, H_2SO_4: -3, H_3PO_4: 2.1) where the acidity strength inversely proportional to the pKa value. Furthermore, the experiments of the reaction time parameters have been performed with different time ranges; 15, 30, 45 and 60 minutes. Thus, sulfuric acid was used in further experiment. Figure 2, 3 and 4 show the amount levulinic acid and formic acid formed from sugars degradation of matured coconut water in interval 15 minutes of reaction times.

Table 3. Average yield of levulinic acid with different concentration of acid catalyst

Concentration of acid catalyst	Yield of levulinic acid, LA (mol %)	Yield of formic acid, FA (mol %)
0.50M H_2SO_4	0.046	0.009
0.75M H_2SO_4	0.052	0.016
1.00M H_2SO_4	0.071	0.028
0.50M HCl	0.046	0.001
0.75M HCl	0.053	0.006
1.00M HCl	0.079	0.011
0.50M H_3PO_4	0.039	0.005
0.75M H_3PO_4	0.046	0.010
1.00M H_3PO_4	0.057	0.013

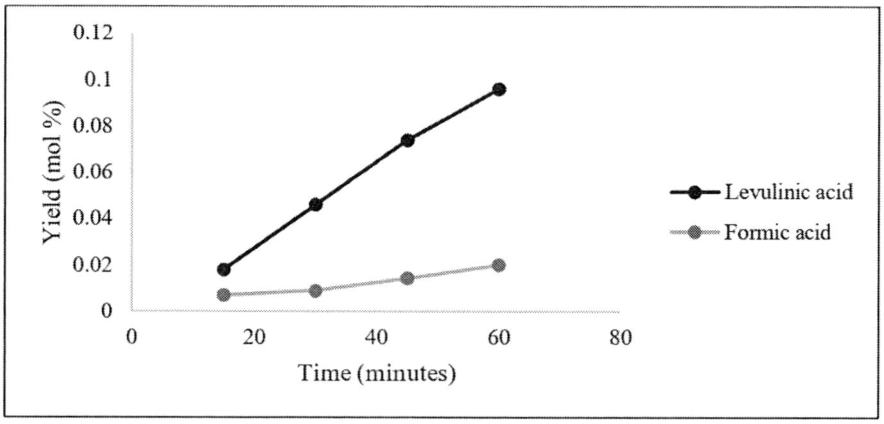

Figure 2. The amount of levulinic and formic acid produced from matured coconut water catalyzed by 0.5 M H_2SO_4

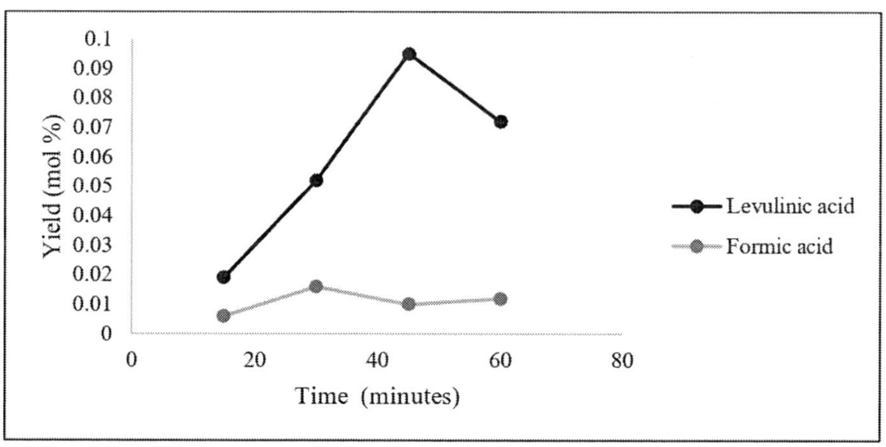

Figure 3. The amount of of levulinic and formic acid produced from matured coconut water catalyzed by 0.75 M H_2SO_4

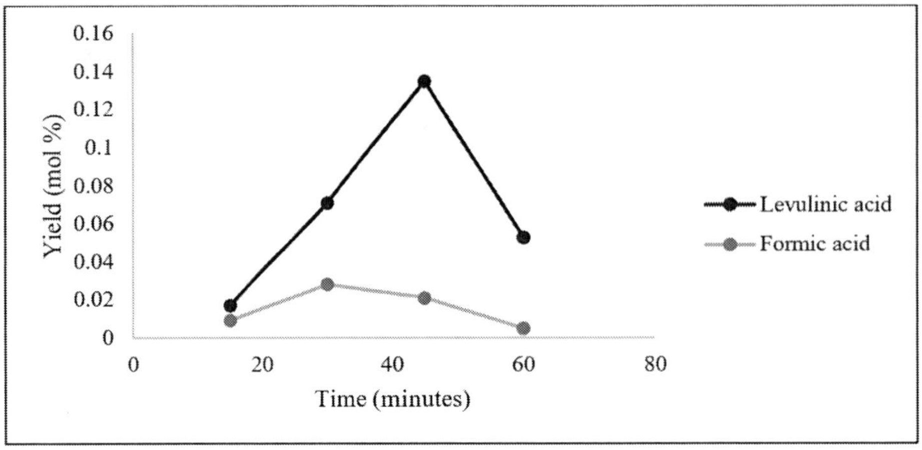

Figure 4. The amount of levulinic and formic acid produced from matured coconut water catalyzed by 1.0 M H_2SO_4

The levulinic acid production increased as the reaction time prolong in these three acid concentrations but after 60 minutes of reaction, a decline pattern was found in both 0.75 M and 1 M H_2SO_4. These finding were however contradicting with Rackemann (2014) who stated that improvement trend of yield will be seen by increasing catalyst concentration and response time. These unpredicted finding might due to error occurred during experimental setup or formation of humin might has begun once the reaction time was extended to one hour.

A simulation model is a mathematical model that tries to resemble a real life system through use of computer software. For the simulation of levulinic acid production, a modelling system was developed using the MATLAB R2012b simulink software. The system has been build base on the equation (7)-(18) that enable the control catalyst concentration input along with the reaction time. The kinetic parameters used in this simulation were regulated based on Girisuta et al. [8] by assuming concentration of HMF is equal to concentration of acid catalyst where it gives the value of $k_{1H} = 0.0029$, $C_{HMF} = 0.11$, aH = 0.88, αH = 1.38 and the concentration of acid is 1.0 M H_2SO_4 (figure 5).

Figure 6 depicts an increasing trend of levulinic acid production for the first half hour consistent with the experimental trend and the production was predicted to increase as it reaches 2 hours of reaction while remain constant by slightly decrease for the rest of reaction time. To validate this model, experiment was repeated with the suggested conditions and percentage of differences was calculated. The differences that less than 20 percent was considered as a good model.

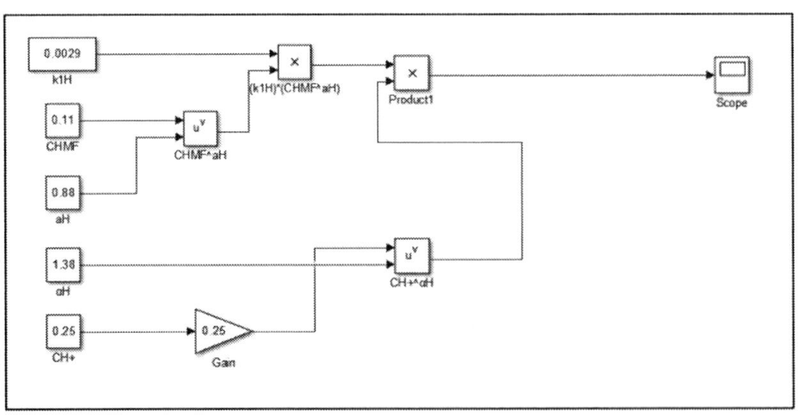

Figure 5. Simulink model of levulinic acid production

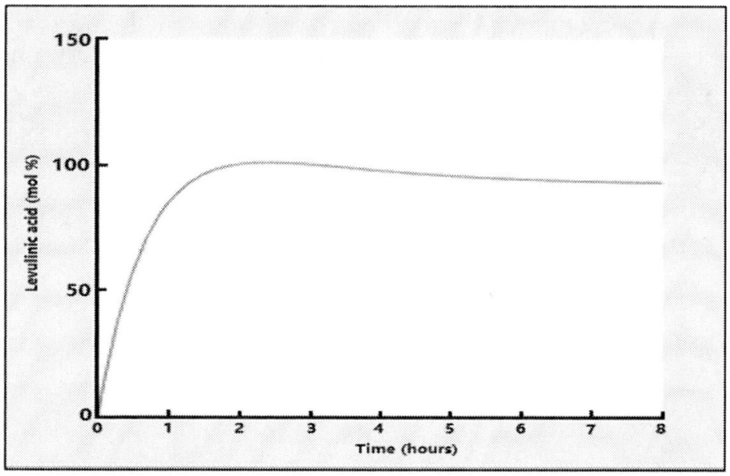

Figure 6. Predicted yield percentage of levulinic acid according to simulation model

4. Conclusion

The 9:1 ratio of matured coconut water to acid catalyst was selected as the best ratio to carry on the sequential experiment. Both HCl and H_2SO_4 are strong acids that gave high levulinic acid amount at operating temperature 100°C and 30 minutes reaction time. Yet, matured coconut water which commonly as agrowaste in Malaysia has the potential to produce levulinic acid. As compared with previous studies by Rackemann and Girisuta, the levulinic acid finding in this research was significantly low. Sophisticated technology is one of the factor that contribute to their high results. Simulation of levulinic acid production systems has been developed using MATLAB R2013b (Simulink) through generating kinetic model from several related equations. Although the simulation and experimental graph were slightly different, some errors might occur during these processes.

References

[1] Frankiewicz A 2016 Overview of 4-oxopentanoic (levulinic) acid production methods – an intermediate in the biorefinery process *Chemik* **70(4)** 206–08

[2] Huber G W, Iborra S and Corma A 2006 Synthesis of Transportation Fuels from Biomass: Chemistry, Catalysts, and Engineering *Chem Rev* **106(2)** 4044–98

[3] Rackemann D W 2014 *Production of Levulinic Acid and Other Chemicals From Sugarcane Fibre* (Queensland: PhD thesis Queensland University of Technology) p 264

[4] Prades A, Dornier M, Diop N and Pain J P 2012 Coconut water uses, composition and properties: a review *Fruits* **67(2)** 87–107

[5] Bozell J J and Petersen G R 2010 Technology development for the production of biobased products from biorefinery carbohydrates—the US Department of Energy's "Top 10" revisited. *Green Chemistry* **12(4)** 539

[6] Watson L J and Connors C G 2008 Furfural - A value adding opportunity for the Australian Sugar Industry *Proceedings of the Australian Society of Sugar Cane Technology* **30** 429–36

[7] Fang Q and Hanna M A 2002 Experimental studies for levulinic acid production from whole kernel grain sorghum. *Bioresource Technology* **81(3)** 187–92

[8] Girisuta B, Danon B, Manurung R, Janssen L P and Heeres H J 2008 Experimental and kinetic modelling studies on the acid-catalysed hydrolysis of the water hyacinth plant to levulinic acid *Bioresource Technology* **99** 8367–75

[9] Tarabanko V E, Chernyak M Y, Aralova S V and Kuznetsov B N 2002 Kinetics of levulinic acid formation from carbohydrates at moderate temperatures. *Reaction Kinetics and Catalysis Letters* **75(1)** 117–26

[10] Chen H, Yu B. and Jin S 2011 Production of levulinic acid from steam exploded rice straw via solid superacid S2O82-/ZrO2-SiO2-Sm2O3 *Bioresource Technology* **102(3)** 3568–70
[11] Meinita M D N, Hong Y K and Jeong G T 2012 Comparison of sulfuric and hydrochloric acids as catalysts in hydrolysis of *Kappaphycus alvarezii* (*cottonii*) *Bioprocess Biosyst. Eng.* **35(1-2)** 123-28

IOP Conf. Series: Materials Science and Engineering **702** (2019) 012011 doi:10.1088/1757-899X/702/1/012011

Aspen Plus simulation and optimization of industrial spent caustic wastewater treatment by wet oxidation method

S Chandraseagar, A H Abdulrazik*, S N Abdulrahman and M A Abdaziz

Faculty of Chemical and Process Engineering Technology, College of Engineering Tecnology, Universiti Malaysia Pahang, Lebuhraya Tun Razak, 26300 Gambang, Kuantan, Pahang, Malaysia

*E-mail: abdhalim@ump.edu.my

Abstract. Spent caustic is a waste generated by petrochemical refineries, which used to eliminate acid components such as hydrogen sulphide (H_2S) and mercaptans from the refined product streams. A study was carried out on simulation and optimization of spent caustic wastewater treatment system by using Aspen Plus based on wet air oxidation (WAO) method. WAO method uses air under elevated temperature and pressure to carry out oxidation in the aqueous phase. Process flow diagram (PFD) of WAO method was drawn and relevant input data was keyed in Aspen Plus according to experimental results on WAO. Both results were validated and the results were match each another as the percentage of allowable error between both the results were less than 5 %. After that, optimization was carried out to determine the optimum ratio of spent caustic feed flow rate to air flow rate and operating temperature of the flash separator unit involved in the process. From the optimization results, it can come to a conclusion that 1: 9.2 is the minimum spent caustic feed flow rate to air flow rate ratio and optimum operating temperature of flash separator unit is 140°C in order for maximum flow rate of CH_3SSCH_3 in offgas stream in this simulation model of spent caustic wastewater treatment system using WAO method.

1. Introduction

The availability of quality water has grown to be quintessential on a daily basis. In recent times, variety of activities that have affected the standard of water were inclining. Sewage from agricultural and industrial activities are viewed as one of the worst effluent to the water [1]. Water pollution is principally coming from industrial and domestic wastewater. Industrial wastewater incorporates high hazardous and toxic chemicals which will incredibly impact human health and aquatic lives. These additionally consist of heavy metal such as sulphide, copper, zinc, iron, cadmium and manganese [2].

Sodium hydroxide (caustic) which generally known with chemical formula NaOH are used in petrochemical and petroleum refineries for the elimination of acid components such as hydrogen sulphide (H_2S), cresylic acids, mercaptans and naphthenic acids from the subtle product streams. When the sodium hydroxide is used in the plant to get rid of those components, the resultant waste is defined as spent caustic. According to Suarez [3], three types of spent caustic are typical from the petroleum industries, and they are from sulphidic, naphthenic and cresylic origins. Sulphidic spent caustic streams are generated when light petroleum fractions are treated with a dilute of 10% caustic solution to convert odorous or objectionable chemical constituents to their salt forms which are then soluble in aqueous solution. This caustic extraction is done to improve color, odor, and oxidation stability of the petroleum product [4]. Naphthenic spent caustic is produced from the caustic scrubbing of kerosene and diesel products which contain high concentrations of polycyclic aliphatic organic compounds such as naphthenic acids. While the cresylic spent caustic is generated when gasoline scrubbing in the fluidized bed catalytic cracking processes with caustic containing high concentrations of organic compounds including phenols and cresols.

Once spent, the caustic solution can be a challenge to handle and dispose as it typically contains pH greater than 12, sulphide concentrations exceeding 2 to 3 wt%, and about 80% of residual alkalinity which comes from excess sodium hydroxide. Constituents in the spent caustic for example sodium bisulphide (NaHS) and mercaptans which can be represented by sulphide are extremely odorous as they have the tendency to cause diseases or even death to living organisms [5]. One of the most common methods used in industries for treating spent caustic is wet air oxidation (WAO) method. Other method included chemical oxidation[6–8]. WAO method can be categorized into three, which are low WAO, medium WAO and high WAO, depending on the reaction operating temperature and pressure. In detail, Clark [9] has compared properties and performances for both WAO and chemical oxidation methods, and has shown that medium and high WAO are better in terms of cost and handling. However, the elimination of pollutants in effluent may need one or combination of treatment techniques depending on the type of compounds and concentration in solution [10]. Like other processes, design and operational analyses of industrial wastewater are challenging without having a process model. In the case of WAO method, the specific challenges are to have an established mass balances for a typical flowsheet and finding the optimal operating conditions. Tian et al. [11] have simulated a scheme to treat organic wastewater produced from Eastman process. Weibin et al. [12] have developed Aspen Plus simulation and optimization of ammonia removal from azodicarbonamide foaming agent (ADC) wastewater for both absorber and stripper sections. Gamba et al. [13] have simulated scrubbing processes to obtain biomethane from municipal sewage sludge-derived biogas using Aspen Plus with emphasis to energy analysis. To the author's knowledge, Aspen simulation of WAO method with optimization of process parameters has yet to be studied. Hence this paper would be detailing the simulation procedures using Aspen plus for the process, as well as to find the optimal ratio values of spent caustic and water flowrates, and flash separator temperature.

2. Methodology

The overall methodology for this work is shown in figure 1. Simulation model of WAO method was developed in Aspen Plus based on the process flow diagram (PFD) obtained from Siemens Water Technologies Corporation [14]. Validation of the simulation results was based on the experimental work by Jonathan [15]. Optimization was carried out for two purposes, firstly, to determine the optimum ratio of spent caustic feed flow rate to air flow rate in order to achieve constant sodium sulfate (Na_2SO_4) flow rate in the treated effluent stream. Na_2SO_4 was the product formed from conversion of NaHS by WAO method as shown in equation (1). Secondly, it was to determine the optimum operating temperature of the flash separator unit involved in the process, as to increase the flow rate of dimethyl disulfide (CH_3SSCH_3) in the off-gas stream. CH_3SSCH_3 was the product formed from conversion of mercaptans by WAO (see equation (2)). Increase flow rate of this compound was desired as it is less harmful compared to mercaptans.

$$NaHS + NaOH + 2O_2 \rightarrow Na_2SO_4 + H_2O \quad (1)$$

$$4CH_3SH + O_2 \rightarrow 2CH_3SSCH_3 + 2H_2O \quad (2)$$

IOP Conf. Series: Materials Science and Engineering **702** (2019) 012011 doi:10.1088/1757-899X/702/1/012011

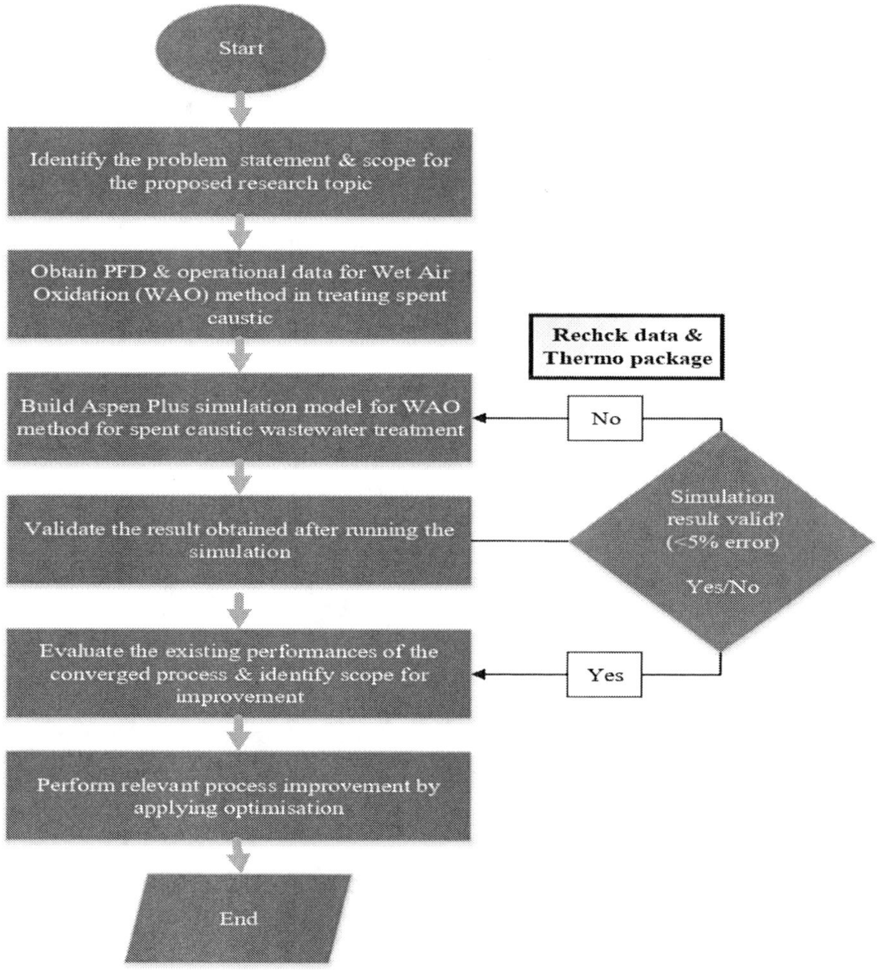

Figure 1. WAO flowchart.

2.1 WAO Process Description

Figure 2 shows the WAO flowsheet which was developed in Aspen Plus software. WAO is the oxidation of soluble or suspended components in an aqueous environment using 21 % O_2 in the air as the oxidizing agent. When air is used as the source of O_2, the process is denoted as WAO process. Oxidation reactions happen at the temperature range between 150 °C and 320 °C (423 K to 593 K), and at pressure range from 10 to 220 bar (1000 kPa to 22000 kPa). Requirement of the operating temperature is determined by the treatment objectives. Operation at higher temperature needs higher pressure to maintain a liquid phase in the system [14].

Rotating equipment such as pump is used to raise the pressure of spent caustic feed stream whereas compressor is used to raise the pressure of air feed stream to the required operating pressures. Once the pressure of spent caustic feed and air feed are raised, both of them are mixed in a mixer before entering heat exchanger for temperature increment with constant pressure. Heat exchangers are routinely employed to preheat the spent caustic feed and air feed mixture before entering the reactor and at the same time recover energy from the reactor effluent. Commonly in industries, the hot mixture outlet from heat exchanger is further heated using a heater to meet the operating condition of the reactor.

After that, the reactor operating temperature and pressure are set based on the compounds in the spent caustic feed stream that is desired to be converted into simpler form. Oxidation reaction occur inside the reactor whereby the reactants which are complex constituents in spent caustic and O_2 in the air react to produce simpler form of products that are soluble in water and is biodegradable. Hot effluent from reactor re-enter into heat exchanger for cooling down and it is further cooled using a cooler before entering flash drum. Flash drum usually operates at low pressure. Thus, high pressure effluent from cooler is reduced to low pressure using a pressure control valve. Liquid effluent consist of several components is partially vaporized in a flash drum at a certain temperature and pressure. The vapour is taken off overhead in the off-gas stream meanwhile the liquid drains to the bottom of the drum in the treated effluent stream whereby it will further treat before the final discharge.

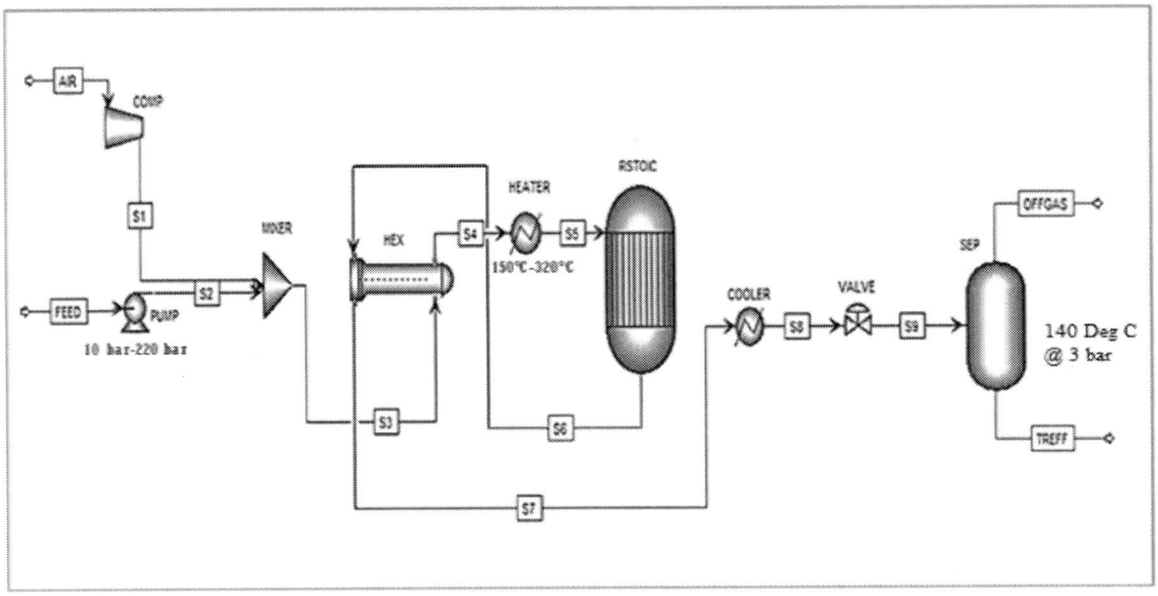

Figure 2. WAO Flowsheet in the Aspen Plus.

2.2 Selection of Blocks and Thermodynamic Property
Blocks used to build this simulation model were pressure changers (pump, compressor and valve), mixer, exchangers (heater, heat exchanger and cooler), reactor (R-stoic), and separators (flash). R-stoic reactor was chosen because stoichiometry of the reactions and fractional conversion of the reactants were known. Thus, by keying the stoichiometry of reactants and products involved in the reaction and entering the desired fractional conversion value of specific reactant in the R-stoic reactor, Aspen Plus would calculate and generate output result of reactants that being converted into products based on the input mass concentration or mass/mole fraction of reactants in the feed. Thermodynamic property applied in this simulation modelling was Soave-Redlich Kwong (SRK), as it was for non-polar and mildly polar mixture, as in the WAO process.

3. Results and Discussion

Aspen Plus simulation was carried out in the direction to validate the data by Jonathan [15], with emphasis on the H_2S and mercaptans concentrations after WAO treatment [15]. Initial concentration value of H_2S was entered into the spent caustic feed stream composition column with assumed flow rate 1000 of litres/hr. Table 3.1 showed the optimal operating conditions and simulation results. From these results have obtained the fractional conversions for NaHS (88.39%) and mercaptans (94.46%) were both less than 5% differences against the experimental data (NaHS, 88.39%; mercaptans, 94.26%). This was required and would be based for further process improvement through optimization.

Table 1. Optimal operating conditions & simulation results attained in Aspen Plus

Optimal Operating Condition			
Temperature	°C	140	
Pressure	bar	3	
Time	minutes	40	
		FEED	OFFGAS
Mass Flow Rate	L/hr	828.1340	46.2177
O_2	L/hr	0	0.0074
N_2	L/hr	0	8.4340
H_2O	L/hr	823.4510	37.1916
CH_3SH	L/hr	0.3682	0.0188
CH_3SSCH_3	L/hr	0	0.3148
NaOH	L/hr	2.1574	4.7021e-15
Na_2SO_4	L/hr	0	0.0009
NaHS	L/hr	2.1574	0.2503

3.1 Optimum ratio of spent caustic feed flow rate to air flow rate
Optimization was done to identify minimum ratio of spent caustic feed flow rate to air flow rate by fixing spent caustic feed flow rate and manipulating air flow rate. Flow rate of Na_2SO_4 which was product produced from the oxidation reaction acts as the responding variable. While spent caustic flow rate was set at 1000 litres/hr, seven simulated runs were done and results were plotted as shown in Figure 3. Based on Figure 3, the optimum spent caustic feed flow rate to air flow rate ratio is 1:9.2 which represented by the dotted lines as Na_2SO_4 flow rate in the treated effluent stream reached steady state beyond this ratio value. Besides that, in order for WAO process occur completely in the reactor and results available without warning, O_2 in air must be in excess. Usually there will be 5 to 21 % of O_2 and some volatile compounds in the off-gas stream from flash drum indicates WAO reaction took place [16].

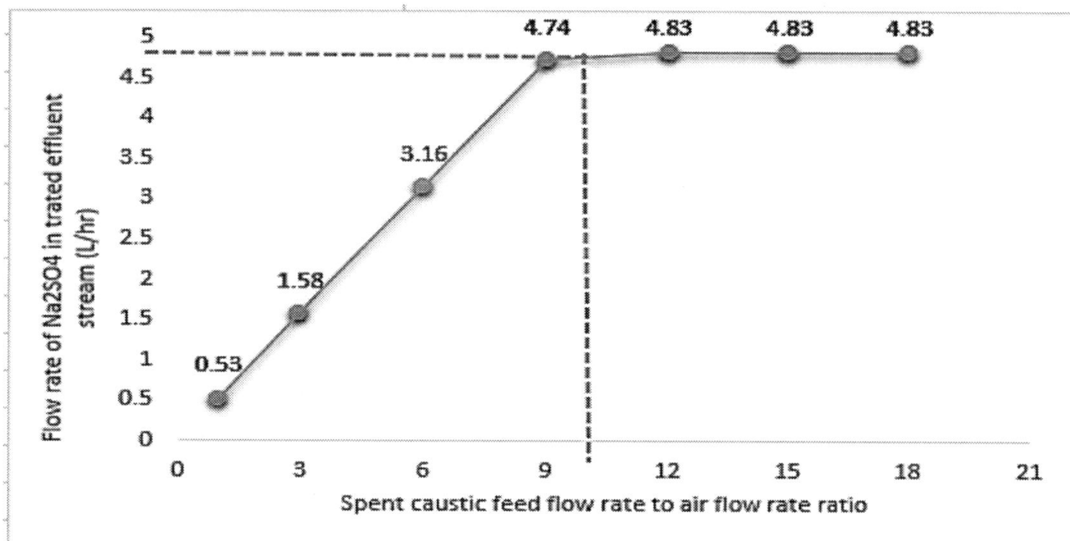

Figure 3. Response of Na_2SO_4 flow rate in treated effluent stream with different spent caustic feed flow rate to air flow rate ratio.

3.2 Optimum operating temperature of the flash separator unit
Optimization was done to determine operating temperature of the flash separator unit involved in the process in order for maximizing the flow rate of CH_3SSCH_3 produced in the off-gas stream. Pressure

of the flash drum was fixed at 3 bar. Simulation results of response in CH_3SSCH_3 flow rate in the off-gas stream by varying the flash drum temperature was observed and plotted in figure 4.

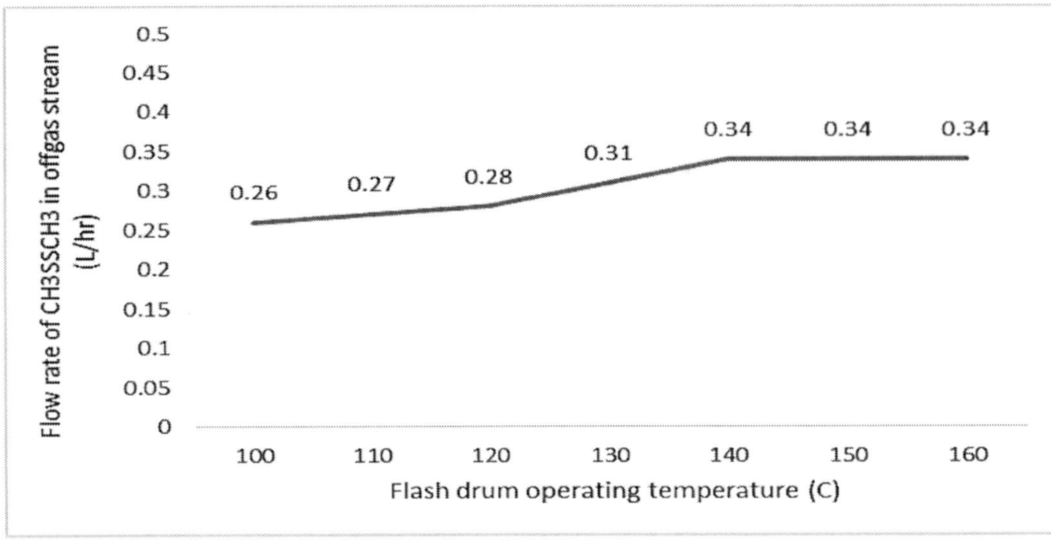

Figure 4. Response of CH_3SSCH_3 flowrate in off-gas stream by increasing flash drum operating temperature.

The flash separator unit separates a component into top product (off-gas stream) and bottom product (treated effluent stream) according to pressure and/or manipulations. In off-gas stream, the components present in vapour phase meanwhile in treated effluent stream, were in liquid phase. Initially operating temperature of the flash drum is set at 100 °C as it is boiling point of water and the response in flow rate of CH_3SSCH_3 is observed. At this temperature, major portion of CH_3SSCH_3 flows in off-gas stream and there is some portion of this compound which about 0.08 L/hr in treated effluent which is not preferred. Therefore, maximum flowrate of this component in vapour phase is desired as it is not hazardous. As the temperature of flash drum was increased beyond 140 °C at 3 bar pressure, flow rate of the compound remains constant at 0.34 L/hr. Therefore, this was the maximum flowrate of CH_3SSCH_3 in the off-gas stream. Moreover, CH_3SSCH_3 presence in the treated effluent stream was almost at zero flowrate and similar happened to mercaptans.

4. Conclusion

This paper has focused on developing Aspen Plus model for both simulation and optimization purposes. Based on the published experimental data and the reported WAO process flow diagram, Aspen Plus simulation model was developed and validated for fractional conversion of NaHS and mercaptans. With the difference less than 5% between the simulation result and the published experimental data, the model was successfully considered to approximate the actual system and provided a base for further improvement. Optimization for the ratio of spent caustic flowrate (fixed) to the air flowrate (varied), and operating temperature of flash separator were then done by simulating for the ranges which the response started to show constant value. It can be concluded that 1: 9.2 was the optimum spent caustic feed flow rate to air flow rate ratio and optimum operating temperature of flash separator unit was 140 °C.

Acknowledgement

This work was funded under the RDU1703170 and RDU1703325 research grants and the authors would like to acknowledge to Faculty of Chemical and Process Engineering Technology, College of Engineering Technology and Universiti Malaysia Pahang (UMP) for the technical assistances and supports.

IOP Conf. Series: Materials Science and Engineering **702** (2019) 012011 doi:10.1088/1757-899X/702/1/012011

References

[1] Elhassadi A 2008 Horizons and future of water desalination in Libya *Desalination* **220** no 1– 3 115–22

[2] Carlsberg Group A S 2019 *Sustainability Report 2018*, ed Carlsberg Group Corporate Affairs (Copenhagen V: Carlsberg A/S) pp 1-62

[3] Suarez F J 1996 Pluses and minuses of caustic treating *Hydrocarb Process* **75**(10)

[4] Tarullo D K 2004 The EU and the WTO: Legal and Constitutional Issues *J. of Int. Economic Law* **5**(4) 941-43

[5] Maugans C and Kumfer B 2005 The use of catalyst to enhance the wet oxidation process *Water Sci. and Technol.* **55**(12) 189–93

[6] Cheremisinoff N P 2017 Groundwater Remediation: A Practical Guide for Environmental Engineers and Scientists *Knovel* pp 262-300

[7] Rosansky S and Chen A 2009 In-situ Chemical Oxidation Current Advancement

[8] Sievers M 2010 Advanced Oxidation Processes *Treatise Water Sci* **4** 377–408

[9] Oil and gas Online 2018 A better method for treating spent caustic (Siemen Energy, Inc: Water Solutions) pp 1-2

[10] Andreozzi R Caprio V Insola A and Marotta R 1999 Advanced oxidation processes (AOP) for water purification and recovery *Catal Today* **53**(1) 51–9

[11] Tian W Cui Z Qin H and Li L 2017 Conceptual design of the Eastman organic wastewater treatment process *Chem Eng Trans* **61** 109–114

[12] Chen W Guo T and Wen L 2016 Simulation and optimization with Aspen- Plus of ammonia removal from azodicarbonamide foaming agent (ADC) wastewater by a stripping process *Beijing Huagong Daxue Xuebao (Ziran Kexueban) J. Beijing Univ Chem Technol Natural Sci Ed* **43**(4) 20–26

[13] Gamba S Pellegrini L A and Langè S 2014 *Energy analysis of different municipal sewagesludge derived biogas upgrading techniques Chem. Eng. Trans.* **37** 829-834

[14] Siemens AG Energy Sector 2013 Zimpro Wet Air Oxidation Systems: The cleanest way to treat the dirtiest water, ed Siemen AG Oil & Gas Division (Erlangen: Siemens AG/ Duisburg: Semen AG/ Wisconsin: Siemens Energy Inc.) pp 1-8

[15] Ketu J 2013 Treatment and disposal of spent caustic at tema oil refinery (TOR) (The Department of Materials Eng: Kwame Nkrumah University of Sci and Technol) pp 1-85

[16] Grover R and Gomaa H 1993 Proven Technologies Manage Olefin Plant's Spent Caustic*Hydrocarb. Process.* **72**(9) 61–70

IOP Conf. Series: Materials Science and Engineering **702** (2019) 012012 doi:10.1088/1757-899X/702/1/012012

Assessment of heat pumping technology in oleochemical fatty acid fractionation

N M Sidek* and M R Othman

Faculty of Chemical and Process Engineering Technology, College of Engineering Technology, Universiti Malaysia Pahang, 26300 Gambang, Pahang, Malaysia.

*E-mail: norulmalakiah.sidek@gmail.com

Abstract. Similar to petroleum industry, major energy consumption in oleochemical plants is also dominated by separation process in order to obtain purified oleochemical cuts. Combination of the distillation column with heat pumping system has emerged as one of the most popular techniques in heat integration. Although heat pumping technology has been proven to be effective in petroleum separation, the research on this technology in oleochemical separation has not yet been discovered. Hence, it would be appealing to investigate the feasibility of integrating heat pump technology with distillation units in hopes of reducing energy usage in the separation of oleochemical products. In this study, two configurations of heat pumping system, namely direct vapor recompression (VRC) and bottom flashing heat pump (BFHP) are simulated in Aspen Plus particularly for fractionation of palm kernel oil (PKO) fatty acid. Proper selection of thermodynamic package is discussed in detail. Only three major components of PKO-based fatty acid are involved in the simulation in order to arrive at simple simulation and easy convergence. Simulation results indicates that both configurations of heat pump can be feasibly integrated with the distillation column. However, to satisfy the heating and cooling requirement, supply of make-up utility is necessary.

1. Introduction

Over the past decades, most research in separation processes has emphasized the use of heat integration approach which not only provides economic benefits but also reduces the greenhouse gas emissions [1]. Principally, the effective separation process involves the supply of heat at an elevated temperature in the reboiler and reject the heat at a low temperature in the condenser [2]. Due to the significant temperature difference between the reboiler and condenser, the overall thermodynamic efficiency of a distillation column is considerably low, which is typically ranging from 5% to 20% [1]. In light of this, distillation has become an utmost priority for general improvement in energy efficiency in most chemical plants.

In the history of energy-efficient distillation development, heat integration has been thought of as a key factor in reducing the energy consumption in the column. The most promising external heat integration approach is the association of heat pumping with distillation unit. The idea of heat-pump assisted distillation (HPAD) lies in the fact that available heat sources and heat sinks within the distillation process may be used for heat recovery. A heat pump is purposely used in an effort to fully utilize the waste heat released from the condenser to assist the evaporation in the reboiler [3]. Of all the available systems, vapor recompression (VRC) is the most popularly known HPAD amongst the researchers. Application of HPAD system would be beneficial to reduce the associated utility costs, by making use of waste heat either from excess vapor at any stages in the stripping section or directly from overhead vapor to aid the evaporation process in the reboiler. There are various possible VRC flowcharts

IOP Conf. Series: Materials Science and Engineering **702** (2019) 012012 doi:10.1088/1757-899X/702/1/012012

can be employed depending on the influencing factors of the process system. Typical VRC configurations in distillation process are direct VRC and bottom flashing heat pump (BFHP).

The research on heat-integrated distillation using HPAD method to date has tended to focus on petroleum processes and very few writers have been able to draw on any systematic research into energy-efficient distillation process in oleochemical plants, let alone the integration of the column with heat pumping and other intensified technologies. The main challenges are coming from the limitations in familiarity and flexibility of oleochemicals which will lead to a complexity in operation; high degree of freedom [4]. Nonetheless, one published literature has successfully developed a model of fatty acid fractionation in a dividing wall column (DWC), closely represented by a four-column configuration [5]. Though the study requires more works on its design controllability, it offers some important insights into integrating the heat integration system to the existing column design.

This present work attempts to demonstrate that the fractionation of palm-based fatty acid in conventional distillation column can be feasibly integrated with the VRC technology. The models are developed using Aspen Plus V10 simulator and the thermodynamic package that is well-suited the process is selected based on the validation process as explained in the methodology section.

2. General process description of fatty acid fractionation

Fractional distillation of oleochemical fatty acid is commonly carried out by means of sequential separation, employing two units of conventional distillation column (CDC) to purify and separate heavy cut (C16-C18), medium cut (C12-C14) and light cut (C8-C10) [5]. The process is operated under vacuum, most of which ranges from 5 to 80 mbar [6]. Due to its heat sensitive characteristic, the chemical decomposition of fatty acids can be avoided if the operating temperature is kept between 180-250°C [7]. The pressure drop within the column should be retained as low as possible, in such a way that the difference between top column pressure and bottom column pressure is 40-60 mbar [7].

3. Principle of distillation column with vapor recompression system

3.1. Direct vapor recompression (VRC)

As illustrated in figure 1, this configuration eliminates the employment of a condenser, as the vapor leaving the top of the column is straight out sent to a compressor. For the purpose of introducing the latent heat of condensation into the reboiler, the overhead stream must be compressed to a point where the temperature is sufficiently high for an efficient heat exchange [8]. In the compression stage, saturated vapor might condense in the compressor. The consequence from this unfavorable situation is that the available latent heat cannot be fully utilized to vaporize the liquid stream in the reboiler [9]. Nonetheless, condensation of the fluid can be avoided by superheating the stream in a superheater prior to compression [8-9]. Large temperature difference indicates high compression ratio, which requires more mechanical works [9-10]. For that reason, the VRC is only considered as cost-effective if the temperature difference is small [10]. The heat transfers from compressed vapor to bottom product occurs through a heat exchanger which operates as a combination of a reboiler and a condenser [11]. In practice, partial condensation might take place during heat transfer process. In view of this, the stream must be further cooled in a trim condenser to a temperature that will not make it vaporized when its pressure is reduced to a column pressure through a throttling valve [9, 12]. The saturated liquid will be then returned to the column, whilst the remaining liquid is discharged as distillate [12]. On the other hand, the saturated vapor in bottom stream is separated from its liquid in the flash drum and recycle to the column whereas the liquid outlet is discharged as bottom product [9, 12].

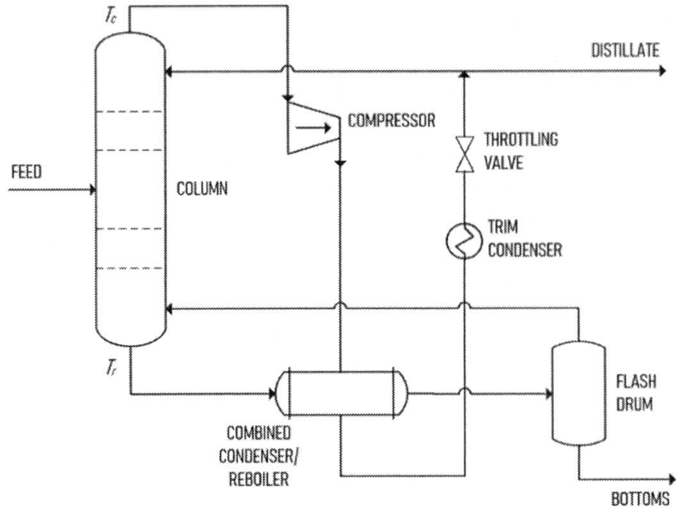

Figure 1. Typical configuration of distillation column with direct VRC system.

3.2. Bottom flashing heat pump (BFHP)

In this heat pump system, the bottom liquid is expanded in a valve to decrease its temperature, bubble point and pressure so that it can be heat exchanged with top vapor stream in a heat exchanger [13]. The bottom stream is flashed until its dew point temperature is much lower that the top vapor temperature [14]. After passing through the heat exchanger, the bottom outlet stream is recompressed in a compressor to the column pressure, increasing the temperature as well. Due to that, the stream needs to be air-cooled so that it can be recycled to the column. Similar to direct VRC system, the vapor is separated from liquid in a flash drum and sent to the column as boil-up, whilst the liquid is discharged as final bottom outlet. In the event that the boil-up is not sufficiently produced, a trim reboiler might be needed to supply the remaining boil-up. For comparison purposes, the condensed top outlet with similar distillate flow rate in the CDC is discharged as a final top product, and the rest is recycled to the column as reflux. Figure 2 shows the standard configuration of BFHP.

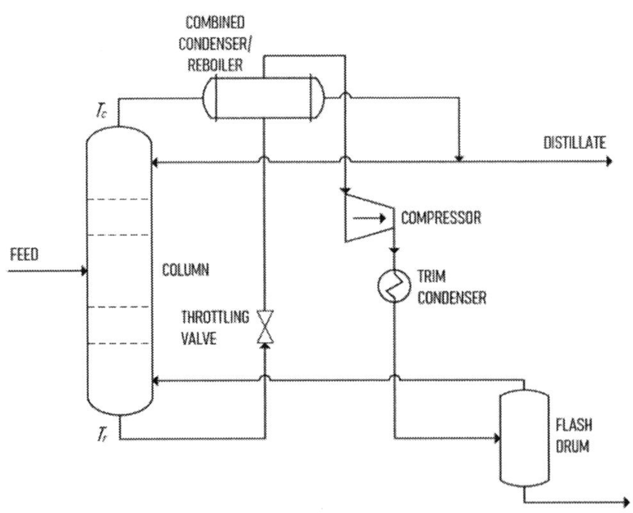

Figure 2. Typical configuration of distillation column with BFHP system.

4. Simulation Methodology

4.1. Feed data selection

In this study, the feed was adopted from Othman and Illner [5] which utilizes industrial palm kernel oil (PKO) based fatty acid. To ease the simulation process, three major compositions of the fatty acid as tabulated in table 1 were used for feed stream input.

Table 1. Major components in PKO based fatty acid.

Component	ID (Aspen Plus)	Formula	Mass fraction	Dipole moments (MUP)
Lauric acid	1-DOD-01	$C_{12}H_{24}O_2$	0.663	2.761
Myristic acid	N-TET-01	$C_{14}H_{28}O_2$	0.227	1.679
Palmitic acid	N-HEX-01	$C_{16}H_{32}O_2$	0.110	1.739

4.2. Selection of thermodynamic model

Many researchers and engineers overlook the importance of choosing an appropriate thermodynamic model prior to carrying out the process simulation in the process simulator. Improper selection will result in a poor performance of the developed process model, or worse, inaccurate result will be generated for the overall simulation. Apparently, there are many thermodynamic models available to choose from and the selection method can be facilitated by using a decision tree [15] which provides appropriate heuristic for better screening options.

As can be seen in Table 1, all three major components of PKO based fatty acid possess dipole moments (MUP) value larger than zero, so as to indicate that the feed to the distillation column is polar fluid. Given that the fatty acid is non-electrolyte system with pressure under 10 bar, it is suggested by the decision tree that the suitable models for this study are NRTL, Wilson, UNIQUAC and UNIFAC. This is inconsistence with previous studies where the thermodynamic model used in fatty acid fractionation were Soave-Redlich-Kwong [5] and Lee-Kesler-Plocker [16]. To determine whether the suggestion from the decision tree is the best choice, further validations for the suggested models as well as the ones used in the published literatures were thoroughly carried out. The validation process using average absolute deviations (AAD) calculations using equation (1) is summarized in table 2.

$$AAD = \frac{\sum_i^n \left| T_i^{exp} - T_i^{pred} \right|}{n} \tag{1}$$

The model predictions results were validated against experimental data from National Institute of Standards and Technology (NIST) available in the Aspen Plus databank. The binary experimental data that could be retrieved from NIST Thermodata Engine (TDE) is only myristic acid/palmitic acid binary system. Under these circumstances, analysis for all listed models was also done for myristic and palmitic acid system to generate more reliable comparison data. The AAD value for each model indicates how much it deviates from the base data. For comparison purposes, the temperature for model predictions was extracted at similar liquid mole fraction of experimental data.

From table 2, it was clearly observed that NRTL, Wilson, UNIQUAC and UNIFAC models acquired considerably small differences which is in agreement with the decision tree. UNIQUAC model has the lowest AAD value which suggests that it is the most suitable model to be used when involving fatty acid components. LK-Plock model, on the other hand, proved the invalidity of previous researches that have employed the particular model for fatty acid process simulations. While SRK model deviations was very much smaller than that in LK-Plock, it is still not the best one to be used.

Table 2. AAD of thermodynamic model predictions from experimental data in Aspen Plus.

Liquid fraction	TDE T_{exp} (K)	Model prediction, T_{pred} (K) NRTL	Wilson	UNIQUAC	UNIFAC	LK-Plock	SRK
0.000	465.150	463.908	463.908	463.908	463.908	457.611	461.960
0.044	563.149	462.752	462.760	462.751	462.659	455.850	460.728
0.160	461.148	459.904	459.921	459.917	459.664	452.124	457.772
0.303	458.147	456.751	456.765	456.790	456.461	448.746	454.605
0.502	454.145	452.926	452.931	452.983	452.689	445.327	450.866
0.629	452.145	450.770	450.779	450.829	450.603	443.606	448.795
0.840	449.644	447.630	447.630	447.654	447.565	441.208	445.769
1.000	447.143	445.546	445.546	445.546	445.546	439.631	443.751
AAD		**1.3095**	**1.3038**	**1.2865**	**1.4469**	**8.3209**	**3.3030**

4.3. Base case: Conventional distillation column (CDC)

The shortcut model in Aspen Plus, DSTWU is usually used to estimate the column performance such as number of stages and reflux ratio, which will later be used for rigorous distillation known as RADFRAC. The column capacity for PKO based fatty acid fractionation in this simulation study was 9000 kg/hr and feed mass compositions were set as tabulated in Table 1. The feed vacuum pressure and temperature throughout this work were taken to be 20 mbar and 180°C, respectively. The estimation of reflux ratio value was determined using DSTWU results and with reflux ratio value of 0.88, 19 stages were required to reach the desired separation. Due to its very low-pressure operation condition, it was assumed that there was no pressure drop across the column. A simulative investigation into the performance of two VRC configurations was carried with comparison to this base case. In all simulations, the feed was introduced to the column at the same condition, and both distillate and bottom product were required as saturated liquid at 20 mbar.

4.4. Distillation column with direct VRC system

For the purpose of arriving at simple process simulation in the simulator, the degree of freedom must be reduced to avoid convergence issues. In this study, the assumption made was that no condensation occurs during compression. The overhead outlet of CDC was directly fed to the compressor with an outlet pressure of 40 mbar, where the temperature was hot enough to supply required heat for bottom liquid heating. The temperature difference was kept within 5-7°C for economical reason. In the heat exchanger, compressed vapor exchanged heat with the bottom liquid stream, where the pressure drops of shell and tube side were neglected. However, the recovered heat was insufficient to satisfy the heating requirement in the reboiler. Thus, a trim reboiler was employed to provide the remaining heat. The throttle valve brought back the stream to its column operating pressure and was then subcooled in the cooler or trim condenser prior to stream splitting. It is very important to make sure that the stream does not vaporize because the reflux stream must return to the column as liquid as well as for comparison purposes, the product streams must be discharged as saturated liquid. In the splitter, mass flowrate of distillate stream was set to be the same as in base case. A portion of liquid is recycled to the top column as reflux. Meanwhile, the heated stream was introduced into the flash drum. The vapor stream was returned to the bottom of the column as boilup. The final product of bottom liquid with similar flowrate as in the CDC was obtained by manipulating the operating temperature of the flash drum. Figure 3 shows the simulated process flow diagram of the system in Aspen Plus. In this simulation, the typical RADFRAC column was substituted with a simple RADFRAC absorber without both condenser and reboiler.

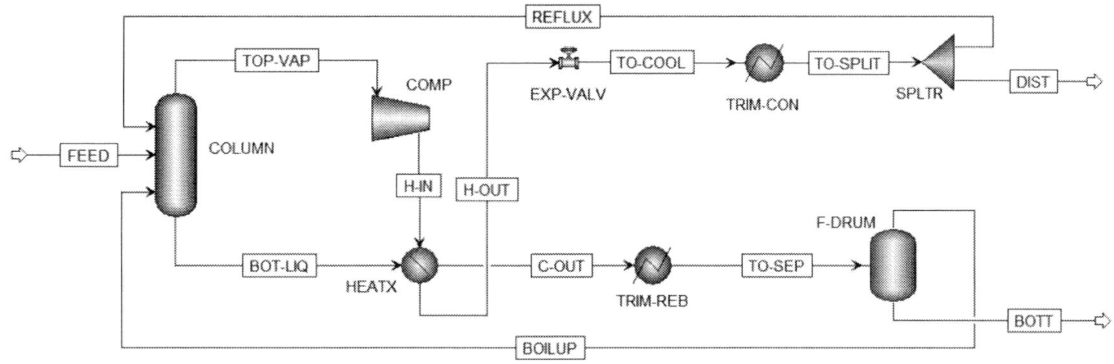

Figure 3: Flowsheet of the direct VRC-assisted distillation column.

4.5. Distillation column with BFHP system

In this configuration, the pressure of the bottom liquid was reduced to 8.5 mbar and the temperature was decreased to 172.38°C to enable heat exchange with the column top vapor stream. Similarly, make-up heat was also required in this system due to insufficient heating duty provided by top column vapor. Hence, additional of a trim reboiler in the flowsheet as can be seen in figure 4. The bottom column outlet entered the compressor to be recompressed to the column operating pressure before being separated in the flash drum. However, a superheater was needed to ensure the stream entering the compressor contained maximum amount of vapor for compression, otherwise the system would not be converged. The recompressed bottom column outlet stream was cooled and sent to the flash drum. Vapor outlet was recycled as boil up and liquid product was discharged as final bottom product. The hot outlet stream of the heat exchanger was straight away divided in the splitter as no cooling needed. Liquid reflux returned to the column while the other portion of liquid was withdrawn as final distillate product.

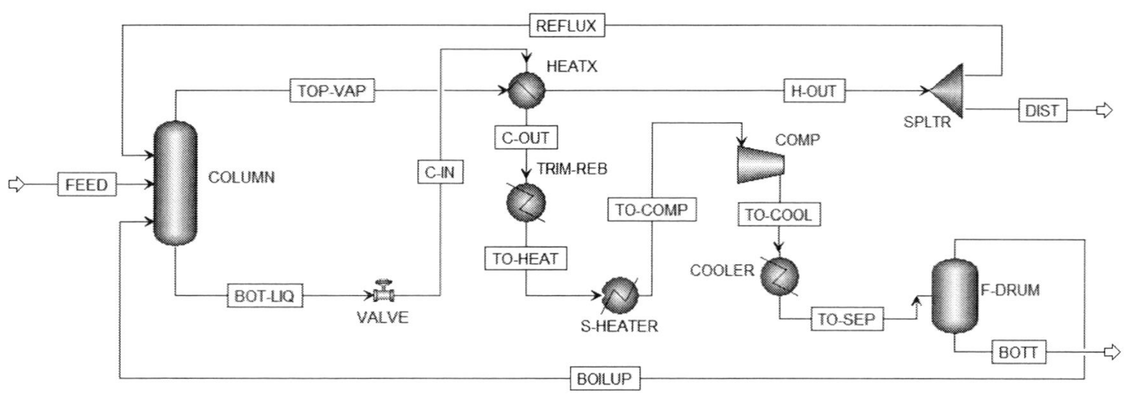

Figure 4. Flowsheet of the BFHP system.

5. Results and discussions

In reviewing the literature, no study was found on the simulation of oleochemical separation using heat pumping scheme. In the current study, comparing the conventional column with two HPAD configurations generates significant differences in energy requirement. The simulation process was straightforward for the VRC model, however for the BFHP, some manipulations of the variables were needed to get the system converged. The process conditions of product streams for all simulations are presented in table 3.

Table 3. Process conditions of product streams.

	CDC	VRC	BFHP
Purity of $C_{12}H_{24}O_2$ in distillate (%)	99.89	99.89	99.78
Distillate temperature (°C)	181.11	180.40	181.50
Bottoms temperature (°C)	195.18	195.17	195.17
Condenser load (kW)	882.27	-	-
Reboiler load (kW)	943.15	-	-
Recovered top vapor heat (kW)	-	822.31	882.26

*(-) Not applicable

As can be seen from the table, both HPAD configurations recorded high purity of the desired product, $C_{12}H_{24}O_2$ in the distillate stream. Product temperatures of integrated columns also showed very minimal differences than in the CDC. Meanwhile, the systems proved an excellent performance in recovering heat from column overhead vapor to aid in heating process of the liquid in the column bottom stream. It is no doubt that an immense amount of energy could be conserved with both HPAD systems. Heat pumping technology, though not completely, but reduced more that 80% of reboiler duty and more than 70% of condenser duty. Table 4 summarises the external heating and cooling duty for direct VRC and BFHP arrangements after the elimination of the condenser and reboiler units as compared to conventional column. The mechanical work in the compressor is also compared in the table for both integrated systems.

Table 4. Summary of relative performance of all simulations.

System	CDC	Direct VRC	BFHP
Cold utility requirements (kW)	882.27	97.56	285.74
Hot utility requirements (kW)	943.15	127.21	425.14
Mechanical work requirements (kW)	-	38.95	76.53

*(-) Not applicable

In direct VRC case, approximately only 11% of cold utility is required than in the CDC. Further cooling was necessary to decrease the vapor fraction of the reflux stream. The hot utility input in this system is coming from the trim reboiler which was purposely used to supply auxiliary heat for bottom liquid reboiling. Even so, 87% of column top vapor heat could be recovered to provide heat for the bottom liquid, which is quite a promising option.

For BFHP case, if we look closely at the utility requirement values, it is apparent that this system requires substantially high external utilities to satisfy the process requirement in comparison to direct VRC. The high heating duty was contributed by two equipment as depicted in figure 4. Even though the recovered heat using this method was 94%, higher than that in the VRC, the superheating of the bottom column outlet has directly escalated the number of hot utility requirements. This situation could not be avoided because the vapor content in the stream after exchanging heat was insufficient to undergo compression, therefore the need to superheat it until vapor fraction has reached 1.00. Due to this, 364.36 kW of energy was demanded only for superheating. On the other hand, the cooling duty for the BFHP as shown in the table is also considerably high due to an increased temperature after the compression.

The elimination of overhead condenser and bottom reboiler implies that the amount of external cooling utility is significantly decreased thus will be reducing more operation costs. However, it is compulsory in heat pumping method to employ a compressor unit. Unlike the conventional operation, the HPAD economic feasibility highly depends on the compression ratio value, because as a matter of fact, higher compression ratio means more mechanical work. Principally, the operation and capital costs for the implementation of the compressor that is driven by electricity is outrageously expensive. Table 4 reported that the compression work in the BFHP is obviously higher than that in the direct VRC. Not only the BFHP system demanded more external utility supply, but it also consumed more electricity. Taken together, the VRC configuration is in favour in this assessment over the BFHP.

6. Conclusions

This study set out to conduct an assessment of heat pumping method in oleochemical fatty acid separation. The results of this study indicate that it is technically feasible to integrate the system with the process. The performance of the direct VRC configuration is apparently better than the BFHP. Although the consumption of external heat source is decreased in the reboiler, along with the reduction of condenser and its utility usage, the HPAD additionally employs a compressor that is operated by electricity, which is far more expensive than the thermal utility. It is a proof that a systematic analysis of the HPAD is necessary from the energetic and economic aspects. There will be certainly a trade-off between operational and capital costs to operate the system. Also, it will be more promising if the simulation process is thermodynamically correct and one of the ways is to properly select the thermodynamic package prior to run the simulation. Further studies, which take the effects of thermodynamic measures to the HPAD system into account, involving all components in the feed stream, will need to be undertaken. This research will serve as a base for future studies in energy efficient oleochemical separation.

Acknowledgement

The authors would like to thank Universiti Malaysia Pahang (UMP) for providing the research fund for this project under grant number RDU1803178.

References

[1] Kiran B, Jana A and Samanta A 2012 *Energy* **41** 443-453
[2] Nakaiwa M, Huang K, Endo A, Ohmori T, Akiya T and Takamatsu T 2003 *Chem. Eng. Res. Des.* **81** 162-177
[3] Bruinsma D and Spoelstra S 2010 *Proc. Distill. Absorption* 21-28
[4] Faessler P, Kolmetz K, Seang K W and Lee S H 2007 *Asia-Pacific J. Chem. Eng.* **2** 315-321
[5] Illner M and Othman M R 2015 *PERINTIS e-Journal* **5** 34-44
[6] Aly G and Ashour I 1992 *Sep. Sci. Technol.* **27** 955-974
[7] Stage H 1984 *J. Oil Fat Industries* **61** 204-214
[8] Navarro-Amorós M A, Ruiz-Femenia R and Caballero J A 2013 *AIChE J.* **59** 3767-81
[9] Felbab N, Patel B, El-Halwagi M, Hildebrandt D and Glasser D 2013 *AIChE J.* **59** 2977-92
[10] Kazemi A, Mehrabani-Zeinabad A and Beheshti M 2017 *Energy* **125** 449-458
[11] Kazemi A, Mehrabani-Zeinabad A and Beheshti M 2018 *Chem. Eng. Process.* **123** 158-167
[12] Diez E, Langston P, Ovejero G and Romero M D 2009 *Appl. Therm. Eng* **29** 1216-23
[13] Kazemi A, Mehrabani-Zeinabad A and Beheshti M 2018 *Appl. Energy* **211** 1261-81
[14] Chew J M, Reddy C C S and Rangaiah G P 2014 Cycle *Chem. Eng. Process.* **76** 45-59
[15] Carlson E 1996 *Chem. Eng. Progress* **92** 35-46
[16] Bono A, Pin O P and Chin P J 2010 *J. Appl. Sci.* **10** 2508-15

IOP Conf. Series: Materials Science and Engineering **702** (2019) 012013 doi:10.1088/1757-899X/702/1/012013

Integrating ILS in dissolving palm fruit cellulose

K N Mansor[1], M A M Halim[1], N Ali[2], W Z W A Rahman[2] and S B Abdullah[1*]

[1] Faculty of Chemical and Process Engineering Technology, Universiti Malaysia Pahang, Lebuhraya Tun Razak, 26300 Gambang, Kuantan, Pahang, Malaysia
[2] D3 Technology Solutions Sdn Bhd, Gebeng Industrial Area Phase II, 26080 Kuantan, Pahang, Malaysia

*E-mail: syamsul@ump.edu.my

Abstract. Ionic liquids (ILs) are introduced to improve the production of palm oil in industry in order to extract more thus lowering the oil losses. The main objective of this study is to investigate the potential of sulfate-based ILs in dissolving palm fruit cellulose. The interpretation study was done in lab scale by dissolving cellulose (that directly extracted from palm fruit) in ILs and percentage of cellulose dissolved is calculated based on weight percentage dissolved. The highest average percentage dissolving of cellulose was 10.67 wt% with condition concentration of sulfate-based ILs is 9000 ppm and the mixing time was two hours. In a nutshell, the higher concentration of ILs and more mixing time will achieve higher percentage of cellulose dissolved.

1. Introduction

World production of palm oil, the most widely traded edible oil, has also seen significant leaps in production and planted areas; production had almost doubled from 1990 to 2001, with Malaysia and Indonesia contributing to most of the increased production. This had been achieved mainly by opening of new land for oil palm plantations [1]. Although the production was increased year by years, but the production counted many losses in term of residue oil that still left during pressing process and slightly in empty fruit bunch though trenching [2]. The highest mean is the oil loss in the pressed fiber and the lowest mean is the oil loss in the wet nut [3].

It can be seen the major of palm oil production was producing about 74 wt% for palm oil-based products, from food to cleaners to cosmetics. Huge production also comes with huge losses of palm oil by considering the amount of oil loss from 5 – 10 wt% during pressing step to be exactly. Hence, oil losses can be reduced during the pressing process by introducing the ionic liquids (ILs) as pre-treatment right before pressing of palm fruit. Perhaps by applying ILs before pressing process will breakup more oil cell within cellulosic structures of fresh fruit bunch for palm oil. Many researchers agreed that pre-treatment was the key to unlock the recalcitrance of lignocellulose for cellulosic structure containing oil palm hence gives proof that using ILs can also help to dissolve the cellulose. In general, a palm oil mill extracts 20 wt% of oil from the fresh fruit bunch while also produces 23 wt% of empty bunch, 15 wt% of fiber, and 12 wt% of nut. Oil losses occur in various by-products, including the fiber, which remains after the mesocarp oil is extracted by a screw press [3-5]. In order to accommodate above palm oil problem, the objective in improving palm oil is to investigate potential sulfate-based ILs in dissolving cellulose whereas directly extracted from palm fruit. Several process parameters have been identified to evaluate the dissolution of palm fruit cellulose in sulfate-based ILs such as temperature, concentration and mixing time.

2. Materials and Methods

2.1. Materials
Cellulose fiber which directly extracted from the palm fruit was supplied by D3 Technology Solutions Sdn. Bhd. Pure 1-butyl-3-methylimidazolium methylsulfate ILs was prepared accordingly in Faculty of Chemical and Process Engineering Technology (FTKKP) Laboratory.

2.2. Breakdown of cellulose and dilution
The cellulose fiber was separated and breakdown into powder before immersing in sulfate-based ILs namely 1-butyl-3-methylimidazolium methylsulfate. But before immersion process is taking place, the ILs was diluted into three different concentrations and the experiment was prepared for three different mixing times. After immersion, drying process was conducted to remove moisture from the sample for weighing purposes.

2.3. ILs dilution process
Pure 1-butyl-3-methylimidazolium methylsulfate ILs was diluted using deionized water to obtain three different concentrations of 5000 ppm, 7000 ppm and 9000 ppm. The total of 6 ml each of diluted ILs were used to immerse cellulose powder for heating with stirring processes [6].

2.4. Sample preparation
Experiment was started with 0.05 g cellulose powder was added into 50 ml beaker contain 6 ml of diluted ILs solution. 150 ml of tap water as water bath was heat up until 70°C using hotplate. Then the solution in the 50 ml beaker was placed into the water bath on the hotplate and the stirrer was adjusted at 800 rpm while the temperature remains at 70°C. The solution was placed in covered beaker to avoid contamination and impurities. The heating and stirring process was carried out for two hours (changes follow the time set). The solution was rinsed with 25 ml of deionized water before filtered using filter paper for 20 min. After that, rinsed process was proceeded to remove the remaining ILs. Then, the filtered sample was dried in an oven with the temperature of 70°C for 24 hours and was put at room temperature for 20 min of cooling process. Bear in mind, the filter paper was weighed before filtering the solution to prevent error in mass change of cellulose. Then, the dried sample was weighed, and the data was recorded [7].

2.5. Dissolution performance
Change in mass of cellulose fiber was calculated after the drying process. The percentage dissolution was calculated using equation (1) below:

$$\text{Percentage dissolution } (\%) = \frac{\text{Weight Before (g)} - \text{Weight After (g)}}{\text{Weight Before (g)}} \; x \; 100 \qquad (1)$$

3. Result and Discussion

3.1. Percentage of dissolution
The percentage of dissolved cellulose fiber at various process parameters was summarized in table 1. The overall result obtained depicts the highest average percentage for dissolved cellulose at 9000 ppm ILs solution for two hours mixing time. In many studies with ILs, it was foreseen that both cation and anion plays great importance and responsible for the dissolution of cellulose especially for sulfate-based ILs [6].

Table 1: Summary of cellulose dissolved percentage

Concentration of ILs (ppm)	Cellulose Dissolved (wt%)
1 hour	
5000	1.4
7000	5.8
9000	8.8
1.5 hours	
5000	4.4
7000	8.4
9000	10.8
2 hours	
5000	7.0
7000	10.4
9000	12.4

3.1.1 Effect of ILs concentration

Based on figure 1 to 3, the ILs solution with concentration of 9000 ppm has better dissolution percentage compared to ILs solution with concentration of 5000 ppm and 7000 ppm. For ILs solution with concentration of 9000 ppm, the average dissolution percentage is 10.67 wt% higher compare to the other two ILs concentrations which merely has the average dissolution percentage of 4.27 wt% and 8.2 wt% for 5000 ppm and 7000 ppm, respectively. As the concentration of the ILs increases from 5000ppm to 9000ppm, the percentage of cellulose dissolved also increased. The reason behind the improvement of electro spinnability of cellulose with the addition of ILs as solvent is that ILs penetrates into the gaps between the chains and the ILs enhancing the chain mobility due to weaker interaction with the chains. As a result, cellulose chains can be easily elongated and fibrillated during the process [7]. This can deduce that increased of concentration ILs brings high dissolution percentage of cellulose because the chain of cellulose become weaker after mixing with solvent [8].

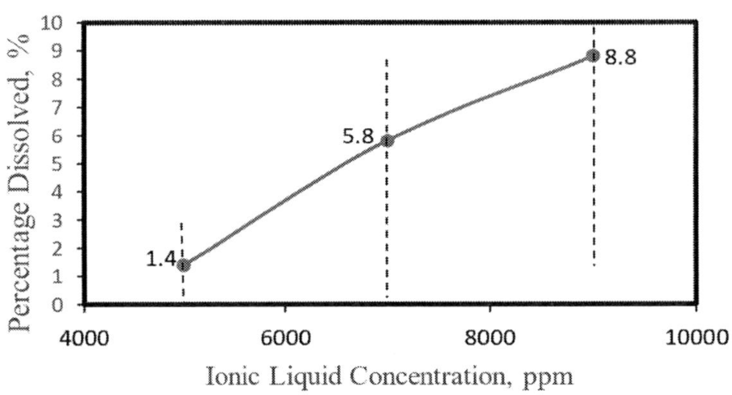

Figure 1: Effect of ILs concentration for mixing time at 1 hour

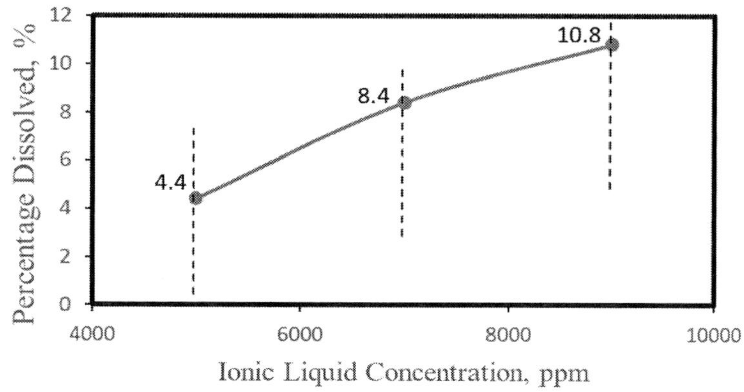

Figure 2: Effect of ILs concentration for mixing time at 1.5 hours

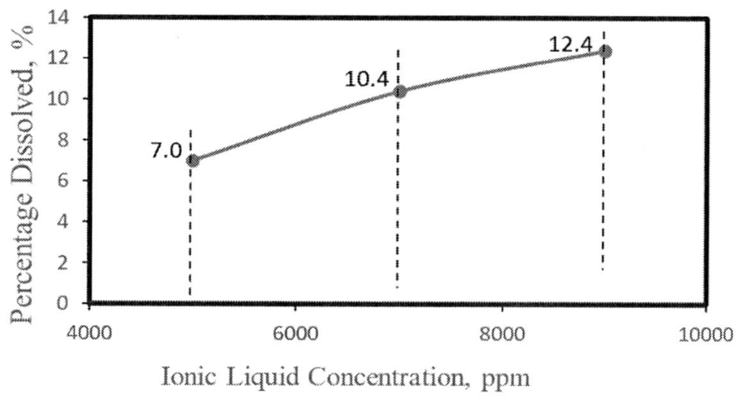

Figure 3: Effect of ILs concentration for mixing time at 2 hours

3.1.2 Effect of mixing time

Mixing time play an important role in determining the rate of cellulose fiber dissolved with ILs whilst mixed up together during allocation time. Based on the graphs shown in figure 4 to 6, mixing time for two hours recorded has the higher rate of cellulose fiber dissolved with the average of 9.93 wt% compared to the one hour with the average of 5.33 wt% and one and half hour with the average percentage of 7.87 wt%. The results showed that increasing of the mixing time for cellulose mixed up with ILs will increase the percentage of cellulose dissolved and the dissolution process generally occurred regarding on the variable of mixing time. Mixing time is important parameter because it significantly shows for substance to react with solvent and how long the time will decide the reaction time. From previous study done in 2017, which had identified the optimum time for the reaction between cellulose and different type of solvent is more than 2 hours at 90°C [9].

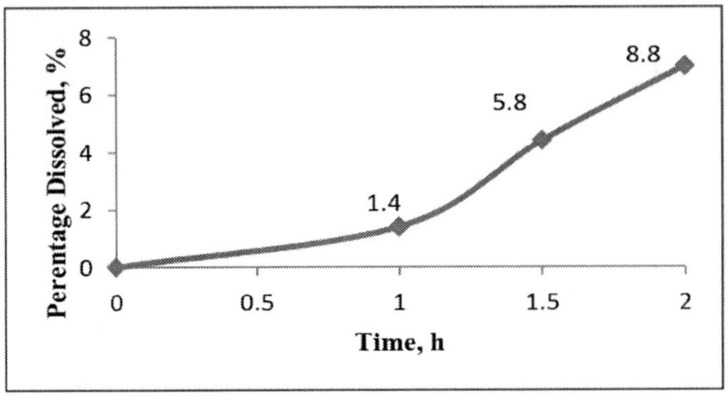

Figure 4: Effect of mixing time for cellulose dissolution at 5000 ppm of ILs concentration

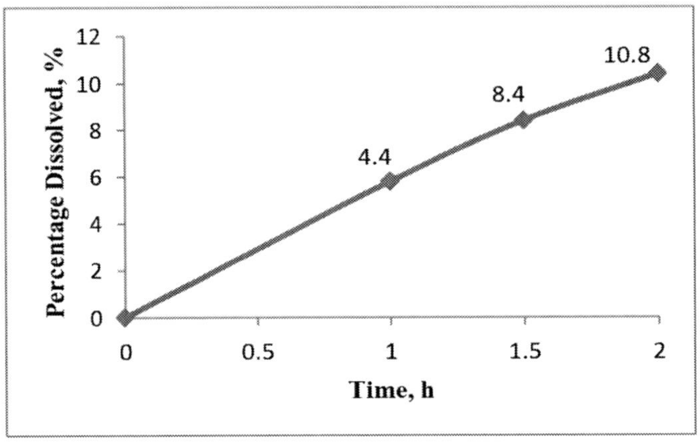

Figure 5: Effect of mixing time for cellulose dissolution at 7000 ppm of ILs concentration

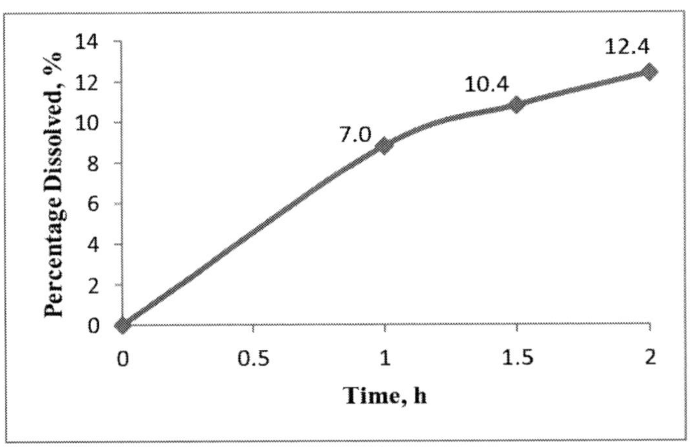

Figure 6: Effect of mixing time for cellulose dissolution at 9000 ppm of ILs concentration

4. Conclusion

In a conclusion, based on the two main parameters which are ILs concentration and mixing time, it shows that the highest percentage of cellulose fiber dissolution occurred for concentration at 9000 ppm during mixing time of two hours. A higher concentration of ILs solution needed for better cellulose fiber dissolution whereas required higher temperature and longer mixing time. The optimum temperature for this experiment was set up at 70°C because sample will evaporate to the atmosphere when conducting above 70°C.

Acknowledgement
The authors would like to thank the Ministry of Higher Education Malaysia under the Fundamental Research Grant Scheme (FRGS/1/2016/TK02/UMP/03/1 - RDU160128) and D3 Technology Solutions Sdn. Bhd. for the research financial/material supports.

References
[1] Inyang J, Munot M A and Shazali S T S, Tanjong S J 2019 A Model to Manage Crude Palm Oil Production System *MATEC Web of Conferences* 255 02001
[2] Lai O M, Tan C P and Akoh C C 2012 *Palm Oil: Production, Processing, Characterization, and Uses* 1st Edition Elsevier
[3] Zulkefli F, Othman N, Syahlan S, Zaini M R and Bakar M A 2017 Fresh Fruit Bunch Quality and Oil Losses in Milling Processes as Factors that Affect The Extraction Rate of Palm Oil *International Journal of Agriculture, Forestry and Plantation* **5** 99-103
[4] González B, Gómez E, Domínguez Á, Vilas M and Tojo E 2011 Physicochemical Characterization of New Sulfate Ionic Liquids. *Journal of Chemical and Engineering Data* **56** (1) 14-20
[5] Azmi S I, Azizan A and Salleh R M 2018 Pretreatment of Oil Palm Frond with Ionic Liquid. *IOP Conference Series: Materials Science and Engineering* 358
[6] Salvador Â C, Santos M and Saraiva J M A 2010 Effect of The Ionic Liquid [bmim]Cl and High Pressure on The Activity of Cellulase *Green Chemistry* **12**(4), 632-35
[7] Isik M, Sardon H and Mecerreyes D 2014 Ionic Liquids and Cellulose: Dissolution, Chemical Modification and Preparation of New Cellulosic Materials *International Journal of Molecular Science* **15**(7), 11922-40
[8] Sashina E S and Novoselov N P 2009 *Effect of Structure of Ionic Liquids on Their Dissolving Power toward Natural Polymers Russian Journal of General Chemistry* **79**(6) 1057- 62
[9] Mohd N, Draman S F S, Salleh M S N and Yusof B 2017 Dissolution of Cellulose in Ionic Liquid: A Review *AIP Conference Proceedings* **1809**(020035).

IOP Conf. Series: Materials Science and Engineering **702** (2019) 012014 doi:10.1088/1757-899X/702/1/012014

Synthesis of optimal biogas production system from multiple sources of wastewater and organic waste

M A Misrol, S R Wan Alwi*, J S Lim and Z A Manan

Process Systems Engineering Centre (PROSPECT), Research Institute for Sustainable Environment (RISE), School of Chemical and Energy Engineering, Faculty of Engineering, Universiti Teknologi Malaysia, 81310 UTM Johor Bahru, Johor, Malaysia.

*E-mail: syarifah@utm.my

Abstract. Biogas is one of the products that can be produced from wastewater or organic waste. Given the input streams for biogas digester can come from multiple sources rich in organic compound, optimization can be performed to maximise the production. In an industrial park with several types of industry, wastewaters and/or organic wastes from the industries or surrounding area can be used as feedstock for biogas production. Compared to conventional industrial park, Eco-Industrial Park (EIP) could be the appropriate platform for implementation as it advocates for industrial symbiosis through materials exchange and minimisation of waste, which are also in-line with the circular economy concept. In this study, a superstructure applicable for an EIP, which consists of multiple biogas sources, biogas digester, biogas upgrading, and biogas demands are developed. The model considers parameters which can affect biogas production quality and quantity such as Chemical Oxygen Demand (COD), Carbon/Nitrogen ratio (C/N ratio), and temperature. It also considers percentage of methane lost, removal efficiency of Carbon Dioxide (CO_2), electricity cost, capital cost of the selected biogas digesters as well as the cleaning and upgrading units. The main objective of the mathematical model is to maximise the profit generation from biogas generation for the EIP centralized biogas production plant. The model is solved by using General Algebraic Modelling System (GAMS) software. Case study conducted shows that anaerobic closed lagoon is selected as digester for the industrial wastewaters, while water scrubber is suggested as the biogas upgrading technology. A yearly profit is estimated at USD13,700,000, which is generated from selling of 279nm^3 of purified biogas. This may suggest that generating biogas from several sources of industrial wastewater and organic waste could be potentially feasible, thus reducing unnecessary pollutant while capturing the carbon source as an additional revenue stream.

1. Introduction

The circular economy concept promotes minimization of waste and fresh resources via the 3R (reduce, reuse, recycle) [1] or 6R (reuse, recycle, redesign, remanufacture, reduce, recover) [2] practice. These would not only improve the plant's process, but also generates additional revenue stream through monetizing the potential carbon source available, which in this case, comes from the wastewater and organic waste.

Instead of being treated via processes that release the methane directly to the atmosphere, the biogas could be tapped as a source of renewable energy. In an industrial park context, a centralized wastewater treatment facility that could produce biogas from several wastewater streams that originated from different industries is identified as one of the potential synergies [3]. Compared to conventional industrial park, the Eco-Industrial Park (EIP) could be the appropriate platform for implementation it advocates the dedicated philosophy i.e industrial symbiosis through materials exchange and minimisation of waste, which are also in-line with the circular economy concept.

However, the wastewater or organic wastes that are sent to the biogas digester may have different characteristics, which are not as homogeneous as they are sourced from an individual plant. This will open opportunity for optimization to be implemented in order to maximize the biogas production [4]. Drobez et al. [5] has conducted a study to optimize biogas production which was sourced from industrial wastewater and other several organic sources from poultry industry, combining heat integration via inclusion of heat exchanger. However, it does not consider COD content and C/N ratio to determine optimal biogas production.

Anaokar et al. [6] has presented a predictive model for urban wastewater based on the Biological Oxygen Demand (BOD) as the most significant parameter. That said, the source for digestion is relatively homogenous. Thongnan et al. [7] has conducted a modeling study regarding co-digestion of farm waste and domestic organic waste, though it does not consider industrial wastewater as one of the feedstocks. Apart from wastewater, organic source from other sources e.g farm animal waste, kitchen waste and food, oil, and grease waste could also be useful as additional feedstock for the biogas production. A study done by Abdeshahian et al. [8] indicates that 4589.49 million m^3 per year biogas can be produced from animal farm waste in Malaysia. Egieya et al. [9] has developed a multi-integer linear program (MILP) to optimize biogas supply from various types of biomass and manure as it also conducted economic and sensitivity analysis.

Study made by Mohtar et al. [10] analyses the economics of biogas plant for usage as bio compressed natural gas (bio-CNG). It considers selection of 6 biogas upgrading technologies in the model. However, specific parameters associated for each technology e.g methane loss and carbon dioxide (CO_2) removal efficiency were not accounted. Angelidaki et al. [11] and Sun et al. [12] have conducted review regarding the biogas upgrading technology as its (the biogas upgrading technology) efficiency in preventing methane loss, removing CO_2, and having lower electricity consumption are regarded as among of important items for attention, which will be considered in this study.

Overall, to the best knowledge of the authors, the study to optimize three important parameters at or prior to the biogas digester inlet, namely COD, C/N ratio, and temperature from different industrial wastewaters and organic waste is yet to be developed. Therefore, the motivation of this study is to develop mathematical model that could provide optimal biogas production from different sources of industrial wastewater and organic waste, at the same time considering selection of biogas upgrading technology, with the main objective is to maximize the profit generation. Considerations in selection of the biogas digester and the biogas cleaning as well as upgrading unit are also performed in the model. Section 2 will describe the superstructure and the mathematical equations used in the model. Information regarding the case study is provided in Section 3, while Section 4 provide the subsequent result and analysis. Section 5 will conclude main takeaway regarding the study conducted.

2. Superstructure and Model Development

2.1. Superstructure

Superstructure of the study is shown in figure 1 as follows:

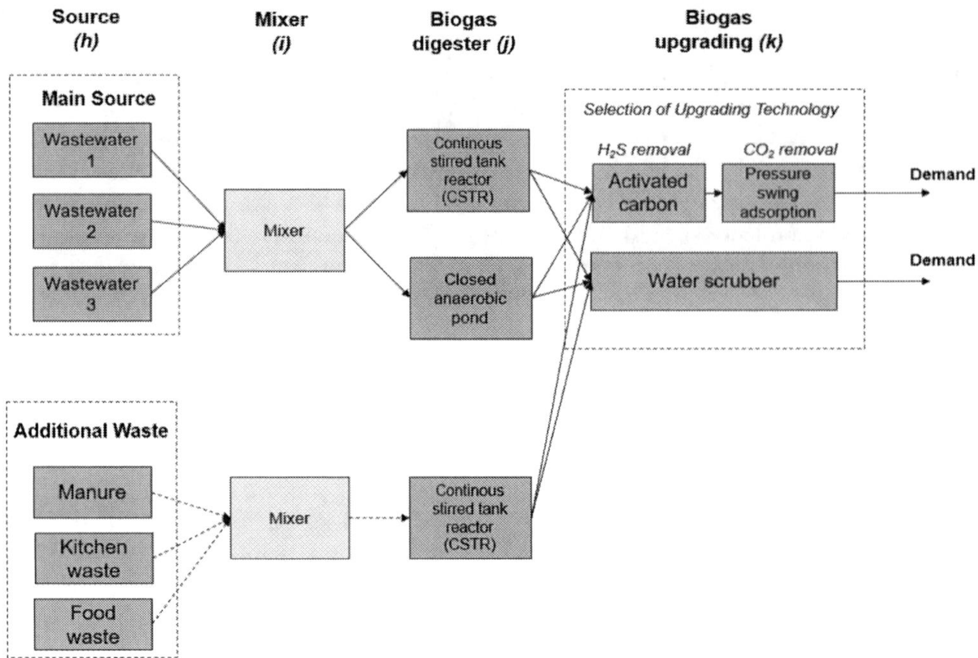

Figure 1. Superstructure of the study

An EIP ecosystem consists of multiple plants from different types of industry. It would have centralized facility to encourage materials exchange among the industries. Hence, in this context, it would have a centralized biogas facility that able to produce biogas from wastewaters, which are sourced from multiple industries. At the same time, it is also open to receive organic sources from surrounding domestic and agricultural activities, such as cow manure, kitchen waste, and food waste in order to maximize the carbon content conversion to biogas. The superstructure shown consists of multiple biogas sources, biogas digester, biogas upgrading, and biogas sets. The wastewater streams sourced each will have different wastewater strength with regards to the pollutant content. Given the adjustment of temperature, C/N and final average COD content are to be made in this study, it is assumed that the sources will not require pretreatment prior to the mixing.

The sources are grouped into two main streams i.e. the main sources and the additional sources. The former has different treatment objectives compared to the latter; it is intended to lower the COD as much as possible while tapping the biogas from the wastewater, whereas main objective for the organic sources is to harness the biogas production instead of leaving it 'as-is' i.e decompose naturally. Plus, COD removal of the additional sources in general is only 60% maximum, while COD removal of the main sources could be as high as 90%. Thus, it is proposed to have different digesters for both type of sources.

There will be two different technology configuration pertaining biogas digestion to be considered i.e continuous stirred-tank reactor (CSTR) and closed anaerobic pond. The selection of biogas purification consists of activated carbon and water scrubber for removal of Hydrogen Sulfide (H_2S), followed by technologies for removal of Carbon Dioxide (CO_2), namely pressure swing adsorption (PSA) and water scrubber. The model considers parameters which can affect biogas quality and quantity as mentioned earlier The model is solved by using General Algebraic Modeling System (GAMS) software. Finally, it is applied to a case study to show its applicability

Prior to be sent into the digesters, the sources streams will be mixed in order to ensure that the average COD content, C/N ratio content (specifically for organic sources only), and temperature are above or within or same the optimal condition for the biogas production. In this study, the optimal boundary of the parameters is shown in Table 1 as follows:

Table 1. Optimal boundary of the parameters in the mixers

Parameter	Optimal Value or Limits
COD content	$\geq 10{,}000\ mg/L$
C/N ratio	≥ 25 and ≤ 35
Temperature	32°C

There are two options regarding biogas digester technology specifically for the wastewater sources i.e CSTR and closed anaerobic pond, while all the organic sources will be digested in the separate CSTR. Biogas generated from both main streams will be sent to the biogas cleaning and upgrading unit as it will have two options of technology i.e. (1) activated carbon combined with PSA or (2) water scrubber. Details of both technologies are further described in Section 3. It is assumed that purified biogas from both options are able to produce methane up to 97% concentration. The demand point is assumed to be the natural gas pipeline with selling price of USD 6.5 per nm^3.

2.2. *Mathematical Formulation*

Main objective of the model is to obtain maximum profit (Pr) from selling of the purified biogas. The list of equations for the model are listed as follows:

Objective function:

$$Pr = Rv - CCostBg - OMCostBg - CCostBgCln - OMCostBgCln - ECostBgCln - MP \quad (1)$$

Rv is revenue from selling of the purified biogas. *CCostBg* stands for total annualized capital cost of the biogas digesters selected and *OMCostBg* is the associated total operation and maintenance cost. *CCostBgCln* is total annualized capital cost of the biogas cleaning and upgrading technology selected, with associated operation and maintenance cost is termed as *OMCostBgCln*. Electricity cost for the biogas digesters is assumed to be included in the *OMCostBg* while electricity cost for the biogas cleaning and upgrading technology is put as *ECostBgCln*. *MP* is the manpower cost.

Constraints:

Source-to-mixer flow rate constraint:

F_h^{source} and F_i^{mix} are the flow rate of the sources and mixers respectively. $B_{h,i}$ is a binary parameter to assign the sources h to the mixer i, where K is a large integer value to limit the flow rate from the sources of the intended mixer. COD content of each sources is C_h^{source} and the resulting average COD content at the mixer is C_i^{mix}. Total COD content at set h and i are COD_h^{source} and COD_i^{mix} respectively. It is required for the average COD value in the mixer to be at more or equal to 10 kg per m^3 in order to obtain higher yield of the biogas production. CN_h^{source} is the C/N ratio of the sources, thus CN_i^{mix} is the C/N ratio in the mixers. Only C/N ratio for organic sources is considered to require optimal C/N ratio as one of the condition to achieve optimal biogas production.

$$K \times B_{h,i} = F_{h,i}^{sourcemix} \quad (2)$$

$$F_h^{source} = \sum_i F_{h,i}^{sourcemix} \quad \forall h \quad (3)$$

$$\sum_h F_{h,i}^{sourcemix} = F_i^{mix} \quad \forall i \quad (4)$$

$$COD_h^{source} = \sum_i F_{h,i}^{sourcemix} \times B_{h,i} \times C_h^{source} \quad \forall h \quad (5)$$

$$\sum_h F_{h,i}^{sourcemix} \times B_{h,i} \times C_h^{source} = COD_i^{mix} \quad \forall h,i \quad (6)$$

$$COD_i^{mix} = C_i^{mix} \times F_i^{mix} \qquad \forall i \qquad (7)$$

$$C_i^{mix} \geq 10 \qquad \forall i \qquad (8)$$

$$F_i^{mix} \times CN_i^{mix} = \sum_i F_{h,i}^{sourcemix} \times CN_h^{source} \qquad ; i = 2 \qquad (9)$$

$$25 \leq CN_i^{mix} \leq 35 \qquad ; i = 2 \qquad \forall i \qquad (10)$$

Temperature balance at the mixers:

T_h^{source} is the temperature of sources and T_i^{mix} is temperature of the mixers. C_p is the specific heating value of water. Total energy that able to be withdrawn from the mixers is termed as $Qtotal$. Temperature of the organic sources are assumed to be at ambient temperature (32°C), thus only the 1st mixer is considered for potential energy withdrawal.

$$F_h^{source} \times T_h^{source} \times C_p = \sum_i F_{h,i}^{sourcemix} \times B_{h,i} \times C_p \times T_h^{source} \qquad \forall i \qquad (11)$$

$$\sum_h F_{h,i}^{sourcemix} \times B_{h,i} \times C_p \times T_h^{source} = F_i^{mix} \times T_i^{mix} \times C_p \qquad \forall h, i \qquad (12)$$

$$Qtotal = F_i^{mix} \times \frac{C_p\left(T_i^{mix} - 32\right)}{3600} \qquad ; i = 1 \qquad (13)$$

Mixer-to-digester flow rate constraint:

Since F_i^{mix} has to be sent to digester(s), flow rate from the mixer is termed as $F_{i,j}^{mixbg}$. F_j^{bg} is the flow rate at inlet of the digester. $B_{i,j}$ is a binary parameter to assign the streams from i to the digester j, where L is a large integer value to limit the flow rate from the mixers to the intended digester. C_j^{bg} is the COD content in the digester which will be used to determine the amount of biogas produced.

$$L \times B_{i,j} = F_{i,j}^{mixbg} \qquad (14)$$

$$F_i^{mix} = \sum_j F_{i,j}^{mixbg} \qquad \forall i \qquad (15)$$

$$\sum_i F_{i,j}^{mixbg} = F_j^{bg} \qquad \forall j \qquad (16)$$

$$C_i^{mix} = \sum_j C_{i,j}^{mixbg} \qquad \forall i \qquad (17)$$

$$\sum_i C_{i,j}^{mixbg} = C_j^{bg} \qquad \forall j \qquad (18)$$

Total production of biogas:

F_j^{bgout} is the flow rate of biogas produced from the digesters. Eff is the production efficiency. $Bgyield$ is the constant value of biogas produced per kg of COD removed, $Bgdensity$ is the biogas density.

$$F_j^{bg} \times C_j^{bg} \times Eff \times \frac{Bgyield}{Bgdensity} = F_j^{bgout} \qquad \forall j \qquad (19)$$

Digester-to-upgrading flow rate constraint:

$F_{j,k}^{bgcln}$ is the flow rate from the digester to the biogas cleaning and upgrading unit. F_k^{cln} is the flow rate of the raw biogas received at the inlet of k. X_k is the conversion yield of respective technology to produce purified biogas at amount of F_k^{clnout}. Revenue from selling of the purified biogas is Rv. R_k and A_k are the selling price in USD and the annual operating hour respectively.

$$F_j^{bgout} = \sum_k F_{j,k}^{bgoutcln} \qquad \forall j \qquad (20)$$

$$\sum_j F_{j,k}^{bgoutcln} = F_k^{cln} \qquad \forall k \qquad (21)$$

$$F_k^{cln} \times X_k = F_k^{clnout} \qquad \forall k \qquad (22)$$

$$Rv = \sum_k F_k^{clnout} \times R_k \times A_k \qquad (23)$$

Capital, Operating and Maintenance Cost estimation:

$CCostBg$ is the capital cost of the selected biogas digesters. CBg is the cost per m^3 of the feed volume. α is the power factor and Af is the annualized factor. Estimated cost for associated operating and maintenance cost for the selected biogas digesters, $OMCostBg$ is multiplication of $OMBg$ which is the percentage value of $CCostBg$.

Both CBg and $CBgCln$ are obtained via calculation based on the cost index formula:

$$CBg \text{ or } CBgCln = BaseCost\left(\frac{I}{I_{base}}\right) \qquad (24)$$

Where $BaseCost$ is cost of the system or unit based in certain years e.g 2007, while I and I_{base} are the current cost index (referred as 2018) and the cost index in year of the $BaseCost$ is referred.

The same formula is principally applied for the biogas cleaning and upgrading. $CCostBgCln$, $CBgCln$, $OMCostBgCln$, and $OMBgCln$ are the capital cost of the selected cleaning and upgrading technology, cost per m^3 of the feed raw biogas volume, associated operating and maintenance cost (excluding electricity cost), and the capital cost percentage value respectively.

$$CCostBg = \sum_j((CBg \times F_j^{bg})^{\alpha})Af \qquad (25)$$

$$OMCostBg = \sum_j OMBg \times CCostBg \qquad (26)$$

$$CCostBgCln = \sum_k((CBgCln \times F_k^{cln})^{\alpha})Af \qquad (27)$$

$$OMCostBgCln = \sum_k OMBgCln \times CCostBg\,Cln \qquad (28)$$

Electricity cost of the biogas cleaning and upgrading:

The electricity cost, $ECostBgCln$, is given by multiplication of the electricity price (EP), cost per nm^3 of raw biogas in USD (EC), annual operation hour (AOH_k)and F_k^{cln}

$$ECostBgCln = \sum_k EP \times EC \times AOH_k \times F_k^{cln} \qquad \forall k \qquad (29)$$

3. Case Study

The produced raw biogas from the digesters is assumed to possess 60% methane, 39.4% CO_2, 0.2% nitrogen gas (N_2), and 0.4% H_2S. Properties of the sources are shown in table 2 follows:

Table 2. Properties of the sources

Sources	Hourly Flow Rate (m^3/h)	COD (mg/L)*	Temperature (°C)	C/N ratio (%)
Wastewater from palm oil mill	20	55,000	80	N/A
Wastewater from food industry	20	1,120	32	N/A
Wastewater palm oil refinery	20	10,000	60	N/A
Cow manure	2	17,400	32	23.6
Food waste	2	682,600	32	28.0
Fat, oil, and grease waste from kitchen waste	2	148,000	32	28.0

*The value is same if the unit is converted to kg/m^3 as per used in the model. Please refer the nomenclature list for details

In this study, density of the wastewater all are assumed at 1 metric ton/m^3. Properties of the biogas digester options are given in table 3 as follows:

Table 3. Properties of the biogas digester technology options

Sources	Base Cost (USD)	Operating and Maintenance Cost (USD)	References
CSTR	1,395,000 (2007)*	2% from capital cost	[13],[14]
Closed anaerobic pond	168,000 (2007)*	15% from capital cost	[13],[14]

*1USD = 3.26MYR in 2007

Properties of the biogas cleaning and upgrading technology are given in table 4 follows:

Table 4. Properties of the biogas cleaning and upgrading technology options

Technology	Methane losses (%)	CO_2 removal efficiency (%)	Base Cost (USD)	Maintenance Cost (USD)	Electricity Consumption (kWh per nm^3 biogas)	References
AC + PSA	2%	98%	1.39 million USD for 500nm^3/h raw biogas input (2009)*	2% from capital cost	0.3	[15],[16]
Water scrubber	1%	95%	1.38 million USD for 500nm^3/h raw biogas input (2009)*	3% from capital cost	0.275	[15]

*1 Euro = 1.11 USD

Other parameters used for in model is shown via table 5 below:

Table 5. Other parameters used in the mathematical model

Parameters	Value
Annual operating hours	8000 hours
Operating years of the system	20 years
Electricity cost	MYR 0.355* per kWh
Selling price of the purified biogas	USD 0.3 per nm^3
Specific heating value of the stream	4.2 kJ per K kg
Biogas production yield	75% for the wastewater streams 60% for the organic sources
Biogas density	1.12 kg/m^3

*1 USD = 4 MYR as current exchange rate

4. Result and Discussion

The non-linear program (NLP) model is run via GAMS version 25.1.2 in a computer with processor capacity of IntelCore i3-8130U 2.2GHz. It was solved using CONOPT solver as the solver provides a locally optimal solution, suggesting that feasible solution is exist for the model. Execution time takes less than 1 second. The average COD content for the wastewater sources combined is 22,000 mg/L. Specific for the organic waste, the average COD content and C/N ratio are 282,000 mg/L and 26.5 respectively. An estimated 1.77MW of energy is able to be withdrawn from the mixer while the adjusting the temperature to the required operating condition. The optimal network selected is shown in figure 2 follows.

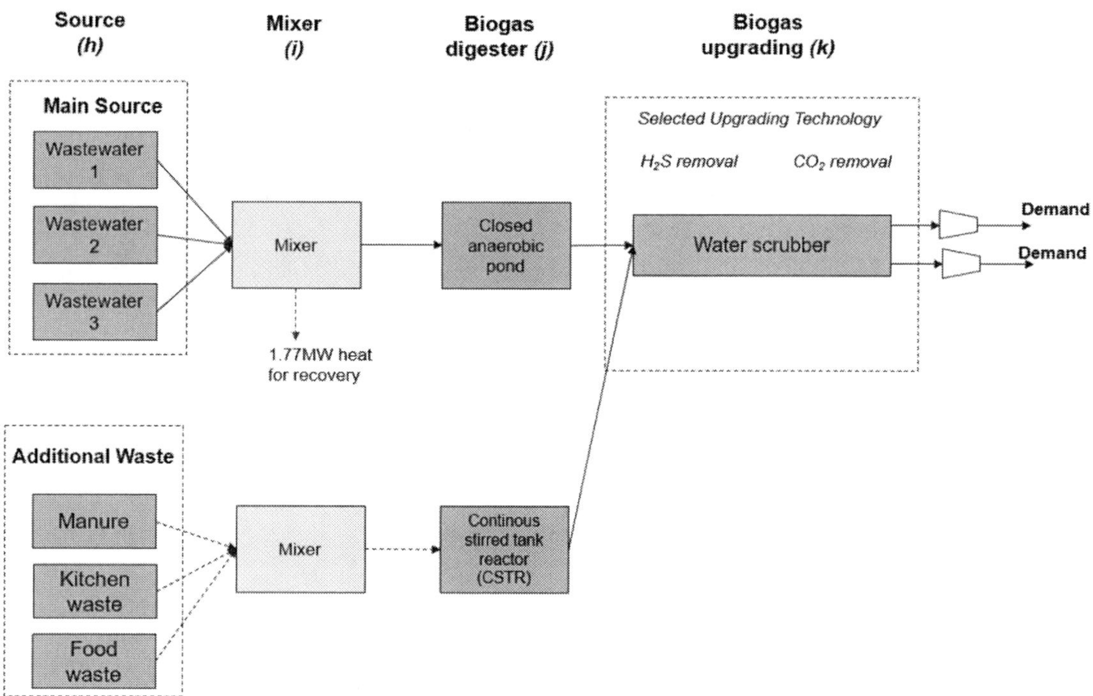

Figure 2. The optimized superstructure

The optimization result suggests closed anaerobic pond as the digester type for the wastewater streams and water scrubber is chosen for the biogas cleaning and upgrading technology. Total annualized capital cost for the biogas digesters and the water scrubber are calculated at USD 31,900 and USD 87,000 respectively. Combined operating cost of the selected biogas digesters and biogas cleaning and upgrading unit is USD 85,200. Electricity cost is USD 88,183. Revenue generated is calculated at USD 669,800 per year while profit obtained is USD 377,600. Translated into local currency, it is equal to MYR 1,510,400 of yearly profit.

IOP Conf. Series: Materials Science and Engineering **702** (2019) 012014 doi:10.1088/1757-899X/702/1/012014

An estimated amount of 452 nm^3 raw biogas will be produced from both digesters as it will be further sent to the water scrubber for cleaning and upgrading. Total annual revenue generated is based on selling of 279 nm^3 of purified biogas.

5. Conclusion

In this study, a mathematical model to provide optimal production of biogas from multiple sources of wastewater and organic waste in centralized EIP site level is proposed. Optimization of the input streams considering the COD, temperature, and C/N ratio has been performed together with selection of the digester and cleaning and upgrading technologies in order to provide optimal condition pertaining the biogas production. The case study conducted shows that the presented superstructure could be possibly developed in real industrial world. That said, the model, which is planned to be further expanded in the next study, will include consideration of pH, piping, compression, and associated logistics costs. It is also to consider opportunity to recycle or treat the effluent and to monetize the solid waste, thus the project feasibility could be comprehensively determined prior to any potential real engineering works.

Nomenclature

Sets

h	Source
i	Mixer
j	Biogas digester
k	Biogas upgrading

Parameters

K	Integer value to limit the flow rate from h to i
L	Integer value to limit the flow rate from i to j
$B_{h,i}$	Binary parameter to assign the sources h to the mixer i
F_h^{source}	Flow rate of source h (m^3/h)
C_h^{source}	COD content of source h (mg/L)
CN_h^{source}	C/N ratio of source h (%)
T_h^{source}	Temperature of source h (°C)
C_p	Specific heat capacity of water (4200 kJ/m^3 K)
Eff	COD removal efficiency (%)
$Bgyield$	Biogas production yield (%)
$Bgdensity$	Biogas density (kg/m^3)
X_k	Conversion yield of respective technology to produce purified biogas (%)
R_k	Selling price of purified biogas (USD per nm^3)
A_k	Annual operating hours (hours)
$BaseCost$	Cost of the base equipment/system (USD)
I_{base}	Cost index of the base equipment/system
I	Cost index of the selected equipment/system
α	Power factor
Af	Annualization factor
	$OMBg$ Percentage of operating and maintenance cost of biogas digester from selected technology (%)
EP	Electricity price (USD)
	EC Electricity factor per capacity of biogas cleaning and upgrading system installed (kWh/nm^3)
AOH_k	Annual operating hours (hour)

IOP Conf. Series: Materials Science and Engineering **702** (2019) 012014 doi:10.1088/1757-899X/702/1/012014

Variables

$F_{h,i}^{sourcemix}$	Flow rate from source h to mixer i (m^3/hour)
F_i^{mix}	Flow rate at mixer i (m^3/hour)
COD_h^{source}	Total COD of sources h (kg)
COD_i^{mix}	Total COD at mixer i (kg)
C_i^{mix}	COD content for each mixer stream i (kg/m^3)
CN_i^{mix}	Cumulative C/N ratio value at mixer i (kg)
T_i^{mix}	Temperature for each mixer stream I (°C)
$Qtotal$	Total energy able to withdrawn from mixer I (MW)
$F_{i,j}^{mixbg}$	Flow rate from mixer i to biogas digester j (m^3/hour)
F_j^{bg}	Flow rate at biogas digester j (m^3/hour)
$C_{i,j}^{mixbg}$	COD content from mixer i to biogas digester j (kg/m^3)
C_j^{bg}	COD content at biogas digester j (kg/m^3)
F_j^{bgout}	Amount of raw biogas generated at j (nm^3)
$F_{j,k}^{bgcln}$	Flow rate from raw biogas generated at j to biogas cleaning and upgrading system k (nm^3)
F_k^{cln}	Flow rate of raw biogas at biogas cleaning and upgrading system k (nm^3)
F_k^{clnout}	Amount of purified biogas produced from k (nm^3)
Rv	Revenue generated from selling of purified biogas (USD)
CBg	Capital cost of biogas digester (USD)
$CBgCln$	Capital cost of biogas cleaning and upgrading system (USD)
$CCostBg$	Annualized capital cost of selected biogas digester (USD)
$CCostBgCln$	Annualized capital cost of selected biogas cleaing and upgrading system (USD)
$OMCostBg$	Operating and maintenance cost of biogas digester (USD)
$OMCostBgCln$	Operating and maintenance cost of biogas cleaning and upgrading system (USD)
$ECostBgCln$	Electricity cost of the biogas cleaning and upgrading system (USD)

References

[1] Wu H, Shi Y, Xia Q and Zhu W 2014 Effectiveness of the policy of circular economy in China: A DEA-based analysis for the period of 11th five-year-plan. *Resources, Conservation and Recycling* **83** 163–175

[2] Jawahir I S and Bradley R 2016 Technological Elements of Circular Economy and the Principles of 6R-Based Closed-loop Material Flow in Sustainable Manufacturing. *Procedia CIRP* **40** 103–108

[3] Sharib S & Halog A 2017 Enhancing value chains by applying industrial symbiosis concept to the Rubber City in Kedah, Malaysia *Journal of Cleaner Production* **141** 1095–1108

[4] Kim H-W, Dong L, Choi A E S, Fujii M, Fujita T and Park H-S 2018 Co-benefit potential of industrial and urban symbiosis using waste heat from industrial park in Ulsan Korea *Resources, Conservation and Recycling* **135** 225–234

[5] Drobež R, Novak Pintarič Z, Pahor B and Kravanja Z 2011 Simultaneous heat integration and the synthesis of biogas processes from animal waste. *Asia-Pacific Journal of Chemical Engineering*, **6**(5) 734–749

[6] Anaokar G S, Khambete A K and Christian R A 2019 Predicting Biogasification Potential of Urban Wastewater Using Multiparameter Aggregated Index of Influent. *Environ. Prog. Sustainable Energy* **38** 13119

[7] Thongnan R, Thongpan H, Rakmak N and Siripatana C 2016 Modeling of anaerobic co-digestion of pig manure and domestic organic waste *Jurnal Teknologi* **78**(5-6) 117-124

[8] Abdeshahian P, Lim J S, Ho W S, Hashim H and Lee C T 2016 Potential of biogas production from farm animal waste in Malaysia *Renewable and Sustainable Energy Reviews* **60** 714–723

[9] Egieya J M, Čuček L, Zirngast K, Isafiade A J, Pahor B and Kravanja Z 2019 Synthesis of biogas supply networks using various biomass and manure types *Computers & Chemical Engineering* **122** 129–151

[10] Mohtar A, Ho W S, Idris A M, Hashim H, Lim J S, Liew P Y, Gabril L H T and Ho C S 2018 Palm Oil Mill Effluent (POME) biogas techno-economic analysis for utilisation as bio compressed natural gas. *Chemical Engineering Transactions* **63** 265–270

[11] Angelidaki I, Treu L, Tsapekos P, Luo G, Campanaro S, Wenzel H and Kougias P G 2018 Biogas upgrading and utilization: Current status and perspectives *Biotechnology Advances*, **36(2)**, 452–466.

[12] Sun Q, Li H, Yan J, Liu L, Yu Z, & Yu X 2015 Selection of appropriate biogas upgrading technology-a review of biogas cleaning, upgrading and utilisation *Renewable and Sustainable Energy Reviews* **51** 521–532

[13] Abas R, Abdullah R and Hawari Y 2013 Economic Feasibility Study on Establishing an Oil Palm Biogas Plant in Malaysia *Oil Palm Industry Economic Journal* **13** 14–21

[14] Malaysian Palm Oil Board (MPOB) 2014 National Key Economic Areas Biogas Capture and Cdm Project.

[15] Hauser M J 2017 *Cost evaluation and life cycle assessment of biogas upgrading technologies for an anaerobic digestion case study in the United States* Master thesis (Norwegian University of Science and Technology)

[16] Bauer F, Hulteberg C, Persson T and Tamm D 2013 *Biogas upgrading – Review of commercial technologies* Swedish Gas Technologies.

IOP Conf. Series: Materials Science and Engineering **702** (2019) 012015 doi:10.1088/1757-899X/702/1/012015

Towards water integration in Eco-Industrial Park: An overview of water recovery from industries

S F Sa'ad, R Zailan, S R Wan Alwi*, J S Lim and Z A Manan

Process Systems Engineering Centre (PROSPECT), Research Institute for Sustainable Environment (RISE), School of Chemical and Energy Engineering, Faculty of Engineering, Universiti Teknologi Malaysia, 81310 Skudai, Johor, Malaysia

*E-mail: syarifah@utm.my

Abstract. Water integration among different industries in Eco-Industrial Park brings a promising opportunity in reducing freshwater consumption by implementing industrial symbiosis concept. Mixing of wastewater from industries might be cost-effective, however, it is not the best option for water reuse since different reuse option requires different water qualities and treatment. Thus, wastewater segregation based on the wastewater source and quality is required. This paper reviews the potential wastewater source from the industry and wastewater recovery prior to be reused. The review is based on the literature survey from the industries, suppliers and guidelines. From the survey, wastewater sources that can be recovered for reuse option are water generated as by-product, wastewater from the cleaning operation and utility operation. These wastewater sources cannot be mixed because each source has different types of contaminant content. Appropriate water quality can be achieved through primary, secondary and tertiary treatment. Advance technology such as membranes are efficient to remove non-desirable compounds from wastewater. The treated wastewater can provide a viable solution as an alternative water source to ease water-related stress and wastewater segregation can reduce the burden of the treatment.

1. Introduction

Freshwater is used vigorously in industries yet the scarcity of freshwater has become a severe global issue recently with increasing of competition and climate change [1]. According to Connor [2], the water demand from the manufacturing sector will increase to 400% by 2030. The growing of water reuse in the industry could help to solve the issues. Water exchange between industries is one of the synergies besides waste and energy intended to minimize freshwater consumption in the Eco-Industrial Park (EIP) [1]. Such a strategy could assist companies through water efficiency and water recycling initiatives within the EIP to save industrial water demand [3-4]. EIP is designed and implemented on the basis of mutual collaboration between multiple industries in the same industrial site to achieve sustainable industries by applying industrial symbiosis concept [5-6]. Industrial symbiosis represents a cyclic flow of wastes from one industry to the other industry as new resources [1]. The option of water reuse in the industry described by the water produced after undergoing necessary treatment with appropriate quality suits with the process requirement. The objectives of water reuse are primarily to fulfil industrial water demand, reduce pressure on freshwater consumption, reduce the cost of energy use and environmental protection. From the environmental view, water reuse also helps to reduce greenhouse gas emissions and prevent wastewater penetration into water bodies.

European Union (EU) water policy and EU policy on resource efficiency in 2015 have embraced safe and cost-effective water reuse as an alternative source of water in the planning for water scarcity

[7]. Water reuse also contributes to the United Nations 2030 Sustainable Development Agenda through its objective of significantly increasing the effectiveness of water use and improve water quality across all sectors by the year 2030. Previous studies have presented an overview of the wastewater management related to EIP development globally. Piadeh et al. [8] have studied the industrial estates in Iran towards the EIP development and stated that reusing of wastewater for industrial consumption through the installation of advanced treatment units in Iran is successful. Through a recent case study by Jia et al. [9], highly efficient water utilization, as well as economic and environmental benefits, were achieved by implementing industrial symbiosis concept for coal chemical industry in Erdos, China. Whereas, in order to develop reuse of industrial water in Port Melbourne, a cost-effectiveness study of five water treatments and industrial reuse options in Port Melbourne's Fishermans Bend industrial area has been done. The research identified that further pilot testing of the quality of input water requires strong financial and continuous stakeholder involvement [3].

Various industries in the EIP like the power plant, oil refinery, and chemical plant require water reuse for different purposes such as cooling, processing, aggregate washing, concrete manufacturing, soil compaction and dust control [7]. The typical source of wastewater for the water integration is mixed wastewater either from the urban wastewater or secondary effluent of wastewater treatment plant (WWTP). Mixing of wastewater from industries might be cost-effective, however, it is not the best option for water reuse [10]. This is due to the different reuse option requires different types of water quality as well as the level of treatment. Rubio-Castro et al. [11] have proposed the water integration model with wastewater source segregation into two different stages of treatment. However, the types of potential wastewater source that can be recovered are not being addressed. Furthermore, the wastewater segregation based on the regeneration unit and not based on the specific wastewater quality. In the real case, there are many types of wastewater with different contaminant content. These wastewaters need to be collected separately and if possible, at each point of the source. In this paper, the potential wastewater sources that can be recovered for reuse option are presented which give an overview to the designers and developers to develop water integration between different industries. Each wastewater source needs to be segregated based on the quality, then recovered based on the reuse option requirement. Different qualities of water reuse will be sold to the industries based on the plant's requirement. Figure 1 shows the illustration of the water integration process between industries that will be presented in the next section of this paper.

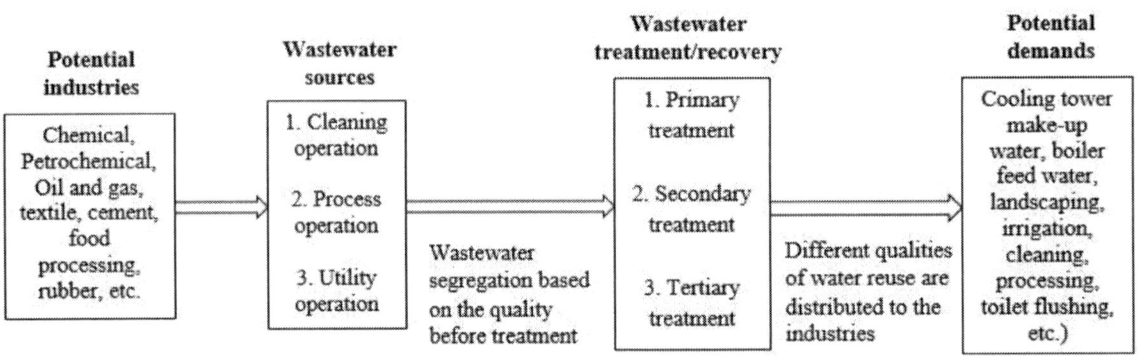

Figure 1. Water integration process between industries.

2. Existing of water integration

The section reviews several existing successful water integrations in the established EIP globally. The most interesting and successful EIP is the Kalundborg, Denmark, which has practised water exchange between tenants, i.e., Asnaes Power Station and Statoil refinery [7]. The power station receives 700,000 m^3 of cooling water each year from Statoil as boiler feed water and the cooling water becomes steam that is returned to Statoil. Furthermore, the power station also utilizes approximately 200,000 m^3 of Statoil's treated wastewater each year for cleaning. The savings recorded about 3 million m^3 of groundwater and 1 million m^3 of surface water per year [7, 12]. One of profitable EIP is the Kwinana

Industrial Area (KIA) in Western Australia, the largest EIP in the world managed to minimize about 6 Gl/year scheme water for the industrial process by replacing it with recycled water. At the Kwinana Water Reclamation Plant (KWRP), about 24 ml/day of secondary treated wastewater from the WWTP at Woodman Point undergoes microfiltration/reverse osmosis to generate high-quality industrial-grade water. Subsequently, about 17 ml/day of recycled water is supplied to Kwinana industrial sites including cogeneration plant [13].

The formation of Chinese EIP had begun since China launched the national EIP demonstration program in 2000 [4]. Tianjin Economic-technological Development Area (TEDA) was instated as one of the National Demonstration EIPs in 2008. The EIP was certified with ISO14001 and adopted its primary principles for decreasing wastewater discharge to the water scarcity problem during the year 2000. According to Yu et al. [14], TEDA was committed to employing the following water integration initiatives:

i. Operated the first wastewater reclamation plant and preceded with two measures: the sewage disposal fee exemption in 2000 and encouragement of the use of alternative water sources in 2002.
ii. In order to encourage companies to enhance their wastewater treatment and use recycled water, the price of recycled water was set 20% cheaper than tap water.
iii. In 2007, TEDA has come out with the regional water cycle system by integrating in-situ regeneration of wastewater, WWTP, recycled water and ecological landscape restoration to provide high-quality water to the EIP tenants.
iv. Almost 25.29% of discharged water was recycled by 2009.

In addition, Shanghai Chemical Industrial Park (SCIP) was endorsed as a national demonstration EIP in 2013. EIP is enthusiastic to reduce pressure on freshwater consumption for its industrial process by downgrading the high water-intensive to less water-intensive sectors (i.e. from oil refinery and raw chemical production to fine chemicals) and practicing water reuse option among EIP tenants. The WWTP accentuated on the quality of wastewater and separated it into four classifications which are organic, inorganic, rainwater and municipal wastewater instead of mixing them together. Furthermore, SCIP also commissioned an artificial wetland experiment to further the treatment of WWTP effluent to be reused before final discharge to Hangzhou bay [15].

3. Potential industries and equipment

The water reuse in industrial sectors can help to minimize the use of potable water and maximize the return on investment. High consumption of water also means a high potential to generate a high quantity of wastewater. As most water will become wastewater except those being consumed by the process, evaporated or lost. Most of the water in the industry is used for cooling purposes and the major water consumption is chemical, steel, and pulp and paper industrial sectors [16]. Water reuse can be used for different types of options according to the requirements of the industries such as cooling, heating, product cleaning, irrigating, washing and others. The potential uses of reused water for industrial sectors are cooling tower make-up water, boiler feed water, stack gas scrubbers, process waters, and construction [17]. Figure 2 shows the global application of water reuse after treatment.

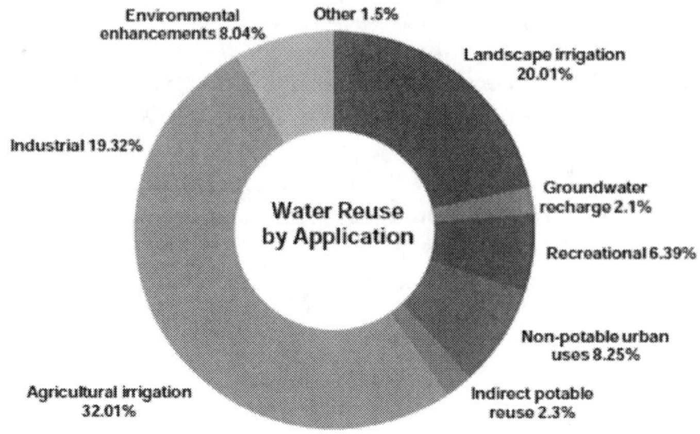

Figure 2. Global application of water reuse after treatment [18].

Landscaping can accept water reuse with a high concentration of nitrogen and phosphorus [10]. Most plants cannot survive the irrigation water with high total dissolved solids (TDS) level of 1000 ppm and the recommended value is 500 ppm [19]. The continual use of high TDS water will cause the minerals and salts to build-up in the soils, which will greatly impair the root growth of plants. Water reuse option for cooling purposes requires lower water quality compared to the other industrial water and it needs the removal of residual organics, suspended solids, ammonia, phosphorus, calcium, magnesium, iron and silica [10]. Meanwhile, boiler, process and cleaning water require the removal of heavy metals and water softening [10]. Table 1 shows the recycled water quality requirement for irrigation and cooling tower. Table 2 shows the recommended water quality standards of boiler feed water for low and medium pressure boiler from American Water Works Association (AWWA), British Standards Specification (BSS) and US Environmental Protection Agency (USEPA) as a benchmark.

Table 1. Recycled water quality requirement for irrigation and cooling tower [18, 20].

Parameter	Irrigation	Cooling Tower
Total suspended solids (TSS)	< 5-35 mg/l	30 mg/l
Biochemical oxygen demand (BOD)	< 5-45 mg/l	30 mg/l
Chemical oxygen demand (COD)	< 20-200 mg/l	-
TDS	< 450-4000 mg/l	-
Heavy metals	< 0.001 mg Hg/l < 0.01 mg Cd/l < 0.02-0.1 mg Ni/l	-
Chlorine residual	0.5-5 mg Cl/l	1 mg Cl/l
Nitrogen	< 10-15 mg N/l	-
Phosphorus	< 0.1-2 mg P/l	-

Table 2. Water quality standards for boiler feed water [21].

Parameter	AWWA	BSS	USEPA
TSS (mg/l)	300-600	100-1500	10
TDS (mg/l)	3000-5000	3000	700
Hardness (mg/l)	0-10	10	0.3
Alkalinity (mg/l)	700	700	350
COD (mg/l)	-	-	5

Reused/Recycled water can be alternative sources of raw water if the requirements are met. Water quality for boiler feed water depends on the pressure of the boiler, where higher pressure requires higher quality water. Since 2000, several Los Angeles refineries have been using recycled water as the main source of boiler feed water [18].

4. Potential wastewater source

According to Rossiter and Rutkowski [22], there are three types of potential sources of wastewater from industries that can be recovered for reuse option. Firstly, wastewater from the cleaning operation to remove impurities such as desalting in oil refining, brown stock washing in pulp mills, solvent extraction and others. Secondly, wastewater generated as a by-product from the process operation such as decanted water from crude oil or water generated by reaction processes. Water as a by-product of crude oil or natural gas extraction is commonly defined as produced water. This produced water contains high saline concentration, TDS and hydrocarbons, which requires high treatment technology [18]. However, the quality of produced water varies widely depending on the geographical location, source of geological formation and type of hydrocarbon. For example, the coalbed methane production has low TDS water sources and this produced water can be reused with very little treatment. The treatment technologies for produced water include oil-water separator, dissolved gas flotation, adsorption and filtration [18].

The third potential wastewater source is wastewater generated by the utility operation such as blowdown from boilers and cooling tower. The blowdown of the boiler is mainly used to regulate dissolved solids in boiler water and to remove loose scale contaminants and sediments. The blowdown frequency depends on the type of boiler, operating conditions of the system and the feed water contaminant levels [20]. The blowdown water could have high TSS, TDS, hardness, alkalinity, bacteria, oil and grease, chemical additives, heavy metals and others [20]. At equilibrium, the quantities of TDS removed by boiler blowdown exactly equal the quantities of TDS present in boiler feed water [23]. Blowdown water from the cooling tower could have high TDS range from 500 to 1300 ppm, depending on the feed water and the operation of the cooling system [19]. The treatment of cooling tower blowdown can optimize the cooling tower consumption, reduce operational costs and water footprint [20]. Table 3 shows the sources of wastewater and contaminants content from different industries. Table 4 shows the blowdown water characteristics of the cooling tower.

Table 3. Source of wastewater and contaminants content from different industries [10].

Industry	Operation	Wastewater content
Discrete parts	Cleaning system	Metals and oily water
	Mop water	Heavy metals
Textiles (assembly)	Compressor operation	Oily water
Automobile parts	Metal plating and cleaning	Heavy metals
Power plant	Cooling system	Heavy metals
	Flue gas desulphurization	Heavy metals, gypsum, fly ash
Seafood processing plant	Cleaning system	Solid particles, nitrogen and phosphorus
Oil refinery	Cooling system	Heavy metals
	Spent soda streams	Phenols, oil and gases
	Main commingled effluent	Oily water
Chemical plant	Cooling system	Heavy metals

All of these three potential sources of wastewater cannot be mixed before treatment and reuse because each wastewater source has different types of contaminant content. Thus, wastewater sources need to be channelled separately to the centralised recovery system by using water header collectors, which transport different quality of wastewater. The segregation of wastewater not only improve the water reuse quality but also reduce the burden of treatment.

Table 4. The blowdown water properties of the cooling tower [24].

Parameter	Range	Average
pH	6-9	7.93
Conductivity (μS/cm)	0.4-7940	1977.12
Alkalinity (ppm CaCO3)	19.2-800	208.42
Hardness (ppm CaCO3)	14.4-1000	426
SO4 (ppm)	200-1400	763.5
Silica (ppm)	25-800	132.5
Chlorine (ppm)	0.5-2102	381.13
Iron (ppm)	0.03-3	1.14

5. Potential wastewater treatment

As water shortages have increased and environmental policies have introduced more stringent blowdown wastewater laws, high technologies for water treatment have been continually created to satisfy the requirements of the industry. The regenerator is used to remove contaminants in the contaminated water so that the water is suitable to be reused [22]. There are three levels of treatment for water recovery which are physical and/or chemical treatment, biological treatment and treatment for removal of nutrient and pathogen [7]. The choice for the specific treatment option depends on the quality requirement. In the United States, all wastewater is required to be treated to secondary levels, at a minimum [18]. Table 5 shows the treatment phase and application for water recovery.

Table 5. Treatment phase and application for water recovery [25].

Classification	Treatment	Application
Preliminary treatment	Equalization	Wastewaters with high variability
	Neutralization	Wastewaters with extreme pH values
	Temperature adjustment	Wastewaters with extreme temperatures
	Screening	Wastewaters containing large solids
	Grit removal	Wastewaters containing significant amount of large and heavy solids
Primary treatment	Sedimentation	Wastewaters containing suspended solids
	Dissolved-air flotation (DAF)	Wastewaters containing oils, fats, suspended solids
Secondary treatment	Activated sludge	Biologically treatable organic wastes 80-90% of BOD and suspended solids removal
	Aerated pond	Biologically treatable organic wastes 80-90% of BOD removal
	Trickling filter	Biologically treatable organic wastes 80-90% of BOD removal Lower operating costs than activated sludge
	Pressure filtration	Wastewaters high in suspended solids 90-95% solids removal

Sedimentation is the primary step to reduce 50-70% of TSS and 25-40% of BOD [26]. Flocculation/DAF system is used to remove suspended solids and oil [10]. Water treatment technologies are usually focused on the secondary treatment in which the biodegradable organic matter and suspended solids are being biologically removed [27]. Activated sludge (AS) is an aerobic slurry to remove soluble organic matters [26]. Organic matters are converted to biomass in a suspended growth reactor by the microorganism in the wastewater. The aerated pond is used for biological treatment and water from this treatment can be used for once-through cooling [10]. Trickling filter (TF) is an attached-growth process of a circular packed bed of media with a biological film of microorganisms, used to remove organic matters [26]. The removal rate is depending on the surface area and the contact time of the stream with the surface.

Filtration and disinfection provide additional removal of pathogens and nutrients [18]. Filtration removes pathogen by physical adsorption or entrapment. Biological filtration is a granular media filter to remove biodegradable constituents such as total organic content (TOC). On the other hand, granular activated carbon (GAC) filtration uses physical and biological processes for contaminant removal. Besides, GAC has adsorption property and can accumulate more microbial biomass on activated carbon media [18]. Disinfection processes kills many classes of microorganisms through the chemical or biophysical method [18]. The most common disinfection method is chlorination. The effectiveness of chlorine disinfection relies on the temperature of the water, pH, degree of mixing, contact time, the presence of interference substances, chlorinating species concentration and form, and concentration of the organisms to be destroyed [18]. In general, bacteria are less chlorine resistant than viruses. Meanwhile, UV disinfection is increasing in use due to energy efficient and low-cost UV technologies. However, UV disinfection performance and efficiency are highly impacted by the presence of particle-associated microorganisms and the UV transmittance (UVT) of the wastewater [18]. Thus, wastewater needs to undergo filtration to remove particles prior to UV disinfection.

Membranes are one of the most widely used technology for the treatment of water and have the ability to remove non-desirable compounds from wastewater [28]. Membrane filtration has high removal capacity and recovery of 50-90% for brackish water [29]. The main disadvantages of the membrane are high cost causing by fouling, chemical demand and high energy consumption [27, 28]. The concept of membrane filtration is the implementation of a semi-permeable filter that allows the water to permeate due to pressure differences and removes certain suspended or dissolved contents. The membrane filtration consists of microfiltration (MF), ultrafiltration (UF), nanofiltration (NF) and reverse osmosis (RO). UF is commonly used for the pre-treatment of RO by providing solid-free feed for the subsequent treatment step [27]. UF can remove TSS, sediment, algae, viruses and bacteria [26]. UF reduces water turbidity up to lower than 0.35 NTU, TSS up to lower than 0.6 mg/l, and TOC and COD up to 30% [27]. However, UF is not an effective barrier for dissolved solids, pH, conductivity, and hardness. Table 6 shows the composition of water before and after the treatment of UF.

Table 6. Composition of water before and after treatment of UF [27].

Parameter	Units	Average UF feed	Average UF filtrate
Turbidity	NTU	2.5-6	0.18-0.25
TSS	mg/l	8	0.3-0.4
BOD	mg/l	7	-
COD	mg/l	44	32-40
TOC	mg/l	18	13.4-23
Ammonia	mg/l	6.31	4-6
TDS	mg/l	899	same
Hardness (CaCO3)	mg/l	343	same
Calcium	mg/l	85	same
Magnesium	mg/l	32	same
Chloride	mg/l	314	same

On the other hand, NF is used to separate solutes with low molecular weight and effective in reducing hardness, dye and heavy metal [30]. NF has been used to remove monovalent and divalent ions in olive mill effluent, pulp and paper wastewater treatment, and oily wastewater treatment [30]. NF is efficient to remove BOD and conductivity up to 80%, calcium and magnesium up to 85%, COD up to 90%, TOC up to 60%, and TDS up to 70% [31, 32, 33]. However, NF membranes do not remove soluble gases such as CO_2 and permeate has low pH resulting from the alkalinity removal from the water [34]. NF has low operating pressure, low energy consumption and higher permeate flux that can replace RO. Besides, NF has high rejection compared to UF for wastewater treatment. NF performance is affected by the operating conditions, feed properties and membrane characteristics. NF characteristics are between UF and RO which is very innovative, economical and environmentally friendly due to low energy consumption and better selectivity. Table 7 shows the application of NF in the industry.

Table 7. Application of NF in the industry [30].

Application	Membrane performance	Wastewater reuse
Rubber industry	COD removal up to 95% Water recovery up to 80-90% Energy cost decrease up to 55%	Permeate can be reused as boiler feed water
Dairy industry	COD removal up to 99% Solids removal up to 93%	Permeate can be reused as water in boiler and cooling tower
Coking industry	COD, hardness, conductivity, fluoride and chlorine removal	Industrial reuse

RO membrane is a dense polymer layer which is used for removal of dissolved compounds and microorganisms in the water with rejection rates higher than 96% [27]. It is very crucial to decrease the particulate matter in the feed water of RO by using UF to prevent the blockage of RO components and fouling. The second RO pass can be used to further treat the RO permeate for the application as cooling water or boilers. Figure 3 shows the comparison of the membrane process based on the types of materials rejected and pore sizes. The recovery rate for the membrane process depends on the quality of the water source, fouling feed rate, operating pressure and the membrane type. Table 8 shows the typical rates of recovery and concentrate flowrates of the feed water for the membrane process.

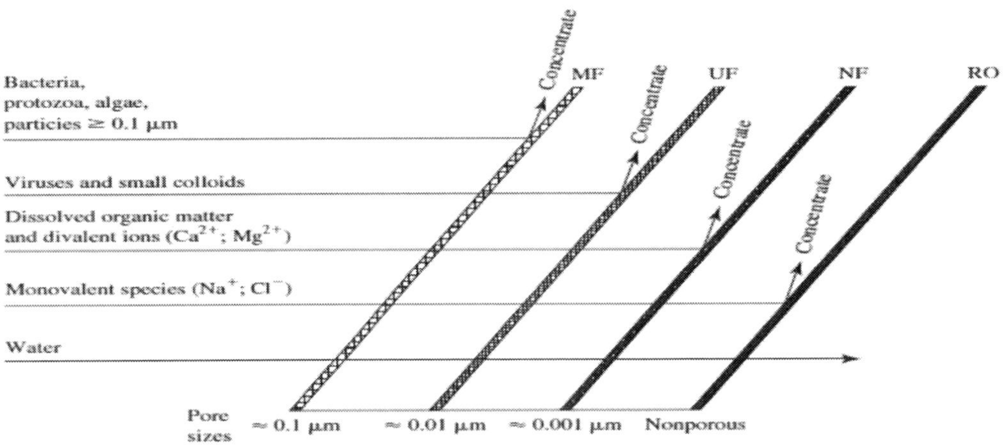

Figure 3. Membrane process comparison [34].

Table 8. Rates of recovery for the membrane process [34].

Process	Recovery rate (%)	Concentrate flow (%)
Microfiltration	85 - 98	2 - 15
Ultrafiltration	85 - 98	2 - 15
Nanofiltration	75 - 90	10 - 25
Brackish water RO	60 - 85	15 - 40
Seawater RO	20 - 50	50 - 80

Ion exchange (IX) is a reversible process in which the exchange of the charged ion in solution with the charged ion attached to the immobile solid particle [26]. The stream has to pass through the cationic and anionic resins for complete removal. The cooling tower of Gaojing Power Plant in China used the multimedia filter (MMF) as a pre-treatment to remove TSS and turbidity, followed by UF and RO [24].

For the application as boiler make-up water, the permeate of the first stage RO was degasified and pH was increased before the second stage RO. The blowdown water requires pre-treatment including filtration, softening, pH adjustment and/or IX [20]. For example, the cooling tower blowdown reuse in Switzerland required MMF, UF, GAC, and IX. The last step, water is demineralized with use of a high recovery close circuit RO. The treated water is used as make-up water for cooling tower.

6. Conclusion

The demand for water in Malaysia and Southeast Asia continues to grow and the natural resources are being impacted by climate change and pollution. Finding alternative and environmental-friendly sources is currently becoming a necessity to arrest challenges of water security. The benefits of wastewater recovery as a second water source cannot be denied. Wastewater emitted from different sources has high potential to be recovered and reused for many purposes in order to reduce freshwater consumption. Wastewater segregation is required before treatment in order to reduce the burden of water treatment as well as to improve the quality of water reuse. Water treatments require adequate technology to produce water that is suitable to be reused. Different reuse option requires different water quality and treatment. Appropriate water quality can be achieved through primary, secondary and tertiary treatment. Advance technology such as membranes treatment are efficient to remove non-desirable compounds from wastewater.

Acknowledgement

The authors would like to thank Universiti Teknologi Malaysia (UTM) for providing the research fund for this project under Vote Number Q. J130000.2509.19H34 and Q. J130000.3509.05G96.

References

[1] Aviso K B 2014 Design of robust water exchange networks for eco-industrial symbiosis *Process Saf. Environ. Prot.* **92** 160-70

[2] Connor R 2015 *The United Nations world water development report 2015: water for a sustainable world* (France: UNESCO)

[3] Giurco D, Bossilkov A, Patterson J and Kazaglis A 2011 Developing industrial water reuse synergies in Port Melbourne: cost effectiveness, barriers and opportunities *J. Clean. Prod.* **19** 867-76

[4] Shi H, Chertow M and Song Y 2010 Developing country experience with eco-industrial parks: a case study of the Tianjin Economic-Technological Development Area in China *J. Clean. Prod.* **18** 191-99

[5] Lowe E A 2001 *Eco-Industrial Park Handbook for Asian Developing Countries* (Oakland: Indigo Development)

[6] Afshari H, Farel R, Gourlia J P and Peng Q 2016 Energy symbioses in Eco-Industrial Parks: models and perspectives *Proc. ASME 2016 Int. Design Engineering Technical Conf. and Computers and Information in Engineering Conf.* (North Carolina: ASME) pp 1-10

[7] EU Water Directors 2016 *Guidelines on Integrating Water Reuse into Water Planning and Management in the context of the Water Framework Directive* (Amsterdam)

[8] Piadeh F, Moghaddam M R A and Mardan S 2014 Present situation of wastewater treatment in the Iranian industrial estates: Recycle and reuse as a solution for achieving goals of eco-industrial parks *Resour. Conserv. Recycl.* **92** 172-78

[9] Jia S, Zhuang H, Han H and Wang F 2016 Application of industrial ecology in water utilization of coal chemical industry: A case study in Erdos, China *J. Clean. Prod.* **135** 20-29

[10] Martin S A, Weitz K A, Cushman R A, Sharma A and Lindrooth R C 1996 *Eco-Industrial Park: A case study and analysis of economic, environmental, technical, and regulatory issues* (Oakland: Indigo Development)

[11] Rubio-Castro E, Ponce-Ortega J M, El-Halwagi M M, Serna-Gonzales M and Jimenez-Gutierrez A 2011 Synthesis of water integration networks in eco-industrial parks *21st European Symposium on Computer Aided Process Engineering* (Greece: Elsevier) pp 1170-74

[12] Harris S, Van Berkel R and Kurup B 2008 Fostering industrial symbiosis for regional sustainable

development outcomes *CRRC 2008* (UK: Queen's University Belfast) pp 1-21
[13] Rosano M and Schianetz K 2014 Measuring sustainability performance in Industrial parks: A Case Study of the Kwinana Industrial Area *Int. J. Sustain. Dev.* **17** 261-80
[14] Yu C, de Jong M and Dijkema G P J 2014 Process analysis of eco-industrial park development–the case of Tianjin, China *J. Clean. Prod.* **64** 464-77
[15] Yune J H, Tian J, Liu W, Chen L and Descamps-Large C 2016 Greening Chinese chemical industrial park by implementing industrial ecology strategies: A case study *Resour. Conserv. Recycl.* **112** 54-64
[16] Magara Y 2017 (cited 7 Nov 2018) Industrial Water Available from: https://www.eolss.net/sample-chapters/c07/E2-19-02-04.pdf
[17] Rimer A E 2018 *Planning for the Distribution of Reclaimed Water* (United States: American Water Works Association)
[18] United States Environmental Protection Agency 2012 *Guidelines for Water Reuse* (Washington: U.S. Agency for International Development)
[19] Alliance for Water Efficiency 2019 (cited 21 July 2019) Blow-Down Water Introduction Available from: http://www.allianceforwaterefficiency.org/blow_down_water_introduction.aspx
[20] Lenntech 2019 (cited 14 June 2019) Water treatment and purification Available from: https://www.lenntech.com/
[21] Jami M S, Amosa M K, Alkhatib M F R, Jimat D N and Muyibi S A 2013 Boiler-Feed and Process Water Reclamation from Biotreated Palm Oil Mill Effluent (BPOME): A Developmental Review *Chem. Biochem. Eng. Q.* **27** 477-89
[22] Rossiter A P and Rutkowski M A (cited 13 Feb 2019) Process integration for wastewater minimization Available from: https://pdfs.semanticscholar.org/b48c/947a19b595e638b89cda379d87ee9274c0e6.pdf
[23] KLM Technology Group 2011 (cited 24 July 2019) Process Design of Water Systems Available from: https://www.klmtechgroup.com/PDF/ess/PROJECT_STANDARDs_AND_SPECIFICATIONS_design_of_water_systems_Rev01.pdf
[24] Olariu R 2015 *Treatment of cooling tower blowdown water-The effect of biodispersant on the ultrafiltration membrane* (Netherlands: Delft University of Technology)
[25] Eludoyin A O 2016 Sustainability and Water Reclamation *Urban Water Reuse Handbook* (Florida: CRC Press Taylor and Francis)
[26] Yang L, Salcedo-Diaz R and Grossmann I E 2014 Water network optimization with wastewater regeneration models *Ind. Eng. Chem. Res.* **53** 17680-695
[27] Niewersch C, Arias A, Gilron J, Sukopova M, Vila S, Gilabert G, Pi J R, Guell I J, Svoijtka J and Martinez X 2016 *Report on innovative membrane technologies and schemes for water reuse* (Spain: DEMOWARE)
[28] Martinez X, Hochstrat R, Miehe U, Casado F, Frijns J, Bortoli J, Denieul M P, Orsoni J, Alcalde L and Jeffrey P 2017 *Final Publishable Summary Report* (Spain: DEMOWARE)
[29] Altman S J, Jensen R P, Cappelle M A, Sanchez A L, Everett R L, Anderson Jr H L and McGrath L K 2012 Membrane treatment of side-stream cooling tower water for reduction of water usage *Desalination* **285** 177-83
[30] Mulyanti R and Susanto H 2018 Wastewater treatment by nanofiltration membranes *IOP Conf. Ser.: Earth Environ. Sci.* **142** 012017
[31] Zulaikha S, Lau W J, Ismail A F and Jaafar J 2014 Treatment of restaurant wastewater using ultrafiltration and nanofiltration membranes *J. Water Process Eng.* **2** 58-62
[32] Bodzek M, Tomaszewska B and Rajca M 2017 Nanofiltration renovation of mineral water *Arch. Environ. Prot.* **43** 51-59
[33] Naidu L D, Saravanan S, Chidambaram M, Goel M, Das A and Babu J S C 2015 Nanofiltration in Transforming Surface Water into Healthy Water: Comparison with Reverse Osmosis *J. Chem.* 1-6
[34] Davis M L 2010 *Water and Wastewater Engineering* (New York: McGraw Hill)

IOP Conf. Series: Materials Science and Engineering **702** (2019) 012016 doi:10.1088/1757-899X/702/1/012016

Superstructure optimization of bioethanol production from corn stover

S S Azizan, N F I Effendy and Z B Handani*

Faculty of Chemical and Process Engineering Technology, College of Engineering Technology, Universiti Malaysia Pahang, Gambang Campus, Lebuhraya Tun Razak, 26300 Kuantan, Pahang, Malaysia

*E-mail: zainatul@ump.edu.my

Abstract. Bioethanol is a liquid fuel that can be produced from various raw materials. The first generation of bioethanol which are starch and sugar has begun to decline as these materials are considered to be the food sources for human as well as animals. However, the rise of production of bioethanol from the second generation which are the lignocellulosic materials has been proven to give a lot of benefits to mankind instead. Various technologies were studied in order to design an optimal processing route for bioethanol production from lignocellulosic biomass. A superstructure on the available process technologies which include the three main technologies, the pretreatment, sugar conversion, separation and purification was first developed before screening it down to a final superstructure. All of the data have been used in this study were collected from open literature and followed by the selection and development of the models required in order to solve the generic models and problems by using an optimization software, the Generic Algebraic Modelling System (GAMS). The optimization results indicate that the optimal processing route in producing bioethanol from lignocellulosic biomass include the technologies of steam explosion, dilute acid followed by fermentation, beer and dehydration of ethanol on zeolite for each process interval with maximum yield of 2238.4 kg/h.

1. Introduction

As the global energy demand keep increasing, the shortage of energy will be a global threat to mankind. The presence of bioethanol can be considered as an important renewable fuel to replace fossil fuels and it is also the most used liquid biofuel in the world. Along with the growth of modernization and industrialization, the demand of bioethanol also has shoot up. In fact, ethanol has long been considered as a suitable alternative to fossil fuels either as a sole fuel in cars with dedicated engines or as an additive in fuel blends with no engine modification requirement when mixed up to 30% [1]. Over the years, the ethanol produced from biomass also has become an increasingly popular alternative to gasoline [2]. As stated by Demirbas [3], unlike gasoline, ethanol is an oxygenated fuel that contains 35% oxygen, which reduces particulate and NO_x emissions from combustion.

Bioethanol can be produced from various raw materials. Sugar and starch based materials such as sugarcane and grains are two groups of raw materials usually used as the main resources for ethanol production while the second group is lignocellulosic materials representing the most viable option for production of ethanol [1]. Due to the agro-ecological conditions, North American and European are the countries that mainly produced bioethanol from starchy materials. Meanwhile, in Brazil, sugarcane is the main feedstock used for bioethanol production. However, in the subsequent years, sugar and starch based materials are not preferable anymore in producing bioethanol as these materials are considered to

IOP Conf. Series: Materials Science and Engineering **702** (2019) 012016 doi:10.1088/1757-899X/702/1/012016

be the source of food. This has led researchers to find another alternative, which are the lignocellulosic materials that are much more promising as feedstock taking into account the viability as well as the cost needed to utilize the feedstock. In many developing countries, food-related feedstock is preferably replaced by non-food raw materials such as sweet sorghum or cassava [4]. Other lignocellulosic materials include agricultural residues or crops such as sugarcane bagasse, corn stover and rice husks.

In general, lignocellulosic biomass can be converted into ethanol through two types of conversion, namely biochemical and thermochemical. Biochemical conversion of biomass involves the technologies of pretreatment followed by hydrolysis to convert the feedstock to simple sugars before these sugars are being fermented to bioethanol. Meanwhile, thermochemical conversion of biomass includes biomass gasification to syngas which are mainly CO and H_2 that can be further synthesized into a wide range of different fuels and chemicals under different catalysts and operating conditions [5]. However, with abundance of technologies available to produce ethanol, the concern of finding the best process route is very significant to produce a high quality fuel grade ethanol. Therefore, in this study, various process technologies available in biochemical conversion were studied in order to synthesis and design an optimal processing route of producing bioethanol from corn stover feedstock through an optimization software, GAMS [6].

2. Materials and Method

2.1. Construction of Systematic Approach for Synthesis of Bioethanol Production

The systematic approach comprises of five main steps and the flow chart of the overall process is shown in figure 1 where the computer-aided support tools are integrated with the workflow in order to deal with the complexity of the synthesis problem. Each step presented in the framework is explained in the following section.

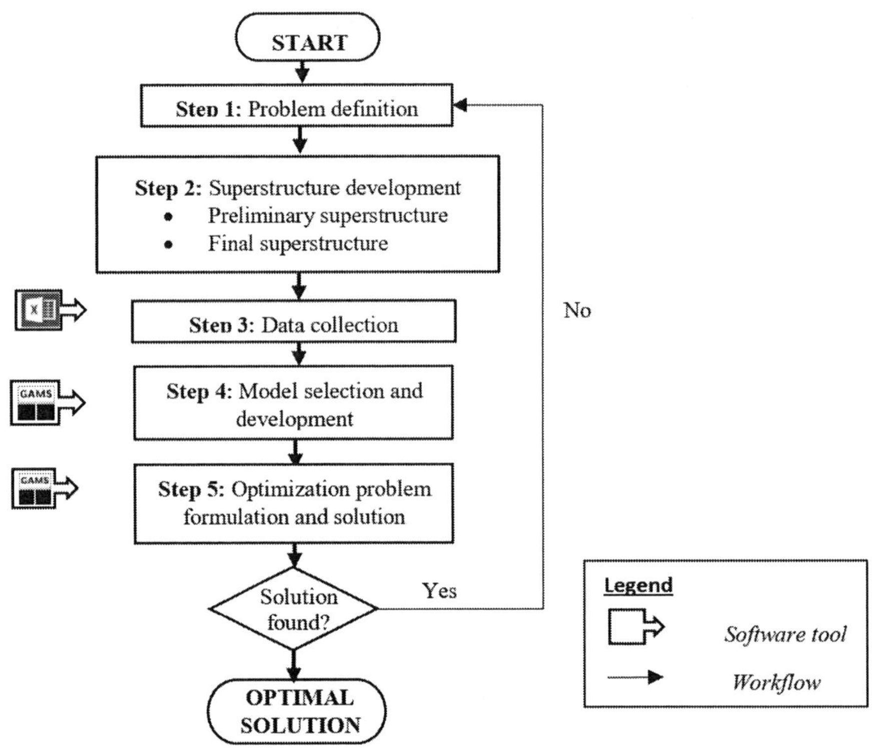

Figure 1. A systematic approach for synthesis of bioethanol production from lignocellulosic biomass.

2.1.1. Step 1: Problem Definition. The objective of the optimization problem is to find the optimal process topology for producing bioethanol through biochemical route. Various alternatives and technologies were listed before finalizing the best process to be used in the production of bioethanol.

IOP Conf. Series: Materials Science and Engineering **702** (2019) 012016 doi:10.1088/1757-899X/702/1/012016

2.1.2. Step 2: Superstructure Development. In this section, the superstructure was developed by listing all the available routes exist in producing bioethanol. The pretreatment methods, sugar conversion technologies as well as separation and purification methods were the factors to be considered in the selection of the process technologies and superstructure development. A prior screening process should be performed for the identification of the technologies to be located in the superstructure and the interconnections between the technological alternatives which defined by the experts. If the data and information for feasible technology was not available, the technology was discarded from the superstructure representation. Each processing step in the superstructure consists of one or more intervals that represented a block. The generic block incorporates generic models to represent alternative processing steps and a simple block used to represent the activities of processing technologies [7-8]. Depending on the step and process technology employed, process intervals can be characterized by a simple process flowsheets of different structure or represented as a series of unit operations which consist of flow mixing, utility dosage, reaction, waste separation, product separation and flow division as shown in figure 2. Table 1 summarizes the available technologies for each conversion step involves biochemical processing routes.

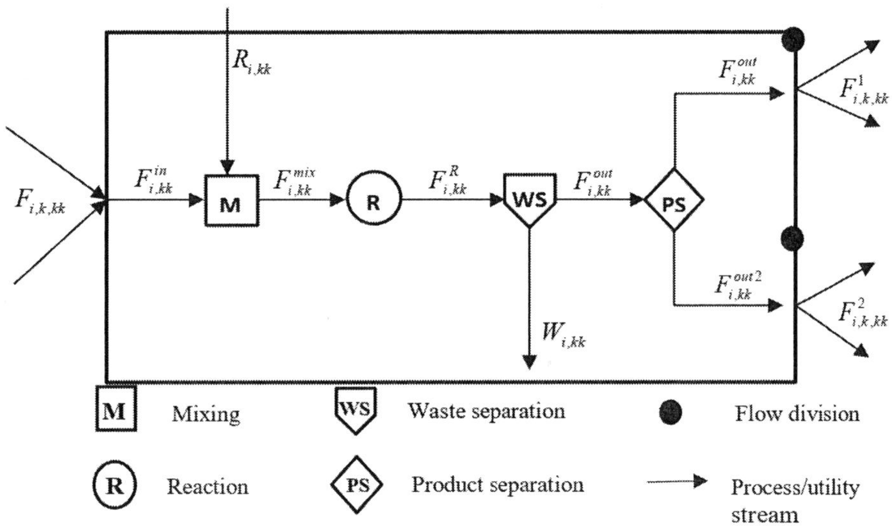

Figure 2. Schematic representation of the generic process interval.

Table 1. Available technologies for conversion of corn stover.

Conversion step	*Process Technology*	*References*
Pretreatment	Steam Explosion	[9]
	Controlled pH	
	Dilute Acid	
	Ammonia Fiber Explosion	
	Ammonia Recycling Percolation	
	Lime	
Hydrolysis	Dilute Acid	[3]
Fermentation	Concentrated Acid	
Separation	Beer column	[10]
Purification	Dehydration of ethanol on zeolite (Rect+Zeo)	[11]
	Dehydration of ethanol on glycerol (Glycer)	[12]
	Dehydration of ethanol on ethylene (Ethyl)	[13]
	Dehydration of ethanol on ionic liquids (ILs)	[14]
	Dehydration of ethanol on silica (Rect+Sil)	[15]

2.1.3. Step 3: Data Collection. For data collection, significant data and information for each possible technology were collected from different resources. All of these data and information then were stored and saved in Microsoft Excel as a database. All relevant information on raw materials as well as the process intervals were also collected. The data and parameters required for the generic data for the intervals include reaction conversion, utility consumption, and split fraction, which are identified by input-output information. The data were obtained from open literature including experimental and pilot plant studies.

2.1.4. Step 4: Model Selection and Development. After defining superstructure and collecting related data, the next task is the selection and development of models describing each of the element of the superstructure as well as the models required for the calculation of the selected objective function (i.e maximum yield) and the performance criteria. In this case, objective function is to obtain maximum product yield as given in equation (1).

$$Yield = \sum_{i,kk} F_{i,kk}^{out} \tag{1}$$

The process models used in this paper are adapted from [7] as presented in the Appendix. Sets of generic equations representing a sequence of processing tasks, which are mixing, reaction, waste removal and product separation, as well as utility consumption were modelled for each interval in the superstructure. Multiple inlets and outlets from the interval were allowed, including recycle streams from downstream intervals and bypasses.

2.1.5. Step 5: Optimization Problem Formulation and Solution. In this last step, the equations, constraint and logical are generally expressed as Mixed Integer Linear Programming (MILP) problem. The optimization problem was solved by employing solver from an optimization software, GAMS. The inputs to the solver were the generic model and an input file with all the necessary problem data such as the raw materials, feed composition, component and model parameters. The outputs from the optimization were the optimal values of the objective function, the corresponding optimization variables and all other process variables.

3. Results and Discussions

The conversion of lignocellulosic biomass to bioethanol in the biochemical conversion platform may go through multiple routes with different available technologies that have different economic performances. In this case study, it is desired to produce 99.5% pure bioethanol from corn stover. The feedstock composition is presented in table 2. It is assumed that the hydrolytic enzymes were purchased from commercial suppliers.

Table 2. The feed composition.

Component	*% Dry Basis*
Lignin	16
Xylan	16
Arabinan	2
Galactan	1
Cellulose	31
Water	24
Ash	10

The objective of the optimization problem is to find the optimal process topology for producing bioethanol through biochemical route that gives the highest bioethanol yield. The superstructure was constructed based on the available of data and information for all viable options of process topology for the bioethanol production from corn stover as shown in figure 3. As can be seen in figure 3, the bioethanol processing network superstructure are composed of 1 raw material (corn stover) 17 process intervals represent different technologies structured in 5 processing steps and 1 product (bioethanol). Also, 34 components and 3 utilities are considered in this problem.

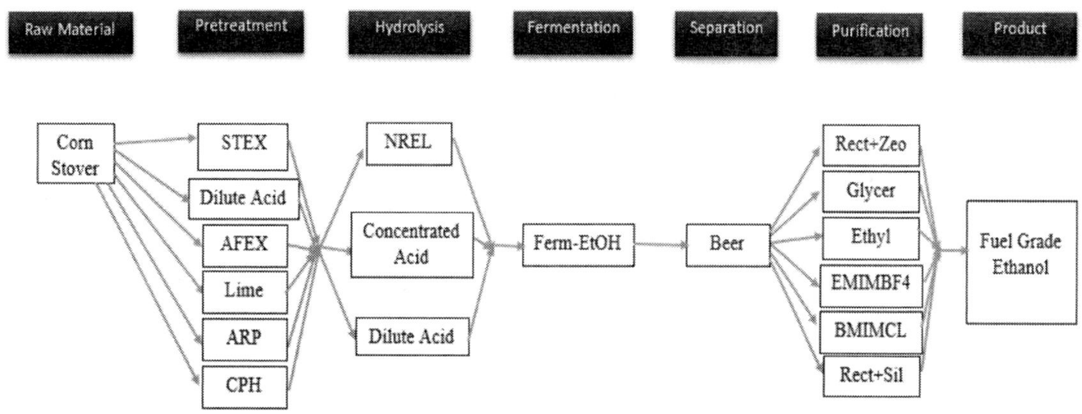

Figure 3. Superstructure of available raw materials and processing routes.

The optimization problem was formulated as MILP and solved using CPLEX/GAMS using GAMS version 23.9.5 [6] to identify an optimal processing route. The case study was performed using a standard computer, equipped with 2.40 GHz Intel (R) Core™ i5-6200U. The superstructure features 987,938 number of variables, and 1,644,456 number of equations as presented in table 3.

Table 3. Statistics optimization problem for bioethanol production process

Number of variables	987,938
Number of equations	1,644,456
Problem type	MILP
Solver	GAMS/CPLEX
Execution time	0.249 seconds

Results for the selected process topology are given in table 4. It is found that steam explosion (STEX) has been selected for the pretreatment step. In STEX process, high-pressure saturated steam is added in which it initiates the hydrolysis reaction. Pressure is then suddenly reduced, exposing the feedstock to an explosive decompression which opens the biomass structure, thus, increasing the accessibility of the enzyme [16]. Meanwhile, for hydrolysis, the results on GAMS showed that the optimized process technology was dilute acid. According to Alvira et al. [17], this method does not only solubilize the hemicellulose which is mainly xylan, but it also converts the solubilized hemicellulose to fermentable sugars. The process route then continued with fermentation and separation before being purified to fuel grade ethanol. In this last step, the optimized route based on GAMS was rectification with zeolite. Abdeen et al. [11] stated that zeolites are proven to be efficient in removing the water from ethanol-water azeotrope due to their small pore size of less than 0.3nm which allows only water to adsorb to the inner large surface area of zeolite. Based on the optimization results obtained from GAMS, the maximum yield that can be achieved for the selected process technologies is 2238.4 kg/h. The optimal processing route is shown in figure 4.

Table 4. Mass flow for the selected process technologies.

Process Technology	*Mass Flow (kg/hr)*
STEX	5175.4
DILA	5175.4
FERM-ETOH	6790.0
FILTR	8090.0
BEER	6896.1
RECT+ZEO	2314.8
FEUL GR ETOH	2238.8

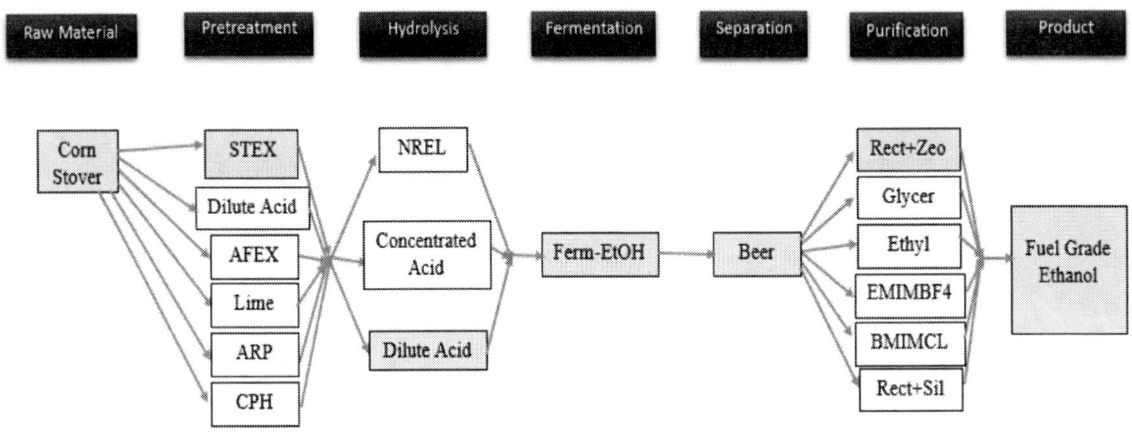

Figure4. An optimal processing route for bioethanol production from corn stover –highlighted the selected process technologies.

4. Conclusion

Based on the results obtained, the selected processing route for pretreatment, hydrolysis, separation and purification process interval were steam explosion, dilute acid followed by fermentation, beer and dehydration of ethanol on zeolite respectively with maximum yield of 2238.4 kg/h. The superstructure-based optimization methodology has been shown to be a useful decision support tool for early stage synthesis and design of bioethanol production from lignocellulosic biomass by screening a number of process technologies.

Acknowledgement
The authors would like to thank Universiti Malaysia Pahang for the financial support (RDU1703181).

Appendix

$$F_{i,kk}^{mix} = F_{i,kk}^{in} + \sigma_{i,kk} CU_{i,kk} \quad \text{(A.1)}$$

$$CU_{i,kk} = \sum_{ii} (\mu_{i,ii,kk} \cdot F_{i,kk}^{in}) \quad \text{(A.2)}$$

$$F_{i,kk}^{R} = F_{i,kk}^{mix} + \sum_{rr,react} (\gamma_{i,kk,rr} \cdot \theta_{react,kk,rr} \cdot F_{react,kk}^{mix} \cdot \frac{MW_i}{MW_{react}}) \quad \text{(A.3)}$$

$$W_{i,kk} = F_{i,kk}^{R} \cdot SW_{i,kk} \quad \text{(A.4)}$$

$$F_{i,kk}^{out} = F_{i,kk}^{R} \cdot (1 - SW_{i,kk}) \quad \text{(A.5)}$$

$$F_{i,kk}^{out1} = F_{i,kk}^{out} \cdot SPr_{i,kk} \quad \text{(A.6)}$$

$$F_{i,kk}^{out2} = F_{i,kk}^{out} \cdot (1 - SPr_{i,kk}) \quad \text{(A.7)}$$

$$F_{i,kk}^{out1} = \sum_{kk} F_{i,k,kk}^{1} \quad (A.8)$$

$$F_{i,kk}^{out2} = \sum_{kk} F_{i,k,kk}^{2} \quad (A.9)$$

$$F_{i,kk}^{in} = \sum_{k} (F_{i,k,kk}^{1} + F_{i,k,kk}^{2}) \quad (A.10)$$

$$F_{i,k,kk}^{1} = F_{i,kk}^{out1} \cdot \xi_{k,kk}^{P} \cdot SM1_{k,kk} \quad (A.11)$$

$$F_{i,k,kk}^{2} = F_{i,kk}^{out2} \cdot \xi_{k,kk}^{S} \cdot SM2_{k,kk} \quad (A.12)$$

$$F_{i,kk}^{out} = \emptyset_{i,kk} \cdot y_{kk} \quad (A.13)$$

$$0 \leq f_{i,kk} \leq y_{kk} \cdot M \quad (A.14)$$

$$\sum_{kk} y_{kk} \leq 1 \quad (A.15)$$

References

[1] Talebnia F, Karakashev D and Angelidaki I 2010 *Bioresource Technology* **101** 4744–53
[2] Binod P, Sindhu R, Singhania R R, Vikram S, Devi L, Nagalakshmi S and Pandey A 2010 *Bioresource Technology* **101** 4767–74
[3] Demirbas A 2005 *Energy Sources* **27** 327-37
[4] Kang Q, Appels L, Tan T and Dewil R 2014 *The Scientific World Journal* 1-13
[5] Ramaswamy S, Huang H –J and Ramarao B V 2013 *Separation and Purification Technologies in Biorefineries* 1st ed. (West Sussex, United Kingdom: John Wiley & Sons, Ltd)
[6] GAMS Development Corporation 2011. GAMS GDX facilities and tools. GAMS Development Corporation, Washington DC, USA.
[7] Quaglia A, Sarup B, Sin Gand Gani R 2012 *Computers and Chemical Engineering* **38** 213-23
[8] Handani Z B, Quaglia A, Gani R 2015 *Computer Aided Chemical Engineering* **37** 875-80
[9] Zhao, X-Q, Zi L -H, Bai F -W, Lin H -L, Hao X -M, Yue G –J and Ho N W 2012 *Adv Biochem. Engin./Biotechnol.* **128** 25–51.
[10] Hamelinck C N, van Hooijdonk G andFaaij A PC 2005 *Biomass and Bioenergy* **28** 384 – 410
[11] Abdeen F R, Mel M, Al-Khatib M and Azmi A 2011 *Proceedings of the 3rd CUTSE International Conference* 312 - 917
[12] Gil D I, Gómez J M and Rodríguez G 2012 *Computers and Chemical Engineering* 39 129–42
[13] Zhang M and Yu Y 2013 *Industrial & Engineering Chemistry Research* **52** 9505−14
[14] Zhu Z, Ri Y, Li M, Jia H, Wang Y and Wang Y 2016 *Chemical Engineering and Processing*, **109** 190-198
[15] LutsT and Katz A 2012 *Catalysis* **55** 84–92
[16] Chiaramonti D, Prussi M, Ferrero S, Oriani L, Ottonello P, Torre P and Cherchi F 2012 *Biomass and Bioenergy* **46** 25-35
[17] Alvira P, Tomás-Pejó E, Ballesteros M and Negro M 2010 *Bioresource Technology* **101** 4851–4861

IOP Conf. Series: Materials Science and Engineering **702** (2019) 012017 doi:10.1088/1757-899X/702/1/012017

Study of natural gas powered solid oxide fuel cell simulation and modeling

A Nafees* and R Abdul Rasid

Faculty of Chemical and Process Engineering Technology, Universiti Malaysia Pahang, 26300 Gambang, Kuantan, Pahang, Malaysia.

*E-mail: anafeespi@gmail.com

Abstract: Solid oxide fuel cell (SOFC) system has been proposed to address the issue of waste gas emission due to gas flaring in oil and gas industry. System has unique advantage of consuming the waste gases and generating electricity as bye product. To analyze and quantify the proposed benefits, a robust cell performance model is highly desirable. A detailed understanding of SOFC component including electrode, electrolyte, interconnect, fuel processing and electrochemical reactions are first step in accurate determination of characteristic performance of the system. For this purpose, a review of modeling philosophies of SOFC system was undertaken in this study. Specifically, SOFC simulation and modeling using commercial software such as Aspen Plus, Aspen Hysys was focused in detail. SOFC models available in literature are either mathematical model or numerical models and ever evolving and improving. SOFC simulation rely on split approach due to absence of built-in module. Authors have developed an Aspen Hysys simulation model using split approach and discussed briefly here. Split approach approximates the SOFC phenomena, thereby inducing error. To overcome this deficiency, authors are developing a MATLAB based user model that can be integrated using 'user unit operation' available in Hysys. Details of the MATLAB program approach is included.

1. Introduction

Fuel cell is an old and proven technology that has been known to the mankind since 1938. Sir William Robert Grove is commonly referred as father of the fuel cell as he firstly demonstrated the concept using water electrolysis and reverse electrolysis that produces the electricity (1). Fuel cell technology has significantly grown ever since and offers promising pathways for inexpensive, safe and clean energy today with application from domestic electricity supplies to automobile to space to air travel and communication. Loosely based on type of electrolyte material used, fuel cell is categorized in different categories namely polymer electrolyte membrane fuel cell (PEM / solid electrolyte), Alkaline fuel cell (AFC / liquid electrolyte), direct methanol fuel cell (DMFC / solid electrolyte), molten carbonate fuel cell (MCFC / liquid electrolyte), Phosphoric acid fuel cell (PAFC /liquid electrolyte), Solid-Oxide Fuel Cell (SOFC / solid electrolyte) and have wide range of application, operating and usage. Among these, a PEM fuel cell is most common and acceptable type of the fuel cell. It consists of proton-conducting polymer electrolyte and precious metals such as platinum containing electrodes. A PEMFC has advantage of low temperature (<100 °C) and reliable operation. But even at a low concentration of carbon monoxide (CO) in ppm, can severely poisons anodes and reduce cell performance. Therefore, requirement of an ultra-pure hydrogen feed is prohibiting its wider application and technology maturity.

IOP Conf. Series: Materials Science and Engineering **702** (2019) 012017 doi:10.1088/1757-899X/702/1/012017

In contrast to PEMFC, a SOFC is quite tolerant to impurities including CO. SOFC utilizes a ceramic based solid electrolyte to transport oxygen-ions between electrodes. SOFC systems employs solid electrode and electrolytes and operation range 500-1000 °C (2). Because the solid materials, it involves only gas and solid only (binary phase). This makes it simpler system than all other type of fuel cells as. Due to its extended operating temperature range, SOFCs are preferable option for various CHP systems beginning sub kilowatt to multi-MW. A summary of the application of various fuel cells is given by Larmini and Dicks [2]. Overall impact of fuel cells becomes significantly more visible and important on CHP systems of any size and scale applications.

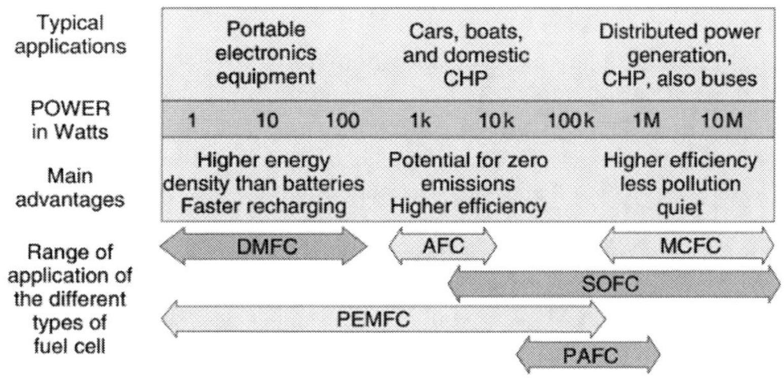

Figure 1. Applications and main advantages of fuel cells of different types, and in different applications [2]

As can be inferred from above figure 1, SOFC has a wider application advantage due to its high efficiency and high temperature of operation. A typical SOFC working principal is shown in Figure 2. The fuel is supplied at anode, while oxidant enters at cathode. Reduction of oxygen and oxidation of fuel takes place. The negative oxygen ion is transferred from cathode to anode. Water is formed as bye product electron is released. The electron then passes through an external load to produce electricity.

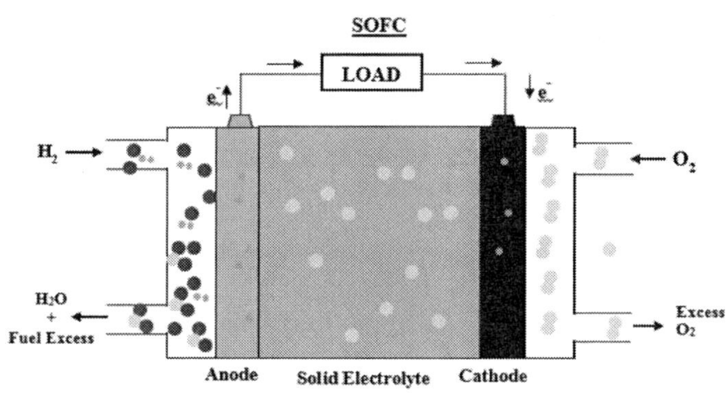

Figure 2. Working principle of a SOFC fuel cell system

A summary of the reactions occuring in the cell given below [2]:

Overall reaction in the cell;

$$CH_4 + 2O_2 \rightarrow 2H_2O + CO_2 \quad (1)$$

Anodic electrochemical reaction;

$$CH_4 + H_2O \rightarrow CO + 3H_2\ (Reformation) \qquad (2)$$

$$CO + H_2O \rightarrow CO_2 + H_2\ (Watergas\ shift) \qquad (3)$$

$$4H_4 + CO_2 \rightarrow CO_2 + 8H^+ + 8e^- \qquad (4)$$

$$CO_2 + 8H^+ + 4O^{-2} \rightarrow CO_2 + 4H_2O \qquad (5)$$

Cathodic electrochemical reaction;

$$2O_2 + 8e^- \rightarrow 4O^{-2} \qquad (6)$$

Due to significantly higher temperature of operation and tolerance, any hydrocarbon fuels can in principle be used with minimum treatment or without rigorous external fuel oxidation and processing that produces hydrogen fuel. This feature of fuel processing elimination makes SOFCs technology attractive. This in turn allow various hydrocarbon fuel choices for fuelling a SOFC system. Notable fuel choices include, hydrogen, associated and non-associated natural gas, methanol, ethanol, biomass and liquid hydrocarbon mixtures such as gasoline, kerosene. Liso et al. [3] using methane, ethanol, methanol, gasoline as fuel option demonstrated the flexibility of a SOFCs performance. As summarized by Weber et al., important SOFC components that helps in understanding of processes includes type and nature of anode and cathode materials, electrolyte, metallic interconnect and sealing, materials [4-5]. A brief overview and review of recent progress is pre-requisite to understand and enable subsequent modeling of phenomena occurring inside the cell.

2. Key SOFC Component and Advancement

A details analysis and understanding of advancement in the major aspects of SOFC system is required to be able to understand the transport and reaction processes.

2.1. Anodic material

SOFC anode typically consist of composite ceramic based material. The requisite characteristic of an anodic material is highly porous material that support ionic transport of ions and electronics. The material should support and facilitate the movement of electron ion phases. Zirconia cermet is a default choice of anodic material and it consist of ceramic and nickel metallic component. There are two main reason why Ni is used as metallic phase in anode. Firstly, it possesses excellent stability and have high electronic conductivity. Secondly, it is highly stable in chemically active environment.

Electrons are released at anode-electrolyte interface boundary travels through the conducting phase. The oxygen ions get transported through the pores of the ceramic phase. Metallic phase material is crafted from catalytic active source and therefore support the heterogeneous reaction of the system. Performance of an electrode directly depend on the composition and structure of the phase. Hence a fine control mechanism is needed to produce a metallic phase of desired activity for electrodes [6]. Ionic phase, metallic phase and gas phase form a network of three-phase boundary. The control of this three-phase bouncy is essential to achieve desired anode characteristics and overall cell electrochemical performance. Anode SOFC material essential meet five characteristics features [7]. Firstly, it should provide active sites for catalytic activity. Secondly, it should have good electronic conductivity. Thirdly, thermal stability should be such as way that its thermal expansion properties are similar to hydrocarbon fuel and other components present in the system. Fourthly, it should have good mechanical stability and have high transport efficiency to / from the three-phase boundary. None of the currently used anode materials fulfil all these requirements. Due to its critical nature, search for an ideal anode material is going at a high pace. Together with cathode, anode is the most preferred aspect and area of current research.

Zirconia stabilized by Nickel / yttria (Ni / YSZ) is currently preferred anode material. Performance degradation with time, coking propensity of Ni for hydrocarbons, low tolerance to sulfur and other acidic compounds and volume uncertainty and instability are some of the major issues with this anode material (8). A number different anode materials involving different metallic phases are being proposed and tested (9-11). For direct hydrocarbon operated systems, Ceria based anode material such as yttria (YSZ) or ceria doped samaria (YDC and SDC), are alternative choices for. As these materials possesses coking suppression abilities, they allow utilization of gaseous hydrocarbons including methane rich fuels with low steam requirement for fuel. Another alternative material with good characteristic features is Perovskite ceramics. A Perovskite ceramic can be described by a generic formula of ABO_3. The alphabet 'A' represents large cations such as Pb, La, Ca and Sr, alphabet B are small cations such as Co, Ti, Zr, Cr, Fe and Ni, As reported by Tao et al, Perovskite ceramics anode are tolerant to carbon deposition as well as poisoning by sulfur compounds (12). Nickel free Perovskite ceramics has comparable electrochemical performance (13, 14).

2.2. Cathodic materials
In the SOFC cell, cathode provides the site for reduction of the oxidant. The reduction releases ions that passes through the electrolyte to anode. Finding a suitable cathode material has been a challenge due to the nature of the requirements. Due to excellent corrosion and oxidation resistance characteristics, noble metals such as Ag, Ru, Rh, Au, Pd, and Pt were preferred choice of materials since the early stages of SOFC development. Due their prohibitive cost, these materials have been losing the application and are being replaced by alternate materials [15-18].

A cathode material should facilitate transportation of oxidant / O_2 / air from the cathode entry channel for reduction at three-phase boundary of cathode and electrolyte. Materials such as strontium-doped lanthanum manganite (LSM) exhibits high catalytic efficiency for the dissociation of oxygen molecules and therefore used a material of choice.

Cathode material typically consist of perovskite materials such as strontium cobalt oxide ($SrCoO_3$) or strontium-doped lanthanum manganite (LSM), which exhibits catalytic efficiency for the dissociation of oxygen molecules. Electrochemical performance of a cathode material depends on aspects such as materials composition, materials doping process, process condition of sintering, material density, surface morphology, and resistivity [19-23]. As search for an ideal cathode material goes on, above factors should be taken into account to develop a new material.

2.3. Electrolyte materials
In a SOFC system, anode and the cathode layers are separated by a solid layer of electrolyte material. The electrolyte typically consists of a ceramic material that redistrict the gas transport, while supporting the ion movement. Any gas transport would lead to thermal oxidation of the fuel resulting in release of heat and loss of fuel. This fuel would have contributed to the electrochemical process and produced electrical energy. Therefore, a desirable electrolyte material should possess excellent ionic conductivity to facilitate movement of oxygen ions from one end of the electrolyte to the other end. As oxidants enters from cathode side, oxygen ions are produced at the interphase of cathode and electrolyte. While fuel enters at the anode side, and these ions are consumed at the interface of anode and electrolyte. As electrons will be produced at the anode and electrolyte boundary by the oxidation of the fuel, the electrolyte material should not possess any electronic conductivity. Even with a low electronic conductivity, electrons will diffuse through the electrolyte layer to the cathode side and might not pass through the external load causing loss of potential.

An electrolyte material should be stable in oxidizing environment existing at the cathode – electrolyte interphase and reducing environmental existing at the cathode – electrolyte interphase. YSZ is currently choice of electrolyte material as it possesses excellent stability in both oxidation and reduction environments present in the system [24-29]. The search for an ideal electrolyte is ongoing and a lot of research studies are being undertaken. Thin film electrolyte (~10 μm) are preferred electrolyte choice of research studies as thickness reduces the ohmic losses.

2.4. Metallic interconnect material
There are two main functions of a SOFC interconnect. Firstly, it provides an electrical contact between the anodes of one individual cell and cathode of the neighboring cell. Secondly, it acts as a physical barrier for the protection of the cathode / anode material of one individual cell from the reduction /

oxidation of the neighboring fuel / air channel (30). Therefore, stability of interconnect materials in oxidizing and reducing environments is pre-requisite and must be fulfilled. Additionally, it provides structural support to the stack material. Cost contribution of interconnects can be significantly in a stack overall cost and depends on system's operating conditions. Sophisticated refractory alloy or ceramic based materials such as lanthanum chromate are preferred materials for interconnect [31, 32].

2.5. Sealing material

Method of sealing of the ceramic components is a major issue for SOFCs as it directly affects the requisite gas tightness of the system. This issue is particularly critical for planar SOFCs and can severely limit the performance. A sealing should possess high thermal and mechanical stability under oxidizing and reducing environment at the typical operating conditions of the system. Thermal and mechanical stability is particularly critical for the SOFC stacks as any considerable thermal expansion difference between the glass and cell components can lead to high residual stress as well as might leads to crack at the sealing interface. This in turn also compromises the cell integrity and electrochemical performance. Sealing material typically consist of glass and glass ceramic. These are preferred choice for a planar SOFCs as it exhibits desired characterize features.

Developing reliable sealing glass has achieved low success due the electrochemical reaction involved ta the interphases. The interfacial reaction among sealing material and fuel, oxidants other components of SOFCs causes performance degradation of stacks. The interaction of glass-ceramics with Fe-Cr alloy interconnect is the main concern area among the various possible interaction and reactions, is of a great concern. As demonstrated by Zhang et al., Nb_2O_5 addition in a borosilicate sealing system improves as it enhances the thermal and mechanical stability by decreasing the interaction among the glass and Fe-Cr metallic interconnect in cells [33]. In ceramic stacks, glass ceramics have been used as sealing materials, but the migration of silica component leads to issues with anode and cathode reactions and performance.

2.6. SOFC poisoning by impurities

High temperature SOFC is tolerant to fuel impurities to carbon monoxide, which is a strong poison for low to moderate temperature operating fuel cells. But SOFC cells are susceptible to sulfur and chloride compounds. Fuel impurities such as H_2S, CH_3SH, COS, Cl^{-2} and iloxane shows poisoning effects on commonly used anode material of Ni-YSZ cermet, thereby having detrimental effect on cell performance [34]. Especially, currently used anode materials have limited tolerance to sulfur compounds that are normally present in many hydrocarbons' fuels including natural gas. The sulfur poising of anode leads performance reduction in conjunction carbon deposition that leads to anode deactivation [12]. One of the reasons for interest in Perovskite ceramics material for anode is its better sulfur tolerance than Ni-YSZ anodes.

3. SOFC Performance

In a typical SOFC system, electrons through the external circuit with an external load due to the potential difference between anode and cathode. However, electrochemical processes occurring inside the system are not simple. These processes are mostly complicated and complex as it involves several intermediates and phases. The charge transfer electrochemistry at the interphase of electrode - electrolyte have been focus of many research, review and studies. The cell operation involves various electrochemical irreversibilities that leads to several potential losses that decrease the net voltage output of the cell. Total resistance of electrodes in the cell comes from the factors such as resistance from internal flow, polarization resistance from concentration gradient, resistance from contact and polarization resistance due to activation. Among these most critical is internal resistance and refers to the resistance offered by the electrode for electron transport. It depends on the electrode material's electronic / metallic phase conductivity and electrode micro - structure and thickness. Contact resistance arises due to imperfect contact of electrodes with electrolyte structure. As resistive losses causing overpotential losses of the cell depends on the local current density, it can be minimized by choice of electrode material as well as by management and control of the micro-structural characteristic's properties at the time of the fabrication [35].

The performance prediction of SOFC is quite difficult due to complex charge transfer phenomenon involved. A voltage vs current performance of a typical SOFC system with operating temperature of approx. 800 ° C is shown in figure 3 (2). As can be inferred from the graph, when higher current density is withdrawn from the cell, its overpotential performance drops from the possible highest voltage. Highest voltage that can be drawn from a SOFC under reversible condition is the open circuit voltage (OCV) or 'no – loss' voltage. Therefore, actual cell voltage will always be less than the OCV voltage as fraction of OCV will be used for driving charge transfer reaction that takes place at the interface of electrodes and electrolyte. The cell potential has a fairly linear relationship at small to moderate current densities but becomes non-linear at higher current densities of the cell. Initially, when current is drawn from the cell, the drop-in voltage is relatively small due to less overpotential required for charge transfer. OCV is the highest voltage that cell can produce, and it is equal to the theoretical OCV value.

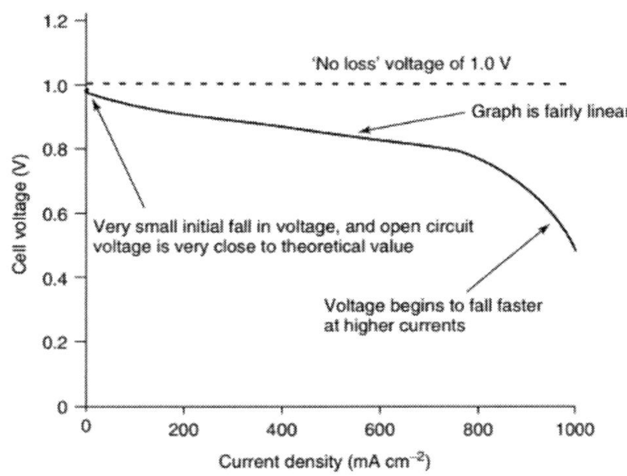

Figure 3. Variation of voltage of a SOFC operating at 800 °C [2]

4. SOFC Cell Modeling

The electrochemical phenomena that occurs in SOFCs is quite complex and some of the phenomena are not understood and clear enough. Therefore, most models of electrochemical phenomenon occurring inside the SOFC system are usually simplified representations of the actual physical and chemical processes. Hence accuracy of these model predictions requires model validation from experimental data over a full range of operations. SOFCs performance its behavior prediction using numerical simulation involving models of the various complex electrochemical processes and their interactions is therefore a relative recent trend in fuel cell research [36]. Theories, philosophies and approach behind modeling of fuel cell is best summarized by Eikerling [37]. This article presents the highlight of the achievement and bottlenecks like developing multiscale and multi-parameter, electrode structure vs functions, model implementation etc. Therefore, a comprehensive model of a SOFC can be developed by incorporating contribution from aspects such as electrochemistry, heterogeneous chemistry, heat and mass transport and unit cells and stacks.

4.1. SOFC Electrochemistry

Activation overpotential (potential needed to drive the charge transfer at electrodes interface), ohmic overpotential (potential needed to overcome the ionic resistance of the electrolyte material), concentration overpotential (potential needed to overcome transport limitation with in porous electrode structure), charge transfer are main resistance factors that needs to be accounted in electrochemistry. Additionally, charge transfer chemistry (reaction pathways, reaction kinetics and mechanism, and rate limiting steps) of the active electrochemical species and Nernst equation modification and solution are integral to the cell electrochemistry.

In a SOFC, electrochemically active species of H_2 and CO are involved in charge transfer process. Amon the two, H_2 oxidation kinetics usually dominates over CO oxidation. Butler - Volmer equation for hydrogen oxidation, equation (7) and oxygen reduction, equation (8) are as follows;

$$i = i_0\left[\exp\left(\frac{(1+\beta_a)F\eta_a}{RT}\right)\exp(-\frac{\beta_c F\eta_a}{RT})\right] \tag{7}$$

and

$$i = i_0\left[\exp\left(\frac{\beta_a F\eta_c}{RT}\right) - \exp(-\frac{\beta_c F\eta_c}{RT})\right] \tag{8}$$

The cell operating potential, results from the collected differences in the electric potential among both phases in the cell [35]. The operating cell potential depends on the overpotential losses and can be written as a function of local current density, I as in equation (9);

$$E_{Cell} = E_{rev} - \eta_a(i) - |\eta_c(i)| - \eta_{ohm}(i) - \eta_{conc}(i) \tag{9}$$

Where η_a and η_c are the activation losses at the anode and cathode side respectively, η_{ohm} is the Ohmic overpotential, and η_{conc} is the concentration overpotential. For the case of hydrogen oxidation E_{rev} is given by Nerst equation in equation (10);

$$E_{rev} = E^o + \frac{RT}{2F}\ln(\frac{P_{H2,a}P_{O2,c}^{1/2}}{P_{H2O,a}}) \tag{10}$$

Where, E^0 is the electro-motoric force (EMF) at standard pressure.

Nerst potential (E_{rev}) is commonly referred as open circuit voltage (OCV). This is the maximum possible potential that can be derived from a cell operating reversibly. Therefore, the Nerst potential is also known as the reversible potential. Therefore, actual potential will always be lower than the Nerst potential during the operation of the cell due to the irreversibilities.

Equation (7), equation (8), equation (9) that are non-linear and coupled with equation (10) can be solved by two different approaches to find the performance of the system. Firstly, it can either be solved for the local current density, *i*, by fixing the operating cell potential, E_{cell}. Secondly, these equations can be solved for cell potential by fixing the average current density. An iterative solution approach will be required.

4.2. SOFC Heterogeneous Chemistry

As Ni-YSZ is predominantly common material of anode, reaction mechanism of CH_4 on Ni is of utmost interest. Even though perovskite materials have exhibited improved characteristic performance but their use as anode material in commercial installation is still not there. For methane or any other hydrocarbon fuel, catalytic reactions will occur inside the anode structure leading to a complex chemical composition of the reactants at the interphase. Intermediate products such CO, CO_2 and H_2O, which are formed at the interphase also actively takes part in the catalytic chemistry, thereby affecting the kinetics of fuel oxidation. Therefore, understanding of the catalytic kinetics is critical for precise cell performance prediction and consumption of fuel. A review and summary of various heterogeneous mechanism of CH_4 over Ni is given by Zhu et al. [38].

4.3. SOFC Heat and Mass Transport

Since SOFC electrode are porous with multi - flow channels, modified form of the general heat and mass conservation equation are needed. The Dusty Gas Model (DGM) can be used to model multi component species transport in porous media. Xu et al., has described the detail equations to calculate DGM diffusion coefficients to calculate the heat and mass transport of the cell [39]. Amiri et al. [40] has developed a mathematical model for entire SOFC transport phenomena. Developed mathematical

IOP Conf. Series: Materials Science and Engineering **702** (2019) 012017 doi:10.1088/1757-899X/702/1/012017

model was for a tubular micro-solid oxide fuel cell and accounted for mass, momentum, heat, and electric charge transfer. The model also included reversible and irreversible heat generation and detailed electrochemical reactions. Experimental data on electrochemical performance and temperature distribution was used to validate the simulation results. Negligible local heating within the electrodes and negligible temperature gradients along the cell's active length are main finding of the study [41].

4.4. SOFC Unit Cells and Stack

In lieu of the complex electrochemical interaction and kinetics, SOFC models are usually simplified representations of the actual physics and chemistry, and accuracy of the model predictions needs to be validated using experimental data. Therefore, numerical simulation of SOFC characteristics employing detailed models of physical and chemistry processes and their interactions is relative new focus of SOFC research.

A complete SOFC system typically includes of a SOFC stack (with variable number of SOFC cells), fuel pre-reformer, fuel ejector, after burner, pre-heater and cogeneration unit. This is done in order to maximize the benefit of high temperature SOFC operation, thereby recovering heat. Above arrangement allow the combined heat hybrid system to reach efficiency up to 80% [42]. All these components are integral and important and have to be considered for the prediction of the cell efficiency and performance of a single or stack of SOFC power generation system [43].

It is desirable to have a SOFC system model comprising of fuel cell stack with its auxiliary components that can predict the cell performance under different conditions and process parameters. Amiri et al. have done outstanding work in this aspect. They have first developed a rigorous SOFC module 43 for a stand-alone SOFC cell. This was then extended to simulate an entire SOFC fuel cell stack system with auxiliary units [44-45]. This work was a good effort to model the stack as it considered factors such as flow maldistribution in the stack manifold, balance of plant (BoP) components and interaction between the stack and the BoP. They have reported another study involving a short stack composed of six micro-tubular fuel cells. This study was mainly focused on stack's electrochemical performance under different electrical connection configurations such as parallel, series and mixed / parallel-series at intermediate temperatures (46). Negligible performance differences were found among different arrangements. A parallel-series configuration was found to best suitable for stack assembly. As can be inferred from above listed studies, fuel cell stack modeling is a work in progress and different aspect and factors needs to be investigated. Zhang et al. [47] have used Aspen Plus A SOFC system may include fuel pre-processors, heat exchangers, turbines, bottoming cycles, etc., all of which can be very effectively modeled in process simulation software. The focus of the study was on optimization of the fuel cell stack and the balance of plant with respect to efficiency and economics to come up with an improved design of a fuel cell system. Therefore, new studies are needed to look for internal SOFC phenomena such as anode blinding by carbon deposition, recycling of depleted fuel, heat and thermal management etc. and analyses at compartment, channel and cell scales. Figure 4 and figure 5 represent schematic of the SOFC stack for counter flow configuration and a single cell for co-current configuration [39].

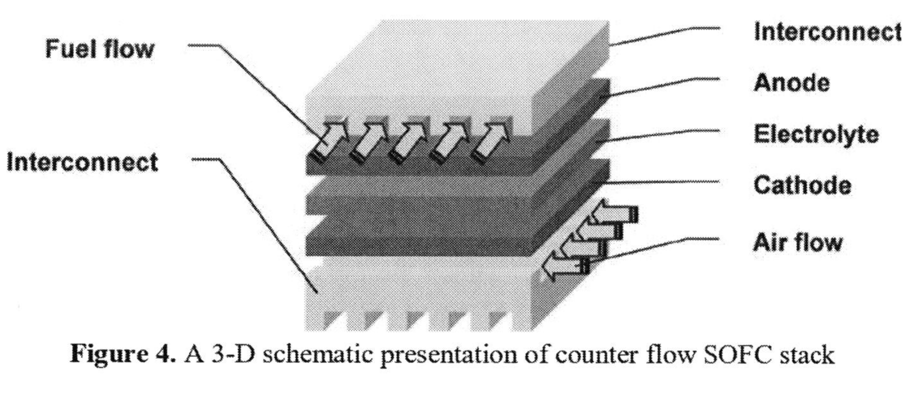

Figure 4. A 3-D schematic presentation of counter flow SOFC stack

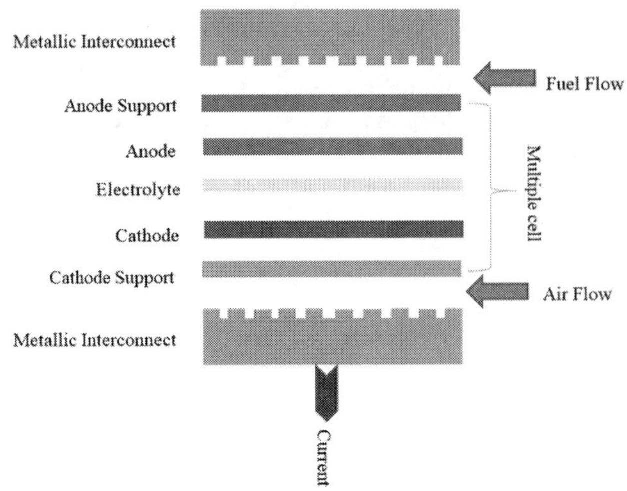

Figure 5. A planer SOFC with co-configuration flow

When natural gas or other hydrocarbons such as diesel, kerosene etc. are used as primary fuel, the fuel is pre-reformed to a certain extent (often completely) prior to feeding to the fuel cell stack. Nevertheless, the anode of the cell also provides ample opportunity for the fuel to undergo reforming. The reforming reactions produce synthesis gas, H_2 and CO, undergoing charge transfer reactions at the electrode and electrolyte interfaces. The reaction products (H_2O CO and CO_2) further replenish the reforming chemistry within the cell. However, the reaction products also dilute the fuel thereby decreasing the Nerst potential as well as the current density of the cell. Generally, the reactants and the products have opposite fluxes within the porous anode. Additionally, Knudson diffusion limitation within the porous electrodes can lead to steep gradients of the species concentration profiles at the electrode-channel interfaces, thereby affecting the chemical kinetics [36].

To predict an accurate performance of a single SOFC or stack, the SOFC model equation needs to be incorporated with models of transport of chemical species in anode and cathode, fluid flow in fuel and air channels, kinetics in anode, heat and mass transport in fuel / air channels and electrode as well as heat transport in entire solid structure of the cell. For this purpose, three-dimensional model, two-dimensional model, one-dimensional model and three-dimensional models of SOFC stack performance are available in literature.

5. Results and Discussion

Background review of the phenomenon happening in SOFC system is presented above. This section describes and summarize the result of SOFC system modeling and simulation. This section is divided in three section and each section focus on experimental data on SOFC performance with simulation data validation, SOFC system simulation using split approach and SOFC simulation using MATLAB model, respectively.

5.1. SOFC Simulation and Modeling

A SOFC system model for a standalone SOFC cell and for an integrated system comprising of fuel cell stack with its auxiliary components needed to model and predict the performance and validation. This task would have been easy by employing one of the commercial process simulations and flowsheeting software such as Promax, Aspen Hysys, Aspen Plus, Pro II, Unisym, CHEMCAD etc. software. However, no in-built SOFC module exists within any of these commercial flowsheet simulation software. Therefore, efforts are being done to come up with a rigorous SOFC model that can be integrated to the standard software.

In literature, many studies have been reported on Simulation of small to medium sized SOFC systems using commercial software. Almost all studies have utilized Aspen Plus software due to inherent advantage such as better solid modeling, stability and accuracy of the results [47-52]. SOFC simulation modeling is a work in progress and provides important bridging to numerical, mathematical and experimental pathways have been employed. The advantage of a simulation modeling is that user can create different case studies and scenarios and achieve a high optimization for the cell.

The authors have used academic version of the Aspen Hysys software (v10) to simulate and model performance of a SOFC fuel cell system that employs waste natural gas as fuel. The details of this simulation model and performance is given here [53]. The study used a split modeling approach to come up with the SOFC fuel performance. Hysys V10 is the version of the software and Peng Robinson property package was employed. A snapshot of the simulation model is shown in figure 6. This simulation was conducted using the simulation parameters listed in table 1. The amount of electricity output for the model was 20197.3 KW. Option of energy integration was explored, and it was found that power generation after energy integration was approximately 18.5% more at 23888.3 KW.

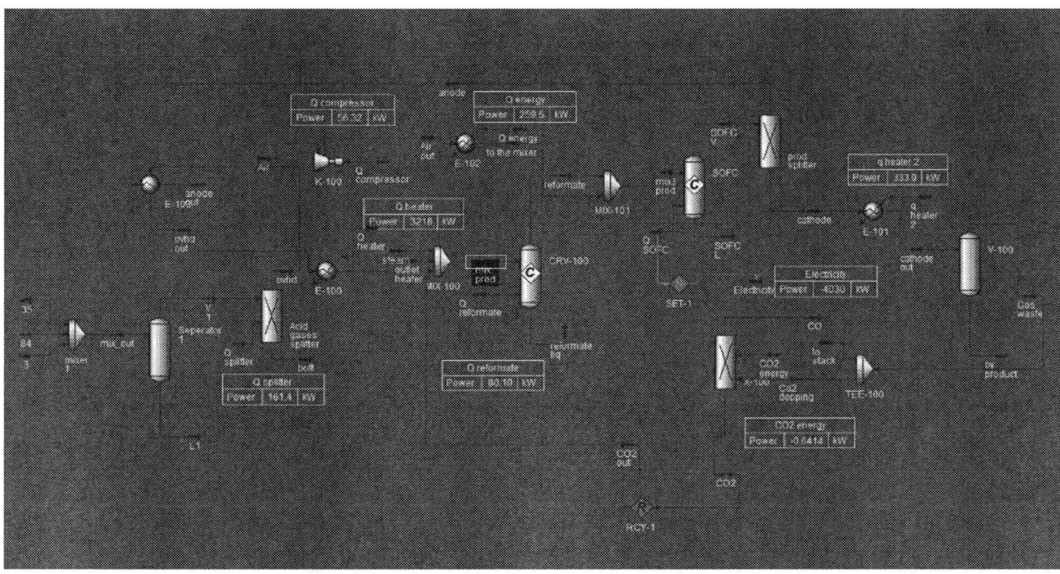

Figure 6. Aspen Hysys simulation model of integrated SOFC fuel cell system

Table 1. Basic parameters for SOFC simulation model

Parameters	Values
Type	A Single SOFC cell
Fluid Package	Peng Robinson
Base parameter	CH_4
Feed	Waste gas (Consisting of methane and other impurities)
Composition	Methane upto 85%
Parameter of interest	Energy / Electricity
Unit	SI
Software version	V10
Nature	Steady state and Dynamic simulation

Subsequently this model was fine-tuned and analysed further. Case studies involving parameters such as fuel and oxidant's flowrate, pressure and temperature, fuel purity, presence of impurities and fuel flexibility. The performance of the model is then compared and validated with the experimental data of a standard SOFC cell. A fair agreement has been obtained with scope for further improvement (53). For this reason, this model is still not the final simulation model as improvement in performance will be undertaken.

5.2. SOFC Simulation with MATLAB SOFC model

The limitation of the present Aspen Hysys simulation model was primarily the absence of a built in SOFC unit operation / module in the design object pellet database of the program, which lead to approximate representation of the SOFC processes to their nearest available unit operations. This impacted the simulation model as electrode were modeled as reactor and electrolytes as splitter and other available unit operations. To overcome this shortcoming, authors are proposing to come up with their own complete SOFC model using MATLAB that can be imported in Aspen Hysys simulation database. To import this model in Aspen Hysys simulation, "user unit operation' option available in the Aspen Hysys design pellet will be used. Figure 7 shows the availability and information needed for the incorporation of a user defined unit operations. The 'user unit operation' can be defined by developing a code with a programming language such as Fortran 90 / 77 or MATLAB. Author choose MATLAB option as it's more applied and stable.

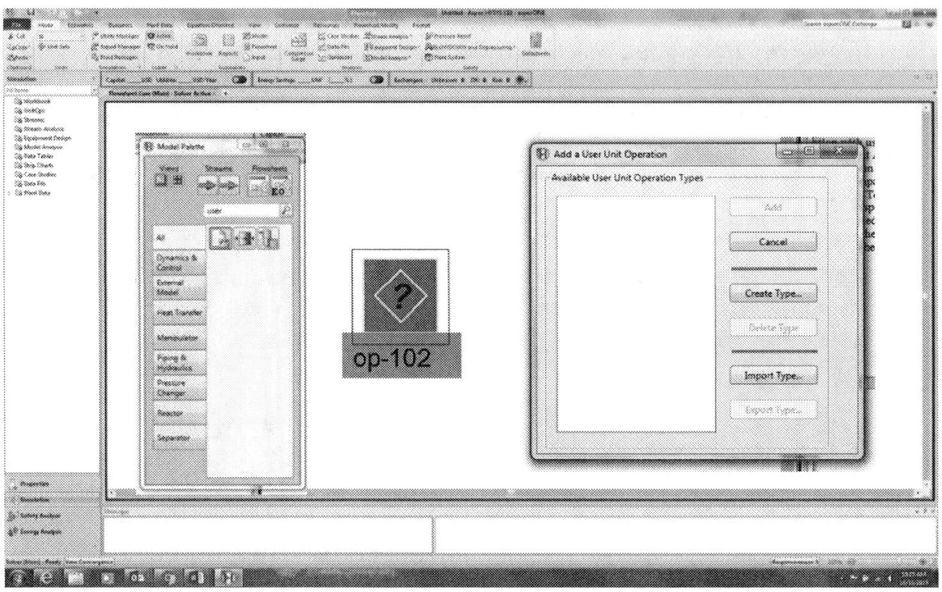

Figure 7. Availability of user unit operation in Aspen Hysys v10

Therefore, in the current study, author is coming up with a 'user defined SOFC model' by coding of the SOFC electrochemistry, heat and mass transport. For this purpose, above described model (equation (7) to (10)) will be solved in a MATLAB by an iterative program. Schematic of the approach that is being followed to come up with a MATLAB program for 'user unit operation' is given in Figure 8. The model works top to bottom approach in either of the two mode namely, voltage or current determination. Basic SOFC details needs to be entered first as well as information about stack cell numbers. Once the feed condition and specifications are defined, the routine will be able to do an iteration to solve and find out the unknown parameters such voltage or current, flowrate of exit streams, energy and heat flow. Troubleshooting and testing of the model is underway and a MATLAB model is not yet available. Once a final MATLAB model is completed, it will be imported in Aspen Hysys simulation to obtain accurate SOFC performance through 'user unit operation'.

A summary of the parameters that is being employed for the MATLAB model / code is given table 2. The final simulation model employing the 'user unit operation' using the MATLAB program is currently being pursued by the authors and final result of the new model will be publish later. Afterwards, benefits can be quantifying and at next stage of future studies, economic feasibility analysis can be conducted.

Table 2. Details of the parameters for MATLAB program model

Parameters	Values
a. Programing tool	
Type	Iterative program
Tool	MATLAB / Simulink program code
MATLAB Version	R2018b (9.5.0.944444)

Parameters	Values
b. SOFC Cell details	
Type	A single SOFC cell
Anode material	Ni - YSZ
Cathode material	GDC/LSCF
Electrolyte	YSZ
Capacity	1 kW
V-I characteristic	As per the reference
c. Assumption for kinetics / reaction mechanism	
Base reaction species	CH_4
Reaction extent	100

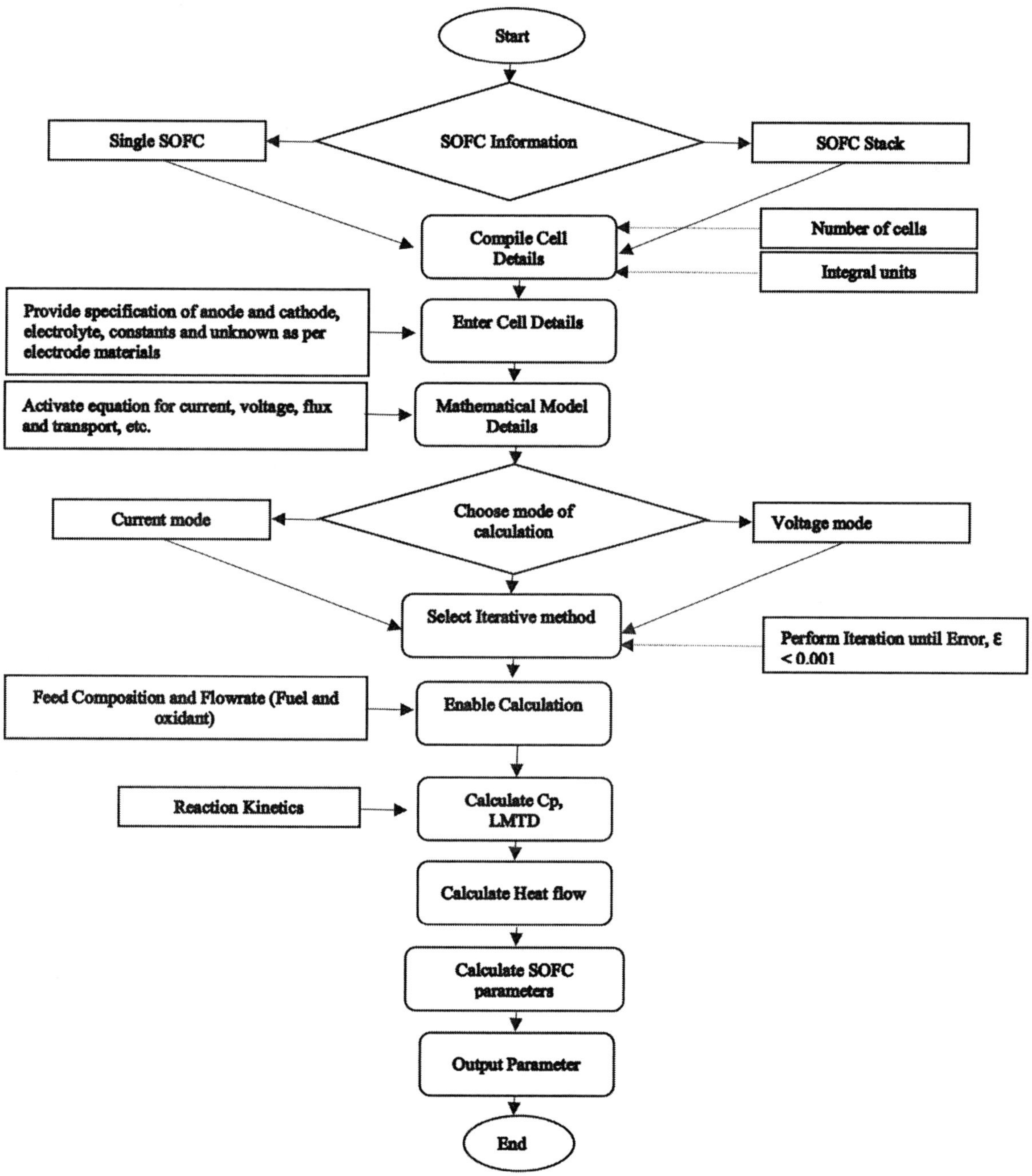

Figure 8. Schematic of the MATLAB program for SOFC performance modeling

5.3. SOFC reaction pathways for MATLAB model

In literature, MATLAB model has been reported for all type of fuel cells including SOFC cell and stacks. But studies are limited to using the hydrogen fuel employing Autothermal reformer or internal reforming that converts the hydrocarbon to hydrogen fuel. Two of the important studies are highlighted here. Boccaleti et al. were the first one to develop a MATLAB based simulation models for a PEMFC and SOFC (54). The study mainly focuses on the advantages and disadvantages of the SOFC cell over PEMFC. Georgis et al. [55] have developed a MATLAB model for a single SOFC system and subsequently extended to the design and operation of energy integrated solid oxide fuel cell (SOFC) systems for in situ hydrogen production and power generation. This study was based on the hydrogen as fuel. In the current study, authors have used the alternative fuel pathways i.e. directly fueled by natural gas. Therefore, reaction kinetic of both studies are altogether different. Authors work will be complimentary to the published work as Georgis et al, study involves the first pathways of fueling the SOFC, while authors have opted for the second pathway of the fueling of SOFC.

5.4. SOFC simulation validation and comparison methodology

To validate and compare the results of the current research, authors are using the open sources available data. There are a number of publications with experimental data of the SOFC cells and their performance and modeling. The research group of Amiri et al. (4, 5) have done extensive publication on SOFC and therefore has been selected by the authors to be the reference experimental data.

This experimental data is then compared with the split approach SOFC simulation and have been found to have a good agreement (53). Scope for further improvement of the simulation via user developed MATLAB model has been found. Now new SOFC simulation using the user. In each of the three cases, performance of the SOFC cell is obtained, i.e. experimental, split based SOFC simulation and user unit operation based SOFC. In particular, cell characteristic performance is obtained by varying the current density of the cell and corresponding voltage measurement. For statistical purpose, the simulation data is taking up to 4 decimal point only. Afterwards, the data will be plotted on a single standard x-y plot using current and potential for each of the three approaches. Percentage difference between from the experimental data will be calculated and any possible disagreement will be analyzed accordingly. Other experimental SOFC data available in open literature will also be used for further validation.

6. Conclusion

Due to energy conversion with high efficiency, solid oxide fuel cells are gaining increasing acceptability as a choice of energy conversion technology with range of industrial and domestic applications. Despite recent advancement in all aspect of SOFC, there is still missing link or incomplete information about exact physio-chemical processes occurring inside the cell.

Therefore, a complete robust SOFC model with universal acceptability is still not developed, either for a single cell or a stack. An accurate SOFC model should incorporate details of transport in electrodes, coupled interactions of heat and mass transport, thermo-catalytic chemistry and electrochemistry. In comparison to H_2 fueled SOFC, natural gas and hydrocarbon fueled SOFC leads to thermo-catalytic chemistry in the anode that affects the electrochemistry due to formation of complex composition of the gas at three phase interfaces. Therefore, development of reliable models and computational tools for the simulation of the SOFC characteristic is complex task.

In this research, authors have reviewed SOFC aspects to come up with an enhanced SOFC simulation and model, which in turn will boost SOFC performance optimization and applications. Technical review of SOFC kinetics pathways that is ore of any SOFC model has been achieved. Accordingly, a MATLAB model of SOFC has been proposed to address the technical gap in the previous SOFC modeling studies. The structure and execution approach of the proposed MATLAB model is included. The details of the integration of this MATLAB model in Aspen Hysys simulation using 'user unit operation' is presented. Result of MATLAB model incorporated SOFC performance will be done in future publication.

Acknowledgment

The authors would like to thank Faculty of Chemical and Process Engineering Technology, FTKKP at University Malaysia Pahang (UMP) for assistance and support for this study.

IOP Conf. Series: Materials Science and Engineering **702** (2019) 012017 doi:10.1088/1757-899X/702/1/012017

References

[1] Milewski J, Swirski K, Santarelli M and Leone P 2011 SOFC Modeling. Advanced Methods of Solid Oxide Fuel Cell Modeling *Green Energy and Technology* 91-200

[2] Dicks JLaA. *Fuel Cell Systems Explained.* 2003 2nd ed. (John Wiley & Sons Ltd.)

[3] Liso V, Cinti G, Nielsen M and Desideri U 2016. Solid oxide fuel cell performance comparison fueled by methane, MeOH, EtOH and gasoline surrogate C_8H_{18}. *Applied Thermal Engineering* **99** 1101-9

[4] Ivers-Tiffee E, Weber A and Herbstritt D 2001 Materials and technologies for SOFC-components. *Journal of the European Ceramic Society* **21**(10-11) 1805-11

[5] Weber A and Ivers-Tiffee E 2004 Materials and concepts for solid oxide fuel cells (SOFCs) in stationary and mobile applications *Journal of Power Sources* **127**(1-2) 273-83

[6] Matsuzaki Y, Tachikawa Y, Somekawa T, Hatae T, Matsumoto H and Taniguchi S 2015 Effect of proton-conduction in electrolyte on electric efficiency of multi-stage solid oxide fuel cells. *Sci Rep.* **5** 12640.

[7] Janardhanan V M, Heuveline V and Deutschmann O 2007. Performance analysis of a SOFC under direct internal reforming conditions. *Journal of Power Sources* **172**(1) 296-307

[8] Janardhanan V M and Deutschmann O 2007 Modeling of solid-oxide fuel cells. Zeitschrift Fur Physikalische Chemie-International *Journal of Research in Physical Chemistry & Chemical Physics* **221**(4) 443-78

[9] Atkinson A, Barnett S, Gorte R J, Irvine J T, McEvoy A J and Mogensen M 2004 Advanced anodes for high-temperature fuel cells. Nat Mater. **3**(1) 17-27

[10] Bae J, Chang I, Kang S, Hong S, Cha S W and Kim Y B 2014 Post-annealing of thin-film yttria stabilized zirconia electrolytes for anode-supported low-temperature solid oxide fuel cells. *J Nanosci Nanotechnol.* **14**(12) 9294-9

[11] Cascos V, Alonso J A and Fernandez-Diaz M T. 2016 Novel Mg-Doped SrMoO(3) Perovskites Designed as Anode Materials for Solid Oxide Fuel Cells. Materials (Basel) **9**(7)

[12] Ding H, Tao Z, Liu S and Zhang J A 2015 High-Performing Sulfur-Tolerant and Redox-Stable Layered Perovskite Anode for Direct Hydrocarbon Solid Oxide Fuel Cells. *Sci Rep.* **5** 18129

[13] Tao S and Irvine J T 2003 A redox-stable efficient anode for solid-oxide fuel cells. Nat Mater **2**(5) 320-3

[14] Tao S and Irvine J T 2004 Discovery and characterization of novel oxide anodes for solid oxide fuel cells *Chem Rec.* **4**(2) 83-95

[15] Choi et al. 2013 Highly efficient and robust cathode materials for low-temperature solid oxide fuel cells: PrBa0.5Sr0.5Co(2-x)Fe(x)O(5+delta). *Sci Rep.* **3** 2426

[16] Li M, Wang Y, Wang Y, Chen F and Xia C 2014. Bismuth doped lanthanum ferrite perovskites as novel cathodes for intermediate-temperature solid oxide fuel cells. *ACS Appl Mater Interfaces* **6**(14) 11286-94

[17] Yang W, Hong T, Li S, Ma Z, Sun C, Xia C and Chen Liquan 2013 Perovskite Sr(1)-xCexCoO(3)-delta (0.05 </= x </= 0.15) as superior cathodes for intermediate temperature solid oxide fuel cells. *ACS Appl Mater Interfaces* **5**(3) 1143-8

[18] Zhang et al. 2015 Enhanced oxygen reduction activity and solid oxide fuel cell performance with a nanoparticles-loaded cathode. *Nano Lett* **15**(3) 1703-9

[19] Ai N, Li N, Rickard W D, Cheng Y, Chen K and Jiang S P 2017 Highly Stable Sr-Free Cobaltite-Based Perovskite Cathodes Directly Assembled on a Barrier-Layer-Free Y2 O3 -ZrO2 Electrolyte of Solid Oxide Fuel Cells *ChemSusChem.* **10**(5) 993-1003

[20] Aniagyei A, Dzade N Y, Tia R, Adei E, Catlow C R A and de Leeuw N H 5 2018 Ab initio investigation of O2 adsorption on Ca-doped LaMnO3 cathodes in solid oxide fuel cells. *Phys Chem Chem Phys.* **20**(45) 28685-98

[21] Azcondo M T, Yuste M, Perez-Flores J C, Munoz-Gil D, Garcia-Martin S, Munoz-Noval A, Orecnch I P, Garcia-Alvarado F and Amador U 2017 Defect Chemistry, Electrical Properties, and Evaluation of New Oxides Sr2 CoNb1-x Tix O6-delta (0</=x</=1) as Cathode Materials for Solid Oxide Fuel Cells *ChemSusChem.* **10**(14) 2978-89

[22] Cascos V, Alonso J A and Fernandez-Diaz M T 2016 Nb(5+)-Doped SrCoO3-delta Perovskites as Potential Cathodes for Solid-Oxide Fuel Cells. *Materials (Basel)* **9**(7).

[23] Suwanwarangkul R, Croiset E, Pritzker M D, Fowler M W, Douglas P L and Entchev E 2007 Modeling of a cathode-supported tubular solid oxide fuel cell operating with biomass-derived synthesis gas. Journal of Power Sources **166**(2) 386-99

[24] Buccheri M A, Singh A and Hill J M 2011 Anode- versus electrolyte-supported Ni-YSZ/YSZ/Pt SOFCs: Effect of cell design on OCV, performance and carbon formation for the direct utilization of dry methane. *Journal of Power Sources* **196**(3) 968-76

[25] Chao C C, Hsu C M, Cui Y and Prinz F B 2011 Improved solid oxide fuel cell performance with nanostructured electrolytes. *ACS Nano* **5**(7) 5692-6

[26] Chen et al. 2019 Advanced Fuel Cell Based on New Nanocrystalline Structure $Gd_{0.1}Ce_{0.9}O_2$ Electrolyte. ACS Appl Mater Interfaces **11**(11) 10642-50

[27] Chen H T, Chang J G, Chen H L, Ju S P 2009 Identifying the O_2 diffusion and reduction mechanisms on CeO2 electrolyte in solid oxide fuel cells: a DFT + U study. *J Comput Chem.* **30**(15) 2433-42

[28] Fu Y-P and Chen S-H 2010 Preparation and characterization of neodymium-doped ceria electrolyte materials for solid oxide fuel cells. *Ceramics International* **36**(2) 483-90.

[29] Zhang Z 2013 The influence of doped iron oxide on the performance of ceria-based solid electrolyte. *J Nanosci Nanotechnol.* **13**(6) 4283-90

[30] Zhu WZ, Yan M 2004 Perspectives on the metallic interconnects for solid oxide fuel cells. *J Zhejiang Univ Sci.* **5**(12) 1471-503.

[31] Liou Y C, Tsai W C, Yen H H and Chang Y C 2015 Preparation and Properties of (YCa)(TiMn)O3-d Ceramics Interconnect of Solid Oxide Fuel Cells. *Materials (Basel).* **8**(7) 4239-48.

[32] Bozzini et al. 2011 In situ X-ray spectromicroscopy investigation of the material stability of SOFC metal interconnects in operating electrochemical cells *ChemSusChem.* **4**(8) 1099-103

[33] Zhang Q, Du X, Tan S, Tang D, Chen K and Zhang T 2017 Effect of Nb_2O_5 doping on improving the thermo-mechanical stability of sealing interfaces for solid oxide fuel cells. *Sci Rep.* **7**(1) 5355

[34] Boldrin P, Ruiz-Trejo E, Mermelstein J, Bermudez Menendez J M, Rami Rez Reina T, and Brandon N P 2016 Strategies for Carbon and Sulfur Tolerant Solid Oxide Fuel Cell Materials, *Incorporating Lessons from Heterogeneous Catalysis Chem Rev.* **116**(22) 13633-84.

[35] Costamagna P, Grosso S, Travis R and Magistri L 2015 Integrated Planar Solid Oxide Fuel Cell: Steady-State Model of a Bundle and Validation through Single Tube Experimental Data. *Energies* **8**(11) 13231-54

[36] Ormerod R M 2003 Solid oxide fuel cells *Chem Soc Rev.* **32**(1) 17-28

[37] Eikerling M 2016 Guest Editorial Topical Issue on "Theory and Modeling of Fuel Cells" Theory and Modeling of Fuel Cells: Extravaganza to Panacea? *Fuel Cells* **16**(6) 663-8

[38] Zhu H Y, Kee R J, Janardhanan V M, Deutschmann O and Goodwin D G 2005 Modeling elementary heterogeneous chemistry and electrochemistry in solid-oxide fuel cells. *Journal of the Electrochemical Society* 152(12) A2427-A40

[39] Xu H, Dang Z and Bai B F 2012 Numerical Simulation of Multispecies Mass Transfer in a SOFC Electrodes Layer Using Lattice Boltzmann Method. *Journal of Fuel Cell Science and Technology* **9**(6)

[40] Amiri S, Hayes RE, Nandakumar K and Sarkar P 2013 Modeling heat transfer for a tubular micro-solid oxide fuel cell with experimental validation. *Journal of Power Sources* **23** 3190-201

[41] Amiri S, Hayes R E and Sarkar P 2018 Transient simulation of a tubular micro-solid oxide fuel cell *Journal of Power Sources* **407** 63-9

[42] Calise F, d'Accadia M D, Vanoli L and von Spakovsky M R 2006 Single-level optimization of a hybrid SOFC-GT power plant. *Journal of Power Sources* **159**(2) 1169-85.

[43] Bessler et al. 2010 Model anodes and anode models for understanding the mechanism of hydrogen oxidation in solid oxide fuel cells. *Phys Chem Chem Phys.* **12**(42) 13888-903

[44] Amiri et al. 2016 Planar SOFC system modeling and simulation including a 3D stack module *International Journal of Hydrogen Energy* **41**(4) 2919-30

[45] Amiri et al. 2015 Solid oxide fuel cell reactor analysis and optimisation through a novel multi-scale modeling strategy *Computers & Chemical Engineering* 78 10-23

[46] Sandhu et al. 2016 Electrochemical performance of a short tubular solid oxide fuel cell stack at intermediate temperatures *Applied Energy* **183** 358-68

[47] Zhang W, Croiset E, Douglas PL, Fowler M W and Entchev E 2005 Simulation of a tubular solid oxide fuel cell stack using AspenPlus ((TM)) unit operation models *Energy Conversion and Management* **46**(2) 181-96

[48] Doherty W, Reynolds A, Kennedy D 2009 Simulation of a Tubular Solid Oxide Fuel Cell Stack Operating on Biomass Syn-gas using Aspen Plus *Solid Oxide Fuel Cells 11 (Sofc-Xi)* **25**(2) 1321-30

[49] Doherty W, Reynolds A and Kennedy D 2009 The effect of air preheating in a biomass CFB gasifier using ASPEN Plus simulation. *Biomass & Bioenergy* **33**(9) 1158-67

[50] Doherty W, Reynolds A and Kennedy D 2010 Computer simulation of a biomass gasification-solid oxide fuel cell power system using Aspen Plus *Energy* **35**(12) 4545-55

[51] Doherty W, Reynolds A and Kennedy D 2010 Simulation of a Tubular Solid Oxide Fuel Cell Stack Operating on Biomass Syngas Using Aspen Plus *Journal of the Electrochemical Society* **157**(7) B975-B81

[52] Hauck M, Herrmann S and Spliethoff H 2017 Simulation of a reversible SOFC with Aspen Plus *International Journal of Hydrogen Energy* **42**(15)10329-40

[53] Nafees A, Rasid R B A, Ideris A 2019 Aspen Hysys Simulation of a Natural Gas Fueled Solid Oxide Fuel Cells TBS

[54] Boccaletti C, Duni G, Fabbri G and Santini E 2006. Simulation models of fuel cell systems

[55] Georgis D, Jogwar S S, Ali S. Almansoori A and Daoutidis P 2011 Design and control of energy integrated SOFC systems for insitu - hydrogen production and powergeneration *Computers and Chemical Engineering* **35** 1691-704.

IOP Conf. Series: Materials Science and Engineering **702** (2019) 012018 doi:10.1088/1757-899X/702/1/012018

Minimizing loop interaction in Multi Input Multi Output (MIMO) system using partial decoupler approach

M S Sulaiman and Z Ahmad*

School of Chemical Engineering, Engineering Campus, Universiti Sains Malaysia, 14300, Nibong Tebal, Penang, Malaysia

*E-mail: chzahmad@usm.my

Abstract. This research focus on controlling Multiple Input Multiple Output (MIMO) process for Reflux (L) – Boil up rate (V) configurations. There are a total of 3 phases in the simulation process. In phase 1, transfer functions and best fit curves are obtained for LV. The best fits are 88.28 % (CV1) and 93.07 % (CV2). The controller for LV configurations was developed. In phase 2, the controllers were tuned and analyze based on the set point tracking and disturbance rejection. In phase 3, partial decouplers were developed and analyzed for a static plus dynamic approach. It is observed that the partial decoupler provides better performance as compared to conventional full decoupler. The performance was proven based on the outcomes from the simulations and calculations. As a conclusion, the partial decoupler is able to minimize the loops interaction and obtain the best set point tracking and disturbance rejection.

1. Introduction

The function of the distillation is to separate feed (F) into multiple products depending on the requirement of each industry. Distillation columns are categorized into 2 types which are batch column and continuous column. Continuous columns are classified into 2 types, binary column (feed contains only 2 components) and multi-component column (feed contains more than 2 components) [1]. The key concept in the distillation column is to maintain the product streams compositions.

This is done by identification, classification (manipulated variables and disturbances) and control of all the parameters that contribute to the change in product composition [2]. Thus, a control system is implemented to accomplish the target. From a control perspective, in a binary distillation column, reflux flow (L), boil up flows (V), distillate (D) and bottom (B) can all be used as control inputs to control the logarithmic purities of top product (X_B), bottom product (X_D) and the levels in the condenser and reboiler [3]. Depending on which control inputs are used to control which outputs, different control configurations for the binary distillation column can be defined. In this study, one control configurations are used which is Liquid-Vapor (LV) control configuration [4]. In LV configuration, reflux rate (L) is used to control the composition of the top product and boil-up rate (V) is used to control the composition of the bottom product.

These research study the interactions of loops particularly between the composition of top and bottom products. These interactions occur because changes in a manipulated variable cause the changes in multiple loops. In other words, the loops are interconnected by a variable. Normally the controlling parameter is selected in pairs and for that, a multiple input multiple output (MIMO) system is required. MIMO system contains multiple loops which, in most cases, interact with one another. These interactions result in the destabilization of a system. A control system is required in order to stabilize and minimizes the loop interactions. Therefore, partial decoupler is introduced in this study to further reduce or minimize the effect of the loop interactions.

2. Methodology

2.1. Modelling

The case study is taken from Skogestad and Postlethwaite [6]. It's a binary mixture with constant pressure, constant relative volatility, and equilibrium on all stages, total condenser, constant molar flows, no vapor holdup and linearized liquid dynamics. However, the main effects important for dynamics and control is there for multi loops control study and in this paper on LV configuration were considered. The column has 40 theoretical stages and separates a binary mixture with relative volatility of 1.5 into products of around 99 % purity [7]. In order to get the model of the system for control and decoupler design, step test analyses were implemented and it runs in the Matlab^{TM} Simulink environment as shown in figure 2.

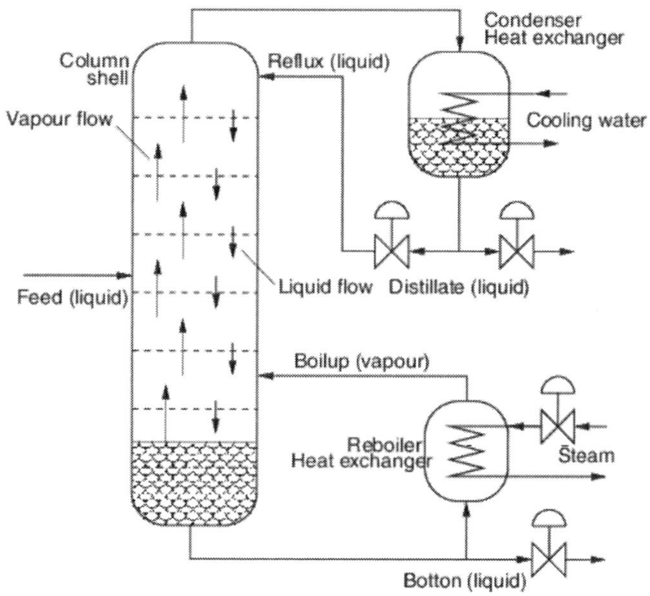

Figure 1. A Binary Distillation Column [5].

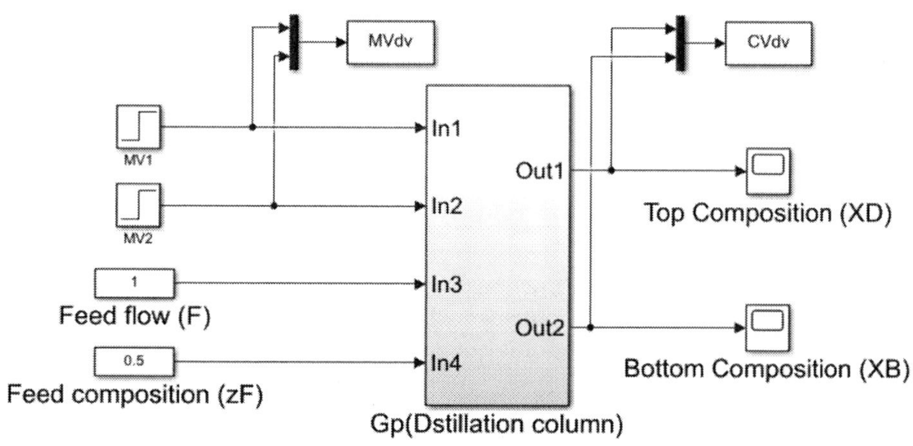

Figure 2. A Step Test Block Diagram configuration.

IOP Conf. Series: Materials Science and Engineering **702** (2019) 012018 doi:10.1088/1757-899X/702/1/012018

System Identification toolbox is selected for transfer function model identification. In the toolbox, the best-fit percentage is displayed for the model prediction for both CV1 and CV2. In this study, the best-fit percentage for either CV1 or CV2 or both is set to higher than 80 %, if the best-fit percentage is lower than 80 %, the step test is repeated with the change of inputs to improve the model prediction.

2.2. Tuning of Controller: Direct synthesis(DS) method

The gain (K_P), time constant (τ_P) and time delay (θ) values obtained from the transfer function model from section 3.1 are used to tune the controllers. The equation to calculateK_C, τ_i and τ_d for direct synthesis method are stated in the table 1.

Table 1. A K_C, τ_i and τ_d Formula for PI and PID Controller

Controller	K_C	τ_i	τ_d
PI	$\tau_P/K_P(\lambda+\theta)$	τ_P	-
PID	$\frac{\tau_P+0.5\theta}{K_P(\lambda+\theta)}$	$\tau_P + 0.5\theta$	$\tau_P\theta/(2\tau_P+\theta)$

2.3. Decoupler

After the controller is tuned, the controlled variables are observed with respect to performance indicator for further improvement by implementing decoupler. For a static decoupler, the value of the gain for the decoupler is obtained through the gain of the process model and for dynamics decoupler, the dynamics part is added to the block as part of the decoupler. The block diagram for static decoupler is shown in Figure 3.

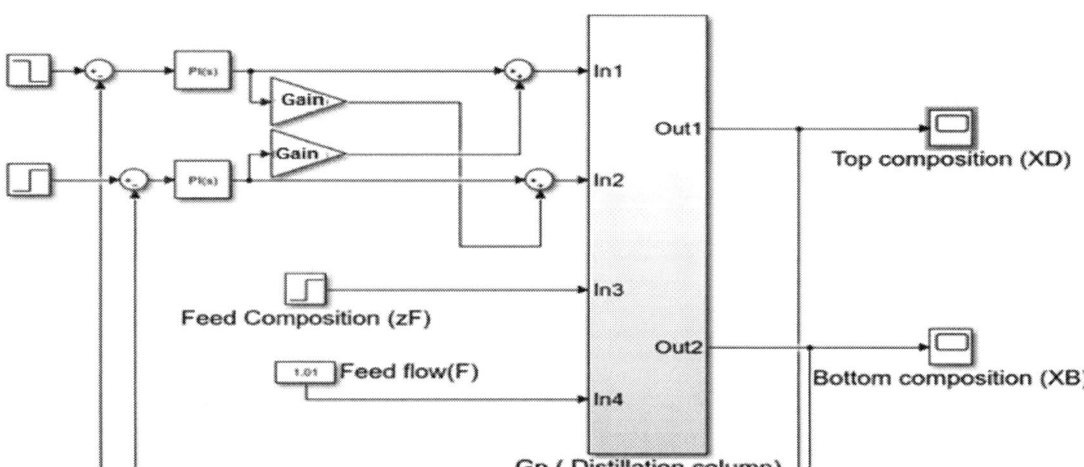

Figure 3. Control System with Static Decoupler.

Furthermore, the full type of decoupler is also compared with the partial top and bottom type decoupler. This is to provide a better insight into the functionality of each decoupler directly to the respective loops. Thus a total of 3 types of decouplers are implemented and compared which are conventional decoupler, partial top decoupler and static partial bottom decouplers. Figure 4 and 5 represent the partial top and bottom decoupler respectively.

IOP Conf. Series: Materials Science and Engineering **702** (2019) 012018 doi:10.1088/1757-899X/702/1/012018

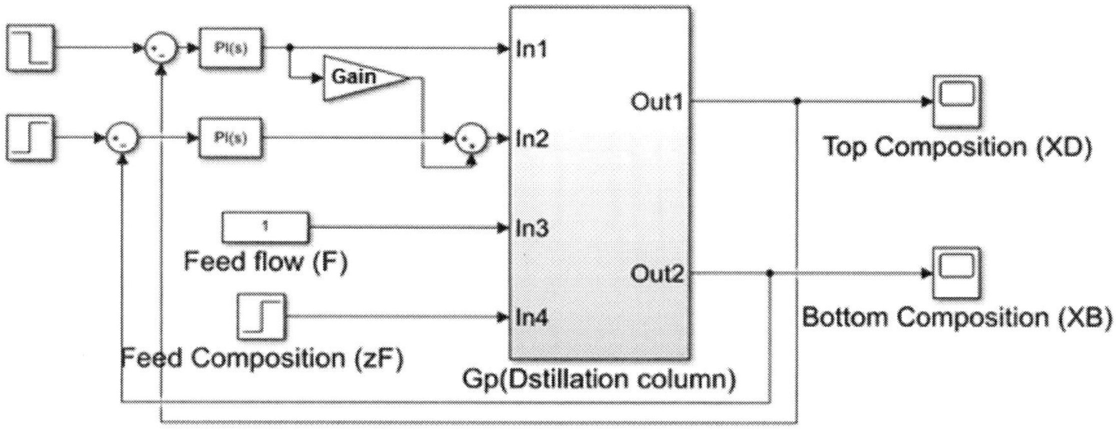

Figure 4. Control System with Static Partial Top Decoupler.

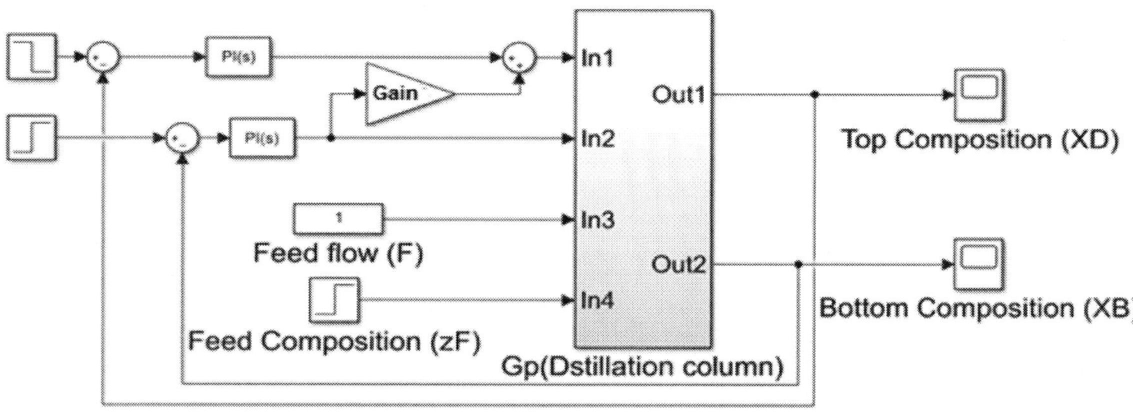

Figure 5. Control System with Static Partial bottom Decoupler.

3. Results and discussions

3.1. Modelling: Step test

Based on the step test result through identification toolbox in Matlab™, after multiple steps testing, the best fit curve for both the top and bottom process (CV1 and CV2) was obtained as shown in Figure 6 and 7 respectively. In those figures, the best fit of 88.28 % and 93.07 %, and the model error within the range of 0.02 to -0.01 and 0.0005 to -0.0007 for CV1 and CV2 were obtained respectively and it is within the valid range [7]. Thus the transfer functions determine from the steps test function were accepted to be used for the tuning of the controller and also for the decoupler design.

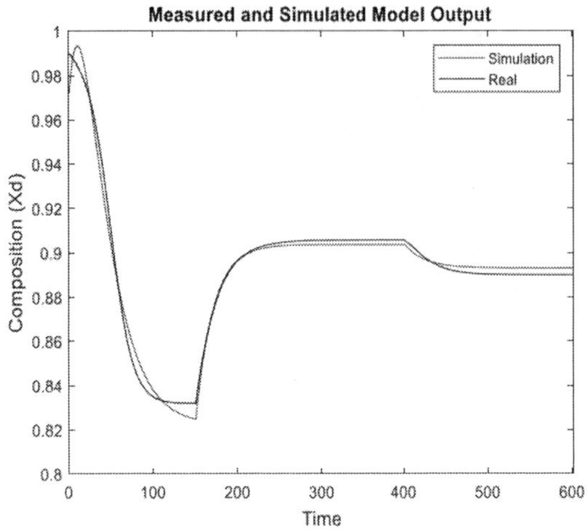

Figure 6. Best Fit Curve of CV1.

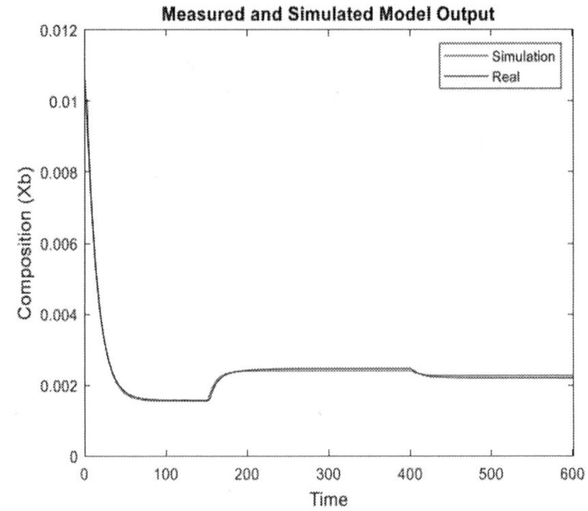

Figure 7. Best Fit Curve of CV2.

3.2. Direct Synthesis (DS) Tuning Method

The set point change with the step time started at 700 seconds with the initial value of the set point of 0.98 and the final value of 0.99 for CV1. In CV2 the set point change from 0.02 to 0.03 at the step time of 350 seconds was implemented respectively. The transfer functions for the distillation column based on the step test result shows in Figure 6 and 7 are shown below:

$$\begin{bmatrix} CV1 \\ CV2 \end{bmatrix} = \begin{bmatrix} \dfrac{0.3254e^{-6s}}{6s+1} & \dfrac{-0.0202e^{-22s}}{26s+1} \\ \dfrac{0.0052e^{-14s}}{6s+1} & \dfrac{-0.0039e^{-14s}}{5.37s+1} \end{bmatrix} \begin{bmatrix} MV1 \\ MV2 \end{bmatrix} \quad (1)$$

The initial DS tuning parameter are shown in table 2 with the λ of 0.5 from the time constant

Table 2. Controller setting for PI using the DS method

Controller	K_C	τ_i
PI (CV1)	3.0-10.0	6
PI (CV2	-83-(-100)	5.37

The controller performances based on DS tuning method for set point tracking are shown in Figure 8 where the initial values of Kc is based on DS method however the final Kc values can be adjusted to get a good set point tracking for each loops. It clearly shows that the control variables were able to follow the set point tracking for both CVs, however due to the loop interaction, the spike was observed when the set point change for both CVs at step time of 350 seconds for loop 1 and 700 seconds respectively for loop 2. It confirms that the loops interaction has significantly occurred in the system. The data analyses for both responses were shown in table 3. Therefore, based on this observation, full conventional and partial decoupler was proposed in this study in the aims of to significantly reduce or minimize the loop interaction in both loops.

Table 3. Set point tracking controller performance

Controller	Rise time/fall time (s)	Overshoot (%)	Settling time (s)
CV1	50	<1	100
CV2	< 1	<1	70

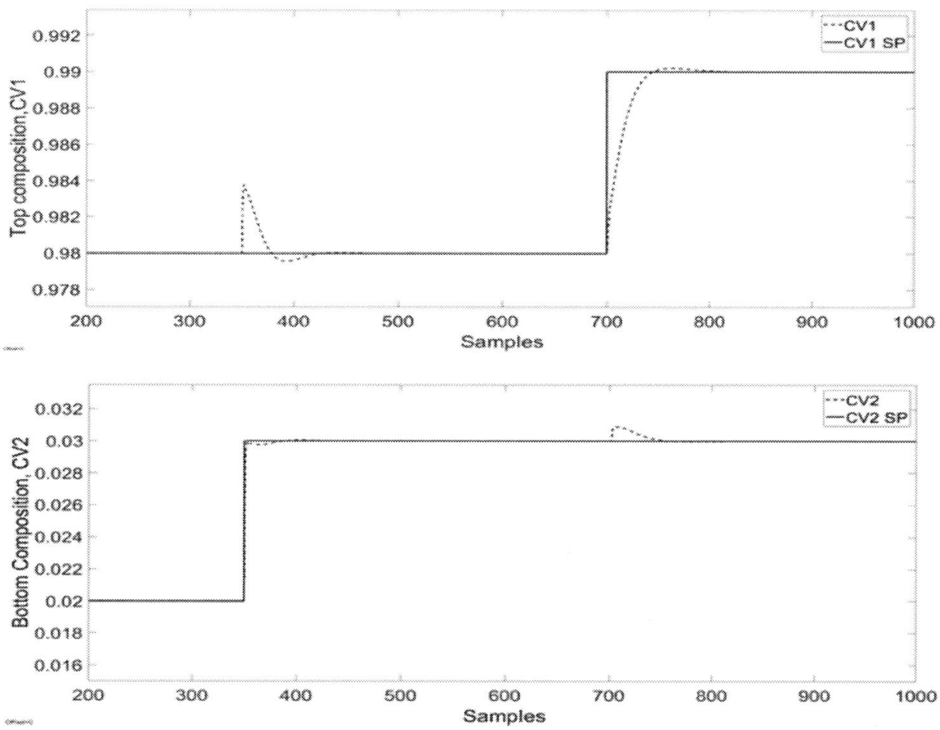

Figure 8. Response curve for top and bottom composition after DS tuning.

3.3. Decoupler

Conventional decoupler is the most common and simple decoupler used to eliminate loop interaction and disturbance. Figure 9 and Figure 10 show the performance of conventional decoupler for CV1 and CV2 as well as after adding the full and partial decoupler to the system. The decoupler values were calculated for both the static and dynamic parts. It shows that a loops interaction occurred due to the magnitude of CV1 and CV2 spike when a new set point was introduced. Set point changed was introduced for CV is at time 700 second from 0.98 to 0.99 while the CV2 set point change was set in time 350 second from an original set point 0.02 to new set point 0.03.

Ideally, when CV1 changed, CV2 should remain at the current set point but due to the loop interaction, the spike was observed in CV2 and vice versa at step time of 350 seconds and 700 seconds respectively. However, even though the decoupler were put in place to eliminate the loop interaction, significant reduction of loop interaction was observed as compared to without decoupler as shown in those figures. It clearly shows that the performances of the controllers were improved by introducing the partial decoupler for both top and bottom respectively. The spikes for both CVs were significantly reduced when full and partial decoupler was added to the control loops. The performances of the CV2 were more less the same as in CV1. This result was supported by the controller performance analysis as shown in table 4. Based on this result, the priority of the loop can be determined and based on that the setting or the design of the decoupler can be made based on that information without compromising the performance of the other loops. In addition to that, it is good to carry out the relative gain analysis (RGA) and sensitivity analysis to determine and reconfirm the priority of each loop for implementation of top and bottom decoupler.

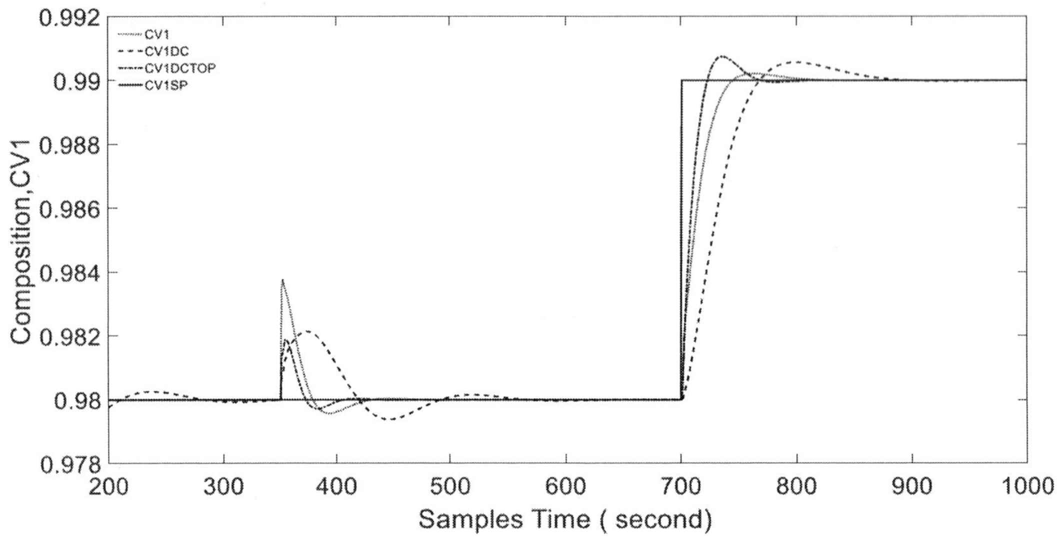

Figure 9. Responses curves for all controllers for CV1.

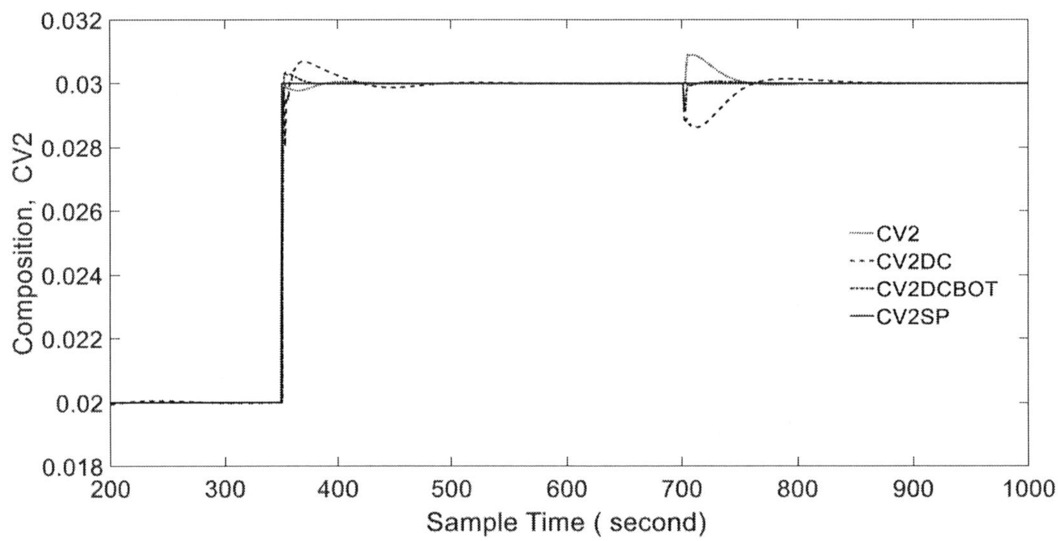

Figure 10. Responses curves for all controllers for CV1.

IOP Conf. Series: Materials Science and Engineering **702** (2019) 012018 doi:10.1088/1757-899X/702/1/012018

Table 4. Overall controller's performances analysis

Criteria	**PI Controller**		**PI controller with partial DC**	
	CV1	CV2	CV1 (Top)	CV2 (Bottom)
Rise time (SP)	50 s	< 1 s	**< 50 s**	< 1 s
Settling time (SP)	100 s	70 second	**< 100 s**	**< 70 s**
Settling time (Interaction)	70 s	60 second	**< 70 s**	**< 60 s**
Overshoot	<1 %	< 1 %	<1 %	< 1 %
Overshoot (Interaction)	3 %	1 %	**< 3 %**	**< 1 %**

4. Conclusions

Implementation of partial decoupler effectively reduces the interactions between control loops. Analysis and simulation results indicated that partial decoupler decoupling control systems with dynamics decoupler are able to reduce the interaction and the same time control the composition at the bottom and the top of the binary distillation column. Ideal static conventional decoupling control yields a less robust performance as compare to partial dynamic decoupling control for the proposed process studied. The performance was proven based on the outcomes from the simulations and calculations. As a conclusion, the partial dynamics decoupler is able to minimize the loops interaction and also able to achieve the best set point tracking for the binary distillation column.

Acknowledgments

The authors would like to thank Universiti Sains Malaysia (USM) for their support through RU research grant 8014100.

References

[1] Ming T T and Costello R C 1997 Types of Distillation Columns. Retrieved from http://www.rccostello.com/distil/distiltyp.htm (Access Date: 15/10/2018).

[2] Gani R and Bek-Pedersen E 2004 Design and Synthesis of Distillation Systems Using A Driving-Force-Based Approach. *Chemical Engineering and Processing: Process Intensification* **43**(3) 251-62

[3] Hori E S and Skogestad S Al-Arfai M A 2006 Self-Optimizing Control Configurations for Two-Product Distillation Columns. *Institution of Chemical Engineers Symposium Series 152 London, UK*

[4] Thone M, Potters M and Baldi S 2016 Control Configurations in Distillation Columns: A Comparative Study *European Control Conference* 37-42

[5] Juan I Y and Graham C 2003 Loop Performance Assessment for Decentralized Control of stable linear systems *European Journal of Control* **9** 118-132.

[6] Skogestad S and Postlethwaite I 2005 *Multivariable Feedback Control Analysis and Design (2nd Edition)* (London: Wiley)

[7] Skogestad S 2019 Matlab Distillation Column Models. Retrieved from https://folk.ntnu.no/skoge/book/Matlab_M/Cola/Cola.Html (Access: 25 May 2019)

IOP Conf. Series: Materials Science and Engineering **702** (2019) 012019 doi:10.1088/1757-899X/702/1/012019

Adaptive PLS inferential soft sensor for continuous online estimation of NOx emission in industrial water-tube boiler

S H Hasnen[1*], H Zabiri[1,2], K K Prakash[2] and T T Mat[2]

[1] Chemical Engineering Department, Universiti Teknologi PETRONAS, Seri Iskandar, Perak, Malaysia.
[2] Group Technical Solutions (GTS), PETRONAS, Kuala Lumpur, Malaysia.

*E-mail: haslindazabiri@utp.edu.my

Abstract. In common industrial application, the use of a linear and static PLS soft sensor for online prediction and monitoring of industrial boiler is often preferred due to its simple and intuitive framework. However, process dynamics and time-variant factors can negatively affect the accuracy and reliability of PLS soft sensor over its long-term application in process industries. In this paper, development of adaptive soft sensor based on dynamic PLS method has been applied to an industrial water-tube boiler for continuous online prediction of Nitric Oxides emission. In the case study, it is found that the adaptive PLS soft sensor which includes lagged measurements of NOx emission in the model input can significantly improve the prediction accuracy and reliability by 72.7% relative to the performance of linear and static PLS soft sensor when tested on online dataset containing gradual and abrupt changes in the process operating conditions.

1. Introduction

Chemical process industries are faced with pressing responsibilities of devising their production policies to adhere accordingly with safety rules and environmental regulations. For process involving combustion, it has been made compulsory for all parties involved to monitor flue gases in the exhaust streams due to stringent air regulations [1,2]. In general practice, the industries rely mostly on hardware analysers, like Continuous Emission Monitoring System (CEMS), to be installed at every air emission release point to give measurement of the exhaust gas composition. Despite so, the hardware systems are complex and costly to install and maintain [3,4]. On the other hand, Predictive Emission Monitoring (PEMS) or soft sensors can offer a few benefits in place of hardware sensors in the industrial applications [5].

Soft sensors are inferential model that use information from easily measured online variables, such as temperature, pressure, flowrate and level, to estimate a target variable(s) that is significantly harder to make continuous measurement in online environment, such as feed or product's composition or concentration. The target variable is often informative of the state of production quality and is therefore used as a reference for estimating and monitoring of quality of an operational production and environmental regulation. These soft sensors application field is further on referred to as on-line prediction [4,6]. The data-driven algorithm applied for soft sensor development are either based on statistical methods or machine learning approaches for linear and non-linear processes. The linear modelling methods like Multivariate Linear Regression (MLR), Principal Component Analysis (PCA) and Partial Least Squares (PLS) are mostly applied in the industries, i.e.- they account for more than 90% of the soft sensors used in chemical plants [5].

Despite the simplicity to develop a linear and static soft sensor model based on the PLS method, its accuracy and validity usually suffer declination when changes happen in the process behaviour. This is so because process dynamics and other time-variant factors have caused the process operating conditions to deviate from the ranges that the trained soft sensor have been initially modelled to [6]. To update a linear and static PLS model with the new process behaviour, it is important to appropriately adapt the model with updating methods that are suitable with the PLS model structure. There are a significant number of adaptive soft sensors that are based on PLS modelling or its extension to dynamic and non-linear form, such as dynamic PLS, kernel PLS, neural network PLS, moving-window PLS, recursive PLS, and etc [7].

Furthermore, the ensemble-model based adaptive methods such as local learning, state-partitioning and Just-in-Time(JIT) have paved ways for PLS to do model regression for each local region defined in the process data, which is found to be more efficient in predicting abrupt changes in the process data, compared to adaptation based on global model structure [8-12]. For instance, a recently published work on highly-overlapped, recursive PLS soft sensor with state partitioning via local variable selection adaptation methods have been applied to industrial dataset of a thermal oxidizer unit for online prediction of NOx emission [2].

In this paper, an industrial case study on the development of adaptive PLS soft sensor based on dynamic PLS method for continuous online prediction of NOx emission in the industrial water-tube boiler derived from actual industrial data of the system is briefly discussed. There are three key areas that have been highlighted in the development of the adaptive PLS soft sensor :- suitable selection of model inputs from available industrial process variables based on process knowledge related to the combustion process and NOx formation mechanisms in the industrial water-tube boiler, PLS model training based on normal operating region of the industrial water-tube boiler system, and consequent adaptation of PLS model with online industrial dataset based on dynamic PLS updating approaches in response to gradual and abrupt changes in the operating regions of the industrial water-tube boiler system.

2. Background

2.1 Partial Least Squares (PLS)

The inferential model is based on PLS model structure, which is essentially a mathematical equation based on the relationships between independent key process variables (X) in the boiler system that are related to the NOx emission (Y) in the boiler and it is derived from real measurements made in the boiler system. The detailed theory and derivation of PLS outer and inner regression models can be found in several influential works by Wold and Geladi [13].

$$Y = Y(mean) + X.W.C + F \tag{1}$$

The PLS basic regression model represented in equation (1) above is the inferential model for prediction of NOx (Y) in which *Y(mean)* is the average value of NOx emission, *W*C* are the loading weights or coefficients for respective key process parameters *(X)*, and *F* is the residual between predicted and actual NOx value from the regression. PLS is a supervised regression method that uses Y variable to look for correlations between systematic variations in X and variations in Y in the process data.

Among the linear regression methods, PLS has rapidly increased in popularity due to several practical reasons. First, the PLS method is better at handling the commonly existed data co-linearity and measurement noise in industrial dataset than the MLR method [7]. Secondly, while the PCA method can only transform the dataset into several latent variables based on only correlations among the process variables, the PLS method can derived latent projections from the dataset transformation as such that the covariation between the process variables to the target variable is maximized [10-11]. This can make the prediction of target variable from the process variables become more accurate. Finally, PLS model has a transparent model structure that can be easily understood by engineers and process operators when performing data analysis and soft sensor development [3,11].

2.2 NOx Emission Formation

The boiler produced harmful emissions from the combustion process, especially NOx and CO gases, of which formations depend mainly on reaction rate in the boiler's combustion chamber. In the case of gas-fired combustion application, the dominant mechanism is mostly attributed to the thermal NOx route, as the range can account from 80 to 95% of the NOx emission [1]. NOX emission is correlated to the actual flame temperature in the combustion zone and several process parameters such as fuel flow, air flow, combustion air temperature, pressure, humidity and more. Hence a data-driven soft sensor can be developed based on these process parameters for the online prediction of NOX emission.

2.3 Case Study : Industrial Water-tube Boiler

The studied industrial boiler is a water-tube type boiler, fired by natural gas. The objective of the boiler is to generate high pressure (HP) steam for plant-wide utilization. The boiler by design can generate maximum 83 tons per hour of steam at 107.9 barg and 515 °C. Fuel gas feed input to the boiler mainly consists of methane. The main parts of the boiler consist of the upper and lower drums, combustion chamber and superheaters. Figure 1 illustrates the schematic diagram of the industrial boiler used in the case study.

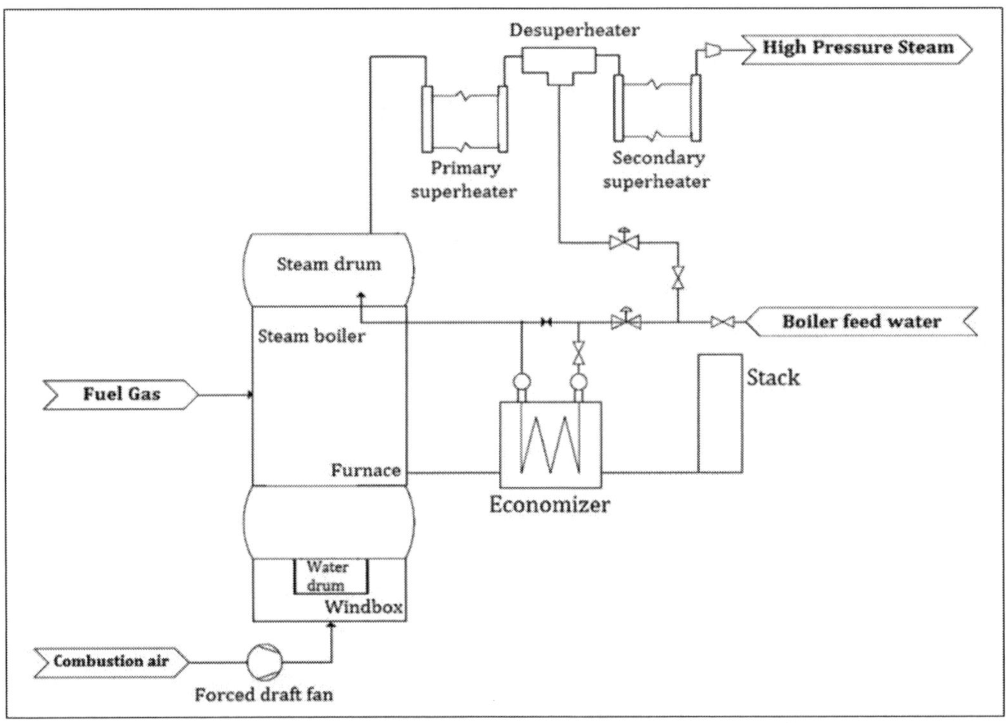

Figure 1. Schematic Diagram of Industrial Process Boiler.

3. Methodology

The procedure for soft sensor development based on PLS for prediction of NOx concentration in the boiler stack is illustrated in Figure 2. In summary, the development of the inferential soft sensor model begins with offline data pre-processing and variable selection. Then it is continued with offline PLS model training and validation analysis development before it is ready for online prediction and online model adaptation studies. All the research works were conducted in the SIMCA-p data analysis tools and the MATLAB R2013a commercial software.

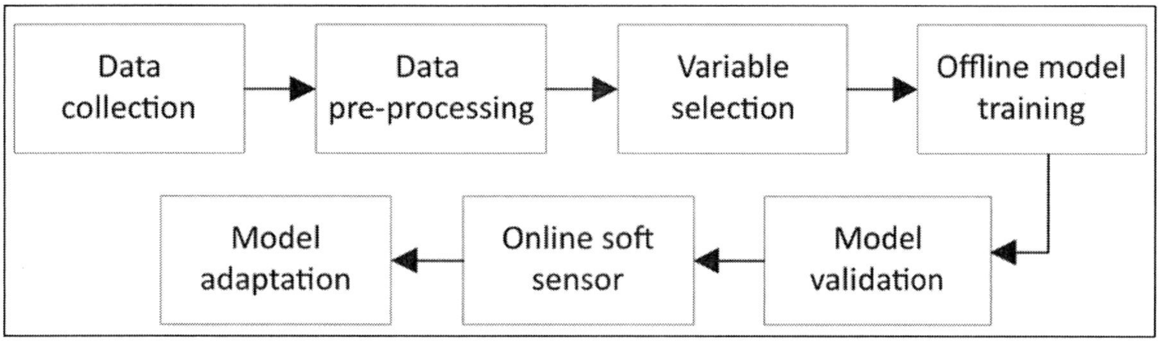

Figure 2. Soft Sensor methodology flowchart.

The model development begins with data collection of samples of available industrial data of the water-tube boiler, which is a set of data consisted of 27 process variables (26 X and 1 Y). Process measurements in the data set were recorded within a duration of 2 weeks and taken at every one-minute sampling threshold. Next, the collected industrial data need to be pre-processed through a series of statistical analysis to filter out missing values and critical outliers. The total numbers of samples after data pre-processing are around 13000, and they needed to be normalized and scaled to zero mean and one-unit variance, before these data were ready to be used in PLS modelling. For data selection, the offline dataset used for model training formed 70% of the whole dataset, while the other 30% of the pre-processed dataset is dedicated for online prediction testing and model adaptation.

Normally the PLS regression performance increase with the amount of input variables used in model training as PLS is relatively capable of handling collinearity and noises such as missing values and incomplete variables. However, incorporating high number of model inputs can implicate the process of model maintenance. Therefore, the theoretical causes for emergence of NOx in the boiler system, as discussed in Section 2.2, are used as the main criteria for the variable selections. Consequently, the key process variables (X) were finalized and then used as the model inputs to predict Y in PLS model training.

The execution of PLS outer and inner regression calculation on a pair of data matrices {X,Y} which has m input variables, p output variables and n samples produce results of the corresponding scores , loadings and regression coefficient matrices which can be represented based on equation (2) as follow :

$$(\boldsymbol{X},\boldsymbol{Y})_{OFFLINE} \xrightarrow{PLS} (\boldsymbol{T},\boldsymbol{W},\boldsymbol{P},\boldsymbol{B},\boldsymbol{Q})_{OFFLINE} \tag{2}$$

The static and linear PLS model's accuracy is prone to deterioration after online implementation due to changes in the process operating conditions. In addition, development of inferential model with static data does not give good representation of the studied chemical process due to dynamic characteristics generally observed in chemical processes. For instance, the future state of Y can be due to cumulating effects of the current and previous operating conditions in the process. To update the offline PLS model with new process behaviours, it is beneficial to use adaptive methods that are suitable with the PLS model structure.

When considering both process dynamics that are represented in the online dataset and a suitable adaptive method for PLS-based model, the selected input variables for model training as well as the previous measurements of NOx emission can be used as new model inputs (Z) for PLS transformation to its dynamic version. The PLS model is converted into a DPLS model by adding lag of X and Y

IOP Conf. Series: Materials Science and Engineering **702** (2019) 012019 doi:10.1088/1757-899X/702/1/012019

variables together with X variables as the new model inputs (Z), notated in equation (3) to train the DPLS model with offline dataset as in equation (4), and consequently validate its adaptability on online dataset as follows:

$$\boldsymbol{Z} = [\, \boldsymbol{X1}_n, \boldsymbol{X1}_{(n-1)}, \boldsymbol{X2}_n, \boldsymbol{X2}_{(n-1)}, \dots, \boldsymbol{Y}_{(n-1)} \,] \tag{3}$$

$$(\boldsymbol{Z}, \boldsymbol{Y})_{OFFLINE} \xrightarrow{DPLS} (\boldsymbol{T}, \boldsymbol{W}, \boldsymbol{P}, \boldsymbol{B}, \boldsymbol{Q})_{OFFLINE} \tag{4}$$

In actual practise, data-driven inferential model can only be updated when a validated or actual measurement of the boiler emission by hardware analyser such as gas chromatography become available. If the actual measurement of NOx emission is available after a sampling threshold of 10 minutes, the PLS model can be extended to dynamic PLS model by including the dataset of X and Y with the lag period of 10 minutes, for example X(n-1) until X(n-10) and Y(n-10), to the initial PLS model inputs accordingly.

4. Results and Discussion

4.1 Variable selection based on process knowledge

From all of 26 process variables that are measured around the industrial water-tube boiler system in Figure 1, there are only 5 of them that are directly associated with the combustion reaction, particularly the NOx formation in the boiler through the thermal NOx route. The list of the key predictors is given in table 1. Data set of the finalized key variables are then gathered together to form a set of model inputs (X) for the development of PLS model. The variable selection analysis helped in making the optimal selection of the model inputs for future works in PLS model development.

Table 1. List of key variables based on combustion theory

Important Variables
x1- Combustion air pressure
x2- Flue gas Economizer Inlet temperature
x3- Combustion air flowrate source 1
x4- Fuel gas flowrate
x5- Fuel Latent Heat Value

4.2 Comparison between PLS model and DPLS model on their adaptive capability for online prediction.

Following the variable selection, the soft sensor model development continues with offline model training of the PLS and DPLS model from offline data, which is consisted of process states and corresponding NOx emission measurements that are within the normal range of the boiler operating conditions. Next, to be able to see capability of the offline PLS and DPLS models to adapt to the changes in the process, the online testing data were chosen such that it contained several process states that were operated outside of the normal range.

The prediction performance of the developed model is calculated based on the Root Mean Square Error (RMSE) statistic parameter to find error between predicted Y and actual Y values. Furthermore, visual interpretation method is also used to help with analysis of the online prediction of NOx emission in the water-tube boiler

IOP Conf. Series: Materials Science and Engineering **702** (2019) 012019 doi:10.1088/1757-899X/702/1/012019

Table 2. RMSE for PLS and DPLS models on online dataset

Model	RMSE
PLS	0.8882
DPLS	0.2480

Two sets of inferential soft sensor models based on PLS and DPLS methods have been developed from the offline dataset and tested on the online dataset to compare both of their accuracy and adaptive capability after significant changes have been made to the industrial water-tube boiler operating conditions. In table 2, the differences between RMSE values calculated from the prediction error made by PLS model and DPLS model relative to the actual NOx emission measurements in the online dataset indicate that the adaptive DPLS model has significantly better prediction accuracy compared to the non-adaptive PLS model by 72.7%.

Furthermore, the visual interpretation of prediction performance of PLS and DPLS models on the online test data as shown in Figure 3 and Figure 4 respectively also reflects that the adaptive DPLS model has very good tracking ability in response to gradual process change (from time 1:2500) and abrupt process change (from time 2501:5000) that can be observed in the actual NOx emission levels in the online dataset. Based on these two criteria to evaluate adaptive capability of the inferential models, the results have proved that DPLS model is significantly more accurate in predicting NOx emission and better at tracking changes in the boiler operating changes for long-term implementation in process industry

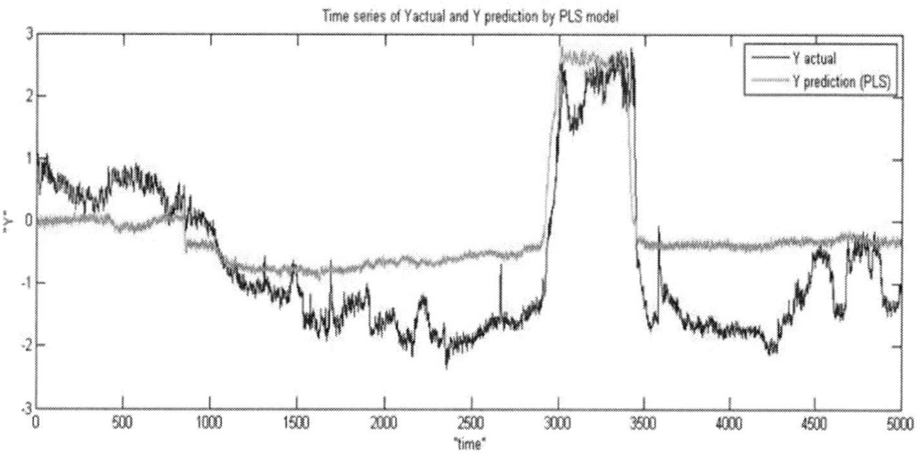

Figure 3. Comparisons between actual NOx trend (blue line) and prediction of NOx emissions (green line) made by the PLS model on the online test data.

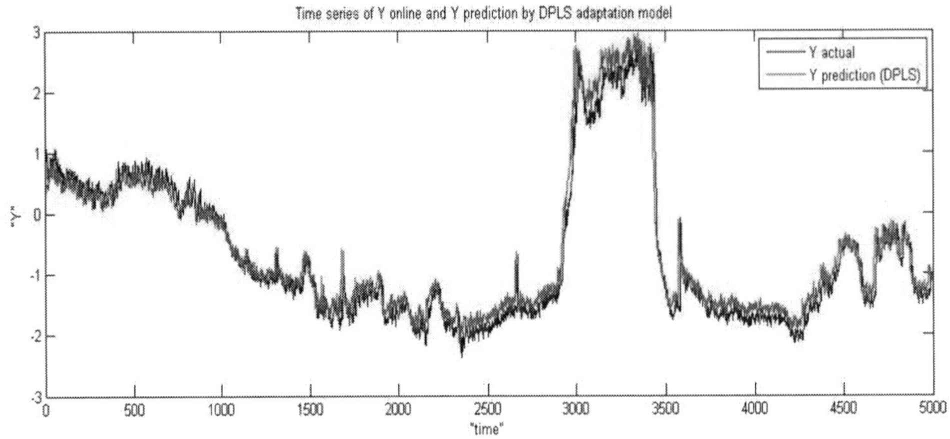

Figure 4. Comparisons between actual NOx trend (blue line) and prediction of NOx emissions (purple line) made by the adaptive Dynamic PLS model on the online test data.

The significant improvement in the prediction and adaptive performance of the DPLS model can be associated to several reasons. First, the information of the previous Y variables helps DPLS to also capture serial correlations between lagged Y(n-10) and the current Y(n) observed in the training dataset, besides than the spatial correlations between X(n)and Y(n) for prediction of future Y(n+1). In addition, statistical analysis such as corresponding regression coefficient and variable importance projection (VIP) of each variable in the DPLS model inputs can be used to determine the most important process variable that can explain majority of the variation observed in NOx emission level.

Upon further inspection of the model regression coefficients and the correlation value, it could be found that lagged of NOx emission measurements significantly dominates other process variables in the DPLS model inputs. Hence, it could also be suggested here that the lagged of NOx emission measurement is responsible for the superior prediction and highly adaptive capability of DPLS model for prediction of NOx emission in the industrial water-tube case study.

5. Conclusion

Inferential model which is the main body of a soft sensor for continuous online prediction of NOx emission has been developed based on actual industrial data of a gas-fired, water-tube boiler. The model inputs for the inferential model development have been selected based on theoretical knowledge related to combustion reaction and thermal NOx formation. Based on the results from the online testing made on the developed inferential models, the PLS model seems not able to properly adapt to the changes in the process operating conditions. On the other hand, the DPLS model is found to be significantly more superior than the PLS model in terms of its prediction performance and adaptive capability for continuous online estimation of NOx emission in the industrial boiler by 72.7%. It has also been suggested that the lagged of NOx emission measurement Y(n-10) could be responsible for the significant advantage in prediction accuracy and highly adaptive capability of DPLS model for continuous prediction of NOx emission in the industrial water-tube case study.

Acknowledgements

The authors would like to thank Group Technical Solutions (GTS) PETRONAS and PETRONAS University of Technology for the industrial data and facilities provided in carrying out this research.

IOP Conf. Series: Materials Science and Engineering **702** (2019) 012019 doi:10.1088/1757-899X/702/1/012019

References

[1] Nielsen JS and Weel M 2015 *First Principle Models for Emission and Performance Monitoring.* Abu Dhabi International Petroleum Exhibition and Conference: Society of Petroleum Engineers.

[2] Poerio DV and Brown SD 2018 Highly-overlapped, recursive partial least squares soft sensor with state partitioning via local variable selection. *Chemometrics and Intelligent Laboratory Systems* **175** 104-15.

[3] Kano M and Fujiwara K 2012 Virtual sensing technology in process industries: trends and challenges revealed by recent industrial applications. *Journal of chemical engineering of Japan.* **12** 167

[4] Kadlec P, Gabrys B and Strandt S 2009 Data-driven soft sensors in the process industry *Computers & Chemical Engineering* **33**(4) 795-814

[5] Souza FA, Araujo R and Mendes J 2016 Review of soft sensor methods for regression applications *Chemometrics and Intelligent Laboratory Systems* **152** 69-79

[6] Kadlec P, Grbić R and Gabrys B 2011 Review of adaptation mechanisms for data-driven soft sensors *Computers & Chemical Engineering* **35**(1) 1-24

[7] Qin SJ 2012 Survey on data-driven industrial process monitoring and diagnosis *Annual Reviews in Control* **36**(2) 220-34

[8] Shao W and Tian X 2015 Adaptive soft sensor for quality prediction of chemical processes based on selective ensemble of local partial least squares models *Chemical Engineering Research and Design* **95** 113-32

[9] Shao W and Tian X 2017 Semi-supervised selective ensemble learning based on distance to model for nonlinear soft sensor development *Neurocomputing* **222** 91-104

[10] Ge Z and Song Z 2010 A comparative study of just-in-time-learning based methods for online soft sensor modeling *Chemometrics and Intelligent Laboratory Systems* **104**(2) 306-17

[11] Kim S, Okajima R, Kano M and Hasebe S 2013 Development of soft-sensor using locally weighted PLS with adaptive similarity measure *Chemometrics and Intelligent Laboratory Systems 124* 43-9

[12] Ni W, Brown SD and Man R 2014 A localized adaptive soft sensor for dynamic system modeling *Chemical Engineering Science* **111** 350-63

[13] Wold S, Sjöström M and Eriksson L 2011 PLS-regression: A basic tool of chemometrics *Chemometrics and Intelligent Laboratory Systems* **58**(2) 109-30

IOP Conf. Series: Materials Science and Engineering **702** (2019) 012020 doi:10.1088/1757-899X/702/1/012020

Temperature control of vacuum dividing wall column – case study on oleochemical fatty acid fractionation

A Alias[1] and M R Othman[2*]

[1] INSTEP, Batu Rakit, 21020 Terengganu, Malaysia
[2] Faculty of Chemical and Process Engineering Technology, College of Engineering Technology, Universiti Malaysia Pahang, 26300 Gambang, Pahang, Malaysia.

E-mail: rizza@ump.edu.my

Abstract. Analysis of oleochemical compositions in distillation column often have large process delays. Inferential control is commonly used by means of stage temperature as the measured variable which provide more responsive in composition control. This work aims to evaluate the performance of temperature control in vacuum dividing wall column (VDWC) for fatty acid oleochemical fractionation. Product purity at 99% used as inferred parameter to determine the temperature. Sensitivity analysis was used to determine the relationship between stage and temperature difference for changes in the manipulated variables. The most sensitive tray was selected and implemented to a Distillate- Side Stream- Boilup (DSV) control configuration in Aspen Dynamics following the work by Othman (2019b). Controller adopted with PI and PID settings using Ziegler-Nichols (ZN) and Internal Model Control (IMC) tuning calculation method. Both methods were compared based on the settling time and overshoot. The best setting was then fine-tuned before tested to set point tracking without any disturbances. From the sensitive analysis, temperature at stage 6, 29 and 34 were selected used as controlled variable which inferred distillate, middle and bottom product purity at 99% respectively. PID controller setting based on ZN method provide the best setup with fastest settling time and smallest overshoot and provide good performance for set point tracking.

1. Introduction

Oleochemical industry particularly in Malaysia mostly uses typical distillation column (DC) for its product fractionation. In process design perspective dividing wall column (DWC) shows very promising alternative to DC which able to reduce around 20% of capital and operating cost [1]. However, one of the potential hurdles for commercial implementation of DWC is the challenges in design, simulate, operation and control [2-3] as well as complexity in operating and controllability due to the introduction of a wall within the column internal [4]. Various research has been conducted for DWC control. One of the most common controlled variable in distillation column is composition. However, composition is difficult to measure. For such variable, inferential control is often implemented which uses easily measure process variables i.e. temperature, pressure and flow to infer more difficult process variables such as compositions and molecular weight. Parrish and Brosilow [5] stated that for higher order and long-dead-time processes, inferential control systems will generally outperform conventional feedback control systems. Because of that, inferential control has excellent performances such as disturbance resisting and set point tracking. However, the application is restricted when strong load disturbance exists or stable control accuracy and response speed are highly required in the system. Besides, it can be much less expensive in terms of capital and operating cost as well as can provide measurement that are not available any other way. According to Ansari and Tadé [6] inferential models which are based on fast and continuously available temperature, pressure and flow measurements reduce the negative impact of the sample intervals and time delays with minimum compromise on accuracy. The resulting continuous and fast response keeps the product qualities on specifications and minimizes the quality giveaway. In order to establish composition inferential control, one needs to know the particular correlation between process variables and product composition. According to Marlin [7], a good inferential control is when inferred variable is closely related to true variable so that controlling inferred controlled variable will maintain true controlled variable close to desired value. The use of tray temperature to infer composition is widely used in distillation column [8]. This paper adopted the criteria in selecting the tray that give largest change in temperature upon changes made at manipulated variable as proposed by Luyben [8].

DWC is according to Nguyen is the integration of two or more different separation units into one single device with one or more vertical partitions in the central section. Dividing wall splits a single column into two parts, which are a pre-fractionator section, and a main column. It uses only one reboiler and one condenser. Various studies have been conducted for inferential control in DWC. For example Wang et al. [9] have investigated temperature inferential control of DWC for separating ethanol, n-propanol, and n-butanol ternary mixture. Yuan et al. [10] has studied inferential temperature control for the process medium of benzene–toluene–o-xylene in DWDC system. Ignat and Woinaroschy [11], has analyzed the controllability of inferential temperature of 4 point control structure for a case study of separation of a ternary nonideal methanol – ethanol – 1-propanol mixture in a DWC. Most of inferential control studies of DWC focuses on petrochemical processes. Study on vacuum dividing wall column (VDWC) inferential control particularly for oleochemical industries however received less attention. Moreover, oleochemical products were analyzed using analytic apparatus i.e. High Performance Liquid Chromatograph HPLC which had large process delay. This practically hindered composition as the controlled variable. The composition of product with greater amount of light key will resulting higher vapor pressure thus lessen heat duty required and give lower temperature at particular stage to keep similar product purity. Hence, temperature tray could be adopted to infer product composition. Therefore, this work aims to evaluate the performance of stage temperature control in vacuum dividing wall column (VDWC) for oleochemical application. Oleochemical fatty acid was used as the case study.

2. Methodology

2.1. Steady state and dynamic modelling

The process under study involves fractionation of oleochemical fatty acids which constitute of three carbon chains namely C10, C12 and C14 with boiling point of 270 °C, 299 °C and 326 °C respectively. To avoid product degradation, the column temperature was operated below 270 °C at pressure between 0.01 to 0.1 bar. The product purity for each streams were set to 99 mole%. Due to the polarity of the fatty acid as well as low operating pressure, non-random two-liquid NRTL thermodynamic model and its variances can be used. In this work NRTL was chosen. For process flowsheeting, four RADFRAC model blocks were used for both steady state and dynamic modelling using Aspen Plus and Aspen Dynamics, respectively. RadFrac is a rigorous model use to simulate all types of multistage vapour-liquid fractionation operations which includes ordinary distillation, absorption and stripping. In Aspen Plus, Radfrac model consisted of 4 column sections which are 1 stripper, 2 parallel absorbers and 1 rectifier to resemble actual 1 dividing wall column. Steady state model used for sensitivity analysis to study the relationship between tray temperature and product composition as well as to determine the tray number to be inferred for controlling the product composition. Dynamic model used for controllability analysis of the inferential control configuration. The control configuration used in the dynamic model was based on Othman [12]. Othman [12] conducted a controllability analysis of VDWC for oleochemical fatty acid fractionation using relative gain array (RGA) and singular value analysis (SVA). From his findings, it was found that Distillate-Side Stream- Boilup (DSV) control configuration was the best 3x3 control pairing due to the low interaction between control loops. Therefore, DSV control configuration was adopted for this work. Figure 1 shows the DSV control configuration.

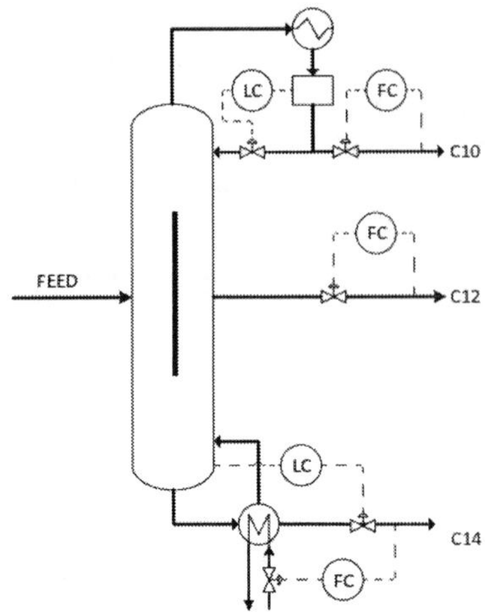

Figure 1. DSV control configuration [12]

2.2. Steady state sensitivity analysis

In this work, sensitivity analysis was applied to establish a column temperature profile for different operating points which was then compared with base case data. Table 1 shows the base case condition. In this work, several operating points were considered by changing the manipulated variables. The changes for distillate rate, side stream rate and reboiler duty at stipulated percentage of deviation are calculated from the base case value in the table. Three conditions were considered (1) ±10% change in distillate flowrate, D (2) ±10% change in middle flowrate, S and (3) ±60 change in reboiler duty, V. The percentage of change for respective manipulated variables are decided based on expected feed composition change from upstream unit resulting the offsetting differences in distillate and middle flow which can be up to 10 % and reboiler duty up to 60 %. The temperature of the most sensitive trays will be the inferred parameters for product composition at 99 % as per Table 1. From the column profile, temperature deviation from the base case (ΔT) were plotted. From this plot, one can determine the most sensitive tray. The identified tray for each product were then implemented in Aspen Dynamics and evaluated for its controllability performance

Table 1. Base case steady state condition.

Parameters	
Flowrate, kg/h	
- Feed	6240
- Distillate	316.11
- Side stream	4405.92
- Bottom	1517.96
Product composition, mol%	
- C10 at distillate	99
- C12 at side stream	99.8
- C13 at bottom	99
Reflux ratio	60.65
Reboiler duty, kW	2275.5

2.3. Tuning and closed loop response

The selected inferential variable from the previous step were added to developed DSV based control configuration by replacing the controlled variables from composition to the designated tray temperature. PID controller was adopted in this work. No measurement delay was included. Ziegler-Nichols (ZN) tuning method is a heuristic method of tuning PID controller which is widely used for tuning. Meanwhile, Internal Model Control (IMC) tuning method is also adopted because of its ability to compensate disturbances and model uncertainty. However, ZN setting result in very good disturbance response for integrating processes, but are otherwise known to result in rather aggressive setting and also give poor performance [13]. On the other hand, IMC setting it result in poor disturbance response for integrating processes but are robust and generally give very good responses for set point changes [14]. The latest current tuning method which is based on IMC known for giving non oscillatory response to setpoint change and load disturbance however it does not address specific application in fatty acid fractionation application compared to ZN which still predominat method in use today. Therefore, in this work both ZN and IMC setting was compared before further fine-tuned. The controller setting was subjected to set point change for distillate, side and bottom product composition. The controller performance was tabled in term of their settling time and overshoot. Tuning with fastest settling time and small overshoot were selected before further fine-tuned.

3. Results and discussions

Sensitivity analysis for distillate, middle and bottom sections are performed individually. For distillate section, Figure 2a gives the temperature deviation for ±10% change in distillate flow. The temperature deviation for increase flow rate was negative, while positive for decrease flow rate. The plot also shows several sensitive trays. Changes in flowrate at particular product stream will register disturbance at temperature column profile. For example greater flow of product resulting heat withdrawal at the column thus create drop in column temperature. However, each stage responds differently depending on the tray location. For flowrate increment tray number 2 to 7 and 28 to 34 were the most sensitive whereas for flowrate decrement tray 2 to 8 and 21 to 31 were the most sensitive. In this work tray 6 were selected as the preferred temperature tray to be inferred to distillate product stream. Note that, For the tray number other than this region, it would be a very poor inferential variable, because the sensor error and low magnitude noise would invalidate any correlation drawn from simulation. Besides, a small temperature deviation indicate that valve saturation can easily occur and operability region could be limited.

Figure 2b gives the temperature deviation for 10% change in the middle stream. The temperature deviation also shows the same pattern as in (a) but with fewer peaks. For flowrate increment tray number 21 to 31 were the most sensitive while tray number 28 to 38 were the most sensitive whereas for decrease in middle flowrate. Tray 29 were selected as the preferred temperature tray to be inferred to middle product stream. Figure 2c gives the temperature deviation for 10% change in the reboiler duty. The temperature deviation was apparent for 60% decrease in reboiler duty with different peak at the stripping section and rectifying section while not so much change for reboiler duty increment. For reboiler duty increment tray number 4 to 8 and 29 and 37 were the most sensitive Tray 34 were selected as the preferred temperature tray to be inferred to middle product stream due its location near the reboiler which quickly affected by changes in reboiler duty. Figure 3a shows updated inferential control loop of DSV based control configuration of the VDWC. The first control loop is using temperature tray no. 6 as controlled variable and distillate flow as manipulated variable. Second loop is measure temperature tray at no. 29 as controlled variable and manipulating middle stream flow as manipulated variable. The third loop is measuring temperature tray at no. 34 while giving the output at steam supply line for reboiler duty regulation. Reflux ratio and column bottom level is held a constant The loop was implemented in Aspen Dynamics as shown in Figure 3b. For PID controller setting, ZN method was compared to IMC. Its performance towards set point tracking were evaluated.

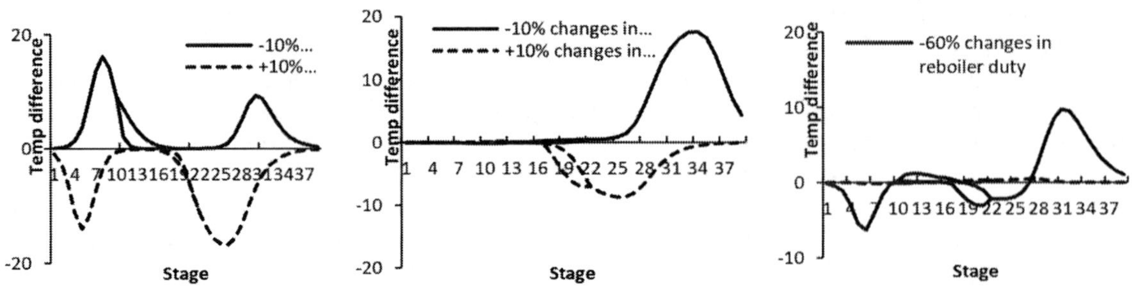

(a) (b) (c)

Figure 2: Stage temperature sensitivity analysis (a) ±10% changes in distillate flow (b) ±10% changes in middle flow (c) ±60% changes in reboiler duty

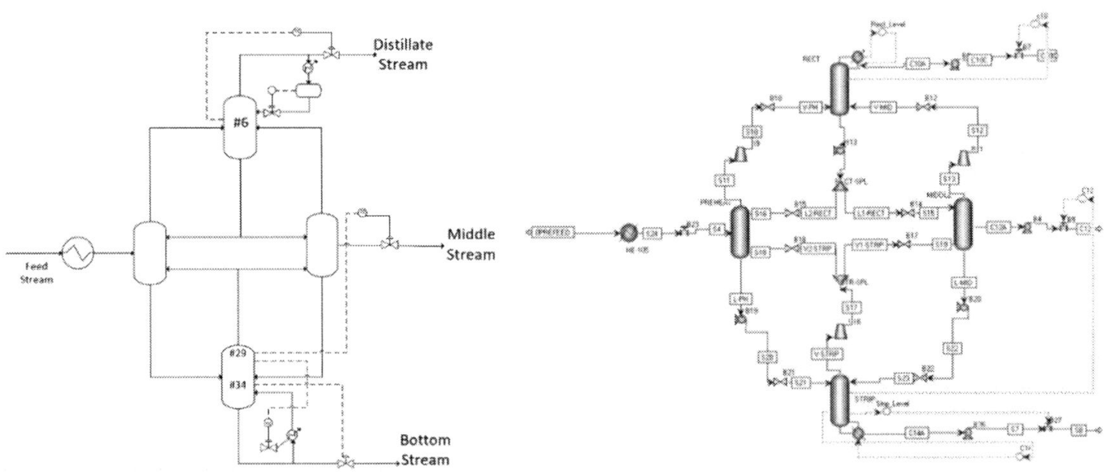

(a) (b)

Figure 3 (a) Inferential control using identified tray in DSV control configuration (b) the corresponding flowsheet in Aspen Dynamics

Table 2 shows the ZN and IMC tuning value as well as its corresponding controller performance for set point tracking. Prior to that open loop step test were performed at 5 % output change to identify process model values such as process gain, time constant and dead time respectively for each control loop which subsequently give the tuning values depending on tuning calculation used either ZN and IMC. Overall, in terms of settling time and overshoot both approach able to compromise and meet the satisfactory target. PID controller based on ZN tuning however provides better performance in term of settling time and overshoot. The settling time shall be short as possible and in ideal case settling time lesser than 10 multiply by dead time, t_d however any value approaching towards that is consider acceptable since no valid data available to be compare with. Hence, PID-ZN setting adopted and further fine-tuned.

Table 2: Control settings for the proposed inferential control in DSV control configuration with the comparison between Ziegler-Nicols and IMC method as well as comparison between PI and PID

Control Loop		Ziegler Nichols		IMC
		PI	*PID	PID
Distillate Stream, C10	Gain	86.87513	**144.7919**	2.350773
	Integral Time	35.47682	**26.63425**	375.1321
	Derivative Time	0	**4.26148**	5.251207
	Settling Time	6.09 hours	**2.64 hours**	3.42 hours
	Overshoot	0.45 %	**5 %**	0.45 %

Control Loop		Ziegler Nichols		IMC
		PI	*PID	PID
Middle Stream, C12	Gain	5.742162	**95.7021**	22.86846
	Integral Time	7150.78564	**2000**	2171.743
	Derivative Time	0	**858.9532**	295.5417
	Settling Time	1983 hours	**321.1 hours**	359.1 hours
	Overshoot	0 %	**0%**	0 %
Bottom Stream, C14	Gain	0.921668	**1.536113**	0.393431
	Integral Time	60.83798	**45.67416**	39.798
	Derivative Time	0	**7.307865**	7.038206
	Settling Time	16.83 hours	**11.62 hours**	30.31 hours
	Overshoot	44.2 %	**36.35 %**	70.7 %

Table 3 and figure 4 shows the tuning values and well as the controller performance for set point tracking after fine-tuned. Both distillate and bottom stream shown fast response with minimal overshoot. However, for middle stream when the settling time is longer than other controller loop. This due to the large amount of C12 which is require longer time for heating. The disturbance rejection test at feed change is not consider in this paper due invalid preset tray temperature set point to keep product quality at stipulated specification of 99 mole %. The new feed changes introduce will alter temperature profile of the column thus require new set point to keep the product quality steadily uninterrupted. The estimator shall be incorporate to predict the new set point upon registration of disturbance at feed flow.

Table 3: After tuning of ZN PID controller for distillate, middle and bottom stream

Control Loop		**Before tuning**	**After tuning**
Distillate Stream, C10	Gain	144.7919	6.6
	Integral Time	26.63425	600
	Derivative Time	4.26148	4.2618
	Settling Time	2.64 hours	1.82 hours
	Overshoot (< 5 %)	5 %	4.75%
Middle Stream, C12	Gain	2521.756	2.2
	Integral Time	35.80275	39
	Derivative Time	5.72844	5
	Settling Time	46.7	50.62 hours
	Overshoot (< 5 %)	12.4 %	0 %
Bottom Stream, C14	Gain	1.536113	80
	Integral Time	45.67416	3
	Derivative Time	7.307865	7.03826
	Settling Time	11.62 hours	1.31 hours
	Overshoot (< 5 %)	36.35 %	3.05%

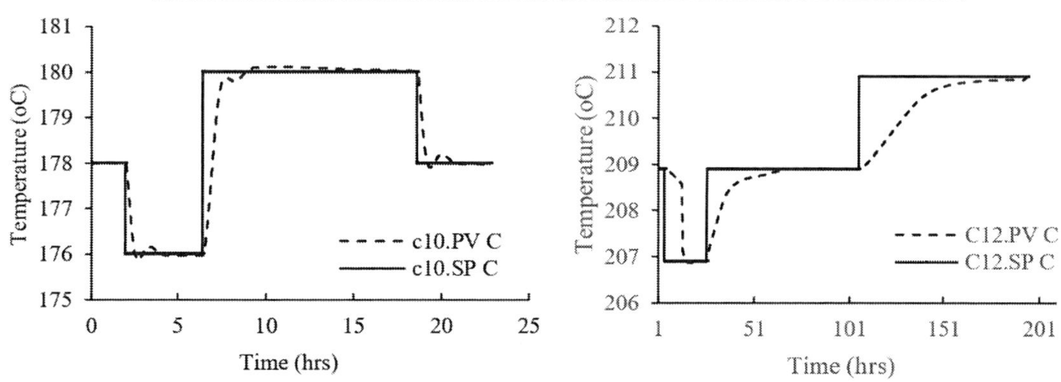

(a) (b)

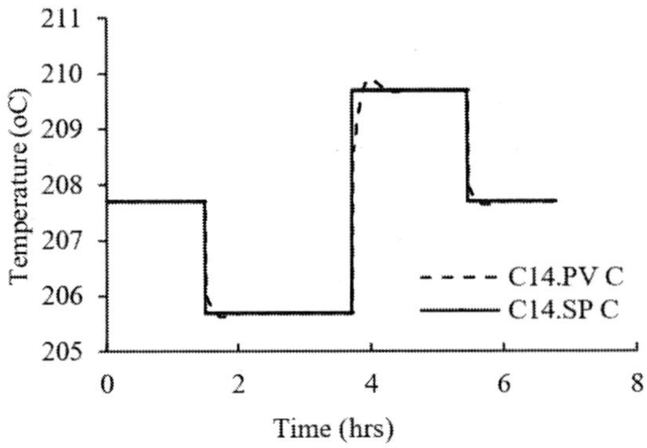

(c)

Figure 4: Set-point tracking for tray temperature change in (a) stage 6 (b) stage 29 (c) stage 34

4. Conclusions

The sensitivity analysis was successfully determining the best tray temperature location for distillate at 6th stage, middle at 29th stage and bottom at 34th stage based on the most temperature deviation from base case. The comparison from two tuning methods; ZN and IMC, with different controller setting of PI and PIC reveal that PID-ZN setting was preferred with good controller response in term of better settling time and minimum overshoot. The fine-tuned has improved the quality of controller response target thus give good set point tracking in the stage temperature.

Acknowledgments

Authors wishing to acknowledge assistance and support by Universiti Malaysia Pahang and Ministry of Education Malaysia through grant number RDU1603148 and FRGS/1/2014/TK05/UMP/02/1.

References

[1] Othman M R and Rangaiah G P 2019a Performance Analysis of Dividing Wall Column Model for Fractionation of Oleochemical Fatty Acid *Journal of Oil Palm Research* In Press

[2] Dejanović I, Matijašević L and Olujić Ž 2010 *Dividing wall column—A breakthrough towards sustainable distilling Chemical Engineering and Processing* **49** 559-80

[3] Yildirim Ö, Kiss A and Kenig E 2011 Dividing wall columns in chemical process industry: A review on current activities *Separation and Purification Technology* 403-17

[4] Kiss A A and Rewagad R R 2011 Energy efficient control of a BTX dividing-wall column *Computers & Chemical Engineering* 2896-2904

[5] Parrish J R and Brosilow C B 1985 Inferential control applications Automatica 2 527

[6] Ansari R M and Tadé M O 2008 On-Line Inferential Models in Nonlinear Multivariate Control Applications *Developments in Chemical Engineering and Mineral Processing* **4**(1–2) 71–83

[7] Marlin T E 1995 Process Control: Designing Processes and Control Systems for Dynamic Performance (McGraw-Hill)

[8] Luyben W L 2006 Evaluation of criteria for selecting temperature control trays in distillation columns *Journal of Process Control* **16** 115–34

[9] Wang J, Yu N, Chen M, Cong L and Sun, L 2017 Composition control and temperature inferential control of dividing wall column based on model predictive control and PI strategies *Chinese Journal of Chemical Engineering* **26**(5) 1087-01

[10] Yuan Y, Huang K, Chen H, Zhang L and Wang S 2017 Asymmetrical temperature control of a BTX dividing-wall distillation column *Chemical Engineering Research and Design* **123** 84-98

[11] Ignat R and Woinaroschy A 2011 Dynamic analysis and controllability of dividing-wall distillation columns using a four points control structure UPB Scientific Bulletin, Series B: Chemistry and Materials Science **73**(4) 71-80

[12] Othman M R 2019b Controllability Analysis of A Vacuum Dividing Wall Column For

Oleochemical Fatty Acid Fractionation. *IOP: MSE*
[13] Skogestad S 2004 Simple Analytic Rules for Model Reduction and PID Controller Tuning *Journal of Process Control* **13** 291-309
[14] Rivera D, Morari M and Skogestad S 1986 Internal Model 4. PID Controller Design Industrial & Engineering Chemistry, Process Design and Development **25** 252-65

IOP Conf. Series: Materials Science and Engineering **702** (2019) 012021 doi:10.1088/1757-899X/702/1/012021

Application of extended method of classes for solving population balance equations and optimization study for crystallization process involving dissolution phenomena

S Z Adnan and N A F A Samad*

Faculty of Chemical and Process Engineering Technology, College of Engineering Technology, Universiti Malaysia Pahang, Lebuhraya Tun Razak, 26300 Kuantan Pahang, Malaysia

*E-mail: asmafazli@ump.edu.my

Abstract. Crystallization is a well-established chemical process for producing high quality of crystals. Several phenomena such as nucleation, crystal growth and dissolution are involved in producing good crystal's quality. Thus modelling of crystallization process is essential for representing crystal quality as well as for process control and optimization purposes. One of the widely approach to represent crystal quality in terms of crystal size distribution (CSD) and to solve population balance equations (PBE) is by using method of classes (MOC). However, MOC is only applicable for the case of nucleation and crystal growth and thus needs to be extended for the case of dissolution. Therefore, the objective of this paper is to extend MOC for the case of dissolution and demonstrate crystallization process involving dissolution in order to achieve desired CSD with minimum fine crystals. The mathematical model of the process is developed and simulated in Matlab. The optimization algorithm is employed to generate set-point trajectory for Proportional-Integral (PI) controller in the case of with and without dissolution phenomena. Based on the simulations, the MOC have shown comparable results with published literature, which indicates MOC has been successfully extended to solve the PBE and representing CSD for size dependent growth rate system.

1. Introduction

Formation of crystals from a supersaturated solution or simply crystallization process is an extremely old unit operation used to produce highly valuable products of pharmaceuticals, bulk chemicals and fertilizer chemicals [1]. These products have to conform certain product specifications with respect to crystal size and shape, purity, percentage of impurities as well as crystal size distribution (CSD) in order to be considered as good quality of crystals. Also, these product properties will determine the types of crystallization process, crystallization mode and types of equipment used for production [1-3]. Therefore, it is necessary to conform to these final specifications in order to avoid unnecessary problem in the downstream processes such as long filtration times and inadequate drying [2-6].

In crystallization process, several phenomena involves in a molecular level such as nucleation, crystal growth, agglomeration, breakage and dissolution. However, in this paper only nucleation, crystal growth and dissolution are discussed in detail as agglomeration and breakage can be neglected due to the weak impact in the solid-liquid systems [7]. Nucleation can be defined as the initial process that occurs in the formation of a crystal from a solution in which small amount of molecules aggregate to become arranged in a pattern characteristic of a crystalline solid, forming a site upon which additional particles are deposited as the crystal grows [8]. Crystal growth refers to the process of progressive addition of the particles to the surface of initial seed crystal causing an increase in crystal size. Crystal growth is a layer-by-layer process, and only occur at the face of the crystal where the material must be transported to that face from the bulk of solution [9]. Crystal growth along with the nucleation are the most common crystallization phenomena that control the final CSD obtained in a system. In addition, the conditions and rate of crystal growth have a significant impact on the purity and the crystal habit [9]. An understanding of both nucleation and crystal growth theory are imperative in the development of industrial crystallization process.

Other crystallization phenomena, which is dissolution is referred as the reverse process of crystallization where dispersed solid that goes into solution is the solute [10]. As dissolution proceeds, the concentration of the solute increases. Given sufficient time at the fixed conditions, the solute will eventually dissolve up to a maximum solubility where the rate of dissolution equal to the rate of crystallization. Then, the solution is incapable of dissolving any solute under these equilibrium conditions. In cooling crystallization, dissolution may be achieved by simply heating up the solution to a specified temperature in which would cause the solute to dissolve. The alternating cycle of high and low temperature allow the crystal particles grown from seed crystals and fine crystals produced by secondary nucleation to dissolve back during reheating cycle. Therefore, dissolution would become beneficial in terms of dissolving fine crystals produced by secondary nucleation if properly designed into the model [4]. However, there is limited available published literature regarding dissolution model and mostly assume that the kinetic model of dissolution is the inverse of crystal growth model [2-4, 8, 11].

Besides, describing crystallization process using model-based approach or population-based modelling is considered as the commonly accepted modelling approach [7]. Generally, crystallization models involve three types of dynamic balance equations which are population, mass and energy. Population balance equation (PBE) is a hyperbolic partial differential equations that is usually coupled with a set of equations describing crystallization phenomena, saturation equilibrium as well as mass and energy balances in the forms of ordinary differential equations (ODEs). This PBE requires complex numerical solution techniques which is characterized by significant computational burden [10]. One of the commonly used PBE numerical solution technique is the standard method of moments [3-4, 12]. The approach of this method is by converting the partial differential equations (PDEs) that represent the population balance into a set of coupled ordinary differential equations (ODEs) for the n-moments considered. Since the ODE solvers are readily available, many researchers use this method to ease the solution of PBE. Another available method to solve PBE is method of classes (MOC). Method of classes permits the partial differential equations (PDEs) to be sectioned along the size domains into finite classes. This means that the method of classes allows estimation of crystal size distribution (CSD) at any point during crystallization process of either size dependent or size independent growth. This could not be achieved by using standard method of moments due to its limited function of n-moments. It is worth noting that method of moments capable of characterizing the crystal product only by mean crystal size or mean size diameter. In order to construct CSD, the standard method of moments needs to be extended or modified which increases the complexity of PBE solution. Thus, in this paper applications of method of classes are studied. Other than that, most of the crystallization work found in the literature used method of classes for the case of nucleation and crystal growth only. There is only one research that used method of classes as a solution for generic PBE in the case of including dissolution phenomena [13]. In their work, crystal growth and dissolution models are developed based on the assumptions of size independent rate but no detailed clarification on the construction of CSD is provided in their work. The relationship between method of classes and concentration or energy balances are also not presented.

Therefore, in this paper the study of seeded batch cooling crystallization is demonstrated using extended method of classes for both without and with dissolution phenomena. Potassium nitrate crystallization is chosen as the case study which is adopted from published literature [3]. The mathematical model of potassium nitrate crystallization using extended method of classes is simulated in Matlab software. For both cases of without dissolution and with dissolution phenomena, specified optimization algorithms are employed to generate the temperature trajectory of the system which is then served as the set-point for Proportional-Integral (PI) controller. The performances for both cases are studied and compared with the published literature [3] especially in terms of the construction of CSD for demonstrating the capability of the extended method of classes.

2. Extension of Method of Classes

In this section, the population balance model for a batch crystallizer with size dependent growth of the one characteristic length (L) adopted from [3] is described as shown in equation (1).

$$\frac{\partial f_n(L,t)}{\partial t}+\frac{\partial G(L,S)f_n(L,t)}{\partial L}=B_{nuc} \tag{1}$$

where $f_n(L,t)$ is the population density function, $G(L,S)$ is the size dependent linear crystal growth rate, B_{nuc} is the nucleation rate and t is crystallization time. This one-dimensional PBE is solved by using extended method of classes as shown in equations (2)-(7) instead of the conventional standard method of moments. The usage of extended method of classes is to demonstrate the capability of this method to incorporate the dissolution terms and to generate the CSD accurately and similarly as the published literature [3]. Please note that dissolution term, D_x is used as the relative supersaturation level proceeds lesser than 0 where the relative supersaturation (S) is applied as $S=(c-c_{sat})/c_{sat}$, where the c is the concentration of the solute and c_{sat} is the saturation concentration. The usage of the dissolution term, D_x is based on the assumption that it is the inverse of crystal growth model.

For $S>0$,

$$\frac{dN_1}{dt}+\frac{G_{x1}}{2\Delta Cl_2}N_2+\frac{G_{x1}-G_{x0}}{2\Delta Cl_1}N_1=B_{nuc}, i=1 \tag{2}$$

$$\frac{dN_i}{dt}+\frac{G_{xi}}{2\Delta Cl_{i+1}}N_{i+1}+\frac{G_{xi}-G_{xi-1}}{2\Delta Cl_i}N_i+\frac{G_{xi-1}}{2\Delta Cl_{i-1}}N_{i-1}=0, 1\le i\le n \tag{3}$$

$$\frac{dN_n}{dt}+\frac{G_x}{2\Delta Cl}N_n+\frac{G_{x1}-G_{x0}}{2\Delta Cl}N_n=0, i=n \tag{4}$$

For $S<0$,

$$\frac{dN_1}{dt}+\frac{D_{x1}}{2\Delta Cl_2}N_2+\frac{D_{x1}-D_{x0}}{2\Delta Cl_1}N_1=B_{nuc}, i=1 \tag{5}$$

$$\frac{dN_i}{dt}+\frac{D_{xi}}{2\Delta Cl_{i+1}}N_{i+1}+\frac{D_{xi}-D_{xi-1}}{2\Delta Cl_i}N_i+\frac{D_{xi-1}}{2\Delta Cl_{i-1}}N_{i-1}=0, 1\le i\le n \tag{6}$$

$$\frac{dN_n}{dt}+\frac{D_x}{2\Delta Cl}N_n+\frac{D_{x1}-D_{x0}}{2\Delta Cl}N_n=0, i=n \tag{7}$$

where N_i is the number of crystals per unit suspension for class i, G_x is the crystal growth rate in length direction x and D_x is the dissolution rate in length direction x.

In addition, the overall mass balance and energy balance for potassium nitrate crystallization are shown in equations (8) and (9) respectively.

$$\frac{dc}{dt}=-\frac{\rho_c k_v V}{m_w}\left(\sum_{i=1}^{i=N}S_{xi}^3\frac{dN_i}{dt}\right) \tag{8}$$

$$\rho V c_p\frac{dT}{dt}=-\Delta H_c\rho_c k_v V\left(\sum_{i=1}^{i=N}S_{xi}^3\frac{dN_i}{dt}\right)-U_1A_1(T-T_w) \tag{9}$$

where ρ_c is the crystal density, k_v is the crystal shape factor, V is the volume of solution, m_w is the mass of solvent, c_p is the heat capacity, T is the temperature of the solution, ΔH_c is the heat of crystallization, U_I is the heat transfer coefficient of internal crystallizer, A_I is the internal area of crystallizer and T_w is the cooling water temperature.

Next, the saturation concentration for potassium nitrate is generated and adapted based on the following polynomial expression as the function of solution temperature [3].

$$c_{sat}=0.129+5.88\times10^{-3}T-1.72\times10^{-4}T^2 \tag{10}$$

Also, secondary nucleation is assumed while the effect of agitation is neglected in the nucleation, crystal growth and dissolution model equations. It has been assumed that the crystal is grown or dissolved based on size dependent rate. The dissolution rate is set to have the same order of magnitude as the growth rate and a first-order dependence on the degree of under saturation is assumed. The models

IOP Conf. Series: Materials Science and Engineering **702** (2019) 012021 doi:10.1088/1757-899X/702/1/012021

for secondary nucleation, crystal growth and dissolution rates in the power law form are shown in equations (11)-(13):

$$B_{nuc} = k_b S^b V \tag{11}$$

For $S > 0$,

$$G_{xi} = k_g S^g (1+\alpha_g L_{xi})^{\beta_g}, i = 1,2,3..n \tag{12}$$

For $S < 0$,

$$D_{xi} = k_d S^d (1+\alpha_d L_{xi})^{\beta_d}, i = 1,2,3..n \tag{13}$$

where k_b, k_g, and k_d are the kinetic coefficients for nucleation, crystal growth and dissolution respectively. b, g and d are the nucleation order, crystal growth order and dissolution order respectively. α_g and β_g are the crystal growth parameters and α_d and β_d are the dissolution parameters.

Other than that, the crystal size distribution (CSD) is calculated based on the following equation (14).

$$f_n(L_{xi}) = \frac{N_i/\Delta Cl + N_{i+1}/\Delta Cl}{2} \tag{14}$$

where L_{xi} is the length of crystal particles and ΔCl is the size of classes where $\Delta Cl = L_{xi} - L_{xi-1}$.

2.1. Operating scenario

The assumed scenario for crystallizer operation is that initially the crystallizer is loaded with a feed solution whose saturation temperature is T_{feed}. At the beginning of the batch process, the crystallizer is cooled to a temperature below saturation temperature and then a crystallizer is loaded with an initial seed distribution which is based on the uniform distribution adapted in [3] where it is characterized by the mass C_s, the average mean size L_s, and the percentage of width w of the distribution. The initial mean characteristic length is set to be 300 μm and targeted to grow to 800 μm as shown in figure 1.

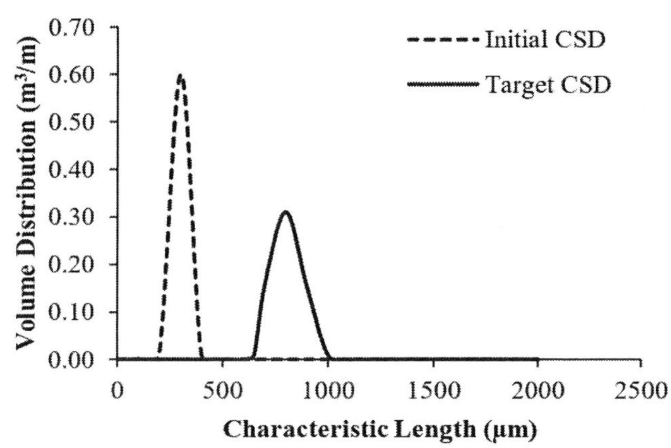

Figure 1. Initial seed of CSD and targeted CSD.

The temperature then is lowered to T_{fin} within the time duration t_{cool}. The specific values of the operating conditions and kinetic parameters are shown in Tables 1 and 2 respectively.

Table 1. Operating conditions and major constraints adapted from [4].

Nomenclature		Values
T_{feed}	Saturation temperature of feed, °C	30.0
T_{fin}	Final temperature, °C	20.0
t_{cool}	Cooling duration, min	120
t_{batch}	Batch time, min	150
C_s	Seed loading ratio, %	0.5
L_s	Seed size, µm	300
w	% Width of distribution, %	10
T_s^{max}	Upper limit of seeding temperature, °C	30.0
T_s^{min}	Lower limit of seeding temperature, °C	27.0
T_{min}	Upper limit of crystallizer temperature, °C	30.5
T_{max}	Lower limit of crystallizer temperature, °C	20.0
$c_0 = c_{sat}(30\ °C)$	Initial feed concentration, g/g – water	0.459
ρ_c	Density, kg/m^3	2.11 x 10^3
k_v	Volumetric shape factor	1

Table 2. Kinetic parameters of crystallizer model adapted from [4].

Parameters	Units	Values
Nucleation parameter (k_b)	#/s/g-water	4.21 x 10^5
Nucleation parameter (b)	-	1.59
Growth parameter (k_g)	m/s	1.0741 x 10^{-4}
Growth parameter (g)	-	1.28
Growth parameter (α_g)	1/m	1000
Growth parameter (βg)	-	0.5
Dissolution parameter (k_d)	m/s	1.19 x 10^{-4}
Dissolution parameter (d)	-	1
Dissolution parameter (α_d)	1/m	1000
Dissolution parameter (β_d)	-	-1.0

3. Applications of Extended Method of Classes

The mathematical model for the potassium nitrate crystallization process is developed in Matlab 2014b software and is solved using backward differentiation formula (BDF) method known as 'ode15s' solver using the initial conditions in Table 1. Two cases of potassium nitrate crystallization without and with dissolution phenomena are demonstrated in the next subsections 3.1 and 3.2 respectively using specified optimization algorithm to achieve desired target CSD.

3.1. Potassium nitrate crystallization without dissolution phenomena

The optimization algorithm as shown in equation (15) is employed to generate the supersaturation set-point for the controller to maintain. This objective function adapted from [14] is set to minimize the error between calculated CSD and targeted CSD in order to obtain desired target CSD as shown in figure 1.

$$F_{obj} = \sum_{i=1}^{N} \left(\frac{f_{n,i} - f_{n,i_{target}}}{f_{n,i_{target}}} \right)^2 \quad (15)$$

Subject to:

$$0 \le \phi \le \phi_{\max}$$
$$t_{\min} \le t_c \le t_{\max}$$
$$c_{tbatch} \le c_{f\,\max}$$

The equation is called as analytical CSD estimator, where N is the number of discretization points, f_{ni} is the predicted CSD obtained from mathematical model (equations (2) to (14)) and $f_{nitarget}$ is the desired target CSD. Φ is the supersaturation set-point, c_{tbatch} is the expected solute concentration at the end of the batch and c_{fmax} is the maximum acceptable solute concentration at the end of the process to achieve required yield. This analytical CSD estimator is developed based on the concept of constant supersaturation level throughout the entire crystallization process. Based on the CSD estimator, in order to obtain desired CSD the supersaturation level that needs to be maintained within 150 minutes is 0.015 g/g. Please note that the crystallization time is fixed at 150 minutes in order to allow for model validation with published literature.

Subsequently, the generated supersaturation set-point from equation 15 is used as the set-point for the PI controller in order to achieve desired CSD. Proportional-Integral (PI) controller is used as controller since it is proven to be feasible for maintaining the concentration at its set-point [14]. The temperature of crystallizer is maintained and controlled at its set-point by manipulating the water inlet temperature of the crystallizer using PI controller. The parameters for PI controller is calculated from Internal Model Control (IMC) tuning method where the proportional gain (K_c) is 33 and integral time (τ_i) is 2.3.

Meanwhile, figure 2 shows the closed-loop simulation results using PI controller for temperature profile of potassium nitrate crystallization with constant supersaturation level at 0.015. It can be seen that the temperature profile in figure 2 (a) drops consistently from 29 °C to 20 °C. Similarly, the solute concentration drops steadily from 0.45 g/g to 0.32 g/g until the end of operational time at 150 minutes. Figure 2 (b) shows there is a constant gap between solute concentration and saturation concentration indicating that the supersaturation level is successfully kept constant to induce high crystal growth and nucleation rate. Nevertheless, sharp drop from 0 to 30 minutes is observed on temperature profile which is mainly due to the compensation of high supersaturation at the beginning of the process in order to keep constant supersaturation level throughout the batch crystallization. It is noted that the solute concentration is operated above the saturation line which indicates that there will be no dissolution for this case study. Both temperature and concentration profiles have similar trend due to concentration profile is temperature dependent as shown in equation (10).

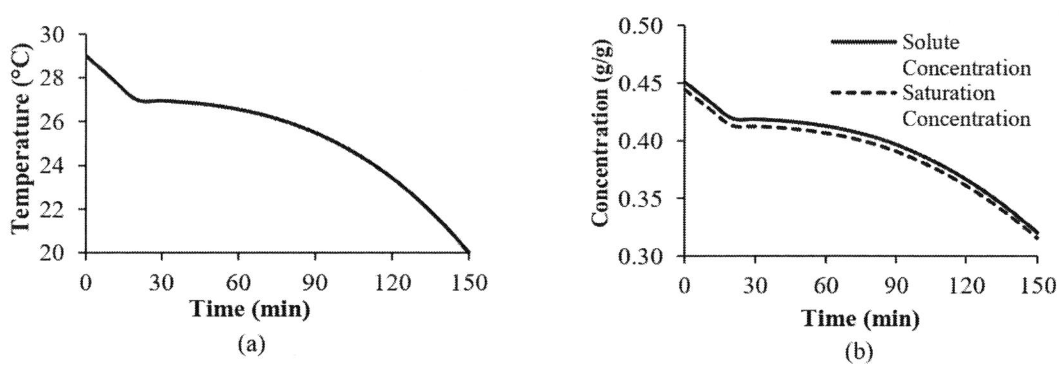

Figure 2. (a) Temperature and (b) concentration profiles without dissolution phenomena.

Next, figure 3 shows the CSD profile of the seed crystals is grown from mean crystal size of 300 μm to 800 μm as the result of specified level of supersaturation. As shown in figure 3, high supersaturation level at 0.015 g/g results in high nucleation and crystal growth rate which contributes to the higher secondary nucleation that distributes below 500 μm although the targeted CSD adapted from [3] is successfully achieved. The distribution of the initial seed is divided into two peaks which is desired CSD target and undesirable peak of secondary nucleation. Large size of crystals has been obtained but the amount of fine crystals is also significant. These fine crystals may cause difficulties in the downstream processing equipment such as filtration and drying whereby longer filtration time and drying is needed [2-6]. The disperse distribution is due to the size dependent crystal growth properties whereby large crystals have high crystal growth rate and vice versa at the selected process conditions. The distribution of the initial seed is decreasing from 0.6 m^3/m to 0.3 m^3/m. This is due to the fact that the low supersaturation would lead to low crystal growth rate that contributes to the low value of the mean crystal size [10]. In summary, desired target CSD is able to achieve but there is significant amount of fine crystals at the end of crystallization process. The simulation of mathematical model of potassium nitrate crystallization process employing method of classes without dissolution case has been successfully performed in this work under closed-loop simulation. The CSD profile obtained through employing method of class is similar to the published literature as shown in figure 3 which proves that the method of classes is capable for achieving similar results for the case of without dissolution.

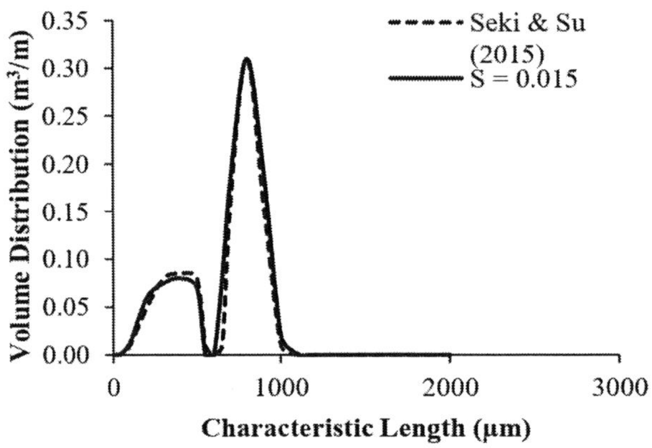

Figure 3. CSD profile based on supersaturation set-point at 0.015.

3.2. Potassium nitrate crystallization with dissolution phenomena

Since the fine crystals in the case of without dissolution step is high despite achieving the desired target CSD, the optimization algorithm adapted from [3] as shown in equation (16) is employed to further minimize the CSD that fall under the size of 500 μm by implying the effect of temperature swing (dissolution steps).

$$F_{obj} = \min_{\theta} f_n(500\mu m) \tag{16}$$

Subject to:

$$\dot{T}_{\min} \le \dot{T}(t) \le \dot{T}_{\max}$$

$$T_s^{\min} \le T(0) \le T_s^{\max}$$

$$T_{\min} \le T(t) \le T_{\max}$$

where the lower limit for the rate of crystallizer temperature change, $\dot{T}_{min}$ and the upper limit for rate of crystallizer temperature change, $\dot{T}_{max}$ are set to be -0.5 °C/ min and 0.5 °C/ min and the values of each T_s and T limits are as shown in the Table 1. The temperature trajectory generated by equation (16) is shown in figure 4 where it is served as temperature set-point for PI controller. The temperature of crystallizer is maintained at its set-point by manipulating the water inlet temperature of the crystallizer. The controller parameters which are calculated from IMC tuning method are set to be 33 for proportional gain (K_c) and 2.3 for integral time (τ_i).

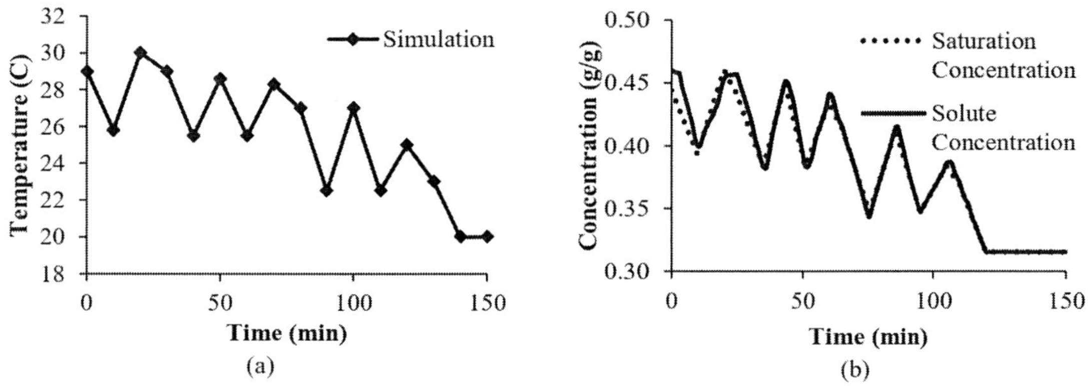

Figure 4. (a) Temperature and (b) concentration profiles for dissolution case.

Figure 4 (a) shows temperature profile of potassium nitrate crystallization process with temperature swing (dissolution) where it can be seen that the solution is repetitively being cooled and heated from 29 °C to 20 °C until the end of the batch crystallization operation. The same trend can be seen from figure 4 (b) where concentration profile of potassium nitrate in this case is zigzagged from 0.46 g/g to 0.32 g/g following the temperature swing profile. The crystallization take place as the process begins at operational time of 0 minute due to the first concentration dropped following the first temperature dropped. As consequence, there are huge distance between solute concentration and saturation concentration whereby this condition promotes nucleation and crystal growth phenomena. The solution is further cooled, heated and re-cooled to create multiple supersaturation conditions where the crystal nuclei start to grow following the crystal growth rate model and dissolve following the dissolution rate model. Due to the crystallization kinetics' properties of size dependent growth, fine crystals have smaller growth rate to be compared with large crystals. Therefore, when dissolution takes place fine crystals would dissolve faster due to the fact that small size of particles have higher surface area than big size of particles. The temperature profile is designed and optimized based on this theory.

Based on figure 5, the supersaturation profile has multiple peaks reacting to the temperature swing profile. As the process frequently proceeds far from the saturation concentration and recurrently become undersaturated, the supersaturation concentration is optimized where the growth of seed crystals is favoured instead of the secondary nucleation of the fine crystals. In each cooling cycle as the solute concentration is higher than saturation concentration, the nucleation and crystal growth rate are further enhanced. This condition fortifies the growth of the crystal particles that did not dissolve when heating cycle and production of new crystals. However, in each heating cycle as the nucleation and crystal growth rate drop, crystal particles are forced to stop growing and dissolve. Small crystals formed by secondary nucleation dissolve faster than those of bigger crystals grown from the seed crystals due to the fact that small size of particles have higher surface area than big size of particles. This disappearance of crystals or dissolution phenomena benefits in reducing the fine crystals. This can be confirmed by figure 6 where the generation of small fines is significantly reduced in the case of operating with temperature swing (dissolution). In overall, the supersaturation profile involves both positive and negative values which indirectly mean that both crystal growth and dissolution take place accordingly.

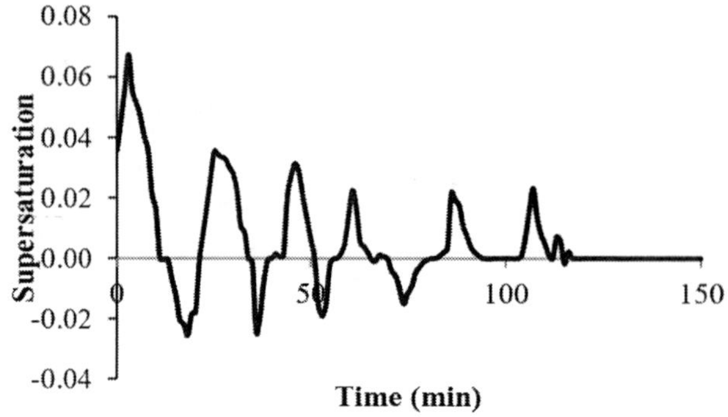

Figure 5. Supersaturation profile for dissolution case.

Other than that, figure 6 (a) shows the generation of crystal fines has been considerably reduced when the process involves dissolution phenomena through multiple dissolution steps. This shows that dissolution enables small crystals to dissolve and accelerate growth for the remaining crystals for the next cooling cycle. Apart from that, the mean crystal size for temperature swing case is larger than that of no dissolution step, from 300 μm of seed mean crystal size grown to approximately 1250 μm of mean crystal size which is 56.2% bigger than targeted mean crystal size. The significant difference of crystal size from the seed crystals clearly explains that the dissolution step provides more room for crystals to grow compare to the case with no dissolution. These findings are also supported by the conclusion made by [3] where the final product CSD is significantly improved and the generation of small fines is considerably reduced. Also, figure 6 (b) shows that the simulation of mathematical model of potassium nitrate crystallization process employing extended method of classes which is performed in this work under closed-loop simulation is successfully implemented. The CSD profile generated using method of classes shows similar results with the method employed by published literature [3].

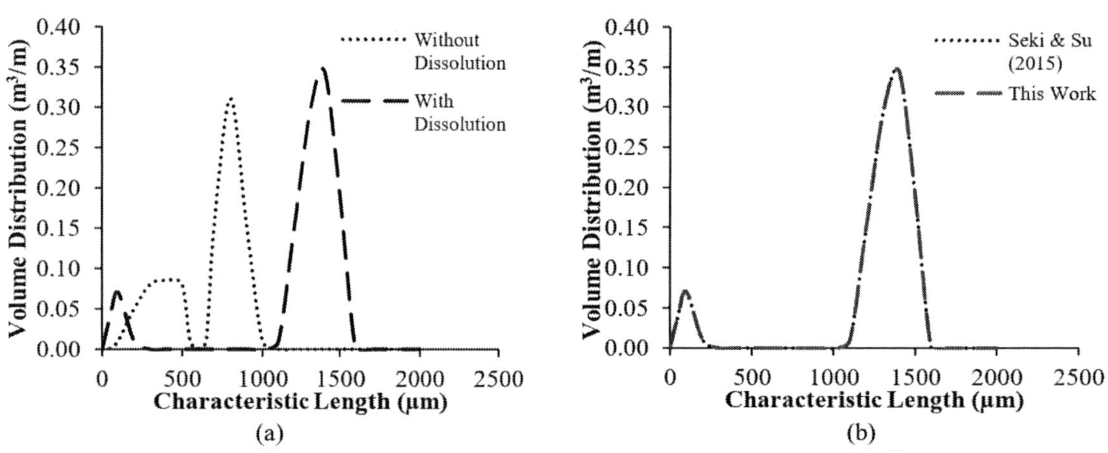

Figure 6. (a) Performance comparison for with and without dissolution phenomena and (b) CSD comparison obtained from extended method of classes and published literature [4].

4. Conclusion

Simulation of seeded batch cooling crystallization for potassium nitrate case study is successfully demonstrated for both with and without dissolution phenomena. In this work, the method of classes has been extended and simulated under closed-loop simulation using PI controller. By using extended method of classes, the CSD profiles obtained from both with and without dissolution case are similar and comparable with experimental data indicating a reliability of the extended method. The desired target CSD is successfully achieved for both cases and better performances with 56.2% larger mean

IOP Conf. Series: Materials Science and Engineering **702** (2019) 012021 doi:10.1088/1757-899X/702/1/012021

characteristic length than target CSD with dissolution case is obtained. Also, the amount of crystal fines has been reduced by employing dissolution steps in the process.

References

[1] Lewis A E, Seckler M M, Kramer H, and Van Rosmalen G 2015 *Industrial Crystallization: Fundamentals and Applications* (Cambridge: Cambridge University Press)
[2] Acevedo D, Jarmer D J, Burcham C L, Polster C S and Nagy Z K 2018 *Che. Eng. Res. Des.* **135** 112–20
[3] Hemalatha K, Nagveni P, Kumar P N and Rani K Y 2018 *Comp. Che. Eng.* **112** 292–303 Seki H and Su Y 2015 *Che. Eng. Sci.* **133** 16–23
[4] Adnan S Z, Saleh S and Samad N A F A 2019 *AIP Conf. Proc.* vol 2124 (United States: American Institute of Physics) p 020042
[5] Lenka M and Sarkar D 2018 *J. Cryst. Growth* **486** 130–36
[6] Zhang D, Liu L, Xu S, Du S, Dong W and Gong J 2018 *J. Cryst. Growth* **486** 1–9
[7] Qamar S, Noor S, Rehman M and Seidel-Morgenstern A 2011 *Comp. Chem. Eng.* **35** 412–22
[8] Myerson A S 2015 *Faraday Discuss* **179** 543-7
[9] Green W G and Perry H R 2013 *Perry's Chemical Engineers' Handbook* (New York: McGraw Hill)
[10] Smith R 2005 *Chemical Process Design and Integration* (New Jersey: John Wiley & Sons)
[11] Samad N A F A, Singh R, Sin G, Gernaey K V and Gani R 2011 *Comp. Che. Eng.* **35** 828–43
[12] Hulburt H M and Katz S 1964 *Chem. Eng. Sci.* **19** 555-74
[13] Mangin D, Garcia E, Gerard S, Hoff C, Klein J P and Veesler S 2006 *J. Cryst. Growth* **286** 121-25
[14] Samad N A F A, Sin G, Gernaey K V and Gani R 2013 *Comp. Chem. Eng.* **54** 8–23
[15] Zahari Z M, Saleh S and Samad N A F A 2017 *Chem. Eng. Trans.* **56** 205-10

IOP Conf. Series: Materials Science and Engineering **702** (2019) 012022 doi:10.1088/1757-899X/702/1/012022

Fractionation of oleochemical fatty acid using vacuum dividing wall column: A controllability analysis

M R Othman*

Faculty of Chemical & Process Engineering Technology, College of Engineering Technology, Universiti Malaysia Pahang, Lebuhraya Tun Razak, 26300 Gambang, Kuantan, Pahang, Malaysia

*E-mail: rizza@ump.edu.my

Abstract. Dividing wall column (DWC) provides a good alternative for oleochemical fractionation. However, the internal configuration and multiple input multiple output (MIMO) system of DWC leads to complexity in operation and control. This work aims to analyze the controllability of fractionating oleochemical fatty acid using vacuum dividing wall column (VDWC). To achieve this, Aspen Plus and Aspen Dynamics were used to develop a rigorous steady state and dynamic model of the column. Five manipulated variables (MVs) were considered namely reflux flowrate (L), distillate flowrate (D), bottom flow rate (B), side-stream flowrate (S) and vapor boilup (V) while controlled variables (CVs) were the product compositions. Pairing of MV and CV to determine the best 3×3 control configuration was performed using relative gain array (RGA) and singular value analysis (SVA). The selected control structure was tested on PID controllers for several regulatory and servo problem. The results of RGA and SVA shows that DSV was the best control configuration. Performance analysis was found to be successful in rejecting the disturbances as well as obtaining good set point tracking. However, distillate and bottom composition shows poor controllability compare to middle composition.

1. Introduction

Dividing wall column (DWC) provides a good alternative for processes using typical distillation (DC) column due to possible saving in both energy and capital cost. Because of its advantages, extensive research has been done and the first industrial application of DWC's was implemented in 1985 by BASF [1]. So far, about 40 DWCs are in operation worldwide and about 30 of them within the BASF group [2]. The potential of DWC is however restricted and not applied to broad range of the separation processes due to the challenges in design, simulate, operation and control [3-4]. In the oleochemical industry particularly in Malaysia, mostly used typical (DC) for its product fractionation. In our recent study, fractionating oleochemical fatty acids using DWC reduces around 20% of capital and operating cost compared to typical DC [5].

The integration of two columns into one shell leads to changes in the operating mode and controllability thus becoming a potential hurdle for commercial implementation of DWC [6]. Moreover, distillation columns exhibit a multiple-input multiple-output (MIMO) system in which manipulated variables (MVs) affect multiple controlled variable (CVs) [7]. To provide good process control in such system it is crucial to consider the probability of pairing CV and MV and applying an effective feedback controller to each selected pair. A good paring will ensure effective rejection of process disturbances as

IOP Conf. Series: Materials Science and Engineering **702** (2019) 012022 doi:10.1088/1757-899X/702/1/012022

well as maintaining the product specifications. Despite various research on DWC, study on oleochemical fractionation using vacuum dividing wall column (VDWC) received less attention. Therefore, this study aims to analyze the controllability of VDWC for fractionation of oleochemical fatty acid. Our scopes of work in paper include development of steady state and dynamic model in Aspen Plus and Aspen Dynamics. Determination of suitable 3×3 control configuration using steady state relative gain array (RGA) and singular value analysis (SVA) and test the performance of the selected controller configuration to disturbances in feed flowrate, feed composition as well as set point change.

2. Methodology

The process under study involves fractionation of oleochemical fatty acids which constitute of three carbon chains namely C10, C12 and C14. The boiling point of C10, C12 and C14 are 270 °C, 299 °C and 326 °C respectively. To avoid product degradation, the column temperature was operated below 270 °C at pressure between 0.01 to 0.1 bar. The feed information is listed in Table 1. The feed comprises mostly of C12 and C10 is the most least. C10 will be fractionated at the distillate stream whereas C12 and C14 will be fractionated at side and bottom stream respectively. The product purity for each streams were set to 99 mole%.

The steady state VDWC model was developed in Aspen Plus. NRTL was chosen as the thermodynamic model due to the polarity of the fatty acid as well as low operating pressure [5]. Instead of typical two or three model configuration, this work employs four RADFRAC model blocks to mimic the four internal and hydrodynamic behavior of an actual DWC internal sections. Furthermore, four column configuration was usually applied for dynamic simulation. The product purity for each streams were set to 99 mole%. Design of the VDWC was based on the work by Othman and Rangaiah [5]. The design parameters are shown in Table 1.

Table 1. VDWC design parameters

Reflux ratio	46.8
Stages (A/B/C/D) [a]	10 / 11 / 11 / 18
Feed stage	5 (at B)
Pressure, mbar	40
Feed flowrate, kg/h	6000
Mass fraction (C10/C12/C14)	0.05 / 0.71 / 0.24
Feed temperature, °C	30

[a] A = Rectifying section, B = Pre-fractionation section, C = Middle section D = Stripping section.

2.1. Analysis tool

DWC implies a multi-input-multi-output (MIMO) control scheme. According to Koko and Barakat [8] there are seven degree of freedom (DOF) of DWC corresponding to seven MVs namely reflux flowrate (L), vapor boilup flowrate (V), side stream flowrate (S), distillate flowrate (D), bottom flowrate (B), liquid split ratio (Rl) and vapour split ratio (Rv). However, Rl and Rv are not suggested to be the MV because it could cause a serious operation and control problem [9-10]. In addition, perfect level control in reboiler and condenser were assumed. This reduces the DOF to three. The controlled variables (CVs) were x_{C10}, x_{C12} and x_{C14} represent mole fraction of distillate, side and bottom stream, respectively. For a 3×3 configuration there are nine possible pairings.

In order to screen suitable pairing of MV-CV, relative gain array (RGA) was applied. RGA has been widely used among the researcher [9-12] to examine the proposed control system in the distillation column and DWC. In RGA, the best control loop pairing was determined by the steady-state gain (K). For a 3×3 system, the steady-state gain matrix is denoted by:

$$K_{ij} = \begin{pmatrix} K_{11} = \left.\frac{\Delta y_1}{\Delta u_1}\right|_{u_{2,3}} & K_{12} = \left.\frac{\Delta y_1}{\Delta u_2}\right|_{u_{1,3}} & K_{13} = \left.\frac{\Delta y_1}{\Delta u_3}\right|_{u_{1,2}} \\ K_{21} = \left.\frac{\Delta y_2}{\Delta u_1}\right|_{u_{2,3}} & K_{22} = \left.\frac{\Delta y_2}{\Delta u_2}\right|_{u_{1,3}} & K_{23} = \left.\frac{\Delta y_2}{\Delta u_3}\right|_{u_{1,2}} \\ K_{31} = \left.\frac{\Delta y_3}{\Delta u_1}\right|_{u_{2,3}} & K_{32} = \left.\frac{\Delta y_3}{\Delta u_2}\right|_{u_{1,3}} & K_{33} = \left.\frac{\Delta y_3}{\Delta u_3}\right|_{u_{1,2}} \end{pmatrix} \quad (1)$$

where y indicates the output and u indicates the input. K_{11} in the matrix indicates the changes on y_1 when u_1 is altered while u_2 and u_3 are constant. K_{12} denotes the changes on y_1 when u_2 is altered but u_1 and u_3 are kept constant and so on. From the steady-state matrix, RGA (Λ) can be calculated using the following relationship:

$$\Lambda = K \otimes (K^{-1}) \quad (2)$$

where $\otimes$ denoted as the element by element multiplication. T denotes the transpose of the steady-state gain matrix. If $\Lambda_{ij} = 0$, it means that y_i does not respond to m_j thus m_j could not be used to control y_i. While $\Lambda_{ij} = 1$ means that y_i only responds to m_j and not interact with other manipulated variables. If $0 < \Lambda_{ij} < 1$ or $\Lambda_{ij} > 1$, then an interaction occurs because more than one y_i react to m_j. Last but not least, if $\Lambda_{ij} < 0$, it means that the interaction exists between the related manipulated and controlled variables is in opposite direction and cause instability. The least interaction, where the arrangement which gives a RGA with diagonal element values close to unity will be selected as the best pairing. For RGA, four possible control configuration schemes were studied namely DB/LSV, LB/DSV, DV/LSB and LV/DSB.

Alternatively, singular value analysis (SVA) was also applied. Singular value analysis (SVA) is an alternative to RGA for design of multivariable control systems. Singular values arise from the decomposition of K:

$$K = W\Sigma V^T \quad (3)$$

where $\mathbf{\Sigma}$ is the diagonal matrix of singular values. W and V are unitary matrices. The columns of W are referred to as the input singular vectors whereas the columns of V are the output singular vectors. The final matrix property of interest in the condition number (CN). If K is non-singular, the CN number of K is a positive number defined as the ratio of the largest and smallest nonzero singular values:

$$CN = \frac{\sigma_1}{\sigma_r} \quad (4)$$

If the CN value is small, then the multivariable effects of uncertainties are not likely to be serious. For SVA, three CVs were manipulated by three out of five MVs with total nine possible pairings. In this work, both RGA and SVA were applied to screen possible MV-CV for the VDWC. RGA was conducted first followed by SVA. The steady state gains for these analysis were obtained through the developed Aspen Plus steady state model. The best possible matching from RGA and SVA analysis will be selected for closed loop performance analysis.

2.2. Dynamic model

In dynamic model, modifications were made to the steady state model. Pressure changes along the column were considered by fitting all four column stages were with Mellapak 350Y packing. Reflux drum and column sump were sized assuming residence time of 5 min and 10 minutes, respectively. Valves, pumps and compressors were added to the dynamic model flowsheet to achieve pressure consistency. Because of the system dependency on pressure, pressure driven mode was selected. The default control loops were reflux drum level control, column sump level control, top column pressure control and bottom column temperature control. Pressure check was made to ensure pressure consistency prior exporting to Aspen Dynamics.

2.3. Closed loop response

The selected controller configuration from the previous step were added to the dynamic model flowsheet. PID controller was adopted in this work. No measurement delay was included. The controller settings were determined using conservative Ziegler-Nichols (ZN) open loop test tuning method. Fastest loop was tuned first and then closed, followed by the second controller tuning while the first controller remains closed. This continues until all control loops were tuned. The controller was tested to several scenarios. First was a regulatory problem in which the process was subjected to feed rate change and change in feed compositions. For the latter three scenarios were considered each with different sets of feed compositions. Second was the servo problem where set point change was introduced to the distillate, side and bottom product composition.

3. Results and discussions

The RGA controllability indices results for all four control configurations is given in Table 2. RGA values with less than 0 were excluded due to caused instability whilst value close to 1 is preferred. The most suitable pairing of MV-CV are D-x_{C10}, S-x_{C12} and V-x_{C14} as their value were close to 1 compared to the other indices. This indicate each MV has a good inner interaction with its correspond CV whilst minimally affect other variables. Table 3 shows the CN results from SVA analysis. Pairing no 7 (D-x_{C10}, S-x_{C12}, V-x_{C14}) and 9 (D-x_{C10}, V-x_{C12}, B-x_{C14}) are having the smallest CN value and therefore are preferred as it shows minimal dependency to other MVs. Pairing 7 is consistent with the 3×3 RGA, but not pairing 9. Therefore, DSV control loop arrangement was selected and the updated controller loop of the VDWC is shown in figure 1(a).

Table 2. RGA steady state controllability indices for 3×3 control problem. The CVs are x_{C10}, x_{C12} and x_{C14}.

Controlled variables, x_i	Manipulated variables, m_j		
	L	S	V
x_{C10}	-3.3070	4.3080	-0.0010
x_{C12}	4.2978	-3.4826	0.1848
x_{C14}	0.0092	0.1746	0.8162
	L	S	B
x_{C10}	-3.0273	4.0208	0.0065
x_{C12}	0.6892	-0.4059	0.7166
x_{C14}	3.3381	-2.6149	0.2768
	D	S	V
x_{C10}	**0.9916**	-0.0949	0.1033
x_{C12}	-0.1153	**1.1151**	0.0002
x_{C14}	0.1237	-0.0202	**0.8965**
	D	S	B
x_{C10}	-13.2389	-3.4140	17.6529
x_{C12}	0.7780	-0.2493	0.4713
x_{C14}	13.4609	4.6632	-17.1242

Table 3. CN values for steady state controllability analysis for 3×3 control problem. The controlled variables are x_{C10}, xC12 and xC14.

Pairing	Controlled variables	Manipulated variables [a]	CN
1	x_{C10}, x_{C12}, x_{C14}	LDS	2765.00
2	x_{C10}, x_{C12}, x_{C14}	LSV	147.84
3	x_{C10}, x_{C12}, x_{C14}	LSB	222.00
4	x_{C10}, x_{C12}, x_{C14}	LDV	139.46
5	x_{C10},x_{C12}, x_{C14}	LDB	18.81
6	x_{C10}, x_{C12}, x_{C14}	LVB	1004.33
7	**x_{C10}, x_{C12}, x_{C14}**	**DSV**	**3.83**
8	x_{C10}, x_{C12}, x_{C14}	DSB	33.03
9	x_{C10}, x_{C12}, x_{C14}	DVB	4.22

[a] In each pairing, the first controlled variables is paired with the first manipulated variable, and so on i.e. in pair 1, x_{C10} is paired with L, x_{C12} is paired with D and x_{C14} is paired with S.

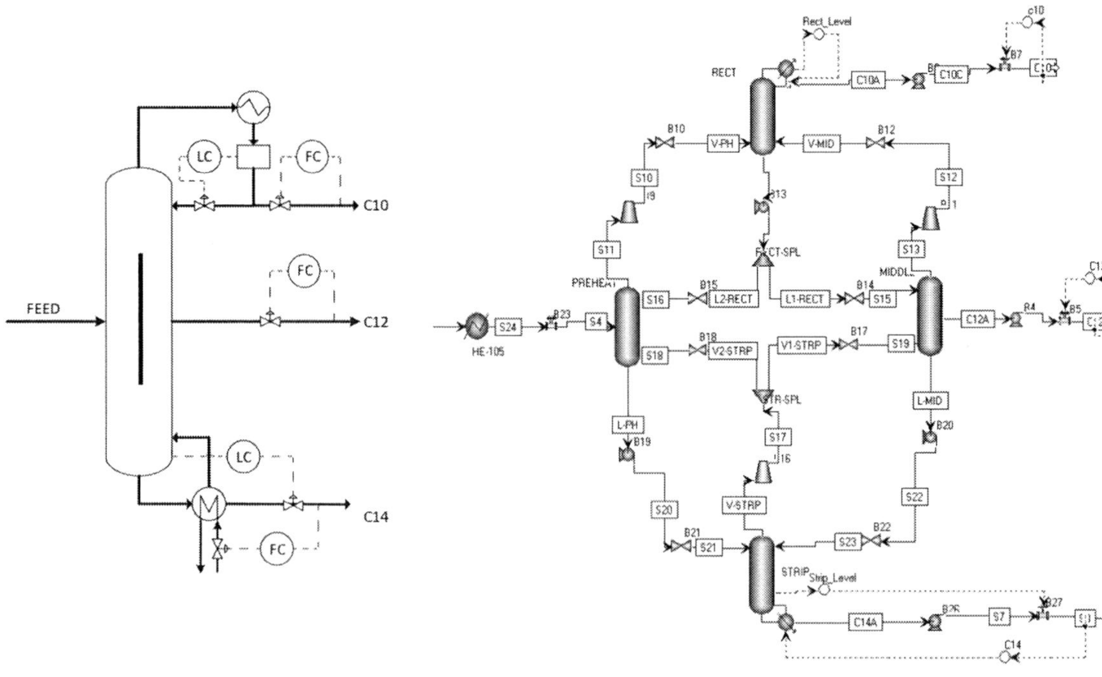

(a) (b)

Figure 1. (a) DSV control configuration (b) Aspen Dynamics VDWC layout for DSV control configuration

The controller loops were added to the dynamic model. Figure 1(b) shows the Aspen Dynamic flowsheet. The dynamic flowsheet layout consists of five control loops corresponding to the DSV configuration as well as level controller for sump and reflux tank. The controller tuning for each controller loop was perform using the Tuning Option. The value is shown in table 4. Time variation of the product composition when subjected to 2% and 4% changes in the feed rate is shown in Figure 2a & 2b, respectively. We see from the simulations that middle composition (C12) provides much better control compared to distillate (C10) and bottom composition (C14). Bottom composition oscillates more than distillate composition. Variation in feed flowrate seems has minimal effect on the middle composition. This is due to the large amount of C12 in the feed stream and minimal effect in the internal flows. Overall, the feed rate change disturbance was able to be rejected.

Figure 2c on the other hand, shows the time profile when the process was subjected to several feed composition variation scenarios namely Scenario 1, 2 and 3. From the results, distillate composition has poor controllability compared to the middle and bottom composition with high overshoot and long settling time. Bottom composition incur some oscillation but has a lower overshoot and faster settling time compare to distillate composition. Since large portion of the feed contain C12, variation of C10 and C12 in the feed affect the internal flows and purity which eventually effect the controllability of the distillate and bottom stream. Middle composition is however unaffected by variation in feed composition with good controllability. Overall, variation of feed stream has minimal effect on middle composition.

Table 4. PID tuning results based on ZN tuning method.

Loop	K	τ_I	τ_D
D	107.4	90.2	8.5
S	111.0	5.6	-
V	4.0	23.1	2.1
Reflux level	1	20	-
Sump level	1	20	-

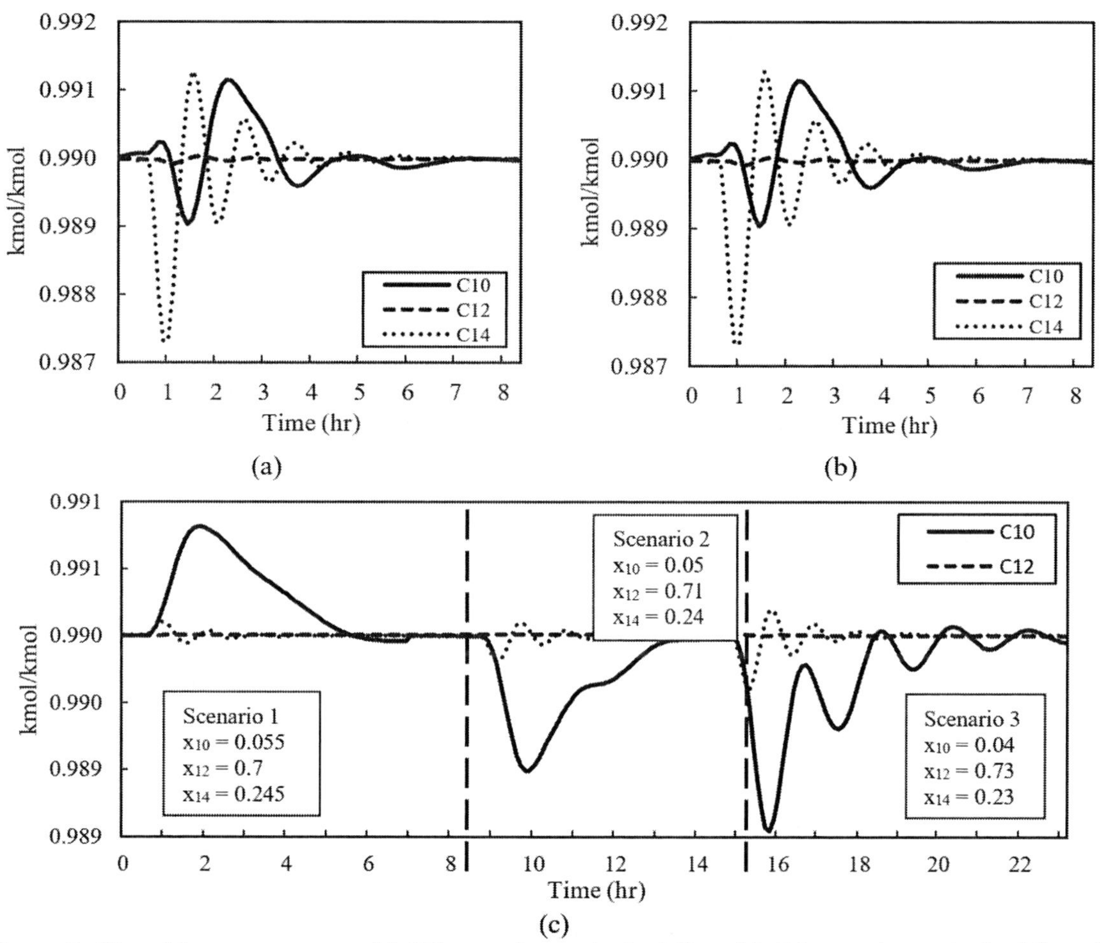

Figure 2. Closed loop response to (a) 2% step change in feed flow (b) 4% step change in feed flow (c) variation in feed composition

Figure 3 shows the result of set point tracking in all product stream with change of 0.001. We can see that distillate and bottom composition control shows some oscillation and settle after almost five hours. Middle composition on the other hand, has very good response with almost no overshoot and fast settling time. This is reasonable and consistent with previous results due to the large amount of C12 in the feed stream and has minimal effect in the internal flows.

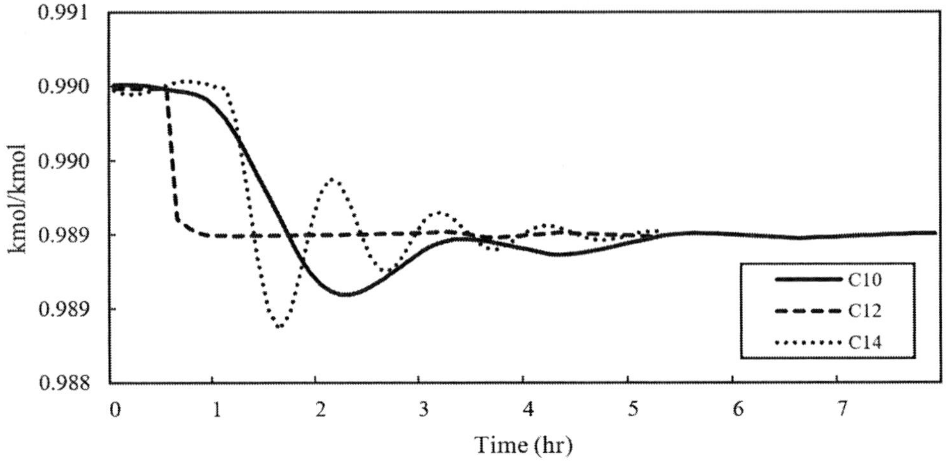

Figure 3. Closed loop response to set point change of 0.001 in distillate, middle and bottom composition.

4. Conclusions

VDWC controllability analysis for fractionating oleochemical fatty acid has been studied in this work. Steady state and dynamic model of the column have been successfully developed using Aspen Plus and Aspen Dynamics. Pairing of 3×3 control configuration was performed using relative gain array (RGA) and singular value analysis (SVA). Both approach resulted in DSV configuration being the most suitable configuration. The configuration was applied to the dynamic model and the controller performance was tested to feed rate and feed composition variations as well as set point change. Performance analysis was found to be successful in rejecting the disturbances as well as obtaining good set point tracking. However, distillate and bottom composition shows poor controllability compare to middle composition mainly due to the composition proportion of the feed.

Acknowledgments

Authors wishing to acknowledge assistance and support by Universiti Malaysia Pahang and Ministry of Education Malaysia through grant number RDU1603148, RDU1803178 and FRGS/1/2014/TK05/UMP/02/1.

References

[1] Jobson M 2005 Dividing wall distillation comes of age *Chemical Engineer* no. **766** pp 30-31

[2] Adrian T, Schoenmakers H and Boll M 2004 Model predictive control of integrated unit operations: Control of a divided wall column *Chemical Engineering and Processing: Process Intensification* **43**(3) 347-55

[3] Dejanović I, Matijašević L and Olujić Ž 2010 Dividing wall column – a breakthrough towards sustainable distilling. *Chemical Engineering and Processing: Process Intensification* **49**(6) 559-80

[4] Yildirim Ö, Kiss A A and Kenig E Y 2011 Dividing wall columns in chemical process industry: a review on current activities *Separation and Purification Technology* **80**(3) 403-17

[5] Othman M R and Rangaiah G P 2019 Performance Analysis of Dividing Wall Column Model for Fractionation of Oleochemical Fatty Acid *Journal of Oil Palm Research* In Press

[6] Kiss A A and Rewagad R R 2011 Energy efficient control of a BTX dividing-wall column *Computers & Chemical Engineering* **35**(12) p2896-2904
[7] Sankaranarayanan D and Deepakkumar G 2014 *Implementing the Concept of Relative Gain Array for the Control of MIMO System: Applied to Distillation Column*. (Coimbatore: Annals University Regional Centre)
[8] Koko I O and Barakat T A 2016 Modelling and Control Analysis of Dividing Wall Distillation Columns *University of Khartoum Engineering Journal* Oct 23 **5**(1)
[9] Mutalib M A and Smith R 1998 Operation and control of dividing wall distillation columns: Part 1: Degrees of freedom and dynamic simulation *Chemical Engineering Research and Design* **76**(3) 308-18
[10] Cho Y, Kim B, Kim D, Han M and Lee M 2009 Operation of divided wall column with vapor sidedraw using profile position control *Journal of Process Control* **19**(6) 932-41
[11] Wang S J and Wong D S 2007 Controllability and energy efficiency of a high-purity divided wall column *Chemical Engineering Science* **62**(4) 1010-25
[12] Strandberg J P 2011 *Optimal operation of dividing wall columns* PhD thesis (Norwegian University of Science and Technology)

IOP Conf. Series: Materials Science and Engineering **702** (2019) 012023 doi:10.1088/1757-899X/702/1/012023

Performance modelling and validation on co-gasification of coal and sawdust pellet in research-scale downdraft reactor

F Z Mansur[1], C K M Faizal[1*], [N] A F A Samad[2], S M Atnaw[3] and S A Sulaiman[4]

[1] Faculty of Engineering Technology, University Malaysia Pahang, 26300 Gambang, Pahang, Malaysia
[2] Faculty of Chemical and Natural Resources Engineering, University Malaysia Pahang, 26300 Gambang, Pahang, Malaysia
[3] College of Electrical and Mechanical Engineering, Addis Ababa Science and Technology University, Ethiopia
[4] Department of Mechanical Engineering, Universiti Teknologi Petronas, 32610 Bandar Seri Iskandar, Perak, Malaysia

*E-mail: mfaizal@ump.edu.my

Abstract. Co-gasification of the fossil fuel with the biomass is considered a very promising clean energy opt-in reduce the greenhouse gas emission. The main objective of this research is to develop a simple and reliable model provided as a preliminary tool to evaluate the performance of the co-gasification of sub-bituminous coal with densified biomass (sawdust pellet, SP). The simulation model using Aspen Plus was validated with the experimental data for minimization of the Gibbs free energy model. Three performance parameter; the calorific value of the syngas (CV_{syngas}), syngas yield (Y_{syngas}) and gasification efficiency (ηGE) were studied along with three different control parameter. The increase of the sawdust pellet blending ratio denoted in a decrease of the CV_{syngas}, Y_{syngas} and ηGE ranged from 3.00-6.00 MJ/Nm^3, 1.20-2.20 Nm^3/kg and 25%-37%, respectively. On the contrary, effect of the gasification temperature at the various blending ratio exhibits an increase for all the performance parameters. In addition, ER_{air} resulted in the decline of the CV_{syngas} from 8.50 to 1.58 MJ/Nm^3 and ηGE from 52 to 15% while vice versa for Y_{syngas}. Furthermore, it is found that the result obtained from the developed model agrees well with the experimental data that have been conducted in replicate.

1. Introduction

Rapidly rising on the issue of depletion of fossil fuels together with the production of greenhouse emissions during the energy production, eventually causing global warming and acid rain [1]. Thus, it has heightened the need to search for promising solution that is renewable, environmentally friendly, sustainable, economically, and lessen the current environmental issues. Recent research has proven that co-utilization of coal and biomass has been a development of sustainable bioenergy network between a renewable and non-renewable resource, especially in the co-gasification to produce syngas and electricity in a sustainable manner [2]. Co-gasification technology aids in reduces potentially the exploitation of a significant amount of conventional coal resources, assists in lower the greenhouse gases (GHG) emissions, but also boost the overall gasification process efficiency [3]. It has been discovered that the co-gasification of these two fuels exhibits synergism reaction that reduces the GHG emission without deprived the energy content of the product gas [4]. Furthermore, biomass characterization and the percentage mixture of the biomass with coal play an essential role as it is directly associated with the fuel gas composition. The co-gasification process between biomass and fossil fuels significantly produce low carbon footprint on the environment and enhance the H_2/CO ratio in the produced syngas, which is essential in liquid fuel synthesis. Another crucial point, it has been discovered that the inorganic matter present in biomass functioning as a catalyst for the coal gasification [5]. Thus, production of the superior gas quality by using coal-biomass blends at different operating condition of temperatures and equivalence ratio of air have gained interest among the researchers [6]. Several numbers of research on co-gasification of the various blending ratio of biomass with the coal have been conducted by with the result indicates that blending coal with biomass eventually enhances the gasification with beyond levels that impossible to be achieved by gasifying these feedstocks alone [7]. Most of the studies were focused on the raw biomass co-gasified with coal; however, the co-gasification on the pre-treated biomass, especially, palletization still lacks. Dafnomilis et al. [8] expressed the opinion that the pre-treated of the biomass in the form of pelletized or otherwise densified resulted in better fuel operability in term of handling, transportation, storage, and feeding compared than raw biomass. Gasification of pellet fuel has widely been applied in the commercial gasification resulted that the syngas composition is much more stable by maintaining the gasification more steady and efficient; the uniform shape and density of the pellet fuels aid in smooth feeding by making less of a biomass bridge and gasification reactions [9]. It has been discovered a number of researches have been found on the co-gasification of biomass pellets/coal conducted with the results promised an efficient production of syngas. Although, the pelletized biomass has been utilized as a co-feed in gasification or combustion system; however, the reason for the improvement in the efficiency of the pelletized case gasification is not apparent [10]. The development of fuel-flexible gasification in the pellet form co-gasified with coal remains a challenge, and the field requires further attention. Hence, this study attempt to simulate co-gasification model of the downdraft fixed bed gasifier with the application of the Aspen Plus® software environment.

Numerous researches have been conducted the modelling of the downdraft gasification on various feedstock either using the agriculture residue or forestry residue. This is due to the application of the software that serves as a suitable alternative to minimize the experimental cost and time [11]. Simulate the gasifier by breaking gasification into the drying zone, the pyrolysis zone, the oxidation zone and the reduction zone as well as considered the heat and mass transfer in the model; this tool is capable of predicting gasification performance effectively. Subsequently, the performance of a gasifier system at the different operating and design parameters which can be validated from the optimal model allows designers to speculate the effects of parameters even without any further experimental data [12]. Keche et al. [13] built the developed model with the different biomass fuels in an atmospheric fixed bed reactor to investigate the syngas composition. A model develops by Gao et al. [14] investigated the production of hydrogen gas from the co-gasification of coal and biomass in the presence of calcium oxide as a sorbent. Co-gasification of the charcoal with empty fruit bunch also being developed by Monir et al. [11] which found out that the highest mole fraction of H_2 and CO occur at 975 °C and 35 bar. Ali et al. [3] develop a simulation model of the rice-husk and coal that indicated the model capable of serving as a reliable benchmark for revamping an existing Egyptian natural gas-based power plant. Kuo and Wu [15] design the co-gasification of the coal with the pre-treated biomass, which is the torrefied woody biomass as the substitutions to the raw woody biomass. The simulation denoted that the utilization of the torrefied woody biomass significantly improves syngas yield. Meanwhile, according to the power generation's view, the co-gasification of coal and torrefied biomass resulted in an optimal input condition in terms of power generation and system efficiency. As far as the authors are aware, there are still limited works on modelling of coal co-gasification with pre-treated biomass, especially densified biomass using ASPEN Plus software.

Therefore, the objective of this study was to investigate the co-gasification of sub-bituminous coal (CL) with sawdust pellet (SP) by modelling and simulate a kinetic free equilibrium model of fixed-bed downdraft gasifier in Aspen Plus. The sawdust pellet (SP) blending ratio, gasification temperature and air equivalence ratio were varied to predict the calorific value of the syngas (CV_{syngas}), syngas yield (Y_{syngas}) and gasification efficiency (ηGE). Furthermore, the results obtained from the experimental measurement was used to verify the simulation results. The results acquired through this study is served for preliminary investigating on gasification performance of the pre-treated biomass co-gasified with coal.

IOP Conf. Series: Materials Science and Engineering **702** (2019) 012023 doi:10.1088/1757-899X/702/1/012023

2. Methodology

2.1. ASPEN Plus simulation model

In modelling the co-gasification process, a kinetic free equilibrium model was developed using ASPEN Plus software (ver. 8.6) by including the major chemical reactions occurring in the gasifier. The co-gasification process was divided into three sub-systems to form a downdraft gasifier system. The drying sub-system is to minimize the moisture content of the feed before being fed into the next reactor. The second sub-system aids in decomposed the feed into volatile components and char. Moreover, a FORTRAN statement was included to specify the yield distribution for each conventional component. Next, the RGibbs sub-system simulated the partial oxidation and gasification process by minimizing Gibbs free energy. In carrying out the modelling, some assumptions were made. The assumptions applied in the model were:

- The process occurs in a steady state with kinetic free and the residence time is not considered.
- Atmospheric pressure was assumed in all sub-system.
- Air was introduced in the RGibbs to enhance the co-gasification process at ambient temperature and pressure.
- The gasification agent mixed homogenous and reacted with feed instantly in the reactor.
- All sulfur produced in the H_2S form; meanwhile no oxide of nitrogen was formed as only NH_3 produced.
- Tars are assumed to be negligible in the syngas.

The gases involved are in compliance with the Peng-Robinson equation of state with Boston-Mathias alpha function (PR-BM) to estimate all physical properties of the conventional and non-conventional components at the multiple phase in the gasification process [16].

2.2. Model development

The simulation model flowsheet used in the developed model is shown in **Error! Reference source not found.**. Firstly, the stream of feed consists the mixture of sawdust pellet and coal, with their blends of a ratio of 0, 25, 50, 75, 100% w/w, respectively were fed into the system. Sawdust pellet blending ratio was defined as the mass ratio of sawdust pellet to the total of biomass and coal, therefore, 0, 100% of sawdust pellet blending ratio refers to the pure coal and pure sawdust pellet, respectively. The feed stream was passed through all block with different reaction temperature. The feedstock was specified as a non-conventional component in Aspen Plus was defined by their ultimate and proximate analysis present in Table 1. In additional, Table 2 provides the operation model that was used in this study. Drying process that removes the residual moisture in the feed was simulated in the 'RStoic' block by including the FORTRAN statement in the calculator block to control the drying operation. After drying, the feed was then decomposed into its components constituent (C, H, O, S and N) by specifying yield distribution in the block 'RYield'. In the 'RYield' block, the yield distribution of the feed was specified by FORTRAN statement in a calculator block into its components. The total yield of volatiles was assumed to be equal to the volatile content of the parent fuel by taken into account the proximate analysis of the fuel. The co-gasification process was simulated in 'RGibbs' block by minimizing the Gibbs free energy assumed the complete chemical equilibrium calculations. The gasifying agent, which is air, was introduced into the block where partial oxidation and gasification reactions take placed. Furthermore, the 'RGibbs' block also capabled of calculating the syngas composition as it can generate light gases such as CO, CO_2, H_2, H_2O, N_2, CH_4, and H_2S. The outlet stream of the 'RGibbs' was passed through the 'Sep' block to separate gases from ash according to the specified splits fractions as desired.

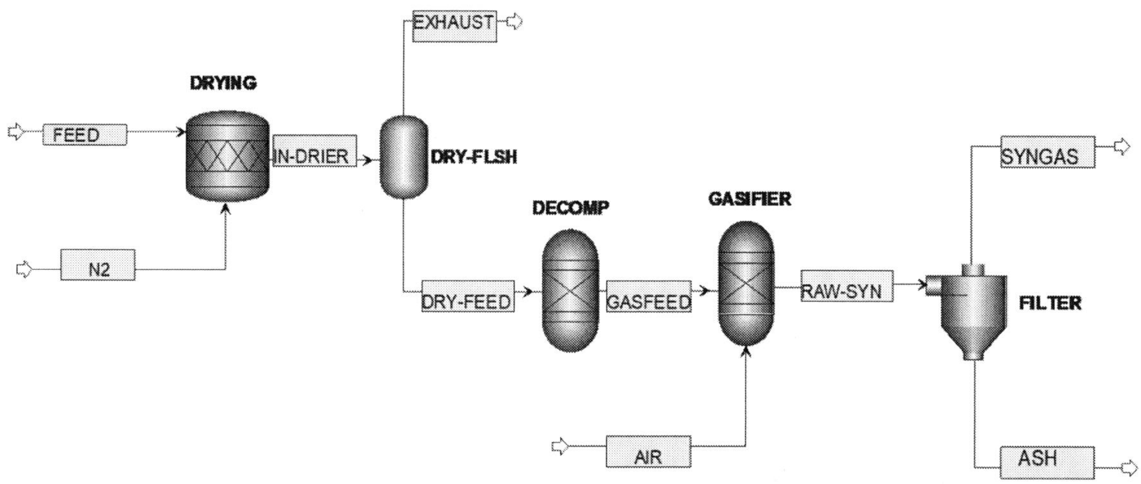

Figure 1. ASPEN Plus simulation model of co-gasification on coal and SP.

Table 1. Proximate, ultimate and calorific value of the coal and SP.

Components	Coal (CL)	Sawdust Pellet (SP)
Proximate analysis (wt %)		
Moisture content	8.18	9.19
Volatile matter	39.79	79.00
Fixed carbon	33.81	10.16
Ash content	18.22	1.65
Ultimate analysis (wt %)		
Carbon (C)	52.58	44.28
Hydrogen (H)	5.90	6.09
Nitrogen (N)	1.49	1.05
Sulphur (S)	1.14	0.28
Oxygen (O)	38.90	48.62
Calorific value (MJ/kg)	20.19 ± 0.082	17.46 ± 0.085

Table 2. List of ASPEN Plus unit operation model.

Aspen Plus Model	Operation	Description	Function
RStoic	Drying	Conversion reactor with known stoichiometry	Reduce the moisture content of the wet feed
RYield	Decomposed	Yield reactor with known products yield	Decomposed non-conventional feed into its element constituents applying FORTRAN statement
RGibbs	Gasifier	Multiphase chemical equilibrium reactor (non-stoichiometry)	Models gas-phase chemical equilibrium and aids in calculating the syngas

Aspen Plus Model	Operation	Description	Function
			composition by minimizing Gibbs free energy
Sep	Separator	Split a stream into two stream or more by specifying split fractions	Separates gas from ash

2.3. Model verification
In order to demonstrate the validity of the proposed model, the experimental measurement of co-gasification of coal with SP displayed in Figure 2 was carried out in a lab-scale electrical downdraft gasifier, as shown in Figure 3.

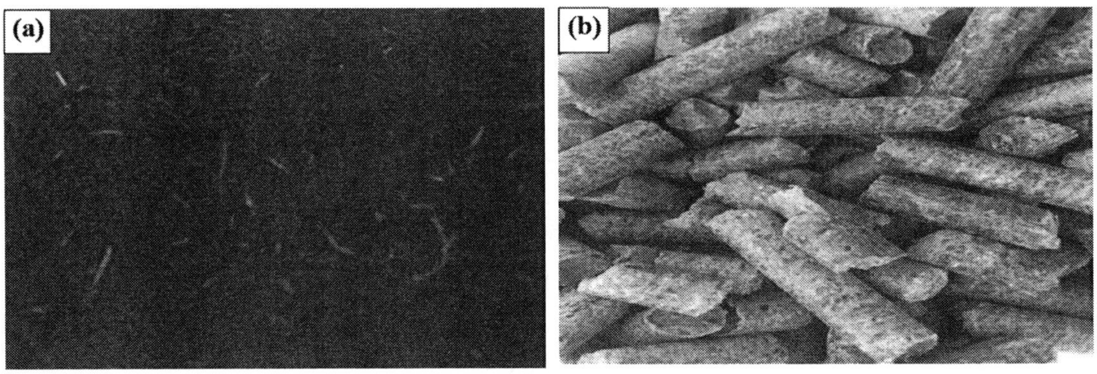

Figure 2. Image of coal (a) and SP (b) that have been used in this study.

The system was custom-fabricated and located in the Biomass laboratory under the Department of Mechanical Engineering of Universiti Teknologi Petronas, Malaysia. The reactor is a cylindrical tube made up of stainless steel class SS316 consist of an internal diameter of 80 mm and 500 mm long. The gasifier was positioned vertically as a function in a free-fall, gravity-fed reactor. About 100g of the feed was then loaded into the gasifier applying "drop chute method" with the different mixture ratio of SP at 0, 25, 50, 75, 100% w/w. The electrical downdraft gasifier was heated with a WATLOW 240 V, 1300 W ceramic-embedded radiant tube heater with a maximum heating temperature of 1000 °C on continuous duty. Furthermore, the operating temperature varying from 650 °C to 850 °C was measured by an external PID controller coupled with a K-type thermocouple mounted on the gasifier reactor. Meanwhile, a stainless steel grate was held at the center inside the gasifier acted as the feedstock holder and also as the reactor bed where the thermal conversions took place. Air as an oxidizing agent was distributed in the gasifier by the compressed air through a 5 mm welded connection situated slightly below the top of the gasifier and controlled by a rotameter. The equivalence ratio of air fixed at 0.20, 0.25 and 0.30 was varied using the airflow rate from 2 L/min - 4L/min. The gasifier has two threaded openings at the top and bottom purposely for the feedstock loading and ash removal for cleaning access, respectively. Meanwhile, gases flowed towards the bottom of the reactor, certifying a downdraft fixed bed configuration [17]. The gases were then flowed to a gas conditioning unit before entering the online gas analyzer for gas composition measurement.

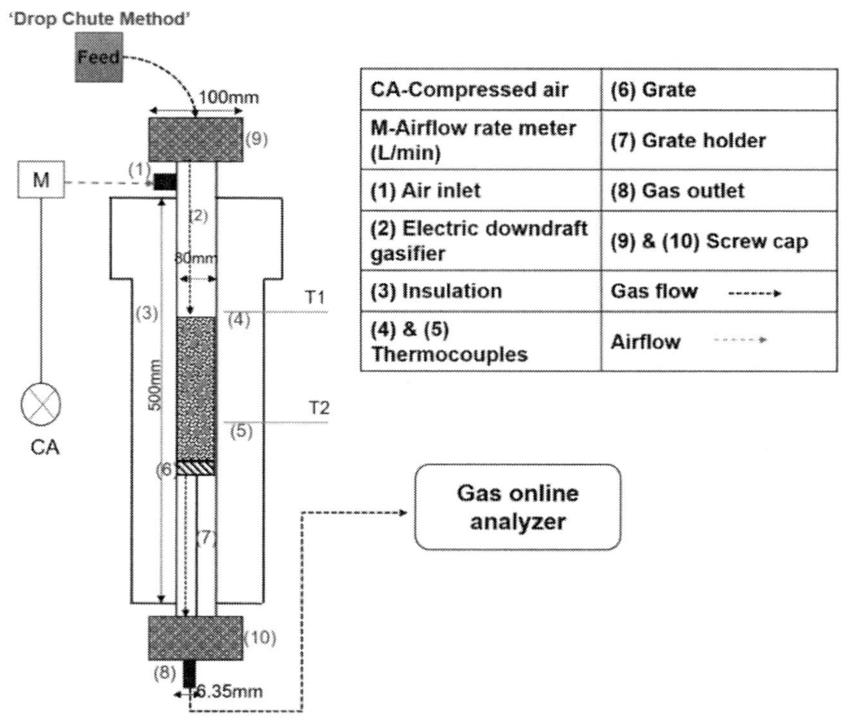

Figure 3. Schematic diagram of laboratory scale electrical downdraft gasifier used for the model validation.

The developed simulation model for co-gasification of CL and SP was used to perform the sensitivity analysis. The effect of the sawdust pellet (SP) blending ratio, gasification temperature and air equivalence ratio (ER) on each syngas composition, the calorific value of the syngas (CV_{syngas}), syngas yield (Y_{syngas}) and gasification efficiency (ŋGE) were investigated. The calorific value of the syngas (CV_{syngas}) was calculated as it is important output parameter that defines the quality of syngas produced from gasification in terms of energy content per unit volume or mass. The calorific value of the syngas (CV_{syngas}) was calculated by taking into account the volume percentage of combustible gas components in the syngas (CO, H_2 and CH_4) produced from the co-gasification experiment with their specific calorific value obtained from the US National Renewable Energy Laboratory (NREL) in the unit of MJ/Nm^3 as per standard value [18]. The equation was expressed in $CVsyngas{=}(V_{CO} \times 12.63) + (V_{CH_4} \times 39.82) + (V_{H_2} \times 12.74))$.

$$CV_{syngas} = (V_{CO} \times 12.63) + (V_{CH_4} \times 39.82) + (V_{H_2} \times 12.74) \quad (1)$$

Where CV_{syngas} is calorific value of the syngas in the unit of MJ/Nm^3 and V is volumetric percentage for each of CO, CH_4 and H_2 obtained from online gas analyzer measurements (%). Meanwhile, the syngas yield (Y_{syngas}) in the unit for each experiment is taken into account the volume of syngas produced per unit mass of feedstock consumed in gasifier by considering the nitrogen balance method that has been proposed and applies by several authors [19]. It was applied by taking into account the continuous exposures of the high temperatures as well as the tar depositions in the measuring equipment cause the inaccuracy of the reading. The calculated value was given in Equation (2).

$$Y_{syngas} = \frac{Q_a \times 79\%}{m_{feed} \times N_2} \quad (2)$$

Where Y_{syngas} described as the syngas yield (Nm^3/kg), Q_a is volume flow rate of air (Nm^3/h), m_{feed} is mass flow rate of the feedstock in the gasifier system (kg/h) and N_2 % is refer to volumetric percentage of N_2 in the dry fuel gas. Furthermore, the gasification efficiency (ηGE) can be calculated either from the cold gas efficiency or hot gas efficiency [20]. It is possible to define the cold gas efficiency as the ratio between the chemical energy leaving the system associated with the cold and tar-free syngas and the chemical energy energy entering the system related to the biomass [21]. Thus, the gasification efficiency was calculated by considering the specific gas production and the energy content of the biomass. The gasification efficiency (ηGE) was calculated using Equation (3)

$$\eta GE = \frac{CV_{syngas} \times Y_{syngas}}{CV_{feed}} x\, 100 \quad (3).$$

Where ηGE is refer to gasification efficiency (%); CV_{syngas} and CV_{feed} is the calorific value of the syngas and feed respectively in the unit of MJ/kg. Meanwhile Y_{syngas} described as syngas yield (Nm^3/kg). The comparison on the predicted data from the simulation model with experimental for the co-gasification of CL and SP were discussed on the gasification performance in term of CV_{syngas}, Y_{syngas} and ηGE. In additional, the root mean square error (RMSE) was calculated using Equation. (4) for each gasification performance at different gasification conditions to measure the error between simulation and experimental.

$$RMSE = \sqrt{\frac{\sum_{i=1}^{n}(P_i - O_i)^2}{n}} \quad (4).$$

Where P, O and n refer to predicted value, observed value and number of dataset respectively.

3. Results and Discussions

A sensitivity analysis was performed in the GASIFIER block by varying the gasification temperature and airflow parameter at different sawdust pellet blending ratio. This was to investigate the effect of sawdust pellet blending ratio, gasification temperature and air equivalence ratio on the gasification performance. In ASPEN Plus, sensitivity tool was used to determine the gasification performance on the varying input parameter. Table 3 shows the input parameters of the temperature and airflow applied in the model.

Table 3. Range of input parameter for operational model

Variable	Type	Block/stream	Variable	Unit	Limits	Increment
Temperature	Block-var	GASIFIER	TEMP	°C	600-1000	50
Air flow	Stream-var	AIR	MASS-FLOW	kg/hr	0.1-0.4	0.025

3.1. Effect of sawdust pellet blending ratio

Figures 4 (a), 4 (b) and 4 (c) present the effect of the sawdust pellet at blending ratio of 0, 25, 50, 75, 100% w/w on the CV_{syngas}, Y_{syngas} and ηGE, respectively. The gasification temperature and the ER_{air} was fixed at 750 °C and 0.25. It can be seen that all the gasification performance increase with the increasing of the sawdust pellet blending ratio from 0 to 50%. The range of the CV_{syngas}, Y_{syngas} and ηGE calculated were 3.00 - 6.00 MJ/Nm^3, 1.00-2.00 Nm^3/kg and 25% - 37%, respectively. It had been discovered that the maximum value of the CV_{syngas} at 5.78 MJ/Nm^3 . Y_{syngas} at 2.00 Nm^3/kg and ηGE at 37% were obtained from the simulation result that occur at 50% of the sawdust pellet blending ratio. However, as the amount of the sawdust pellet blending ratio increase to 75%, all of the gasification performance was dropped down averagely 30%. A similar trend had been found out by Seo et al. [22] that denoted the increasing of the Y_{syngas} at all temperature together with an increase in biomass ratio is due to the transfer of hydrogen radicals in biomass to coal that resulted in higher decomposition of coal. It is noted that the suggested minimum blend 40% pine chips to 60% Sabero refuse coal by Pan et al. [23] with the value at 1.78 Nm^3/kg were quite similar to those for the highest yield of the syngas obtained from this study when assessing influence of the biomass blending ratio on the Y_{syngas}. In term of the RMSE value, both of the CV_{syngas} and Y_{syngas} are in the range of the 0 to 1.60 that is relatively low and generally well satisfactory with the experimental result. Hence, this denoted that the purpose model was validated and reliable. In contrast, the RMSE value of the ηGE was in the range of the 0 to 21, in which the highest was recorded at 50% of the sawdust pellet blending ratio. This deviation might be due to the equilibrium condition that was applying in the gasification model that eliminated insignificant reaction between the coal and biomass [24].

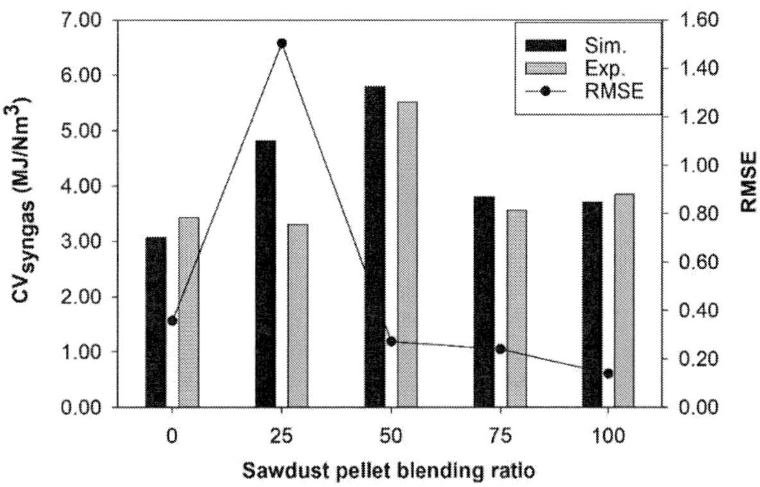

(a)

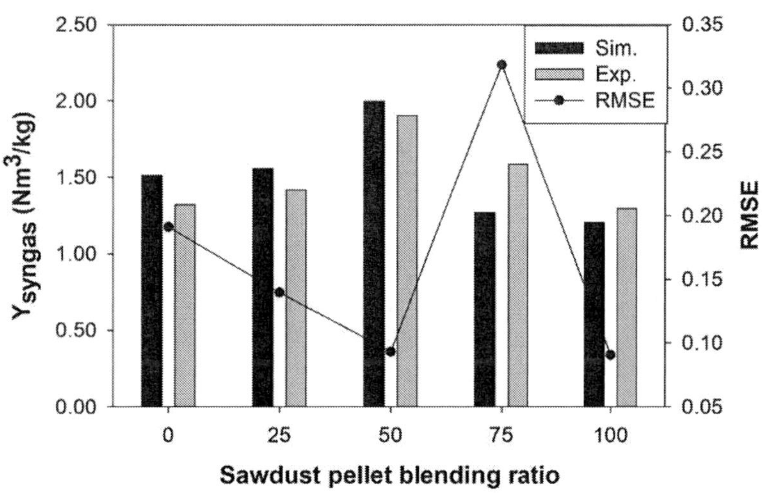

(b)

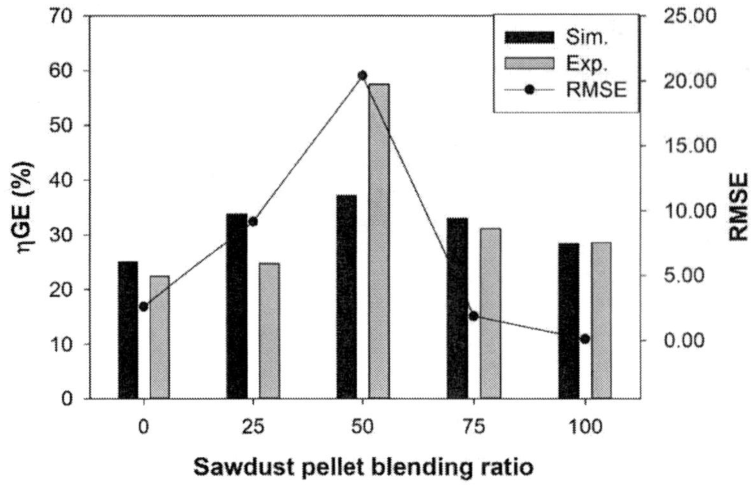

(c)

Figure 4. Effect of the sawdust pellet blending ratio with calculated RMSE value on (a) CV_{syngas} (b) Y_{syngas} and (c) ηGE at gasification temperature and E_{air} fixed at 750 °C and 0.25.

3.2. Effect of gasification temperature

The influence of the gasification temperature from 650 to 850 °C on various sawdust pellet blending ratio with the ER_{air} fixed at 0.25 on the gasification performance is illustrated in Figures 5 (a), 5 (c) and 5 (e) show the CV_{syngas}, Y_{syngas} and ηGE as a function of gasification temperature, respectively. Meanwhile, the RMSE value for every gasification performance in term of CV_{syngas}, Y_{syngas} and ηGE are illustrated in Figures 5 (b), 5 (d) and 5 (f), respectively. Altogether, it can be seen that the gasification temperature at the various sawdust pellet blending ratio exhibits an increase for all the performance parameters. The range of the CV_{syngas} was from 2.00 to 6.00 MJ/Nm^3, and Y_{syngas} was ranged from 1.00 to 2.00 Nm^3/kg. Furthermore, ηGE was ranged from 18% to 37%. The maximum of each of the gasification performance occurs at 50% of the sawdust pellet blending ratio. It can be concluded that the optimum blending ratio for the sawdust pellet with coal was at 50%. A study conducted by Masnadi et al. [25] assesses that the increased of the CV_{syngas} is associated with higher gasification temperature resulted in the endothermic gasification reactions [26]. Complementary to this higher gasification temperature, more heat losses of the system and eventually improved the gasification process on the syngas production. Meanwhile, increasing of the gasification temperature enhance the release of gaseous product from the pyrolysis, steam reforming, gasification and cracking reactions inside the gasifier and contribute to the high total amount of Y_{syngas} [27]. These results were also attributed by several researchers that state the influence of temperature on Y_{syngas} in co-gasification [28]. Considering rising of the gasification temperature improved the endothermic char reactions in the gasifier, it can be concluded that the increase of the Y_{syngas} can be expected due to the increasing concentration of gaseous product [29]. It can be seen that the increase of ηGE as the gasification temperature increases is mainly due to the rise in CV_{syngas}. As previously mentioned, the lower RMSE value indicates the least error between simulation and experimental. It can be seen from the RMSE value for each the gasification performance parameter against the gasification temperature is lower than 20. Consequently, the model is suitable to serve as a preliminary for the co-gasification of coal with pellets.

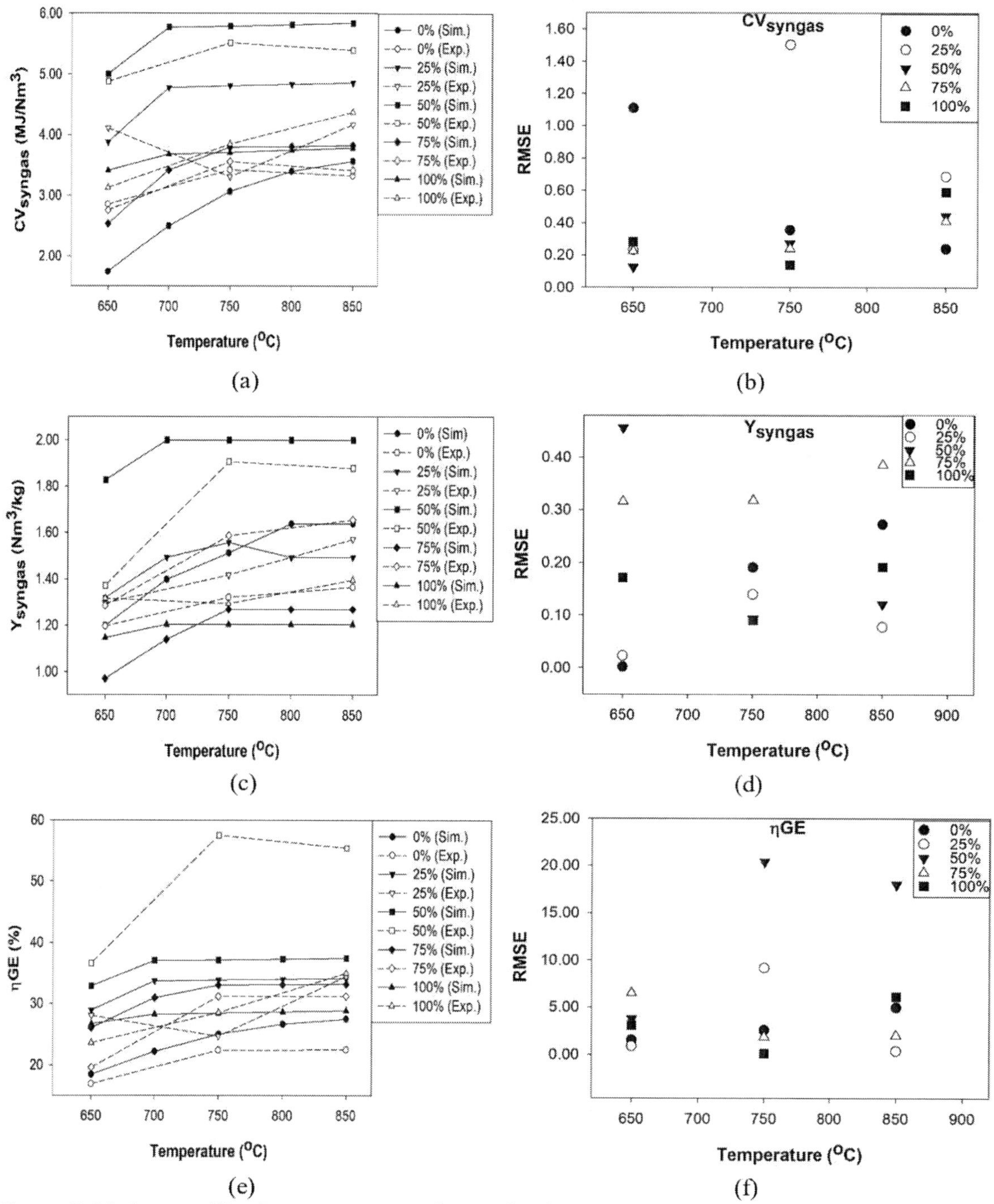

Figure 5. Various gasification temperature of ER_{air} fixed at 0.25 on (a) CV_{syngas} (c) Y_{syngas} and (e) ηGE with the respectively calculated RMSE (b), (d) and (f) on the different SP blending ratio.

3.3. Effect of air equivalence ratio (ER_{air})

CV_{syngas}, Y_{syngas} and ηGE plot are presented in Figures 6 (a), (c) and (e), respectively for co-gasification of sawdust pellet at various blending ratio testing at gasification temperature fixed at 750 C. On the other hand, Figures 6 (b), (d) and (f) exhibited the calculated RMSEE value for each CV_{syngas}, Y_{syngas} and ηGE respectively. It can be seen that the CV_{syngas} and ηGE gradually decreased as the ER_{air} increased to 0.4. The CV_{syngas} and ηGE were ranged from 1.6 to 8.4 MJ/Nm^3 and 14% to 51%, respectively. Both of

IOP Conf. Series: Materials Science and Engineering **702** (2019) 012023 doi:10.1088/1757-899X/702/1/012023

the highest of the CV_{syngas} were recorded at pure sawdust pellet achieved at 8.369 MJ/Nm^3 and 51%, respectively. This can be predicted as the nature of the pellet form enhance the energy density per unit volume, uniformity and defined structure of fuels thus possess higher stability without depending on the critical variation of time [9]. In additional for others feed, increasing the ER_{air} contribute to the higher airflow rate resulted to the lower heating values for syngas and significantly reduced the gasification process efficiency [7,30]. This is believed to occur due to the ER_{air} is related to the airflow rate, therefore increasing of the airflow rate resulted in the shorter residence time of the feed to undergoes reactions (Basu, 2010; Yan *et al.*, 2018). Inversely, increasing the ER_{air} value, the value of the Y_{sngas} also increased. The range of the Y_{syngas} at different sawdust pellet blending ratio is Upadhyay et al. [33] stated that the total Y_{syngas} is mainly associated with the fuel and air consumption rates. The study conducted on the co-gasification of lignite and sawdust briquette was found that the higher gas yield was reached at 2.99 Nm^3 /kg that obtained at the high ER_{air} of 0.386. Commonly, it has been stated that for effective downdraft gasification, the ER is between 0.2-0.4 (Basu, 2010). For the RMSE value, as previously described, the RMSE value both for the CV_{syngas} and Y_{syngas} were lower ranged from 0 to 2.3. The RMSE value for ηGE was ranged from 0 to 32 that is quite high, which calculated at 50% of the sawdust pellet blending ratio. This might be due to the kinetic reaction that takes place in the gasifier during the experimental measurement.

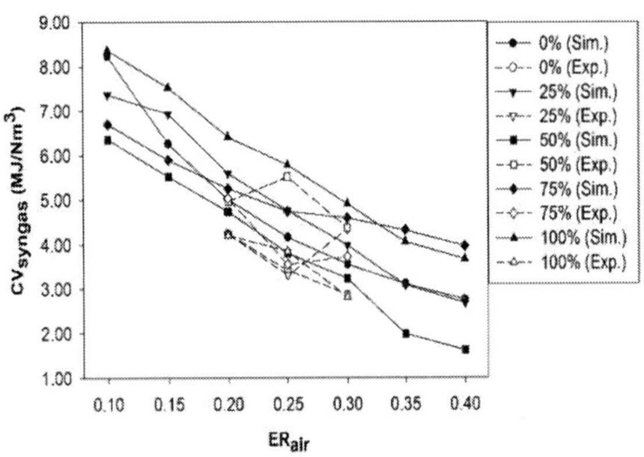

(a)

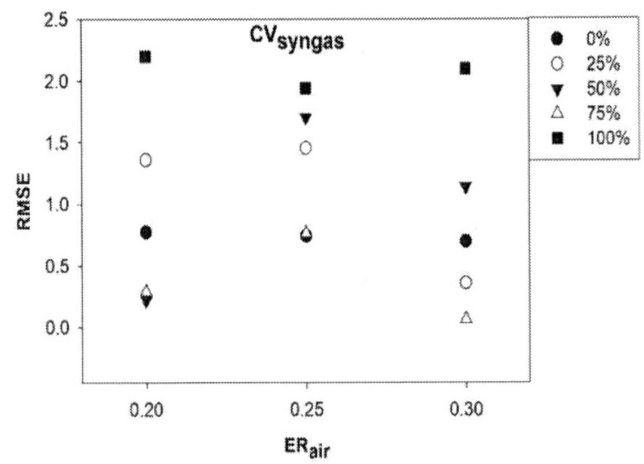

(b)

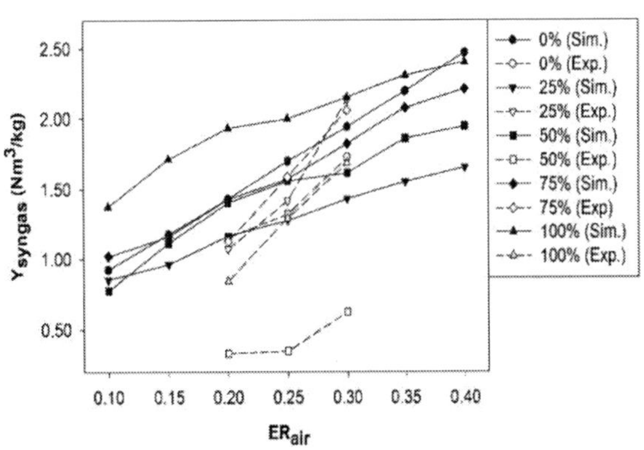

(c)

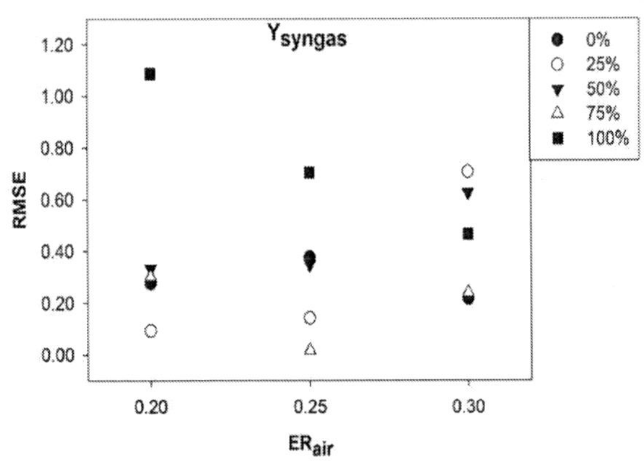

(d)

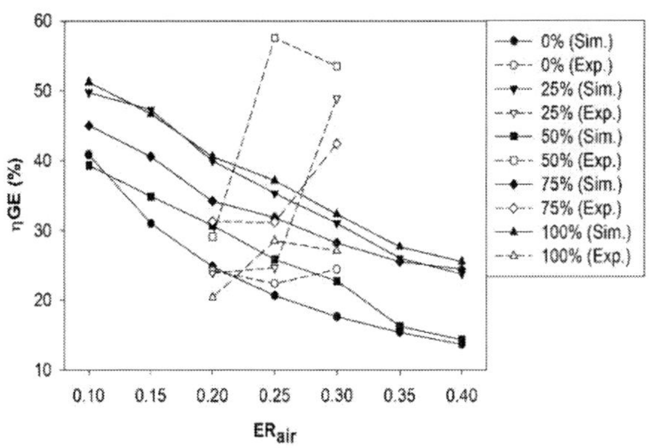

(e)

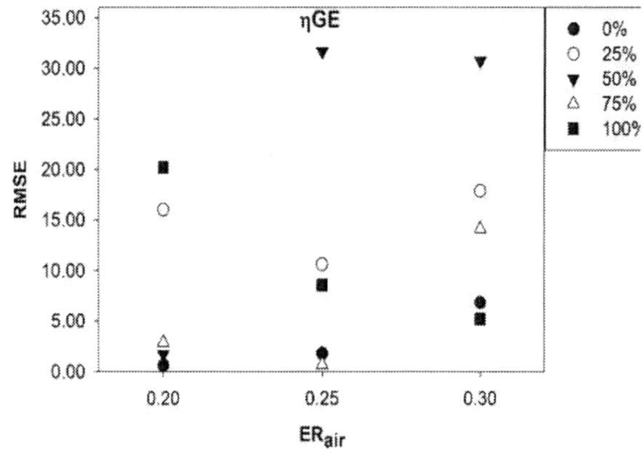

(f)

Figure 6. Influence of ERair at the various sawdust pellet blending ratio from 0, 25, 50, 75, 100% w/w at gasification temperature 750 °C on the (a) CV_{syngas} (c) Y_{syngas} and (e) ηGE with the respectively calculated RMSE (b), (d) and (f).

4. Conclusions

Simulation modelling on the co-gasification of the coal and sawdust pellet was developed using the Aspen Plus software. The effect of the sawdust pellet blending ratio, gasification temperature and E_{air} on the gasification performance are investigated. The result shows that 50% of the sawdust pellet blending ratio in the co-gasification possess the maximum of the CV_{syngas}, Y_{syngas}, ηGE at 5.84 MJ/Nm^3, 2.00 Nm^3/kg and 37%, respectively. Increasing of the gasification temperature are parallel increasing the gasification performance of the co-gasification. In additional, the sensitivity results indicate that the higher ER_{air} contribute to the lower value of the CV_{syngas} at 1.58 MJ/Nm^3 and ηGE at 14.52% that occur at 50% of the sawdust pellet blending ratio. Meanwhile, for the RMSE value; CV_{syngas} and Y_{syngas} shows relatively low value calculated at 0-2 and 0-1.5 indicated that the proposed model could be adopted to measure the gasification performance. In contrast, RMSE value on the ηGE is calculated at 5-32 due to the equilibrium state assume in the model.

Acknowledgement

This project was funded by the Ministry of Higher Education Malaysia and Universiti Malaysia Pahang through the FRGS/1/2014/TK06/UMP/02/6 and PGRS190394, respectively. The authors would also like thanks Universiti Teknologi Petronas for providing full facilities and assistance during the research work.

References

[1] Alauddin Z A B Z, Lahijani P, Mohammadi M, Rahman A and Mohamed A R 2010 Gasification of lignocellulosic biomass in fluidized beds for renewable energy development: A review *Renew. Sustain. Energy. Rev.* **14** 2852-62

[2] Paiman M E S, Hamzah N S, Idris S S, Rahman N A and Ismail K 2018 Synergistic effect of co-utilization of coal and biomass char: An overview. *IOP Conf. Ser. Mater. Sci. Eng.* **358**

[3] Ali D A, Gadalla M A, Abdelaziz O Y and Ashour F H 2016 Modelling of coal-biomass blends gasification and power plant revamp alternatives in Egypt's natural gas sector. *Chem. Eng. Trans.* **52** 49-54

[4] Ciferno J P and Marano J J 2002 Benchmarking Biomass Gasification Technologies for Fuels, Chemicals and Hydrogen Production *National Energy Technology Laboratory* (U.S Department of Energy)

[5] Satyam N V, Aghalayam P and Jayanti S 2016 Synergetic and inhibition effects in carbon dioxide gasification of blends of coals and biomass fuels of Indian origin. *Bioresour. Technol.* **209** 157-

65
[6] Sarker S 2016 Thermochemical Gasification of Local Lignocellulosic Biomass via Fixed-Bed and Fluidized-Bed Reactors (University of Agder)
[7] Valdés C F, Chejne F, Marrugo G, Macias R J, Gómez C A, Montoya J I et al 2016 Co-gasification of sub-bituminous coal with palm kernel shell in fluidized bed coupled to a ceramic industry process. *Appl. Therm. Eng.* **107** 1201-9
[8] Dafnomilis I, Lodewijks G, Junginger M and Schott D L 2018 Evaluation of wood pellet handling in import terminals *Biomass and Bioenergy* **117** 10-23
[9] Yoon S J, Son Y Il, Kim Y K and Lee J G 2012 Gasification and power generation characteristics of rice husk and rice husk pellet using a downdraft fixed-bed gasifier *Renew. Energy* **42** 163-7
[10] Pradhan P, Mahajani S M and Arora A 2018 Production and utilization of fuel pellets from biomass: A review *Fuel Process Technol.* **181** 215-32.
[11] Monir M U, Abd Aziz A, Kristanti R A and Yousuf A 2018 Co-gasification of empty fruit bunch in a downdraft reactor: A pilot scale approach *Bioresour. Technol. Reports* **1** 39-49
[12] Ismail T M and El-Salam M A. 2017 Parametric studies on biomass gasification process on updraft gasifier high temperature air gasification *Appl. Therm. Eng.* **112** 1460-73.
[13] Keche A J, Gaddale A P R and Tated R G 2015 Simulation of biomass gasification in downdraft gasifier for different biomass fuels using ASPEN PLUS *Clean Technol. Environ. Policy.* **17** 465-73
[14] Gao W, Yan L, Tahmoures M and Asgari Safdar A H 2017 Hydrogen production from co-gasification of coal and biomass in the presence of CaO as a sorbent *Chem. Eng. Technol.* 1-13
[15] Kuo P C and Wu W 2015 Design of co-gasification from coal and biomass combined heat and power generation system. *Energy Procedia* **75** 1120-5
[16] Yu M M, Masnadi M S, Grace J R, Bi X T, Lim C J and Li Y 2015 Co-gasification of biosolids with biomass: Thermogravimetric analysis and pilot scale study in a bubbling fluidized bed reactor *Bioresour. Technol.* **175** 51-8
[17] Susastriawan A A P, Saptoadi H and Purnomo 2017 small-scale downdraft gasifiers for biomass gasification: A review *Renew. Sustain. Energy Rev.* 76 989-1003
[18] Basu P. 2010 *Biomass Gasification and Pyrolysis Handbook* Kidlington (Oxford: Academic Press)
[19] Atnaw S M, Sulaiman S A and Yusup S. 2011 Downdraft gasification of oil palm fronds *Trends Appl. Sci. Res.* **6** 1006-18
[20] Kumar A, Sharma A and Bhandari P 2014 *Biomass Gasification and Syngas Utilization* Lijun W Sustain. Bioenergy Prod. (CRC Press)
[21] Inayat M, Sulaiman S A, Kumar A and Guangul F M 2016 Effect of fuel particle size and blending ratio on syngas production and performance of co-gasification *J. Mech. Eng. Sci.* **10** 2188-200
[22] Seo M W, Goo J H, Kim S D, Lee S H and Choi Y C 2010 Gasification characteristics of coal/biomass blend in a dual circulating fluidized bed reactor *Energy and Fuels* **24** 3108-18
[23] Pan Y G, Velo E, Roca X, Manyà J J and Puigjaner L 2000 Fluidized-bed co-gasification of residual biomass/poor coal blends for fuel gas production. *Fuel* **79** 1317-26
[24] Ghassemi H and Shahsavan-markadeh R 2014 Effects of various operational parameters on biomass gasification process: A modified equilibrium model *Energy Convers. Manag.* **79** 18-24.
[25] Masnadi-Shirazi M S 2014 Biomass/fossil fuel co-gasification with and without integrated CO_2 capture The University of British Columbia
[26] Adeyemi I, Janajreh I, Arink T and Ghenai C 2017 Gasification behavior of coal and woody biomass: Validation and parametrical study *Appl. Energy* **185** 1007-18.
[27] Patel B and Narnaware S 2018 Co-gasification of coal / biomass blends and effects of parameter variation on the performance of gasifier : A review. *Int. J. Adv. Res. Innov. Ideas Educ.* 94-100.
[28] Fermoso J, Arias B, Plaza M G, Pevida C, Rubiera F, Pis J J et al. High-pressure co-gasification of coal with biomass and petroleum coke *Fuel Process Technol.* **90** 926-32
[29] Ahmed A M A, Salmiaton A, Choong T S Y and Wan Azlina W A K G 2015 Review of kinetic and equilibrium concepts for biomass tar modeling by using Aspen Plus Renew. *Sustain. Energy Rev.* **52** 1623-44

IOP Conf. Series: Materials Science and Engineering **702** (2019) 012023 doi:10.1088/1757-899X/702/1/012023

[30] Azargohar R, Gerspacher R, Dalai A K and Peng D Y 2015 Co-gasification of petroleum coke with lignite coal using fluidized bed gasifier *Fuel Process Technol.* **134** 310-6
[31] Basu P. 2010 *Biomass Gasification and Pyrolysis: Practical Design and Theory* Kidlington (Oxford: Academic Press)
[32] Yan L, Cao Y, Li X and He B 2018 Characterization of a dual fluidized bed gasifier with blended biomass/coal as feedstock. *Bioresour. Technol.* **254** 97-106
[33] Upadhyay D S, Sakhiya A K, Panchal K, Patel A H and Patel R N 2019 Effect of equivalence ratio on the performance of the downdraft gasifier-An experimental and modelling approach. *Energy* **168** 833-46

IOP Conf. Series: Materials Science and Engineering **702** (2019) 012024 doi:10.1088/1757-899X/702/1/012024

Influenced of Ni loading on SBA-15 synthesized from oil Palm ash silica for syngas production

H A Razak[1], N Abdullah[1], H D Setiabudi[1,2], C S Yee[1,2] and N Ainirazali[1*]

[1]Faculty of Chemical and Process Engineering Technology, College of Engineering Technology, Universiti Malaysia Pahang, 26300 Gambang, Kuantan, Pahang, Malaysia
[2]Centre of Excellence for Advance Research in Fluid Flow, Universiti Malaysia Pahang, 26300 Gambang, Kuantan, Pahang, Malaysia

*E-mail: ainirazali@ump.edu.my

Abstract. The nickel loading (1-3wt.%) effect on SBA-15 prepared via sol-gel method for dry reforming of methane (DRM) was investigated. Silica source of sodium silicate-oil palm ash (Na_2SiO_3-OPA) was obtained from extraction of OPA using mass ratio Na_2SiO_3/P123~2.9. The SBA-15 was synthesized via hydrothermal treatment reflux route at treatment temperature of 80 °C. A vertical stainless steel fixed-bed reactor was applied for catalytic DRM process at 800 °C, atmosphere pressure and ratio CH_4:CO_2 is 1:1 for 30 h. Characteristic of structure SBA-15 was successfully analyzed by TEM, indicates to the positive role of Na_2SiO_3-OPA as silica sources. XRD, BET and FTIR results revealed that 2Ni/SBA-15 exhibited high surface area (460.89 m^2/g) and pore diameter (6.53 nm), small NiO particle (14.71 nm) which assist good metal dispersion and led to better metal-support interaction. The increase of Ni loading from 1 to 3wt.% results in lowering the surface area and pore volume of catalyst as more Ni particles tend to agglomerate and positioning on the SBA-15 pores. The performance of Ni/SBA-15 catalyst in DRM decrease with the trend of 2Ni/SBA-15>3Ni/SBA-15>1Ni/SBA-15 whereby the H_2/CO followed the order of 2Ni/SBA-15≈3Ni/SBA-15>1Ni/SBA-15.These studies revealed the Ni loading influences the metal-support interaction and significantly affect the H_2/CO production.

1. Introduction

Carbon dioxide (CO_2) and methane (CH_4) are the major contributor to the greenhouse gases (GHG_S) which cause the heat entrapment in the atmosphere that led to the global warming [1]. Methane can be transformed into syngas (H_2 and CO) via three reactions which are steam reforming (SRM), dry reforming (DRM) and partial oxidation (POM). In SRM, the process was conducted in a highly endothermic reaction for considerable syngas yields that making it as a major energy consumer. Meanwhile, in POM, methane was converted into syngas with a lower H_2/CO ratio~2 with one step reaction that may affect the reaction control when high space velocities were applied [2]. Thus, DRM was gained great attention since its acknowledged as a dominant route to convert major GHG_S (CO_2 and CH_4) into syngas (H_2 and CO) compared to SRM and POM [3]. Syngas production with ratio (H_2/CO) 1.0 has been recognized in petrochemical industries as its can directly been used in production of liquid hydrocarbon via Fisher-Tropsch process [4]. Unfortunately, commercialize of DRM facing a major obstacle on the reactor blockage causes by continuous carbon formation in the system which disturbing the reaction process. Thus, researchers struggled in developing a reforming catalyst with anti-coke, anti-sintering and long life-time for better DRM process in future.

Even though noble metal based (Pd, Pt, Rh, Ru) catalysts exhibited high catalytic activity and good resistance to carbon deposition for DRM, metallic nickel (Ni) seems more attracted to the researchers due to its readily available and inexpensive compared to the noble metals [5]. Unfortunately, Ni based catalysts susceptible to fast deactivated due to the coke deposition on the active site and sintering at high temperature. Maria et al. [6] has been claimed that the selection of proper support and synthesis of method metal loading might affect the metal dispersion onto the framework that led to the high catalytic performance for DRM. It is acknowledged that ordered mesoporous materials act as a heterogenous catalyst due to its structural features as a support. On the other hand, the mesoporous materials represent high surface area and large pores that greatly contribute the dispersion of active sites and to improving mass transfer [7,8]. Among the mesoporous materials, catalysts such as MCM-41, MCM-48, MSN, Mesoporous Silica Barbara Amorphous (SBA-15) has been considered owing to their properties that contained high surface area (600-1000 m^2 g^{-1}), tunable pore dimeter (5-30 nm) and thicker pore walls (3-6 nm) [9–11]. As a matter of fact, SBA-15 possess highly ordered two-dimensional hexagonal structure with narrow pore size that be able to restrain the mobility species of metal sintering [12].

Generally, there are two common types of silica precursor that had been used for synthesis SBA-15 including tetraethyl orthosilicate ($SiC_8H_{20}O_4$, TEOS) and sodium silicate (Na_2SiO_3) [13]. Unfortunately, to prepare large amount of SBA-15 using commercial silica precursor required high cost and harmful to human kind. Thorough studies were investigated to extract silica from agricultural waste such as corn cob ash (CCA), rice husk (RH), sugarcane bagasse ash (SCBA), and oil palm ash (OPA) [14–17]. OPA was discovered as the most promising waste materials to generate high purity of silica (>60%) that normally dumped in open field without any profitable return [18]. The overall production of OPA every year was estimated around 2.6 million in Malaysia and the method of disposal normally will affect to the environment [19]. Thus, oil palm ash (OPA) waste has been used as an alternative silica sources to synthesis SBA-15 support by using reflux method.

Besides of selection suitable catalyst support, the metal loading method and the percentages of metal loading are significantly affecting the performance of the catalyst including metal particle size, metal dispersion, and metal-support interaction. Regarding of the types metal loading method, sol-gel (SG) method has been chosen due to its ability to produce uniform and homogenous dispersion with less amount of metal loading [20]. Thus, the optimum metal loading was determined to interpret the effect on the catalyst characteristic behavior and its performance. Thus, the main focus in these studies to investigate the influence of Ni loading on the SBA-15 synthesized from OPA catalysts prepared via SG methods towards DRM reaction.

2. Materials and Method

2.1 Synthesized of Na_2SiO_3-OPA

The oil palm ash (OPA) waste was supplied by Felda Lepar Hilir Palm Oil Mill, Gambang, Pahang. In order to remove unspent coarse and biomass residue, OPA was calcined in an electric furnace at 600 °C for 6 h. OPA was then pre-treated with 2M HCl at 80 °C for 3 h to eliminate the acid-soluble impurities and contaminants present. The resultant solution of HCl-OPA was filtered and washed several times with deionized water. The obtained solid residue was allowed to dry in an oven at 110 °C for 12 h followed by calcination at 600 °C for 3 h and was abbreviated as AOPA.

The reflux method was chosen to extract silica (SiO_2) from AOPA by mixing with 2.5 N NaOH under constant stirring at 120 °C for 3 h. The resultant suspension of sodium silicate (Na_2SiO_3) was cooled at room temperature followed by filtration and the solution obtained was assigned as Na_2SiO_3-AOPA. The amount of SiO_2 content was detected using Inductive coupled plasma-optical emission spectrometry (ICP-OES, APHA 3010) before being used for synthesis of SBA-15 support.

2.2 Catalysts preparation

The synthesis of SBA-15 was carried out using hydrothermal treatment reflux which has been declared by syahida et al. [21]. Briefly, 4.9 g of triblock poly (ethylene glycol)–block-poly (propylene glycol)–block-poly (ethylene glycol) ($EO_{20}PO_{70}EO_{20}$, Pluronic P123, molecular weight = 5800, Aldrich) was dissolved in an aqueous solution of 2M hydrochloric acid (HCl, 37 wt.%) under constant stirring at 40

°C for 1 h. The Na_2SiO_3-OPA was added dropwise, pH~1 after the homogenous solution was obtained. The solution was then vigorously stirring for 24 h to form white sediment before being transferred into a hydrothermal reflux at 80 °C for 4 h. The resulting white sediment was cooled at room temperature followed by filtration, washed with D.I water, pH~7. The obtained solid product was allowed to air-dried overnight at 110 °C followed by calcination using muffle furnace at 550 °C for 3 h.

By referring the preparation technique metal loading for sol-gel method as mentioned by Xiaohua [22], an appropriate amount of Ni salt precursor, ($Ni(NO_3)_2.6H_2O$) was added slowly in SBA-15. Subsequently, 1.9 g citric acid and 1.6 mL ethylene glycol were mixed at room temperature to form a sol before being added into the suspension. The resultant green gel was then heated slowly at 80 °C until all the water fully evaporated. Thereafter, the solid residue was dried in air at 110 °C overnight followed by calcination at a heating rate of 5 °C min^{-1} up to 550 °C for 3 h. Similar procedure was repeated by varying percentages of Ni loading at 1,2, and 3 wt.% and were assigned as 1NS, 2NS, and 3NS, respectively.

2.3 Catalyst characterization
The preparation of Ni-based SBA-15 catalyst was characterized by the following technique. X-ray Fluorescence (XRF, Model S8 Tiger) was conducted to determine the chemical composition of OPA and AOPA meanwhile the amount of SiO_2 was detected using Inductively Coupled Plasma Optical Emission Spectrometry (ICP-OES, APHA 3010). Powder X-ray diffraction (XRD, Philips X' Pert MPD) was employed to analyse the crystalline phase identification, using monochromatic Cu-K∝ radiation (λ) of 1.5405 $\dot{A}$ at 15 mA and 30 kV. The crystallite size (D_{NiO}) was estimated by applying the Scherrer equation:

$$D_{NiO} = \frac{0.9\lambda}{\beta \cos\theta} \tag{1}$$

where λ represents the X-ray wavelength corresponding to the Cu-K∝ radiation (0.15405 nm), β is the full-width at half maximum (FWHM) and θ is the angle of diffraction corresponding to peak broadening. The specific surface area was measured by Brunauer-Emmett-Teller (BET) using AUTOSORB-1 model ASAP-2020 instrument. The pore volume (V_p) and the averages pore diameter (D_p) were obtained from Barrett-Joynet-Halenda (BJH) methods, respectively. Fourier transform infrared spectroscopy (FTIR, Nicolet Avatar 370 DTGS) was carried out using KBr compression method to determine the surface functional group of catalysts. The FTIR spectra were taken in a wavelength between range 400 to 1300 cm^{-1} with 32 scanning times. Thermogravimetric thermal analysis (PL-TGA) was employed for spent catalyst to determine the amount of carbon deposition under a mixture of air (20% O_2, 80% N_2) with heating rate 10 °C min^{-1} up to 900 °C.

2.4 Catalytic activity in the dry reforming of methane
The catalytic performance of the synthesized catalysts was performed using a vertical stainless-steel fixed bed reactor with an inner diameter of 11 mm. Approximately, 0.2 g of catalyst was loaded in the centre of reactor by quartz wool was placed at the both ends and it was subjected by H_2 reduction (H_2= 50 mL min^{-1}) at 700 °C. Then, the reaction was allowed by total flowrate 50 mL min^{-1} (STP) with feed mixture of CH_4 (99.9%) and CO_2 (99.9%) with a molar ratio 1:1 at 800 °C. The product gases were determined via Agilent Gas Chromatography (AGILENT 6890 N) which was equipped with Supelco molecular sieve 13x and Agilent Hayasep DB packed column with thermal conductivity (TCD). The following equation were used to determine the conversion of CO_2 (2), conversion of CH_4 (3), and molar ratio of H_2/CO (4):

$$CO_2\ Conversion,\ X_{CO_2}(\%) = \frac{F_{CO_2(in)} - F_{CO_2(out)}}{F_{CO_2(in)}} \times 100\% \tag{2}$$

$$CH_4\,Conversion, X_{CH_4}(\%) = \frac{F_{CH_4(in)} - F_{CH_4(out)}}{F_{CH_4(in)}} \times 100\% \quad (3)$$

$$Molar\ ratio\ of\ \frac{H_2}{CO} = \frac{F_{H_2}}{F_{CO}} \quad (4)$$

3. Results and Discussion

3.1 OPA and AOPA characterization

The chemical composition of OPA and AOPA was obtained using XRF analysis. The percentage of SiO_2 was the highest compared to other chemical component as observed in Figure 1. It is obviously seen that after HCl treatment, the composition of SiO_2 in OPA was increased from 47.71 % to 85.46 %. This was due to the process of acid treatment was probably related with the removal alkaline residue from the raw OPA. Based on the previous studied, Irfan Khan et al. [23] reported the amount of silica content become higher (55%) after being treated with HCl which was related with the removal of metallic impurities in the OPA material. The XRF analysis has been proven that AOPA can be as alternative low-cost material of SiO_2. Thus, the pre-treatment of OPA become significant before being extracted as Na_2SiO_3-OPA.

The optimum condition for extraction of Na_2SiO_3-OPA via hydrothermal reflux process at 100 °C for 3 h. Inductively Coupled Plasma Optical Emission Spectrometry (ICP-OES) analysis revealed that the amount of SiO_2 content in Na_2SiO_3-OPA was 61,021 ppm.

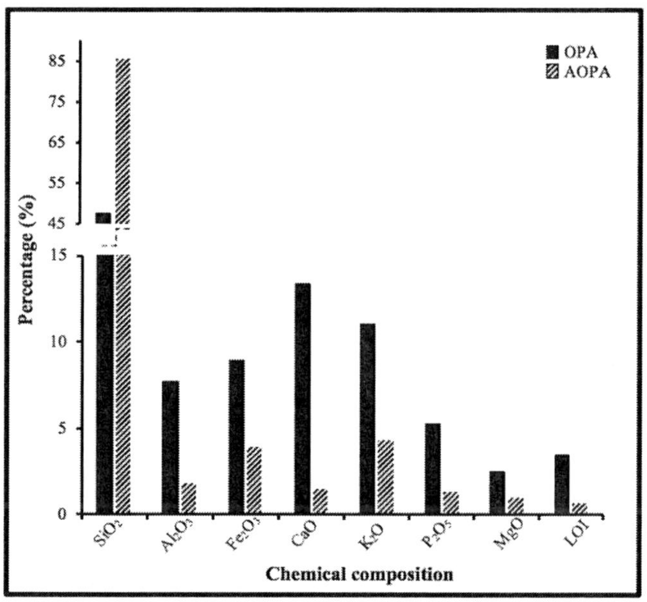

Figure 1. Percentage of chemical composition of OPA and AOPA

3.2 Characterization of catalysts

The wide-angle of XRD in the range 2θ=10°–80° (Figure 2) was analyzed to determine the presence of silica oxide of SBA-15 and metal crystallites of Ni supported on SBA-15 (NS). All diffractograms exhibited similar broad diffraction peak at range 2θ=15°–35° which centered at 22.6°. The observed peak was assigned to the characteristics of the amorphous SiO_2 frameworks of the SBA-15 support [24]. In fact, the peak demonstrated that SBA-15 from waste OPA was successfully synthesized similar as commercial SBA-15. In addition, there are three peaks at 2θ= 38.52°, 45.48°, and 64.93° were observed for NS catalysts, corresponding to the face-centered cubic crystalline nickel oxide presence in the sample. These peaks are become more intense by increasing nickel species (1-3 wt.%) due to the stronger accessibility and incorporation between NiO particles with SBA-15 support [23]. The absence of NiO peaks for 1 wt.% Ni loading probably due to the less existence of Ni active sites that led to the undetected respective signals during the XRD measurement. Nurul et al. [25] declared that there are no peak was observed at 1Ce/SBA-15 catalyst due to the complete incorporation of 1wt.% Ce species into the SBA-15 support. Additionally, it can be obviously seen that the peaks for 2NS was broadened and less intense compared to the 3NS indicating less accumulation of NiO crystallites located onto the SBA-15 support as evidence by TEM analysis [26]. Similar occurrence was also reported by Sidik et al. [27] for Ni/MSN. They found that the peaks intensities appear for 15 wt.% was higher than 10 wt.% due to the large Ni particles that effect the aggregation of Ni species inside the support. Therefore, it can be point out that exceed of metal loading led to less dispersion of metal species due to the agglomeration of Ni crystals.

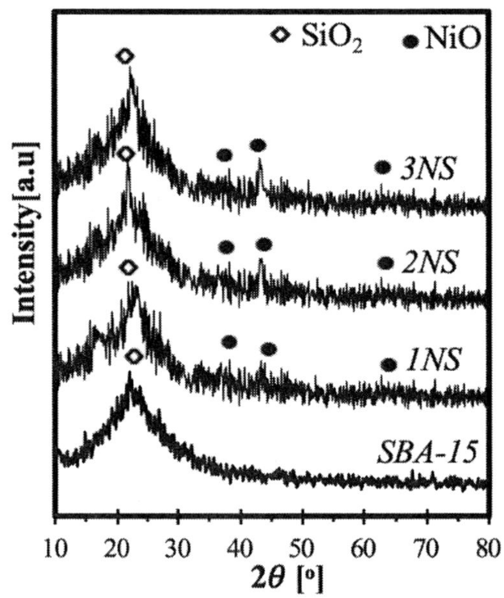

Figure 2. XRD patterns of SBA-15, 1NS, 2NS, and 3NS

The textural and physical properties of the SBA-15 and different percentages of Ni loading were displayed in Table 1. The highest BET surface area (S_{BET}) and pore diameter presented of bare SBA-15 were 524.7 m^2/g and 0.77 cm^3/g. The introduction of Ni loading (1-3 wt.%) led to the decreased of S_{BET} ranging from 490.54 m^2/g to 426.42 m^2/g due to the blockage of the pores with Ni species [28]. The change in the BET surface area was related with the different percentage of Ce species loaded into SBA-15 framework has been studied by Nurul *et al.* [29]. The results proved that the increment of Ce loading was attributed to the incorporation of Ce active sites with SBA-15 support. Additionally, it can be obviously seen the NiO crystallites sizes were 10.09, 14.71, and 16.61 nm for 1, 2, and 3 wt.% of NS that synthesized by using SG method. This result proved that SG method can generated smaller size of Ni that formed strong interaction between metal species and supports due to uniform dispersion on SBA-15 supports follows the order 2NS>3NS>1NS as evidence by XRD analysis. Previous literature by Radlik et al. [30] also was reported that NiO crystallites larger than ~12 nm easier to susceptibility of coke formation that lead to deactivation of catalyst. Apart from this, Hassani et al. [31] reported that high specific surface area was related with the reactants adsorption step which provide more active species that attributed to enhance the catalytic performance. It can be seen that, there is no significant different for the average pore volume and pore diameter which is proven that Ni species favorably dispersed outside the SBA-15 pores.

Table 1. Physical properties of SBA-15, 1NS, 2NS, and 3NS

Catalyst	Nickel content (%)	BET surface area (m^2/g)	Pore volume (cm^3/g)	Pore diameter (nm) [a]	d_{NiO} [b] (nm)
SBA-15	0	524.74	0.77	6.42	-
1NS	1	490.54	0.74	6.48	10.09
2NS	2	460.89	0.72	6.53	14.71
3NS	3	426.42	0.71	6.56	16.61

[a] Obtained from Barret-Joyner-Halenda (BJH) desorption method
[b] NiO crystallite size was calculated from XRD using Scherer equation.

According to Figure 3, TEM analysis for bare SBA-15 synthesized from OPA was conducted in order to investigate the structure and morphology of the catalyst. As it can be observed, SBA-15 (OPA) showed relatively a highly ordered hexagonal with three-dimensional network. Additionally, the apparent of hexagonal array was closely related with the formation of Si-O-Si framework influenced by appropriate amount of Na_2SiO_3-OPA and surfactant. Abdullah et al. [32] has been declared that the amount of silica to surfactant ratio was correlated with the morphology of SBA-15 framework. SBA-15 synthesized from waste OPA has been remarkable to create similar patterns as a typical feature of SBA-15 with the optimum amount of Na_2SiO_3-OPA/surfactant~ 2.9 [10, 32].

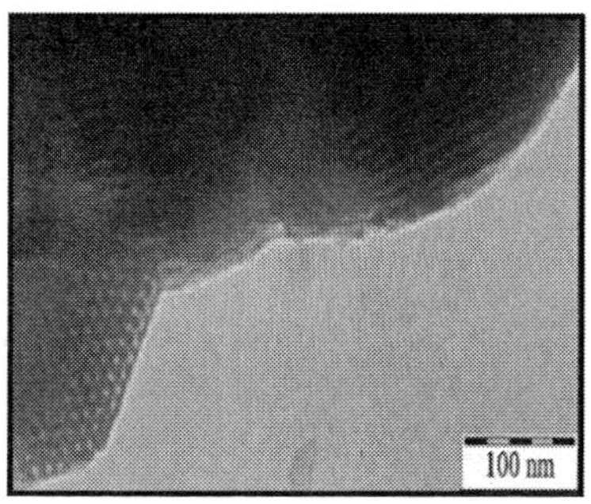

Figure 3. TEM image of SBA-15(OPA)

The distribution of 2NS on the SBA-15 support was illustrated in Figure 4. It can be clearly seen that the structure of SBA-15 was retained after insertion of 2 wt.% of Ni species. The circle dark points attributed to the Ni active sites that were well scattered onto the SBA-15 support. The well-dispersed of 2NS was corresponding with the FTIR analysis which less intense resulting good metal-support interaction of Si-O-Ni. In fact, this observation was parallel with the Ni particle size in an average (14.71 nm) as evidence by XRD analysis.

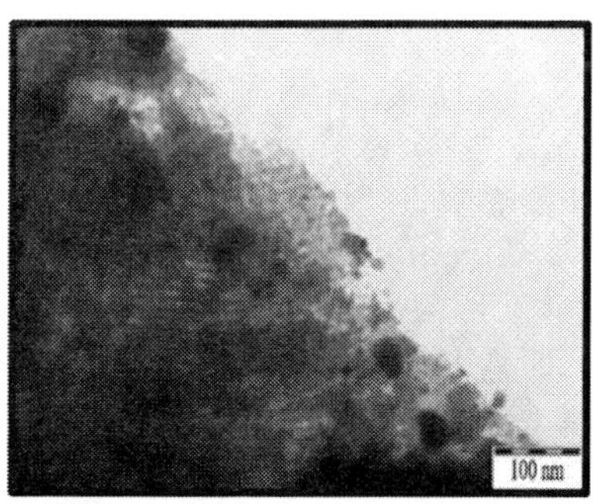

Figure 4. TEM images of 2NS with NiO particle size distribution

The functional groups present on the surface of pure SBA-15 and NS were detected via FTIR spectroscopy analysis in the range between 1300-400 cm^{-1} as shown in Figure 5. The broad absorption band at approximately 1060 and 801 cm^{-1} were corresponded to the asymmetric and symmetric of Si-O-Si stretching vibrations, respectively [5,33]. Moreover, the vibration frequency peaks at 961 and 510 cm^{-1} were attributed to the Si-O stretching vibration of Si-OH and bending vibration of framework Si-O-Si [32,34]. It can be seen that the bands observed at peak 961 cm^{-1} was reduced for all the samples after introduction of Ni into SBA-15 support. The peak of 961 cm^{-1} becomes enveloped in the band 1060 cm^{-1} probably due to the incorporation of metal ions in the silica framework, suggesting the replacement of hydrogen atoms of O-H to O-Ni to produce Si-O-Ni bond [34]. It can be observed that 2NS was less enveloped compared to the 3NS and 1NS suggesting the perfection of the incorporation Si-O-Ni. Based on the previous literature, Tomer et al. [35] revealed that the presence of Ag in SnO_2/SBA-15 was decreased the peak intensity at 961 cm^{-1} that related with the substitution of Si-O-H to form Si-O-M. Additionally, the bond observed at 1060 cm^{-1} for 3NS was weaker at the higher of Ni loading which corresponding to the partial failure of the silica network probably due to the excessive of metal ions in the Si-O-Si formation [36]. Thus, according to the results obtained, it can be concluded that the metal support interaction in the silica framework was following the order of 2NS>3NS>1NS which attribute to the appropriate amount of metal ions with strong Ni support interaction able to reduce carbon formation and prevent catalyst deactivation.

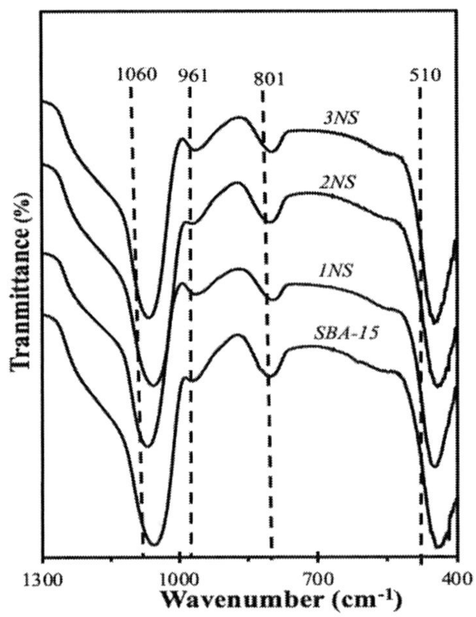

Figure 5. FTIR spectra of KBr in the range of 1300 - 400 cm^{-1} for SBA-15, 1NS, 2NS, and 3NS

3.3 Methane dry reforming reaction

The effect of different metal loading by using SG method on the catalytic performance for CH_4 conversion, CO_2 conversion and H_2/CO ratio were displayed in Figure 6. The presence of 1 wt.% of Ni loading on SBA-15 promotes the conversion of CH_4 and CO_2 which are 84.45% and 73.50%. However, the promotion was limited due to the less amount of active sites for the reaction. The performance of catalytic activity gradually increased with further increasing of Ni content 1-2 wt.%. It can be seen that 2NS exhibited the highest catalyst performance with CH_4 and CO_2 conversion of 90.26% and 93.92% due to the smaller NiO species and strong interaction as evidence by XRD, BET and FTIR. Additionally, the performance of catalytic reaction at different percentage of Ni loading synthesized by SG method followed the order of 2NS>3NS>1NS, meanwhile the average H_2/CO ratio followed the order of 2NS≈3NS>1NS. Similar observation for Ni/MgO, the catalytic performance gradually increased from the range (2-10 wt.%) yet, further increasing to 20 wt.% led to the decreased catalytic activity. This result closely related with the deposition of carbonaceous materials [37]. As suggested by Abdullah et al. [32] the enhancement in the catalytic activity was related with Ni particles that distributed uniformly on the surface SBA-15 which related with the size of Ni species. As observed, no remarkable increase of catalytic performance with further increase in 3Ni wt.% due to the aggregation of metal species in SBA-15 support as evidence by XRD. The lowest catalytic performance for 1NS was probably due to the smallest NiO crystallite sizes that led to the Ni oxidation and accessibility of sintering. Moreover, the influence of higher percentage of Ni content on SBA-15 support is one of the vital factor that lead to carbon deposition and metal sintering has been declared by Yang et al. [38].

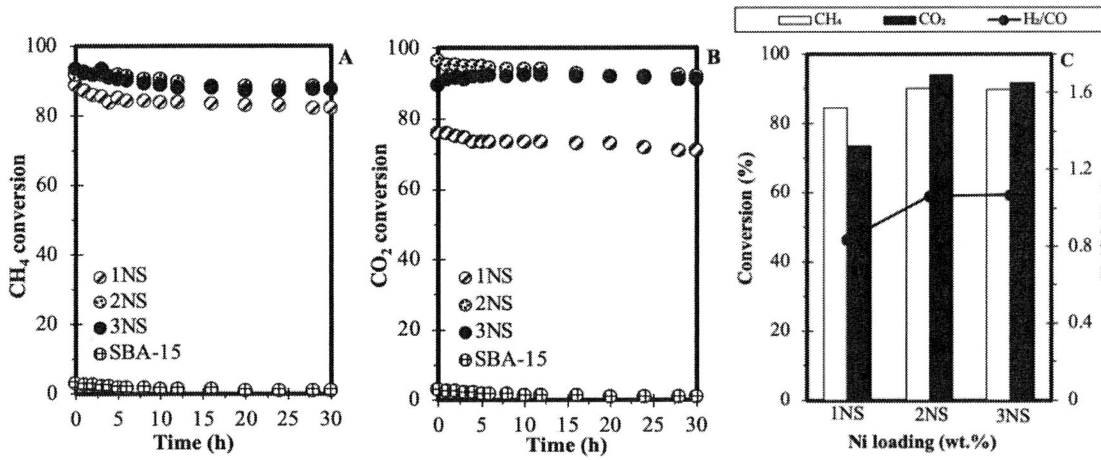

Figure 6. (A) CH_4 conversion, (B) CO_2 conversion, and (C) H_2/CO ratio of 1NS, 2NS, and 3NS. [Reaction condition m_{cat} =0.2 g, reactant flowrate = 25 mL/min, $CH_4:CO_2:N_2$ = 1:1:1, T=800 °C, P= 1 atm, t= 30 hr]

4. Conclusions

According to the present studies, a series of Ni loading with different percentage (1, 2, and 3) wt.% were loaded on SBA-15 catalyst on DRM has been investigated. The different percentage of Ni loading greatly influenced the metal distribution in the catalyst support that affect the performance of the catalytic activity. the overall characterization analyses showed that metal support interaction followed the order of 2NS>3NS>1NS, whereas the catalytic activity of catalysts followed the order of 2NS>3NS>1NS. Based on the results obtained, 2NS possessed the best catalytic activity with CH_4 and CO_2 conversion of 90.3% and 93.9% which is indicate smaller NiO particle species led to the well dispersion and strong metal support interaction as confirmation by TEM, XRD, BET, and FTIR. Thus, it can be concluded that SBA-15 synthesized from waste OPA obviously can generate similar catalytic performance as typical feature of SBA-15 with low amount of Ni loading.

Acknowledgement

The authors acknowledge the financial supports from Research Acculturation Grant Scheme Universiti Malaysia Pahang (RDU170330 & RDU1803184) and Postgraduate Grant Research Scheme (PGRS180387).

References

[1] Takahashi R, Sato S, Sodesawa T and Tomiyama S 2005 *Appl. Catal. A Gen.* **286** 142
[2] Schmal M, Brackmann R, Alves R M B and Giudici R 2019 *Int. J. Hydrogen Energy* **44** 8166
[3] Aramouni N A K, Touma J G, Tarboush B A, Zeaiter J and Ahmad M N 2018 *Renew. Sustain. Energy Rev* **82** 2570
[4] Razali S M, Asikin-mijan N and Sivasangar S 2018 *Int. J. Hydrogen Energy* **32** 186
[5] Omoregbe O, Danh H T, Nguyen-Huy C, Setiabudi H D, Abidin S Z, Truong Q D and Vo D V N 2017 *Int. J. Hydrogen Energy* **42** 11283

[6] Grzybek M, Rønning M and Da P 2018 **4** 312
[7] Karam L and El Hassan N 2018 *J. Environ. Chem. Eng.* **6** 4289
[8] Roblero J G, Pola-Albores F, Valenzuela M A, Rojas-García E, Ríos-Valdovinos E and Valverde-Aguilar G 2019 Int. J. Hydrogen Energy **42** 1854
[9] Taherian Z, Yousefpour M, Tajally M and Khoshandam B 2017 Int. J. Hydrogen Energy **42** 24811
[10] Chong C C, Abdullah N, Bukhari S N, Ainirazali N, Teh L P and Setiabudi H D 2019 *Int. J. Hydrogen Energy* **44** 20815
[11] Setiabudi H D, Chong C C, Abed S M, Teh L P and Chin S Y 2018 *J. Environ. Chem. Eng.* **6** 745
[12] Wang N, Chu W, Zhang T and Zhao X S 2012 *Int. J. Hydrogen Energy* **37** 19
[13] Wang J, Fang L, Cheng F, Duan X and Chen R 2013 *J. Nanomater* 1
[14] Okoronkwo E A, Imoisili P E, Olubayode S A and Olusunle S O O 2016 *Adv. Nanoparticles* **5** 135
[15] Vaibhav V, Vijayalakshmi U, Roopan S M 2015 *Spectrochim. Acta - Part A Mol. Biomol. Spectrosc* **139** 515
[16] Embong R, Shafiq N and Kusbiantoro A 2016 *ARPN J. Eng. Appl. Sci.* **11** 7304
[17] Razak H A, Abdullah N, Gasang J, Setiabudi H D, Yee C S and Ainirazali N 2019 *IOP Conf. Ser. Earth Environ. Sci.* **220** 012.
[18] Ismail M, Lau S K, Majid Z and Ismail M E 2009 *Asia Pacific Struct. Eng. Constr. Conf*
[19] Alsubari B, Shafigh P and Jumaat M Z J 2016 *Clean. Prod.* **137** 982
[20] Yao D, Yang H, Chen H and Williams P T 2018 *Appl. Catal. B Environ* **239** 565
[21] Bukhari S N, Chong C C, Teh L P, Vo D V N, Ainirazali N, Triwahyono S, Jalil A A, H.D. Setiabudi 2018 *Int. J. Hydrogen Energy* **1** 13
[22] Li X, Agarwal V, Liu M and Rees W S 2000 *J. Mater. Res.* **15** 2393
[23] Irfan Khan M, Azizli K, Sufian S, Man Z and Khan A S 2015 *RSC Adv.* **5** 20788
[24] Li D, Zeng L, Li X, Wang X, Ma H and Assabumrungrat S 2015 *J. Gong, Appl. Catal. B Environ.* **176** 532
[25] Ainirazali N, Ainun N, Abghazab N, Setiabudi H D and Yee S Y 2017 *J. of Science and Tech.* **10** 1
[26] Takeishi K and Akaike Y 2016 *Appl. Catal. A Gen.* **510** 20
[27] Sidik S M, Triwahyono S, Jalil A A, Aziz M A A, Fatah N A A and Teh L P 2016 *J. Chem. Eng.* **13** 71.
[28] Setiabudi H D, Lim K H, Ainirazali N and Chin S Y 2017 *J. Mater. Environ. Sci.* **8** 573
[29] Ainirazali N, Setiabudi H D and Dai-Viet N 2017 *J. Malaysian Cat.* **2** 41
[30] Radlik M, Adamowska-Teyssier M, Krztoń, K. Kozieł A, Krajewski W, Turek W and Da Costa P 2018 *Comptes Rendus Chim.* **18** 1242
[31] Hassani Rad S J, Haghighi M, Alizadeh Eslami A, Rahmani F and Rahemi N 2016 *Int. J. Hydrogen Energy* **41** 5335
[32] Abdullah N, Chong C C, Razak H A, Ainirazali N, Chin S Y and Setiabudi H D 2018 *Mater. Today Proc.* **5** 21594
[33] Chong C C, Owgi A H K, Ainirazali N, Chin S Y, Setiabudi H D 2018 *Mater. Today Proc.* **5** 21644–21651.
[34] Bukhari S N, Chin S Y, Setiabudi H D and Vo D V N 2017 *J. Environ. Chem. Eng.* **5** 3122
[35] Tomer V K, Devi S, Malik R, Nehra S P and Duhan S 2016 *Microporous Mesoporous Mater.* **219** 240
[36] Ahmed R A, Pang Y X, Olea M and Hodgson S N B 2012 **81** 71
[37] Lin L I, Lu-ming Z, Yu-hua Z and Jin-lin L I 2015 *J. Fuel Chem. Technol.* **43** 315
[38] Yang W, Liu H, Li Y, Wu H and He D 2016 *Int. J. Hydrogen Energy* **41** 1513

IOP Conf. Series: Materials Science and Engineering **702** (2019) 012025 doi:10.1088/1757-899X/702/1/012025

Electro-oxidation of renewable glycerol to value added chemicals over phosphorous-doped Pt/MCNTs nanoparticles

M S Ahmad[1*], C K Cheng[1, 2], H R Ong [3] and H Abdullah[1], M R Khan[1, 2]

[1]Faculty of Chemical and Natural Resources Engineering, Universiti Malaysia Pahang, Lebuhraya Tun Razak, 26300 Gambang, Pahang, Malaysia.
[2]Centre of Excellence for advanced Research in Fluid Flow (CARIFF), Universiti Malaysia Pahang, Lebuhraya Tun Razak, 26300 Gambang, Pahang, Malaysia.
[3]Faculty of Engineering and Technology, DRB-HICOM University of Automotive Malaysia, Malaysia

*E-mail: chemist019@gmail.com

Abstract. In the present work, Pt/MCNTs and P-dopedPt/MCNTs catalysts were synthesized via hydrothermal method and utilized for glycerol oxidation reaction (GOR). The catalysts were physio-chemically characterized by Brunauer-Emmett-Teller (BET), X-ray diffraction, field emission scanning electron microscopy (FESEM) and transmission electron microscopy (TEM) to investigate pore size, pore volume, structure and morphology of the catalysts. A remarkable reduction in BET surface area was found for P-dopedPt/MNCTs which 1.56 time less than the sole MNCTs is confirming the incorporation of phosphorous into the pore channels of Pt/MCNT. The TEM analysis of P-dopedPt/MCNTS demonstrated the uniform distribution of Pt nanoparticles over multiwalled carbon nanotubes with an average particle size of 6.23 nm. The activity and stability of the catalysts were analysed for GOR using cyclic voltammetry (CV) and chronoamperometry (CA). The CV testing showed the higher current density for P-doped Pt/MCNTs (28.33 A/m^2) which is 2 times higher than the Pt/MCNTs (14.20 A/m^2) current density. The synthesized products were analysed by high performance liquid chromatography (HPLC) analysis which demonstrated the selective synthesis of tartronic acid as a major product with maximum yield. It was found that the P-dopedPt/MCNTs have higher selectivity for tartronic acid and glyceric acid as compared to its counterpart.

1. Introduction

Rapid and overwhelming demands for alternative green fuels to cope with upcoming energy crisis leading towards the utilization of low cost, non-toxic and environment friendly raw materials such as glycerol as fuel of choice [1]. Glycerol produced as a by-product during biodiesel production via esterification of fatty acids of oil and fats [2]. Excess of this by-product founds its applications in electricity production and synthesis of valuable organic compounds by promoting C-C bond cleavage and ultimately complete oxidation to CO_2 [3-5]. Historically, glycerol oxidation was extensively carried out over noble metal catalysts such as Pt, Pd, Ru, Ir and Rh, due to their excellent catalytic properties, although the main disadvantage of Pt and Pd is their deactivation at high reaction time [6]. To overcome this problem, the distribution of metals particles over active supports proved to be a promising technique to enhance the overall activity and selectivity of the catalysts. Among them, carbon nanotubes (CNTs) as inert support drawn continuous attention over past few decades. The unique properties such as high conductivity, enlarged surface area, unique 1 D structure and chemical stability made MCNTs highly suitable for research and industrial applications [7-8].

The physiochemical properties of MCNTs can be tuned either by surface functionalization or by doping with metallic or non-metallic heteroatoms such as nitrogen (N), phosphorous (P), sulphur (S), boron (B) and cobalt (Co) [9-10]. Hu et al. [11] studied the electro-oxidation of glycerol over CeO_2 supported Pt catalysts, resulting the tartronic acid yield of 40%. Carrettin et al. [12] reported the production of dihydroxy acetone selectivity ranging in between 10-80% over charcoal-supported Pt catalysts within pH 2-3, with the incorporation of Bi and Pt with a glycerol conversion of 80%. Similarly, under mild condition, 1 % Au-supported charcoal or 1% Au-supported graphite, resulted synthesis of glyceric acid with selectivity of 100 % [13]. Similar results have been reported by Gomes et al. [14] which synthesized glyceric acid with a selectivity of 55% and 77% with glycerol conversion of 90% for Pt/C and Pd/C, respectively [14]. The chemical doping of graphene and carbon nano tubes with substitutional dopants such as P, B, N and Si atoms can control the size, degree of dispersion and stability of catalyst in order to improve the activity of catalyst [15].

Herein, we report the facile synthesis of Pt/MCNTs and P-dopedPt/MCNTs catalysts via hydrothermal method and utilized for GOR. The physical properties of the catalysts were analysed by using BET, XRD, and TEM. While the electro-chemical studies have been done by conducting cyclic voltammetry (CV) and chronoamperometry (CA) measurements. High performance Liquid chromatography (HPLC) analysis demonstrated the selective synthesis of mesoxalic acid.

2. Experimental

2.1. Chemical reagents and materials

Hexachloroplatinic (IV) acid hexahydrate ($H_2PtCl_6.6H_2O$), was purchased from Merck, Germany. While glycerol anhydrous, phosphorous pentoxide (P_2O_5), hydrazine and Nafion® 117 solution (for ink preparation) were purchased from Sigma-Aldrich, USA. The chemicals are in analytical reagent grade, so used without any further purification.

2.2. Catalyst synthesis

The hydrothermal method was adopted to synthesize the P-doped Pt/MCNTs and Pt/MCNTs catalyst by using hydrazine as reducing agent. In brief, 5 mL of $H_2PtCl_6.6H_2O$ aqueous solution was mixed with 450 mg of phosphorous pentoxide (aqueous solution) for P-dopedPt/MCNTs catalysts. The resulting mixture was put on sonication for 60 min at ambient temperature. The colour of the mixture was light yellow (for both the catalysts) representing that the Pt is in unreduced form. Afterward, hydrazine was added drop wisely to reduce the Pt and was put on sonication for more 30 min. Thereafter, the reduced mixture was then transferred to a hydrothermal reactor and 800 mg of activated MCNTs was added into it. The whole assembly was then kept in an oven at 120 °C temperature for 5 h. The residue was taken out from cooled vessel and centrifugal washing was carried out thrice with distilled water followed by drying at 100 °C in a vacuum oven. After complete drying, the final sample was grinded using a piston mortar and stored for further physiochemical studies

2.3. Physiochemical Characterization

The X-ray diffraction patterns of P-dopedPt/MCNTs and Pt/MCNTs were taken using a Rigaku Miniflex diffractometer, with the Cu Ka radiation (1.54 Å) X-ray generator at 30 kV and 15 mA The Scherer's equation ($d = 0.9\ \lambda/\beta \cos\theta$) was used to determine the crystallite size of the catalyst for 111 peak

The Bruauner, Emmett, Teller (BET) analysis was used to measure the specific surface area, pore size and pore volume of the catalysts. A Micrometrics system (Model ASAp-2000) was used to analyse the sample using a vacuum pressure of 6.67 Pa, and a temperature of 77 K. The degassing of the samples was carried out at 473.15 K for 5 h.

The quantitative measurement of the size distribution, particle size, and morphology of the catalysts was carried out by direct imaging of nanoparticles. A Tecnai G2 20S-TWIN, with accelerating voltage of 220 kV. TEM was employed to examine the structural properties of the electrocatalysts. The sample for transmission electron microscopy (TEM) was prepared by suspending a 0.1 mg of catalyst powder in an isopropyl alcohol solution followed by sonication for an hour.

2.4. Electrochemical Characterization of Catalyst

A three electrode assembly along with a Pt counter electrode, Ag/AgCl (sat. KCL) as reference electrode and catalyst pasted glassy carbon (GC) as the working electrode were used to conduct the electrochemical studies. A microcomputer-controlled Potentiostat/Galvanostat NOVA 1.10 was used to perform CV and CA analysis. An oxygen free (N_2 bubbled) 0.5 M H_2SO_4 solution was used as electrolyte to perform the CV for electrochemical surface area (ECSA) measurement using equation 1.

$$\mathrm{ECSA} = \frac{Q_{Pt}}{W_{Pt}\ S\ C} \tag{1}$$

where Q_{Pt} is the area calculated by integration of the total charge corresponding to reduction peak of PtO, W_{Pt} is Pt content deposited on glassy electrode (~ 0.157 mg based on the 0.3 μL ink used), S is the charge density for reduced PtO monolayer: 210 $\mu C/cm^2$ and C is the scan rate: 0.05 V/s.

While, 0.5 M glycerol and 0.5 M KOH solution was employed as electrolyte for GOR analysis. Prior to each CV test, the electrolyte solution was bubbled with N_2 for 60 min to remove the excess oxygen. The CV curves were obtained at a scan rate of 0.05V/s under inert environment. The catalyst stability against poisoning intermediates was measured by performing CA analysis at -0.28 V for 3600 s in alkaline medium. The samples were also taken during CA analysis for product separation and quantification through HPLC.

2.4.1. Catalyst ink and working electrode preparation

A catalyst ink was prepared for working electrode preparation for each electrochemical study. In a typical procedure, a 0.02 g of fine catalyst powder was weighted in a glass vessel and mixed with 0.15 mL of Nafion® 117 (5% solution and 240 mL of isopropyl alcohol. The mixture was put on stirring overnight in order to completely homogenize. Afterwards, a 0.3 μL of ink was taken using micropipette and transferred onto a cleaned GC electrode (~0.8 cm^2 area), followed by drying at 80 °C in a vacuum oven overnight. The electrode surface was cleaned by cyclizing through CV within a potential window of (-0.2 V -1.4 V @ 0.01 V/s) and (-0.8 V -0.4 V @ 0.05 V/s) for ECSA and GOR analysis, respectively.

2.5. Product distribution analysis by High performance Liquid chromatography (HPLC)

An Agilent 1260, furnished with ultraviolet (210 nm) and refractive index detector (RID). The Phenomenex HPLC column, Model: Rezex ROA® (4.6 mm × 250 mm, 5 mm) was used to perform the chromatographic analysis for GOR products. The column was kept at 65 °C and the wavelength employed was 210 nm (maximum absorbance for organic compounds). The RID detector was kept at 35 °C. The sample injection volume was 10 μL and eluent flow was set at 0.7 mL/min. The mobile phase was 3 mmol H_2SO_4. The possible GOR products were compared with standard samples for identification. The product selectivity was calculated using equation (2).

$$\mathrm{Selectvity} = \frac{n_j}{\Sigma n_j} \times 100\% \tag{2}$$

where n_j is the number of moles of individual product and $\sum n_j$ is the sum of all the products obtained during HPLC analysis. The identification of the products was carried out by comparison with standard samples.

3. Results and discussion

3.1. X-ray diffraction analysis (XRD)

The X-ray diffraction patterns of MCNTs, Pt/MCNTs, and P-dopedPt/MCNTs are shown in the figure 1. The diffraction peaks at 2θ of 43.41° and 53.86° were found for MCNTs. While the diffraction peaks for Pt/MCNTs and P-dopedPt/MCNTs were located at 2θ of 40.00°, 46.24°, 67.85° and 39.886°, 46.22°, 67.83°, respectively corresponding to (111), (200) and (220) lattice planes.

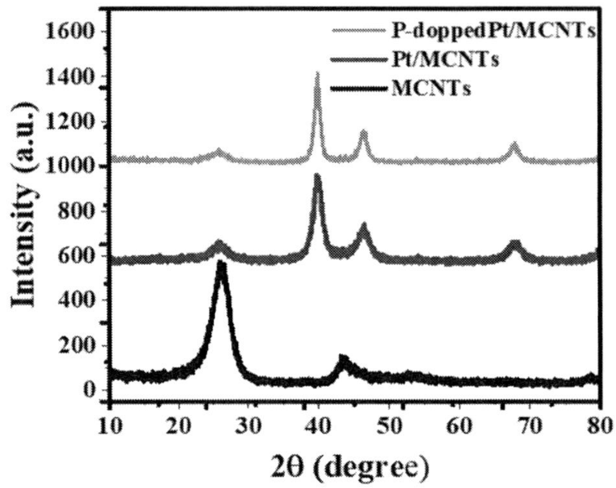

Figure 1. XRD analysis of MCNTs, Pt/MCNTs and P-dopedPt/MCNTs.

Regardless of the catalyst, a strong peak at ca. 2θ ~ 26° was shown for all the catalysts, which corresponds to MCNTs graphitic structure. A typical face centered cubic lattice structure was illustrated from the peak positions for Pt, and Pt alloy catalysts (COD: 1011107). Besides, the XRD pattern were similar for both the catalysts. There was no typical peak detected for phosphorous. In addition, the intensity of these peaks were slightly increased after the addition of phosphorous. Moreover, a negative peak shift (2θ = 0.02) has been observed for P-dopedPt/MCNTs for 220 lattice plane. This revealed that the lattice size of the P-dopedPt/MCNTs catalyst expanded due to incorporation of P into pore channels of Pt/MCNTs. These findings are further confirmed by the calculation of average crystallite size for Pt/MCNTs and P-dopedPt/MCNTs catalysts. The average crystallite sizes of 4.27 nm and 6.24 nm were calculated (for 220 lattice plane) for Pt and P-dopedPt respectively. The increase in average crystallite size for P-dopedPt justified the above statement and also confirmed the incorporation of P and alloy formation with Pt. In one of the recent studies, it was found that the incorporation of Pd into Pt increased the particle size and the crystallite size [16]. A similar explanation has been reported by Ref. [37] for the negative peak shift for incorporation of P into Pt/MCNTs [17]. The d- spacing was also calculated for all the catalysts and values were 0.13801 nm, and 0.13805 nm for Pt/MCNTs and P-dopedPt/MCNTs respectively. The XRD crystallite size for P-dopedPt/MCNTs is line with TEM particle size (6.23 nm).

3.2. BET analysis

The N_2 adsorption and desorption curves for all the materials (MCNTs and electrocatalysts) are presented in figure 2.

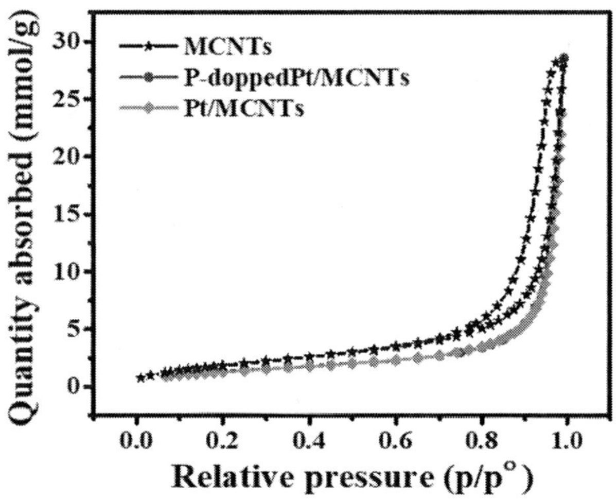

Figure 2. BET analysis of MCNTs, Pt/MCNTs and P-dopedPt/MCNTs.

While all the calculated results for specific surface area and pore size are shown in table 1.

Table 1. Physical properties of MCNTs, Pt/MCNTS and P-dopedPt/MCNTs.

Catalyst	[a] Specific Surface Area (m^2/g)	[a] Pore Volume (cc/g)	[b] Crystallite size (nm)	[c] Crystallite size (nm)
MCNTs	179.30	1.04	-	-
Pt/MCNTs	112.39	0.90	4.27	3.45
P-dopedPt/MCNTs	114.43	1.03	6.24	6.23

[a] Calculated form BET analysis.
[b] Calculated form XRD analysis.
[c] Calculated form TEM analysis.

The Pt/MCNTs, P-dopedPt/CNTs and MCNTs, followed the type-IV adsorption-desorption isotherms characteristics [18]. The MCNTs showed the highest specific surface area as compared to Pt/MCNTs and P-doped/MCNTs which is in an order of 1.59 and 1.56 times respectively. In one of the recent studies by Ref. [19] explained that the reduction in pore size distribution in comparison to MCNTs is dependent to the functional groups along the catalyst surface. While the decrease in total pore volume is through anchoring the functional groups [19]. Likewise, Ref. [20] obtained an increase in a BET surface area for acid (HCL, HF and HNO_3) treated MCNTS and explained this is due to the removal of the inorganic constituents after acid treatment [20]. These results demonstrated that the reduction in surface area of metal catalysts compared to MCNTs can be attributed to the blocking of pores and channels of MCNTs and in agreement with the published literatures [21-23].

3.3. Transmission electron microscope analysis (TEM)

The morphology, particles size and distribution of the metal nanoparticles were revealed by TEM analysis as shown in figure 3.

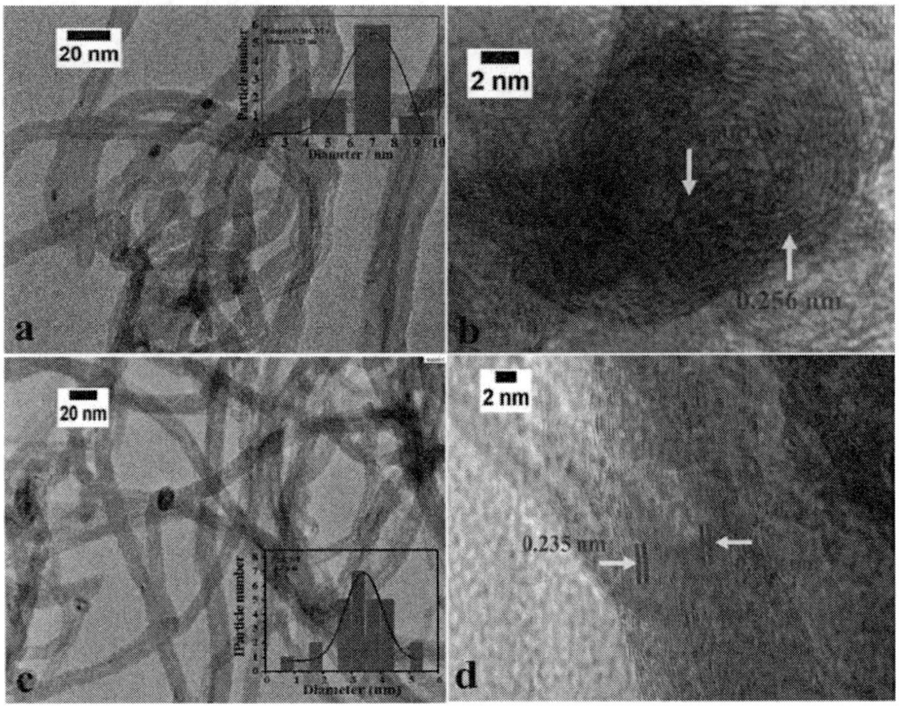

Figure 3. TEM images (a&b); P-dopedPt/MCNTs, and (c&d); Pt/MCNTs, with histograms (insight).

An average particle diameter of 3.45 nm and 6.23 nm was measured using image J software for Pt/MCNTs and P-dopedPt/MCNTs respectively. All the catalysts shown the uniform particle distribution while some of aggregation was also observed for P-doped Pt/MCNTs catalyst which could be attributed to the fast reduction of Pt [24]. The TEM results are in close agreement with XRD data showing almost the similar crystallite size for P-dopedPt/MCNTs (see table 1). In addition, the average atomic spacing (d-spacing = 0.276 nm) was calculated using ImageJ software (figure 3b) for the doped catalyst showing a slight increase which might be due to higher crystallite size (in consistent with XRD).

3.4. Electrochemical characterization

3.4.1. Cyclic Voltammetry (CV)

The electrocatalytic activities of Pt/MCNTs and P-dopedPt/MCNTs were measured in 0.5 M H_2SO_4 and 0.5 M glycerol/0.5 M KOH electrolytes for ECSA and GOR respectively at a scan rate of 0.05 V/s (figures 4(a) and 4(b)).

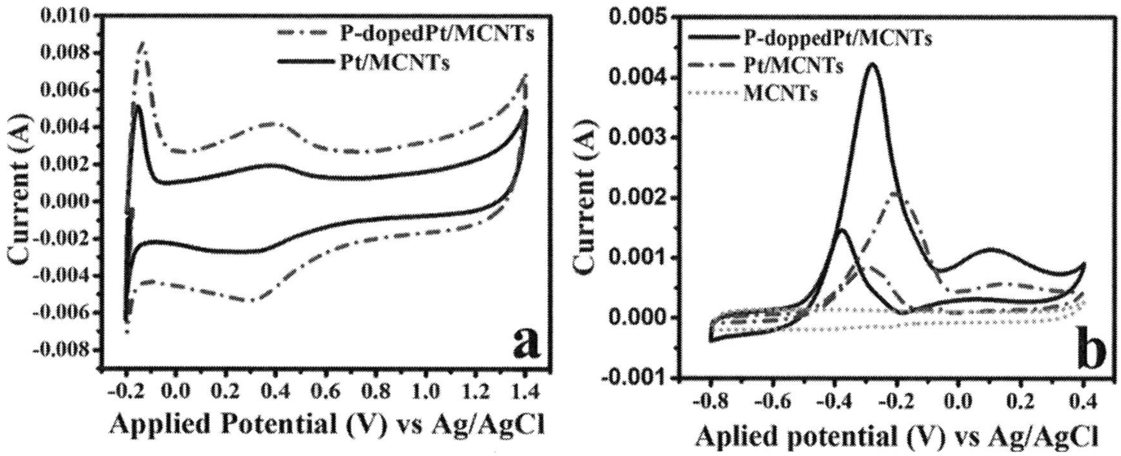

Figure 4. CV curves of Pt/MCNTs and P-dopedPt/MCNTs: (a) ECSA calculation in N_2 saturated 0.5 M H_2SO_4 @ 0.05V/s, (b) glycerol oxidation reaction in 0.5 M glycerol/0.5 M KOH @ 0.05V/s.

As shown in figure 4a, irrespective of the nature of the catalyst, three prominent electrochemical areas are represented in CV profile for both catalysts including double layer capacitance, PtO oxide formation (anodic seep) and PtO oxide reduction (cathodic sweep). Moreover, the formation of surface oxides for P-dopedPt/MCNTs observed at 0.40 V while 0.34 V for Pt/MCNTs, which is 0.06 times, less than the doped catalyst. In contrary, the reduction of these oxides occurs at (0.32 - 0.0 V) and (0.34 - 0.00 V) respectively vs Ag/AgCl for P-doped Pt/MCNTs and Pt/MCNTs.

In addition, the presence of only forward peaks (anodic) in both sweeps demonstrated the irreversibility of the GOR [25-27]. The P-dopedPt/MNCTs showed a higher current density (28.33 A/m^2) as compared to Pt/MCNTs (14.20 A/m^2) which is 1.99 folds higher than the counterpart, indicating the higher GOR activity for P-dopedPt/MCNTs. The ECSA was calculated for both the catalysts and values are presented in table 2.

Table 2. ECSA calculation for Pt/MCNTs and P-dopedPt/MCNTs in 0.5 M KOH @ 50 mV/s.

Catalyst	*I, (A/m^2)	$^{\#}$ECSA, (m^2/g)	$^{\Delta}$I, (A/m^2)
P-dopedPt/MCNTs	28.33	670.95	0.83
Pt/MCNTs	14.20	548.19	0.13

* Current density, measured from CV in 0.5 M glycerol/0.5 M KOH @ 0.05 V/s.
\# Electrochemical surface area, measured from CV in 0.5 M H_2SO_4 @ 0.05 V/s.
Δ Current density, measured from CA in 0.5 M glycerol/0.5 M KOH.

While the respective CV curves are shown in figure 4(b). The P-dopedPt/MNCTs demonstrated 1.22 times higher ECSA as compared to Pt/MCNTs catalyst. The higher ECSA for P-dopedPt/MNCTs ascribed to the better dispersion of metal nanoparticles over multi walled carbon nanotubes support [28-29]. It also demonstrated the high availability of effective surface area for the conversion of glycerol to valuable products. Moreover, the incorporation of P into Pt lattice can also enhance the CO poisoning resistance of the catalyst due to higher oxidizability of the catalyst due availability of excess oxygen-containing species [16, 30].

Since the GOR is carried out in an alkaline medium (pH ~ 13) which also facilitates the rate of glycerol oxidation reaction either through deprotonation of the hydroxyl group or desorption of the acid (H^+ ions) from the metal resulting to suppress the rate of CO poisoning [31-33]. Thus a higher current density, lower poisoning rate and high ECSA for P-doped catalyst in comparison to Pt/MCNTs account for the higher activity of the P-doped catalyst for GOR.

3.4.2. Chronoamperometry analysis (CA)
The durability of the as prepared catalysts was measured by performing the CA tests at – 0.28 V for 3600 s (figure 5) and current densities were calculated and presented in Table 2. Regardless of the catalysts, current density abruptly decayed at the initial stage of the reaction while it becomes almost stable at the intermediate stage and finally becomes constant near the completion of reaction. This CA trend revealed that the rate of GOR is higher at the start and intermediate state of reaction indicating the high availability of active sites facilitating the conversion of glycerol to products. While a decrease at final stage clearly demonstrated the poisoning of the reaction sites due to adsorbed oxygenates (CO & CO_2). In addition, it was also found that the current density for P-doped/Pt/MCNTs is 6.38 times higher than the Pt/MCNTs representing the high rate of GOR due to incorporation of P. While a slight reduction at the final stage can be ascribed due to pore blocking of the catalyst [34, 35]. These results are in agreement with the findings reported by Ref. [34] for bimetallic PtPd catalyst for ethanol electro-oxidation [36].

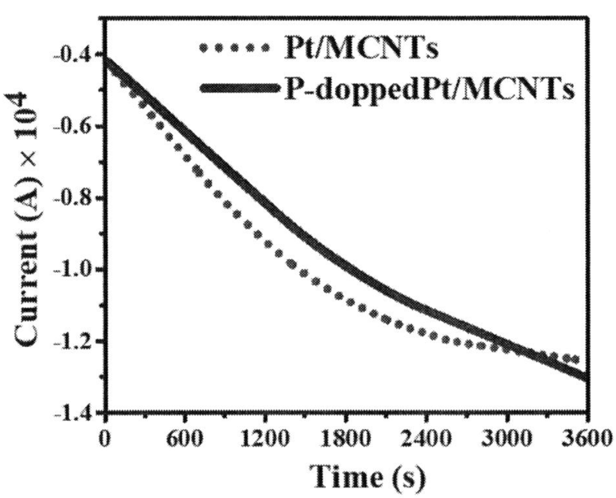

Figure 5. CA curves for Pt/MCNTs and P-dopedPt/MCNTs in 0.5 M glycerol/0.5 M KOH.

3.5. Products Analysis by HPLC
The sample solution from glycerol oxidation reaction is collected during CA test at regular interval of 15 min and products were separated by applying HPLC. Since the glycerol is polyol with complex structure and provides a range of reactive sites over primary and secondary carbons. Thus the complete oxidation of glycerol to organic products (GTO) is susceptible primarily due to slow kinetics and the high energy barrier of C-C bond breaking. In addition, the possibility of complete oxidation is also less due to bulkiness (hydroxyl groups) of the nucleophilic centre providing stearic hindrance to attacking nucleophiles [37]. That reason, partial oxidation of the glycerol produced glyceraldehyde (GALD), tartronic acid (TAT), glyceric acid (GLY) and mesoxalic acid (MOXA) in their salt form but described as acid for simplicity in the paper.

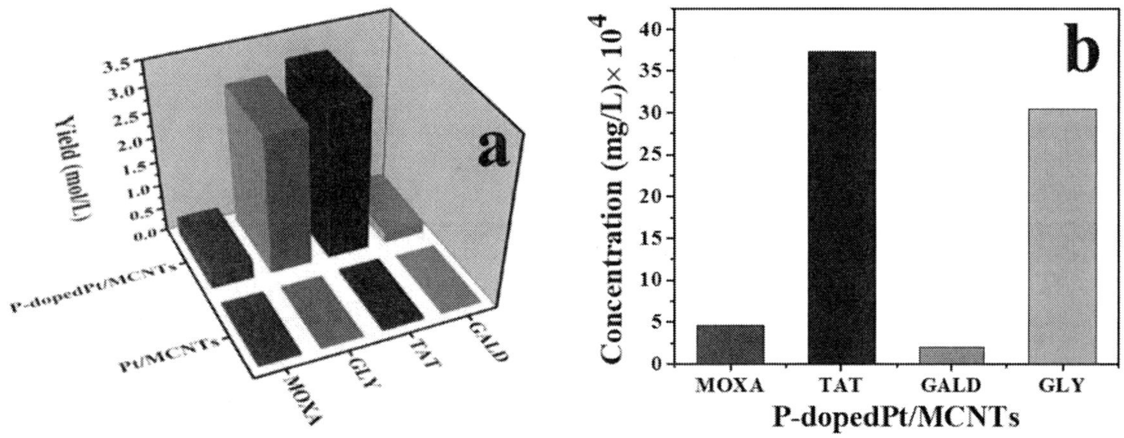

Figure 6. (a) Product distribution for Pt/MCNTs and P-dopedPt/MCNTs in 0.5 M glycerol/0.5 M KOH, (b) product selectivity for P-dopedPt/MCNTs.

Figure 6(a) presented the products distribution for P-dopedPt/MCNTs and Pt/MCNTs. It can be clearly seen that the nature of the products is similar for both the catalysts indicating that the route of reaction is unaltered even after the incorporation of P. while the product yield was enhanced for P-doppedPt/MCNTs. The tartronic acid (TAT) and glyceric acid (GLY) was separated as the major oxidation product while small quantities of mesoxalic acid (MOXA) and glyceraldehyde were also found. In addition, TAT and GLY was found in high yield for P-dopedPt/MCNTs as compared to Pt/MCNTs. This could be explained due to the higher ECSA and stability of the doped catalyst provided due to the incorporation of the P resulting an enhancement in rate GOR.

Furthermore, the selectivities of all the products during GOR of P-doppedPt/MCNTS catalyst were calculated and showed in figure 6b. It is worth mentioning, that TAT have the highest selectivity (~ 47%) which is 1.2 times higher than GLY. While GLY has the second highest selectivity around 40%. The higher yield and better selectivity of the catalyst can ascribe to role of P doping by reducing the accumulation of carbonaceous intermediates and increasing the OH^{-1} availability. The Ref. [24] studied the detailed mechanism for formation and oxidation of carbonaceous intermediates (CO & CO_2) through involvement of hydroxyl ions.

The enhancement in yield of desired C3 compounds acid can be attributed to the strong adsorption of the glycerol molecules with exposed P-dopedPt/MCNTs reactive sites along with uniform distribution of nanoparticles over activated MCNTs support. These results conclude that the incorporation of non-metal P into channels of Pt/MCNTs can remarkably enhance the activity, stability and selectivity of the desired products while keeping the glycerol oxidation pathway unchanged.

4. Conclusions

Highly dispersed P-dopedPt and Pt nanoparticles supported over multiwalled carbon nanotubes (MCNTs) were prepared to study the possible products synthesis from GOR. Through doped P and selectively modified synthetic process of the electrocatalysts, interaction of the non-metal with MCNTs supported noble metal catalyst was researched. The electrochemical analysis suggested the high activity and stability for the P-dopedPt/MCNTs catalyst in comparison to undoped Pt/MCNTs catalyst. In addition, a higher product yield and selectivity of TAT and GLY were also obtained through selective GOR. The enhancement in C3 products yield along with high current density (28.33 A/m^2) ECSA (670.95 m^2/g) and better stability can be attributed to the incorporation of P and strong interaction between metal particles and support. Therefore, P-dopedPt/CNTs can served as an efficient selective catalyst for the valuable products synthesis from electro-oxidation of glycerol at ambient temperature.

Acknowledgements
The authors thank Universiti Malaysia Pahang, Malaysia for providing financial assistance for this project under PGRS-180335.

References
[1] Hunsom M and Saila P 2013 *Int. J. Electrochem. Sci.* **8** 11288-300
[2] Zhou C H C, Beltramini J N, Fan Y-X and Lu G M 2008 *Chem. Soc. Rev.* **37** 527-49
[3] Arechederra R and Minteer S 2009 *Fuel Cells* **9** 63-9
[4] Painter R M, Pearson D M and Waymouth R M 2010 *Angew. Chem. Int. Ed.* **49** 9456-9
[5] Xin L, Zhang Z, Wang Z and Li W 2012 *Chem. Cat. Chem.* **4** 1105-14
[6] Kostecka M, Kowalska D, Kozłowska M and Kowalski B 2013 *J. Oleo Sci.* **62** 893-900
[7] Machado B F, Oubenali M, Axet M R, Nguyen T T, Tunckol M, Girleanu M, Ersen O, Gerber I C and Serp P 2014 *J. Catal.* **309** 185-98
[8] Brandner A, Lehnert K, Bienholz A, Lucas M and Claus P 2009 *Top. Catal.* **52** 278-87
[9] Wepasnick K A, Smith B A, Schrote K E, Wilson H K, Diegelmann S R and Fairbrother D H 2011 *Carbon* **49** 24-36
[10] Su D S, Perathoner S and Centi G 2013 *Chem. Rev.* **113** 5782-816
[11] Hu W, Lowry B and Varma A 2011 *Appl. Catal. B: Environ.* **106** 123-32
[12] Carrettin S, McMorn P, Johnston P, Griffin K, Kiely C J and Hutchings G J 2003 *Phys. Chem. Chem. Phys.* **5** 1329-36
[13] Lakshmanan P, Upare P P, Le N-T, Hwang Y K, Hwang D W, Lee U-H, Kim H R and Chang J-S 2013 *Appl. Catal. A: Gen.* **468** 260-8
[14] Gomes J F, Gasparotto L H and Tremiliosi-Filho G 2013 *Phys. Chem. Chem. Phys.* **15** 10339-49
[15] Tang Y, Yang Z, Dai X, Ma D and Fu Z 2013 *J. Phys. Chem. C* **117** 5258-68
[16] Carvalho L L, Tanaka A A and Colmati F 2017 *J. Solid State Electr.* 1-11
[17] Yang G, Chen Y, Zhou Y, Tang Y and Lu T 2010 *Electrochem. Commun.* **12** 492-5
[18] Tamilarasan P and Ramaprabhu S 2014 *J. Mat. Chem. A* **2** 14054-63
[19] Kohantorabi M, Giannakis S and Gholami M R 2019 *Appl. Catal. A: Gen.*
[20] Wang S and Lu G M 1998 *Carbon* **36** 283-92
[21] Miguel S R, Heinen J C, Castro A A and Scelza O A 1989 *React. Kinet. Catal. L.* **40** 331-5
[22] Torres G C, Jablonski E L, Baronetti G T, Castro A A, de Miguel S R, Scelza O A, Blanco M D, Pen˜a Jiménez M A and Fierro J L G 1997 *Appl. Catal. A: Gen.* **161** 213-26
[23] Hwang S-W, Rather S-u, Naik M-u-d, Soo C S and Nahm K-S 2009 *J. Alloy. Compd.* **480** L20-L4
[24] Zhou W J, Song S Q, Li W Z, Zhou Z H, Sun G Q, Xin Q, Douvartzides S and Tsiakaras P 2005 *J. Power Sources* **140** 50-8
[25] Yang G, Zhou Y, Pan H-B, Zhu C, Fu S, Wai C M, Du D, Zhu J-J and Lin Y 2016 *Ultrason. Sonochem.* **28** 192-8
[26] Hong W, Shang C, Wang J and Wang E 2015 *Energy Environ. Sci.* **8** 2910-5
[27] Li S, Lai J, Luque R and Xu G 2016 *Energy Environ. Sci.* **9** 3097-102
[28] Cha B-C, Jun S, Jeong B, Ezazi M, Kwon G, Kim D and Lee D H 2018 *J. Power Sources* **401** 296-302
[29] Bharti A and Cheruvally G 2018 *Int. J. Hydrogen Energ.* **43** 14729-41
[30] Arroyo-Gómez J J, Barrera D, Castagna R M, Sieben J M, Alvarez A E, Duarte M M E and Sapag K *Chem. Cat. Chem.*
[31] Geraldes A N, Silva D F, Silva J, Souza R F, Spinacé E V, O Neto A, Linardi M and Santos M C 2014 *J. Braz. Chem. Soc.* **25** 831-40
[32] Frota Jr E F, de Barros V V S, de Araújo B R, Purgatto Â G and Linares J J 2017 *Int. J. Hydrogen Energ.* **42** 23095-106
[33] Geraldes A N, da Silva D F, e Silva L G d A, Spinacé E V, Neto A O and dos Santos M C 2015 *J. Power Sources* **293** 823-30

[34] Dutta A and Datta J 2013 *Int. J. Hydrogen Energ.* **38** 7789-800
[35] Dutta A, Mondal A and Datta J 2015 *J. Power Sources* **283** 104-14
[36] Datta J, Dutta A and Biswas M 2012 *Electrochem. Commun.* **20** 56-9
[37] Zhang Z, Xin L and Li W 2012 *Appl. Catal. B: Environ.* **119** 40-8

IOP Conf. Series: Materials Science and Engineering **702** (2019) 012026 doi:10.1088/1757-899X/702/1/012026

A kinetic model for the photocatalytic reduction of CO_2 to methanol pathways

H Abdullah[1*], M M R Khan[1], Z Yaakob[2] and N A Ismail[1]

[1]Faculty of Chemical and Process Engineering Technology, College of Engineering Technology, Universiti Malaysia Pahang, Lebuhraya Tun Razak, 26300, Gambang, Pahang, Malaysia.
[2]Department of Chemical and Process Engineering, Faculty of Engineering and Built Environment, Universiti Kebangsaan Malaysia, 43600, Bangi, Selangor, Malaysia.

[*]E-mail: hamidah@ump.edu.my

Abstract. Carbon dioxide (CO_2) is one of the greenhouse gases that contribute to global warming. CO_2 could be converted to valuable products such as hydrocarbons through the photocatalytic process. The aim of this research was to develop the kinetic model for the photocatalytic reduction of CO_2 to methanol (CH_3OH) in liquid phase reaction using cerium oxide-titanium dioxide (CeO_2-TiO_2) catalyst. The Langmuir-Hinshelwood approach was used in developing rate laws for the catalytic reaction using the catalytic reaction mechanism proposed. The catalytic reaction mechanism is about the adsorption of reactant (CO_2 dissolved in the liquid phase), the reaction on catalyst surface and desorption of product. The experimental kinetic data were evaluated in the Polymath 6.1 software. In this study, two types of mechanism are proposed whereas one is considered the carbon monoxide (CO) oxidation while the other is not. Based on the model fitting, it was found that the model considers the CO oxidation is fitted well with the experimental data represents that the oxidation reaction of intermediate product, CO is the rate-determining step in the photocatalytic reduction of CO_2 to CH_3OH in liquid phase reaction.

1. Introduction

The technologies of reducing CO_2 emission is important to tackle the challenge of global climate change. Reduction of CO_2 to hydrocarbons using catalyst and photon as an energy which is called photocatalytic reaction has the potential to be parts of renewable energy as well as to reduce CO_2 emission. Titanium dioxide (TiO_2) catalyst has gained a lot of interest as potential photocatalyst due to its non-toxicity, low cost and excellent photochemical properties [1]. However, it has limitations such as has wide band gap which makes it only active under UV light and has high electron/hole pair recombination rate as compared to the rate of a chemical reaction [2]. Both factors lead to low activity and yield of the product. Therefore, various research has been studied in order to design the ideal catalyst for photocatalytic reduction of CO_2 to hydrocarbons such as doping TiO_2 with anions [3], doping TiO_2 with cation [4], coupling TiO_2 with another semiconductor [5] and etc. Koci et al. [11] found that modification of TiO_2 by doping with silver (Ag) could enhance the yield of methane (CH_4) and CH_3OH under visible light irradiation mainly caused by the narrow band gap and decreases of electron-hole recombination rate [6]. TiO_2 modified by doping with gold (Au) also improved the photocatalytic activity of CO_2 reduction due to the efficient electron-hole separation and surface plasmon response of Au-nanoparticles [7]. Among them, modifying TiO_2 with cation is widely investigated [8]. Cation as rare earth elements, instead of transition and noble metals, have caught the interest of researches because it is abundant, non-toxic and inexpensive. Moreover, it has a 4f electron configuration that contributes to lower the band gap and reduce the high electron/hole recombination rate problem [9].

However, the physical properties of the catalyst are not enough to describe the net photocatalytic activity since the photocatalytic reduction of CO_2 has a multistep process such as CO_2 adsorption, activation and dissociation of the C-O bond [10]. Moreover, the photocatalytic reduction of CO_2 to CH_3OH also has various pathway [11]. Therefore, understand the kinetics of a photocatalytic reaction is important in order to determine the rate -determining step and the right pathway. Besides, kinetic study is vital to develop the photocatalytic system as well as practical application in the future.

Thus, the objective of this study is to develop a kinetic model based on the Langmuir- Hinshelwood model using CeO_2-TiO_2 catalyst in order to determine the kinetic parameters and to provide fundamental insights on the possible reaction mechanism.

2. Materials and Method

2.1. Materials

The CeO_2-TiO_2 (2 wt % CeO_2 loading) catalyst used was prepared as per the reported procedure [12]. The sodium hydroxide (NaOH) with 99 % purity was purchased from R & M Chemicals. Purified carbon dioxide (99.98 %) was supplied by Siad Technical Gases, Malaysia.

2.2. Kinetic Measurement

The photocatalytic reduction of CO_2 was performed according to the procedure reported elsewhere[13]. The CO_2 with 0.4 L/min was bubbled through the 300 ml of 0.1 M NaOH solution in the reactor. Then, 0.3 g CeO_2-TiO_2 catalyst was dispersed in the reactor. The solution was agitated under the visible light irradiation for 6 hours. The gas and liquid samples were taken every 1 hour and analyzed using a gas chromatography (GC-Agilent Technologies 6890N) with a thermal conductivity detector (TCD) and flame ionization detector (FID), respectively.

The experimental kinetic data, which is the concentration of CH_3OH for each interval in 6 hours reaction, were evaluated using non-linear regression method (Levenberg-Marquardt) in the Polymath 6.1 software to estimate the unknown kinetic constants.

3. Results and Discussion

Based on the previous result [12], the CH_3OH was observed as the main product in the reaction. The CH_3OH is possibly produced by the reduction of CO_2 in gas and CO_2 which dissolved in water and form carbonate as equations 1 to 3. The carbonate formed was supported by the result shown in figure 1. It shows that the pH dropped drastically from pH 12.4 to 6.87 in the first 10 minutes and then remain stable. The carbonic acid (H_2CO_3) , ion bicarbonate (HCO_3^-) and ion carbonate (CO_3^{2-}) were formed and in equilibrium as in equation 4 when a CO_2 is dissolved in solution [14].

$$CO_2 + 6H^+ + 6e^- \rightarrow CH_3OH + H_2O \tag{1}$$

$$H_2CO_3 + 6H^+ + 6e^- \rightarrow CH_3OH + 2H_2O \tag{2}$$

$$CO_3^{2-} + 8H^+ + 6e^- \rightarrow CH_3OH + 2H_2O \tag{3}$$

$$CO_2(aq) + H_2O(l) \leftrightarrow H^+(aq) + HCO_3^-(aq) \leftrightarrow 2H^+(aq) + CO_3^{2-}(aq) \tag{4}$$

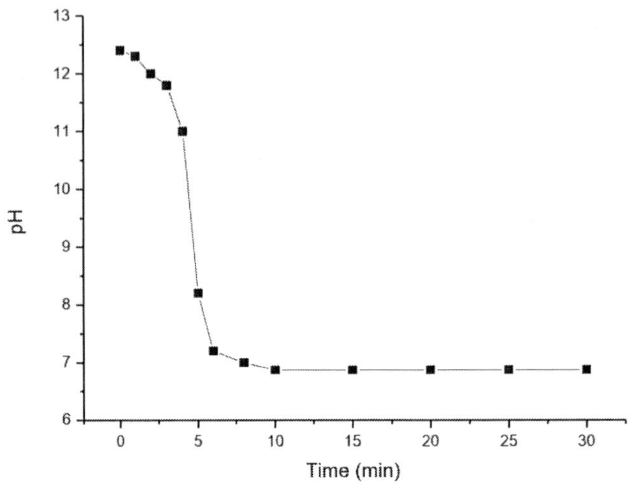

Figure 1. pH as function of time in NaOH solution under the bubbling of CO_2.

Sasirekha et al. [15] proposed that the CH_3OH was formed from the intermediate product (CO) as equation (5) [15]. The details proposed mechanisms are as equations 6 to 12. Besides, the using of NaOH in the reaction also contribute to the formation of CH_3OH. The oxidation of OH ions in NaOH solution could form OH radicals, thus supplying more OH radicals for CH_3OH formation. Tseng [17] reported that the CH_3OH yield from the photocatalytic reduction of CO_2 increases after adding NaOH in solution [16-17].

$$CO_2 \rightarrow CO \rightarrow CH_3OH \quad (5)$$

$$H_2O_{adsorbed} + h^+ \xrightarrow{catalyst} {}^{\bullet}OH + H^+ \quad (6)$$

$$H^+ + e^- \xrightarrow{catalyst} {}^{\bullet}H \quad (7)$$

$$CO_2 + e^- \xrightarrow{catalyst} {}^{\bullet}CO_2^- \quad (8)$$

$${}^{\bullet}CO_2^- + {}^{\bullet}H \xrightarrow{catalyst} CO + OH^- \quad \text{(CO as an intermediate product)} \quad (9)$$

$$CO + e^- \xrightarrow{catalyst} {}^{\bullet}CO^- \quad (10)$$

$${}^{\bullet}CO^- + {}^{\bullet}H \xrightarrow{catalyst} C + OH^- \quad (11)$$

$$C + {}^{\bullet}H \xrightarrow{catalyst} {}^{\bullet}CH + {}^{\bullet}H \xrightarrow{catalyst} CH_2 + {}^{\bullet}H \xrightarrow{catalyst} {}^{\bullet}CH_3 + {}^{\bullet}OH \xrightarrow{catalyst} CH_3OH \quad (12)$$

Thus, the proposed mechanisms with a combination of the Langmuir Hinshelwood (L-H) model were used in this study as the basis in developing the kinetic model. The reactants and products were assumed to be adsorbed on the same active sites as equation (13) [18-19].

IOP Conf. Series: Materials Science and Engineering **702** (2019) 012026 doi:10.1088/1757-899X/702/1/012026

$$r_{reduction} = k_{reduction} I^{\alpha} \frac{K_{H_2O}C_{H_2O}K_{CO_2}C_{CO_2}}{(1+K_{CO_2}C_{CO_2}+K_{H_2O}C_{H_2O}+K_{CH_3OH}C_{CH_3OH}+K_{CO}C_{CO}+K_{O_2}C_{O_2})^2} \quad (13)$$

The rate of reduction equation can be deduced as equation (14) by using this assumption:

i) the reaction rate is proportional to the fraction of the surface covered by CO_2
ii) H_2O, CH_3OH, and O_2 are weakly adsorbed over the catalyst
iii) CO_2 and CO are moderately adsorbed. However, the reaction was observed to be very fast at the beginning and become slow after a certain time of reaction. Besides, CO was not observed. Thus, it indicates that the CO formed is not effectively desorbed from the catalyst surface and accumulated ($K_{CO}C_{CO} >> 1 + K_{CO_2}C_{CO_2}$)

$$r_{reduction} = k_1 \frac{K_{H_2O}C_{H_2O}K_{CO_2}C_{CO_2}}{(K_{CO}C_{CO})^2} \quad (14)$$

where k_1 is the photoreduction rate constant affected by temperature and light intensity ($k_1 = r_{reduction} I^{\alpha}$) and $K_{CO_2}, K_{H_2O}, K_{CO}$ are the ratios of the rate constant for adsorption of CO_2, H_2O, and desorption of CO, respectively. Since H_2O was in great excess and CO_2 was continuously feed into the reactor, the concentration of CO_2 and H_2O can be assumed constant. Thus, the rate of reaction becomes as per equation (15).

$$r_{reduction} = k_2 \frac{1}{(K_{CO}C_{CO})^2} \quad (15)$$

where $k_2 = k_1 K_{H_2O}C_{H_2O}K_{CO_2}C_{CO_2}$

Based on equation (15), it can be said that the rate of formation of products is dependent on CO conversion and/or desorption over the catalyst surface. Tan et al. [18] proposed that the CO also sometimes undergoes partial oxidation with oxygen [18, 20]. They suggested that the CO oxidation undergoes dissociative adsorption process. The rate of photo-oxidation of CO could also be evaluated by L-H model as equation (16).

$$r_{oxidation} = k_3 \frac{K_{CO}C_{CO}\sqrt{K_{O_2}C_{O_2}}}{(1+K_{CO}C_{CO}+\sqrt{K_{O_2}C_{O_2}})^2} \quad (16)$$

where k_3 is the kinetic constant of oxidation while K_{CO} and K_{O_2} are the ratios of rate constants for adsorption and desorption of CO and O_2, respectively. Equation 16 is simplified to equation 17 by assuming that all active sites are occupied by CO due to the CO being strongly bound to the catalyst surface $(K_{CO}C_{CO} >> 1)$

$$r_{oxidation} = k_3 \frac{\sqrt{K_{O_2}C_{O_2}}}{K_{CO}C_{CO}} \quad (17)$$

In this study, two types of mechanism are proposed. One is considered the CO oxidation while the other is not. The rate of CH_3OH formation that does not consider the CO oxidation is per equation (15) while the net rate of reduction by considering the CO oxidation is shown in equation (18).

$$r_{netreduction} = k_2 \frac{1}{(K_{CO} C_{CO})^2} - k_3 \frac{\sqrt{K_{O_2} C_{O_2}}}{K_{CO} C_{CO}} \quad (18)$$

Both equations can be simplified to equations (19) and (20) by assuming that the concentration of CO (C_{CO}) and O_2 (C_{O_2}) is proportional to the concentration of CH_3OH (C_{CH_3OH})

$$\frac{dC_{CH_3OH}}{dt} = k_4 \frac{1}{(C_{CH_3OH})^2} \quad (19)$$

$$\frac{dC_{CH_3OH}}{dt} = k_4 \frac{1}{(C_{CH_3OH})^2} - \frac{k_5}{\sqrt{C_{CH_3OH}}} \quad (20)$$

where $k_4 = \frac{k_2}{K_{CH_3OH}{}^2}$ and $k_5 = k_3 \frac{\sqrt{K_{O_2}}}{K_{CO}}$

Equations 19 and 20 were integrated and arranged as equations (21) and (22), respectively.

$$t = \frac{(C_{CH_3OH})^3}{3k_4} \quad (21)$$

$$t = k_6 \lg(1 - k_7 (C_{CH_3OH})^{3/2}) + k_8 (C_{CH_3OH})^{3/2} \quad (22)$$

where $k_6 = \frac{-2k_4}{3k_5{}^2}, k_7 = \frac{k_5}{k_4}, k_8 = \frac{-2}{3k_5}$

$k_4, k_6, k_7 .k_8$ are the kinetic constant related to reaction rate constants, adsorption-desorption ratio constants and light intensity. The unknown kinetic constants were estimated with a non-linear least square regression method (Levenberg-Marquardt) using Polymath 6.1 software. The CeO_2-TiO_2 performance has been explored in the proposed kinetic model. Table 1 shows the summary results of the processing of kinetic data using the method of non-linear regression. The experimental data of photocatalytic reduction of CO_2 to CH_3OH were compared with the data calculated according to both equations. The profiles of CH_3OH formation for both proposed models and experimental data are shown in figure 2. It is clearly seen that model 2 fit well with the experimental data when compared to model 1. Moreover, the correlation coefficients in table 2 also show that the model 2 is suitable for CH_3OH formation rate using CeO_2-TiO_2 catalyst in liquid phase reaction. Thus, it is confirmed that CO is produced as an intermediate product, the CO was oxidized with O_2, and the reaction of CO to CH_3OH is the rate-determining step in the photocatalytic reduction of CO_2 to CH_3OH in liquid phase reaction. It is different if the reaction is occur in the gas phase reaction whereas the rate-determining step is the adsorption of the reactants and desorption of products over the catalyst surface [21].

Table 1. The results of the kinetic data by non-linear regression.

Equation	Constant	95 % interval of reliability	Correlation coefficient	Variance
(21) (Model 1)	$k_4 = 0.00034$	$\pm 8.745 \times 10^{-5}$	0.805	0.912
(22) (Model 2)	$k_6 = -0.078$	± 0.00035	0.991	0.064
	$k_7 = -3.020$	± 0.08060		
	$k_8 = -0.078$	$\pm 5.627 \times 10^{-6}$		

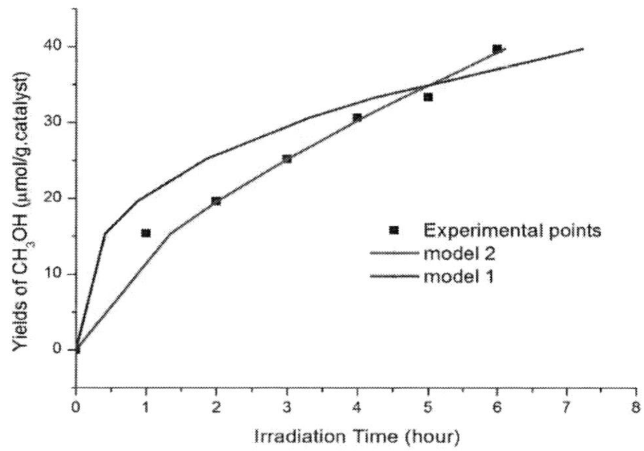

Figure 2. Comparison of model fitting with the experimental data for the formation of CH_3OH on the CeO_2-TiO_2 catalyst.

4. Conclusions

The experimental data obtained in this research were used to validate the model and it was fitted very well with the proposed model that considered the CO oxidation occur in the reaction. Thus, a kinetic model simulating the CH_3OH formations by photocatalytic reduction of CO_2 in NaOH solution on CeO_2-TiO_2 catalyst has been developed in this study and the kinetic constants have been determined with correlation coefficient of 0.991.

Acknowledgement

Funding from Universiti Malaysia Pahang (RDU 1603127) is gratefully acknowledged.

References

[1] Anandan S, Ikuma Y and Niwa K 2010 *Solid State Phenom.* **162** 239–60
[2] Ola O and Maroto-Valer M M 2015 *J Photochem. Photobiol. C Photochem. Rev.* **24** 16–42
[3] Patil S B, Basavarajappa P S, Ganganagappa N, Jyothi M S, Raghu A V and Reddy K R 2019 *Int. J. Hydrogen Energy* **44 (26)** 13022–39.
[4] Ambrozova N, Reli M, Sihor M, Kustrowski P, Wu J C S and Koci K 2018 *Appl. Surf. Sci.* **430** 475–87.
[5] Biswas M R U D, Ali A, Cho K Y and Oh W-C 2018 *Ultrason. Sonochem.* **42** 738–46.
[6] Koci K, Mateju K, Obalova L, Krejcikova S, Lacny Z, Placha D, et al. 2010 *Appl. Catal. B Environ.* **96 (3–4)** 239–44.
[7] Tahir M, Tahir B and Amin N A S 2015 *Appl. Surf. Sci.* **356** 1289–99.
[8] Daghrir R, Drogui P and Robert D 2013 *Ind. Eng. Chem. Res.* **52 (10)** 3581–99.

IOP Conf. Series: Materials Science and Engineering **702** (2019) 012026 doi:10.1088/1757-899X/702/1/012026

[9] Tong T, Zhang J, Tian B, Chen F, He D and Anpo M 2007 *J. Colloid Interface Sci.* **315 (1)** 382–8.

[10] Liu L and Li Y 2014 *Aerosol Air Qual. Res.* **14 (2)** 453–69.

[11] Koci K, Obalova L, Solcova O 2010 *Chem. Process Eng.* **31** 395–407.

[12] Abdullah H, Khan MR, Pudukudy M, Yaakob Z and Ismail N A 2015 *J Rare Earths* **33 (11)** 1155–61.

[13] Michalkiewicz B, Majewska J, Kądziołka G, Bubacz K, Mozia S and Morawski A W 2014 *J CO_2 Util.* **5** 47–52.

[14] Zhong H, Fujii K, Nakano Y and Jin F 2015 *J. Phys. Chem. C* **119 (1)** 55–61.

[15] Sasirekha N, Basha S and Shanthi K 2006 *Appl. Catal. B Environ* **62 (1–2)** 169–80.

[16] Tseng I-H and Wu JC-S 2004 *Catal. Today* **97(2–3)** 113–9

[17] Tseng I-H, Chang W-C and Wu J C S 2002 *Appl. Catal. B Environ.* **37(1)** 37–48.

[18] Tan S S, Zou L and Hu E 2008 *Catal. Today* **131(1–4)** 125–9.

[19] Tahir M and Amin N S 2013 *Appl. Catal A Gen* **467** 483–96.

[20] Tan S S, Zou L, Hu E 2006 *Catal. Today* **115(1–4)** 269–73.

[21] Delavari S, Amin N A S 2015 *Appl. Energy* **162** 1171-85.

IOP Conf. Series: Materials Science and Engineering **702** (2019) 012027 doi:10.1088/1757-899X/702/1/012027

Characterization of a novel terpolymer containing maleic anhydride-*co*-stearyl methacrylate-*co*-benhely acrylate for wax inhibition application

I Elganidi, B S Elarbe, N Abdullah* and N Ridzuan

[1]Faculty of Chemical and Process Engineering Technology, College of Engineering Technology, University Malaysia Pahang, 26300 Gambang, Pahang, Malaysia.

*E-mail: yatiabdullah@ump.edu.my

Abstract. Wax deposition in petroleum pipelines is deliberated as one of the most severe operational issues, and extraordinary efforts have been made to remediate and prevent this flow assurance problem. Using, polymeric pour point depressants (PPD) as a wax control strategy to improve the fluidity of waxy crude oil and reduce the pour point have received significant attention. The purpose of the current research is to synthesize novel terpolymers which used as PPD and viscosity improvers. The synthesis process occurred *via* polymerization of stearyl methacrylate (SMA), benhely acrylate (BA), and maleic anhydride (MA) monomers using free radical polymerization method in toluene as a solvent, and (1 wt%) benzoyl peroxide (BZP) as initiator. The chemical structure of the prepared terpolymer was characterized by FTIR and NMR. Also, the crystallization behavior of paraffin wax was investigated by differential scanning calorimetry (DSC). Experimental investigations furnish that the synthesized PPD showed high yield (87%) at the concentration (1:1:1) of BA-*co*-SMA-*co*-MA. From FTIR and ^{1}HNMR characterization results observed that maleic anhydride moiety, alkyl chain moiety, and methacrylate moiety are existing in the prepared terpolymer. Furthermore, DSC analysis showed that the glass transition temperature of the prepared terpolymer is weakly stated over a wide temperature range.

1. Introduction

The crude oil is a complex compound that consists of different hydrocarbons such as resins, asphaltenes, aromatics, and paraffin waxes [1-3]. Usually, in the grilling area or at high temperatures, the paraffin crystals stay liquefied in the crude. However, when the temperature decreases beneath the wax appearance temperature (WAT) the fluidity of crude oil becomes weak due to starting the precipitation of paraffin crystals in the crude and also on the inner wall of the pipeline [4-5]. The crystallization process of paraffin wax may lead to severe issues in the storage, production, and transportation due to loss in the flowing capability of crude oil as well as rising in the crude oil viscosity and some cases lead to blockage in the pipeline [6-7]. Paraffin wax could be successfully spread by using a little number of additives that are known as pour point depressants (PPD) [8-9]. There are different postulated mechanisms for such additives to form slight and regular shaped wax crystals, for instance, nucleation, adsorption, and co-crystallization which lead to increase in wax solubility [10-12]. Notably, there are no universal additives that can be effective for all types of crude oil. Consequently, the best wax inhibitor performance should be investigated for every well on an individual basis, and exact compatibility with the raw petroleum must be present. Many methods have used to mitigate and prevent the issue of wax deposition, and some of these techniques are chemical treatment, thermal, electromagnetic, mechanical, bacterial. Nevertheless, chemical treatment practices using wax inhibitor has shown the best result to mitigate the wax precipitation in the oil and gas sector [5, 13-14]. Moreover, treating the wax deposition issue using polymeric additives as chemical treatment is considered as the most valuable and economical method to inhibit the wax precipitation. These polymeric compounds consist of the polar section which is chargeable for the modification of wax crystals morphology, as well as, contain hydrocarbon chain which extends the interaction between the additives and paraffin [6, 15]. A considerable number of sorts of polymeric additives have been created and utilized as PPDs to impact the formation manner of paraffin crystals [12, 15-17]. However, in recent decades close attention has been made on the methacrylate group and maleic anhydride polymers due to the highly polar functional groups that contain [15, 18].

Moreover, Dao in (2017) has proved that stearyl methacrylate and behenyl acrylate with vinyl acetate using AIBN as initiator gives high efficiency to reduce pour point temperature [19]. Furthermore, many other researchers have proven that maleic anhydride with other different monomers shows good results in reducing the pour point and yield stress of crude oil [20-21]. The goal of this study is to synthesize a newly terpolymer which working as a pour point depressant (PPD) using maleic anhydride with stearyl methacrylate and behenyl acrylate *via* free radical polymerization method to inhibit the wax precipitation in pipelines and reservoir formation. These three monomers integrated a new terpolymer which has not yet been invented by other researchers in wax inhibition field. Consequently, the significant effort of this research is to characterize the newly synthesized polymeric additives by confirming the chemical structure and rheological characteristics of the prepared terpolymer.

2. Experimental section

2.1. Chemical Materials

The material used in this study are stearyl methacrylate (SMA) (96%), behenyl acrylate (BA) (99%), maleic anhydride (MA) (<99%), which obtained as analytical reagents from Aladdin chemicals and used as received, as well as, Xylene (≥98.5%) from Sigma- Aldrich as solvent in which the monomers dissolved. Also, benzoyl peroxide (75%) from Sigma- Aldrich used as an initiator of the reaction after the recrystallizing process from methanol.

2.2. Reaction yield

Typically, the efficiency of the chemical reaction in the polymeric synthetic is determined by the percentage of yield (%) [22]. The percent yield figures out by determining the amount of the monomers in the stoichiometry which known as theoretical yield. Whereas, the actual yield measured by determining the amount of the polymer after the chemical reaction carried out, and the percent yield can be measured using Equation 1[23].

$$\textit{Percent Yield} = (\textit{Actual Yield}/\ \textit{Theoretical Yield}) * 100 \qquad (1)$$

2.3. Preparation of Terpolymer

The polymerization reaction conducted in a 200 mL three-necked glass flask which was equipped with a magnetic stirrer and thermometer, as shown in Figure 1. The produced terpolymer prepared by polymerizing of stearyl methacrylate, maleic anhydride, and behenyl acrylate monomers at the concentration (1:1:1) mole ratios under free radical polymerization method. The polymerization is ideally directed with constant stirring (400 rpm) at 80°C for five hours in the presence of 50 mL of toluene as a solvent and benzoyl peroxide (1 wt%) as an initiator under a nitrogen environment. The obtained terpolymers were precipitated in excess of methanol, then filtered and dried under vacuum at 40°C for 12 h.

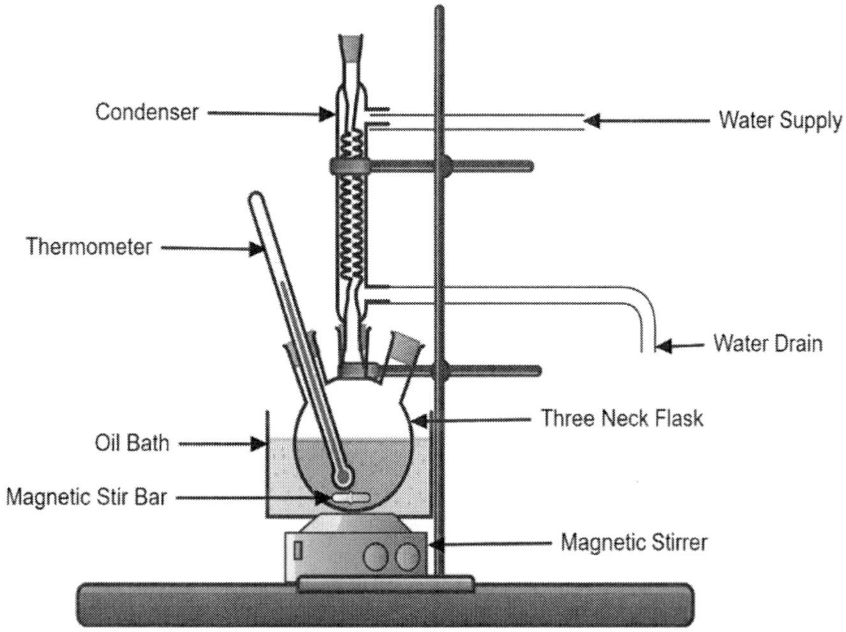

Figure 1. Experimental apparatuses.

2.4. Characterization of Prepared Terpolymer

To approve the chemical structure of the synthesized terpolymer Fourier transform infrared spectrometer (Nicolet iS5 FT-IR spectrometer) has been used. Also, H-NMR was conducted to determine unknown substances as well as the sample purity using Bruker ultra-shield plus 500MHz. Differential scanning calorimetry (DSC) was utilized to explore the response of polymeric additives to heating and also characterize the physical properties of the prepared terpolymer using TA Instruments Q1000 Differential Scanning Calorimeter.

3. Results and Discussion

3.1. Fourier Transform Infrared (FTIR) Spectroscopy

The chemical structure of the prepared terpolymer was characterized by FTIR spectrometry. Figure 2 demonstrates the FTIR spectra of the prepared BA-*co*-SMA-*co*-MA copolymer and clearly illustrates that the influential absorption band at 2916 and 2849 cm^{-1} attributed to stretching vibration of C-H bonds of methyl and methylene group as the similar findings that have been indicated previously by Darvishi et al. [24]. The figure also shows characteristics of C=O stretching peaks of maleic anhydride at 1781 cm^{-1} and sharp absorption peak of the characteristic C=O at 1728 cm^{-1} for stearyl methacrylate as similar observation by Song et al. [25]. The bending vibration peaks of the C-H bonds appear at 1466 and 1379 cm^{-1}, and weak asymmetrical stretching vibration peaks for the epoxy group appears at 835 and 959 cm^{-1}. As well, the typical medium absorption peak of characteristic -CH_2- in long-chain alkyl are distinguished at 720 cm^{-1}, and these results are in line with previous findings by M. A. Zhou et al. [18]. Therefore, the result represents that maleic anhydride moiety, alkyl chain moiety, and methacrylate moiety are existing in the prepared terpolymer and absenteeism of bands between 1640-1680 cm^{-1} which typify C=C group designate the completion of polymerization and this result in line previous reported by other researchers [26].

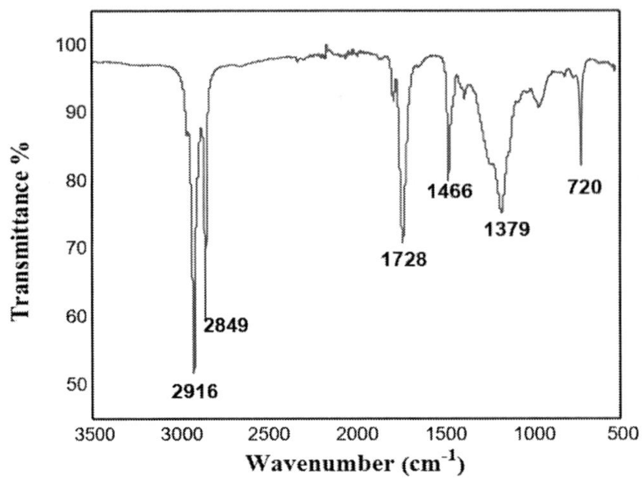

Figure 2. FTIR spectra of (BA-*co*-SMA-*co*-MA) terpolymer.

3.2. Proton Nuclear Magnetic Resonance (^1H NMR)

Further characterization utilizing ^{1}HNMR analysis has been conducted to accumulate more information on the chemical structure of the prepared terpolymer (BA-*co*-SMA-*co*-MA). Figure 3 represents the ^{1}HNMR spectra of the synthesized polymer (1:1:1-mole ratio of BA-*co*-SMA-*co*-MA). The chemical shifts for terpolymer showed a peak at 0.86 ppm was due to methyl of methacrylate chain, whereas the other peaks which ranged between 1.1-1.6 ppm characterized for methylene protons of acrylate chain these results are in accordance with conclusions stated by observation by Upadhyay et al. [27]. Besides, the spectra of the proton showed chemical shift at 3.7 ppm, and 4.1 ppm were due to the group of methylene which adjacent to an oxygen atom (-OCH_2) of methacrylate and behenyl acrylate respectively, a similar pattern of findings was obtained by Song et al. [25]. Nevertheless, the peaks of the proton of maleic anhydride polymeric state are not observable as the similar results reported by Mingan in 2015 [18]. These results designate that the alkyl chain moiety and methacrylate moiety are existing in the prepared terpolymer and same results have also been confirmed by FTIR analysis.

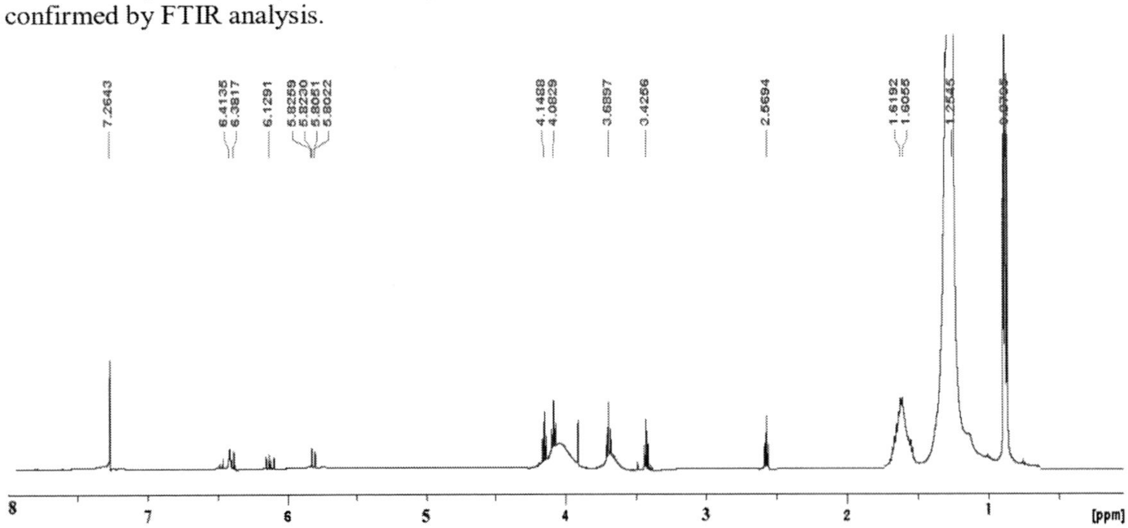

Figure 3. ^{1}HNMR spectra of the prepared terpolymer.

3.3. Differential Scanning Calorimeter (DSC)

Phase transitions and thermal properties for the prepared polymer are significant for the utilization of polymeric materials. Therefore, the thermal properties of the prepared terpolymer have been characterized using DSC. As observed from DSC analysis data in figure 3, the prepared terpolymer relies upon the sort and structure of monomers and these results are consistent with what has been found in the previous literature by Karažija et al. [28]. Also observed from figure 3, at the temperature -60 C° to approximately -10 C° the terpolymer in the glassy state, while at the temperature ranged from -10 C° to 40 C°, the glass transition of the synthesized terpolymer is observed. With additional heating, the crystalline phase of terpolymer melts. Furthermore, the temperature of the glass transition of synthesized terpolymer is weakly stated over a wide temperature range, and without apparent completion surpasses into an endothermic peak as a similar conclusion that obtained by Liu et al. [29]. This is because of the relaxation process in the glass transition phase, as reported previously by Beiner et al. [30]. Consequently, for more precise determination and comprehensive examination of phase transition, additional methods ought to be used as well.

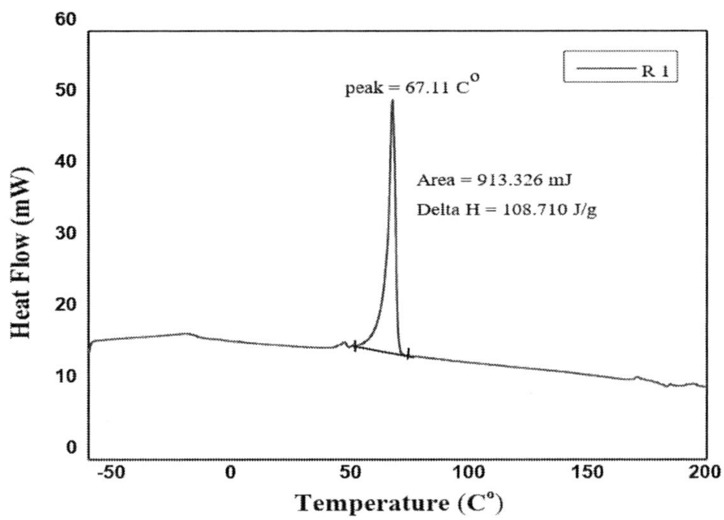

Figure 4. DSC curve of the synthesized terpolymer.

4. Conclusion

A newly synthesized terpolymer has been characterized using FTIR, NMR, and DSC. The synthesis process occurred *via* polymerization of stearyl methacrylate, benhely acrylate, and maleic anhydride monomers using free radical polymerization. The chemical structure of the prepared terpolymer has been confirmed using FTIR (Fourier transform infrared spectroscopy), and NMR (Nuclear Magnetic Resonance). Experimental investigations furnish that the synthesized PPD has high yield 87% at the concentration (1:1:1-mole ratio of BA-co-SMA-co-MA). From FTIR and ^{1}HNMR results observed that maleic anhydride moiety, alkyl chain moiety, and methacrylate moiety are existing in the synthesized terpolymer. Furthermore, DSC analysis reported that additional methods ought to be used as well to obtain precise determination and comprehensive investigation of phase transition.

Acknowledgments
Authors wishing to thank the Faculty of Chemical and Process Engineering Technology at University Malaysia Pahang for the financial support of this research using PGRS 1903103.

References

[1] Chen W, Zhang X, Zhao Z and Yin C 2009 *Fluid Phase Equilibria* **280** 9–15
[2] Duarte L, Filgueiras P, Silva S, Dias J, Oliveira L and Castro E 2016 *Fuel* **181** 660–9
[3] Kasumu A, Arumugam S and Mehrotra A 2013 *Fuel* **103** 144–7
[4] Jung K, Chun B, Park S, Lee C and Kim S 2011 *J. Appl. Poly.Sci.* **120** 2579–86
[5] Aiyejina A, Chakrabarti D, Pilgrim A and Sastry M 2011 *Int. J. Multiphase Flow* **37** 671–94
[6] Al-Sabagh A, Eldin M, Morsi R and Elsabee M 2009 *J. Petro. Sci. Eng.* **65** 139–46
[7] Yao B, Li C, Yang F, Zhang Y, Xiao Z and Sun G 2016 *Fuel* **184** 544–54
[8] Ghosh P and Das M 2014 *J. Petro. Sci. Eng.* **119** 79–84
[9] Han S, Song Y and Ren T 2009 *Energy & Fuels* **23** 2576–80.
[10] Chen W, Zhao Z and Yin C 2010 *Fuel* **89** 1127–32
[11] Fang L, Zhang X, Ma J and Zhang B 2012 *Industrial & Eng. Chem. Research* **51** 11605–12
[12] Liu T, Fang L, Liu X and Zhang X 2015 *Fuel* **143** 448–54
[13] Elbanna S, Abdelrhman A, Alhussaini A and Khalil S 2017 *Petro. Sci. Tech.* **35** 1047–54
[14] Jennings D and Breitigam J 2009 *Energy & Fuels* **24** 2337–49
[15] Wu Y, Ni G, Yang F, Li C and Dong G 2012 *Energy & Fuels* **26** 995–1001
[16] Cao K, Zhu Q, Wei X and Yao Z 2015 *Energy & Fuels* **29** 993–1000
[17] Lü Y, Zhang X and Yao G 2011 *Energy & Fuels* **25** 2115–8
[18] Zhou M, He Y, Chen Y, Yang Y, Lin H and Han S 2015 *Energy & Fuel* **29** 5618–24
[19] Than D, Chuong T and Tuy D 2017 *AIP Conf. Proc.* **1878** p 020002
[20] Soliman E, Elkatory M, Hashem A and Ibrahim H 2018 *Fuel* **211** 535–47
[21] Xu J, Xing S, Qian H, Chen S, Wei X and Zhang R 2013 *Fuel* **103** 600–5
[22] Isacgarcía J, Dobado J, Calvoflores F and Martínezgarcía H 2016 *Academic Press* **8** 29–43.
[23] Sheldon R 2012 *Chem. Society Reviews* **41** 1437–51
[24] Darvishi A, Zohuriaan M, Marandi G, Kabiri K, Bouhendi H and Bakhshi H 2013 *Designed Monomers and Polymers* **16** 79–88
[25] Song Y, Ren T, Fu X and Xu X 2005 *Fuel Processing Tech.* **86** 641–50
[26] Atta A, Elghazawy R, Morsy F, Ali M and Elmorsy A 2013 *Global J. Sci. Res. Chem.* **13** 21–7
[27] Upadhyay M, Saha D and Ghosh P 2018 *J. of Scientific & Industrial Research* **77** 652–6
[28] Karažija T, Vidović E and Jukić A 2013 *Polymer Eng. & Sci.* **53** 2299–307
[29] Liu G, Zhang L, Yao Y, Yang L and Gao J 2003 *J. Appl. Poly. Sci.* **88** 2891–6
[30] Beiner M 2001 *Macromolecular rapid communications* **22** 869–95

IOP Conf. Series: Materials Science and Engineering **702** (2019) 012028 doi:10.1088/1757-899X/702/1/012028

Screening of the factors for novel pour point depressant copolymer synthesis to improve the copolymer yield

B S Elarbe, N Ridzuan*, K Yusoh, I Elganidi and N Abdullah

Faculty of Chemical and Process Engineering Technology, College of Engineering Technology, Universiti Malaysia Pahang, Lebuhraya Tun Razak, 26300 Kuantan, Pahang, Malaysia

*E-mail: norida@ump.edu.my

Abstract. Pour point depressants (PPDs) commonly used in the oil and gas industry as a chemical wax inhibitor to control the temperature of the crude oil below their wax appearance temperature (WAT) points, and to improve the flowability of crude oil. In this study, different conditions of Stearyl Acrylate-Behenyl Acrylate (SABA) copolymer were synthesized by free-radical solution polymerization method at different range conditions of polymerization. Four factors considered during the synthesize of the copolymer are the mass ratio of monomers (1:1, 1:2, 1:3, 2:1, and 3:1 wt%), reaction temperature from 60 to 100 °C, the concentration of initiator from 0.5 to 2.5 wt% and reaction time from 5 to 9 hour. The results showed that the best conditions to obtain the highest yield of SABA copolymer were at the mass ratio of monomers of 1:1 (w/w), the reaction temperature of 90 °C, the reaction time of 7 h, and concentration of initiator at 1 wt%.

1. Introduction

In the previous years, copolymer synthesis with active functional groups as side alternatives of the significant chain was an active field of study in polymer science. In this case, the preparation of monomers consisting of a functional group is one of the most effective methods implemented for synthesizing copolymers [1].

In particular, one of the long-chain alkyl groups is stearyl acrylate (SA) is a category of acrylic monomer that has a good bond with nonpolar solvents, monomer finds out diffuse usage to synthesis pour point depressants. However, poly stearyl acrylate tends to be in a crystalline state. Thus, long alkyl side chains can crystallize despite an amorphous backbone that reduces oil absorbency. Regardless of various lower undesirable influences, the process of copolymerization with other monomers is used to obtain amorphous copolymers [2].

In the polymerization study, the monomer ratio, reaction temperature, reaction time, and concentration of initiator are crucial factors from the chemical industrial point of view, which contributes to improving the yield of the copolymer [3-4], particularly in tailor-made copolymers with desirable physical and chemical properties. Monomer ratios display some information about the relative reactivity of monomer pairs, clarification of copolymer synthesis, and monomer series distribution. [5, 6].

In last years, many studies were carried out on acrylic monomers because they form the tremendous multilateral category of monomers that used to form acrylate polymers based on acrylic acid structure, which contents of a carboxylic acid and a vinyl group [7-8]. Most researchers [4-5, 9-

12] have used acrylate and methacrylate monomers to produce PPD copolymers because of their very reactive double bonds at the vinyl group. Consequently, these copolymers can be utilized as a chemical inhibitor PPD to improve the flowability of crude oil and to reduce wax deposition into the pipeline. Substantially, acrylic and methacrylic ester groups are among the chemicals utilized for this goal since they show excellent capacity either as flow modifiers of crude oil or as wax deposition inhibitors [8].

In other words, pour point depressants (PPDs) can be widely applied as a chemical wax inhibitor to transport crude oil at a temperature below their wax appearance temperature points, widely improving safety and economics. In addition, PPD is a polymeric compound created fundamentally of hydrocarbon chains that interact with wax crystals to change wax morphology and preventing the aggregation stage [8, 13].

Recently studies revealed that PPD copolymers could decrease the viscosity of crude oils during transportation [14-17]. Styrene-maleic anhydride copolymer (PSMA) was synthesized by free-radical solution polymerization. It was found that the yield value and the dynamic viscosity have been decreased for the same sample and conditions [18]. Octadecyl acrylate–maleic anhydride copolymer was prepared by free-radical solution polymerization, According to the results, four types of chemical additives have shown excellent efficiency as PPDs and flow improvers in CQ crude oil with low asphaltene content [19]. Oleic acid-maleic anhydride copolymer (POMA) has been prepared by free-radical solution polymerization of various molar ratios. The obtained data showed that all chemical additives have a positive impact on the dynamic viscosity and Bingham values if they compared with the original results of the waxy crude oil [20].

Apart from that, one of the main problems facing the transportation of crude oil through the pipeline is high viscosity crude oil, which requires effective and expensive ways to transfer crude oil. Therefore, the higher viscosity of crude oil leads to an increase in pumping pressure from the ground, difficulties in transportation and refining operations.

Up till now, there are many monomers that have synthesized with stearyl acrylate to form a copolymeric particle such as styrene [9], glycidyl methacrylate [5], maleic anhydride [19, 21], and vinyl acetate [22]. Meanwhile, these monomers used as a chemical inhibitor to improve the flowability of crude oil. Therefore, for extending the application of polymerization, we have inserted a novel monomer that is behenyl acrylate (BA) to synthesized with (SA) for the formation of highly stable copolymeric particles. This study aims to synthesize and screening linear stearyl acrylate-behenyl acrylate copolymers at different conditions to obtain the optimum conditions and highest yield.

2. Experimental

2.1. Materials

Two monomers were used in this study; include stearyl Acrylate (SA, 97% purity) (purchased from Sigma Aldrich) and behenyl acrylate (BA, 95% purity) which was purchased from Aladdin chemicals. Other chemicals such as benzoyl peroxide 99% were used as the initiator, toluene 99% as solvent and methanol; all the chemicals were bought from Sigma Aldrich.

2.2. Experimental set up for synthesis of copolymer

Stearyl Acrylate-Behenyl Acrylate copolymer (SABA) was synthesized through the free-radical solution polymerization technique by utilizing different mass ratios between SA and BA (1:1, 1:2, 1:3, 2:1, and 3:1 wt%). The reaction was conducted in a 250 mL three-neck round-bottomed flask equipped with a thermometer, magnetic stirrer, reflux condenser, and nitrogen gas inlet. Figure 1 show the experimental apparatus. A mixture of monomers was dissolved in 50 mL of toluene under constant stirring at a range of temperature from 60 to 100 °C for 5 to 9 h of reaction time and a nitrogen atmosphere for 30 min. At this time, benzoyl peroxide (BPO) was dissolved in suitable toluene used as an initiator at a range between (0.5-2.5% w/w of a total of monomers). The solution of benzoyl peroxide was added drop by drop every 15 min during the first hour of the reaction. After the reaction had been completed, it was kept at room temperature; then, the precipitation of copolymer solution was initiated through the addition of methanol to reaction mixture alongside with rapid stirring.

Thereafter, the copolymer was filtered through a Buchner funnel and washed three times with methanol. The obtained product was dried in a vacuum oven at 40 °C for 12 h with the final state of SABA copolymer being a white powder.

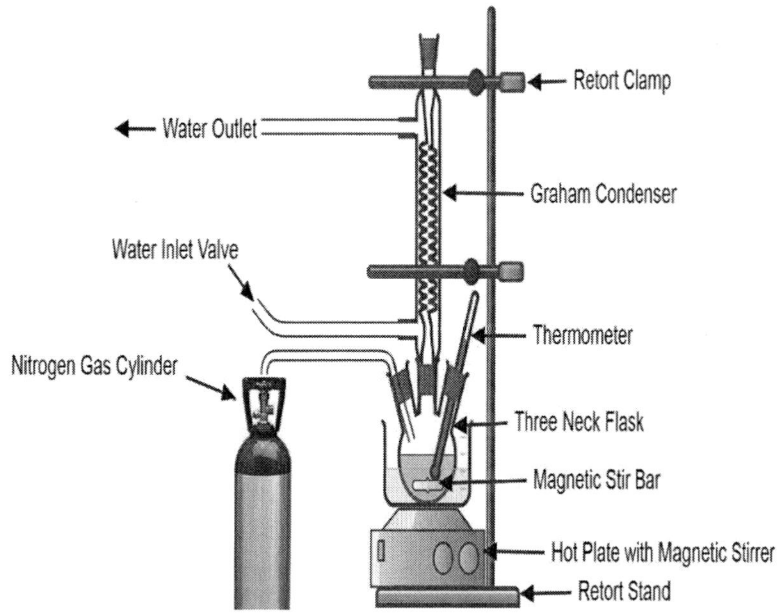

Figure 1. The experimental apparatus.

3. Results and discussion

3.1. Effect of SA: BA ratio on the yield

In this study, monomer ratios are critical factors in the polymerization, especially in tailor-made copolymers. Monomer ratios display information about the reactivity of monomer pairs, clarification of copolymer composition, and monomer sequence distribution [20]. Figure 2 illustrates the relationship between the mass ratio of stearyl acrylate (SA) and behenyl acrylate (BA) on the yield. As seen in Figure 2 there are slightly variations in the yields of the copolymers when the polymerization was conducted at different mass ratios. When the mass ratio of monomers was equal 1:1 (50 wt% to 50 wt%), the yield was high around 95.83%. However, if the mass ratio of stearyl acrylate monomer increases more than the behenyl acrylate monomer or vice versa such as (2:1, 1:2, 3:1, and 1:3), the yield gradually decreases from 95.83% to 80%. This is a clear indication that an equal amount of stearyl acrylate (SA) and behenyl acrylate (BA) can give an optimal copolymer yield. Thus, the highest yield of mass ratio was found at 95.83% of the yield at 1:1 ratio [23].

IOP Conf. Series: Materials Science and Engineering **702** (2019) 012028 doi:10.1088/1757-899X/702/1/012028

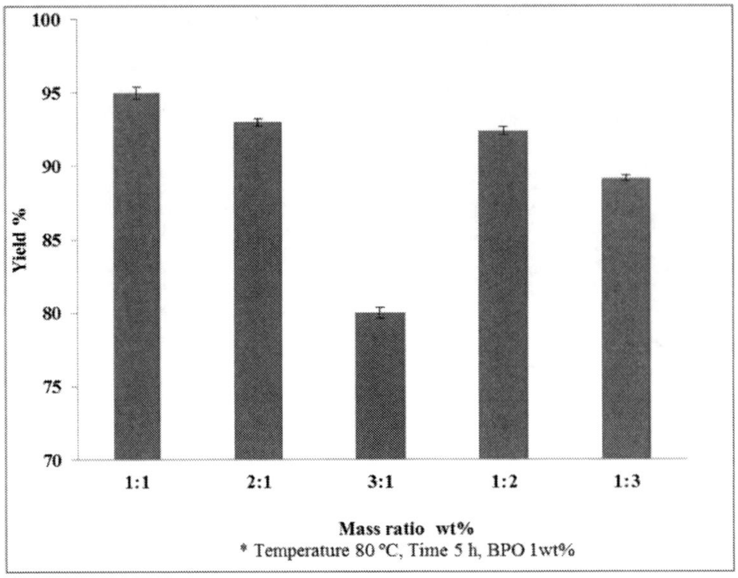

Figure 2. Relationship between the mass ratio of monomers and yield.

3.2. Effect of reaction temperature on the yield

A series of tests were conducted to observe the effect of temperature on the yields of the copolymer. The temperatures used during the polymerization were increased by 10 °C increments between 60 and 100 °C, as shown in Figure 3. From Figure 3, the yield of the copolymer increased with increasing the reaction temperature during the polymerization. However, at a temperature of 90 °C, the yield of the copolymer was slightly decreased from 92.5% at 90 °C to 90% at 100 °C. Therefore, the optimum polymerization reaction to completed and obtained the highest yield of copolymer was found 92.5% of the yield at 90 °C.

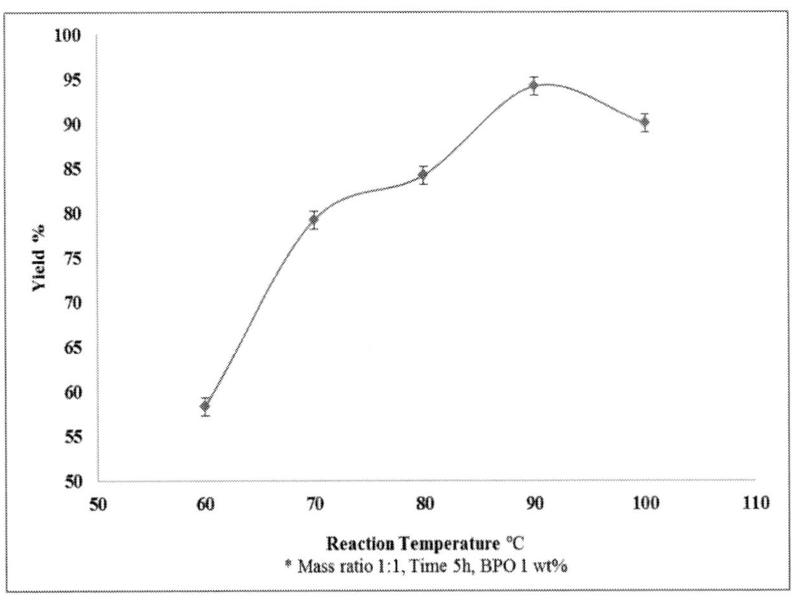

Figure 3. Relationship between reaction temperature and yield.

3.3. Effect of concentration of initiator on the yield

In order to understand the effective concentration of the initiator, a chain of the reaction was conducted at 90 °C whereby the initiator concentration fell between 0.5 and 2.5% wt of BPO, as shown in Figure 4. From Figure 4 the copolymer yields initially increased with increasing concentration of initiator from 80% to 95%. The highest yield was 95% at 1% of BPO. Nonetheless, beyond 1% of BPO, the copolymer yield started declining progressively until the yield remained rather similar at different concentration of BPO 1.5%, 2%, and 2.5%, respectively. At this point, the yield of copolymer remained unchanged at 94.16%. Therefore, this is an indicator for completion of the polymerization process at 1% of BPO, which is the highest yield. In terms of reaction yield, the highest yield of copolymer was 95% at 1% of BPO.

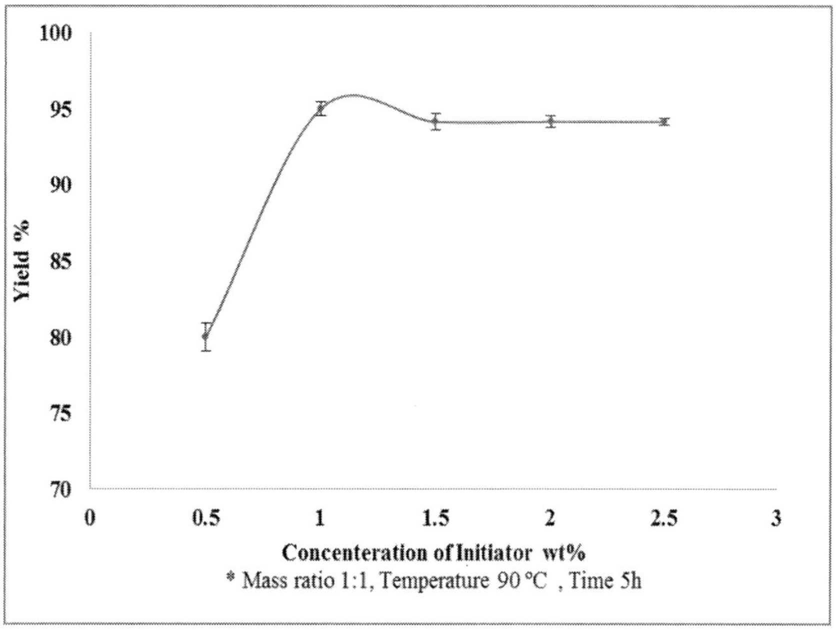

Figure 4. Relationship between the concentration of initiator and yield.

3.4. Effect of reaction time on the yield

As refer to Figure 5 the changing of the reaction time from 5 h to 9 h had a negligible influence on yield loss. Consequently, the result of copolymer yield exhibited a declining from 98.16% to 95.83% with increased reaction time from 7 to 9 hours as similarly reported by another researcher [24]. The copolymer yield showed a significant performance at the beginning of the process and attained the highest yield at 7 h. Meanwhile, there was a decline in the copolymer yields with increasing time. Therefore, this clearly showed that the polymerization process was completed at 7 h. In terms of the reaction yield, the highest yield polymerization of reaction time at 7h was 98.16%.

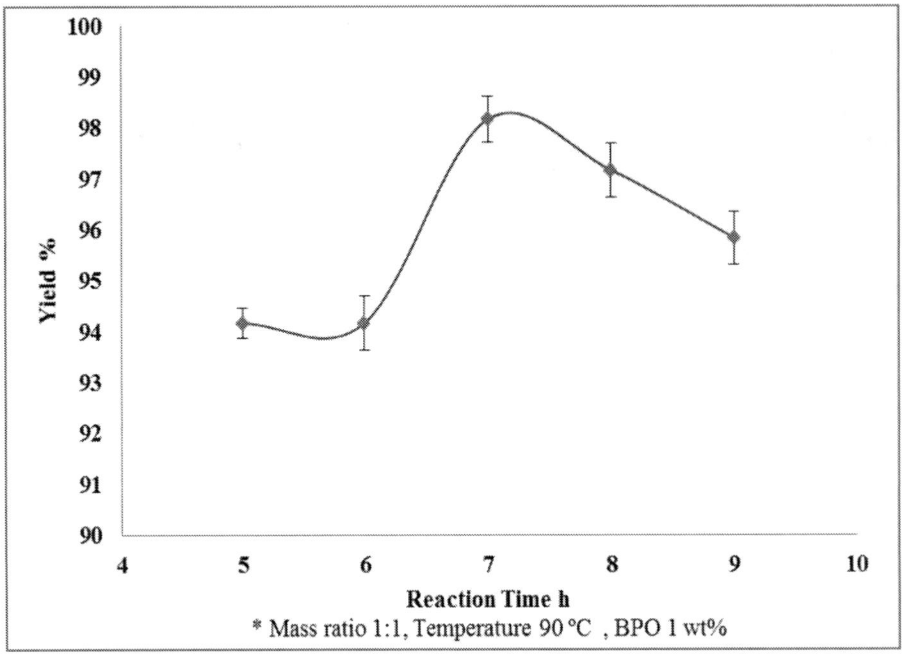

Figure 5. Relationship between reaction time and yield.

4. Conclusion

This study had investigated the screening of factors for synthesizes a novel pour point depressant copolymer. The best operating condition the obtain the highest yield of SABA copolymer is at a mass ratio of monomers is 1:1 (w/w), the reaction temperature of 90 °C, the reaction time of 7 h, and concentration of initiator of 1 wt%. Measuring viscosity will discuss further in the next study.

Acknowledgments

The authors wish to thank the Faculty of Chemical and Process Engineering Technology, College of Engineering Technology (Universiti Malaysia Pahang) to support this study financially under grants post graduate research scheme (UMP.5/26.10/03/PGRS1903102) and RDU170352. We also acknowledge the contributions of Petronas Penapisan Terengganu, Malaysia, for providing the samples of crude oil used in this study.

References

[1] Rodriguez F, Cohen C, Ober CK and Archer L 2014 *Prin. polym. sys. CRC Press*
[2] Farag RK 2008 *Int. J. Polym. Mater.* **57** 189
[3] Kang C, Crockett R and Spencer ND 2016 *Polym. Chem.* **7** 302
[4] Goswami PK, Kashyap M, Das PP, Saikia PJ and Handique JG 2019 *J. Disper. Sci. and Tech.* **1** 9
[5] Darvishi A, Zohuriaan Mehr MJ, Marandi GB, Kabiri K, Bouhendi H and Bakhshi H 2013 *Desi. Monoms. and Polyms.* **16** 79
[6] Singhal R and Gupta K 2016 *Polym. Plas. Tech. and Eng.* **55** 54
[7] Ballard N and Asua JM 2018 *Progs in Polym. Sci.* **79** 40
[8] Yang F, Zhao Y, Sjöblom J, Li C and Paso KG 2015 *J. Disper. Sci and Tech.* **36** 213
[9] Zhu X, Gu Y, Chen G, Cheng Z and Lu J 2004 *J. appl. polym sci.* **93** 1539
[10] Ghosh P, Hoque M, Karmakar G and Das MK 2017 *Int. J. Ind. Chem.* **8** 197
[11] Yang F, Cheng L, Liu H, Yao B, Li C, Sun G and Zhao Y 2018 *Energy & Fuels.* **32** 373
[12] Yang F, Paso K, Norrman J, Li C, Oschmann H and Sjöblom J 2015 *Energy & Fuels.* **29** 1368
[13] Fang L, Zhang X, Ma J and Zhang B 2012 *Ind. & Eng. Chem. Res.* **51** 11605

[14] Sun H, Lei X, Shen B, Zhang H, Liu J, Li G and Wu D 2018 *J. energy chem.* **27** 1198
[15] Chala GT, Sulaiman SA and Japper-Jaafar A 2018 *J. Non-Newtonian Fluid Mechanics.* **251** 69
[16] Wang F, Zhang D, Ding Y, Zhang L, Yang M, Jiang B and Zhang S, Ai M, Liu G, Zhi S and Huo L 2011 *Chinese sci. bull.* **56** 14
[17] Zhu L, Wang Y, Wang S, Huo T, Jing X, Li A and Xia D 2018 *J. Petrol. Sci. and Eng.* **163** 37
[18] Al-Sabagh A, El-Din MN, Morsi R and Elsabee M 2009 *J. Petrol. Sci. and Eng.* **65** 139
[19] Wu Y, Ni G, Yang F, Li C and Dong G 2012 *Energy & Fuels* **26** 995
[20] Al-Sabagh A, El-Hamouly S, Khidr T, El-Ghazawy R and Higazy SA 2013 *J. Disper. Sci. and Tech.* **34** 1585
[21] El-Ghazawy RA, Atta AM and Kabel KI 2014 *J. Petrol. Sci. and Eng.* **122** 411
[22] Yao B, Wang L, Yang F, Li C and Zhao Y 2016 *Energy & Fuels.* **31** 448
[23] Zhu L, Jing X, Wang S, Chen J, Xia D and Li A 2015 *Petrol. Sci. and Tech.* **33** 452
[24] Zhu L, Wang S, Huang C, Tian Y, Yang X, Jing X and Li A 2018 *IOP Conf. Ser: Earth and Environ. Sci.* **108** 022032

IOP Conf. Series: Materials Science and Engineering **702** (2019) 012029 doi:10.1088/1757-899X/702/1/012029

Treatment of Palm Oil Mill Effluent (POME) using Membrane Anaerobic System (MAS)

M S Hybat[1], N H Abdurahman[1*], Z H Yasmeen[1], Z Jemaat[1] and N H Azhari[2]

[1] Faculty of Chemical and Process Engineering Technology, College of Engineering Technology, Universiti Malaysia Pahang, 26300 Kuantan, Pahang, Malaysia.
[2] Faculty of Pure and Applied Science International University of Africa Khartoum, Sudan

*E-mail : nour2000_99@yahoo.com

Abstract. The directly discharged of Palm Oil Mill Effluent (POME) into the river causes environmental hazards due to the biochemical oxygen demand (BOD) and high Chemical Oxygen Demand (COD). The main issue of the traditional methods for POME treatment is that are not suitable for high suspended solid wastewaters and their gas production is less efficient at a high treatment volume. Therefore, in this study Membrane Anaerobic System (MAS) was used as alternative effective method for treating POME. The result obtained from the six steady states with concentration ranging from 11,048 to 15,700 mg/L mg/L for the MLSS, while the MLVSS recorded a concentration range between 10,540 and 17,600 mg/L. Moreover, the COD removal efficiency and HRT recorded from 94 to 97% and 150 to 10 days, respectively. Also, the coefficient of microorganism yield, decay rate, and the produced methane gas were obtained as 0.52g VSS/g COD, 0.31 day-1 and 0.182 to 0.564 l/g, respectively. Kinetic equations from Monod, Contois, and Chen, & Hashimoto were employed to describe the kinetics of POME treatment at organic loading rates ranging from 0.5 to 13 kg COD/m3/day. POME were characterized using Transform Infrared (FTIR) spectroscopy, in the regions of 900 to 1740 and 2800 to 3400 cm-1. Scanning Electron Microscope (SEM), the obtained result confirmed the potential of Membrane Anaerobic System (MAS) for the efficient treatment of POME effluent.

1. Introduction

Palm oil mill effluent is usually associated with palm oil processing which uses large volume of wastewater at high COD, BOD, and high temperatures, 80 to 90 °C [1]. In the recent years, the palm oil industry has been growing rapidly in Malaysia. Palm oil has been the most produced and consumed vegetable oil in the world, and it has been widely used in the field of cosmetics, hygiene, and food products due to its high oxidative stability, affordable price, and efficient production. However, Palm Oil Mill Effluent (POME) is immersive of Chemical Oxygen Demand (COD) and Biochemical Oxygen Demand (BOD) at high level which is discharged into the river and land before treatment at 75–85°C. Previous studies reported that POME had high contents of organic matter melanoid which are generated from the fresh fruit bunches during the sterilization process, and carotene pigment as well as other compounds such as polyphenol compounds, polyalcohol, etc. The physicochemical characteristics of raw POME which was used in this current work was thick with a dark brownish colour and oily with a viscous texture. Its obnoxious odour is usually dangerous to both aquatic life and human beings [2].
Traditional methods which were adopted in order to treat the problem of waste effluent (POME) could not obtained methane. Thus, the main drawback of these traditional methods is that it is not suitable at

high Organic Loading Rate (OLR), at which these methods were rated [3-4]. In addition, POME has low PH which is hard to be treated by conventional methods [5-6]. The traditional method of anaerobic digestion is the regular primary treatment of POME [7-8] and it was found that ponding system for POME treatment adopted more than 85% of the POME, due to its low capital and operating costs [9]. However, this system has disadvantages such as large land area requirement and long retention time [10]. Therefore, this study intends to use membrane separation techniques to overcome the above-mentioned problems.

MAS has shown economic viability, good ability in recycling high biomass solids and has long solid retention time that leads to degrade all particulate matter. In addition, the proposed method is effective for separating high biomass solids from digester. The kinetic equations in table 1 were found to be applicable to the anaerobic treatment of palm oil mill effluent.

Table 1. Mathematical expressions of specifics substrate utilization rates for known kinetic model

Kinetic Model	Equation 1	Equation 2
Monod	$U = \frac{k\,S}{k_s + S}$	$\frac{1}{U} = \frac{K_s}{K}(\frac{1}{S}) + \frac{1}{k}$
Contois	$U = \frac{U_{\max} \times S}{Y(B \times X + S)}$	$\frac{1}{U} = \frac{a \times X}{\mu_{\max} \times S} + \frac{Y(1+a)}{\mu_{\max}}$
Chen & Hashimoto	$U = \frac{\mu_{\max} \times S}{Y\,K\,S_o + (1-K)\,S\,Y}$	$\frac{1}{U} = \frac{Y\,K\,S_o}{\mu_{\max}\,S} + \frac{Y(1-K)}{\mu_{\max}}$

2. Materials and Methods

In this study, the sample were collected from a near local palm oil mill in Lepar Hilir, Kuantan, Malaysia. Temperature ranged from 75 °C to 90 °C and the wastewater was stored in a cooled room at 4 °C before it was used in this experimental study. An effective volume of 60-litre of row pome was treated in the laboratory digester by using MAS consisting of a Cross-flow Ultra-Filtration membrane (CUF) apparatus, pump, and anaerobic reactor shown in figure 1. By gravity flow, the feed system was designed to provide continuous addition of feed volume from feeder tank. The type of membrane of this design was used in the UF membrane module which had a Molecular Weight Cut-Off (MWCO) of 200,000, a tube diameter of 1.25 cm, and an average pore size of 0.1 μm. The length of each tube was 30 cm. The total effective area of the two membranes was 0.024 m^2. The maximum operating pressure on the membrane was 55 bars at 70 °C, and the pH ranged from 2 to 12. In this study, the operating pressure was maintained between 1.5 and 2.5 bars by manipulating the gate valve at the retentate line after the CUF unit.

IOP Conf. Series: Materials Science and Engineering **702** (2019) 012029 doi:10.1088/1757-899X/702/1/012029

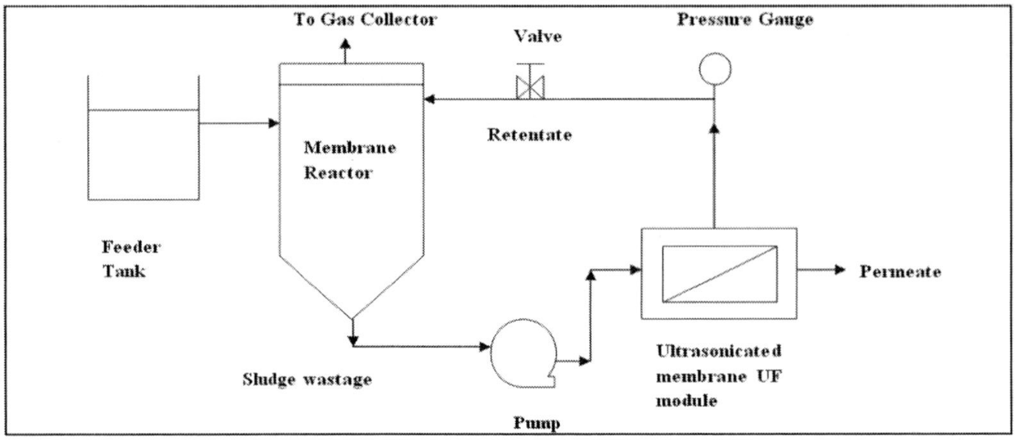

Figure 1. Experimental set-up

2.1. Analysis

Samples were analysed for Chemical Oxygen Demand (COD), Mixed Liquid Suspended Solids (MLSS were determined by using dichromate reflux (HACH Water Analysis method)), Total Suspended solid (TSS), Volatile Suspended Solids (VSS), Substrate Utilization Rate (SUR), Specific Substrate Utilization Rate (SSUR), and pH were determined by the pH meter.

2.2. Bioreactor operation

The performance was evaluated under six steady states as running digester at constant temperature, feed input, output, and gas. The process was evaluated by Organic Loading Rates (OLR) between 0.5 and 13kg COD/m3/d, alongside COD concentrations ranging from 72,100 to 84,400mg/L the condition of steady state was considered acceptable in this study when the operating and control parameters were within ± 10% of the averages. The gas volume was measured daily by using A 20-litre water displacement bottle. J-tube gas analyzer was used to perform this analysis, as shown in figure 1 above. The method assumed that the biogas produced CO_2 and CH_4 gases, and then the sodium hydroxide solution (NaOH) was added to absorb CO_2.The remaining volume was methane gas (CH_4).

3. Result and discussion

The applicability of three substrate utilization kinetic models (Monod, Contois, and Chen, and Hashimoto) were tested and tabulated in table 2 linear relationship and the coefficients. The six steady states were achieved when feed COD, TSS, VSS, PH reactor, and permeate were within increased 10% variation of each average the data for six steady states were summarized in Table 3. In this study the steady states were taken between 9 and 13 days to achieve steady state after each new organic loading rate during the six steady states, the organic loading rate increased from 0.5 to 13kg/COD/m3/d corresponding to the highest influent COD which was recorded at the sixth steady state (84,800mg/L), the steady state conditions with influent COD concentration of 72,100-84,800 mg/L and pH within the optimal working range for anaerobic digesters (6.7-7.8) achieved good performance of MAS.

In addition, the Volatile Suspended Solids (VSS) fraction in the reactor rose to 90.6% of the MLSS. This indicates that the long solid retention times and SRT of MAS are beneficial to the anaerobic contact process over the conventional process because of the ability to work well in different SR. Also, it predicted the decomposition of the suspended solids and the conversion to (CH_4). Table 2 shows that the MLVSS concentration was 14226 mg/L, whereas MLSS concentration was 15700 mg/L; this result was low due to high amount of suspended solids contents in the POME. At six steady states the MAS signified 95% COD removal during an effluent COD of 84800. This value is better than that which was reported by [11].

Thus, this study observed six steady states with an Organic Loading Rate (OLR) of 13 kg COD/m$^{3/}$daily. The highest influent COD was recorded at the sixth steady state (84800mg/L). The value was better than those reported in other studies on anaerobic POME digestion [12,13]. It also found that the Volatile Suspended Solids (VSS) fraction in the reactor increased to 90.6% of the MLSS. This indicates a conversion of the suspended solids to (CH_4). The Hydrolytic Retention Time (HRT) for the first steady state was too high. This may be attributed to very little amount of feeding to the system in the first steady state, S1.

However, the three kinetic models by Monod and Chen and Hashimoto demonstrated an acceptable relationship (R2 > 90%) for the membrane anaerobic system treating POME, as shown in figures 2 to 4. Based on the result of this study, the three kinetic models by Monod and Chen, and Hashimoto showed a better performance, implying that digester organic loading rates quantify and rationalize the digester operation.

Table 2. Results of the application of three known substrate utilization models

Model	Equations	R^2 (%)
Monod	$U^{-1} = 2025\ S^{-1} + 3.6$	0.99
	$K_s = 498$	
	$K = 0.350$	
	$\mu_{Max} = 0.259$	
Contois	$U^{-1} = 0.306\ X\ S^{-1} + 2.78$	0.9993
	$B = 0.111$	
	$\mu_{Max} = 0.344$	
	$a = 0.115$	
	$\mu_{Max} = 0.384$	
	$K = 0.519$	
Chen & Hasimoto	$U^{-1} = 0.0190\ S_0 S^{-1} + 3.77$	0.9998
	$K = 0.006$	
	$a = 0.006$	
	$\mu_{Max} = 0.277$	
	$K = 0.374$	

Table 3. Summary of results (SS: steady state)

Steady State (SS)	1	2	3	4	5	6
COD feed, mg/L	72,100	73,400	75,400	78,900	81,700	84,800
COD permeate, mg/L	1250	1420	1865	2393	3412	3608
Gas production (L/d)	280	290	313	352	370	399
Total gas yield, L/g COD/d	0.26	0.33	0.45	0.69	0.77	0.83
% Methane	70	71.3	68	67.8	70.1	68
Ch_4 yield, l/g COD/d	0.182	0.235	0.306	0.467	0.539	0.564
MLSS, mg/l	11,048	12,410	13,750	14,680	14,800	15,700
MLVSS, mg/L	9,0593	10,450	12,054	13,054	13,262	14,226
% VSS	82	84.2	87.6	88.9	89.6	90.6
HRT, d	150	73.0	25	13.1	11.6	10.0
SRT, d	200	100	70	50.0	30.0	15.0
OLR, kg COD/m^3/d	0.5	1	3	6	7	10
SSUR, kg COD/kg VSS/d	0.152	0.182	0.274	0.427	0.586	0.664
SUR, kg COD/m^3/d	0.472	0.8997	3.3028	5.6657	7.7753	9.4528
Percent COD removal	97.0	96.7	96.2	95.7	95.6	95

IOP Conf. Series: Materials Science and Engineering **702** (2019) 012029 doi:10.1088/1757-899X/702/1/012029

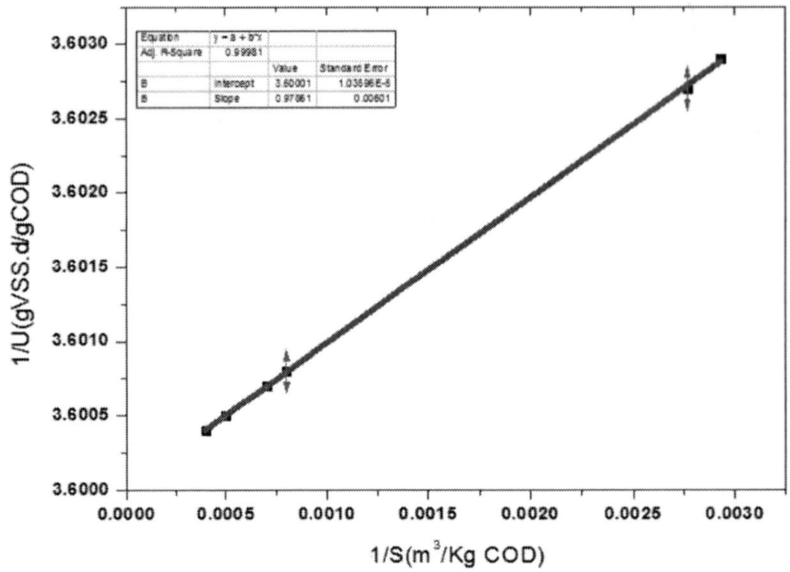

Figure 2. The Monod model

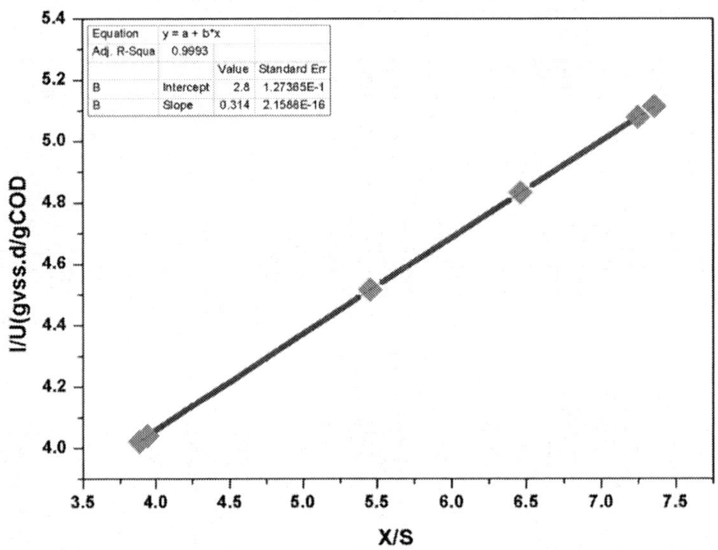

Figure 3. The Contois model.

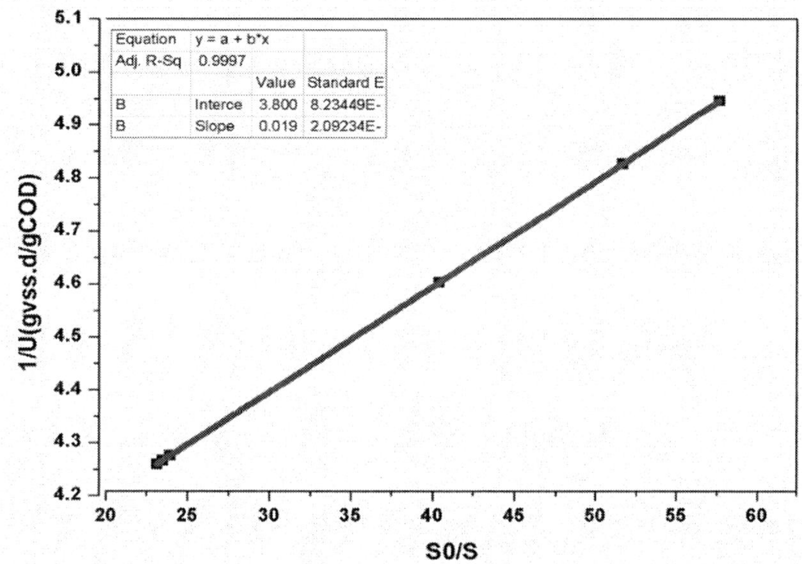

Figure 4. Chen and Hashimoto model

3.1. Gas Production and Composition

To achieve accurate and good result, and to prevent the failure of the performance of anaerobic digester, factors such as pH, temperature, mixing, organic loading rate and nutrient availability must be controlled. Methanogenic activity will decline with the pH in the system of digester perversities from the optimum level (6.8 - 8). In addition, to reduce the resistance of mass transfer and obtain better contact between substrate and microbes by mixing, as shown in figure 5, the observed methane content decreased from 0.26 to0.81/g COD/d while the OLR decreased from 0.05 to 13 kg COD m^3 because the acid bacteria grew more than methanogenic bacteria. In addition, figure 6, showed COD removal efficiency from 95% to 97% increasing, while the HRT increased from 10 to 150 days due to the increasing in contact time between waste water and granular sludge and enhanced COD removal.

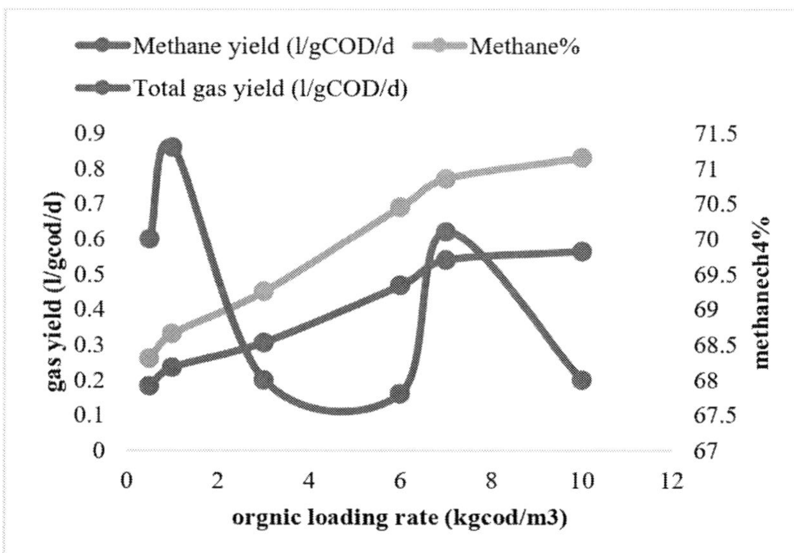

Figure 5. Organic loading rate (kg COD/m^3/day)

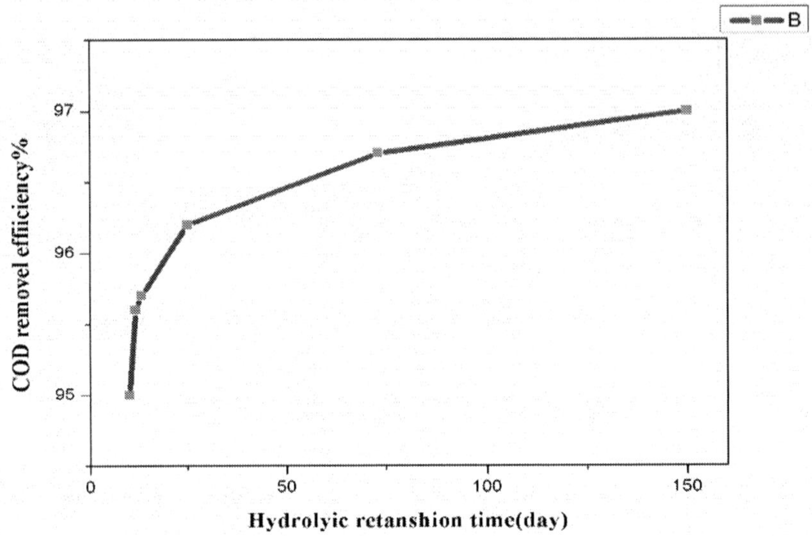

Figure 6. COD removal efficiency of IUMAS under steady state conditions with various hydraulic retention times

3.2. Characterization of POME

Characterization is a tool which has to do with the development, control and optimization of the wastewater processes by the characterization of the activated sludge. The physical and chemical properties of wastewater (POME) were characterized using various analytical techniques such as Scanning Electron Microscope (SEM) and Fourier Transform Infrared (FTIR) spectroscopy.

3.3. Elemental composition and spectrum analysis of the POME (SEM-EDXS) micrographs imaging of POME

The SEM showed the morphology of untreated and treated POME while the EDXs showed the propositional elemental composition spectrum demonstrated by the micrograph imaging and elemental analysis of untreated and treated POME. Figures 7 shows the micrographic imaging of the treated and untreated POME. Thus, Figure 7 (b) showed irregular disperse rod- shape and rough surface image in the bacteria cell of the composite for untreated POME. In this figure also, an external density can be noticed in the hall surface, where this observation is similar to [14]. In addition, the surface showed pore indicating that the gas bubbles were generated in the process inside the reactor. The study used the reactors at high loading rates, which justified the greatest quantity of morphologies observed when compared to [15]. However, in a complex network the appearance of the microorganisms was like noodles and they were connected to each other where Bello et al., has a similar observation [16]. On the other hand, Figure 7 (a) presents that the observed treated POME had smooth surface and had close pores compared with figure 8, which may indicate the interparticle bridging and effective degradation of the particles. Obviously, the tow images can be presented and observed in various textural features between the networks of structures.

SEM-EDX was employed to further ascertain the composition of the formed elemental peak along with others. Figure 8 reveals the elemental composition of the treated and untreated POME. Table 4 shows the major elements observed during the SEM experiment such as carbon and oxygen. In addition, Saphira et al., 2018, they traced amount of phosphorus, sulphur, potassium and calcium that appeared in this result [17]. The high concentration of carbon and low concentration of sulphur is an indication that the sludge can be used as biomass energy source for power generation. The presence of a significant quantity of carbon provided active surface for the attachments of the organic pollutants to the surface of the composite. Over all, the difference in elementals composition peak indicated in the POME treatment further affirmed the biodegradation potential of the model. In row POME, the result presented a concentration of NH3eN which is an evidence of the importance of the treatment of POME before it was discharged into the river or land to avoid the eutrophication phenomenon. The toxic elements such as cd2 and pb2 in row pome are hazardous in the water and to human health [18].

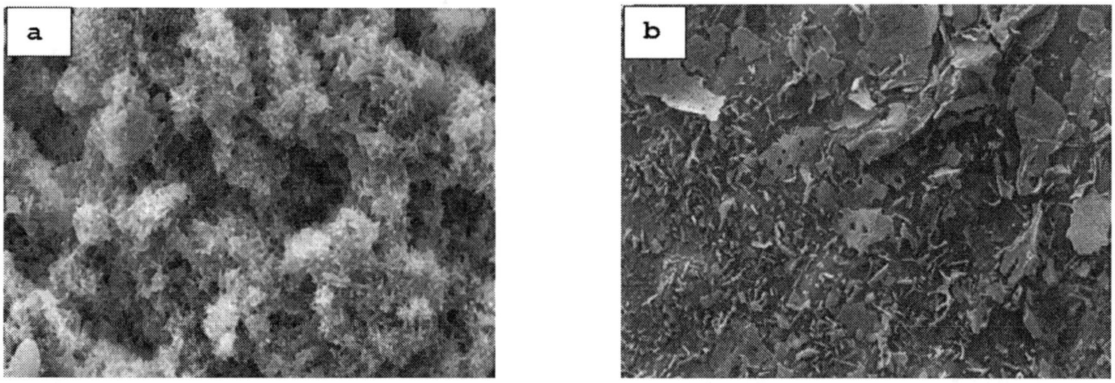

Figure 7. Scanning Electron Microscope (SEM) (a) before treated 500XK POME (b) treated POME

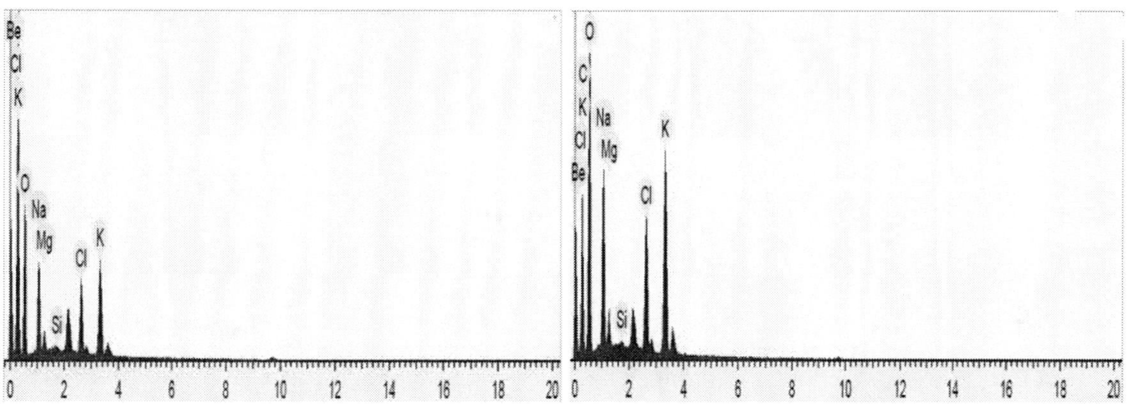

(a) (b)

Figure 8. Elemental composition (a) after treatment (b) before treatment

Table 4: Characteristics of POME before and after treatment

Element	Weight %Before	Weight % After
Oxygen	48	54
Sodium	18	12.637
Magnesium	4.421	2.415
Silicon	0.567	0.299
Chlorine	14.2	9.527
Cadmium	0.231	0
Potassium	17.858	15.583

3.4. Fourier Transform Infrared (FTIR) spectroscopy

The Fourier Transform Infrared (FTIR) spectroscopy was used to discover the functional groups which were presented in the POME samples. FTIR works on the concept of absorption, or emission of solid liquid, or gas. In addition, FTIR has in a wide range ability to collect high resolution data process. Also, FTIR at times, can measure intensity over a narrow range of wave lengths. In this study, FTIR spectra of POME before and after treatment are shown in figure 9. The spectra of row POME in Figure 9(a) showed that the broad spectrum absorption observed at $3391.1m^{-1}$ was designated to OH stretching vibrations in intra molecular H [18]. On other hand, figure 9(b) shows that the Pome after treatment showed that a broad absorption band at 3391.cm^{-1} was reduced to 3250 cm^{-1} due to the completed digestion process of POME. However, due to the consumed component of lipid and fat, the shoulder peak 2855 cm^{-1} disappeared in figure 9(b) after treatment. In addition, the band attributed to C-O-C vibrations in esters was obtained at 1088.75, having the same result as [19]. The absorption spectrum between 1130 and 1260 cm^{-1} POME after treatment were similar for lignin [18]. In addition, two absorptions band appear between1449 and 1417cm-1 related to the methyl asymmetric CH, designated to the effective degradation of the particle in the model [20].

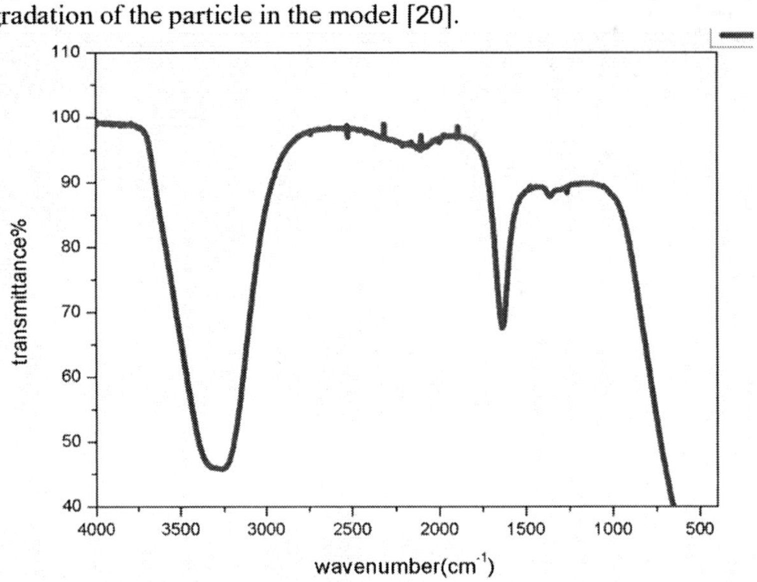

(a)

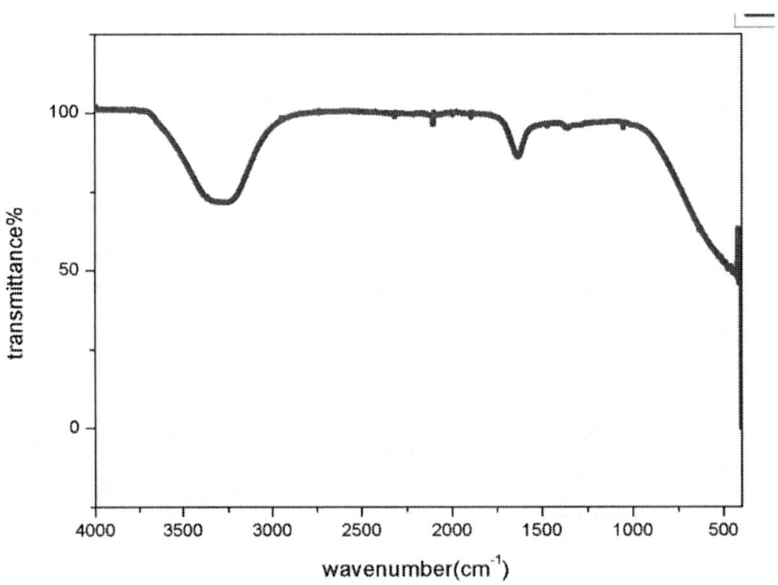

(b)

Figure 9. FTIR for POME (a) before treatment (b) after treatment

4. Conclusion

MAS has been found to be a successful alternative treatment system that achieved high COD removal efficiency (94%-97%) in a short period of time and the recovery of methane CH_4 for energy would be an additional valuable of the process (67.8-70.3%). Therefore, MAS is a good alternative for treating high strength wastewaters. The proposed kinetic equations have been found to be applicable to the anaerobic treatment of Palm Oil Mill Effluent (POME). MAS has been able to operate at a high SRT and has been quite tolerant of variations in influent COD loadings. The SEM characterization and FTIR analysis supported the success of modification with the increase in a peak at C−H group, C=C, O-H and the presence of C=O.

Acknowledgments

This study was kindly supported by the postgraduate student scholarship grant (PGRS 180373) of Universiti Malaysia Pahang.

Nomenclatures

COD	chemical oxygen demand (mg/l)
OLR	organic loading rate (kg/m^3/d)
CUF	cross flow ultra-filtration membrane
SS	steady state
SUR	substrate utilization rate (kg/m^3/d)
TSS	total suspended solid (mg/l)
MLSS	mixed liquid suspended solid (mg/l)
HRT	hydraulic retention time (day)
SRT	solids retention time (day)
SSUR	Specific substrate utilization rate (kg COD/kg VSS/d)
MAS	Membrane Anaerobic System
MLVSS	mixed liquid volatile suspended Solid (mg/l)
VSS	volatile suspended solids (mg/l)
MWCO	molecular weight Cut-Off
BLR	biological loading rate
U	specific substrate utilisation rate (SSUR) (g COD/G VSS/d)
S	effluent substrate concentration (mg/l)
S_o	influent substrate concentration (mg/l)
X	micro-organism concentration (mg/l)
μ_{max}	Maximum specific growth rate (day^{-1})
K	Maximum substrate utilisation rate (COD/g/VSS.day)
K_s	Half velocity coefficient (mg COD/l)
X	Micro-organism concentration (mg/l)
B	specific microorganism decay rate (day^{-1})
Y	growth yield coefficient (gm VSS/gm COD)
T	time

References

[1] Yee T L, Rathnayake T and Visvanathan C 2019 Performance Evaluation of a Thermophilic Anaerobic Membrane Bioreactor for Palm Oil Wastewater Treatment *Membranes (Basel)* 9(4) 55

[2] Mohammed R R and Chong M F, 2014 Treatment and decolorization of biologically treated Palm Oil Mill Effluent (POME) using banana peel as novel biosorbent *Journal of Environmental Management* **132** 237-49

[3] Zaki Y H, Nour A H and Mohamed H S 2018 Application of Membrane Bioreactor in Wastewater Treatment *International Journal of Innovative Research and Scientific Studies* **1**(1)

[4] Nour A H, Zaki Y H, Mohamed H S, Rassem H H 2019 The Potentials of an Integrated Ultrasonic Membrane Anaerobic System (IUMAS) in Treating Sugar Cane Wastewater *Indones. J. Chem.* **19**(3) 804-10

[5] Poh P E, and Chong M F 2009 Development of anaerobic digestion methods for palm oil mill effluent (POME) treatment. *Bioresource Technology* **100**(1) 1-9.

[6] Abdurahman N H, Rosli Y M, Azhari N H, and Bari H A 2015 The potential of ultrasonic membrane anaerobic systems in treating slaughterhouse wastewater *Water Reuse and Desalination* **5**(3) 293-300

[7] Liew W L, Kassim M A, Muda K, Loh S K, and Affam A C 2015 Conventional methods and emerging wastewater polishing technologies for palm oil mill effluent treatment: A review *Journal of Environmental Management* **149** 222-35

[8] Rupani P F, Singh R P, Ibrahim M H, and Esa N 2010 Review of current palm oil mill effluent (POME) treatment methods: vermicomposting as a sustainable practice *World Applied Sciences*

[9] Borja-Padilla R and Banks C J 1993 Thermophilic semi-continuous anaerobic treatment of palm oil mill effluent *Biotechnology Letters* **15**(7) 761-66

[10] Ng W J, Wong K K and Chin K K 1985 Two phase anaerobic treatment kinetics of palm oil wastes *Water Sci. Technol.* 19(5) 667-69

[11] Abdurahman H N, Yunus R M, Azhar H N 2011 Development of a membrane anaerobic system (MAS) for palm oil mill effluent (POME) effluent *Desalination* 266(1-3) 208-12

[12] Fakhru'l-Razi A and Noor M J M M 1999 Treatment of palm oil mill effluent (POME) with the membrane anaerobic system (MAS) *Water Sci. Technol.* **39** (10-11)159-63

[13] EDDY N O, ABECHI S E, Paul A M E H, and Ebenso E E 2013 GCMS, FTIR, SEM, physiochemical and rheological studies on Albizia zygia gum *Walailak Journal of Science and Technology (WJST)* **10**(3) 247-65

[14] Fia F R, Borges A C, Matos A T D, Duarte I, Fia R and Campos LC D 2010 Development of biofilm in anaerobic reactors treating wastewater from coffee grain processing *RBEAA-Agriambi* **14**(2) 210-17

[15] Luo J, Im J H, Mayer M T, Schreier M, Nazeeruddin M K, Park N G, and Grätzel M 2014 Water photolysis at 12.3% efficiency via perovskite photovoltaics and Earth-abundant catalyst *Science* **345**(6204) 1593-96

[16] Bello O S, Adegoke K A and Akinyunni O O 2017 Preparation and characterization of a novel adsorbent from Moringa oleifera leaf *Appl. Water Sci.* **7**(3) 1295-1305

[17] Saphira M R., Daud Z, Ismail N, Ahsan A, Ab Aziz N A, Ndah M and Hijab M 2018 Locally Derived Activated Carbon From Domestic, Agricultural and Industrial Wastes for the Treatment of Palm Oil Mill Effluent. Nanotechnology in Water and Wastewater Treatment: Theory and Applications p 35.

[18] Matias M C, De La Orden M U, Sánchez C G and Urreaga J M 2000 Comparative spectroscopic study of the modification of cellulosic materials with different coupling agents *J. Appl. Polym. Sci.* **75**(2) 256-66

[19] Shak K P Y, and Wu T Y 2014 Coagulation–flocculation treatment of high-strength agro-industrial wastewater using natural Cassia obtusifolia seed gum: treatment efficiencies and flocs characterization *Chemical Engineering Journal* **256** 293-305

[20] Korbag I, and Saleh S M 2014 Extraction of Lignin from Paper Industry Journal of Applied Engineering Research 9(23)19421-28

IOP Conf. Series: Materials Science and Engineering **702** (2019) 012030 doi:10.1088/1757-899X/702/1/012030

Interaction study between M-Phenylenediamine, Trimesoyl Chloride and Nylon 66 using Molecular Dynamic Simulation

W Z A W Jusoh [1], S A Rahman[1*], A L Ahmad [2] and N M Mokhtar[3]

[1]Faculty of Chemical and Process Engineering Technology, College of Engineering Technology, Universiti Malaysia Pahang, 26300 Gambang, Pahang, Malaysia
[2]Pusat Pengajian Kejuruteraan Kimia, Kampus Kejuruteraan, Universiti Sains Malaysia, Universiti Sains Malaysia Pulau Pinang, 11800, Pulau Pinang, Malaysia.
[3]Faculty of Engineering Technology, Universiti Malaysia Pahang, Lebuhraya Tun Razak Gambang,Kuantan, 26300, Pahang, Malaysia.

Email: *sunarti@ump.edu.my

Abstract. The thin film composite (TFC) membrane is a porous membrane layered by a polyamide active layer. In the experimental work, the polyamide active layer prepared by the interaction between monomers, m-phenylenediamine (MPD) and trimesoyl chloride (TMC) which deposited on a membrane support. Considering the possibility of producing a loose TFC membrane and affect the separation performance, this study enlightens the interaction between monomers with Nylon 6 6 (N66) support membrane using simulation - molecular dynamic (MD). Different simulation system has been investigated with varying the ratio molecule of MPD to TMC where, TFC1 (1:1), TFC2 (3:1) and TFC3 (3:2) to the tertiary system of N66/MPD/TMC. The simulation was set for a COMPASS force field, where the equilibrium phases in a micro-canonical (constant volumes and total energy) (NVE) followed by run-production stage (constant pressure and temperature ensembles) (NPT) ensembles. The temperature and pressure were set at 323.15 K and 1atm respectively. The polyamide TFC was found to be formed by the main interactions between nitrogen (N) from MPD group to the carbon, (C) from TMC main chain at ranges of 4.25Å. Overall interaction can be observed between N from TFC layer with O atom from the substrate layer. All of the intermolecular interactions take place at 3.25Å distance. The best interaction in N66 to MPD and TMC system to the weakest is in the arrangement of TFC2 > TFC3 > TFC1. The highest intensity was obtained by the TFC2, indicating the best ratio of 3:1 of MPD molecules to TMC molecule with support polymer when fully polymerized occurred. This study suggests that ratio N66 to MPD and TMC in TFC2 is the best to interaction between TFC layer and substrate for fabrication of the TFC membrane.

1. Introduction

Molecular Dynamic (MD) simulation is employed to simulate the interfacial binding in relation to the molecular structures at the atomic level. It is widely used in the chemical process, especially for polymers. MD is a device to compute the various parameters such as diffusivity, compatibility of the materials blended, and the density of the polymers [1-2]. In MD, the force field represents the interaction between the molecular forces established within the system studied. The common force fields used in the MD are the condensed-phase optimized molecular potentials for atomistic simulation studies (COMPASS), Assisted Model Building with Energy Refinement (AMBER99), Consistent Valence Force Field (CVFF), etc. [2-3]. COMPASS was created based on the empirical parameterization techniques and state-of-the-art ab outlined the relationship between distance and the magnitude of the forces existing.

The COMPASS force field utilized to compute the structural, chemical, and physical characteristics of various series of molecules of organic and inorganic material [4-5]. It is very common to determine the interatomic interactions of organic as well as inorganic materials [6]. Compared with the other force fields, the COMPASS is more relevance to study the compatibility between aqueous and organic monomers to form the TFC membrane [7-8]. Sukitpaneenit and Chung used COMPASS force field in MD to study the compatibility of the polymer and mixed solvent systems [9]. Zhang et al. [10] on the other hand, calculate the binding energies between hydroxyapatite (HA) various polymers of polyethylene (PE), polyamide (PA) and polylactic acid (PLA) using the similar force field.

It is feasible to modify the pristine polymer to the thin film composite (TFC) membrane to enhance the separation performances [11-12]. Study the interaction between the two layers of membrane or two materials in the membrane by implementing the MD simulations have been done by Cho and Sung, Zhang et al. and Fereidoon et al. [5,13-14]. Polyamide (PA) TFC is easily prepared by interfacial polymerization (IP) flash reaction. It has both mechanical and performance strength which ideal for both commercial and laboratory study. The IP processes were manipulated in many MD studies to assist them recognize the materials properties before carried out in the real work [15–18]. Many researchers have demonstrated TFC films performance formulated by m-phenylenediamine (MPD) and trimesoyl chloride (TMC) on a various hydrophilic substrate [19–21]. Unfortunately, the hydrophilic lead to the weak mechanical strength of produced TFC membrane [22]. According to Shi et al., the stability between layers are strongly depending on the well interfacial interactions between the thin layer and the polymers [23]. Huang and McCutcheon applied the combination of MPD and TMC on the nylon 6, 6 (N66) as the support membrane and produced hydrophilic - high mechanical strength balance of TFC membrane [24-25]. In this study, MD is utilized to simulate the molecular interaction and to find the binding strength between monomers and the substrate membrane.

2. Methodology

The pairs of organic and aqueous monomers were analysed using RDF in MD for the shortest distance for the interaction to happen as well as the magnitude of the attraction forces. The strongest bonding between organic and aqueous monomers is associated with the stability of the thin film when deposited on the substrate membrane. The substrate chooses for the reasons which hydrophilic and mechanical strength owned by N66 membrane. N66 with the molecule formula $C_{12}H_{20}N_2O_2$ was used as platform for the aqueous and organic monomers, MPD and TMC respectively to react and form polyamide active layer. MPD is aromatic structure with two amine groups while TMC consist of three acyl chloride groups attach to a benzene ring. The molecular structures of the monomers with different number of functional groups and chain length were presented in figure 1.

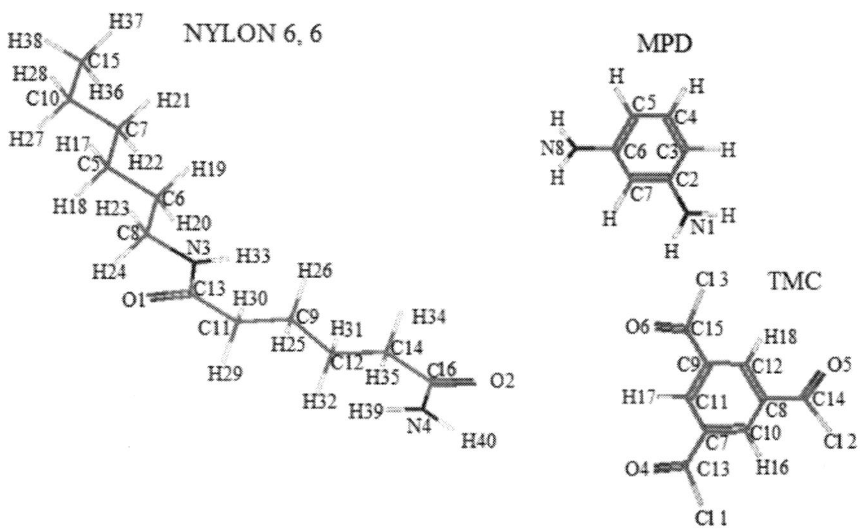

Figure 1. The repeat unit of monomers chain molecules for N66, MPD and TMC.

There is various composition suggested for the organic and aqueous monomers for the IP reactions to form the TFC membrane. MD simulations provide discernment about the process at a molecular state and to evaluate the intermolecular interface between various monomers composition with the substrate membrane. The number of molecules for each monomer was set as shown in table 1. The systems involved of the various composition of molecules namely for N66, 50 molecules, MPD molecules set from 10, and 30 while for TMC was at 10 and 20.

Table 1. Ratio of monomers molecules for the interaction study with support membrane in molecular dynamic simulation.

No.	Name	Number of molecules, N66:MPD:TMC	Equilibrated cell size : A x B x C [Å]
1.	N66	50	26.66 × 26.66 × 26.66
2.	TFC1	50: 10:10	28.83 × 28.83 × 28.83
3.	TFC2	50: 30:10	30.63 × 30.63 × 30.63
4.	TFC3	50: 30:20	31.41 × 31.41 × 31.41

The simulation setting was described according to the Ewald summation, COMPASS force field and velocity verlet algorithm for the function of equilibrium and production phase function, energy interatomic potential calculation and time integrator algorithm calculation respectively. The force field of the bulk phase represented as 3D box as shown in figure 2. The equilibrium phase was performed in micro-canonical, NVE of constant number molecules, volumes, and total energy ensembles, subsequently by constant number molecules, pressure, and temperature (NPT) ensembles for 1000ps each ensemble. NVE was choosing in the annealing step to keep the total energy in the system constant. Meanwhile, NPT ensemble in the production step is applied for the temperature coupling which maintains a fixed pressure rather than a steady volume. The same condition implies for both NVE and NPT ensembles where T and P were set at 323.15 K and 1atm respectively. Then, the final simulation trajectories were evaluated.

IOP Conf. Series: Materials Science and Engineering **702** (2019) 012030 doi:10.1088/1757-899X/702/1/012030

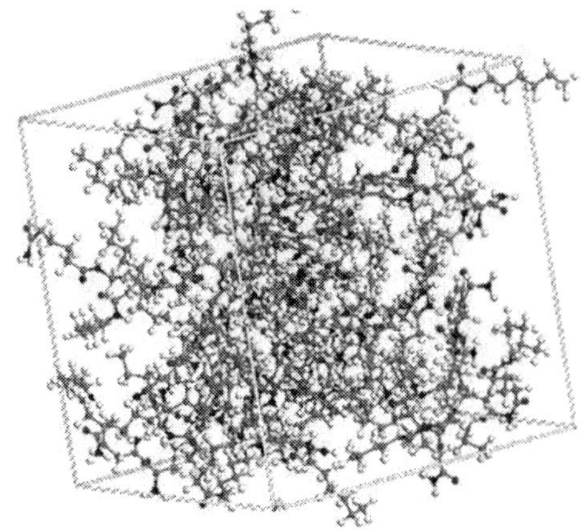

Figure 2. 3D boxes representing the simulation of the N66: MPD: TMC (50:10:10) system.

The simulation started by insert the molecular structures for the geometry optimization, followed by create the box to represent the system where the reaction take place and then minimizing the box. The equilibrium phase was performed in microcanonical, NVE ensembles followed by run-production stage by ensembles NPT. The electrostatics forces were calculation by Ewald summation (convergence parameter of 10^{-6}) in the periodic system. The simulation was run for a 1000 picoseconds (ps) with the time step set 1 femtosecond (fs) for each run and the simulation was periodic in three dimensions (Li et al., 2009). The total simulation time must be 1000ps or exceed than that in order to ensure a complete interaction (Yang et al., 2005). The simulation run-production stage then continues by NPT with the same condition for total 2000 ps with the time step of 1 fs each run. The final simulation trajectory was analyzed by RDFs.

3. Results and discussion

3.1. Interaction within the N66.

There are two main interaction forces can be identified using RDF analysis from the MD simulations which are intra- and intermolecular interactions. Basically, intra-molecular interaction is the repulsive forces between adjacent atoms in a molecule. Meanwhile, intermolecular interaction is the attractive forces among the atoms from different nearby molecules to form larger molecular structure. As for N66, the discussion only involved the intra-molecular interaction as illustrated in Figure 3. The main repulsive forces are within N66 polymer is between the hydrogen (H) and nitrogen atoms (N). These atoms potentially are the active atoms in a N66 functional group to act and react with the other compound introduced.

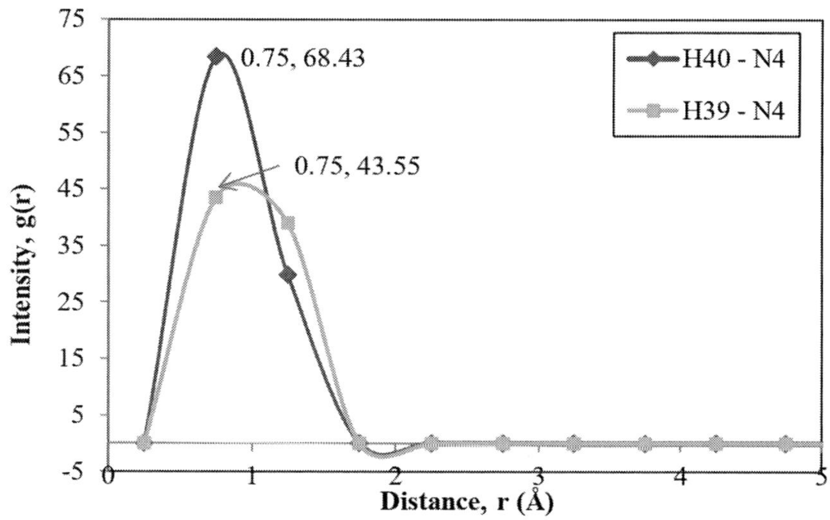

Figure 3. The intra-molecular interaction within N66 chain.

3.2. General interaction of amine group and organic group.

The molecular interaction is important to determine the efficiency of the active thin sheet on the support membrane by avoiding for a loose TFC membrane formation. The COMPASS force field successfully demonstrated the magnitude of binding forces with the contribution to a firm TFC membrane. TFC membrane is produced by the reaction between amine groups (MPD) with the organic group (TMC) on the polymeric support surfaces. The cross-linking interaction can be detected between nitrogen (N) from the aquoues monomer group with the carbon (C) from the acyl group to form the TFC membranes [26]. Figure 4 illustrates the main interaction existing within the N66 support with the TFC membrane. The TFC membrane then interacted with the support membrane for completing the deposition process where the main interaction by the hydrogen bond of oxygen (O) from N66 membrane support with the chloride (CL) from acyl main chain and with the N from amine molecule.

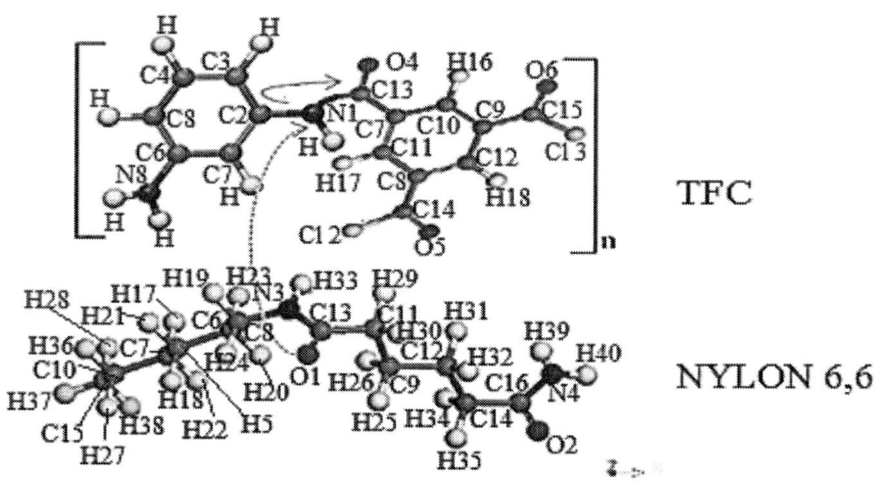

Figure 4. The main interaction between TFC membranes with the N66 support.

3.3. The interaction between amine and acyl group for formation of active layer.

Different simulation system have been investigated with varying the ratio molecule of MPD to TMC where, TFC1 (1:1), TFC2 (3:1) and TFC3 (3:2) to the tertiary system of N66/MPD/TMC in Figure 5.

All the main interactions to form TFC membranes were functioned by N atom from MPD to the C from TMC during IP process at radii of 4.25Å. The best interaction system to the weakest is in the arrangement of TFC3 > TFC1 > TFC2. The highest intensity was obtained by the TFC3 which indicated the best ratio of 3:2 of MPD molecules to TMC molecule. This finding is in agreement with Shen et al., which showed ratio 3:2 of MPD to TMC was the best composition [27]. The ratio is referred as fully polymerized its monomer composition where the polymerization reaction was completed. Two TMC group would attracted with three amide group generated less monomers excess on the membrane surface. Low amount of TMC (TFC1 and TFC2) reduces the IP rate on the membrane surface as the limited reactants. The ratio 1 to 1 of MPD to TMC system found to form a reaction-aggregation Muscatello et al., [26]. If the system has more reactants, there will be more unreacted amine and acyl chloride groups respectively. This condition would lead to thicker membrane formation as amine and carboxyl groups [28].

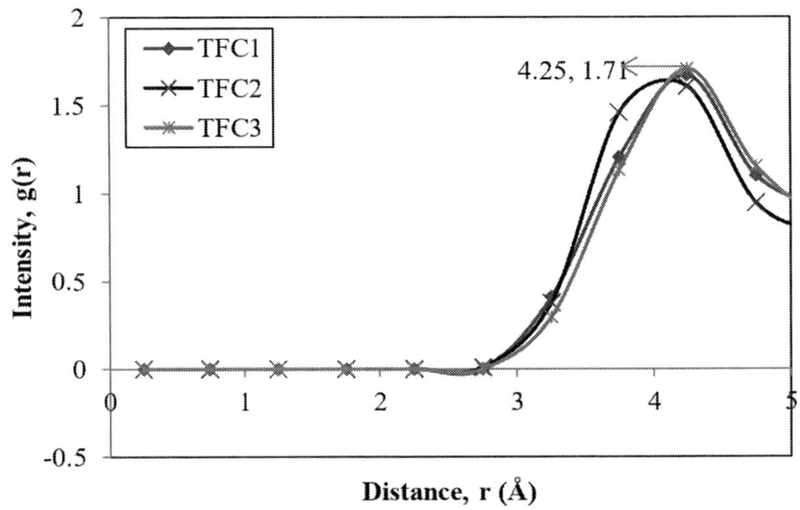

Figure 5. The interaction of nitrogen (N1) from the MPD group with the carbon (C13) from TMC group.

3.4. The interaction between active layer with support membrane, N66.

The significant interaction occurred between TFC layers and the N66 membrane can be observed attributed by the N from TFC layer with O atom from the substrate layer as shown in Figure 6. All of the intermolecular interactions take place at 3.25Å distance. The intensity of interaction forces was discovered in the range from 1.35 Å to 1.93Å which indicate strong interaction. Huang & McCutcheon claimed that TFC formed inside the core of the N66 support [25]. Small pores membrane support consist of the high surface area of substrate which draws the TMC solution to diffuse inside the membrane. This is because of the low diffusion rate of MPD solution from the pore attributed by good hydrogen - bonding interactions between MPD and the N66 membrane pore walls. The TFC in the pores provides a longer effective film thickness for water permeation [29].

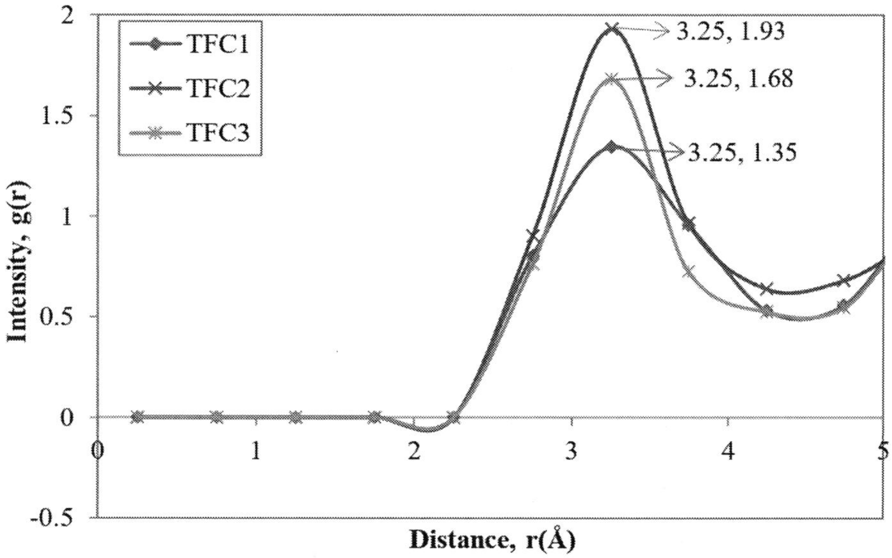

Figure 6. The interaction of atom N1 from TFC (amine group) with the O1 from N66 group.

The ratio of MPD to TMC observed to have affected on the interaction forces from 3:1 (TFC2) > 3:2 (TFC3) > 1:1 (TFC1). More amount of MPD in the system would increase the interaction between TFC layers with the support. Increasing the MPD in real experimental work can be carried out by increasing the MPD concentration or MPD immersion time. This is believed to occur due to higher MPD in the TFC layer and therefore increase the interaction with the carbonyl group of N66 membrane [30]. Result agree to corroborate with the study by Huang et al. [31]. Compare the similar ratio of N66 to MPD to TMC but at larger cell size (between TFC1 to TFC3), the interaction between N66 to TFC observed to be stronger in the smaller cluster size than the bigger size. The higher interaction distance obtained from this simulation may be contributed by the lower amount of molecules in the simulation system [32]. However, the interaction distance still below than 5.0Å [33].

4. Conclusion

Understanding the fundamental of the interaction between two layers at the atomic state by performing the molecular dynamic simulations is much more relevance. The current study, complete atomistic molecular dynamics simulations, explore the effects of cross-linking degree on TFC membrane and the interaction bonding between monomers and polymers. The main interactions to form TFC membrane were by N from amine to the C from TMC at radii of 4.25Å where shown the best by TFC3 followed by TFC1 and TFC2. The highest intensity indicated the best ratio of 3:2 of MPD molecules to TMC molecule when fully polymerized occurred. However, for deposition of MPD and TMC on N66, the result shows 3:1 of amount of MPD to TMC possess higher interaction. The interaction between the N from TFC layer at 3.25Å with the O atoms from the support membrane molecule. In the meantime, simulation study suggests that the larger cell size of TFC3 yield weaker interaction compared smaller cluster size TFC1.

Acknowledgements

The study was funded through grants PGRS180323 and RDU1803113.

References

[1] González M A 2011 *Collect. SFN* **12** pp 169–200.
[2] Xiang Y, Liu Y, Mi B and Leng Y 2013 *Langmuir* **29** pp 11600–11608 .

[3] Ding M, Szymczyk A, Goujon F, Soldera A and Ghoufi A 2014 *J. Memb. Sci.* **458** pp 236–244
[4] Taylor Fu Y, Liao L, Yang L, Lan Y, Mei L and Liu Y 2013 *Mol. Simul.* **39** pp 415–422.
[5] Zhang H, Lu X, Leng Y, Fang L, Qu S, Feng B, Weng J and Wang J 2009 *Acta Biomater.* **5** pp 1169–1181.
[6] Asche T S, Behrens P, and Schneider A M 2017 *J. Sol-Gel Sci. Technol.* **81** pp 195–204.
[7] García J L S, Bastarrachea M I L and De Jesús Aguilar Vega M 2013 *Macromol. Symp.* **325–326** pp 120–124.
[8] Sun H, Jin Z, Yang C, Akkermans R L C, Robertson S H, Spenley N A, Miller S, Todd S M 2016 *J. Mol. Model.* **22** pp 1–10.
[9] Sukitpaneenit P, and Chung T S 2009 *J. Memb. Sci.* **340** pp 192–205.
[10] Zhang X, Cahill D G, Coronell O and Mariñas B J 2009 *J. Memb. Sci.* **331** pp 143–151.
[11] Ong Y T and Tan S H 2016 *Chem. Eng. Res. Des.* **109** pp 116–126.
[12] Shi G M and Chung T S 2013*J. Memb. Sci.* **448** pp 34–43.
[13] Fereidoon A, Aleaghaee S and Taraghi I 2015 *Comput. Mater. Sci.* **102** pp 220–227.
[14] Cho H W and Sung B J 2017 *Soft Matter* **13** pp 1190–1199.
[15] Gao W, She F, Zhang J, Dumée L F, He L, Hodgson P D and Kong L 2015 *J. Memb. Sci.* **487** pp 32–39.
[16] Sridhar S, Smitha B, Mayor S, Prathab B and Aminabhavi T M 200, *J. Mater. Sci.* **42** 9392–9401.
[17] Huang Y H, Chao W C, Hung W S, An Q F, Chang K S, Huang S H, Tung K L, Lee K R, and Lai J Y 2012 *J. Memb. Sci.* **417–418** pp 201–209.
[18] Xiang Y, Liu Y, Mi B and Leng Y 2014 *Langmuir* **30** pp 9098–9106.
[19] Zuo J, Wang Y, Sun S P and Chung T S 2012 *J. Memb. Sci.* **405–406** pp 123–133.
[20] Zhang Y, Le N L, Chung T S and Wang Y 2014 *Chem. Eng. Sci.* **118** pp 173–183.
[21] Zuo J, Hua D, Maricar V, Ong Y K and Chung T S 2017 *J. Appl. Polym. Sci.* **135** pp 1–7.
[22] Singh P S, Joshi V, Trivedi J J, Devmurari C V, Rao A P and Ghosh P K 2006 *J. Memb. Sci.* **278** pp 19–25.
[23] Shi M, Zhang Y, Cheng L, Jiao Z, Yang W, Tan J, Ding Y 2016 *J. Phys. Chem. B* **120** pp 10018–10029.
[24] Huang L. and McCutcheon J R 2014 *J. Memb. Sci.* **457** pp 162–169.
[25] Huang L. and McCutcheon J R 2015 *J. Memb. Sci.* **483** pp 25–33.
[26] Muscatello J, Müller E A A, Mostofi A A A and Sutton A P P 2017 *J. Memb. Sci.* **527** pp 180–190.
[27] Shen M, Keten S and Lueptow R M 2016 *J. Memb. Sci.* **506** pp 95–108.
[28] Harder E, Walters D E, Bodnar Y D and Faibish R S 2009 *J. Phys. Chems.* **113** pp 10177–10182.
[29] Jimenez-Solomon M F, Gorgojo P, Munoz-Ibanez M and Livingston A G 2013 *J. Memb. Sci.* **448** pp 102–113.
[30] Peng M, Li L, Xiong J, Hua K, Wang S, Shao T 2017 *Coatings* **7** 123.
[31] Huang L, Bui N N, Meyering M T, Hamlin T J and McCutcheon J R 2013 *J. Memb. Sci.* **437** pp 141–149.
[32] Shen M, Keten S and Lueptow R M 2016 *J. Memb. Sci.* **509** pp 36–47.
[33] Ding M, Ghoufi A and Szymczyk 2014 *Desalination* **343** pp 48–53.

IOP Conf. Series: Materials Science and Engineering **702** (2019) 012031 doi:10.1088/1757-899X/702/1/012031

Development of keratin based hydrogels for biomedical applications

M S B Husain, A Gupta*, B Y Alashwal

Faculty of Chemical and Process Engineering Technology, College of Engineering Technology, Universiti Malaysia Pahang, 26300 Gambang, Pahang, Malaysia.

*E-mail: arungupta10@gmail.com

Abstract. Synthesis of hydrogel was developed by using keratin protein extracted from a chicken feather. Further, polyvinyl alcohol (PVA), polyvinylpyrrolidone (PVP), and starch have been commonly using for biomedical applications hydrogels due to their non-toxicity and, water solubility. The first sample, (KS-50) was mixed keratin, PVA/PVP and starch, the second sample, (K-50) was mixed the total weight of the (KS-50) without starch. KS-50 and K-50 were synthesized through cyclic freezing/thawing technique. Moreover, characteristics of the hydrogels, scanning electron microscopy (SEM) examined clearly exhibited porous structures, and the chemical structure of the compound was investigated by Fourier transform infrared spectroscopy (FTIR). The swelling degree of KS-50 exhibited slightly more pore size compared with K-50 hydrogels.

1. Introduction

The hydrogels are generally polymeric networks seen in cross-linked structures. They can be distinguished by their indistinct viscoelastic performance, hydrophilic nature. Elevation-water content facilitates the transfer the metabolic waste, nutrients, and oxygen [1]. Due to the presence of physical or chemical crosslinks like crystallites networks are collected from copolymers and homopolymers insoluble [2]. Modulating the polymer concentration and methods of synthesis allows tuning of specific hydrogel properties like biodegradability, swelling behavior, and porosity [3]. The hydrogels have been applied as scaffold materials the controlled release of drugs, proteins, and other applications [4-6]. Also, hydrogels are attractive for biomedical applications because of their diffusion properties/good water sorption biocompatibility [7].

Keratin turns into one of the most significant biomaterials through commercially available biomacromolecules due to its relatively good biocompatibility, excellent mechanical properties, high physicochemical, and good formatting-film ability [8-9]. Feather keratins from chicken feathers, that are considered to be the essential waste in the industry of poultry, because of coloring-dyeing and perming the control of the quality of the feather keratin is easier than hair keratin [10-11]. The featherweight of pure keratin is 90% feathers waste are a source of amino acids, protein, and nitrogen [12-[3]. Due to a great number of disulfide bonds, the keratin is a group of cysteine-rich structural proteins established in the cells of vertebrates that display high-rise mechanical strength [14-16]. Keratin has been utilized in several biomedical applications because of its biodegradability and biocompatibility [17-18].

Starch is one of the cheapest and most abundant polysaccharides. Also, the starch is a very cheap polysaccharide due to every starch derivatives are biodegradable and biocompatible polymers that will be utilized biomedical applications [19-21].

PVA based hydrogel is one of the famous polymer gel because of its good biocompatibility and utilized in several numbers of biomedical applications, such as, implants, dressings in wounds management, contact lenses, and drug delivery devices [22]. It has raised attention due to that it's water-soluble, non-toxic, and biocompatible properties [23-25]. PVP is one of the most common water-soluble, biodegradable, biocompatible, and low toxicity synthetic polymers [19, 24]. It is utilized in many biomedical applications and separation processes to raise the hydrophilic character of the mixed polymeric materials [26].

This paper presents a methodology for the development of hydrogels, composed of keratin protein, PVA, PVP, and Starch by using physical crosslinking by the freeze-thawing method for potential biomedical applications.

2. Materials and Methods

2.1. Materials

The chicken feathers have taken from the chicken plant at Jaya Gading, Kuantan, Malaysia. Polyvinyl alcohol (PVA), Polyvinylpyrrolidone (PVP), Sodium hydroxide (NaOH), and Hydrochloric acid (HCl) were purchased from R&M, chemicals, Kuala Lumpur, Malaysia, and, starch was purchased from Chemmart Asia Sdn. Bhd.

2.2. The keratin protein Extraction from chicken feather

Keratin protein solution was prepared in the chemical engineering lab at University Malaysia Pahang by cleaning feathers according to previously studied methods. The 100g of cleaned, dried and blended chicken feathers were added in 1L of (1N) sodium hydroxide solution in a 2L of the conical flask. The temperature of the mixture of the solution was 50°C, pH was 12 and the mixture was continuously stirred for 4h and cool to room temperature. After that, the solution was filtered through filter stainless steel then centrifuged at 10,000rpm for 10min [27], [28]. The keratin protein was collected carefully and then was filtered through filter paper and of the solution was adjusted to pH 7 with HCl (2N) and stored in the laboratory bottle for the synthesis of hydrogels.

2.3. Preparation of Polyvinyl Alcohol (PVA) Solution

A mass of 10g PVA powder was weighed by using the analytical balance. Then, the powder was added into 100ml distilled water with stirring until PVA dissolved completely. In the same time, the temperature was kept constant in the range of 70- 80°C [29]. The stirring process was done on a hot plate stirrer and continuously stirred until all the solutes dissolved thoroughly [30].

2.4. Preparation of Polyvinylpolypyrrolidone (PVP) Solution

A mass of 10g PVP powder was weighed by using the analytical balance. Then, the powder was added into 100ml distilled water with stirring until PVP dissolved completely. The stirring process was done manually by using a stirring rod for about 15 minutes [31].

2.5. Preparation of hydrogel film

The first sample, the keratin hydrogel (KS-50) was mixed with 50mL of keratin, 30mL of PVA 10mL of PVP, and 10g of starch with continuous stirring at 60 °C for 30min. The second sample, keratin hydrogel (K-50) was the total weight of the (KS-50) without starch and under stirred at 60°C for 30min. The proper amount of this mixture (15mL) were poured in plastic Petri dishes, followed by freezing at −20°C for 8h and thawing for 6h at 25°C for 3 continuous cycles to form hydrogel for further analysis [32], [33].

2.6. Scanning Electron Microscope (SEM)

Hitachi's Tabletop Electron Microscope (TM3030) was selected for carrying out the analysis related to the molecular structure of the study of the hydrogel samples [25].

2.7. Fourier Transform Infrared Spectroscopy (FTIR)
The Perkin-Elmer Model 1000 Series FTIR device was utilized to select spectra of the hydrogel. The frequency scope of the spectra was between 4000 cm^{-1} to 500 cm^{-1}. The all data obtained for hydrogel was through the utilizing of the software of FTIR [25].

2.8. Degree of the Swelling ratio
The hydrogel ability for fluid uptake was measured utilizing an electronic balance by immersion in phosphate-buffered saline (PBS). The hydrogels pieces were weighed (W_d) and put in PBS at 37 °C, and pH 7.4. The swollen hydrogel discs were measured at 10, 20, 30, 40, 50, 60, 1440 minutes, the surface was wiped gently with blotting paper to eliminate surface-adsorbed fluid, and the hydrogels weighed again (W_w). The degree of swelling defines the ability for fluid uptake by the following equation (1) [34]:

$$\text{The Swelling Degree } (\%) = \left[\frac{W_w - W_d}{W_d}\right] \times 100\% \quad (1)$$

Where W_d is the initial weight of the disc at time 0, and W_w is the weight of the disc at time t.

3. Results and Discussion

3.1. SEM analysis
SEM micrographs of the top view of the KS-50 (A) and K-50 (B) smooth with no particular structure. Figure 1, freeze-dried KS-50 and K-50 hydrogels clearly exhibited porous structures. Pore sizes can affect the physicochemical characteristics of the outcoming KS-50 and K-50 hydrogels, like the degree of swelling [35, 26].

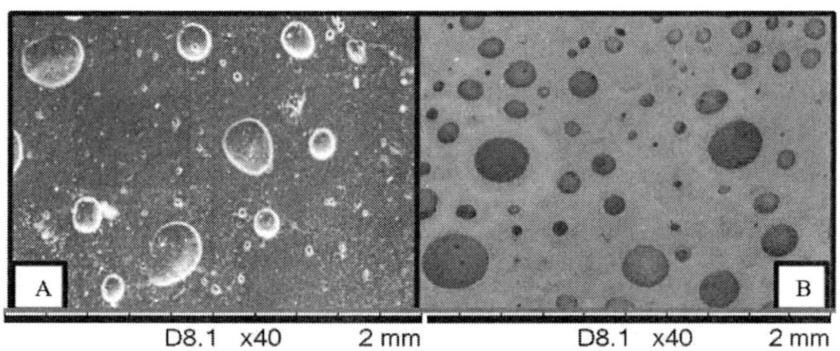

Figure 1. The SEM of KS-50 and K-50 keratin hydrogel samples.

3.2. FTIR analysis
FTIR measurement was utilized to recognize the chemical structure of the hydrogel and the results are shown in figure 2. The attained results validate the presence of keratin chemical structure when compare to the previous study in which keratin extracted from the chicken feathers [28]. The prepared hydrogels demonstrated transmission bands of the peptide bonds (-CONH) and are known as Amide A (3298cm^{-1}) ascribed to stretching vibration of -N-H and-O-H, Amide I (1651cm^{-1}) which are related to C=O stretching bonds, Amide II (1427cm^{-1}) is for N-H bending and C-H stretching, Amide III (1252cm^{-1}) derived from C-N stretching and N-H bending [36], [37]. The closeness of hydrophilic and hydrophobic moieties which was confirmed by FTIR makes them soluble in both aqueous and organic medium. This made it essential for medication framework [38]. The vibration bands from 950cm^{-1} recognized to alcohol. The presence of more -OH gatherings contributed to hydrogen bonding in the hydrogel. Thus, FTIR spectra the similar peaks demonstrated the presence of chemical groups in hydrogels that they can improve their physical and chemical properties for various for potential biomedical applications.

IOP Conf. Series: Materials Science and Engineering **702** (2019) 012031 doi:10.1088/1757-899X/702/1/012031

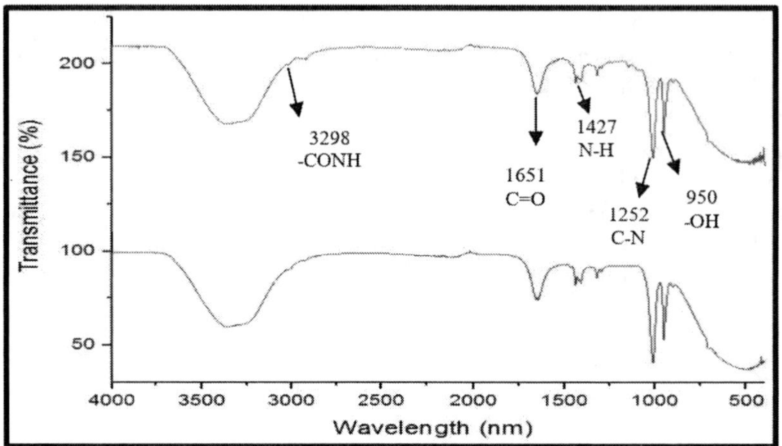

Figure 2. The FTIR spectra KS-50 and K-50 of keratin hydrogel samples.

3.3. Equilibrium degree of swelling

The swelling percentage of the KS-50 has increased only slightly with the time compared with the K-50 and the equilibrium swelling was reached at 50minutes for both KS-50 and K-50 as shown in figure 3. The size of pores is the major factor that controls the swelling degree of the hydrogel [39]. The sample could rapidly restore to the original state after being pressed, suggesting a sponge-like characteristic [40]. For most of the keratin hydrogels, the swelling properties are effective for biomedical applications.

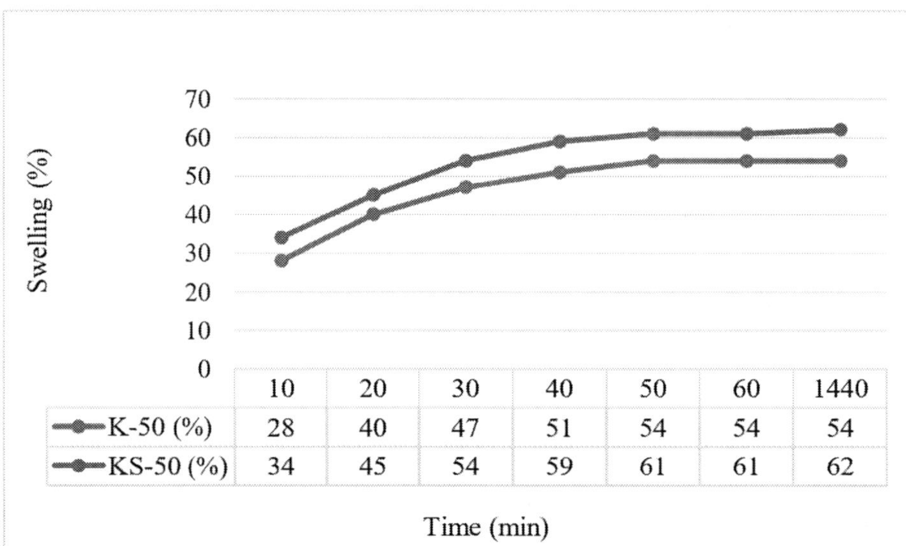

Figure 3.3 The swelling KS-50 and K-50 of keratin hydrogel samples.

4. Conclusion

In summary, the synthesis of keratin-based hydrogel was successfully developed using PVA/PVP/starch. Although the analysis of the results described good properties of hydrogel (K-50), the desired development was seen in the presence of starch (KS-50). FTIR spectra showed that peaks appeared in the hydrogel designating that they provide the macromolecular composition of the keratin, and SEM exhibited a large of pore sizes in the smooth surface. Moreover, the swelling percentage of the KS-50 has increased only slightly compared with the K-50 and the samples have developed like a hydrophilic sponge. However, the developed keratin hydrogels have a future in several biomedical applications.

Acknowledgments
Authors are thankful to University Malaysia Pahang (UMP) for financial support (Financial Aid (PGRS) number: 180334) and providing facilities.

References

[1] Ågren, M ed 2016 Wound Healing Biomaterials -Volume 2: Functional Biomaterials (Woodhead Publishing)
[2] Peppas N A, Hilt J Z and Thomas JB 2007 Nanotechnology in therapeutics: Current technology and applications (Horizon Scientific Press)
[3] Hemshekhar M, Thushara R M, Chandranayaka S, Sherman L S, Kemparaju K and Girish K S 2016 Emerging roles of hyaluronic acid bioscaffolds in tissue engineering and regenerative medicine *International Journal of Biological Macromolecules* 917-928
[4] Hennink W E and van Nostrum C F 2012 Novel crosslinking methods to design hydrogels *Advanced Drug Delivery Reviews* 223-36
[5] George A, Augustine R and Sebastian M eds 2014 Diabetes Mellitus and Human Health Care: A Holistic Approach to Diagnosis and Treatment (Apple Academic Press Canada)
[6] Koehler J, Brandl F P and Goepferich A M 2018 Hydrogel wound dressings for bioactive treatment of acute and chronic wounds *European Polymer Journal* 1-11
[7] Spanoudaki A N N A, Fragiadakis D A N I E L, Vartzeli-Nikaki K A L L I O P I, Pissis P O L Y C A R P O S, Hernandez J C R and Pradas M M 2006 Nanostructured and nanocomposite hydrogels for biomedical applications (Springer Dordrecht) pp 229-40
[8] Xu S, Zhang C, Zhang A, Wang H, Rao H and Zhang Z 2016 Fabrication and biological evaluation in vivo of an injectable keratin hydrogel as filler materials Journal of Bioactive and Compatible Polymers 179-190
[9] M M S B Husain, A Gupta, and B Y Alashwal, and S Sharma 2018 Synthesis of Hydrogel Using Keratin Protein from Chicken Feather Proceedings of ICCEIB2018 **193** 16
[10] Sharma S, and Gupta A 2016 Sustainable management of keratin waste biomass: applications and future perspectives Brazilian Archives of Biology and Technology **59**
[11] Wang J, Hao S, Luo T, Cheng Z, Li W, Gao F, Guo T, Gong Y and Wang B 2017 Feather keratin hydrogel for wound repair: preparation, healing effect and biocompatibility evaluation Colloids and Surfaces B: Biointerfaces **149** 341-50
[12] Kshetri P, Roy S S, Sharma S K, Singh T S, Ansari M A, Prakash N and Ngachan S V 2019 Transforming chicken feather waste into feather protein hydrolysate using a newly isolated multifaceted keratinolytic bacterium Chryseobacterium sediminis RCM-SSR *Waste and Biomass Valorization* 1-11
[13] Alashwal B Y, Bala M S, Gupta A, Sharma S and Mishra P 2019 Improved properties of keratin-based bioplastic film blended with microcrystalline cellulose: A comparative analysis *Journal of King Saud University-Science* In Press
[14] Reichl, S, Borrelli M and Geerling G 2011 Keratin films for ocular surface reconstruction Biomaterials **32**(13) 3375-86
[15] Reichl S, 2009 Films based on human hair keratin as substrates for cell culture and tissue engineering Biomaterials **30**(36) 6854-66
[16] Yua H M, Gupta A, Guptab R and Husaina M S B 2017 Investigation of bioplastic properties developed from acrylate epoxidized soybean oil through ring opening polymerization process *Journal of Chemical Engineering and Industrial Biotechnology* 29-41
[17] Vasconcelos A and Cavaco-Paulo A 2013 The use of keratin in biomedical applications *Current Drug Targets* **14**(5) 612-9
[18] Rouse J G and Van Dyke M E 2010 A review of keratin-based biomaterials for biomedical applications *Materials* 999-1014
[19] Kamoun E A, Chen X, Eldin M S M and Kenawy E R S 2015 Crosslinked poly (vinyl alcohol)

hydrogels for wound dressing applications: A review of remarkably blended polymers *Arabian Journal of Chemistry* 1-14
[20] Sharma S Sharma D and Kanwar K 2015 Technology refinement for micropropagated Aloe vera L.: a miracle plant *Research in Plant Biology* **5**(4) 1-10
[21] Haesler E 2017 Evidence summary: Wound management low resource communities-Aloe vera for wound healing *Wound Practice & Research: Journal of the Australian Wound Management Association* 115
[22] Kokabi M, Sirousazar M and Hassan Z M 2007 PVA–clay nanocomposite hydrogels for wound dressing *European Polymer Journal* 773-81
[23] Swaroop K, Francis S and Somashekarappa H M 2016 Gamma irradiation synthesis of Ag/PVA hydrogels and its antibacterial activity *Materials Today: Proceedings* 1792-98
[24] Husain M, Gupta A, Alashwal B, and Sharma S 2018 Synthesis of PVA/PVP Based Hydrogel for Biomedical Applications: A Review *Energy Sources, Part A: Recovery, Utilization, and Environmental Effects* **40**(20) 2388-93
[25] Ramakrishnan N, Sharma S, Gupta A, and Alashwal B Y 2018 Keratin based bioplastic film from chicken feathers and its characterization *International Journal of Biological Macromolecules* **111** 352-358
[26] Abd El-Mohdy H L and Ghanem S 2009 Biodegradability, antimicrobial activity and properties of PVA/PVP hydrogels prepared by γ-irradiation *Journal of Polymer Research* **16**(1)
[27] Bao Y, Ma J and Li N 2011 Synthesis and swelling behaviors of sodium carboxymethyl cellulose-g-poly (AA-co-AM-co-AMPS)/MMT superabsorbent hydrogel *Carbohydrate Polymers* 76-82
[28] Sharma S, Gupta A, Chik S M S B T, Kee C Y G and Poddar P K 2017 Dissolution and characterization of biofunctional keratin particles extracted from chicken feathers *IOP Conference Series: Materials Science and Engineering* **191** 12013
[29] Imtiaz N, Niazi M B K, Fasim F, Khan B A, Bano S A, Shah G M, Badshah M, Menaa F and Uzair B 2019 Fabrication of an Original Transparent PVA/Gelatin Hydrogel: In Vitro Antimicrobial Activity against Skin Pathogens *International Journal of Polymer Science*
[30] Pan H, Fan D, Cao W, Zhu C, Duan Z, Fu R, Li X and Ma X 2017 Preparation and characterization of breathable hemostatic hydrogel dressings and determination of their effects on full-thickness defects Polymers **9**(12) 727
[31] Adjei F K, Osei YA, Kuntworbe N and Ofori-Kwakye K 2017 Evaluation of the Disintegrant Properties of Native Starches of Five New Cassava Varieties in Paracetamol Tablet Formulations *Journal of Pharmaceutics*
[32] Fahmy A, Kamoun E A, El-Eisawy R, El-Fakharany E M, Taha T H, El-Damhougy B K and Abdelhai F 2015 Poly (vinyl alcohol)-hyaluronic acid membranes for wound dressing applications: synthesis and in vitro bio-evaluations *Journal of the Brazilian Chemical Society* 1466-74
[33] Hassan C M and Peppas N A 2000 Cellular PVA hydrogels produced by freeze/thawing *Journal of Applied Polymer Science* 2075–79
[34] Ponrasu T, Veerasubramanian P K, Kannan R, Gopika S, Suguna L and Muthuvijayan V 2018 Morin incorporated polysaccharide–protein (psyllium–keratin) hydrogel scaffolds accelerate diabetic wound healing in Wistar rats *RSC Advances* 2305-2314
[35] Park M, Kim B S, Shin H K, Park S J and Kim H Y 2013 Preparation and characterization of keratin-based biocomposite hydrogels prepared by electron beam irradiation *Materials Science and Engineering: C* 5051-57
[36] Kamarudin N B, Sharma S, Gupta A, Kee C G, Chik S M S B T and Gupta R 2017 Statistical investigation of extraction parameters of keratin from chicken feather using Design-Expert *3 Biotech* **7**(2) 127
[37] Xu H, Cai S, Xu L and Yang Y 2014 Water-stable three-dimensional ultrafine fibrous scaffolds from keratin for cartilage tissue engineering *Langmuir* 8461-70
[38] Roberts M J, Bentley, M D and Harris J M 2012 Chemistry for peptide and protein PEGylation *Drug Delivery* 116–127
[39] Khan S, Ullah A, Ullah K and Rehman N U 2016 Insight into hydrogels Designed Monomers and

Polymers 456-478
[40] Feng R, Fu R, Duan Z, Zhu C, Ma X, Fan D and Li X 2018 Preparation of sponge-like macroporous PVA hydrogels via n-HA enhanced phase separation and their potential as wound dressing *Journal of Biomaterials Science, Polymer Edition* 1463-1481

IOP Conf. Series: Materials Science and Engineering **702** (2019) 012032 doi:10.1088/1757-899X/702/1/012032

The efficiency of membrane anaerobic system (MAS) in treating sugarcane mill effluent (SCME)

Z H Yasmeen, N H Abdurahman*, M S Hybat and R M Yunus

Faculty of Chemical and Process Engineering Technology, College of Engineering Technology, Universiti Malaysia Pahang, 26300 Kuantan, Pahang, Malaysia.

*E-mail: nour2000_99@yahoo.com

Abstract. In recent times, the environmental issues have been on an alarming rate mostly for public authorities, societies, and industrial establishments. Sugarcane mill effluent (SCME) which contain high miscellaneous pollutants such as the chemical oxygen demand (COD), biological oxygen demand (BOD), total suspended solid (TSS) and the volatile suspended solid (VSS). However, the conventional method has the merit of COD reduction from SCME, but is limited in capturing the methane gas produced. In this study the membrane anaerobic system (MAS) was used as a cost-effective replacement method for treating SCME. In this study, six steady states were reached as part of the kinetic study with concentration ranging from 6312 to 14043 mg/l. for the mixed liquor suspended solids (MLSS) and 4790 to 12887 mg/l for mixed liquor volatile suspended solids (MLVSS). The three kinetic equations (Monod, Contois and Chen and Hashimoto) were used to explain the kinetics of SCME treatment at organic loading rates ranging from 0.5 to 13 kg COD/m^3 /day. The removal efficiency of COD was from 94.2% to 93.9% and produced methane gas (CH_4) 75.4%. The scanning electron microscopy (SEM) coupled with energy-dispersive X-ray spectroscopy (SEM/EDX), and Fourier transforms infrared spectroscopy (FTIR) were subsequently used to quantitatively and qualitatively determine the content of the biomass before and after treatment.

1. Introduction

Industrial wastewater treatment is one of the great worldwide challenging environmental issues. Due to a large amount of water used in a sugarcane milling operation, there are many contaminants in the sugar cane mill effluent. This water contains high BOD and COD level and hence cannot be discharged directly as it has standard disposable limit according to the Department of Environmental (DOE). The COD and BOD level for untreated SCME is around 16000 to 1500 mg/L respectively which shows that it has higher decay of microorganisms and particles (organic and inorganic material) thereby constituting pollution to the water body. if not treated properly [1, 3]. The pollutant capacity could be the main source of environmental pollution with a direct effect on human health if they are not treated perfectly[4]. There are many types of wastewater such as POME, sugarcane, and slaughter mill effluents. The anaerobic digestion was found by researchers to be the most appropriate for effluents treatment which contains a high organic carbon concentration [1, 5]. In engineering, the anaerobic digestion of methanogenic involves an organic matter decay which include anaerobic microorganism's with different types [6,7]. One of the basic requirements in anaerobic digestion is the absence of oxygen molecules to decompose the biomass which includes the organic and inorganic materials [7-8]. The intermediate component should be converted into basic products, such as methane (CH_4) and carbon dioxide (CO_2) which involves the biological transformation of the organic substrate. This transformation happens under the anaerobic requirement initiated by the hydrolysis, acidogenesis, acetogenesis, and methanogenesis [8-9]. However, the anaerobic digestion process has a major role for the decomposition of organic matter to produce methane gas in the electricity generation and save fossil energy [10-11]. These processes belong to the conventional SCEM treatment method which takes a very long period of time and large area for treatment. Anaerobic bioreactors with the high rate, have been used in the laboratory-scaled wastewater treatment like in the up-flow anaerobic sludge blanket (UASB) reactor [7]. Recently, AD has become a good source of treatment of wastewater effluent r treatment in the world [12]. AD technology is employed in renewable energy recapture of biogas and in the treatment of a huge diversity of waste from industrial process, wastewater treatment plants, agriculture, and household [13]. In this study the membrane anaerobic system was used to treat SCME. Economically, the use of MAS has the capacity to overcome high chemical oxygen demand and easily capture the methane gas. In addition, estimation of the dynamic process parameters shall be carried out using three kinetic models (Monod, Contois, Chen and Hashimoto) for growth the specific substrate [14] as presented in table 1.

Table 1. Mathematical Substrate utilization rates for kinetic model

Kinetic Model	Equation 1	Equation 2
Monod	$U = \dfrac{k\,S}{k_s + S}$	$\dfrac{1}{U} = \dfrac{K_s}{K}(\dfrac{1}{S}) + \dfrac{1}{k}$
Contois	$U = \dfrac{U_{max} \times S}{Y(B \times X + S)}$	$\dfrac{1}{U} = \dfrac{a \times X}{\mu_{max} \times S} + \dfrac{Y(1+a)}{\mu_{max}}$
Chen & Hashimoto	$U = \dfrac{\mu_{max} \times S}{Y\,K\,S_o + (1-K)\,S\,Y}$	$\dfrac{1}{U} = \dfrac{Y\,K\,S_o}{\mu_{max}\,S} + \dfrac{Y(1-K)}{\mu_{max}}$

2. Materials and Methods

2.1. Configuration of MAS

The anaerobic digester with volume capacity of 200 L was used for the treatment of the crude sugar cane wastewater. The design is included the feeder conservation tank (to pass the food to microorganisms), a pressure gauge (to control the pressure that ranges from 1-2 bar). The MAS consists of a cross-flow ultra-filtration membrane (CUF), the centrifugal pump, and an anaerobic reactor. The MAS ultrafiltration membrane is characterized with a cut-off (MWCO) of 200,000 molecular weight, a 1.25 cm tube diameter, 0.1 µm pore size, and 30 cm tube length. The membrane house area was 0.024 m². The membrane highest pressure was 55 bars at 60 °C, the pH range started from 4.5 to 0 7.5. The reactor was made from PVC and the internal diameter is measured at 45 cm shown in figure 1.

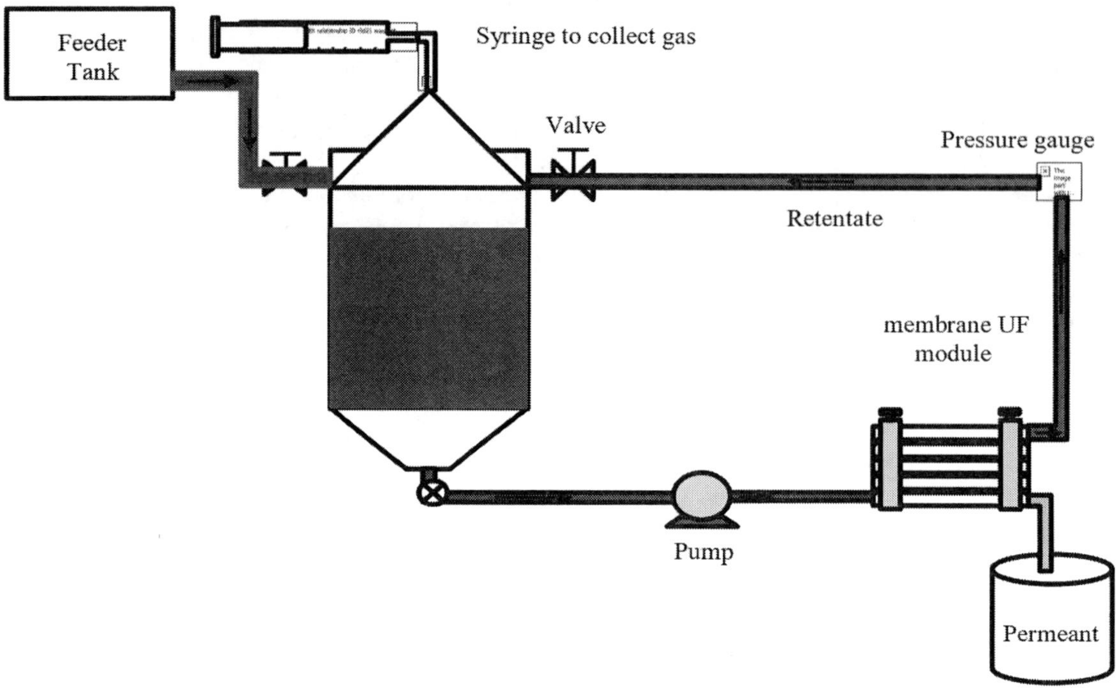

Figure 1. Experimental set-up

2.2. Analytical Testing
The parameters analyzed were TSS, BOD, COD, VSS The HACH colorimetric digestion method (Method 8000, HACH Company, and Loveland, CO, USA) was used for COD measurement. On the other hand, the influence of mixed liquor suspended solids (MLSS) and mixed liquor volatile suspended solids (MLVSS) was carried out by drying the sample at 105 and 550 ± 50 °C. To determine the CH_4, the Gas chromatography was employed with 200 cm × 0.3 cm stainless steel column which is packed with 30–60 mesh active carbon, and thermal conductivity detector.

2.3. Sugarcane (SC) Wastewater
Raw samples of the sugarcane wastewater were obtained from in Central Sugars Refinery located in Selangor, Malaysia. The sugarcane wastewater was characterized with higher COD and BOD, fishy, a fermented odour and with a dark brownish colour which disappears after treatment into a light brown colour. The total suspended solids are made up of carbohydrates, Ca, K, Mg, and organic matter [15]. To therefore evaluate the quality characteristics of the water after SCME treatment, the sample was subjected to some physicochemical characterization such as SEM, EDX, FTIR Which invariably explored the removal mechanism [16].

3. Results and Discussion

3.1. The operation of Bioreactor
The performance of MAS was estimated under six steady states at different CPD concentration as presented in Table 2. These range from 7541 to 23822 mg/L and the organism feeding (OLR) from 0.5 to 13 kg COD/m^3/day. To achieve good result, condition must be provided for operating and control parameters which must be within ± 10% of the average value. Pulled the gas from the top of reactor used a syringe at the end of an elastic tube to easily collect the gas. Then, added sodium hydroxide solution (NaOH) to biogas produced CO_2 and CH_4 only in order to absorb CO_2. The biogas was produced as the organic material degrades by methanogens bacteria in the absence of oxygen. Skorek and

Włodarczyk [16] reported that the biogas contains about 50-75% methane (CH_4), 25–45% carbon dioxide (CO_2) and fewer amounts of the other gases. Then, add sodium hydroxide solution (NaOH) to biogas produced CO_2 and CH_4 only in order to absorb CO_2

Table 2. Summary of results (Six steady state)

Steady state (SS)	**1**	**2**	**3**	**4**	**5**	**6**
COD feed, mg/L	7541	10377	15200	18631	20431	23822
COD permeate, mg/L	440	661	897	903	1200	1455
Gas production (L/day)	200.5	230	280	320	340	381
Total gas yield, L/g COD/day	0.31	0.38	0.42	0.51	0.56	0.63
% Methane	66.4	69.4	68.4	70	73.6	75.4
CH_4 yield, l/g COD/day	0.21	0.26	0.29	0.357	0.41	0.48
MLSS, mg/L	6312	8236	10532	12964	13639	14043
MLVSS, mg/L	4790	7087	9922	10465	11753	12887
% VSS	0.76	86.04	94.21	80.72	86.17	91.77
HRT, day	150.82	103.77	30.4	23.3	18.57364	18.32462
SRT, day	300	250	180	30	20.6	11
OLR, kg COD/m 3 /day	0.5	1	5	8	11	13
SSUR, kg COD/kg VSS/day	0.225	0.659	4.663	7.953	12.164	15.735
SUR, kg COD/m3/day	0.047	0.093	0.470	0.760	1.035	1.221
COD removal (MAS)	94.2	93.6	94.1	95.1	94.1	93.9

3.2 Performance of the Membrane Anaerobic System (MAS) Performance

The data achieved from the evaluated MAS performance is presented in Table 3. The application of MAS process was under different COD influent concentrations, as well as HRTs. From Eq. 2, the coefficients models were obtained using a linear relationship as shown in Table 1. The average of the model's coefficients is summarized in table 3. Furthermore, the important value of the coefficients models was brief in the same table.

Table 3. Results of the application of three known SU models

Model	Equation	R^2(%)
Monod	$U^{-1} = 2025\,S^{-1} + 3.61$ $K_s = 498$ $K = 0.350$ $\mu_{Max} = 0.259$	99.8
Contois	$U^{-1} = 0.306\,X\,S^{-1} + 2.78$ $B = 0.111$ $u_{Max} = 0.344$ $a = 0.115$ $\mu_{Max} = 0.384$ $K = 0.519$	99.3
Chen & Hashimoto	$U^{-1} = 0.0190\,S_o\,S^{-1} + 3.77$ $K = 0.006$ $a = 0.006$ $\mu_{Max} = 0.277$ $K = 0.374$	99.8

The MAS showed the best performance due to the capacity of the membrane ultrafiltration that cleans the surface where the molecules are collected and bacteria from SCME. One of the important parameters to produce methanogenesis bacteria is pH. This was found to be steady to the optimal points ranging from 6.7–7.8 for the reaction's anaerobic digesters when the experimental achieved the steady-state conditions, it found that the influent COD concentrations of 7541 -23822 mg/L. At the six steady states The VSS decreased to became 65.08 due to the presence of high suspended solids in the sugarcane wastewater mill effluent which can lead to the reduction in the value obtained. The highest value of VSS reached 70.05% in the reactor. The results have shown the long solid retention time, SRT of MAS facilitated the decay of the suspended solids and their subsequent conversion to methane (CH_4), these findings reported by [7]. At this stage, the performance of MAS achieved 99.8% removal efficiency. Increasing the OLRs that indicated to increase SUR and SSUR. This suggests that there is a multiplication in the population of bacteria within the MAS [26]. Hence, the increase in the concentration of the substrate might have led to the significant increment in the specific substrate utilization rates, and SSURs. This is an indication that the increasing amount of COD influent supplied to the reactor was accounted for by the amount of COD consumed by the bacteria populations [3]. The results obtained was applied in the famous three modules shown in Figure 2 to 4. MAS can be applied and treat SCME efficiently. The Monod and Chen & Hashimoto model verified an excellent relationship ($R^2 > 99\%$) for the treatment of sugarcane wastewater using an anaerobic membrane system as explained in Figure. 2,3,4.

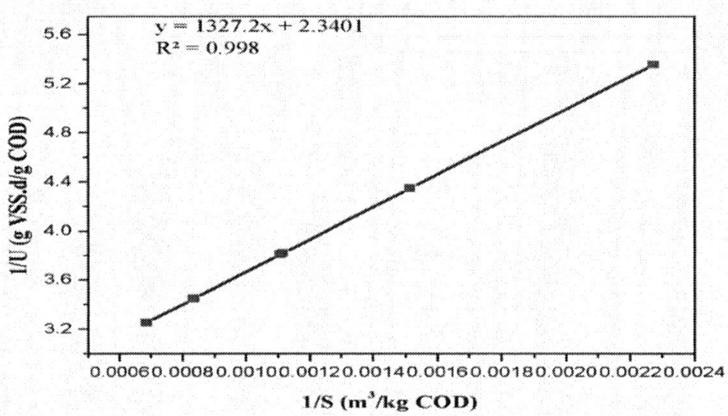

Figure 2. The Monod model

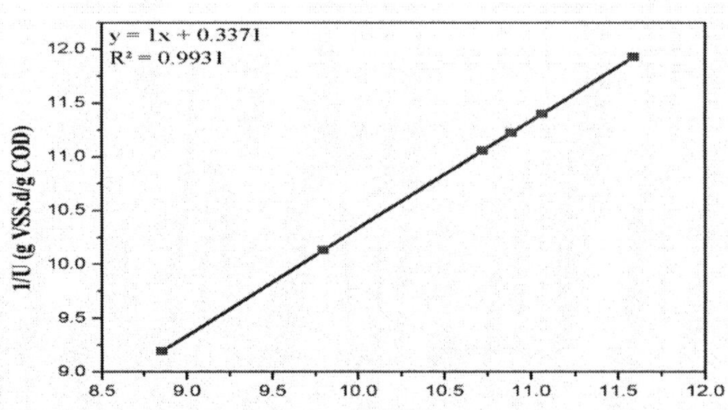

Figure 3. The Contois model

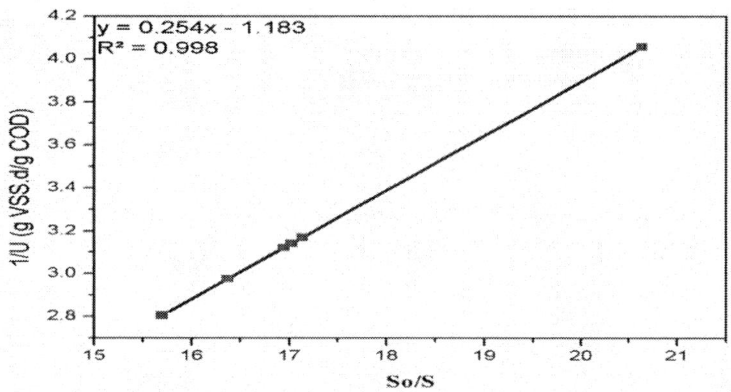

Figure 4. The Chen and Hashimoto model

The Monod and Chen and Hashimoto models performed better, implying that the digester performance should consider OLR. These two models suggested that the predicted permeate COD concentration (S) as a function of influent COD concentration (S_o). In the Contois model, however, S is independent of S_o. The best fit of these three models (R^2 > 99%) in this the study showed that the MAS process has an ability of handling sustained organic loads between 0.5 and 13 kg m^3/d.

3.3 The relationship between COD and HRT

The SCME exhibited higher COD until 25000 mg/l. For that need treatment before discharge to the river. This is estimated in the expression of COD as depicted in Figure 5. This illustrated that the COD removal efficiency increased as HRT increased from 18.32 to 150.82 days at ambit of 94%–93%. Otherwise, COD% decreased for lower HRT of 103.77 the COD was 93.6% this was due to the washout phase as showed in Table 2. Moreover, in the reactor process, while the biomass concentration increased, the washout phase was expected to happen. This may be indicated too high OLR with low HRT. The increase OLR create the VFA which in turn leads to a reduced COD concentration. The ultrafiltration (UF) membrane influx rates, usually exhibit a strong effect on the hydraulic retention times and also could determine the amount volume of the influent (SCME), fed into the reactor.

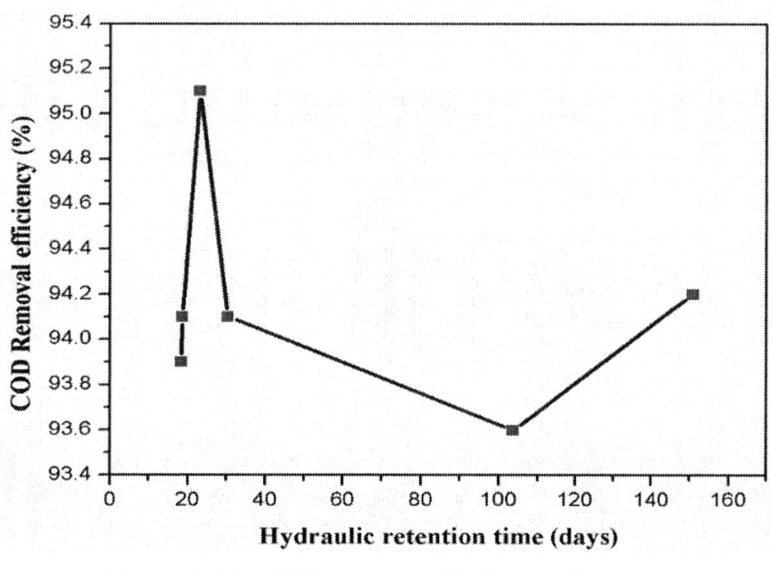

Figure 5. The COD removal efficiency of MAS HRT

3.4 Characterization of biomass

3.4.1 Scanning electron microscope (SEM)

The SEM was characterized by providing very high resolution for image more than the previous analyses. This technique could be connected with the X-ray emission which lead to an electron beam of high energy microanalysis to identify the elemental distribution of biomass structure (anaerobic processes) and morphology [17-18]. It also studied the efficiency of the treatment before and after the biological treatment and the existence of some bacteria that might be responsible for producing biogas [19].

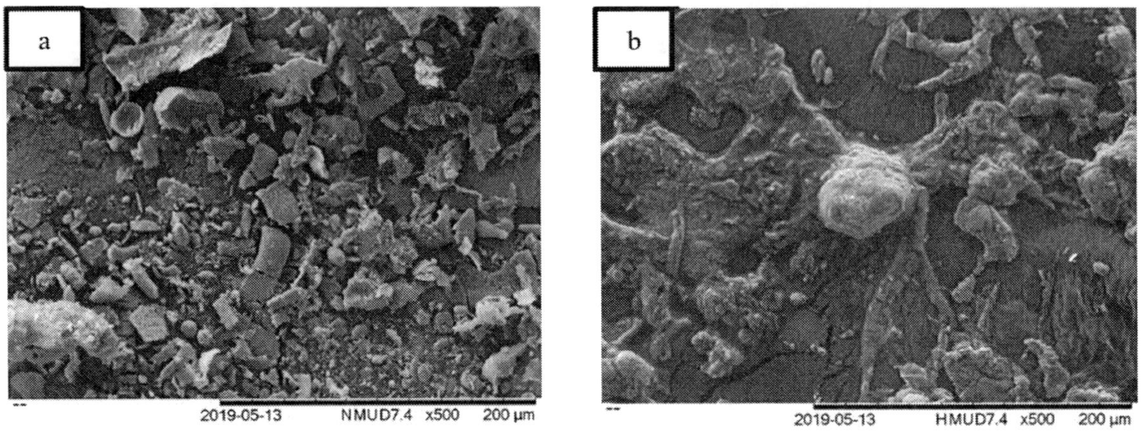

Figure 6. Scanning Electron Microscope (SEM), (a) Before treatment 500 XK (b) After treatment 500KX.

Biomass formation was observed using SEM, under specific conditions after the four reactions process which is responsible for the aggregation of bacteria on the surface of the substrate and produces CH_4 [20] in figure 6 (a) Observation on the surface of the substrate with SEM almost covered with the methanogenesis bacteria. The bubbles gas and the gaps between the particles indicated the presence of methanogenesis bacteria in the substrate of AD. The microorganism appeared noodle-like and they linked to each other in a network shown in figure 6 (a) [21]. This provided information about the various stages of anaerobic digestion biomass development, demonstrating the non-homogeneous distribution of biomass in figure 6 (a) before treatment [19] . The surface morphology after treatment (b) appeared less rough and corrugated and the distance between the particle increase and this shows the presence of high organic material captured inside the membrane. Furthermore, some microorganism are expected to have died during it arrived to steady state in the surface of biomass [22].

3.4.2 Energy dispersive X-ray (EDX) analysis

Energy-Dispersive X-ray spectroscopy (EDX) analysis proved to be one of the characterization methods, which the breakdown of organic compounds in the wastewater. EDX spectrum explains the morphology of SCME before and after the treatment, changing in the structure progression as a result of biochemical digestion of organic admixture inside the anaerobic reactor [23-24]. The EDX spectrum also revealed degradation of some organic material as well as a breakdown and damage of biological material in the membrane. It can be seen from the EDX data he chemical content composition of the adsorbent used is shown in figure 7. before treatment. Figure 8. After treated some elements decrease and the other one disappeared due to digestion during the anaerobic process, shown that in table 3.

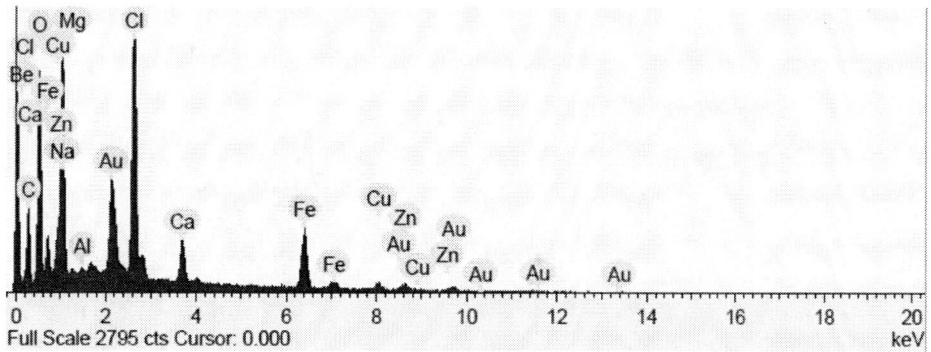

Figure 7. EDX spectra of influent of SCME after operation for 14 days.

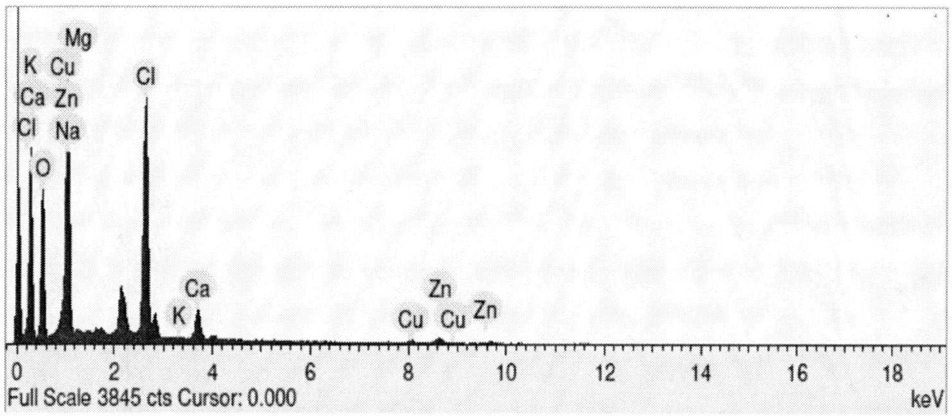

Figure 8. EDX spectra of permeate of SCME after operation for five hours

Table 4. Elemental Composition

Element	Weight (%) Before	Weight (%)After
Carbon	21.699	_
Oxygen	27.771	38.209
Sodium	9.321	24.404
Magnesium	0.177	0.370
Aluminum	0.143	_
Chlorine	11.199	23.098
Calcium	2.276	_
Iron	10.382	_
Copper	3.807	_
Zinc	3.825	_

3.4.3 Fourier Transform Infrared Spectroscopy (FTIR) Analysis

The FTIR spectroscopy mechanism was used to elucidate the surface functional groups and to obtain an inherent interaction of the groups with other parts of the molecule found in both untreated and treated SCME samples. The FTIR spectra of the SCME were recorded on an instrument (Nicolet IS5 FTIR) in the range of 400–4000 cm^{-1} with a resolution of 4 cm^{-1} [23]. The sugarcane mill effluent bands are illustrated in Figure 9. For carboxylate (COO-) and hydroxyl (OH-) groups a strong adsorption band was observed from 3500 cm^{-1} to 3000 cm^{-1}, with a peak maximum at 3302 cm^{-1}. This can be assigned to bonded OH groups [24]. The bands 2100-2245 cm $^{-1}$ which represents the C≡C stretch in alkynes group. The peak of 1642 cm^{-1} indicates OH group in water. Broadband in the range between 1585 and 1600 can be assigned to C–C stretch (in-ring) aromatics [25]. Also, the bands from 1040 cm^{-1} to 1250 cm^{-1} represent carboxylic acids group. After treatment some functional group disappear that indicate to organic material and small particles sedimentation on the membrane and digested during the four anaerobic processes inside the reactor [26].

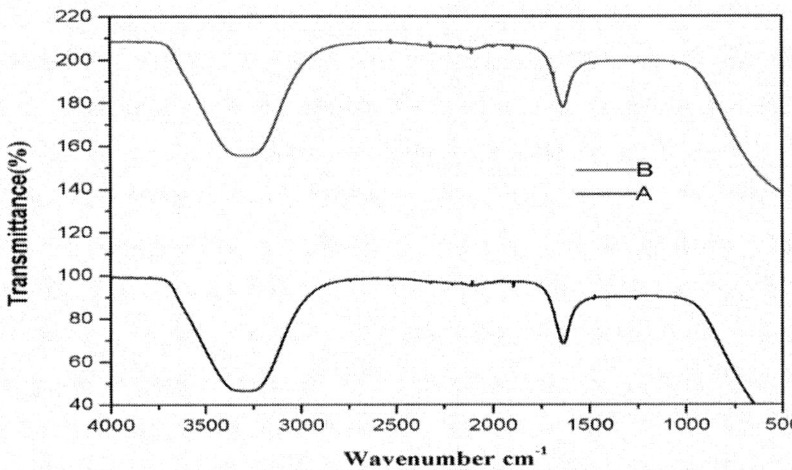

Figure 9. FTIR spectra of sugarcane wastewater mill effluent (wavenumber 4000 to 400 cm^{-1}): (A) Before treatment. (B) after treatment

4. Conclusion

The membrane anaerobic system (MAS) seemed to be an improved biological treatment reactor system, for the treatment of raw SCME. The reactor volumes are considered to be smaller than the conventional digester required volumes. This in a very short period of time has the capacity to achieved high COD removal efficiency up to 95.1%. The gas production, and the methane concentration in the gas, was satisfactory amount and, therefore, both could be considered. The amount of methane gas produced is an additional energy source appropriate for use at the sugarcane mill effluent. In MAS, the preliminary data on anaerobic digestion at 36°C showed the potential of the proposed technology, which substantially reduce the SCME wastewater pollution load. A significant quantity of energy (methane 75.4%) could be recovered by the MAS process, which could be used for electricity generation and fuel purposes. The infrared spectrometric, Fourier transform analyses indicated an improvement of the sugarcane wastewater mill effluent after treatment due to the decrease in the absorbance bands of carboxylate (COO-) and hydroxyl (OH-) groups. The morphological observation by scanning electron microscopy (SEM) for SCME before and after treatment, proved to be having bubbles from methanogenesis bacteria which in turn leads to the production of CH_4. MAS showed that the proposed technology has a potential to substantially decrease the pollution load of SCME wastewater.

Acknowledgements
The authors gratefully acknowledge the University Malaysia Pahang (UMP) for the research project financial support under the grant (PGRS 180332).

Nomenclature

COD	chemical oxygen demand (mg/L)
OLR	organic loading rate ($kg/m^3/d$)
CUF	cross flow ultra-filtration membrane
SS	steady state
SUR	substrate utilization rate ($kg/m^3/d$)
TSS	total suspended solid (mg/l)
MLSS	mixed liquid suspended solid (mg/L)
HRT	hydraulic retention time (day)
SRT	solids retention time (day)
SSUR	Specific substrate utilization rate (kg COD/kg VSS/d)
MAS	Membrane an aerobic System
MLVSS	mixed liquid volatile suspended Solid (mg/L)
VSS	volatile suspended solids (mg/L)
MWCO	molecular weight Cut-Off
BLR	biological loading rate
U	specific substrate utilization rate (SSUR) (g COD/G VSS/d)
S	effluent substrate concentration (mg/L)
So	influent substrate concentration (mg/L)
X	micro-organism concentration (mg/L)
μ_{max}	Maximum specific growth rate (day^{-1})
K	Maximum substrate utilization rate (COD/g/VSS.day)
K_s	Half velocity coefficient (mg COD/L)
X	Micro-organism concentration (mg/L)
b	specific microorganism decay rate (day^{-1})
Y	growth yield coefficient (gm VSS/gm COD)

References

[1] Borja R, Banks, C J and Sánchez, E 1996 Anaerobic treatment of palm oil mill effluent in a two-stage up-flow anaerobic sludge blanket (UASB) system *Journal of Biotechnology* **45**(2) 125-35
[2] Abdurahman, N H, Azhari, N H and Rosli, Y M 2013 Ultrasonic membrane anaerobic system (UMAS) for palm oil mill effluent (POME) treatment *Int. Perspect. Water Qual. Manag. and Pollut. Control* 107-21
[3] Nour A H, Zaki Y H, Mohamed H S, Rassem H H 2019 The Potentials of an Integrated Ultrasonic Membrane Anaerobic System (IUMAS) in Treating Sugar Cane Wastewater *Indones. J. Chem.* **19**(3) 804-810
[4] Trikoilidou E, Samiotis G, Bellos D and Amanatidou E 2016 Sustainable operation of a biological wastewater treatment plant *IOP Conference Series: Materials Science and Engineering* **161**(1) 012093
[5] Zhu M, Hu X, Ye M, Shi F and Liu J 2019 Examples and Analysis of Tobacco Wastewater Treatment Engineering IOP Conference Series: Earth and Environmental Science **233**(5) 052040
[6] Côté C, Massé, D I and Quessy S 2006 Reduction of indicator and pathogenic microorganisms by psychrophilic anaerobic digestion in swine slurries Bioresource technology **97**(4) 686-91
[7] Abdurahman N H, Rosli Y M and Azhari N H 2011 Development of a membrane anaerobic system (MAS) for palm oil mill effluent (POME) treatment Desalination **266**(1-3) 208-12
[8] Guerrero L, Omil F, Mendez R and Lema J M 1999 Anaerobic hydrolysis and acidogenesis of wastewaters from food industries with high content of organic solids and protein Water Research **33**(15) 3281-90
[9] De Mes T Z D, Stams A J M, Reith J H and Zeeman G 2003 Methane production by anaerobic

digestion of wastewater and solid wastes *Bio-methane & Bio-hydrogen* 58-102

[10] Vo T T Q, Xia A, Wall D M, and Murphy J D 2017 Use of surplus wind electricity in Ireland to produce compressed renewable gaseous transport fuel through biological power to gas systems *Renewable Energy* **105** 495-504

[11] Zaki Y H, Nour A H and Mohamed H S 2018 Application of Membrane Bioreactor in Wastewater Treatment *International Journal of Innovative Research and Scientific Studies* **1**(1)

[12] Thomas P, Soren N, Rumjit N P, James JG and Saravanakumar MP 2017 Biomass resources and potential of anaerobic digestion in Indian scenario *Renewable and Sustainable Energy Reviews* **77** 718-30

[13] Achinas S, Li Y, Achinas V and Euverink GJW 2019 Biogas Potential from the Anaerobic Digestion of Potato Peels *Process Performance and Kinetics Evaluation Energies* **12**(12) 2311

[14] Abdullah A L, Idris A, Ahmadun F R, Baharin B S, Emby F, Noor M M M and Nour A H 2005 A kinetic study of a membrane anaerobic reactor (MAR) for treatment of sewage sludge *Desalination* **183**(1-3) 439-45

[15] Sahu O P and Chaudhari, P K 2015 The characteristics, effects, and treatment of wastewater in sugarcane industry *Water Quality, Exposure and Health* **7**(3) 435-44

[16] Skorek A and Włodarczyk R 2018 The Use of Methane in Practical Solutions of Environmental Engineering *Journal of Ecological Engineering* **19**(2)

[17] Lazarova V and Manem J 1995 Biofilm characterization and activity analysis in water and wastewater treatment *Water Research* **29**(10) 2227-45

[18] Linares-Hernández I, Barrera-Díaz C, Roa-Morales G, Bilyeu B and Ureña-Núñez F 2007 A combined electrocoagulation–sorption process applied to mixed industrial wastewater *Journal of Hazardous Materials* **144**(1-2) 240-48

[19] Garcia-Mesa J J, Delgado-Ramos F, Muñio M M, Hontoria E and Poyatos J M 2012 Comparison of activated sludge technologies by particle size analysis *Water, Air, & Soil Pollution* **223**(7), 4319-31

[20] Liu Y, Zhu Y, Jia H, Yong X, Zhang L, Zhou J, Cao Z, Kruse A and Wei P, 2017 Effects of different biofilm carriers on biogas production during anaerobic digestion of corn straw *Bioresource Technology* **244** 445-51

[21] Kim, J O, Kim Y H, Ryu J Y, Song B K, Kim I H and Yeom S H 2005 Immobilization methods for continuous hydrogen gas production biofilm formation versus granulation *Process Biochemistry* **40**(3-4) 1331-37

[22] Kumar A, Sengupta B, Kannaujiya M C, Priyadarshinee R, Singha S, Dasguptamandal D and Mandal T 2017 Treatment of coke oven wastewater using ozone with hydrogen peroxide and activated carbon *Des. Water Treat.* **69** 352-65

[23] Sevda S, Dominguez-Benetton X, Vanbroekhoven K, De Wever H, Sreekrishnan T R and Pant D 2013 High strength wastewater treatment accompanied by power generation using air cathode microbial fuel cell *Applied Energy* **105** 194-206

[24] Bala, M S, Azoddein A M, Asdash, M I and Alshwal, B Y 2018 Comparative Study for Activation of Hydrogen Peroxide by Chemical Reagents (Fe^{2+}, Al^{3+}) to Reduce COD in Petrochemical *Wastewater Journal of Engineering and Technology* **9**(2)

Characterization of dehydrated keratin protein extracted from chicken feather

B Y Alashwal, A Gupta* and M S B Husain

Faculty of Chemical and Process Engineering Technology, College of Engineering Technology, Universiti Malaysia Pahang, 26300 Gambang, Pahang, Malaysia.

*Email: arungupta10@gmail.com

Abstract. Keratin is one of the most important biomaterials due to its relatively good bio-compatibility, high physicochemical with excellent mechanical properties and good film coordination ability. Chicken feathers are a major waste in poultry industries however, 90% of pure keratin can be obtained from the total featherweight. Chicken feathers are abundant of keratin protein which can be considered as a suitable protein source and used as a component in the cosmetic and pharmaceutical industry. The aim of this study is to extract keratin protein from chicken feather source. The chemical structure of the keratin protein was examined by Fourier transform infrared spectroscopy (FTIR). The results showed the presence of, Amide I-III bands, which gave critical information on the protein conformation and alteration in backbone structure of the keratin., Surface morphology analysis was described by Scanning electron microscopy (SEM). The extracted keratin was found to present aspheric shaped, small microspheres with smooth surface. Meanwhile the X-Ray Diffraction results showed the crystalline form of the chicken feather dehydrated keratin. The diffraction peak was indexed for the β-sheet crystalline structure of keratin. In conclusion, the characterization of keratin extracted from chicken feathers was successfully performed using the abovementioned analyses.

1. Introduction

Nowadays, the increase in poultry products consumption led to an increase in waste about by 8.5 billion tons of feather, that is produced from 24 billion from the consumption of chicken annually. The disposal of feather waste has become a global environmental problem due to the traditional and costly feathers disposal strategies [1]. On the other hand, feathers are cheap, eco-friendly as an alternative available abundantly as a natural source of protein (keratin) that it is utilized in many applications such as cosmetic, biomedical and others [2-4].

Keratin from feathers of chicken has included some feature compared to other keratin like the feather keratin of fibrous can extend almost to 6 % and get breaking, but hair keratin can extend to double of length [5-6]. It is famous that feathers are essentially made of a structural keratin protein (>90%), rich in cysteine, and hydrophobic residues that enhances crosslinking by disulfide bonds and includes a variety of predominantly cystine, amino acids, lysine, serine, and proline [7-8].

Extraction of keratin from sheep wool has also been a source of extraction of keratin protein to be used in various cosmetics and biomedical products. According to Fan et al. [9] provided convincing results with up to 72% solubility in the keratin extraction method of wool using L-cysteine. The previous studies have shown that the extract keratin from chicken feathers proved that they produce a total protein mass of up to 53% [4-5]. Several studies reported that increasing protein concentration can contribute to reducing protein degradation during the drying process net gain in kidney protein content [10]. The reduction of keratin is a multi-step, long-lasting method in that sulphitolysis is a key reaction. The disulfide bond are disrupted by sulfite to (Keratin − Cys − S −) cysteine thiol (reduced keratin) and (Keratin −Cys − SSO_3^-) cysteine-S-sulphonate residue (Bunte salt) [11]:

$$\text{Keratin} - \text{Cys} - \text{S} - \text{S} - \text{Cys} - \text{Keratin} + SO_3^{2-} \rightarrow \text{Keratin} - \text{Cys} - S^- + \text{Keratin} - \text{Cys} - SSO_3^-$$

Keratins are classed into two types of Alpha (α) keratins and beta (β) keratins [12]. (α) keratin proteins are found in all vertebrates, while (β) keratin proteins are found exclusively in feathers of birds and reptiles [13]. In this study, feather keratin was extracted from chicken and convert to powder and the drying under vacuum conditions was first developed for keratin protein biomedical, pharmaceutical and cosmetics applications.

2. Materials and Methods

2.1. Materials

Chicken feathers were taken from the chicken was obtained from Balok Poultry Farm Sdn. Bhd. Malaysia, Malaysia. Sodium hydroxide and hydrochloric acid were obtained from Sigma-Aldrich.

2.2. Extraction of keratin protein from chicken feather

Preparation of the feather keratin was based on the previously studied techniques which put 50 g feather with 1 L sodium hydroxide into the conical flask. The mixture was kept at 50 °C and continuously stirred for 5 hours. Then, the mixture was filtered by filter paper and the biomass waste was removed by centrifugation at 10,000 rpm for 10 min [14, 16-17].

2.3. Protein neutralization

The filtrate solution collected earlier was placed in a beaker and stirred. 2 N hydrochloric acid was added drop wise to the solution for making neutral pH. The solution was again centrifuged at 10,000 rpm for 5 minutes and filtered using cellulose filter paper with Grade 1 from Whatman to ensure it is particle free then stored for the further analysis.

2.4. Total mass concentration of Keratin

The content of the keratin protein from the 50 g feather per letter was determined by dehydrating 5 ml of the solution at 104 °C for 24 h in the oven. The powder keratin was 0.24 g and the total mass concentration of keratin was calculated from equation (1):

$$Mass\ concenteration\ of\ keratin = \frac{m}{V} \times 100\% \tag{1}$$

where m is the weight of the solute after evaporation and V is the weight of the initial mixture.

The mass concentration of keratin as powder was 4.8g/mL from 1L keratin.

2.5. Dehydration of keratin

Under vacuum condition, a 50 mL of keratin solution was poured in Pyrex glass dish and dried in the vacuum oven for 20 hours, at 60 °C and 20-inch Hg.

2.6. Characterization of keratin protein

Characterization of keratin was performed by FTIR, SEM and, XRD analyses as shown below:

2.6.1 Fourier transform infrared spectroscopy (FTIR)

FTIR spectra of keratin were read from 400 to 500 cm^{-1} utilizing a Perkin-Elmer Model 1000 Series FTIR spectrophotometer. All data obtained for keratin was analysed using the software of FTIR [16].

2.6.2 Scanning electron microscopy (SEM)

The surface images of the keratin protein powder sample were investigated by SEM using Hitachi TM3030Plus [17].

2.6.3 X-ray diffraction (XRD)

The keratin protein in crystal structures was examined by X-ray diffractometer a RIGAKU, (Miniflex II, Japan. The models were collected with a radiation source of the Cu Kα. The crystallinity index (Crl) of the samples was described by equation (2) [18]:

$$Crl(\%) = \left(\frac{A_{Crystal}}{A_{Total}}\right) * 100 \qquad (2)$$

Where, ($A_{Crystal}$) is the sum of the areas below the crystalline diffraction peaks and (A_{Total}) describes the total area below the diffraction curve 2θ ranging from 5 to 50°.

3. Results and Discussion

3.1. Fourier transform infrared spectroscopy (FTIR)

The FTIR measurement was utilized to identify the chemical structure of the keratin as shown in figure 1. In FTIR spectra an absorption band was scrutinized in the range of 3274-3287 cm^{-1}. The result indicates that the band at 3274.15 cm^{-1} correspond to the peptide bonds (–CO–NH–). The wavenumbers at 1633, 1559 and 1392 cm^{-1} are the waves of the peptide bonds identified as amides I, II and III which provide important information on the keratin protein structure and alteration in backbone composition of the keratin protein. thus, Amide I, is the combination of α-helix and β-sheet and amide III can be attributed to β-sheet structure [19]. As shown, the Amide I region can be deconvoluted to provide more information about the structure presents. at wavenumber 1126 cm^{-1} was described (-OH) the groups of carboxylic acids and the (N-H) band of transmission at 750 to 500cm^{-1}. Hence, FTIR proved the appearance of amino acids like and threonine, cysteine, and glutamine in the keratin sample [17]. Therefore, it can be concluded that the extracted keratin possesses more β-sheet structures as compared to α-helices.

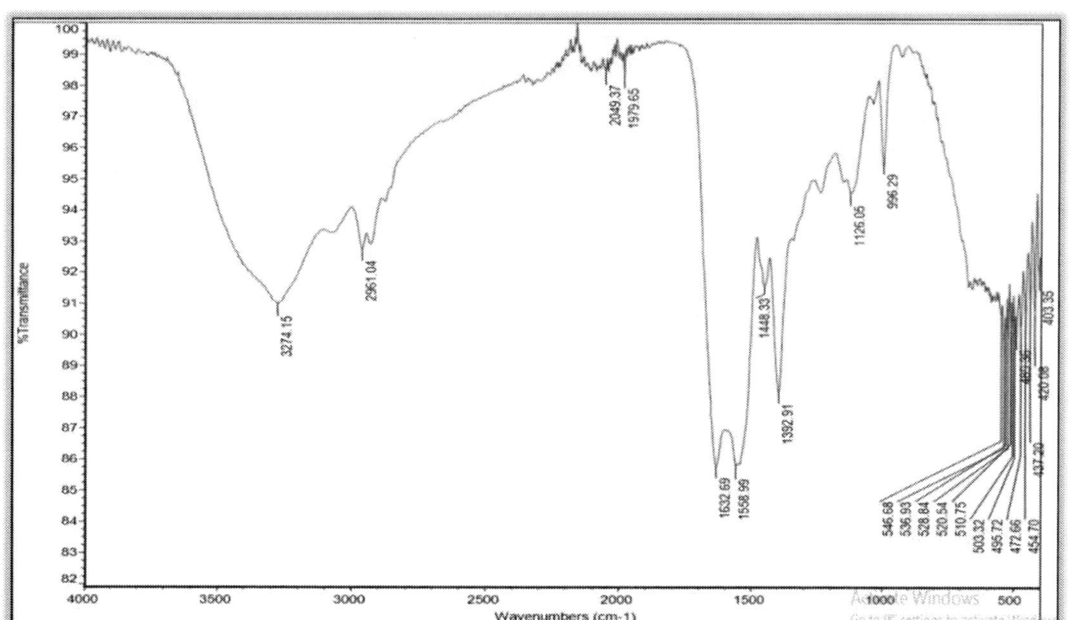

Figure 1. The FTIR spectra of keratin

3.2. Scanning electron microscopy (SEM)

The morphological characteristic of the keratin displays the highly porous structure in the surface with multitude features, like the specimen explained a horny layer, and very thick. The SEM image in figure 2 confirmed that keratin seemed as small particles in aggregates form. The extracted keratin presents aspheric shaped, small microspheres with smooth surface, The surface morphology of these microparticles is similar to the keratin particles synthesized from wool and feather keratin in previous studies [20]. The dried keratin consisted of globular, tightly packed microparticles and randomly arranged microstructures.

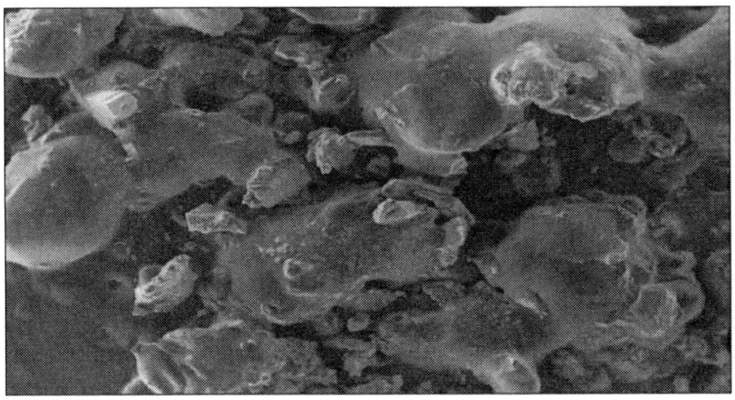

Figure 2. The SEM of keratin

3.3. X-ray diffraction (XRD)

The XRD results determined that the extracted keratin was mainly presented in a semi-crystalline form so, that its aqueous state maintained the crystal. XRD analysis of keratin molecules in figure3 showed a strong peak at 32.75 degrees, which determined the presence of keratin nanoparticles [21]. The dominant peaks at $2\theta = 9\text{-}10.76°$ and at 19-32.75° were allocated to α-helix and β -sheet, respectively [22]. The keratin nanoparticles show the diffraction properties of α-helix that appear at $2\theta = 10.76$ ° and β-sheet at $2\theta = 20.50$ ° [23].The peak ranges from 5 degrees to 46 degrees, with the width of an amorphous structure [24]. From the above results, it can be concluded that keratin possesses two types of crystal structure α and β. However, keratin contains a greater amount of β-sheet. Although keratin molecules are also slightly amorphous, the peaks have not yet changed the formation of nanoparticles that have shown that the crystal form of keratin is retained in the formulation of nanoparticles which showed in a previous study that after the formation of nanoparticles keratin remained unchanged [6].

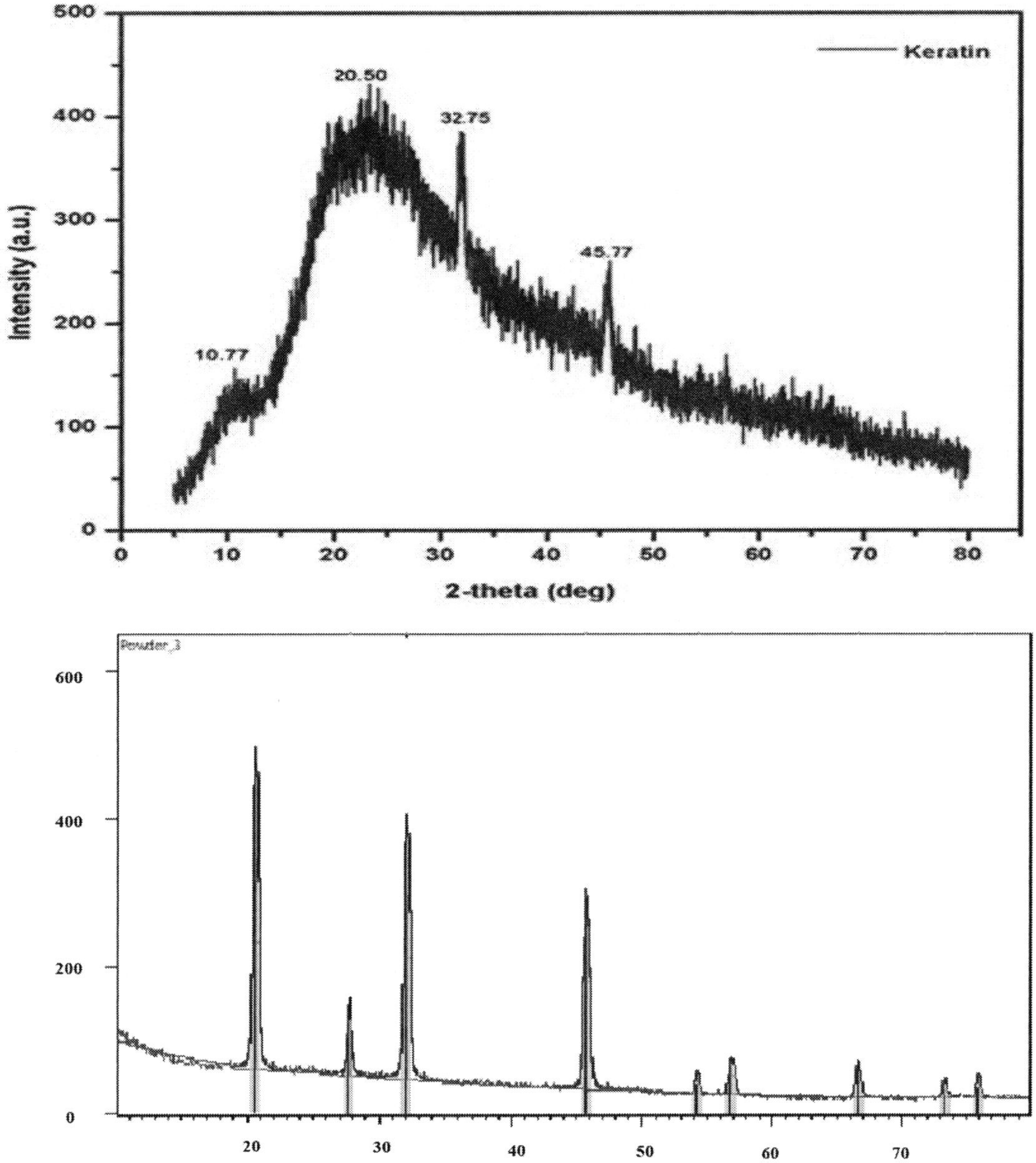

Figure 3. The XRD of keratin

4. Conclusions

In a conclusion, the study aimed to efficiently extract keratin from the biomass waste of chicken feathers and dry the keratin using vacuum oven conditions and obtained success. The fine particles of keratin were studied for morphology and crystalline areas and the appropriate functional groups were analyzed using different assays. Keratin extract, with sequential properties of distinct amino acids, stable properties, and concentration of its mass, has excellent potential for use in future studies and commercial applications.

Acknowledgements
Authors are thankful to University Malaysia Pahang (UMP) for financial support (Financial Aid (PGRS) number: 18033) and providing facilities.

References
[1] Husain M, Gupta A, Alashwal B and Sharma S 2018 Synthesis of PVA/PVP Based Hydrogel for Biomedical Applications: A Review Energy Sources, Part A Recovery, Utilization, and Environmental Effects 2388-93
[2] Husain M S B, Gupta A and Alashwal B Y and Sharma S 2018 Synthesis of Hydrogel Using Keratin Protein from Chicken Feather Proceedings of ICCEIB 2018 **193** p 16
[3] Alashwal B Y, Bala M S, Gupta A, Sharma S and Mishra P 2019 Improved properties of keratin-based bioplastic film blended with microcrystalline cellulose: A comparative analysis *Journal of King Saud University-Science* In Press
[4] Gupta A, Kamarudin N B, Kee C Y G and Yunus R B M 2012 Extraction of keratin protein from chicken feather *Journal of Chemistry and Chemical Engineering* **6** (8) 732
[5] Navone L and Speight R 2018 *Understanding the dynamics of keratin weakening and hydrolysis by proteases* (Public Library of Science) p e0202608
[6] Ma B, Qiao X, Hou X and Yang Y 2016 Pure keratin membrane and fibers from chicken feather *International Journal of Biological Macromolecules* 614–621
[7] Sharma S, Gupta A, Chik S M S, Kee C G, Mistry B M, Kim D H and Sharma G 2017 Characterization of keratin microparticles from feather biomass with potent antioxidant and anticancer activities *International journal of biological macromolecules* 189–196
[8] Fraser R D B and Parry D A D 2017 *Filamentous structure of hard β-keratins in the epidermal appendages of birds and reptiles in Fibrous Proteins: Structures and Mechanisms* (Springer, Cham) 231–252
[9] Fan J, Lei T D, Li J, Zhai P Y, Wang Y H, Cao F Y and Liu Y 2016 High protein content keratin/poly (ethylene oxide) nanofibers crosslinked in oxygen atmosphere and its cell culture *Materials & Design* 60–67
[10] A Shavandi, Bekhit A E-D A, A Carne, and Bekhit A 2017 Evaluation of keratin extraction from wool by chemical methods for bio-polymer application *Journal of bioactive and compatible polymers* 163–177
[11] Sinkiewicz I, Śliwińska A, Staroszczyk H and Kołodziejska I 2017 *Localization of Alpha-Keratin and Beta-Keratin (Corneous Beta Protein) in the Epithelium on the Ventral Surface of the Lingual Apex and Its Lingual Nail in the Domestic Goose (Anser Anser f. domestica) by Using Immunohistochemistry and Raman Microspectros* (The Anatomical Record 300)
[12] Skieresz-Szewczyk K., Jackowiak H, Buchwald T and Szybowicz M 2017 *Localization of Alpha-Keratin and Beta-Keratin (Corneous Beta Protein) in the Epithelium on the Ventral Surface of the Lingual Apex and Its Lingual Nail in the Domestic Goose (Anser Anser f. domestica) by Using Immunohistochemistry and Raman Microspectros* (The Anatomical Record 300)
[13] Fraser R D B and Parry D A D 2017 *Filamentous structure of hard β-keratins in the epidermal appendages of birds and reptiles in Fibrous Proteins: Structures and Mechanisms* (Springer, Cham) 231–252
[14] Bao Y, Ma J and Li N, 2011 Synthesis and swelling behaviors of sodium carboxymethyl cellulose-g-poly (AA-co-AM-co-AMPS)/MMT superabsorbent hydrogel *Carbohydrate Polymers* 76-82
[15] Sharma S, Gupta A, Chik S M S B T, Kee C Y G and Poddar P K, 2017 *Dissolution and characterization of biofunctional keratin particles extracted from chicken feathers* (IOP Publishing) 12013
[16] Ramakrishnan N, Sharma S, Gupta A and Alashwal B Y 2018 Keratin based bioplastic film from chicken feathers and its characterization *International Journal of Biological Macromolecules* **111** 352-58
[17] Kamarudin N B, Sharma S, Gupta A, Kee C G, T Chik S M S B, and Gupta R 2017 *Statistical*

investigation of extraction parameters of keratin from chicken feather using Design-Expert (3 Biotech) 127

[18] Idris A, Vijayaraghavan R, Patti A F, and MacFarlane D R 2014 Distillable protic ionic liquids for keratin dissolution and recovery *ACS Sustainable Chemistry & Engineering* 1888–94

[19] Wang Y-X, and Cao and X.-J 2012 Extracting keratin from chicken feathers by using a hydrophobic ionic liquid *Process Biochemistry* 896–899

[20] Barone J R, Schmidt W F and Gregoire N T 2006 Extrusion of feather keratin *Journal of Applied Polymer Science* 1432–1442

[21] Ayutthaya S I N, Tanpichai S and Wootthikanokkhan J 2015 Keratin extracted from chicken feather waste: extraction, preparation, and structural characterization of the keratin and keratin/biopolymer films and electrospuns *Journal of Polymers and the Environment* 506–516

[22] Song K, Xu H, Xie K, and Yang Y, 2017 Keratin-Based Biocomposites Reinforced and Cross-Linked with Dual-Functional Cellulose Nanocrystals *ACS Sustainable Chemistry & Engineering* 5669–5678

[23] Esparza Y, Ullah A and Wu J 2017 Preparation and characterization of graphite oxide nano-reinforced biocomposites from chicken feather keratin *Journal of Chemical Technology & Biotechnology* 2023–2031

[24] Yin X-C, Li F-Y, He Y-F, Wang Y, and Wang R-M, 2013 Study on effective extraction of chicken feather keratins and their films for controlling drug release *Biomaterials Science* 528–36

IOP Conf. Series: Materials Science and Engineering **702** (2019) 012034 doi:10.1088/1757-899X/702/1/012034

Preparation and characterization of grease formulated from waste transformer oil

S N A Mohd Sofi, M A Abd Aziz*, N S Anang Japar, N W Abdu Rahman, A R Abdulhalim and M Y Mohd Yunus

Faculty of Chemical & Process Engineering Technology, Universiti Malaysia Pahang, 26300 Gambang, Pahang, Malaysia

*E-mail: maizudin@ump.edu.my

Abstract. Grease is known as a lubricating material in a form of solid to semisolid mixture which is produced with the combination of base oil (80% - 95%), thickener (2% - 15%) and additive (0% - 10%). The limitation of soap grease performances as well as the volatile oil prices worldwide has led the interest in seeking alternative grease formulation. This research is focusing on the formulation of non-soap based grease from waste transformer oil (WTO) by two types of non-soap based thickeners, i.e. fumed silica and bentonite. The grease is prepared using different ratios of WTO and non-soap thickener ranging from 90:10 to 65:35. The physical and chemical properties of the formulated grease were evaluated using the ASTM International grease testing standard – for consistency and oil separation, and using the thermalgravimetric analyzer (TGA) and Fourier Transform Infrared Spectroscopy (FTIR) – for thermal stability and composition identification. It was found that the grease consistencies were directly proportional to the amount of thickener and the oil separated from the grease will affect the grease stability. Greases were found thermally stable at temperature as high as 150 °C. From the overall test, the best formulation was the grease formulated using bentonite with the ratio of 80:20 with NLGI grade 2 consistency.

1. Introduction

Grease is a solid to semisolid mixture which is produced with the combination of base oil, thickener, and additive with the range ratio of 80% to 95%, 2% to 15% and 0% to 10% respectively [1]. Grease also is one of the oldest forms of lubricating material and in early 2000, grease was made from environmentally friendly and biodegradable materials. The formulated grease provides satisfactory performance in most of the industrial applications. The important factors that affecting the properties and characteristics of grease are the amount and type of thickener used, the viscosity and physical characteristics of the oil, and the additives. The grease is fundamentally designed to separate moving parts, decreasing the friction between two moving parts to minimize wear and also reducing the heat generation [2]. Variation ratio of base oil, thickener and formulation process, producing different grease properties and specialty. The Association of Southeast Asian Nations (ASEAN) countries, is further contributing to the growth of the lubricant consumption as a whole and grease in particular, in the region [3]. Over the next two years, the compound annual growth rate for industrial grease demand will be a measly 0.3 percent, bringing global consumption to 691,000 tons in 2019 [4]. There are several reasons to get involved in this field of formulating the non-soap based grease from waste transformer oil (WTO).

The first reason is characteristic of soap-based, secondly is economy and availability of raw material which is petroleum and lastly is environment.

Non-soap grease usually reserved for specialist application and it is costly than soap greases [5] thus, this type of grease is not commonly used compared to soap greases. However, it has distinct properties that most soap greases do not have, unless with the presence of selected grease additive. For example, non-soap greases are thermally stable thus, it is able to withstand wider temperature range. Some of the non-soap greases also have excellent water resistance as some of the non-soap thickeners is water insensitive and corrosion resistance. Normally, such properties are apparent in most soap greases only by the help of grease additive for enhancing the grease properties or adding the desired properties. In general, non-soap thickener is known for its key characteristics which are excellent performances over high temperature, water resistance, and good thermal and oxidative stability [6].

The petroleum is classified as one of the non-renewable resources and the oil prices are currently unstable worldwide. This study is providing information on the possibility of formulating alternative grease from waste oil generated by the power industry (waste transformer oil). This could possibly help in reducing the pollution caused by the waste oil – due to illegal disposal [7], as well as in reducing the overall cost of grease production [8].

Due to these reasons, the distinctive non-soap thickener properties, waste oil-related pollution, and the volatile oil prices, this study aimed to investigate the potential of converting the waste transformer oil (WTO) into valuable end products which are non-soap based grease. In order to achieve this, different ratios of WTO and thickener were being tested and analyzed using chemical and physical analysis to ensure the suitability of the grease produced.

2. Materials and methodology

All materials used, the process of preparing the grease, and the characterization of grease were discussed in the following subchapters.

2.1. Materials

The formulation of grease was based on the varies ratio of three main components which are base oil, thickener, and additives. In this study, waste transformer oil (WTO) was used as base oil. Initially the oil was pre-treated in order to eliminate the impurities. For the thickener, two different types of non-soap based thickener were used (bentonite and fumed silica). Thickeners are molecules, polymers or particles that are partially soluble in lubricating fluid. Different types of thickener will affect the grease performance mainly in dropping point, water-resistance and grease consistency.

2.2. Pre-treatment of waste transformer oil

This process is one of the important steps in this study to ensure that the waste transformer oil was free from any impurities and in a good condition before went through the grease formulation process. Two major steps involved in this process were filtering and heating. The filtering process was done by applying the vacuum filter to ensure the waste oil was free from any impurities or particulates such as solids and inorganic material [9]. After that, the waste oil went through the second pre-treatment process which was heating or dewatering process. Waste oil was heated at high temperatures up to 100 °C with continuous stirring for 2 hours to eliminate the water content.

2.3. Preparation of grease

Table 1 and Table 2 show the ratio of the non-soap based grease formulation for two different types of thickener namely, bentonite and fumed silica. The formulation consists of 4 major steps. Initially, the waste transformer oil was pre-treated process. After that, waste transformer oil and thickener are weighted according to the ratios. The waste oil will be heated at temperature 120 °C while stirring using the homogenizer to ensure the oil is free from water. Then, thickener will be added slowly into the solution and the speed of homogenizer need to be adjusted for the homogenizing process for at least 1 hour.

Table 1. Bentonite grease formula.

WTO (%)	Bentonite (%)
90	10
85	15
80	20
75	25
70	30
65	35

Table 2. Fumed silica grease formula.

WTO (%)	Fumed silica (%)
94	6
92	8
90	10

2.4. Characterization of grease
The formulated grease will undergo several physical testing such as consistency test, oil separation test, thermogravimetric analysis and also FTIR analysis in order to identify the characteristic of grease itself.

2.4.1. Consistency. Consistency test (ASTM D-217) is defined as the degree to which a plastic material resists deformation under the application of force. In the case of lubricating greases, this is a measure of the relative hardness or softness and has some relation to flow and dispensing properties. Harder and softer grease will be depending on the grease consistency number or NLGI (National Lubricating Grease Institute) class and also will produce different characteristics of grease itself. Grease will release some of lubricating oil when the mechanical stress had been applied. This oil is squeezed out of the thickener matrix and will be delivered to the lubrication points. The point of the oil release is to provide lubrication, film formation and as well as wear prevention that will make sure the optimum performance of the equipment. The higher the penetration number, the softer the texture of grease.

2.4.2. Oil separation. Oil separation test (ASTM D-1742) also known as stability test upon storage which basically to determine the ability of a grease to separate oil during storage and to also the stability of the grease formulated. An oil separation test monitors the separation of oil from grease at elevated temperatures for a defined period of time. The container is then left at room temperature for one month. The tendency of oil to separate during storage can be an important characteristic. When the base oil begins to separate from other raw materials in grease, the remaining material may change in consistency and potentially affect the ability of the grease to function as designed. In general, the lesser the amount of oil separated, the better the stability and the longer the lifetime of the formulated grease.

2.4.3. Thermogravimetric analysis. Thermogravimetric analysis (TGA) is categorized in the family of thermal analysis techniques used to characterize a wide variety of materials [10]. TGA measures the amount and rate which is the velocity of change in the mass of a sample as a function of temperature or time in a controlled atmosphere [11]. The results are used usually in order to determine the thermal or oxidative stabilities of the materials as well as their compositional properties. This analysis has the capability to predict grease high-temperature life and designed to be a comprehensive method of oil bleeding. Currently, many standard analysis methods are available to identify the thermal stability of grease, but none of them provides comprehensive data related to the oil separation and evaporation of greases within a short time [11]. Next, this analysis also able to applied to select materials for certain end-use applications, predict product performance and improve product quality. Details that can be gained from this analysis such as thermal stabilities, oxidative stabilities, estimation product lifetime and moisture and volatiles content [12]. The analysis can be carried out in an inert atmosphere such as nitrogen or a reactive atmosphere such as oxygen. In general, a few milligrams of the sample were weighed and heated under controlled conditions with a heating rate. The higher the temperature for decomposition of waste oil, the higher the thermal stability of formulated grease.

2.4.4. FTIR analysis. Fourier Transform Infrared Spectroscopy (FTIR) is the analysis to analyze the functional group that exists inside the grease and oil. The FTIR will provide specific information on the molecular structure and chemical bonding of the grease that it is useful to identify the organic and inorganic material involve in the substance. The principle of FTIR is based on there being molecules present in a lubricant that absorb infrared light to different degrees because of their chemical structure. The sample of grease is applied to an attenuated total reflectance (ATR) cell. In the contact zone, the grease will expose to the infrared light and the infrared spectrum will show the corresponding wavenumber based on the absorption of infrared light.

3. Result and discussion

3.1. Grease consistency

The consistency of the grease is basically depending on the result according to the National Lubricating Grease Institute (NLGI) class. This is a measure of the relative hardness (NLGI class:6) or softness (NLGI class:000) and has some relation to flow and dispensing properties. whether the grease is soft or hard, it does not mean that the grease cannot be used. The stiff greases usually used in very high speed and low load applications [13]. The higher the NLGI numbers, the higher the consistency and the grease is suitable for the application of journal bearings with low speed, high operating temperature and to avoid any water washout. On the other hand, the lower the NLGI numbers, the lower the consistency and the grease is usually used for low speed rolling element bearings and cold temperature operation [13].

Tables 3 and 4 show the results gained on the consistency class of each formulated grease. Greases that categorized under NLGI grades of 0 to 1 were bentonite grease with ratio 85:15 and fumed silica grease with ratio 94:6. This type of grease has a very soft to soft texture and suitable for the application of centralized lubrication systems with low temperatures. The grease which is categorized under NLGI class of 2 to 3 is bentonite with ratio 80:20, 75:25, 70:30 and fumed silica grease with ratio 92:8. This type of grease has medium soft to medium texture. Lastly, grease with NLGI class of 4 are bentonite grease with ratio 65:35 and also fumed silica grease with ratio 90:10.

Table 3. Bentonite grease consistency and oil separation.

WTO (%)	Bentonite (%)	Consistency (NLGI)	Oil separation (%)
90	10	-	0
85	15	0 – 1	0
80	20	2	0
75	25	3	0
70	30	3	0
65	35	3 – 4	0

Table 4. Fumed silica grease consistency and oil separation.

WTO (%)	Fumed silica (%)	Consistency (NLGI)	Oil separation (%)
94	6	-	0
92	8	3	0
90	10	3 – 4	0

The commercial greases available nowadays have a consistency ranging from NLGI grade 1 – 4. Most commercial greases are commonly formulated with NLGI grade 2 consistency because, at this consistency, the grease has the most grease-like semi-solid appearances. The present study found that only bentonite grease with ratio of 80:20 is having such consistency. Thus, bentonite grease with ratio of 80:20 is selected as the best grease formulation.

3.2. Separation of oil from grease

The oil separation test was carried in order to identify the tendency of oil to be separated from grease during the storage in accordance with ASTM D-1742 [14]. An oil separation test will monitor the separation of oil from grease at elevated temperatures for a defined period of time. The rate of oil released from the grease will increase with time and vary based on the temperature at which it is stored [15]. When the base oil starts to separate from other raw materials in grease, the remaining material may change in consistency and potentially affect the ability of the grease to function as designed. The sample of formulated grease is stored in an encloser is then leave at room temperature for a month [9].

Tables 3 and 4 show that all formulated greases have no oil separation. Both types of thickeners used which categorized as organo-clay and standard silica have an oil keeping ability and outstanding heat resistance which needs a longer time to do the oil separation process [16].

For bentonite grease with a ratio of 90:10, the inexistence of separated oil was due to the liquid-like appearance of the grease for which indicated that no grease structure is formed. At this ratio, the thickener is dispersed and dissolved in the WTO. This finding is due to the insufficient thickener to form 3D thickener networks to absorb the WTO into the thickener structure [17]. However, the grease structure is forming with the presence of sufficient thickener and able to hold the oil thus resulting in no oil separation over time.

3.3. TGA of grease

According to figure 1, the sample of formulated grease has high thermal stability as the grease is able to operate at higher temperatures. The TG curve presented two weight loss steps, which are at temperature 194.08 °C and 380.44 °C respectively. The WTO started to decompose at temperatures above 150 °C and the quality loss of WTO occurs when the temperature is raised to 150 °C under normal pressure. The loss rate of WTO is gradually increased accompanied by temperature rise. The decomposition accelerated when the temperature keeps increasing until more than 200 °C. Generally, the first decomposition happened at temperature range 190 °C – 360 °C due to the elimination of low molecular weight of sample. The WTO completely decomposed at a temperature 380 °C. This result shows that most of the volatile components in WTO will be vaporized at a temperature above 200 °C and almost all of the oil had vaporized at 490 °C. Thus, WTO can be used at operating temperatures up to 150 °C without undergoing significant vaporization. Other than that, the grease formulated is categorized under medium volatile as the product able to withstand at more than 150 °C and does not evaporate.

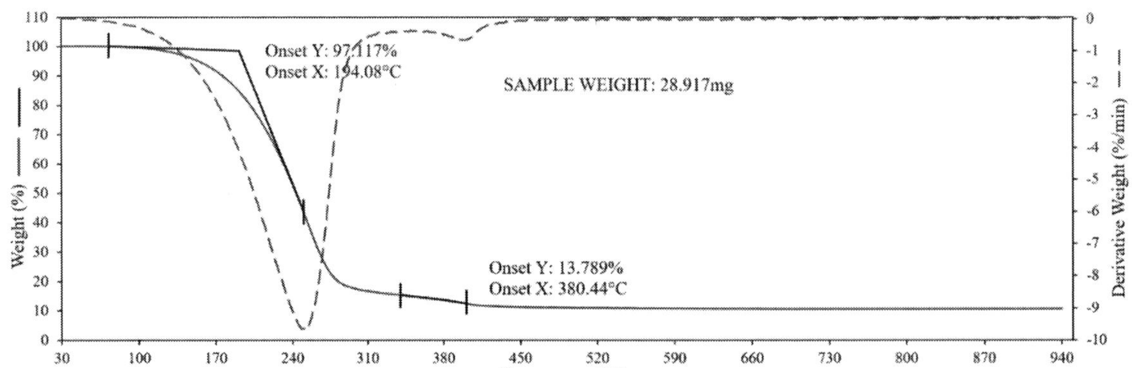

Figure 1. TGA of bentonite grease with 80% WTO.

3.4. FTIR characterization

Figures 2 and 3 show the FTIR peaks of bentonite and fumed silica greases, respectively. Both greases have peaks at wavenumber between 400 cm^{-1} to 600 cm^{-1} and 900 cm^{-1} to 1500 cm^{-1} which represent a C-O functional group, and at 2700 cm^{-1} to 3000 cm^{-1} indicating the C-H functional group.

IOP Conf. Series: Materials Science and Engineering **702** (2019) 012034 doi:10.1088/1757-899X/702/1/012034

According to the analysis, the intensity of absorbance peaks at 2922 cm^{-1} and 2854 cm^{-1} increase with aging due to the adsorption process of mineral oil to the paper surface during aging. The major degradation of the aging mechanism in mineral oil is oxidation. Oxidation will decompose the hydrocarbon molecules into other substances such as hydroperoxide, alcohol, aldehyde, ketones and also esters.

The oxidation reaction of mineral oils also will produce carboxylic acids (R-COOH) that bind with alcohol in order to generate esters. During thermal aging process, the carboxylic acids generated from the oxidation process of mineral oil can react with hydroxyl groups on the cellulose forming a cellulose graft polymer via a condensation reaction. Other than that, Bentonite which is acts as a thickener were found at the wavenumber 1034 cm^{-1} in figure 2 [18].

From figure 3, the wavenumber produced at 3000 cm^{-1} represents the fumed silica thickener content in the grease sample [19]. The other objective of this analysis is to determine the water content in the sample and the wavenumber for water is at 3600 cm^{-1}. Both of the sample have lesser water content which proved that the grease is a good thermal ability product.

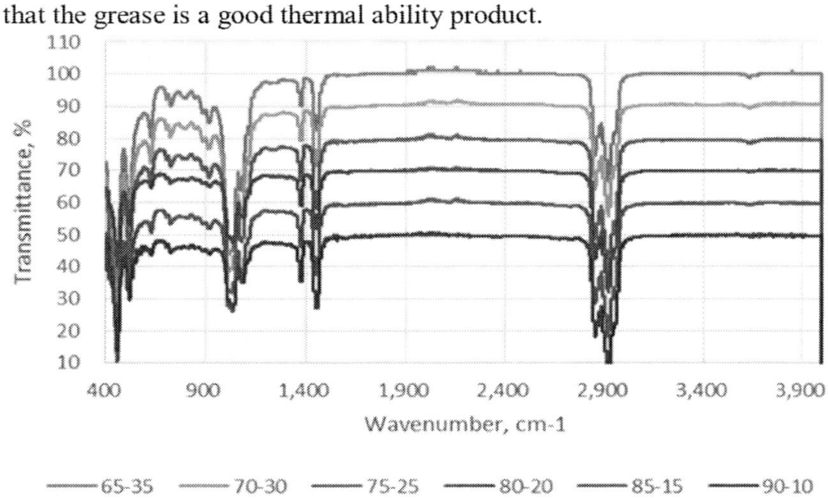

Figure 2. FTIR spectrums of bentonite greases.

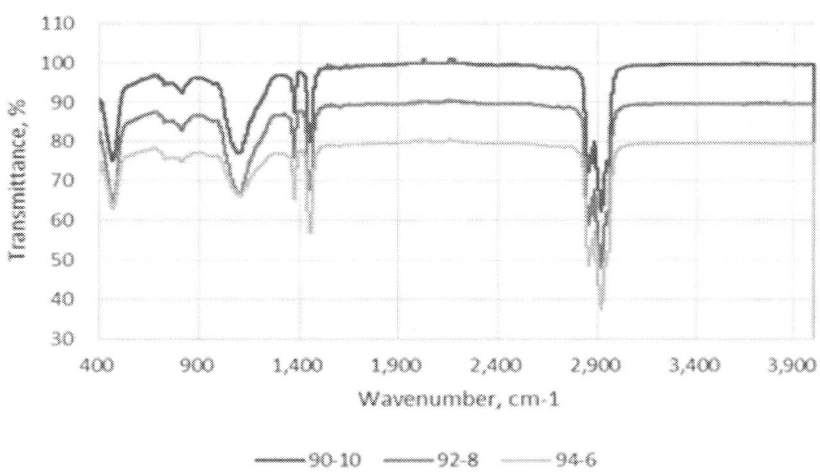

Figure 3. FTIR spectrums of fumed silica greases.

4. Conclusion

In this research, grease from WTO as the base oil was formulated successfully by using two different types non-soap thickener which is bentonite and fumed silica. The best grease formulation is bentonite grease with ratio of 80:20 (WTO: bentonite) because it has the consistency of NLGI grade 2 that is comparable to most commercial greases. The other formulated greases are classified as NLGI grade 3 – 4 grease with no oil separation during storage. The WTO and the greases can be exposed to operating conditions as high as 150°C or above, respectively, without substantial evaporation or degradation. It is proven that the WTO could be used as alternative base oil in grease formulation and at the same time, reducing the dependency of lubricant towards conventional mineral oil.

References

[1] Mang T 2014 *Encyclopedia of Lubricants and Lubrication* ed T Mang (Springer-Verlag Berlin Heidelberg)
[2] Donley E 2012 *Handbook of Advances in Additive Lubricants and Grease Technology* (Auris Reference)
[3] Mordor Intelligence 2019 Grease Market - Growth, Trends, and Forecast (2019 - 2024) *Mordor Intell.*
[4] Vietsea 2016 Grease Demand Forecast to Stagnate *Vietsea Co.*
[5] McGrory S 2001 *Plant Engineer's Handbook* ed R K Mobley (Elsevier Science)
[6] Rizvi S Q A 2008 *A Comprehensive Review of Lubricant Chemistry, Technology, Selection, and Design* (ASTM International)
[7] Mensah-Brown H 2015 Re-refining and recycling of used lubricating oil: An option for foreign exchange and natural resource conservation in Ghana *ARPN J. Eng. Appl. Sci.* **10** 797–801
[8] Kholijah A M S, Yeung S L C, Sazwani S and Yunus R M 2012 Production of High Temperature Grease from Waste Lubricant Sludge and Silicone Oil *J. Appl. Sci.* **12** 1171–5
[9] Japar N S A, Aziz M A A and Razali M N 2019 Formulation of fumed silica grease from waste transformer oil as base oil *Egypt. J. Pet.*
[10] Sichina W J 2000 Characterization of Polymers Using TGA *Perkin Elmer Instruments*
[11] Rhee I S 2009 *Prediction of High Temperature Grease Life Using a Decomposition Kinetic Model* (Warren, MI)
[12] Polini A and Yang F 2017 5 - Physicochemical characterization of nanofiber composites *Nanofiber Composites for Biomedical Applications* ed M Ramalingam and S Ramakrishna (Woodhead Publishing) pp 97–115
[13] Mortier R M, Fox M F and Orszulik S 2011 *Chemistry and Technology of Lubricants* (Springer Netherlands)
[14] ASTM D1742-06 2013 Standard Test Method for Oil Separation from Lubricating Grease During Storage
[15] Ludwig L G 2012 Storing Grease to Avoid Bleed and Separation *Mach. Lubr.*
[16] Abdulbari H A, Abid R T and Mohammad A H A 2008 Fume Silica Base Grease *J. Appl. Sci.* **8** 687–91
[17] Razali M N, Aziz M A A, Hamdan W N A W M, Salehan N A M and Rosli M Y 2017 Synthesis of Grease from Waste Oils And Red Gypsum *Aust. J. Basic Appl. Sci.* **11** 154–9
[18] Holtzer M, Bobrowski A and Grabowska B 2011 Montmorillonite: A Comparison of Methods for Its Determination in Foundry Bentonites *Metalurgija* **50** 119–22
[19] Voronin E F, Gun'ko V M, Guzenko N V, Pakhlov E M, Nosach L V, Leboda R, Skubiszewska-Zięba J, Malysheva M L, Borysenko M V and Chuiko A A 2004 Interaction of poly(ethylene oxide) with fumed silica *J. Colloid Interface Sci.* **279** 326–40

IOP Conf. Series: Materials Science and Engineering **702** (2019) 012035 doi:10.1088/1757-899X/702/1/012035

Properties of microcrystalline cellulose prepared from oil palm empty fruit bunch at different sulfuric acid concentration and hydrolysis temperature

J Sukaimi, W N H Wan Abdullah, M A F Supian, S S Jamari, N A Mohd Azman, S Mohamad and M F Ali*

Faculty of Chemical and Process Engineering Technology, College of Engineering Technology, Universiti Malaysia Pahang, 26300 Gambang, Pahang, Malaysia

*Email: mohdfaizalali@ump.edu.my

Abstract. Microcrystalline cellulose (MCC) is a biodegradable crystalline powder that normally isolated through acid hydrolysis process of cellulose. However, the implication of high concentration of acid, such as more than 5M sulfuric acid (H_2SO_4) in MCC production is concerned. Therefore, this work studies the effect of low acid concentration (1.5, 2.5 and 3.5 M) and hydrolysis temperature (40, 60 and 80 °C) for the MCC isolation from oil palm empty fruit bunch (EFB) assisted with steam explosion process cellulose fibers. Yield, crystallinity and chemical properties of the isolated EFB-MCC were discussed and compared with the commercial MCC. The crystallinity result shows the range of 72 – 77% for the isolated EFB-MCC. The highest crystallinity (77%) achieved at 3.5M H_2SO_4 at 80°C, similar to the crystallinity of the commercial MCC. As for the yield, the trend is decreasing severely as the increment of temperature, however small reduction of yield detected at different H_2SO_4 concentration. While for the chemical properties, FTIR spectra illustrated consistent wavenumbers detection for the EFB-MCC, commercial MCC and exploded EFB-Cellulose. It can be concluded that the optimum condition for the acid hydrolysis process is 1.5M H_2SO_4 at 60 °C with the middle range of crystallinity (74.7 %) and yield (82.4 %).

1. Introduction

The palm oil industry is one of the key economic contributors in Malaysia, respectively [1]. In 2018, the Malaysian crude palm oil output leaped by 12.5% to 43 million tons compared to the previous year's [1, 2], and was estimated to increase by 4 -5% in 2019 [2]. A large amount of palm oil waste would be generated during palm oil extraction oil process [3]. Empty fruit bunches (EFB) is the main solid waste produced, estimated by 25 wt%, from the oil extraction in the palm oil mill [2-3]. Conventionally, this waste is directly disposed of either by mulching and composting in the plantation site or landfills [4]. This lead to the abundant of unused EFB, which not only consuming massive valuable area, triggering severe environmental pollutions but also a huge waste of natural resources [5]. Therefore, reuse and recycling of EFB for the conversion towards the value-added product is an effective opportunity for economic and environmental benefits [4-5].

Microcrystalline cellulose (MCC) is a fine, odorless, and whitish crystalline powder, normally utilized in many industry areas such as stabilizer, binder, and filler in food and medical tablets, emulsifier in certain dairy products, viscosity regulator, and as reinforcement agents in polymer composite development [6]. MCC produced as α-cellulose undergo partial hydrolyzation and depolymerization and leaving mostly crystal regions in the final structure [7]. In term of properties, MCC is a non-toxic material, biodegradable, biocompatible, high mechanical strength, high surface area and crystallinity [6,7]. MCC is highly preferred as it has cheap resources due to its availability and abundance in nature [8]. Depending on the method, techniques, conditions and sources of the cellulose, the structures and features of the MCC could be displayed differently [9].

MCC has been successfully extracted from various sources like cassava bagasse [10], wood cotton waste [11], tea waste [12], bamboo [13], pomelo peel [14], and roselle fibers [15]. Different methods have been developed as to produce MCC such as physical treatment: microwave irradiation [16], high-pressure homogenization [17]; chemical treatment: alkaline and acid hydrolysis [18,19], carbonaceous catalyst [20] or biological treatment: enzymatic hydrolysis [21]. Several studies reported that hydrolysis condition such as temperature, acid concentration, type of acid, and duration give major effects on structures and properties of produced MCC [21-22]. However, by using a huge volume of concentrated acid and high temperature will lead to high installment cost and most crucially causing severe environmental issues [17,21]. Despite many research conducted to investigate the optimum condition to prepare MCC by means to avoid all the problems above, yet acid hydrolysis is the best way of producing MCC [22].

In this study, EFB-MCC was prepared from exploded EFB-Cellulose by the acid hydrolysis method at different sulfuric acid (H_2SO_4) concentration and temperature. The yield, crystallinity and chemical properties of the EFB-MCC have been discussed and compared with commercial MCC. Even though works on the hydrolysis process and characterization of MCC from EFB has been reported, however, no research reported on the isolation of MCC from EFB-Cellulose that has been pre-treated via steam explosion process.

2. Experimental

2.1 Materials

The EFB fibers were collected from the LCSB Lepar Oil Palm Mill (Pahang, Malaysia). The EFB fibers were firstly undergoing steam explosion pretreatment process before chemical treatments (alkaline and bleaching) to isolate the Exploded EFB-Cellulose. The pre-treatment and extraction procedures were described in details in the Supian et al. [23]. The chemical used for the EFB-MCC isolation was sulfuric acid (H_2SO_4) with purity 95% from Fisher Scientific, USA. The commercial MCC, Avicel® PH101, Fluka by Sigma Aldrich, USA, was used as the reference.

2.2 Methods

2.2.1 Preparation of EFB-MCC. The EFB-MCC was isolated from the Exploded EFB-Cellulose through acid (H_2SO_4) hydrolysis process for 30 minutes with constant agitation in the 10 wt% consistency (solid to liquid ratio). Different H_2SO_4 concentrations were used; which were 1.5 M, 2.5 M, and 3.5 M, during the hydrolysis process at different temperatures; 40, 60 and 80 °C. The mixture was then filtered at room temperature and washed repeatedly with water. The produced EFB-MCC was then dried in an oven at 60°C until constant weight.

2.2.2. Percentage Yield of MCC. The percentage yield of EFB-MCCs was calculated using equation (1), given; M_{MCC} = mass of isolated EFB-MCC (g), M_C = mass of exploded EFB-Cellulose (g).

$$Yield\ (\%) = \frac{M_{MCC}}{M_C} \times 100 \tag{1}$$

2.2.3. Characterization. Selected sample of the produced EFB-MCCs were analyzed which are EFB-MCC hydrolyzed with constant concentration; 1.5 M at different temperature; 40, 60 and 80 °C and with constant temperature; 80°C at different concentration 1.5 M, 2.5 M and 3.5 M. The crystallinity of the EFB-MCCs was determined using Broker D8 Advance X-ray Diffractometer (XRD) with Ni-filtered CuK_α radiation. The scanning rate is 2°min^{-1} within a 2θ angle range from 5° to 40°. The crystalline index (*CrI*) illustrated in Equation (2) by applying an established Segal method, where I_{002} is the intensity of both amorphous and crystalline regions; I_{am} is the intensity of amorphous regions.

$$CrI\ (\%) = \frac{I_{002} - I_{am}}{I_{002}} \times 100 \quad (2)$$

Fourier Transform Infrared (FTIR) was conducted using Thermo Nicolet FTIR spectrometer to detect the presence functional groups in the samples. The spectrometer is equipped with a DTGS detector by direct scanning with fitted universal ATR accessory in the wavenumber range of 500 to 4000cm^{-1}.

3. Result and Discussion

3.1 Crystallinity analysis

MCC is typically characterized by the degree of crystallinity; based on grades with a typical value range from 55 – 80 %, as determined by XRD analysis [24]. The XRD patterns of EFB-MCC at different concentrations of acid sulfuric and temperatures are presented in figure 1. The x-ray diffraction pattern of all MCC-EFB shows similar trends with the 2θ values were found at 16, 18, 22.6 and 22.9°. It is reported by De Menezes et al. [25] that the typical XRD patterns for lignocellulosic are displayed at the crystal lattice of cellulose I at 2θ values of 15, 16, 22.6, and 34°.

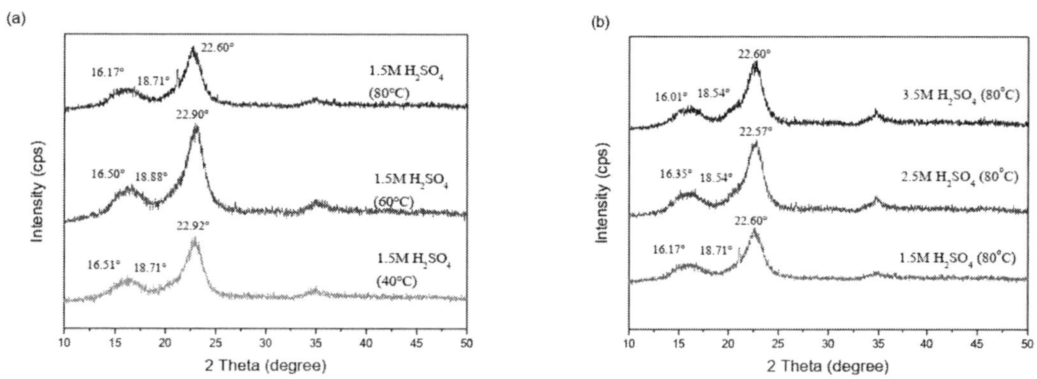

Figure 1. The XRD patterns of EFB-MCC (a) 1.5 H_2SO_4 at a different temperature, (b) different H_2SO_4 concentration at 80°C.

The crystallinity data for the isolated EFB-MCC and commercial MCC are tabulated in Table 1. Analyses showed that the crystallinity index of EFB-MCC isolated at a concentration of 3.5M and 80°C is the highest, which is at 77.03% and between the range of commercial MCC (76 -79%). The higher crystallinity of MCC related to the removal of amorphous regions of cellulose which stimulates the hydrolytic cleavage of glycosidic bonds that leads to rearrangement of cellulose molecules [26]. The percentage of crystallinity for other samples is slightly lower (~5%) than the EFB-MCC isolated at a concentration of 3.5M and 80°C, indicating no significant effect of the crystallinity of MCC from different acid concentration and hydrolysis temperature. Based on the data, the manipulation of concentration and hydrolysis temperature does not offer a significant effect on the crystallinity of the EFB-MCC.

Table 1. The crystallinity of EFB-MCC and commercial MCC.

Sample	Concentration (M)	Temperature (°C)	Crystallinity (%)	Reference
Commercial MCC	n/a	n/a	76 - 79	[24]
EFB-MCC	1.5	40	73.02	Experimental Data
		60	74.71	
		80	73.78	
	1.5	80	73.78	
	2.5		76.75	
	3.5		77.03	

*** n/a is not available**

3.2 Yield of EFB-MCC

Table 2 shows the percentage yield of the isolated EFB-MCC through different H_2SO_4 concentration and hydrolysis temperatures. Generally, the yield obtained is reducing as the concentration of H_2SO_4 increase along with the hydrolysis temperature. Compared to the 1.5M H_2SO_4, the greater influence of hydrolysis temperatures can be identified at a higher acid concentration as the yield reduction is around 35% difference from the temperature of 40 to 80 °C. The highest yield was recorded at 99.3% for the EFB-MCC hydrolyzed with 1.5M H_2SO_4 at 40°C.

This result is in agreement with the yield obtained by previous research [27] where the yield decreased when the temperature increased from 30°C to 40°C. The increment of hydrolysis temperature will promote the hydrolysis which will affect the crystalline region orientation [27] as a result, the cellulose tends to transform into glucose substances [28]. However, a minimal difference in yield is reported with the increment of acid concentration at a constant temperature. Besides, a similar trend is illustrated with the percentage of crystallinity recorded. Thus, it is important to obtain an optimum yield with good crystallinity of MCC.

Table 2. The yield of EFB-MCC at different acid concentrations and hydrolysis temperatures.

Concentration (M)	Temperature (°C)	Yield (%)
1.5	40	99.3
	60	82.4
	80	73.0
2.5	40	85.0
	60	79.7
	80	53.4
3.5	40	79.6
	60	86.6
	80	50.9

3.3 Functional groups analysis

The FTIR spectra of commercial MCC, EFB-MCC at different concentration and Exploded EFB-Cellulose are shown in Fig. 2. No major changes in the FTIR spectra can be detected following hydrolysis of cellulose with acids of different concentrations. The major detected regions are found at several wavenumbers such as 3500 – 3200 cm^{-1} for hydroxyl bonds, 2900 – 2800 cm^{-1} for C-H bonds, 1020 – 970 cm^{-1} for C-O-C bonds.

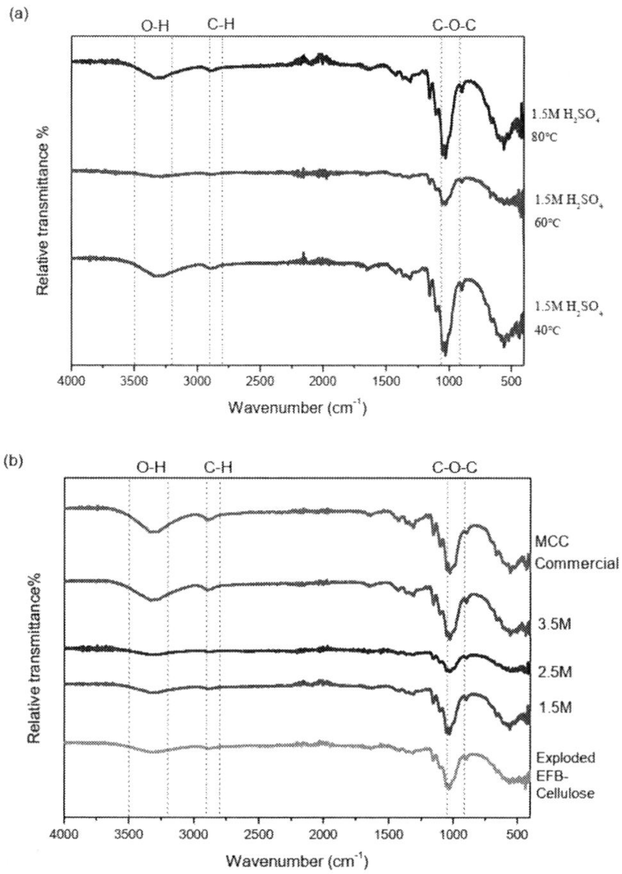

Figure 2. FTIR spectra of EFB-MCC (a) 1.5 H2SO4 at a different temperature, (b) different H2SO4 concentration at 80°C, Exploded EFB-Cellulose and MCC commercial.

The spectra obtained revealed that the similarities between all regions indicating that all isolated samples have similar chemical compositions. These results are also comparable with the spectra reported by Mohamad Haafiz et.al [24], which in their work compared the OPEFB-MCC, OPEFB-pulp and commercialize MCC (C-MCC). The effect of hydrolysis conditions is not affecting the chemical bonding on the prepared samples.

4. Conclusion
The MCC is successfully isolated from the exploded EFB-Cellulose, at different acid concentration and hydrolysis temperature. The XRD data shows that all the samples can be classified as crystalline cellulose material, with the range of crystallinity of 72 – 77%. The highest crystallinity (77 %) value is recorded at 3.5M H_2SO_4 at 80 °C; however, it recorded the lowest yield (50.9 %). Based on the yield obtained, different process temperature gives more influence to the hydrolysis process compare to the H_2SO_4 concentration. While the chemical bonds between the samples show that no changes on the EFB-MCC structure compared to the exploded EFB-Cellulose and commercial MCC. Based on these results, it can be concluded that the optimum condition for the acid hydrolysis process is 1.5M H_2SO_4 at 60 °C with the middle range of crystallinity (74.7 %) and yield (82.4 %).

5. Acknowledgments
The authors would like to be obliged to Universiti Malaysia Pahang for providing laboratory facilities and financial support under the Internal Research Grant Scheme project no. RDU1703182.

References

[1] Nambiappan B, Ismail A, Hashim N, Ismail N, Shahari D N, Nik Idris N A, Omar N, Mohamed Salleh K, Mohd Hassan N A and Khushairi A 2018 *J. Oil Palm Res.* **30** 13-25
[2] Ayoub M and Abdullah A Z 2012 *Renew. Sust. Energ. Rev.* **16** 2671-2686
[3] Reijnders L and Huijbregts M A J 2008 *J. Clean. Prod.* **16** 477-482
[4] Singh R P, Ibrahim M H, Esa N and Iliyana M S 2010 *Rev. Environ. Sci. Bio.* **9** 331-344
[5] Foo K Y and Hameed B H 2009 *J. Hazard. Mater.* **172** 523-531
[6] Zango Z U and Imam S S 2018 *J. Nanosci. Nanotechno.* **8** 1-6
[7] Khattab M M, Abdel-Hady N A and Dahman Y 2017 *Cellulose nanocomposites: Opportunities, challenges, and applications. In Cellulose-Reinforced Nanofibre Composites* (Woodhead Publishing) p 483-516
[8] Siqueira G, Bras J and Dufresne A 2010 *Polymers-Basel.* **2** 728-765
[9] Ng H M, Sin L T, Tee T T, Bee S T, Hui D, Low C Y and Rahmat A R 2015 *Compos. Part B-Eng.* **75** 176-200
[10] Teixeira E D M, Curvelo A A, Corrêa A C, Marconcini J M, Glenn G M and Mattoso L H 2012 *Ind. Crop. Prod.* **37** 61-68
[11] Xiong R, Zhang X, Tian D, Zhou Z and Lu C 2012 *Cellulose* **19** 1189-1198
[12] Zhao T, Chen Z, Lin X, Ren Z, Li B and Zhang Y 2018 *Carbohyd. Polym.* **184** 164-170
[13] Visakh P M, Thomas S, Oksman K and Mathew A P 2012 *Compos. Part A-Appl. S.* **43** 735-741
[14] Liu Y, Liu A, Ibrahim S A, Yang H and Huang W 2018 *Int. J. Biol. Macromol.* **111** 717-721
[15] Kian L K, Jawaid M, Ariffin H and Karim Z 2018 *Int. J. Biol. Macromol.* **114** 54-63
[16] Peng H, Li H, Luo H and Xu J 2013 *Bioresource Technol.* **130** 81-87
[17] Lee S Y, Chun S J, Kang I A and Park J Y 2009 *J. Ind. Eng. Chem.* **15** 50-55
[18] Wang N and Ding E Y 2004 *Acta Polym. Sin.* **6** 925-928
[19] Tang L G, Hon D N S, Pan S H, Zhu Y Q, Wang Z and Wang Z Z 1996 *J. Appl. Polym. Sci.* **59** 483-488
[20] Hu L, Li Z, Wu Z, Lin L and Zhou S 2016 *Ind. Crop. Prod.* **84** 408-417.
[21] Brodeur G, Yau E, Badal K, Collier J, Ramachandran K B and Ramakrishnan S 2011 *Enzyme Research*, 787532
[22] Moon R J, Martini A, Nairn J, Simonsen J and Youngblood J 2011 *Chem. Soc. Rev.* **40** 3941-3994
[23] Supian M A F, Mohd Amin K N, Jamari S S and Mohamad S 2019 *J. Environ. Chem. Eng.* In Press
[24] Mohamad Haafiz M K, Eichhorn S J, Hassan A and Jawaid M 2013 *Carbohyd. Polym.* **93** 628-634
[25] De Menez A J, Siqueira G, Curvelo A A S and Dufresne A 2009 *Polymers-Basel.* **50** 4552-4563
[26] Trache D, Khimeche K, Mezroua A and Benziane M 2016 *J. Therm. Anal. Calorim.* **124** 1485-1496
[27] Tan X Y, Hamid S B A and Lai C W 2015 *Biomass Bioenerg.* **81** 584-591
[28] Kusumattaqiin F and Chonkaew W 2015 *Macromol. Symp.* **354** 35-41

IOP Conf. Series: Materials Science and Engineering **702** (2019) 012036 doi:10.1088/1757-899X/702/1/012036

Three different types of tea as surfactant in liquid exfoliation of graphite: green tea, black tea and oolong tea

F S Ismail[1,*], K Yusoh[1], A S Zainal Abidin[1], A H Abdullah[1] and Z Ismail[2]

[1] Faculty of Chemical and Process Engineering Technology, Universiti Malaysia Pahang, Lebuhraya Tun Razak, 26300 Kuantan, Pahang, Malaysia

[2] Faculty of Manufacturing Engineering,Universiti Malaysia Pahang, Lebuhraya Tun Razak, 26600 Pekan, Pahang, Malaysia

*E-mail: fadwa_wawa@yahoo.my.

Abstract. Liquid phase exfoliation (LPE) with surfactant-assisted is a method to produce graphene that used water and surfactant as solvent. Tea is one of possible surfactants that can be used in LPE of graphite as it reduced the surface tension energy of water (72.75 mJ m^{-2}) to ~40 – 50 mJ m^{-2} which is suitable level to exfoliate graphite. Three different types of tea were used as surfactant through one-step production of graphene which were green tea (GT), black tea (BT) and oolong tea (OT). The exfoliation of graphite in each tea solutions were successful as the UV spectra of supernatant samples show peak at 270 nm which indicates the C – C bond of graphene flakes. Morphology analysis (AFM) of resulted graphene samples show that graphene sheets have sharp edges caused by sonication process. The effect of sonication time was being investigated in this work. Green tea has higher exponent factor of time (1.715) compare to black tea (1.033) and oolong tea (0.762) due to high quantity of small molecules of polyphenols present in green tea solution. Thus, it is better to used GT as surfactant in LPE of graphite compare to BT and OT.

1. Introduction

Tea (*Camellia sinensis*) is one of most consumed beverages by over two-third of the world population after water. In the past 2000 years, tea which originated from China has gained the world's taste [1]. Mostly, tea being consumed as green tea (GT), black tea (BT) and oolong tea (OT) as beverages. 80% of tea are processed as black tea, meanwhile 20% of tea manufactured as green tea and 2% of tea as oolong tea [2]. Those type of tea are produced varies manufacturing process. Green tea unfermented tea that produced by drying and steaming the fresh leaves without oxidation process. It has more catechins than black tea and oolong tea. It also composed of certain minerals and vitamins which increase the antioxidant potential [1]. Black tea is fermented tea. The fresh leaves are left to wither and caused the moisture content in the leaves to be reduced. Then, the withered leaves are crushed and rolled to initiate fermented process of the tea polyphenols and produce black tea [3]. Whereas, oolong tea is manufactured by rolling and firing the tea leaves in short time to end the oxidation process and dry the leaves. Oolong tea is also known as semi fermented tea [3]. Tea has been used as reducing agent to produce graphene from graphene oxide [4]. Rather as reducing agent, tea also has been proven to be one of possible surfactant to exfoliate graphite in LPE method. In Ismail et al. [5] research, graphene has been successfully produced by using black tea as surfactant in LPE of graphite assisted by kitchen mixer. It also proof that polyphenols that contain in tea acts as stabilization agent for graphene in water. Thus, most of graphene sheets does not clumping back to form graphite.

Graphene is two-dimensional (2D) of carbon allotropes building block. It is a promising material due to its own peculiar electric properties as well as outstanding mechanical and thermal properties. Graphene had been used as nanofiller in the composite matrix to enhanced the properties of bulk composite [6]. The production methods of graphene can be divided into bottom-up and top-down approach. Bottom-up process such as chemical vapor deposition (CVD), epitaxial growth and reduction of CO produce high quality of graphene. However, the cost of production for these process is extremely expensive and the yield production is low. Meanwhile, separation or exfoliation of graphite and graphite derivatives is used in top-down process to produce graphene or modified graphene sheets. Liquid phase exfoliation (LPE) is one of top-down process that has potential to produce large scale of graphene with low production cost. LPE of graphite can be divided into two solvent-assisted and surfactant-assisted exfoliation. Most of researches about solvent-assisted exfoliation method used solvents which are toxic to organ and irritant to eye such as N-methyl-2-pyrrolidone, N,N-dimethylformamide and ortho-dichlorobenzene [7]. Thus, the another alternative method is to used water as solvent in LPE method. Surfactant is needed to tune the water surface energy (72.75 mJ m^{-2}) to a suitable level for exfoliation of graphite (~ 40 mJ m^{-2}) [8]. Direct LPE of graphite in water-surfactant solution to produce dispersed graphene has attracted more attention as it is a simplest, non-toxicity and eco-friendly method.

There are three steps involves in LPE process which are dispersion of graphite in a solvent, exfoliation of graphite and purification of graphene [8]. There were several methods that been used to aided the exfoliation of graphite in LPE such as high-shear mixing, high-pressure exfoliation, vortex-fluid exfoliation, kitchen mixer and sonication [5]. In this research, graphite has been exfoliated in solvent aided by ultrasonication to extract individual layers [9]. In Ciesielski et al. [8] graphite had been successfully exfoliated to produce graphene sheets by using sonication method. During ultrasonication, sound is transmitted through solvent and created cavitation bubbles [10]. Formation, growth and implosive of these cavitation bubbles induce the exfoliation of graphite in solvent. When the graphite was exfoliated into graphene sheet, surfactant diffused into newly exposed interfaces to stabilize the graphene sheets from aggregation [11]. Sonication process can be carried out by a probe-type ultrasonic homogenizer. In the previous research, tea was used as an agent to reduce graphene oxide to graphene [12]. Rather than two step production of graphene, graphene was produced directly from industrial graphite using LPE method and three types of tea (green tea, black tea and oolong tea were used as surfactant solution in this paper.

2. Materials and Methods

2.1. Materials
Industrial graphite flakes (332461) was purchased from Sigma Aldrich (Malaysia), commercial black tea leaves (TeaBoh brand), green tea leaves (TeaBoh brand) and oolong tea leaves (Da Hong Pao).

2.2. Tea concentration
The brewing process of each tea solution must be kept in constant condition because the concentration of tea polyphenols can be effected greatly by the effect of temperature, time taken for brewing process and volume of ultrapure water. In this work, 5 g of tea was measured by analytical balance and was put into a beaker with 200 ml of ultrapure water. The solution with tea was heated with a hotplate equipped with thermal control at 120 °C for 30 minutes. After being heated, the solution was filtered using small sieves.

2.3. Exofoliation of graphite
Graphene was prepared by liquid exfoliation of graphite in water-surfactant solution aided with probe-type ultrasonic homogenizer at ~70 W. 2 g of graphite was sonicated in 200 ml of water-tea solution for 90 minutes. After sonication, the resultant supernatant was collected in tubes (50ml) and centrifuged at 5000 rpm for 30 minutes. After centrifugation, 15 ml of upper dispersant was collected from the tube which is graphene. Ultra-pure water was used to wash graphene for 6 times and it was filtered by using filter paper. Finally, graphene was dried in vacuum oven that set at 100 °C for 24 hours.

IOP Conf. Series: Materials Science and Engineering **702** (2019) 012036 doi:10.1088/1757-899X/702/1/012036

2.3. Graphene's characterization
Ultra-violet spectroscopy was used to measure the spectroscopy of blank tea solutions and exfoliated graphene in the resulting supernatant. The resulted supernatants after centrifugation and washing process are concentrated. Thus all supernatant samples were diluted with ultrapure water for the dilution factor of 10. The absorbance measurement for diluted supernatant samples and blank tea solutions were determined by using *Shidmazu UV Spectroscopy* from 800 nm to 200 nm. For morphology analysis, *Park Systems Atomic Microscopy (AFM)* was used to obtained images of graphene sheets in supernatant samples. Imaging was done in tapping mode using silicon cantilevers with spring constant of 3 to 10 N m^{-1} and resonance frequency of 110 to 190 kHz at ambient temperature and pressure with a scanning raster rate 1 Hz. Images were analyzed using *Molecular Imaging* software and properties profiles were obtained from topography images that have been levelled by first-order plane subtraction.

3. Results and Discussion

3.1. Characterization of graphene

3.1.1. Ultra-violet spectroscopy. After preparation of three different teas (BT, GT and OT) solutions in approximately same condition, the UV spectrum of those blank tea solutions were taken as shown in figure 1. Based on figure 1, solid line indicates UV spectrum for green tea solution reached the maximum value at 283 nm. Meanwhile, dashed line which represent UV spectrum of oolong tea reached the maximum value at 311 nm and uv spectrum of black tea solution (small dashed line) at 341 nm. The differences of maximum values was due to the different concentration of polyphenols present in each tea solution [13]. Thus, BT has higher polyphenols content than GT and OT solutions.

UV spectra of resulting supernatant were taken after being washed for 6 time with ultrapure water as shown in figure 2. Based on figure 2, all three spectra have shown UV peak at 270 nm. Durge et al. [14] claims that the UV peak at 270 nm (cyan region) indicates the C-C bond of graphene flakes. Thus, those peaks in figure 2 indicates successful exfoliation of graphite in each tea solution to produce graphene flakes. There is intense peak at 200 nm (yellow region) for G_{GTea} spectrum compare to G_{BTea} and G_{OTea} spectrum. This peak is attributed for remaining of unbound tea polyphenols in collected sample solution [9]. Green tea contains high concentration of simple polyphenols (catechins) than complex polyphenols. Whereas, black tea and oolong tea undergoes oxidation process that change simple polyphenols (flavonoids) into more complex structure of polyphenols (theaflavins and thearubigins) [1]. In reasonable explanation, most of simple polyphenols in green tea solution have tendency to stay in supernatant sample after it being centrifuge because it has less density than complex polyphenols.

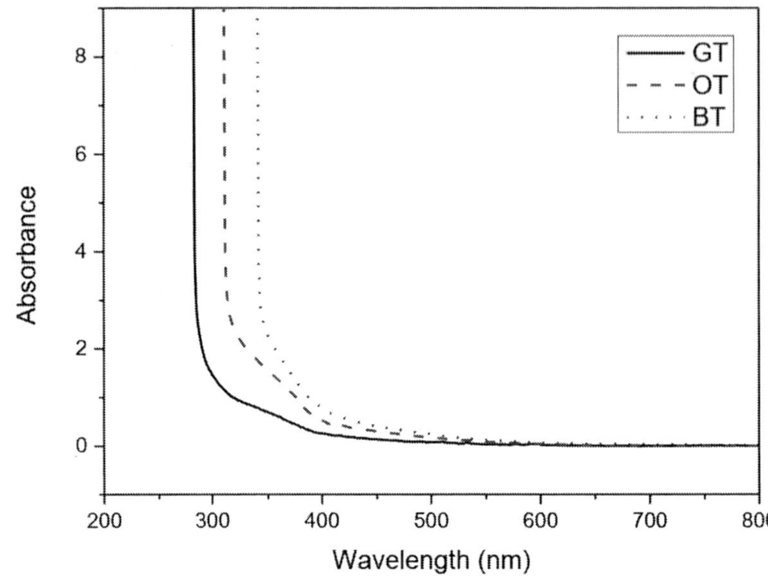

Figure 1. UV spectrum of blank black tea (BT) solution, green tea (GT) solution and oolong tea (OT) solution

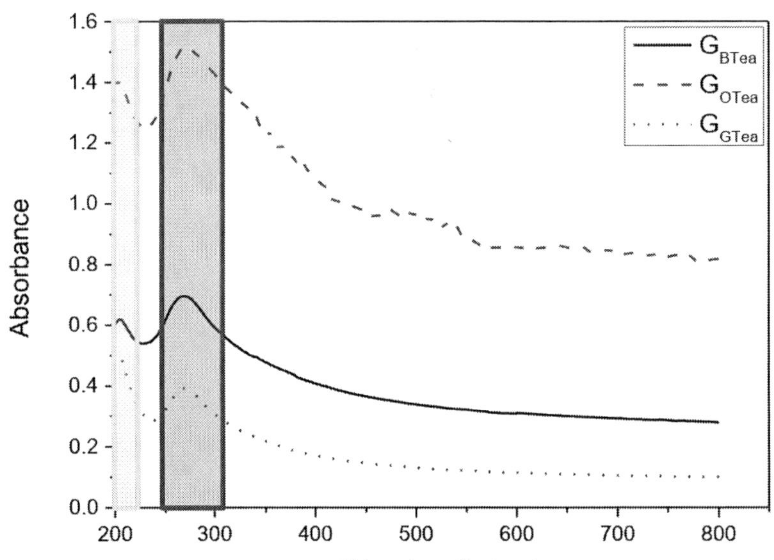

Figure 2. UV spectra of washed G_{Gtea}, G_{Otea} and G_{Btea} supernatant.

3.1.2. Atomic force microscopy (AFM)

The morphology study of those graphene was investigated by using *Atomic force microscopy (AFM)*. Figure 3 shows AFM images of G_{Gtea}, G_{Otea} and G_{Btea}. In figure 3, white color of flakes represent graphene sheet and brown color is for substrates. As shown in figure 3, there were many graphene sheets been distributed in those three samples. All three AFM images for G_{Gtea}, G_{Otea} and G_{Btea} in figure 3 shows the graphene sheets have sharp edge caused by the effect of fragmentation during sonication process by homogenizer. The same effect of sonication on the graphene sheets had been seen in Durge et al. [15] research work. The lateral sizes of graphene sheets in figure 3 are different than each other. Thus graphene that produce using LPE of graphite does not have uniform lateral size unlike chemically modified graphene [16]. Using bare eyes, graphene sheets in figure 3(a) and figure 3(c) have larger lateral size compare to figure 3(b). This may be due to sonication effect in sample for figure 3(c) is high compare to figure 3(a) and figure 3(b). Based on Botas et al. [17] study, the lateral sizes obtained from AFM images for graphene oxide sheets is decreased when the power of sonication and sonication time increased [17].

(a)

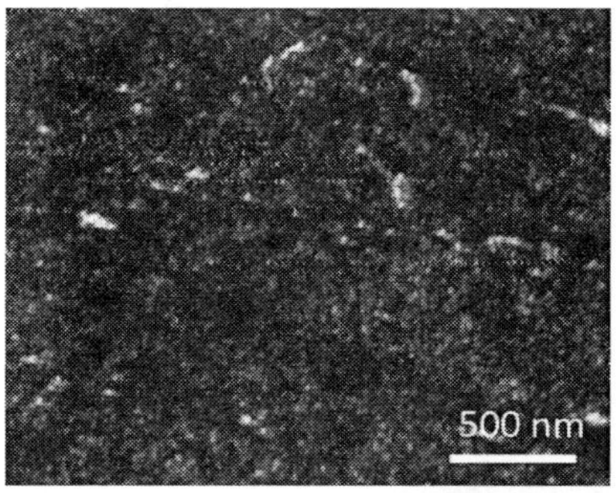

(b)

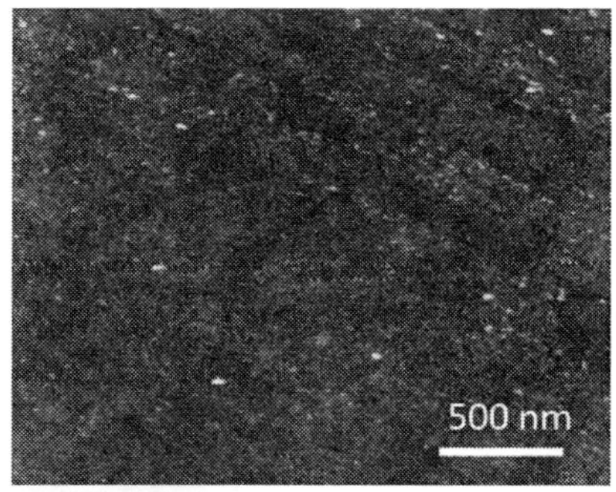

(c)

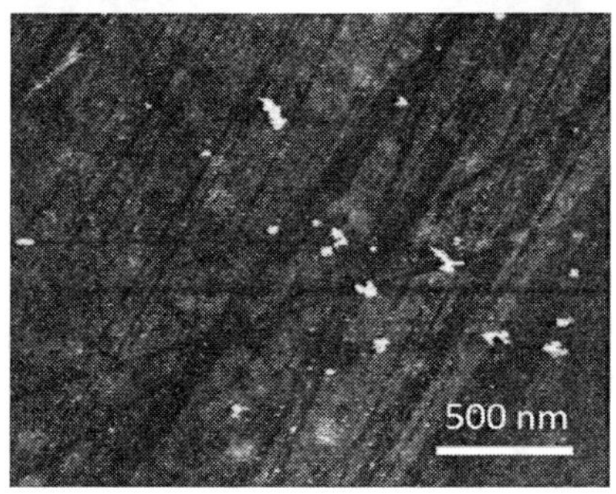

Figure 3. AFM images of (a) G_{Gtea}, (b) G_{Otea} and (c) G_{Btea}.

3.2. Effect of sonication time. The efficiency of graphene production can be improved by investigate several crucial parameters. One of important parameter that usually being investigated is production time [18]. Thus, the effect of sonication process was determined in this research work. 200 ml of BT, OT and GT solutions were prepared and mixed with 2 g of graphite. The duration of sonication time was varied between 15, 30, 45, 60, 75, 90 and 105 minutes. As the result, figure 4 shows the relationship between sonication time and concentration of G_{Gtea}, G_{Otea} and G_{Btea}. In all three relationship graphs on figure 4 shows that the yield concentration of graphene are increased as sonication time increased. Those data are consistent with other previous graphene production research [19]. This shows that sonication can be used to exfoliate graphene and as the time of sonication increased, the yield of graphene production also increased. Figure 5 shows log-log graph of G_{Gtea}, G_{Otea} and G_{Btea} against time. The exponent factor of time as function of graphene concentration can be determined by the gradient of log-log graph of graphene concentration against time [20]. Based on figure 5, G_{Gtea} has higher exponent factor of time (1.715) compare to G_{Otea} (1.033) and G_{Btea} (0.762). Based on UV spectrum in figures 2, GT has high concentration of polyphenols attached to graphene compare BT and OT. In addition, GT has simpler and smaller component of polyphenols. Small molecules of polyphenols acts as efficient surfactant because its can easily attach and locate between the layers of graphite to facilitate the exfoliation of graphene [6]. Small molecules of surfactants also packed tightly and increase its density on the graphene surface so that it can prevent graphene from aggregating and accumulating [21]. Thus

IOP Conf. Series: Materials Science and Engineering **702** (2019) 012036 doi:10.1088/1757-899X/702/1/012036

higher concentration of graphene will be produced using GT as surfactant in LPE of graphite for fixed sonication time compare to BT and OT.

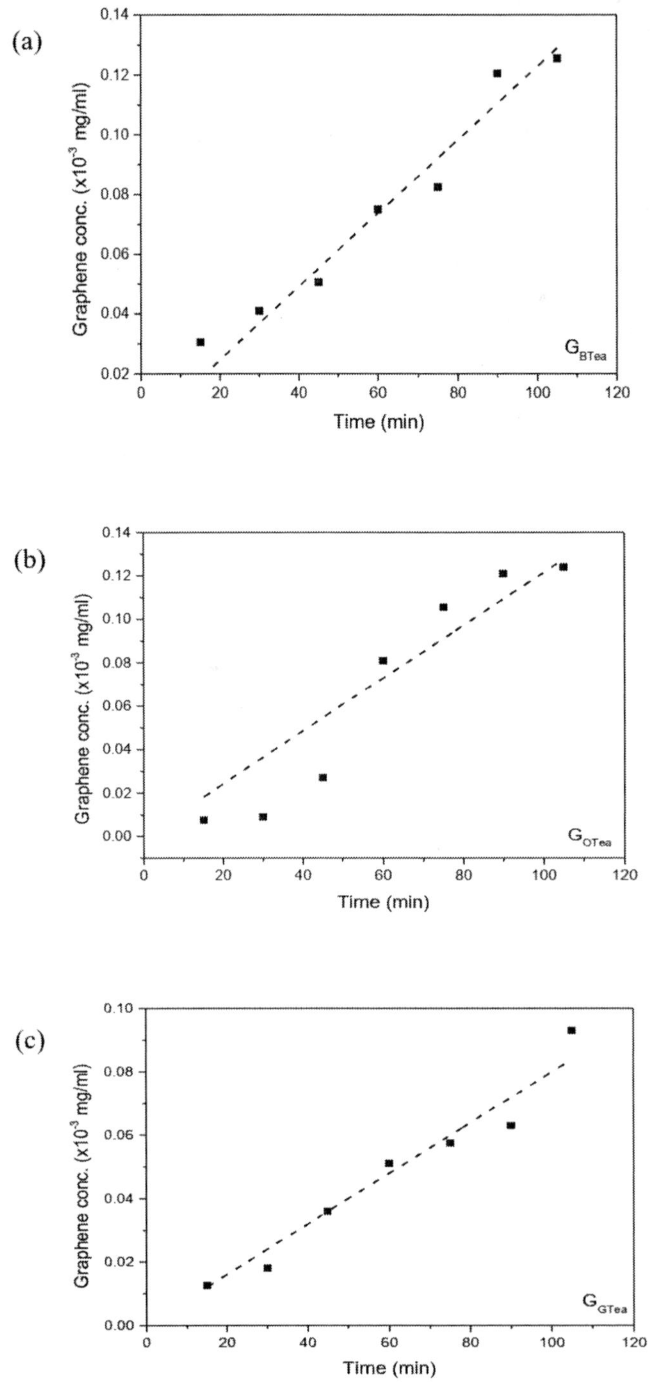

Figure 4. Relationship between time and yield concentration of (a) G_{BTea}, (b) G_{OTea} and (c) G_{GTea}.

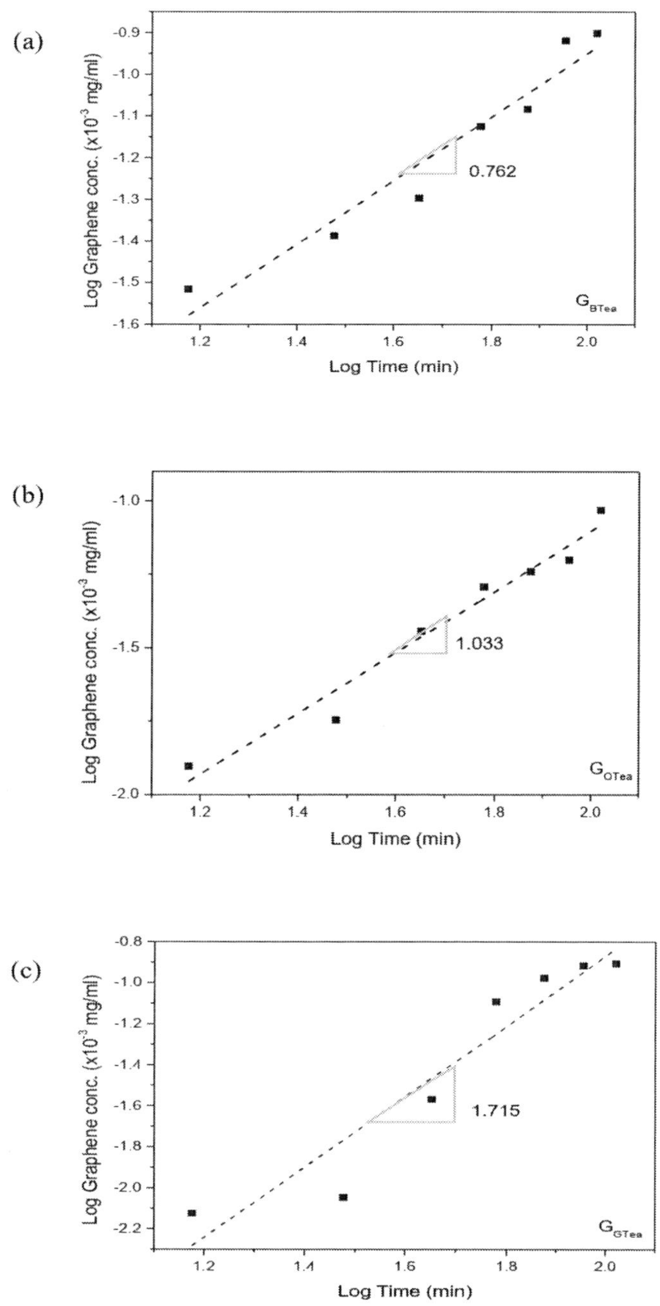

Figure 5. Log-log graph of yield concentration of (a) G_{BTea}, (b) G_{OTea} and (c) G_{GTea} against time.

4. Conclusion

In conclusion, this work has shown that direct exfoliation of graphite in GT, BT and OT solution can produce a few layers graphene. The presence of graphene after exfoliation of graphite in each tea solutions has been determined by using UV spectroscopy and the morphology study on those graphene was conducted by AFM. The effect of sonication time on the production of graphene using three different tea solutions as surfactant is discussed in this work. As the time taken for graphite to sonicated in surfactant solution increased, the yield of graphene concentration is increased. GT has shown highest exponent factor of time (1.715) than BT and OT. This is due to small components of polyphenols that present in green tea. Thus, it is better to used green tea compare to black tea and oolong tea as surfactant in LPE of graphite in order to get high yield of graphene.

Acknowledgements
This work was supported by a grant from the Ministry of Higher Education Malaysia under FGRS scheme in 2016 (Grant RDU160149).

References

[1] Cabrera C, Artacho R and Gimenez R 2006 Beneficial effects of green tea - A review *J. Am. Coll. Nutr.* **25**(2) 79–99
[2] Kushwaha U K and Kushwaha U K S 2016 Health Benefits of Black Rice *Black Rice* **4**(2) 151–83.
[3] Harbowy M E and Balentine D A 2007 Tea Chemistry *CRC. Crit. Rev. Plant Sci* **16**(5) 415–80.
[4] Krane N 1993 Preparation of Graphene Selected Topics in Physics : Physics of Nanoscale Carbon *Growth Lakel* **4**(4) 1–5.
[5] Ismail Z, Abu Kasim N F, Abdullah A H, Abidin A S Z, Ismail F S and Yusoh K 2017 Black tea assisted exfoliation by kitchen mixer allow one-step production of graphene *Materials Research Express* **4** 11
[6] Abidin A S Z, Yusoh K, Jamari S S, Abdullah A H, and Ismail Z 2016 Enhanced performance of alkylated graphene reinforced polybutylene succinate nanocomposite *AIP Conf. Proc.* **1756** 1–6
[7] Cui X, Zhang C, Hao R, and Hou Y 2011 Liquid-phase exfoliation, functionalization and applications of graphene *Nanoscale* **3** 2118-26
[8] Ciesielski A and Samorı P 2014 Graphene via Sonication Assisted Liquid-phase Exfoliation *Chem Soc Rev* **43** 381-98
[9] Ismail Z and Yusoh K 2016 Facile method for liquid-exfoliated graphene size prediction by UV-visible spectroscopy *AIP Conference Proceedings* **1756** 070002
[10] Ginsburg E, Kinsley M D, and Quitral A 1998 The power of ultrasound *Adm. Radiol. J* **17**(5) 17–20
[11] Uysal Unalan I, Wan C, Trabattoni S, Piergiovanni L and Farris S 2015 Polysaccharide-assisted rapid exfoliation of graphite platelets into high quality water-dispersible graphene sheets *RSC Adv.* **5**(34) 26482–90
[12] Ban F Y, Majid S R, Huang N M, and Lim H N 2012 Graphene Oxide and Its Electrochemical Performance *Int. J. of Electrochem. Sci.* **7** 4345–51
[13] Atomssa T and Gholap A V 2011 Characterization of caffeine and determination of caffeine in tea leaves using uv-visible spectrometer *African J. Pure Appl. Chem.* **5**(1) 1–8
[14] Durge R, Kshirsagar R V, and Tambe P 2014 Effect of sonication energy on the yield of graphene nanosheets by liquid-phase exfoliation of graphite *Procedia Eng.* **97** 1457–65
[15] Durge R, Kshirsagar R V and Tambe P 2014 Effect of Sonication Energy on the Yield of Graphene Nanosheets by liquid-phase exfoliation of graphite *Procedia Engineering* **97** 1457-65.
[16] Park S and Ruoff R S 2019 Chemical methods for the production of graphenes *Nature Nanotechnology* **4** 45–47
[17] Botas C, Perez-Mas A M, Alvarez P, Santamaria R, Granda M, Blanco C and Menendez R 2013 Optimization of the size and yield of graphene oxide sheets in the exfoliation step (Mexico: Instituto Nacional del Carbon, CSIC) *Carbon* **63** 576-78
[18] Paton K R 2014 Scalable production of large quantities of defect-free few-layer graphene by shear exfoliation in liquids *Nat. Mater.* **13**(6) 624–30.
[19] Fan J, Shi Z, Ge Y, Wang J, Wang Y and Yin J 2012 Gum arabic assisted exfoliation and fabrication of Ag–graphene-based hybrid *J. Mater. Chem.* **22**(27) 13764
[20] Lotya M, King P J, Khan U, De S and Coleman J N 2010 High-Concentration, Surfactant-Stabilized Graphene Dispersions *ACS Nano* **4**(6) 3155–62
[21] Zhang K 2017 Direct exfoliation of graphite into graphene in aqueous solution using a novel surfactant obtained from used engine oil *J. Mater. Sci* **53** 2484-96

IOP Conf. Series: Materials Science and Engineering **702** (2019) 012037 doi:10.1088/1757-899X/702/1/012037

Thermal stability of melt-blended poly (lactic acid) (PLA)/ polyamide 66 (PA66)/ graphene nanoplatelets (GnP)

W Alhadadi[1], A Almaqtari[1], F Hafidzah[1], M Bijarimi[1*], M S Z Desa[1], H Merzah[1], E Normaya[2] and M Norazmi[2]

[1]Faculty of Chemical and Process Engineering Technology, College of Engineering Technology, Universiti Malaysia Pahang, Lebuhraya Tun Razak, 26300 Gambang, Kuantan, Pahang, Malaysia.
[2]Department of Chemistry, Kulliyah of Science, International Islamic University, Kuantan, Malaysia

*E-mail: bijarimi@ump.edu.my

Abstract. In recent times the blending of different biodegradable materials to produce nanocomposites with improved properties has been a significant point of discourse by many researchers. This study reports a melt blend of poly (lactic acid) (PLA), polyamide (Nylon 66) and graphene nanoplatelets (GnP). The thermal stability and chemical functional characteristics of the nanocomposites produced were succinctly investigated using thermogravimetric analysis (TGA), differential scanning calorimetry (DSC) and Fourier transform infrared (FTIR). The blending was carried out by using a twin-screw extruder at a revolution speed and time of 50 rpm and 20 minutes, respectively. The composition of GnP in the PLA/PA66 was varied from 0 – 5 wt.%. The results showed that the increased of the GnP concentrations improved the thermal stability of the nanocomposites. It was also observed that a strong interaction between the polymeric materials as demonstrated by the FTIR analysis. The use of a low concentration of graphene in PLA/PA66 blend has been demonstrated to improve the thermal stability of the nanocomposites.

1. Introduction

Poly (lactic acid) is classified as biopolymer thermoplastic utilized from renewable resources and often used as a substitute to a petroleum-based polymer. As for now, PLA has gained extensive research due to high-strength, high-modulus, eco-friendly, biocompatibility, processability and 25–55% less energy in PLA production [1]. Despite all the advantages, PLA is lacking in term of reactive side chain groups, less rate of degradation, mostly poor in toughness and hydrophobicity [2]. Therefore, blending with a tough polymer is capable to overcome all these drawbacks and improve the functionality of PLA [3].

Polyamides-66 (PA66) or known as nylon-66 is widely used in engineering thermoplastics, with various applications. PA66 is classified as a semi-crystalline thermoplastic polymer which possesses excellent properties such as low density, processability, good strength, solvent resistance, self-lubricating, and good abrasion resistance [4]. Thus, PA66 is a good candidate for blending with PLA which requires high strength properties. PA66 has been blended with polypropylene and other polyolefins [5]. Novello et al. [6] also investigated the blend of PA66 with polyethylene terephthalate (PET) and short glass fiber (SGF) and found that the blend showed a good impact, tensile and flexural properties. Like any other polymer blends, the binary blending of PLA/PA66 normally showed incompatible as both of the polymers are brittle and resulted in poor mechanical properties. In order to overcome immiscibility issues between PLA with PA66, the addition of nanofiller is required to achieve the desired properties.

Currently, graphene is well-known high surface area nanofiller and it possess excellent mechanical properties such as stiffness and strength [7-8] due to its unique structure which consist of sp2 hybridized of one-carbon atoms packed in a honeycomb-like lattice [9]. Several studies have been conducted in the small capacity of graphene loaded in PLA blends. Bijarimi et al. [10] had studied the various capacity of GnP loaded in the PLA/PA6 blends with 0.5, 1.0, 3.0 and 5.0 wt%. They reported that the 5% GnP has resulted in increasing the thermal and mechanical properties of PLA/PA6 blend. On the other hand, Chieng et al. [11] had prepared plasticized PLA-based nanocomposites by adding various concentrations of graphene nanoplatelets (GnP) and analyzed the mechanical, thermal properties and crystallinity. The results shown that the addition of 3% of GnP in the reinforced plasticized PLA (p-PLA) increased the elongation at break and tensile strength of the nanocomposites by 60.6% and 26.5, respectively. The thermal analyses displayed that increasing the GnP content triggers a significant increment in thermal stability but not the crystallinity. In this study, the effect of various concentrations of GnP loaded in PLA/PA66 blends were investigated in terms of thermal stability and chemical characteristics by using thermogravimetric analysis (TGA), differential scanning calorimetry (DSC) and Fourier transform infrared (FTIR).

2. Materials and methods

2.1. Materials

The polylactic acid (IngeoTM -Biopolymer 2002D-grade) was provided by Unit Technology Ltd, China. It has density, melt flow index and melting temperature of 1.24 g/cm^3, 4–8 g/10 min and 160-170°C, respectively [12]. Polyamide66 (PA66) was supplied by DuPont Co. Ltd. It has a density and melting temperature of 1.14 g/cm^3 and 263°C respectively. M-grade GnP was supplied from XG-Sciences (US Michigan) it has a surface area between 120 to 150 m^2/g, the average thickness between 6-8 nm and the density 2.2 g/cm^3 [13].

2.2. Sample composition and blending procedure

In this study, a thermally stable material was fabricated by blending different proportion of PLA, PA66, and GnP. The blend ratio was fixed at 60/40 PLA/PA66 for all samples by increasing the concentration of GnP in PLA/PA66 nanocomposites. The blends were mixed manually using a glass beaker and then prepared by using twin-screw extruder. The temperature profile at the feed section was 290°C, decreasing to 253°C at the head of the die zone. The rotation speed of the screw fixed at 50 rpm/min. The compound is blended twice for each blending compositions and blended slowly to ensure homogenous blend [12]. The dimension of the die used were of 1 mm thickness and 20 mm width. The total amount of weight of each blend is 100 grams. Table 1 depicts the sample composition and concentration.

Table 1. Sample composition and concentration.

Composition	PLA (wt.%)	PA66 (wt.%)	GnP
PLA	100	0	0
PA66	0	100	0
PLA/PA66	60	40	0
PLA /PA66 /GnP 0.5	60	40	0.5
PLA /PA66 /GnP 3	60	40	3
PLA / PA66/GnP 5	60	40	5

2.3. Characterization

The thermal stability of the material under heat was determined as well as the chemical surface. The samples with an average weight of 1-2 mg were characterized by using differential scanning calorimetric (DSC) with a nitrogen atmosphere. The nanocomposites thermal properties were examined via utilization of DSC as well. The sample was prepared in the aluminum pan with pierced lids and heated up to 300°C at 10°C / min steady rate. The glass transition temperature, melting and crystallization temperatures were analyzed and recorded. DSC was used to identify the crystallinity of the polymer through the polymer related heated (fusion). The change in sample mass during heating was evaluated by using thermogravimetric analysis (TGA). Chemical reactions or physical transformations can cause mass decompositions while the sample heating. Mettler Toledo TGA/SDTA 851e apparatus was used to evaluate the thermal stability of PLA and blends. Samples with average weight of (6 ±7 mg) were added to the alumina crucibles. As a reference, an empty alumina crucible was employed. The temperature, weight and derivative of the sample and heat flow were identified after heating from ambient temperature at increasing rate of 20°C per minute.

The Fourier transform infrared (FTIR) was used to analyze the chemical properties of nanocomposites blends. The infrared (IR) spectra were recorded using a Spectrum 400 FT-IR and spectrometer by with 4 cm^{-1} resolutions and 10 scans in the 4000–500 cm^{-1} spectra region with resolution over 33 scans [14].

3. Results and discussion

3.1. Thermal analysis

The DSC curves of PLA, PLA/PA66 and PLA/PA66/GnP nano-composites are shown in figure 1. The glass temperatures of PLA/PA66/GnP nano-composites for different weight per cent of 0.5, 3 and 5 wt.% were found not to change extensively with the addition of GnP. The result obtained for each of the polymeric material blends, therefore, revealed that there is no difference in the heat flow for PLA/PA66 and PLA/PA66GNPs. The glass transition temperature (T_g), crystallization temperature (T_c) and melting temperature (T_m) values derived could not be estimated.

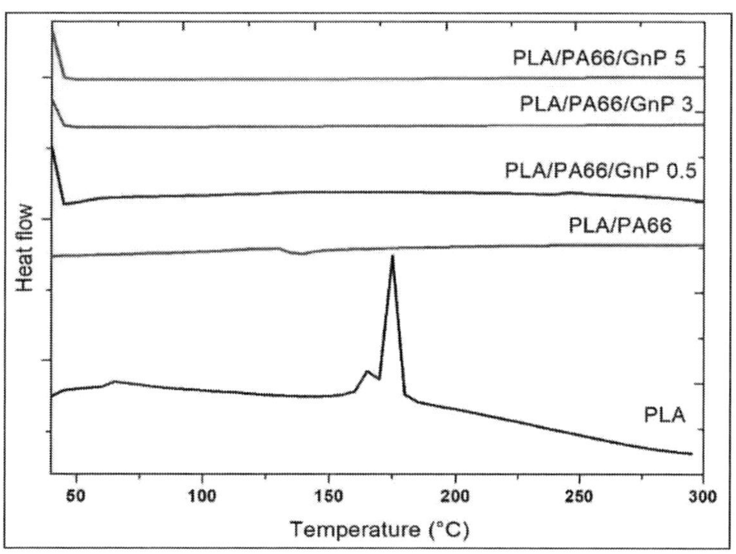

Figure 1. DSC graph of PLA, PA66 and PLA/PA66/GnP nanocomposites.

Figure 2(a) depicts the degradation temperature of PLA, PA66, PLA/PA66 and PLA/PA66/GnP nanocomposites. It was noticed that the incorporation of graphene in PLA/PA66 matrix significantly enhanced the thermal stability due to the good dispersion of GnP in the nanocomposites as shown in the TGA plot. However, the degradation and weight loses at $T_{5\%}$, $T_{10\%}$, $T_{25\%}$ and $T_{50\%}$ for PLA, PA66, PLA/PA66 and PLA/PA66/GnP nanocomposites are summarized in table 2. The pristine PLA blend shows an initial mass loss of 5% at temperature of 299.77°C and loss of 50% at temperature of 346.85°C. On the other hand, the PLA/PA66/GnP 3 blend shows an initial mass loss of 5% at temperature of 324.74°C and loss of 50% at temperature of 407.03°C. In the derivative thermogravimetric (DTG) curve, maximum degradation temperature for all blends can be seen in figure 2(b). It was observed that there were two peaks in the DTG curves which indicate the PLA and PA66 degraded individually due to the immiscible blend of the two polymers.

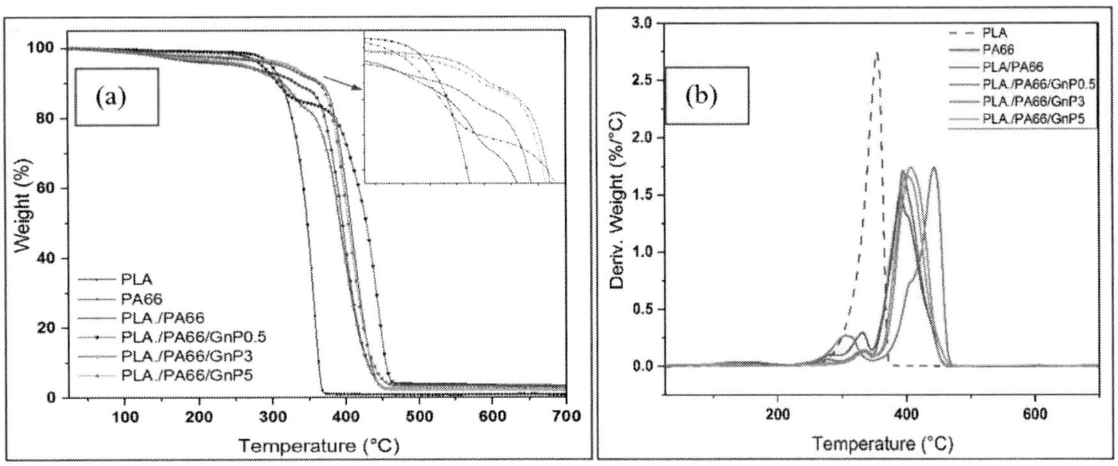

Figure 2. (a) TGA graph of PLA, PA66, PLA/PA66 and PLA/PA66/GnP and (b) DTG graph of PLA, PA66, PLA/PA66 and PLA/PA66/GnP.

Table 2. The degradation of weight loses at $T_{5\%}$, $T_{10\%}$, $T_{25\%}$ and $T_{50\%}$.

Sample	$T_{5\%}$	$T_{10\%}$	$T_{25\%}$	$T_{50\%}$
PLA	299.77	313.21	331.99	346.85
PA66	258.31	331.97	377.51	395.85
PLA/PA66	265.03	314.26	369.45	392.21
PLA/PA66/GnP0.5	287.54	308.2	398.09	428.42
PLA/PA66/GnP3	324.74	366.23	390.97	407.03
PLA/PA66/GnP5	313.39	363.02	387.72	403.86

3.2. Functional group chemical analysis

In this study, the functional group characterization of the sample was observed using the Fourier transform infrared spectra of PLA, PA66, PLA/PA66 and PLA/PA66/GnP as illustrated in figure 3. From the PLA spectrum, four central regions were identified and this includes the –CH stretching at 2900–3010 cm^{-1}, –C=O stretching at 1680–1817 cm^{-1}, C–H bending at 1340–1485 cm^{-1} and –C–O stretching at 1000 –1240 cm^{-1}. Despite the increment of GnP loadings in the blend systems, the characteristic peaks for the PLA were conspicuously dominant. The observation conducted on other polymeric materials shows the presence of peaks which indicated that chemical interaction was observed in the polymer configuration [15]. This indicated a strong interaction between the polymeric materials blended as supported by Chieng et al. [11].

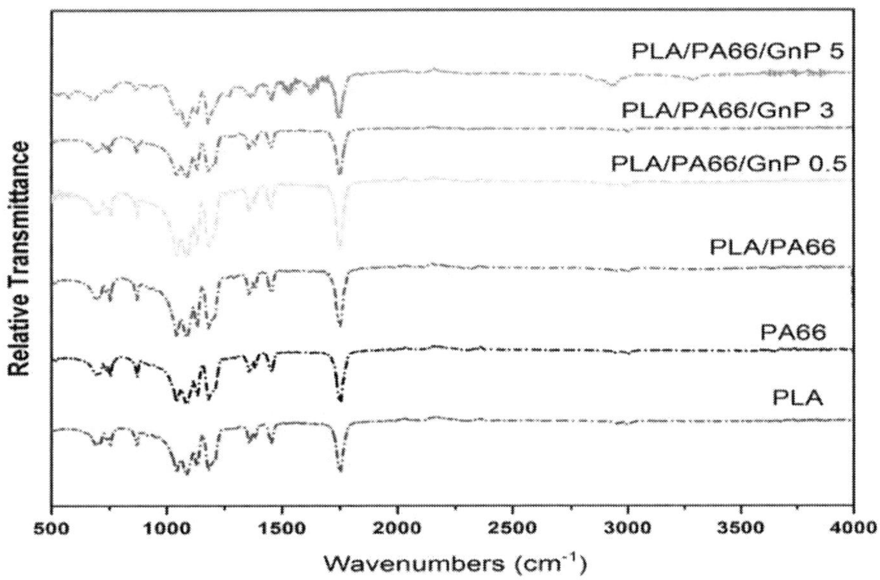

Figure 3. FTIR of PLA, PA66 and PLA/PA66/GnP nanocomposites.

4. Conclusions

PLA/PA-66 blends with different loadings of GnP were prepared by melt blending method and their thermal and chemical properties were investigated. The thermal analysis by thermogravimetric analysis (TGA) indicated that there was a significant improvement of thermal stability in the blends containing graphene as the loadings of GnP was increased. DSC thermogram of PLA/PA-66 also exhibited a reduction in cold crystallization temperature to zero after the addition of graphene nanoparticles. Based on both analyses it was shown that 3% of GnP loaded obtained the optimum thermal stability. Moreover, the results from FTIR also confirmed the interaction between graphene and PLA/PA66 phases. From this study, it can be concluded that the thermal stability of PLA/PA66 could be enhanced by incorporating GnP and suitable for industrial application.

Acknowledgements

The authors wish to acknowledge Universiti Malaysia Pahang (UMP) for the financial assistance provided under the internal research grants (RDU180351) and FRGS/1/2019/TK05/UMP/02/1.

References

[1] Farah S, Anderson D G and Langer R 2016 *Adv. Drug Deliv. Rev.* **107** 367-92
[2] Bayer I S 2017 *Materials (Basel)* **10**
[3] Bijarimi M, N Shahadah, Azizan R, Said N, Waleed A, Zakir M M and Jamiluddin J 2019 *Indonesian J. Chem.* 1–6
[4] Lingesh B V, Ravikumar B N and Rudresh B M 2016 *Indian J. Advances in Chem. Sci.* **4(2)** 168–71
[5] Lingesh B V, Rudresh B M and Ravikumar B N 2014 *Procedia Mat. Sci.* **5** 1231-40
[6] Novello M V, Carreira L G and Canto L B 2014 *Mat. Res.* **17** 1285-94
[7] Ahmad S R, Xue C and Young R J 2017 Mat. Sci. Eng.: B **216** 2-9
[8] Norazlina H and Kamal Y 2015 *Polymer Bulletin* **72** 931-61.
[9] Zhang H, Xing W, Li H, Xie Z, Huang G and Wu J 2019 *Adv. Indus. Eng. Polymer Res.* **2** 32-41
[10] Bijarimi M, Amirul M, Norazmi M, Ramli A, Desa M S Z, Desa M D A and Abu Samah M A 2019 *Mat. Res. Express* **6** 055044
[11] Chieng B W, Ibrahim N A, Wan Yunus W M Z, Hussein M Z and Loo Y Y 2014 *J. Thermal Anal. Calorimetry* **118** 1551-9

[12] Bijarimi M, Ahmad S and Alam A K M M 2016 *Polymer Bulletin* **74** 3301-17
[13] Kashi S, Gupta R K, Kao N and Bhattacharya S N 2017 *Rheology and physical characterization of graphene nanoplatelet/poly (butylene adipate-co-terephthalate) nanocomposites*
[14] Bijarimi M, Ahmad S and Rasid R 2012 *Mechanical, Thermal and Morphological Properties of PLA/PP Melt Blends*
[15] Huang J C, He C B, Xiao Y, Mya K Y, Dai J and Siow Y P 2003 *Polymer* **44** 4491-9

IOP Conf. Series: Materials Science and Engineering **702** (2019) 012038 doi:10.1088/1757-899X/702/1/012038

Synthesis of alumina from aluminium can waste to be applied as catalyst support for biodiesel production

A A Bin Mokaizh[1], N Wirman[1] and J H Shariffuddin[1,2*]

[1] Faculty of Chemical & Natural Resources Engineering, Universiti Malaysia Pahang, Lebuhraya Tun Razak 26300 Gambang, Kuantan, Pahang, Malaysia
[2] Centre of Excellence for Advanced Research in Fluid Flow, Universiti Malaysia Pahang, Lebuhraya Tun Razak 26300 Gambang, Kuantan, Pahang, Malaysia

*E-mail: junhaslinda@ump.edu.my

Abstract. Abundant of aluminum beverage cans are normally discarded after use have caused considerable land pollution and environmental problems. This research is therefore aimed to synthesize alumina from aluminum can waste which is one of the most common kind of waste. The objective of this research is to synthesize and characterize alumina produced from aluminum can waste, and to be applied as catalyst support in the biodiesel production. In this study, the alumina from aluminum can waste was produced via Sol-gel method by varying the aging time. Characterization of alumina was performed by using FTIR, XRD, BET, and SEM-EDX. The synthesized alumina was used as catalyst support for potassium nitrate catalyst to be applied in biodiesel production by using transesterification reaction of cooking oil. The biodiesel produced was analyzed by using gas chromatography-mass spectrometry (GCMS) and FTIR. The experimental results revealed that the alumina powder synthesized at room temperature have high surface area which are suitable to be used as catalyst support of producing biodiesel. In conclusion, it has been demonstrated that it is possible to produce alumina from aluminum can waste that can be used as catalyst support for biodiesel production. From the GCMS and FTIR results, it was proven that biodiesel is produced.

1. Introduction

Waste management in Malaysia has become a challenging undertaking in recent years due to rapid economic development and population growth, land scarcity and poor infrastructure and expertise. Some of the current trends in managing solid waste include reengineered process which helps to reduce quantities of emissions, strategies to maximize current levels of waste utilization and development of high value products [1]. Aluminum found in solid waste includes can, wrapping foil, car engines and body parts, while food packaging and automobile production are being two of many possible sources in municipal solid wastes. Recycling aluminium can waste save precious natural resources, energy, time and money [2].

Alumina is one of the most important ceramic oxides which have vast technological and industrial applications. It is used in many applications such as to contain molten metal, as insulators in spark plugs, and others applications such as dental and medical use [3]. Alumina has also been applied as catalyst, catalyst support and adsorbent. For example, alumina is used to support calcium oxide (CaO) base catalyst for the transesterification of palm oil in biodiesel production [4]. Activated alumina also act as adsorbent in removal of nickel ions from industrial plating effluents.

Various sources can be used to produce biodiesel such as from waste cooking oil, plastics, and vegetable oil, animal fats. For example, by using waste cooking oil that has many disposal problems like water and soil pollution, human health concern; it can be used as an effective and cost efficient feedstock for biodiesel production [5]. Preparation of biodiesel from cooking oil and methanol by transesterification reaction using the heterogeneous catalyst will increase the reaction yield and minimize the methanol molar ratio because huge amount methanol required for reaction will involve an additional cost for the separation of biodiesel [6]. Thus, this research is aimed to utilize the aluminum can waste from the municipal solid waste (MSW) to produce alumina and to be applied as catalyst support for biodiesel production.

2. Materials and Method

2.1. Materials
Aluminum can waste was collected from Mercu Resolution Sdn Bhd landfill site located in Sungai Ikan, Terengganu. The chemicals potassium hydroxide (KOH) (≥85%) and sulfuric acid (H_2SO_4) (≥ 98%) were purchased from Sigma-Aldrich. Deionized water employed in this study was from Millipore Elix 5, USA. Materials needed for the production of biodiesel were cooking oil that was obtained from local market; methanol (≥99.8%) and potassium nitrate (≥99.0%) were purchased from Sigma-Aldrich.

2.2. Methods
Methods used in this study were divided into three stages where the first stage is synthesizing γ-alumina from can waste, second stage involved on the incorporation of the synthesized alumina as catalyst support in KNO_3 and the third stage is the application of the synthesized catalyst KNO_3/Al_2O_3 in the biodiesel production.

2.2.1 Synthesis of mesoporous γ-alumina
The mesoporous γ-alumina was synthesized via sol-gel method. Aluminum can waste was cut into small pieces and dissolved in potassium hydroxide solution. Then, the mixture was filtered and adjusted the pH at 9.0 using sulfuric acid. The solution was stirred until it became gel-like. The mixture was aged by fixing the reaction temperature at room temperature and varying the aging time at 6 hours and 24 hours. After that, the mixture was separated using centrifugation. Finally, the aluminum hydroxide gel was put into the oven at 80°C for half day for drying process and calcined into furnace at 500°C. Alumina synthesized at reaction temperature (room temperature) and aging times (6 – 24) hours was denoted as AL-6-30 and AL-24-30 respectively. The physicochemical properties of the synthesized γ-alumina were investigated by using different characterization techniques. γ-alumina was characterized by using X-Ray Diffraction (XRD) (RIGAKU MINIFLEX II X-ray diffractometer) to determine the crystalline size, structure and phase identification, Fourier Transform Infrared Spectroscopy (FTIR) (Nicolet iS5 Model Spectrum100) to obtain the functional group of sample powders while Brunauer-Emmett-Teller (BET) to measure the specific surface area, pore size, pore volume and pore size distribution of the sample powders while Scanning Electron Microscopy(SEM) (Benchtop Scanning Electron Microscope TM3030Plus) was carried out to inspect the surface morphology whereas Energy-dispersive X-ray (EDX) (SwiftED3000) was used to determine the elemental compositions that present in the catalyst.

2.2.2 Synthesis of catalyst for biodiesel production
The catalyst that was synthesized in this study was KNO_3 on alumina (KNO_3/Al_2O_3) which is also known as base heterogeneous catalyst. The aqueous potassium nitrate loaded on alumina and stirred for 4 hours at 85 °C and 250 rpm by using impregnation method. The purpose of stirring is to make sure the catalyst well distributed on support. The catalyst was further dried for 16 hours at 120°C. Prior to each reaction, the catalyst was calcined at 500°C for 5 hours in furnace. The catalyst prepared to KNO_3 loading amounts of 35 wt%, was designated as 35% KNO_3/Al_2O_3. The characterization of catalyst was investigated by using FTIR, XRD, SEM, EDX, and BET.

2.3. Application of alumina as catalyst support in biodiesel production

The method of conducting the production of biodiesel was based on Zabeti et al. [4] with some modifications [4]. Transesterification reaction was performed in three neck glass batch type reactor. 150 mL of cooking oil and 56.80 mL of methanol were poured into the flask. Then, the reaction was initiated by adding catalyst with a desired amount. The mixture was heated at reaction temperature of 65 °C and stirred with an optimum speed of 600 rpm for 5 hours. After the reaction completed, the catalyst was separated by centrifuged for 10 min. The liquid was put into separating funnel and left it overnight until the liquid separated into two phases. Biodiesel was analyzed using the result by Gas Chromatography-Mass Spectrometry (GC-MS) and Fourier Transform Infrared Spectroscopy (FTIR).

3. Results and discussion

3.1. Characterization of synthesized alumina

Fourier transform infrared (FTIR) analysis of the alumina was done in the wavenumber ranging from 400 to 4000 cm^{-1}. The FTIR spectra of AL-6-30 and AL-24-30 are as indicated in figure 1. In this section, the effect of aging time on the physicochemical of synthesized alumina is studied through the characterization of powder. FTIR spectra of the synthesized samples have been shown in Figure 4.1. The appearance of peaks indicates the presence of a better crystallized phase. The bands of AL-6-30 were around 613 cm^{-1} and 1102 cm^{-1} while the bands of AL-24-30 were around 614 cm^{-1} and 1104 cm^{-1}. The band around 613 cm^{-1} for sample AL-6-30 and the band at 614 cm^{-1} for sample AL-24-30 were assigned to ν-AlO_6. While the band around 1102 cm^{-1} for sample AL-6-30 and 1104 cm^{-1} for sample AL-24-30 were assigned to Al-O-Al. The peak corresponding to 1101 cm^{-1} is assigned to Al-O-Al symmetric bending stretching vibrations. The peaks in the region of 500–750 cm^{-1} are assigned to ν-AlO_6, whereas the shoulder at 890 and the line at 800 cm^{-1} are assigned to ν-AlO_4 [7]. There is no peak around 3500 cm^{-1} and 1600 cm^{-1} which shows that both samples of synthesized γ- Al_2O_3 are fully dried from water. This is supported by Rajaeiyan and Bagheri-Mohagheghi (2013) where intense bands centered around 3500 cm^{-1} and 1600 cm^{-1} are assigned to O–H stretching and bending modes of adsorbed water due to high tendency of γ-Al_2O_3 for adsorbing water [8]. Based on the graph, 6 h and 24 hours aging time gave same intensity of Al-OH at peak of 614 cm^{-1}. Both of aging time FTIR analysis shows that larger intensity of aluminum hydroxide and aluminum oxide in the sample.

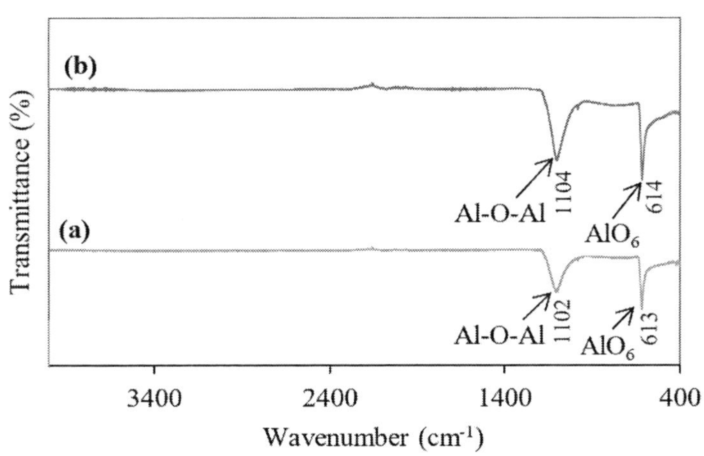

Figure 1. FTIR spectra comparing the effect of aging time on synthesis of alumina; (a) AL-6-30 (b) AL-24-30

Figure 2 shows the X-ray diffraction patterns of the synthesized alumina AL-6-30, AL-24-30 and AL-commercial. For both AL-6-30 and AL-24-30 all the observed peaks were identified as aluminum oxide, or alumina (Al_2O_3) crystalline structure. The XRD patterns of mesoporous alumina in wide angle region suggested the phase of γ-Al_2O_3 at 2θ are 30.2°, 31.2° with broad peaks, 43.0°, and 67.0° [JCPDS card 10-0425] [9]. In the XRD results, there were boehmite, corundum and gamma alumina identified in the synthesis of alumina. Boehmite is an aluminum oxide hydroxide (γ-AlO(OH)) mineral, a component of the aluminum ore bauxite. It is dimorphous with diaspore. The structural collapse of boehmite occurs, after hydrogen transfers and water extraction [10]. Then, through an aluminum migration process, the γ-alumina characteristics appear. Corundum is a crystalline form of aluminum oxide, typically containing traces of iron, titanium, vanadium and chromium [11]. The XRD patterns of synthesized alumina AL-6-30 and AL-24-30 in small angle, which is 1–10° have no XRD peaks. This implies that this mesoporous alumina has no long-range order pore structure. For alumina synthesized at room temperature, as indicated in Figure 2, they appeared to be less crystalline in structure with broader peaks. AL-6-30 had higher intensity for the observed peaks compared to AL-24-30 which showed that AL-6-30 had more crystalline structure. For AL-commercial, the observed peaks were identified as an aluminum oxide, or alumina (Al_2O_3) crystalline structure. The peaks of Al_2O_3 were at 37.2°, 45.92°, 60.48°, and 66.7°.

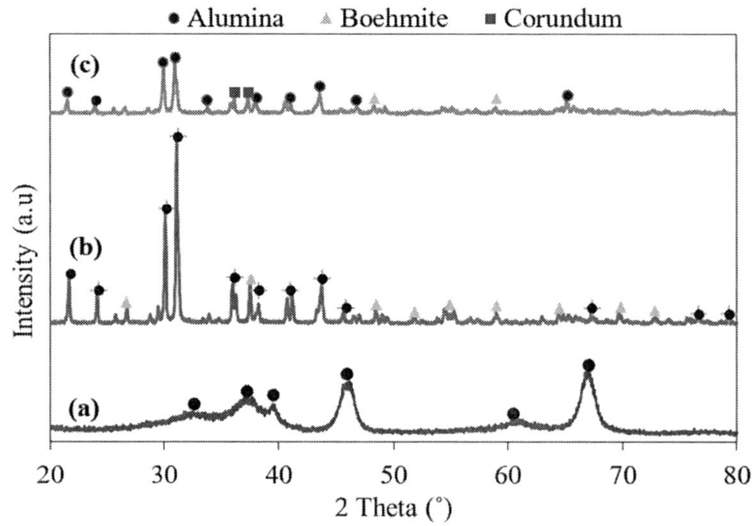

Figure 2. XRD pattern comparing the effect of reaction time on synthesis of alumina; (a) AL-commercial (b) AL-6-30 and (c) AL-24-30

Figure 3 shows the SEM-EDX spectra of AL-6-30, AL-24-30 and AL-commercial with average weight and atomic percentage of elements that present on the samples. From Figures 3(a) and 3(b), it can be observed that the solid morphologies for both AL-6-30 and AL-24-30 were similar, with heterogeneous particle sizes. Both of the synthesized alumina had non-spherical shape. As it is observed, some particles with disordered shape were obtained by joining very small crystallites during the calcination [12]. AL-6-30 and AL-24-30 appeared to be more disordered compared to the surface morphology of AL-commercial. The red circle in the figures represent the spot where EDX sample were taken. Analyzing the spectra for AL-6-30 and AL-24-30 in (a) and (b) it was noted that the powders are mainly composed of oxygen, aluminum, sulphur and potassium. Oxygen has the highest weight and atomic percentage in all samples followed by aluminum. In Figure 3(b) for AL-24-30 shows that aluminum was produced higher when used 24 h aging time. The presence of potassium in AL-6-30 and AL-24-30 was due to the usage of potassium hydroxide solution in dissolving the aluminum waste cans. There were traces of sulphur in the spectra of Figure 3 as suphuric acid was used for pH adjusting. These traces of potassium and sulphur can be eliminated by having sufficient rinsing with deionized water during the formation of aluminum hydroxide gel. Regardless the reaction time, all synthesized aluminas had similar trend in the composition for each element.

IOP Conf. Series: Materials Science and Engineering **702** (2019) 012038 doi:10.1088/1757-899X/702/1/012038

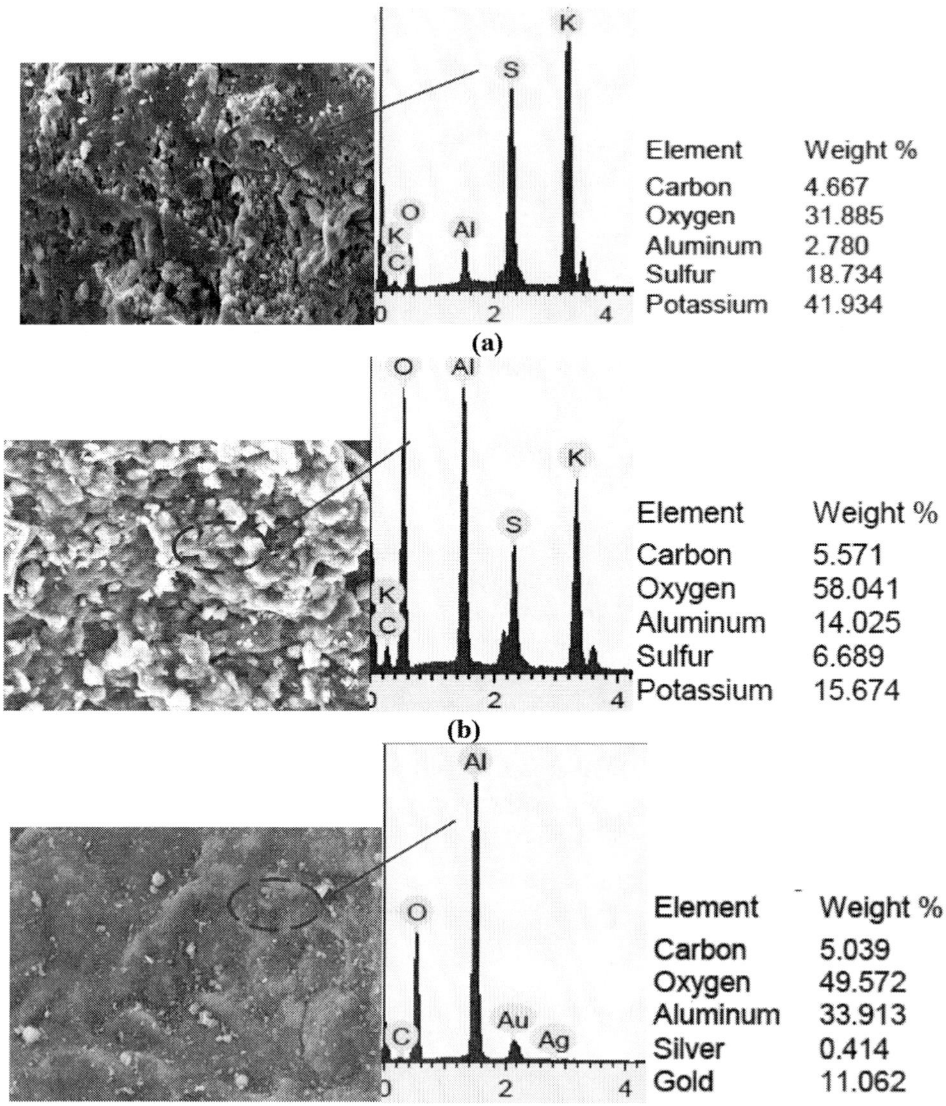

(c)

Figure 3. EDX comparing the effect of reaction time on synthesis of Alumina; (a) AL-6-30 (b) AL-24-30 (c) AL-commercial

From Table 1, it is noted that the Specific Surface Area (SSA) of AL-6-30 and AL-24-30 were 6.5264 m^2/g and 4.3072 m^2/g, showing low surface areas. The pore volume for AL-6-30 was 0.0125 cm^3/g while the pore volume for AL-24-30 was 0.0090 cm^3/g. The AL-commercial has SSA of 176.6364 m^2/g and pore volume of 0.3988cm^3/g. The AL-commercial had high surface area compared to synthesized alumina. The surface area analysis is consistent with the XRD analysis. AL-6-30 and AL-24-30 which had less crystalline structure in XRD had lower surface areas. This indicates that aging time or reaction time is an important factor to obtain high surface area of the products. It has been reported that porous materials having a surface area value above 100 m^2/g is considered having a high surface area and is a potential material for catalyst support [13]. But with low surface area still can used as a catalyst support for biodiesel production.

Table 1. Summary of specific surface area (SSA), pore volume and adsorption average pore diameter of AL-6-30, AL-24-30 and AL-commercial by BET comparing the effect of reaction time on synthesis of Alumina

Sample	**Specific Surface Area (SSA)[a], m^2/g**	**Pore Volume [b], cm^3/g**	**Adsorption Average Pore Diameter (4V/A by BET), nm**	**Crystalline Size [c], nm**
AL-6-30	6.5264	0.0125	7.4083	65.27
AL-24-30	4.3072	0.0090	9.2205	42.54
AL-commercial	176.6364	0.3988	8.9207	5.38

[a] Calculated by Brunauer-Emmett-Teller method

[b] Determined by Barret-Joyner-Halenda (BJH) method

[c] Determined by X-ray peak in XRD using Scherrer formula

3.2. Characterization of catalyst KNO_3/Al_2O_3

Figure 4 shows the FTIR spectrum of the KNO_3/AL-commercial and KNO_3/AL-24-30 catalyst. All samples showed an absorption band at around 1360 cm^{-1} to 1390 cm^{-1} that could be attributed to the vibration of N-O of KNO_3, while this N-O absorption band was not observed on alumina without loaded KNO_3. Further, the intensity of 1384 cm^{-1} band grew as the amount of loaded KNO_3 rose [14]. Also, physically adsorbed water on the solid base was observed at around 1636 cm^{-1} assigned to the bending mode of O–H. With the increase of calcination temperature, the O–H absorption band became weaker and, finally disappeared. On the surface of fully hydroxylated alumina K^+ ion replaced the protons of isolated hydroxyl groups to form Al-O-K groups which were considered to be the active basic sites. There are no bands ascribed to NO_2 (1550, 1320 cm^{-1}), were detected, demonstrating that the resultant of KNO_3 decomposition was K_2O instead of KNO_3.

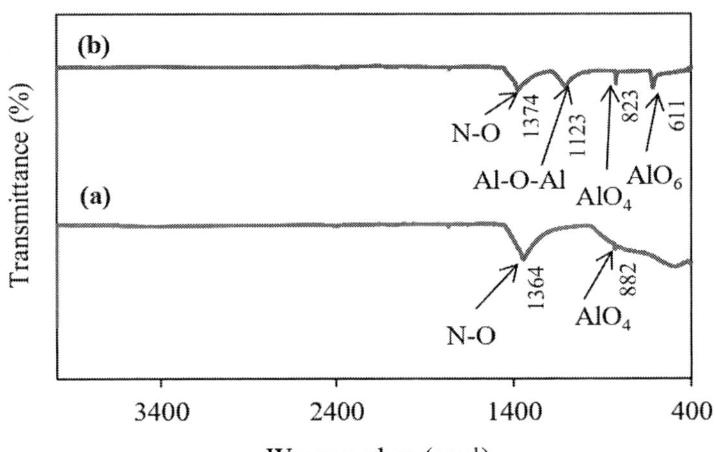

Figure 4. FTIR spectra of (a) KNO_3/AL-commercial (b) KNO_3/AL-24-30

Figure 5 shows the XRD patterns of KNO_3/AL-commercial and KNO_3/AL-24-30 samples. According to Wenlei Xie, 2006 on the patterns of 35% KNO_3/Al_2O_3 sample calcined at 773 K, new diffraction lines appeared with 2Θ of 31°, 39°, 55° and 62°, which were assigned to K_2O. This K_2O phase was probably the cause of the high catalytic activity and basicity of the catalysts because a close relationship was seen between the K_2O phase that appeared and the basicity or the catalytic activity and indicating the good dispersion of KNO_3 on alumina. As the amount of loaded KNO_3 exceeded, the characteristic XRD peaks of KNO_3 appeared on the XRD pattern and their intensities grew with increasing amount of loaded KNO_3, suggesting that a residual phase of KNO_3 remained on the composite. In the XRD results, there were corundum, potassium, potassium nitrate, potassium dinitramide and niter identified in the synthesis of KNO_3/Al_2O_3. The present of corundum which is a crystalline form of aluminum oxide means there has alumina in the catalyst. Potassium dinitramide (KN_3O_4) contains excess oxygen and, therefore, may be also used as an oxidizer in various compositions [15]. Niter is the mineral form of potassium nitrate, KNO_3 also known as saltpeter.

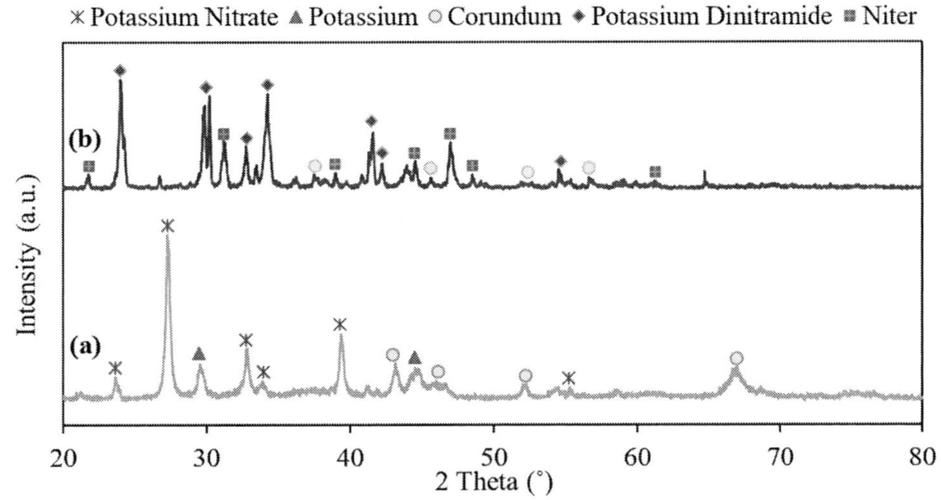

Figure 5. XRD pattern of (a) KNO_3/AL-commercial (b) KNO_3/AL-24-30

3.3. Characterization of catalyst KNO_3/Al_2O_3

Figure 6 shows the FTIR analysis for the comparisons of biodiesel by using catalyst KNO_3/AL-commercial and KNO_3/AL-24-30. Both spectra show similar bands. Complete depletion of the band near 2900 cm^{-1}, assigned to the alkene C-H stretch is observed for the deposits. IR bands in the region of 1160-1200 cm^{-1} for O-CH_3 stretching, clearly demonstrated the transformation of cooking oils into biodiesel [16]. Bands were around 1137 cm^{-1} and 1457 cm^{-1} were assigned to (C-O vibrations) which is the additional ester bands. From the spectra, the band was around 1734 cm^{-1} for both sample were assigned to v(C=O). According to Aryee et al. [17], band observed in the 1750-1720 cm^{-1} range is attributed to the ester v(C=O) absorption of triacylglycerols and is indicative of the extraction of a small amount of oil into the methanol phase [17]. IR bands in the region of 1370-1400 cm^{-1} are for O-CH_2 groups in glycerol contain of triglycerides, diglycerides, and monoglycerides [16].

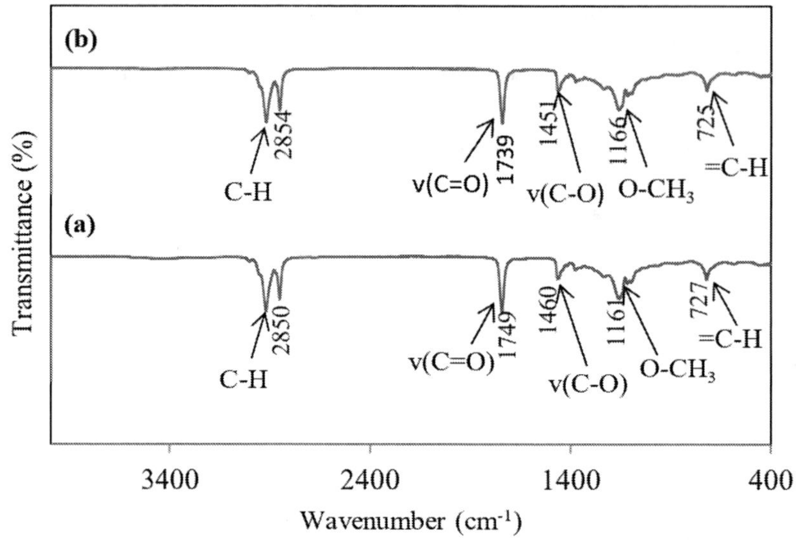

Figure 6. FTIR spectra of biodiesel by using (a) KNO_3/AL-commercial (b) KNO_3/AL-24-30

In this study, GC has been used to identify methyl esters according to their retention time which shows the mass spectrum in Figure 7(a) for biodiesel produced yang using catalyst KNO_3/AL-24-30 and Figure 7(b) for biodiesel produced yang using catalyst KNO_3/AL-commercial. Both figure show the similir mass spectrum of methyl ester in producing biodiesel from cooking oil. The red circle in the figures represent the peak of components in biodiesel. The chemical composition of biodiesel carried out using the GC-MS analysis led to the identification of five major components. From Figure 7(a) the identified components of biodiesel are at retention time 33.788 which is 9-Octadecenoic acid (E), (24.93%) represent the main constituent. n-Hexadecanoic acid, (22.67%) was the second major constituent at retention time 31.504 detected followed by 9-Octadecenoic acid (Z) methyl ester (1.96%) at retention time 33.788, Hexadecanoic acid, methyl ester (1.78%) at retention time 30.622 and 9,12-Octadecadienoic acid (Z,Z) (0.73%) at retention time 40.281. From Figure 7(b) shows that Oleic acid, 14-Pentadecenoic acid (17.96%) represent the main constituent at retention time 34.51, n-Hexadecanoic acid (11.35%) was the second major constituent at retention time 31.397 detected followed by Hexadecanoic acid methyl ester (2.78%) at retention time 30.622, 9-Octadecenoic acid (E), methyl ester (2.58%) at retention time 33.783 and 9,12-Octadecadienoic acid (Z,Z) (1.73%) and retention time 40.281.

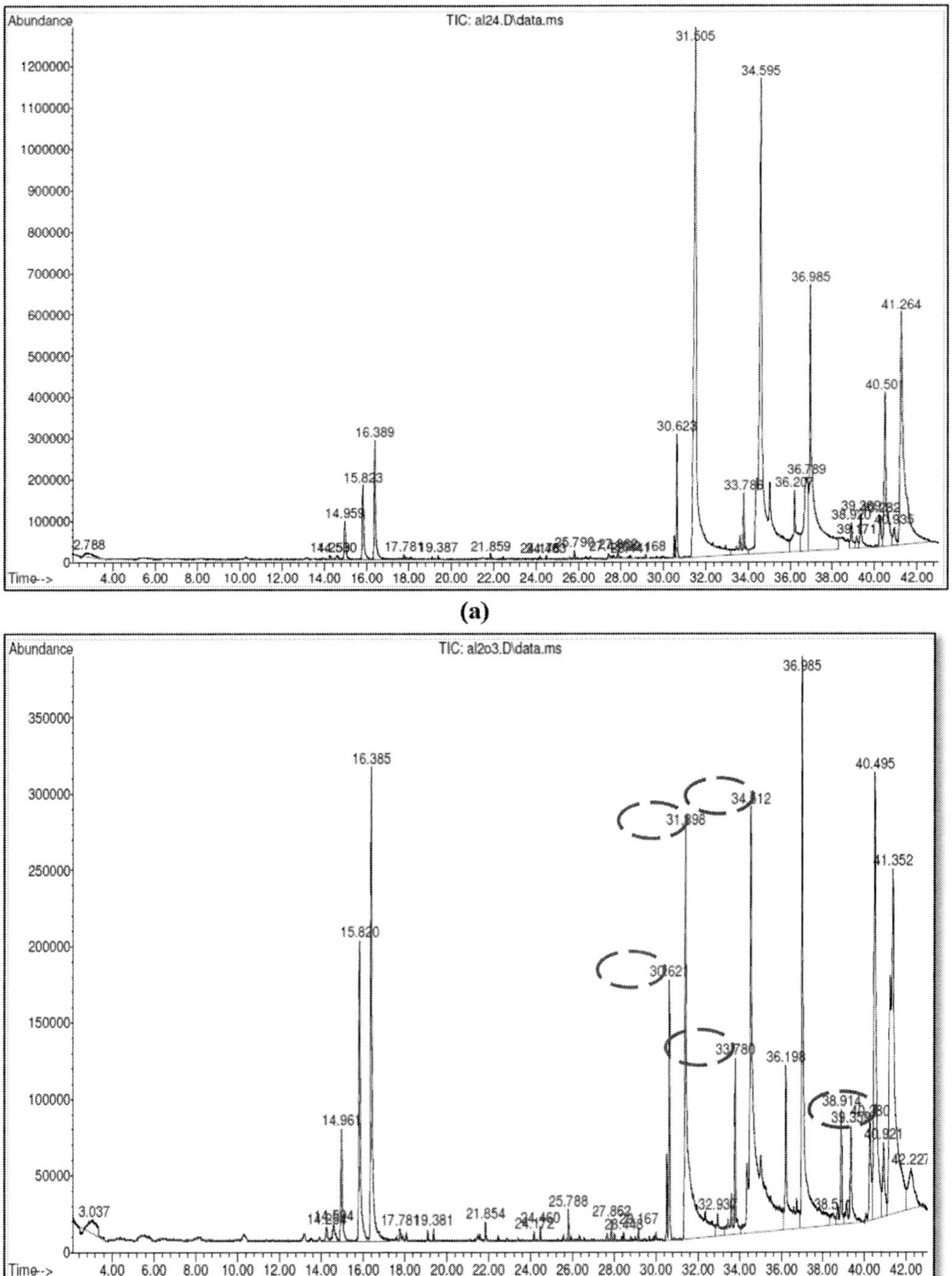

Figure 7. GC-MS analysis on biodiesel by using catalyst; (a) KNO_3/AL-24-30 (b) KNO_3/AL-commercial

4. Conclusion

The experimental results show that γ-alumina was successfully prepared from aluminum can waste at different reaction time which are 6 hours and 24 hours. The result was proven through the characterization of alumina by FTIR, XRD, SEM, EDX and BET analyses. According to the mechanism, a coherent skeleton of γ-alumina inherited from the original boehmite (AlOOH) and corundum as the crystallized alumina also have in the alumina sample. Based on XRD and BET analysis results, alumina synthesized at room temperature with reaction time of 24 hours (AL-24-30) was found to have the best condition compared to alumina with reaction time of 6 hours (AL-6-30). From the analysis, we decided to use AL-24-30 as a catalyst support for biodiesel production.

For the biodiesel production, from the FTIR analysis shows that the band and spectra for biodiesel is clearly demonstrated the transformation of cooking oils into biodiesel. In addition, results from GC-MS analysis have proven that biodiesel components have been produced. Overall, this study indicates that the aluminum can waste are able to be converted to alumina and then can be further used as a catalyst support for biodiesel production.

Acknowledgement

The authors would like to acknowledge financial support for this work from Universiti Malaysia Pahang (UMP) internal grant (RDU180326), Faculty of Chemical & Natural Resources Engineering, UMP.

References

[1] Kumar S, Kumar R and Bandopadhyay A 2006 Innovative methodologies for the utilisation of wastes from metallurgical and allied industries *Resour. Conserv. Recycl.* **48** 301–14

[2] Kumar S and Putnam V 2008 Cradle to cradle: Reverse logistics strategies and opportunities across three industry sectors *Int. J. Prod. Econ.* **115** 305–15

[3] Dixit S, Kashyap S, Kailas S V. and Chattopadhyay K 2018 Manufacturing of high strength aluminium composites reinforced with nano tungsten particles for electrical application and investigation on in-situ reaction during processing *J. Alloys Compd.* **767** 1072–82

[4] Zabeti M, Daud W M A W and Aroua M K 2010 Biodiesel production using alumina-supported calcium oxide: An optimization study *Fuel Process. Technol.* **91** 243–8

[5] Stephen J L and Periyasamy B 2018 Innovative developments in biofuels production from organic waste materials: A review *Fuel* **214** 623–33

[6] Vakh C S, Bulatov A V., Shishov A Y, Zabrodin A V. and Moskvin L N 2014 Determination of silicon, phosphorus, iron and aluminum in biodiesel by multicommutated stepwise injection analysis with classical least squares method *Fuel* **135** 198–204

[7] Hosseini S Y and Khosravi-Nikou M R 2016 Synthesis and characterization of nano-sized γ-Al2O3 for investigation the effect of temperature on catalytic dehydration of methanol to dimethyl ether *Energy Sources, Part A Recover. Util. Environ. Eff.* **38** 914–20

[8] Rajaeiyan A and Bagheri-Mohagheghi M M 2013 Comparison of sol-gel and co-precipitation methods on the structural properties and phase transformation of $γ$ and $α$-Al2O3 nanoparticles *Adv. Manuf.* **1** 176–82

[9] Chotisuwan S, Sirirak A, Har-Wae P and Wittayakun J 2012 Mesoporous alumina prepared from waste aluminum cans and used as catalytic support for toluene oxidation *Mater. Lett.* **70** 125–7

[10] Rani G and Sahare P D 2015 Effect of Temperature on Structural and Optical Properties of Boehmite Nanostructure *Int. J. Appl. Ceram. Technol.* **12** 124–32

[11] Simonet C, Fritsch E and Lasnier B 2008 A classification of gem corundum deposits aimed towards gem exploration *Ore Geol. Rev.* **34** 127–33

[12] Wu Y S, Ma J, Hu F and Li M C 2012 Synthesis and Characterization of Mesoporous Alumina via a Reverse Precipitation Method *J. Mater. Sci. Technol.* **28** 572–6

[13] Kyriakogona K, Giannopoulou I and Panias D 2017 *Extraction of Aluminium from Kaolin: a Comparative Study of Hydrometallurgical Processes*

[14] Xie W, Peng H and Chen L 2006 Transesterification of soybean oil catalyzed by potassium loaded on alumina as a solid-base catalyst *Appl. Catal. A Gen.* **300** 67–74

[15] Sinditskii V P, Chernyi A N, Shmelev D S, Egorshev V Y, Filatov S A, Matveev A A and

IOP Conf. Series: Materials Science and Engineering **702** (2019) 012038 doi:10.1088/1757-899X/702/1/012038

Milekhin Y M 2016 Combustion of potassium dinitramide and its binary mixtures with nitroester binder *Combust. Flame* **172** 206–14
[16] Mumtaz M W, Adnan A, Anwar F, Mukhtar H, Raza M A, Ahmad F and Rashid U 2012 Response Surface Methodology: An Emphatic Tool for Optimized Biodiesel Production Using Rice Bran and Sunflower Oils *Energies* **5**
[17] Aryee A N A, van de Voort F R and Simpson B K 2009 FTIR determination of free fatty acids in fish oils intended for biodiesel production *Process Biochem.* **44** 401–5

IOP Conf. Series: Materials Science and Engineering **702** (2019) 012039 doi:10.1088/1757-899X/702/1/012039

CO_2 removal via promoted potassium carbonate with glycine: solubility measurement at elevated pressure

F Isa[1,2], H Zabiri[1,2*], H and M A Shaarif[1,2]

[1] Chemical Engineering Department, Universiti Teknologi PETRONAS, 32610 Seri Iskandar, Perak, Malaysia.
[2] Institute of Contaminant Management, CO_2 Research Center (CO_2RES), Universiti Teknologi PETRONAS, 32610 Seri Iskandar, Perak, Malaysia.

*E-mail: haslindazabiri@utp.edu.my

Abstract. Potassium carbonate promoted with glycine is considered as potential candidates for CO_2 removal process as it demonstrated less energy usage during the regeneration process. Hence, in order to model the process at upscale condition, the loading of CO_2 into the promoted solvent is measured at temperature (30-60 ^{0}C) over the elevated pressure (10 bar) using high pressure solubility cell. The equilibrium solubility data is measured for 15wt% potassium carbonate + 3wt% glycine and 20wt% potassium carbonate +5 wt% glycine. This work aims to study the solubility of CO_2 into the promoted solvent at higher operating pressure, thus the vapor liquid equilibrium (VLE) can be derived and further facilitate the development of simulation for CO_2 absorption process at the upscale condition. Results show that the CO_2 is less favor to absorb into the promoted solvent at high temperature since the process is exothermic, thus increasing temperature would decrease the tendency of the absorption process. The comparative analysis also showed that solubility is improved at higher pressure as greater force assist to alleviate the surface stress of the solution.

1. Introduction

In natural gas treatment, CO_2 is one of the major contaminants that highly affected the gas quality standard. Thus, the various effort has been introduced to remove CO_2 before they can be transferred into the pipeline. The most commonly matured process is chemical absorption via a conventional fast reacting solution such as MEA, DEA. Nonetheless, the high regeneration energy that will severely increase the uptake of electricity up to 70-80 % [1] has led to many global researches to explore another potential solvent. Aside from amine based solvent, potassium carbonate has also long been studied as chemical absorption and their drawback in a slow kinetic reaction has been eliminated with the addition of various promoter. One of such promoters is amino acids and in the recent year, there has been renewed interest in using glycine as it has the same functional group as amine which leads to same reaction rate as amine but resistance to oxidative degradation, non-volatile and less toxicity [2].

Previous studies [2, 3] have reported various kind of finding related to CO_2 removal using promoted potassium carbonate but mostly focused on experimental works that emphasized on removal efficiency, rate enhancement and solubility. Realizing that the possibility of CO_2 removal using promoted potassium carbonate required a lot of trial and error to find the optimum condition, thus modelling and simulation is a must. Hence, in order to establish an accurate simulation model, a correct physical property and vapor Liquid Equilibrium (VLE) data are very necessary.

In Mondal et al. [4], a new VLE data in aqueous sodium glycinate (5-25 %) is measured and then correlated using Kent Eisenberg model. Hence, a carbamate hydrolysis equilibrium constant and amine deprotonation equilibrium constant are regressed. Eventhough, a model successfully predicted the experimental data but the operating condition is only up to medium pressure (2-6 bar). Another study [5] have also conducted VLE study for the absorption of CO_2 into a glycine promoted potassium carbonate solution over a range of temperatures and in this work ENRTL model has been used to model the data. The comparison between promoted potassium carbonate and pure potassium carbonate indicated that glycine can improve the CO_2 absorption but still the data is only suitable for post combustion process only. When designing or simulating an absorption column associated with chemical reaction, certain key data is required such as physical, thermal and transport properties, vapor liquid equilibrium (VLE) and reaction rate data. Unlike amine-based solvent, the information related with promoted potassium carbonate with glycine is not widely presented and available in the literature. Hence, a few studies have attempted to investigate the physical properties of this solvent but still the data are limited to post combustion operating condition. Hence, this study aims to study the solubility of the promoted solvent at higher operating pressure, thus the CO_2 absorption process at the upscale condition can be model and simulate accurately.

2. Research methodology

2.1 Chemical reaction

Understanding the chemical reaction in the of CO_2 in promoted potassium carbonate is very necessary before any process can be developed or modelled. The chemical absorption in the process can be divided into two main mechanisms. The solvent mechanism and promoter mechanism that taking at parallel reaction in the same time. The common solvent mechanism is listed from (R1) until (R5), the hydrolysis of potassium carbonate is defined in (R1) and (R2) shows the water ionization. Then, bicarbonate formation and carbonate formation are defined in (R3) and (R4). Then, the overall reaction is represented in (R5). The precipitation of potassium salt might occur at high concentration and considered as hydrated potassium carbonate ($K_2CO_3.1.5H_2O$) is listed in (R6) but considered negligible in this reaction.

$$K_2CO_3 + H_2O \leftrightarrow 2K^+ + HCO_3^- + OH^- \quad \text{(R1)}$$
$$2H_2O \leftrightarrow H_3O^+ + OH^- \quad \text{(R2)}$$
$$CO_2 + OH^- \leftrightarrow +HCO_3^- \quad \text{(R3)}$$
$$HCO_3^- + OH^- \leftrightarrow H_2O + CO_3^{2-} \quad \text{(R4)}$$
$$K_2CO_3 + H_2O + CO_2 \leftrightarrow 2KHCO_3 \quad \text{(R5)}$$
$$K_2CO_3.1.5H_2O \leftrightarrow 2K^+ + CO_3^{2-} + 1.5H_2O \quad \text{(R6)}$$

The promoter reaction is defined from (R7) -(R9). Based on the observation in Lee et al. [5], glycine is not only enhancing the kinetic reaction of the solvent but the glycine itself is capable to capture CO2 to form glycine carbamate. In aqueous solution, glycine may present in term of protonated, anionic and zwitterionic.

$$-Gly + H_3O^+ \leftrightarrow -Gly^+ + H_2O \quad \text{(R7)}$$
$$-Gly + CO_2 + H_2O \leftrightarrow -GlyCOO^- + H_3O^+ \quad \text{(R8)}$$
$$-GlyCOO^- + H_2O \leftrightarrow Gly + HCO_3^- \quad \text{(R9)}$$

2.2 CO_2 solubility measurement

Potassium carbonate and glycine with purity (99%) are obtained from Sigma Aldrich. Two different concentration of samples are prepared based on the following ratio; PC:Gly, 15wt% : 3wt% and 20wt%: 5wt%. The solvent concentration as suggested in literature must be less than 40 wt% as higher concentrations will lead to precipitation in the column [6]. The experiments for solubility measurement are carried out at temperature (30-60 °C) at 10 bar using pressure solubility cell (SOLTEQ BP-22). The equilibrium cell consists of 2 mixing vessel that labelled as EC 1 and EC 2 with a volume of 50mL. Initially, N_2 purged and the vacuum condition is formed before 7 mL of sample is injected into the equilibrium cell via a metering pump. The temperature is adjusted at the required condition and maintained until it stabilized. Then, the initial pressure reading in the equilibrium cells are recorded. Then, CO_2 gas is compressed into a pressure vessel (PV) until the desired operating pressure is achieved. Once the temperature in the cell maintained and pressure is stable, the valve is fully open and CO_2 is transferred to the mixing cell. The pressure will be rapidly increased and the valve will be fully closed once there is no change is observed after 5s. Sudden pressure drop is observed and the attainment of equilibrium is confirmed when there is no further change of pressure at least 2 h in equilibrium cell at a constant temperature. This procedure is repeated for each sample at different temperature. The simplified diagram of solubility cell is shown in figure 1.

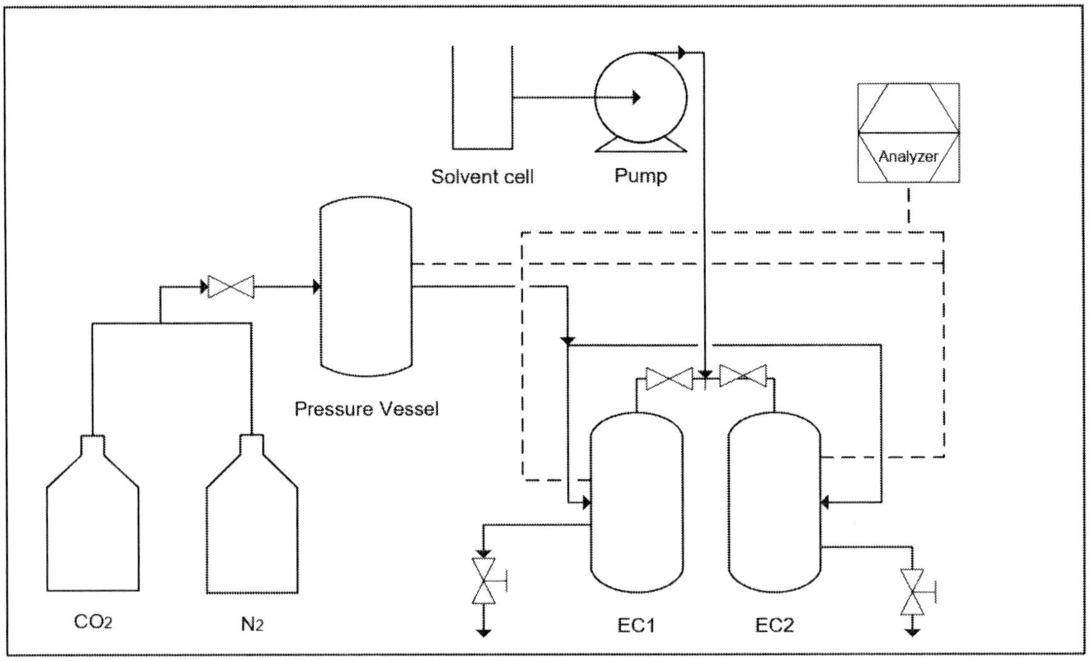

Figure 1. High pressure gas solubility cell (SOLTEX).

Based on [7], mole of CO_2 is obtained using equation (1).

$$n_{CO2} = \frac{V_T}{RT}\left(\frac{P_i}{z_1} - \frac{P_f}{z_2}\right) \tag{1}$$

V_T denotes the volume of mixing cell, Z are the compressibility factor that calculated using Peng rebingson equation of State (PR-EOS) while R is the real gas constant and T is referred to the operating temperature.The loading (α) is calculated as mole of CO_2 per mole of promoted solvent using equation (2) The number of moles for the promoted potassium carboante and glycine is calculated using equation (3).

$$\alpha = \frac{n_{CO_2}^{l}}{n_{PC+Gly}} \tag{2}$$

$$n_{PC+Gly} = \frac{\rho V_l m_{PC+Gly}}{M_{PC+Gly}} \tag{3}$$

ρ denotes the density of solvent, V_l as the liquid volume in cell, m_{PC+Gly} is the mass fraction of the solvent and M_{PC+Gly} is the molecular weight .

3. Results and discussions

Figure 2 shows the pressure reduction for both solvents concentrations at different temperatures (30-60 ^{0}C). The pressure reduction is continued until the promoted solvent is saturated with CO_2 gas and the equilibrium pressure is attained once no further reduction of pressure is observed after 2 hours of operation. According to Mondal et al. [4], the equilibrium pressure is described as a difference of total pressure and vapor pressure while the equilibrium is achieved for at least 1 hour at the constant temperature. However, this study has extended the equilibrium attainment time to ensure the accuracy is improved. It is interesting to note that the attainment time is different for each set of concentration and operating condition but the record is not presented in this paper as the main aim only to observe the loading value.

Pressure decay is recorded as it can give the overview of CO_2 loading and based on the early observation, the CO_2 is not favored to absorb into this promoted solvent at high temperature and this is true based on the Le Chatlier principle [8]. As in nature, CO_2 absorption process is exothermic, thus increasing temperature would decrease the tendency of the absorption process. Besides, it is noticed that higher concentration of solvent (20 wt% PC + 5 wt% Gly) leads to higher pressure drop as the amount of active solvent is higher thus increase the tendency of the absorption process. Initially, the absorption is very vigorous once CO_2 gas is introduced to the equilibrium cell and after sometimes, the absorption process is slowing down until it reached the steady state.

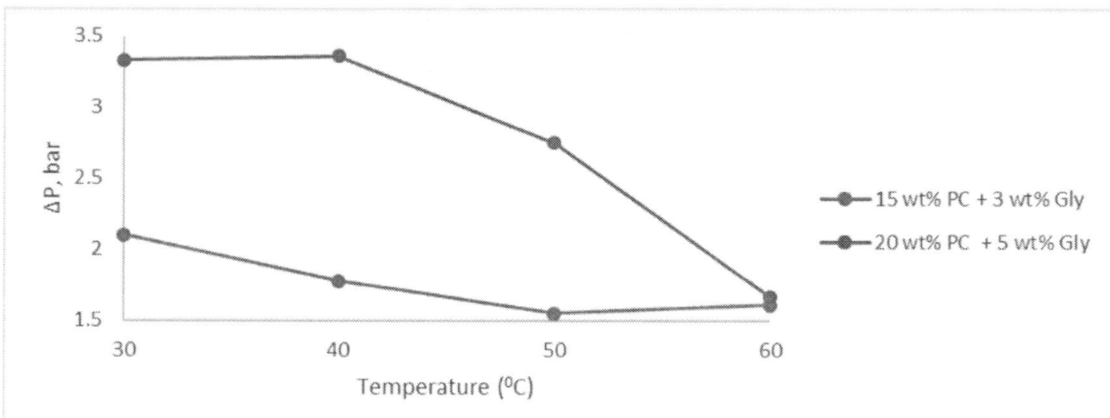

Figure 2. Pressure reduction of CO_2 against temperature

The loading of mole CO_2 over mole promoted solvent is calculated via measuring the difference between initial and final pressures (equilibrium pressure) as shown in figure 3.

At constant elevated pressure (10bar), the results manifested that the loading is decreased with the rise of temperature. The comparative analysis in table 1 shows that similar trend is also observed in [7] at same operating pressure but the solubility test is conducted using a pure potassium carbonate solution.

Table 1. Comparison of CO_2 loading (α) at 10 bar

Temperature (^{0}C)	15 wt% PC + 3 wt% Gly	20 wt% PC + 5 wt% Gly	20 wt% PC (1)
30	0.75	0.96	nil
40	0.62	0.94	1.25
50	0.52	0.75	0.93
60	0.53	0.44	0.79

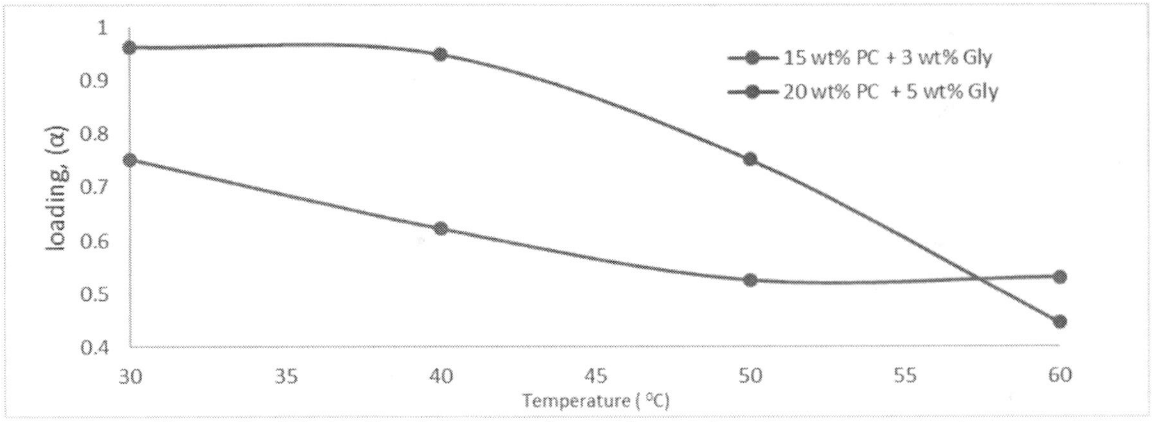

Figure 3. CO_2 loading in promoted potassium carbonate versus temperature

The higher the concentration of potassium carbonate and glycine used (20wt% PC + 5 wt% Gly), the greater CO_2 loading is observed. This is expected as more active solvent present in the solution, the absorption will be improved, thus the loading calculation is also increased. The aim value for loading is 1 as it demonstrated that one mole of CO_2 is absorbed by one mole of solvent. The loading value for 15wt% PC + 3 wt% Gly has also shown a promising value but the calculated value is not able to reach the aim of the experiment.

Based on the literature [7], a pure K_2CO_3 yield a better solubility based on loading calculation but the time taken to reach the equilibrium during the solubility test is much longer than the current promoted solvent. Hence. the addition of promoter is helpful to compensate for the difference in term of mass transfer rate. This is true as [5] has also agreed that using the promoter itself will greatly enhance the reaction kinetic but blended solvent between potassium carbonate and glycine may provide a mean of balancing kinetics and capacity against the solvent cost.

4. Conclusions

The promoted potassium carbonate has good potential for CO_2 absorption. Nevertheless, further investigation at higher operating pressure at various concentration is an essential before the accurate VLE model can be derived. Hence, a scale up condition for the overall process can be designed and simulated at wider operating range.

Acknowledgements

The authors are thankful to Universiti Teknologi PETRONAS and research grant (FRGS-015MA0-036) for financial and technical support in the completion of this research study.

References

[1] Li X, Wang S, Chen C 2013 Experimental study of energy requirement of CO_2 desorption from rich solvent *Energy Procedia.* **37** 1836-43

[2] Thee H, Suryaputradinata Y A, Mumford K A, Smith KH, da Silva G and Kentish SE 2012 A kinetic and process modelling study of CO_2 capture with MEA-promoted potassium carbonate solutions *Chemical Engineering Journal* **210** 271-9

[3] Smith K, Lee A, Mumford K, Li S, Thanumurthy N and Temple N 2015 Pilot plant results for a precipitating potassium carbonate solvent absorption process promoted with glycine for enhanced CO_2 capture *Fuel Processing Technology* **135** 60-5

[4] Mondal B K, Bandyopadhyay S S and Samanta A N 2015 VLE of CO_2 in aqueous sodium glycinate solution–New data and modeling using Kent–Eisenberg model. *International Journal of Greenhouse Gas Control* **36** 153-60

[5] Lee A, Mumford K A, Wu Y, Nicholas N and Stevens G W 2016 Understanding the vapour–liquid equilibrium of CO_2 in mixed solutions of potassium carbonate and potassium glycinat. *International Journal of Greenhouse Gas Control* **47** 303-9

[6] Haghtalab A and Gholami V 2019 Carbon dioxide solubility in the aqueous mixtures of diisopropanolamine+ l-arginine and diethanolamine+ l-arginine at high pressures *Journal of Molecular Liquids* 111064

[7] Bohloul M, Sadeghabadi M A, Peyghambarzadeh S and Dehghani M 2017 CO_2 absorption using aqueous solution of potassium carbonate: Experimental measurement and thermodynamic modeling *Fluid Phase Equilibria* 447 132-41

[8] Shaikh M S, Azmi M S, Bustam M A and Murshid G 2014 Study of CO_2 solubility in aqueous blend of potassium carbonate promoted with glycine *Applied Mechanics and Materials* 19-23.

Simulation of pressurized water scrubbing process for biogas purification using Aspen Plus

N Abu Seman and N Harun*

Faculty of Chemical & Process Engineering Technology, Universiti Malaysia Pahang, 26300 Gambang, Pahang, Malaysia

*E-mail: noorlisa@ump.edu.my

Abstract. Biogas produced from anaerobic digestion process has a major amount of biomethane ($bioCH_4$), carbon dioxide (CO_2), and traces amount of impurities. It must undergo biogas cleaning first to remove the impurities depending on the requirement of the applications. Pressurized water scrubbing is the most applicable method for removing CO_2 because of its simplicity, low cost, and low toxicity. This study aims to develop a pressurized water scrubbing process model for biogas purification using Aspen Plus software. Several process parameters such as absorber column pressure and liquid to gas (L/G) ratios were varied to investigate their effects on $bioCH_4$ purity and percentage of CO_2 removal. The simulated results show good agreement with previous work and agreed well with the typical standard concentration of $bioCH_4$ of over 95% to be used as vehicle fuels or in a gas grid.

1. Introduction

Attaining solution to the depleting source of fossil fuel and environmental issues due to the use of fossil fuel requires long term possible actions for sustainable development. Renewable energy is a promising option as an effective and efficient way to solve these issues. Biogas is one of the renewable energy resources that has received attention worldwide as it can accommodate the depleting resources of fossil fuel. This renewable energy has a continuous supply of resources such as municipal solid waste, agricultural waste, industrial waste, and kitchen waste. It definitely seems like a promising option to solve the problem of fossil fuel shortage and environmental issue due to usage of fossil fuel and waste management. In terms of environmental issue, biogas engine emission of CO_2 is less than fuel from fossil fuel sources. The burning of fossil fuels over the past century has dramatically increased the levels of CO_2 that traps heat in our atmosphere. The implications of the increasing levels of these greenhouse gases are a serious matter. Concerns about the depletion of fossil fuels, energy security, and emission of greenhouse gas have initiated renewable energy studies.

Biogas is produced from the biological process of the decomposition of organic material in the absence of oxygen [1]. The breakdown or decomposition of organic material is a process called anaerobic digestion. Anaerobic digestion involves four phases, namely hydrolysis, acidogenesis, acetogenesis, and methanogenesis [2]. Anaerobic digestion process for biogas production has become one of the important sources of renewable energy to generate electricity, heating, and transportation fuel, and the digestate can be used as fertilizer. According to Yang et al. [2], biogas produced from anaerobic digestion process has a high amount of biomethane ($bioCH_4$), carbon dioxide (CO_2), and significant traces amount of impurities such as hydrogen sulfide, hydrogen, nitrogen, oxygen, and ammonia. These impurities can cause corrosion and toxicity and reduce the heating value of biogas. Biogas purification or upgrading has become an important element in the development of the biogas

IOP Conf. Series: Materials Science and Engineering **702** (2019) 012040 doi:10.1088/1757-899X/702/1/012040

industry. Due to the impurities in biogas, raw biogas cannot be used directly. Biogas cleaning method may require the removal of a large amount of impurities depending on the requirement of the applications [3]. For instance, the typical standard concentration of bioCH_4 is over 95% to be used as vehicle fuels or in gas grid [4].

Pressurized water scrubbing is the most applicable method for removing CO_2 because of its simplicity, low cost, and low toxicity [1]. The pressurized water scrubbing unit comprises compressor, absorber, flash, and stripper. An absorber that operates at high pressure between 0.8 and 1.2 MPa is used to increase the partial pressure of CO_2 [5], which increases the absorption rate of CO_2 and the attainment of the high purity of bioCH_4 in gas form, leaving the top of the absorber. Since CO_2 is more soluble in water than bioCH_4, it is an advantage to separate CO_2 and bioCH_4 with high efficiency [2]. Flash vessel, operates at low pressure between 0.11 and 0.2 MPa is used to depressurize water coming from the absorber [5,6], to ensure a major portion of gases is desorbed to be recycled into the absorber. Stripper, which uses air as stripping agent operates under ambient pressure to regenerate water from CO_2-rich stream and desorb CO_2 from the water. According to Petersson and Wellinger [7], pressurized water scrubbing method is mainly applied in Germany and Sweden for biogas purification for transportation fuel and gas grid. Despite the advantages of biogas upgrading to bioCH_4 using pressurized water scrubbing technology mentioned before, this process faces multiple challenges in the real-life process. The percentage of CO_2 removal and the purity of bioCH_4 content produced are determined by several parameters such as temperature, pressure, and the amount of fresh water [8]. The main concern of this process is the high water demand [2]. Although water is assumed to be recycled continuously, the amount of evaporated water needs to be replaced, and it requires high water demand to increase the absorption capacity in the column. The performance of this process can be improved by operating it at optimum process conditions. Experimental work to investigate the optimum process condition will be costly and time-consuming.

Therefore, this study aims to apply a simulation approach to elucidate the effect of different operating parameters on the purification of biogas to remove impurities. Simulation approach can greatly reduce the cost of the process before real-life operation by optimizing the operating parameters involved. Aspen Plus software was applied in this study to develop a biogas purification using pressurized water scrubbing method. The model was compared against data reported by Cozma et al. [8].

2. Process simulation model

The process simulation model for high-pressure water scrubbing process was developed using Aspen Plus software. The steps to build a simulation using Aspen Plus cover the selection of the property method, the list of component setup, and the creation of a process flowsheet in the simulation environment. The thermodynamic properties package used in this simulation is the nonrandom-two-liquid (NRTL) method because most of the components used in this process simulation are polar mixtures and suitable for pressure up to 10 bar.

The composition of raw biogas in FEED stream is tabulated in Table 1. The process flow sheet of the pressurized water scrubbing process for biogas purification is depicted in Figure 1. The process conditions and input data applied in this study were obtained from Cozma et al. [8].

Table 1. Biogas input specifications [8]

Components	Mole composition (%)
Methane	60
Carbon dioxide	38.9
Hydrogen sulphide	300 ppm
Nitrogen	0.5
Oxygen	0.5

The biogas enters the bottom of the absorber after being pressurized to 10 bar in compressors (COMP1, COMP2) while pressurized water is fed from the top of the absorber. The PRODUCT gas, which contains a high amount of bioCH_4, leaves through the top of the absorber and the CO_2-enriched water leaving the absorber is sent to the flash column. The flash column pressure is set at 3 bar to minimize bioCH_4 losses. The gas released from the flash column containing CO_2, bioCH_4, hydrogen sulfide, nitrogen, oxygen, and water is mixed with raw biogas in the mixer and recycled through the inlet of the compressor, COMP2. The liquid solution leaving the flash column is transferred to the desorption column (STRIPPER) where it meets a counter-current flow of hot air. CO_2 and hydrogen sulfide are separated from water at atmospheric pressure and temperature of 1 atm and 20°C, respectively. Water is recycled back to the top of the absorber.

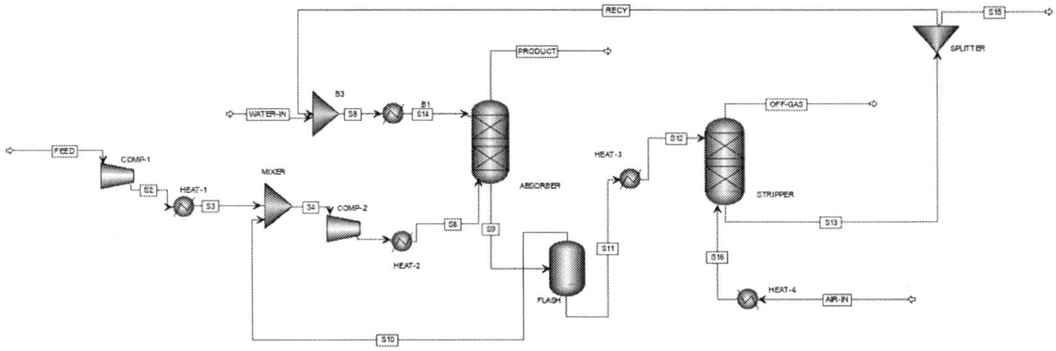

Figure 1. Flow sheet of pressurized water scrubbing process in Aspen Plus

3. Results and discussion

The FEED stream had a total of 604.6 kg/h biogas and 5000 kg/h water [8]. The process condition and column specification are presented in Table 2. The amount of bioCH_4 purified in the PRODUCT stream and the percentage of CO_2 removal in the OFF-GAS stream are considered as the main outputs from this simulation. Table 1 displays the comparison between the results obtained by Cozma et al. [8] and the present study. The percentage of CO_2 removal and bioCH_4 purity in this study was slightly higher than those in Cozma et al. [8]. This is due to the absorber column pressure and the number of theoretical stages chosen were slightly higher than those reported in Cozma et al [8], to minimize the differences of bioCH4 purity and percentage of CO_2 removal between this study and Cozma et al. [8]. The product gas leaving the absorber mainly contained bioCH_4 with 97.6 mole % and traces amount of CO_2, hydrogen sulfide (H_2S), oxygen (O_2), and nitrogen (N_2). The purity of bioCH_4 in the PRODUCT stream is suitable for gas grid and vehicle fuel which require more than 95 mole % bioCH_4 purity. The percentage of CO_2 removal obtained from this study was slightly higher than that from Cozma et al. [8] which is 99.9 mole %. This model shall further be used for sensitivity analysis to study the effect of absorber pressure and liquid-to-gas ratio (L/G) on bioCH_4 purity and percentage of CO_2 removal.

Table 2. Comparison base-case simulated results

Parameters	Cozma et al. (2014)	This study
$P_{absorber}$ (bar)	8	10
$T_{absorber}$ (°C)	20	20
$T_{stripper}$ (°C)	20	20
P_{flash} (bar)	3	3
$P_{stripper}$ (bar)	1	1
Number of theoretical stages (absorber)	11	18
CO_2 removal (%)	**99.12**	**99.93**
Product gas (mole fraction)		
CO_2	0.00397	0.000338
bioCH_4	**0.965**	**0.976**
H_2S	0.000094 ppm	0.000088 ppm
N_2	0.0209	0.0106
O_2	0.00563	0.0092

Parameters	Cozma et al. (2014)	This study
H_2	0.000902	-
H_2O	0.00319	0.00348
Off gas (mole fraction)		
CO_2	0.171	0.1065
bioCH_4	0.000677	0.01151
H_2S	38 ppm	82 ppm
N_2	0.635	0.688
O_2	0.168	0.160
H_2	0.193 ppm	-
H_2S	0.0234	0.03381

3.1. The effect of absorber pressure

The absorber column provides intimate contact of the biogas with water so that the CO_2 molecules can be transferred from the gas phase into the liquid phase. The analysis of various pressures setting for absorber was conducted to investigate its effect on bioCH_4 purity and CO_2 removal efficiency. Figure 2 depicts the bioCH_4 composition and percentage of CO_2 removal for pressure setting between 5 and 20 bar. The purity of bioCH_4 and CO_2 removal efficiency is increasing along with the pressure of absorber. Pressurized water scrubbing is a physical absorption process that requires high pressure to increase the solubility of CO_2 in water [9,10]. At 10 bar, the bioCH_4 purity and CO_2 removal efficiency remained constant and did not affect the performance of CO_2 absorption. Higher pressure up to 10 bar gives better performance of CO_2 absorption. However, increasing the pressure absorber will result in higher power consumption and cost. Hence, the optimum pressure for absorber for this process is 10 bar.

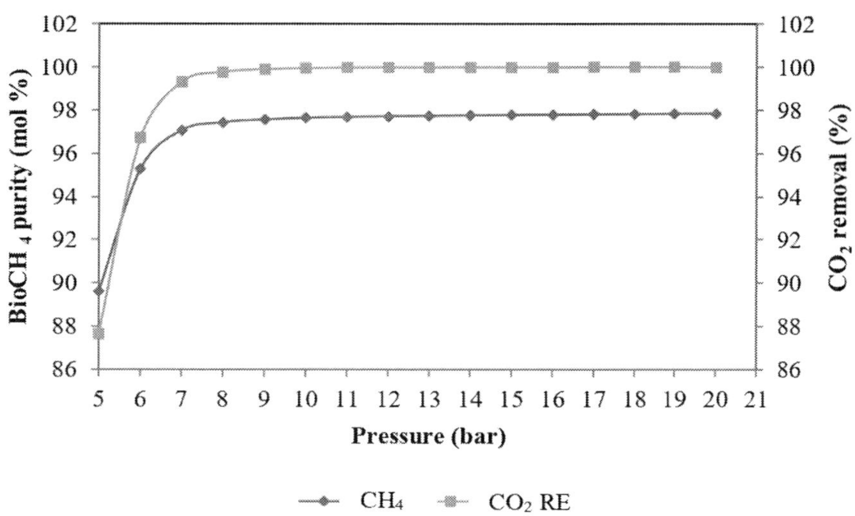

Figure 2. Effect of absorber pressure on bioCH_4 purity and CO_2 removal

3.2. Effect of L/G ratio

The flow rate of feed biogas was kept constant at 604.6 kg/h, and only the flow rate of water was varied. The effect of L/G (water flow into the absorber with biogas inlet flow) ratio on bioCH_4 purity and CO_2 removal is illustrated in Figure 3. The CO_2 removal efficiency increases along with the ratio of L/G. A higher flow rate of water increases the area for mass transfer process between gas and liquid, which increases the mass transfer coefficient since more water molecules are available to dissolve CO_2 [10]. Hence, a higher amount of solvent used increases the rate of gas absorption in the solvent. At the ratio of 10, the CO_2 removal is nearly complete at 99.9%. The L/G ratio of 10 shows a slight decrease in bioCH_4 purity in the product gas. It might be due to some of bioCH_4 dissolved in water as the ratio is high. Hence, at the ratio of 10, the bioCH_4 purity slightly decreases to 97.5% from 97.6% at the ratio of 9.

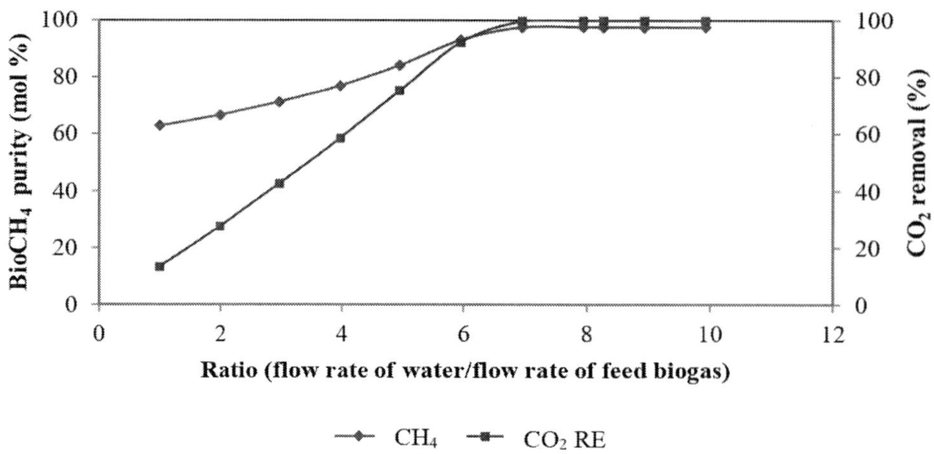

Figure 3. The effect of L/G ratio on bioCH_4 purity and CO_2 removal

4. Conclusions

A simulation of biogas purification to produce bioCH_4 was carried out using Aspen Plus software. The simulation to purify biogas gave a result of 97.6 mole % bioCH_4 purity and 99.9 mole % CO_2 removal. It was pointed out that the influence of absorber column pressure and liquid to gas (L/G) ratio significantly affected bioCH_4 purity and percentage of carbon dioxide removal. Increasing the absorber column pressure up to 10 bar increased the solubility of CO_2 in water and purification of bioCH_4. As the flow rate of fresh water was increased, more CO_2 dissolved in water, and it increased the bioCH_4 purity and CO_2 removal efficiency. It shows that more carbon dioxide can be removed in a high amount of fresh water.

Acknowledgement

The authors would like to thank Universiti Malaysia Pahang (UMP) for the financial support through the research grant RDU1703238.

References

[1] Voicu G, Ungureanu N and Paraschiv G 2017 *Methods of Biogas Purification – a Review* 403–408

[2] Yang L, Ge X, Wan C, Yu F and Li Y 2014 Progress and perspectives in converting biogas to transportation fuels *Renewable and sustainable energy reviews* **40** 1133–1152

[3] Ohenoja M and Sorsa A 2015 Simulation as a Tool for Evaluating Biogas Purification Processes *Proc. Of the 56TH SIMS* (Sweden: ResearchGate) pp 55–61

[4] Nock W, Walker M, Kapoor R and Heaven S 2014 Modelling the water scrubbing process and energy requirements for CO_2 capture to upgrade biogas to biomethane *Industrial and Engineering Chemistry Research* **53** 12783–12792

[5] Wylock C and Budzianowski W 2017 Performance evaluation of biogas upgrading by pressurized water scrubbing via modelling and simulation *Chemical Engineering Science* **170** 639–652

[6] Kapoor R, Subbarao P M V and Vijay V K 2019 Integration of flash vessel in water scrubbing biogas upgrading system for maximum methane recovery *Bioresource Technology Reports* **7** 100251

[7] Petersson A and Wellinger A 2009 Biogas Upgrading Technologies – Developments and Innovations *IEA Bioenergy Task 37 – Energy from biogas and Landfill Gas* **37**

[8] Cozma P, Wukovits W, Mămăligă I, Fredl A and Gavrilescu M 2014 Modelling and simulation of high pressure water scrubbing technology applied for biogas upgrading *Clean Technologies and Environmental Policy* **17**

IOP Conf. Series: Materials Science and Engineering **702** (2019) 012040 doi:10.1088/1757-899X/702/1/012040

[9] Bashar M 2018 Biogas Quality Improvement Using Water Wash and Phosphorus Recovery as Struvite in Jones Island Wwtp *Theses and Dissertations.* 1746

[10] Kapoor R, Subbarao P M V, Vijay V K, Shah G, Sahota S, Singh D and Verma M 2017 Factors affecting methane loss from a water scrubbing based biogas upgrading system *Applied Energy* **208** 1379–1388

IOP Conf. Series: Materials Science and Engineering **702** (2019) 012041 doi:10.1088/1757-899X/702/1/012041

Preliminary Fractional Factorial Design (FFD) study using incorporation of Graphene Oxide in PVC in mixed matrix membrane to enhance CO_2/CH_4 separation

K R Raj, A R Sunarti*

Faculty of Chemical and Process Engineering Technology, College of Engineering Technology Universiti Malaysia Pahang, 26300 Kuantan, Pahang, Malaysia.

*E-mail: sunarti@ump.edu.my

Abstract. Mixed matrix membranes (MMMs) comprising of where polyvinyl chloride (PVC) and graphene oxide (GO) were employed in the membrane fabrication in order to improve the membrane characteristics. In this study of gas separation using mixed matrix membrane, gases such as CO_2 and CH_4 were used. The objectives of this research were to synthesize and develop MMMs PVC/GO, and also to carry out screening to determine best factors condition for membrane to function at high selectivity. The development of MMMs PVC/GO was based on 5 factors. Based on the screening tests, the run which was made up of factors PVC 20%, GO 4%, 1 bar of pressure during gas permeation, NMP solvent, and immersion time of 300s, shows to have the highest selectivity at 26.09 which was above the upper bound lines in the Robeson's Upper bound plot. Fractional Factorial Design (FFD) was applied to study the minimalization of influenced factors during MMMs preparation. Based on the FFD run, the R-squared (R^2) was 1.00. Factors that had less impact on the maximum selectivity were the time membrane immersed in water bath (s), and also gas pressure during gas permeation test.

1. Introduction

Polymer membranes are used commercially to separate air, to remove carbon dioxide from natural gas, and to remove hydrogen from mixtures with nitrogen or hydrocarbons in petrochemical processing applications. The fundamental parameters characterizing membrane separation performance are the Permeability, P, and the selectivity, $\alpha_{A/B}$. Gas selectivity is the ratio of permeability of two gases (P_A/P_B), where P_A is the permeability of the more permeable gas and P_B is the permeability of the less permeable gas in the binary gas pair. Polymers with both high permeability and selectivity are desirable. Higher permeability decreases the amount of membrane area required to treat a given amount of gas, thereby decreasing the capital cost of membrane units. A rather general trade-off relation has been recognized between permeability and selectivity: Polymers that are more permeable are generally less selective and vice versa.

The increasing global demand for energy-efficient separations in carbon capture has prompted international actions on searching for novel, high-performance separation membranes. Polymeric membranes offer advantages over inorganic membrane for their ease fabrication and low cost [1-2]. Unfortunately, the separation performances of polymeric membranes are limited by a trade-off between permeability and separation selectivity which is called the Robeson upper limit [3-4]. Therefore, mixed matrix membranes (MMMs), with hybrid advantages of both inorganic and polymeric membranes, are developed combining polymeric materials with well-selected fillers. So far, a variety of inorganic materials, such as zeolites [5], silica [6], carbon nanotubes [7] and carbon molecular sieves [8] can be used as fillers of MMMs. However, owing to the intrinsic differences between the inorganic and polymeric phases, interface voids and rigidified polymer can easily appear in MMMs with inorganic fillers. Therefore, the synthesis of perfect MMMs with high separation performances is rather challenging. Mixed matrix membranes (MMMs) are considered a new-generation membrane for gas purification applications and have become a focus for research and development in both academic and industrial interests due to their unique properties combining inherent characteristics of polymer and inorganic fillers [9].

Throughout the years there have been a number of researches done related to gas separation using mixed matrix membrane. In a research done by Adams et al. [10], Under both mixed feed conditions at 35 °C, substantial improvements in overall separation performance were observed. At low CO_2 partial pressures, CO_2 permeability roughly doubled with a nearly 50% increase in selectivity versus pure PVA under the same conditions. For the high CO_2 partial pressure feed, CO_2 permeability remained effectively unchanged with a 63% increase in selectivity versus pure PVA. Surprisingly, the performance of these PVA based MMMs approached the properties of current "upper bound" polymers. In another study done by Bao-Sheng et al. [11], GO loading increases from 1 to 3 wt.%. Based on the interaction of the GO and the polymer segmental chains, an increase in free volume of the polymer matrix resulted from the disruption of the polymer chains packing also demonstrates an ascending trend of the permeability. However, with 4 wt.% GO incorporated, the d-spacing of MMMs decreases slightly due to the poor dispersion of GO at higher loadings. A study done by Farahani et al. [12] shows that the effect of polymer concentration and solvent showed by the membranes fabricated using 15 wt% of PES concentration possessed greater fouling resistance and water flux compared to those of fabricated using 18 wt% of PES concentration. Also, membranes fabricated using DMA exhibited a more porous structure with considerably greater water flux as compared with those of fabricated using NMP as the solvent. Mixed matrix membranes are prepared in a dope solution where it is then heated at low heat for it to be casted on the casting machine via NIPS method [13]. However, NIPS method uses water in terms of water immersion of the produced membrane where the time for the membrane to be immersed is still not fixed. Ahmad et al. [14] study shows that MMM prepared with NIPS method shows the immersion time of the membrane in water for 60 minutes in the first immersion and left for another 24 hours for completing the process of removing excess chemicals. However, Farahani et al. [15] stated for the fabrication of NH2-MWCNT/P84 MMMs, the immersion time of the MMMs in water is around 2 days for removing excess chemicals. In another study by Qianyu et al. [16] shows the immersion time is only 1 hour for the fabrication of PVDF-AC MMMs using NIPS method.

Experimental designs are commonly performed in the study of empirical relationship, in terms of a mathematical model, between one or more measured responses and a number of variables or factors where these designs and modelling normally carried out in Design Expert software [17]. Experimental design and mathematical modelling techniques are mathematical tools normally used to optimize a process. Traditional methods of optimization involved changing one independent variable while fixing the others at a certain level. Experiment design techniques were developed to allow the gathering of maximum process information with reduced number of experiments. Experimental design techniques usually depend on empirical model structure in order to interpret experimental data and provide optimum process conditions. For the experimental design and response surface modelling for mixed matrix membrane by Kusworo et al. [18], mixed matrix membrane of PI/PES-Zeolite 4A was used for CO_2 separation. This research carried out experimental design via full factorial design (FFD) and central composite design (CCD) were systematically performed to investigate the main factor of fabrication parameters and the relationship with the mixed matrix membrane performance. A response surface methodology and central composite design were used in optimization experiments and iterative regression analysis to determine the maximum gas permeability and selectivity. Hence, the dominating factors that were likely to be the most important and influential could be diagnosed in order to optimise flat sheet mixed matrix membrane formation process. The effects and interactions of total solid/polymer concentration (X1), composition of polymer blending (X2) and zeolite content (X3) on carbon dioxide permeability and CO_2/CH_4 selectivity for PI/PES-zeolite mixed matrix membranes were investigated.

IOP Conf. Series: Materials Science and Engineering **702** (2019) 012041 doi:10.1088/1757-899X/702/1/012041

The experimental value and predicted responses for 16 trials runs. Another paper by Éva et al. [19], shows the full factorial design (FFD) performance based on the selected DOE procedure to evaluate the effect of every combination of the selected factors which are concentration of absorbent (X1), liquid volume flow (X2), and gas volume flow (X3) and hydrogen sulfide content of gas mixture (X4) with removal efficiency as its response. After the experimental work is finished and all the data is collected and systemized, a statistical modeling closes the procedure to compress the extracted information about the system applying the response surface methodology (RSM).

This study aims to prepare asymmetric PVC flat sheet MMMs using GO as a filler. A total of 5 factors have been selected to determine the to evaluate the significant factors towards the performance of the membrane during gas permeation test. Screening method via full factorial design (FFD) was used to determine the factor which provides the highest effect percentage. Factors used in this study are weight ratio of PVC, weight ratio of GO, pressure used during gas permeation (bar), time of membrane immersed in water (s), and type of solvent used. In this study, the response for the FFD would be selectivity of the MMMs.).

2. Methods

The materials that was used in this experiment were PVC with a melting temperature of 160°C and also GO powder. Solvents such as DMF with 99.5% purity and NMP with 99.5% purity were used in this study. The boiling points of DMF and NMP are 153°C and 202°C respectively. There two gases used for gas permeation test were CO_2 and CH_4 gases. This research shows the usage of NIPS method for the fabrication of MMMs and gas permeation using MMMs for the separation of CO_2 and CH_4.

2.1. Fabrication of PVC/GO MMMs

The asymmetric flat sheet MMMs were fabricated using the NIPS technique [20]. Using the combination of PVC powder, GO powder, and NMP or DMF solution that formed a dope solution, the PVC/GO was stirred and dissolved in the solvents respectively. The dope solutions were prepared according to ratios of PVC, GO, and NMP or DMF, where for NMP or DMF will be constant at 80% of weight ratio of the solution. The dope solution was stirred for about 5 hours continuously at 130°C for NMP mixed dope solution and 125°C for DMF mixed dope solution, to ensure homogeneity. After completely dissolving, the dope solution was left and stirred for another 4 hours in order to remove trapped gas bubbles. In the production of asymmetric MMMs, phase inversion is considered to be a crucial method as it mainly focuses on polymer membrane which mainly goes through this method. Dry/wet phase inversion method is a process where the casted MMMs was left to vaporize and it was immediately immersed into a nonsolvent medium, normally water is used. The phase separation between solvent NMP or DMF, and nonsolvent such as water component, forms the membrane product. This is also assisting in the process of vaporization and polymer clotting. Using a glass plate or a polyester nonwoven fabric, the degassed dope solution was coated at room temperature at a knife gap of 0.3mm. After 10s of exposure in air, the casted film would be immediately immersed into a coagulation. The precipitated membrane was taken out of the coagulation bath and rinsed with running water to remove the residual solvent. The MMMs was then dried at room temperature.

2.2. Screening via Full Factorial Design (FFD)

Software usage, that is Design Expert is crucial for the evaluation the significant factors. Based on earlier journal studies, many factors affect the performance of the MMMs in gas separation process. A 2 Level Factorial Design has been carried out for this study where 5 factors have been chosen to determine the effects of the MMMs upon the performance of the MMMs in gas separation. These factors are weight ratio of PVC (A), weight ratio of GO (B), pressure used during gas permeation (bar) (C), type of solvent used (D), and time of membrane immersed in water (s) (E). From these 5 factors, 4 of these factors are numeric factors and another 1 is a categoric factor. The weight ratio of PVC was divided into 2 ratios which are 0.2% and 0.15%, whereas for GO the weight ratio was divided into 2 ratios also which are 0.02 and 0.04. The pressure that was set for gas permeation were 1 bar and 3 bar. The time the MMMs was immersed in the water was set to two different time which are 30s and 60s. Solvents that were used

IOP Conf. Series: Materials Science and Engineering **702** (2019) 012041 doi:10.1088/1757-899X/702/1/012041

in this study were NMP and DMF. Table 1 shows a brief explanation of the factors used in the screening process.

Table 1. Insertion of factor names and values according to types of factors in FFD

	Name	Units	Type	Low	High
A	Weight Ratio PVC	-	Numeric	0.15	0.20
B	Weight Ratio GO	-	Numeric	0.02	0.04
C	Pressure	bar	Numeric	1	3
D	Type of solvent	-	Categoric	NMP	DMF
E	Time of membrane in water	S	Numeric	300	600

Response values are inserted after taking note of conditions according to 5 factors with a total of 32 runs through experimental runs of experiment. From there, the R-squared will be examined where if it is more than 0.8 (>0.8), it is for bioprocess such as microbial fermentation and many more, while if it is more than 0.9 (>0.9), it is for chemical process. Coefficients will be checked, where for positive coefficients shows positive effect, while negative coefficients show negative effect. Final equation in terms of coded factors will be formed. Interaction graph for each factor interacting will be formed.

2.3. Gas Permeation Test using CO_2 and CH_4 gases

This study involves gas permeation test with a specific result from the performance of the MMMs. It is also to study whether the MMMs can separate the two gases without being ruptured. A gas permeation device was used in this study. Figure 1 shows the entire set up of the gas permeation experiment. Before the MMMs was put inside the device. The flat sheet MMMs was cut into 5cm in diameter circles. Pressure was set to according the runs provided from the screening table where some runs uses 1 bar of pressure while the rest were 3 bar of pressure. When the valve was opened for air flow at specific pressures, gases enter the device and through the MMMs for permeation and finally reaching the burette as gas bubbles. Time was started once the burette pipe is opened. Time was taken until it reached 0 ml reading at the top of the burette.

Figure 1. MMMs inside Gas Permeation Device.

Given below are equation (1) and equation (2), where equation (1) that was used for the calculation of the permeability of the MMMs, while equation (2) is the MMMs selectivity. Selectivity determines the ability of the MMMs to separate 2 gases, which were CH_4 and CO_2.

$$\frac{P}{l} = \frac{V}{At\Delta p} \quad (1)$$

$$\alpha_{CO2/CH4} = P_{CO2}/P_{CH4} \quad (2)$$

Where P is permeability ($cm^3/s.\ cm^2.cm\ Hg$), α is selectivity, l is membrane skin layer thickness (cm), Q is measured volumetric flow rate (cm^3/s), A is Effective membrane area (cm^2), ΔP is pressure differences across the membrane (cm Hg). Equation (2) is the selectivity of membrane to separate 2 gases. P_{CO2} represents the permeability of CO_2 whereas, P_{CH4} represents the permeability of CH_4.
[1 Barrer = 1X10-10 $cm^3/s.cm^2.cm$ Hg]. Permeability should be converted into Barrer unit.

3. Results and Discussions

3.1. Screening using 2-Level Factorial Design

As stated earlier, MMMs has been produced according to 5 factors with specific criteria. These 5 factors have been inserted into the Design Expert software via 2-Level Factorial Design to achieve the experimental design table (table 2). From table 2, a total of 32 runs consisting of 5 factors in random arrangement has been distributed in the table. Each run has its own specific conditions according to all 5 factors introduced earlier. Experimental run was carried out via preparation of MMMs and gas permeation. Both permeability values from different types of gases were calculated to obtain the selectivity value which would then be added into the response section of the experimental table. Table 3 and 4 are the raw data and permeability values for each CO_2 and CH_4 gases respectively. A research done by Narenderan et al., six factors made up of volume of extracting solvent, the volume of dispersant/eluting solvent, extraction time, salt concentration, the flow rate of the sample, the volume of sample solution were considered in this study to determine the most significant factors. The results obtained were evaluated by ANOVA at 5 % significance level and it was observed that the factors, volume of dispersant/eluent solvent, volume of extraction solvent and flow rate of the sample showed statistical significance ($p = 0.05$) which was further investigated using CCD. The other factors (extraction time, salt effect and breakthrough volume in SPE) were not found to be significant ($p > 0.05$) [21].

Based on table 2, the highest selectivity is by Run 26 and the lowest selectivity is by Run 20. By comparing these 2 runs, we can see that in both Run 26 and Run 20 has the weight ratio of PVC and GO which are 20% and 4% respectively. However, when both runs are compared in factor 3 which is the pressure during gas permeation test, Run 26 used 1 bar of pressure whereas Run 20 used 3 bar of pressure. Another difference can be seen is in the categoric factor where Run 26 uses NMP whereas Run 20 uses DMF to prepare dope solution. Another difference was in the categoric factor where run 26 uses NMP whereas run 20 uses DMF to prepare dope solution. As stated earlier in Chapter 2 on the effects of solvent upon membrane performance, a study by Imtiaz Ali et al. shows PVDF homopolymer was blended with PVDF-co-HFP copolymer used different solvents, namely NMP, THF, and DMF solvents, were used to fabricate blended PVDF flat sheet membranes without the introduction of pore forming agent, via NIPS technique. Permeability of the membranes increased with the increase in overall content of PVDF. Mixed-solvents significantly improved permeability of membrane. This is because mixed-solvents undergo sonication where it is to ensure high degree of filler dispersion. [22]. The fifth factor was also the same in terms of the time of membrane immersed in water where both runs used 600s for immersion of MMMs inside water before left for drying. Table 3 is the calculation of selectivity based on the permeability values of both CO_2 and CH_4 gases from all 32 runs.

IOP Conf. Series: Materials Science and Engineering **702** (2019) 012041 doi:10.1088/1757-899X/702/1/012041

Table 2. Experimental Run Table (Design Expert using FFD)

Run	Factor 1 A: Weight Ratio PVC	Factor 2 B: Weight Ratio GO	Factor 3 C: Pressure (bar)	Factor 4 D: Type of Solvent	Factor 5 E: Water Contact of MMMs (s)	Response Selectivity
1	0.20	0.02	3.00	NMP	600.00	25.30
2	0.15	0.04	1.00	NMP	600.00	25.11
3	0.15	0.02	1.00	DMF	600.00	23.19
4	0.15	0.02	1.00	NMP	600.00	24.00
5	0.20	0.04	1.00	DMF	300.00	22.99
6	0.15	0.04	1.00	DMF	300.00	25.23
7	0.15	0.02	1.00	DMF	300.00	23.34
8	0.20	0.04	3.00	NMP	300.00	24.07
9	0.15	0.02	3.00	NMP	300.00	23.55
10	0.15	0.04	1.00	DMF	600.00	24.07
11	0.15	0.04	3.00	DMF	600.00	25.76
12	0.15	0.04	3.00	NMP	300.00	24.97
13	0.20	0.04	1.00	DMF	600.00	25.19
14	0.20	0.04	3.00	NMP	600.00	23.49
15	0.20	0.02	1.00	NMP	300.00	25.77
16	0.15	0.04	3.00	NMP	600.00	24.52
17	0.15	0.02	3.00	NMP	600.00	23.00
18	0.20	0.02	3.00	DMF	300.00	23.24
19	0.15	0.02	3.00	DMF	300.00	24.41
20	0.20	0.04	3.00	DMF	600.00	22.87
21	0.15	0.02	1.00	NMP	300.00	23.78
22	0.20	0.02	1.00	DMF	600.00	23.90
23	0.20	0.04	1.00	NMP	300.00	26.09
24	0.15	0.04	3.00	DMF	300.00	25.40
25	0.20	0.02	1.00	NMP	600.00	22.97
26	0.20	0.04	1.00	NMP	600.00	23.09
27	0.15	0.04	1.00	NMP	300.00	24.41
28	0.20	0.04	3.00	DMF	300.00	23.70
29	0.20	0.02	1.00	DMF	300.00	24.33
30	0.15	0.02	3.00	DMF	600.00	25.06
31	0.20	0.02	3.00	NMP	300.00	23.07
32	0.20	0.02	3.00	DMF	600.00	25.99

Table 3. Permeability and Selectivity for CO_2 and CH_4 gases.

Runs	Permeability of CO_2 gas. P_{CO2}	Permeability of CH_4 gas, P_{CH4}	Selectivity (P_{CO2}/P_{CH4}), α
1	455669.61	18009.58	25.30
2	409900.34	16319.89	25.11
3	406972.21	17548.78	23.19
4	501087.10	20876.28	24.00
5	577517.57	25115.46	22.99
6	396705.47	15718.76	25.23
7	614722.38	26331.06	23.34
8	625567.95	25987.80	24.07
9	320354.56	13600.91	23.55
10	248502.67	10321.25	24.07
11	282851.37	10978.46	25.76
12	385499.58	15437.89	24.97
13	225246.54	8938.56	25.19
14	502819.15	21403.29	23.49
15	361823.06	14035.35	25.77
16	711494.30	29005.30	24.52
17	375186.35	16310.81	23.00
18	508741.18	21887.27	23.24
19	422382.98	17302.83	24.41
20	488666.65	21365.93	22.87
21	188337.73	7919.30	23.78
22	274432.91	11479.57	23.90
23	419568.87	16078.45	26.09
24	470500.31	18518.60	25.40
25	256931.77	11184.80	22.97
26	225481.85	9764.62	23.09
27	647419.36	26515.10	24.41
28	434687.22	18336.20	23.70
29	321829.98	13225.03	24.33
30	405962.36	16196.31	25.06
31	633181.11	27438.35	23.07
32	408772.15	15726.84	25.99

Figure 2 shows the Robeson's plot of CO_2/CH_4 using MMMs. The dotted round circles are the experimental runs done during the screening phase of this study. All 32 runs have achieved high performance membrane as all are above the present and prior upper bound lines of the graph. A study done by Fang et al. shows membranes homopolymer poly (PFMMD) and copolymers poly (PFMMD-co-PFMD) with varied compositions was tested with N_2/CH_4, H_2/CH_4, He/CH_4, and CO_2/CH_4 gas separations where each upper bound, the points lie near the line hence, showing a promising performance from the membrane. Even though it does not show high performance, even more impressive results are obtained when PFMD is added to produce PFMMD–PFMD copolymers. These solution cast copolymers have He/CH_4 selectivity/permeability combinations that place them well above the 2008 upper bound [4], with selectivities higher than even the melt-pressed Hylon AD60 [23].

IOP Conf. Series: Materials Science and Engineering **702** (2019) 012041 doi:10.1088/1757-899X/702/1/012041

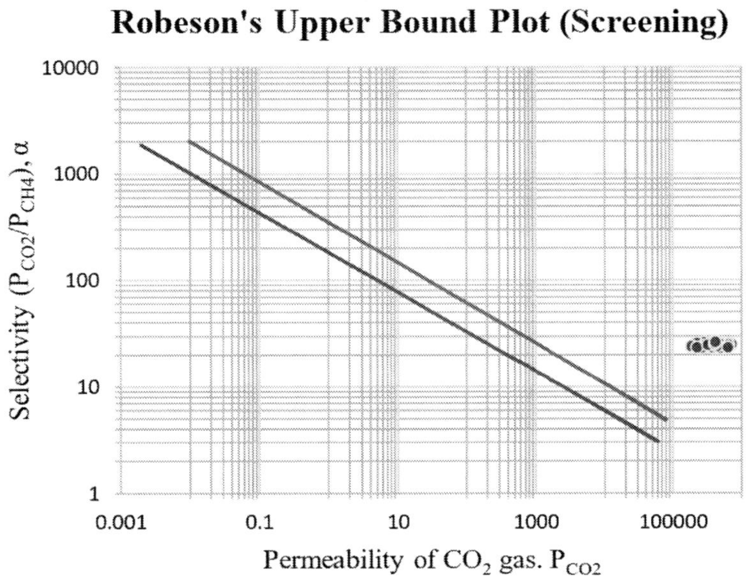

Figure 2. Robeson's Plot of CO_2/CH_4

3.2. Effect Lists of Factors and Pareto Chart

Even though all MMMs from the experimental runs are considered to be high performance via the Robeson's plot (figure 4), there are still more to study in terms of the factors affecting each run and how do they relate to each other. After calculating the permeability and selectivity of each runs, the experimental run table is further proceeded into the effects list of each factors individually and includes the interactions between factors. Table 6 shows the effect these factors individually and interactions between factors throughout the experiment and how they affect the experimental run.

Based on Table 4, as stated earlier where the 5 factors are weight ratio of PVC (A), weight ratio of GO (B), pressure used during gas permeation (bar) (C), type of solvent used (D), and time of membrane immersed in water (s) (E) interact among each other and contribute to each run in a certain percentage. When all 5 factors are compared individually, factor B shows the highest percentage of contribution at 3.66 %, and factor E shows the least percentage of contribution at 0.0072 %. When 2 factors interact, factor AB shows the highest at 14.83 and interaction of factor AE shows the least at $9.63X10^{-4}$ %. A study by Yanbin Cui et al., states that graphene layers in the polymer matrix are capable of producing a tortuous path, which acts as a barrier for gases. A high tortuosity leads to superior barrier properties and lower permeability of PNCs. The influence of the intrinsic properties of these fillers (graphene and its derivatives) and their state of dispersion in polymer matrix on the gas barrier properties of graphene/PNCs is discussed [24]. Hence this shows GO does affect the performance with the most significance.

Table 4. Effect List of Factors and Percentage Contribution (%)

Term	Percentage Contribution (%)
A - Ratio PVC	1.39
B - Ratio GO	3.66
C - Pressure	0.083
D - Type of Solvent	0.22
E - Water contact	0.072
AB	14.83
AC	3.74
AD	2.23
AE	9.63E-04
BC	1.39
BD	0.64
BE	2.15
CD	5.46
CE	6.34
DE	5.74

The effect of these 5 factors were analyzed with selectivity, where the Pareto chart is shown in figure 3. It is clearly seen that the interaction between weight ratio of PVC (A) and weight ratio of GO (B) shows the highest effect against selectivity, which is 93.11 while the lowest one is interactions of pressure used during gas permeation (bar) (C), type of solvent used (D), and time of membrane immersed in water (s) (E) (CDE), which is 24.86. From an individual standpoint, factor B shows the highest effect against selectivity at 46.54, while factor E shows the lowest effect against selectivity which lies slightly above the t-value limit line. The orange colored of bar shows that the effect possessed positive effects while the blue one has negative effects. A study done by Mah et al. [25] shows the Pareto chart where the bar lengths are proportional to the absolute value of the estimated effects, which helps to compare relative importance of the effects. The value of the Student's test parameter for $p = 0.05$ (95 % confidence level) and seven degrees of freedom (df) was 2.36. Thus, a t-value for the model coefficient which surpasses the critical value of 2.36 is considered to be statistically significant over the range of analytical response at the 95% confidence level. Four of the independent variables, namely TEOA concentration, TMC concentration, reaction time, and curing were statistically significant. Hence, these factors had major influence on the response within the limits of studied levels except pH of the aqueous solution. The interaction effect between TMC concentration and curing was the only two-way interaction effect that is statistically significant. The insignificance of effects does not mean that these factors are unimportant, but just implies a little influence on response. From the pareto chart, it is shows that more factors lead to negative effects compared the ones that lead to positive effect, however since all of the effect value of factors are all higher than the limits which is at 4.30, hence all factors are considered significant.

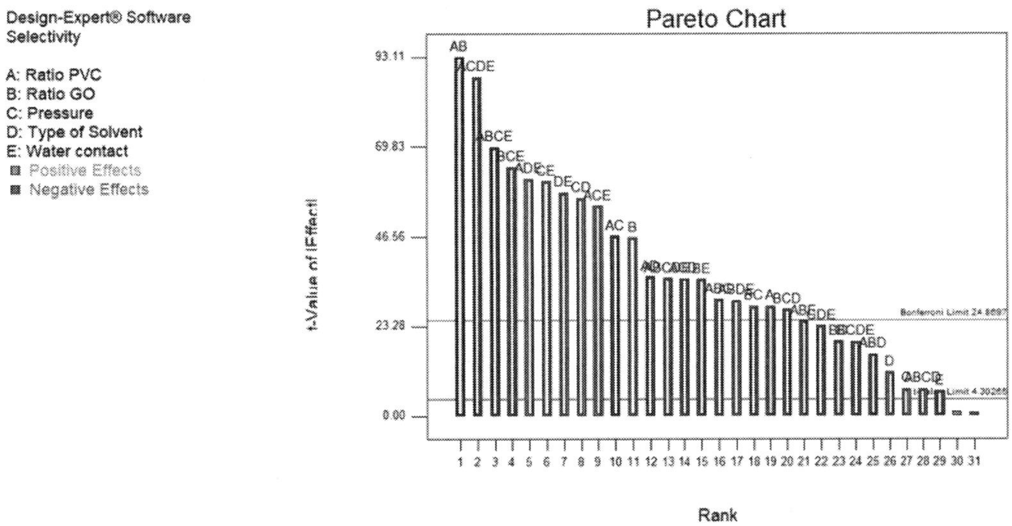

Figure 3. Pareto Chart from MMMs using 5 factors in FFD

3.3. Analysis of Variance (ANOVA) and Empirical Model Analysis

ANOVA is a statistical technique that separates the total variation in a set of data into component parts correlated with specific sources of variation. It is commonly for the purpose of testing hypothesis on the parameters of the model. In this experiment, ANOVA is used to test the statistical significance of the ratio of mean square variation which caused by regression and to test the mean square residual error. By using significance of factor (SOF) and R-squared (R^2) test, this research outlook two statistical point of view to analyse and asses the model of the experimental data. The goodness-of-fit is validated when the coefficient R^2 has the tendency to be closer to unity and when predicted is in agreement with the adjusted R^2. The R^2 for the analysed model must be more than 0.9 [26]. Table 5 show the result for SOF and the interaction for the selectivity of CO_2/CH_4 with the value of R^2. The R^2 for the analysed model must be more than 0.9. In table 5, the original value of R^2, adjusted R^2 and predicted R^2 are demonstrated after neglecting the insignificance terms of the design model. The predicted R^2 is 0.9912 is in reasonable agreement with adjusted R^2 because the value varied by 0.0088. In other words, the design model of the experiment is in an accurate description of experimental data which indicated the relationship between the variables and response data. The model result shows the tendencies for the model to form linear regression fit which showed that the experimental research range is adequate.

Table 5. R^2 statistic for the fitted model

Model Source	Selectivity of CO_2/CH_4
Std. Dev.	0.023
Mean	24.25
Coefficient of variation	0.096
R-Squared	1.0000
Adj R-Squared	0.9995
Pred R-Squared	0.9912
Adeq Precision	142.90

3.4. Interaction Factors, Model Graphs and Effect on Response (Selectivity)

One interaction factor only gives effect to the response (selectivity) and sometimes lack of interaction among another factor's information. However, based on earlier analysis and calculation, there are still interactions between other factors. Multiple interaction factors which requires further in-depth analysis and a better understanding can be seen from figure 4. Multiple interaction factors can sometimes relate two or more interaction factors for one response.

Based on figure 4, these multiple interaction factors can be divided into 5 parts. For the first part, the 3D graph is a plot for multiple interaction factors AB and response (selectivity), which are made up of both weight ratios of PVC and GO interaction factors. Other multiple interaction factors such as ABC and, ABD are also included in this graph section because 3D graph requires 3 axes where they are made up of factor A, B and response, so the rest of the factors which are linked to factors A and B which are also included in this part. From the 3D graph of part 1, the selectivity increases with an increase in weight ratio of PVC and GO. Further analysis shows the contour of 3D graph for part 1 where, more greenish color can be seen in the section where the weight ratio of PVC and GO are higher, and at that same greenish area was where the selectivity were higher. The second part of the table shows the multiple interaction factors AC and response (selectivity), which are made up of factors weight ratio of PVC (A) and pressure used during gas permeation (bar) (C). Other multiple interaction factors are also included in the AC to response plot, such as ACD and ACE, but as stated earlier, 3D graph requires 3 axes where they are made up of interaction factors A, C and response, so the rest of the factors which are linked to factors A and C which are also included in this part. The 3D plot in part 2 shows that even though there is an increase in weight ratio of PVC, an increase in pressure will reduce the selectivity. The contour of the 3D plot for part 2 shows that the bottom is right very yellowish because that area shows the highest response (selectivity), where in that area, the weight ratio of PVC is higher and pressure is low, hence, achieving higher selectivity compared to the upper part of the contour where the greenish area shows a higher pressure and lower selectivity, at increasing weight ratio of PVC. Part 3 shows the 3D plot of multiple interactions of BC and response (selectivity). Interaction factors BC are weight ratio of GO (B) and pressure used during gas permeation (bar) (C). In part 3, other than BC, the multiple interaction factors that come under this model graph are BCD and BCE. Only interaction factors B and C are in the 3D plot axes with response (selectivity) whereas the rest of the interaction factors are only linked to the stated interaction factors which B and C. Based on the 3D plot in part 3, an increase in GO effects the increase in selectivity. However, increase in pressure at low GO weight ratio shows lower selectivity. This can be seen in the contour plot of part 3, where the top left corner of the plot is blueish in color where that area has low selectivity. Even though the pressure is high, but when the weight ratio of GO is increased, the selectivity increases which can be seen on the top right of the plot where it is a bit greenish in color showing that the selectivity is much higher. The highest number of selectivity can be found at the bottom right of the plot where the plot looks slightly yellowish and it is at lower pressure and higher GO weight ratio. Part 4 of figure 4 shows model graph for the multiple interaction factors of BE with response (selectivity), interaction factors B and E are, weight ratio of GO (B) and time of membrane immersed in water (s) (E). Based on the 3D plot in part 4, at higher time of membrane immersed in water, the selectivity decreases. At the lowest weight ratio of GO, selectivity increases as time of membrane immersed in water decreases. Selectivity shows the highest at high GO weight ratio and low time of membrane immersed in water. From the contour plot of part 4, at low GO weight ratio, the top left corner shows high time of immersion but low selectivity. However, as selectivity does increase when time of immersion is reduced and weight ratio is increased where at the bottom right of the plot, the yellowish area represents high selectivity, high GO weight ratio and low time of immersion. Finally, part 5 is model graph for the multiple interaction factors of CE and response (selectivity). The interaction factors involved are pressure used during gas permeation (bar) (C) and time of membrane immersed in water (s) (E). From the 3D plot of part 5, at higher time of immersion in water and lower pressure, selectivity is low. At low pressure and time of immersion in water, selectivity is at its highest in the 3D plot. Based on the contour plot of part 5, the bottom left shows yellowish in color because that area shows the highest selectivity at low pressure and low time of immersion. On the top right of the plot where it is slightly blueish in color, selectivity is at its lowest when pressure and time of water immersion at its highest.

In another study related to screening of MMMs using DoE can be seen in Back et al. research. In that research, seven-channel capillary membranes were fabricated in a steam–dry–wet spinning process, while varying the composition of the polymer solution and the process temperatures in a three-level fractional factorial linear screening design. The polymers PVDF was the chemically resistant main polymer and PVP was added as hydrophilic co-polymer. In the model graph with interaction factors, it showed that the concentration of the main polymer PVDF and the molecular weight of the co-polymer

PVP showed linear relations with both permeability and retention. The obtained membranes may be suitable for micro/ ultra-filtration and, together, demonstrate the merits and limitations of DoE for multi-channel membrane screening. The permeability could be increased using sodium hypochlorite post-treatment, although retention was slightly compromised [26].

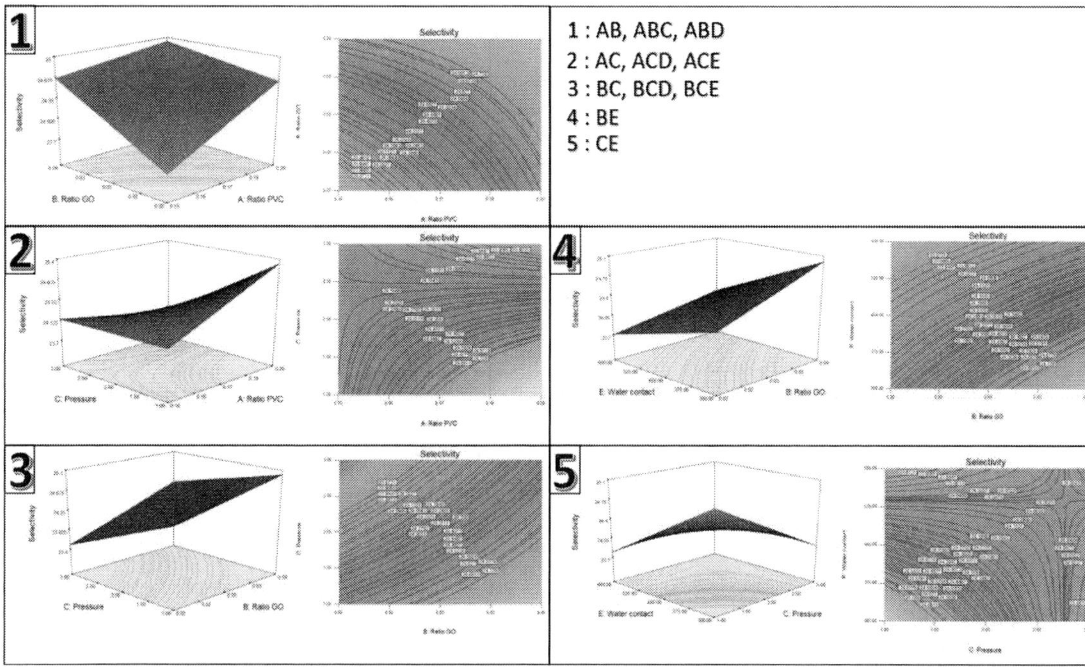

Figure 4. Model Graphs using Interaction Factors from Screening (3D Plot)

4. Conclusion

Based on this study, MMMs with factors of weight ratio of GO and PVC shows the most contribution and highest in the effect list. The interaction between both factors also show the highest in terms of contribution. Type of solvent which are NMP and DMF also show the third most contribution. Based on the 3D plot without considering the categoric factor which is type of solvents, weight ratio PVC and GO show to have the highest selectivity when both have the higher weight ratio, followed by lower pressure and shorter time of membrane immersed in water. From the screening table, run 23 shows the highest selectivity while run 20 shows the lowest selectivity at 26.0951 and 22.8713. In both runs, run 23 has lower pressure at 1 bar while run 20 is carried out at 3 bar of pressure. Run 23 uses solvent NMP while run 20 uses DMF. Run 23 immersed MMMs at 600s in water while run 20 at 300s in water. However, both runs use weight ratio PVC of 20% and weight ratio GO of 4%. Hence, factors weight ratio of PVC, weight ratio of GO and type of solvents contribute to the highest selectivity and also shows highest response during interaction among factors.

Acknowledgement

The authors would like to thank Universiti Malaysia Pahang for funding this project using university grant RDU1803113.

References

[1] Jusoh N, Yeong Y F, Chew T L, Lau K K and Shariff A M 2016 Current development and challenges of mixed matrix membranes for CO_2/CH_4 separation *Separ. Purif. Rev.* 45 321-44
[2] Nasir R, Mukhtar H, Man Z and Mohshim D F 2013 Material advancements in fabrication of mixed-matrix membranes *Chem. Eng. Technol.* **36** 717-27
[3] Robeson L M 1991 Correlation of separation factor versus permeability for polymeric membranes

J. Membr. Sci. **62** 165-85

[4] Robeson L M 2008 The upper bound revisited *J. Membr. Sci.* **320** 390-400

[5] Goh P S, Ismail A F, Sanip S M, Ng B C and Aziz M 2011 Recent advances of inorganic fillers in mixed matrix membrane for gas separation *Separ. Purif. Technol.* **81** 243-64

[6] Khan A L, Klaysom C, Gahlaut A, Khan A U, Vankelecom I F J 2013 Mixed matrix membranes comprising of Matrimid and -SO3H functionalized mesoporous MCM-41 for gas separation J. *Membr. Sci.* **447** 73-79

[7] Vu D Q, Koros W J and Miller S J 2003 Mixed matrix membranes using carbon molecular sieves - I. Preparation and experimental results *J. Membr. Sci.* **211** 311-34

[8] Ebadi Amooghin A, Mashhadikhan S, Sanaeepur H, Moghadassi A, Matsuura T and Ramakrishna S 2019 Substantial breakthroughs on function-led design of advanced materials used in mixed matrix membranes (MMMs): A new horizon for efficient CO_2 in separation *Progress in Materials Science* **102**

[9] Yong W F, Li F Y, Xiao Y C, Li P, Pramoda K P, Tong Y W and Chung T S 2012 Molecular engineering of PIM-1/Matrimid blend membranes for gas separation *J. Membr. Sci.* **407–408** 47-57

[10] Ryan T A, Lee J S, Bae T -H, Jason K W, Johnson J R, Christopher W J, Sankar N, and William J K 2011 CO_2–CH_4 permeation in high zeolite 4A loading mixed matrix membranes *Journal of Membrane Science* **367**(1–2) 197-203

[11] Bao-Sheng G, Wang T, Hai-Xiang S, Gao W and Hao-Ru Z 2018 Preparation of mixed matrix membranes based on polyimideand aminated graphene oxide for CO_2 separation *Polymer Advanced Technology* **29**(4) 1334-43

[12] Farahania M H, Rabieeb H and Vatanpour V 2019 Comparing the effect of incorporation of various nanoparticulate on the performance and antifouling properties of polyethersulfone nanocomposite membranes *Journal of Water Process Engineering* **27** 47-57

[13] Ahmad A L, Ramli W K W, Fernando W J N and Daud W R W 2012 Effect of ethanol concentration in water coagulation bath on pore geometry of PVDF membrane for membrane gas absorption application in CO2 removal *Separation Purification Technology* **88** 11-18

[14] Ahmad N A, Mohd Noh A N, Leo C P and Ahmad A L 2017 CO_2 removal using membrane gas absorption with PVDF membrane incorporated with POSS and SAPO-34 zeolite *Chemical Engineering Research and Design* **118** 238-47

[15] Farahani M H, Dan H and Tai-Shung C 2018 Cross-linked mixed matrix membranes (MMMs) consisting of amine-functionalized multi-walled carbon nanotubes and P84 polyimide for organic solvent nanofiltration (OSN) with enhanced flux *Journal of Membrane Science* **548** 319-31

[16] Qianyu L, Shaobin H, Yongqing Z and Shuaifei Z 2018 Comparing the antifouling effects of activated carbon and TiO_2 in ultrafiltration membrane development *Journal of Colloid and Interface Science* **515** 109-18

[17] Nele M, Vidal A, Bhering D L, Pinto J C and Salim V M M 1999 Preparation of high loading silica supported nickel catalyst: Simultaneous analysis of the precipitation and aging steps *Applied Catalysis A: General* **178**(2) 177-89

[18] Kusworo T D, Ismail A F and Mustafa A 2015 Experimental Design and Response Surface Modelling of Pi/Pes-Zeolite 4a Mixed Matrix Membrane for CO_2 Separation *Journal of Engineering Science and Technology* **10** 1116 – 30

[19] Molnár É, Varga T, Rippel-Pethő D, Bocsi R, Bobek J and Horváth G 2017 Optimization of the Sodium Hydroxide–Assisted Hydrogen Sulfide Selective Removal from Natural Gas *Chemical Engineering Communications* **204**(8) 974–81.

[20] Rabiee H, Farahani M H and Vatanpour V 2014 Preparation and characterization of emulsion poly (vinyl chloride)(EPVC)/TiO2 nanocomposite ultrafiltration membrane *J. Membr. Sci.* **472** 185–193

[21] Narenderan S T, Meyyanathan S N and Karri V V S R 2019 Experimental design in pesticide extraction methods: A review *Food Chemistry* **289**

[22] Ali I, Bamaga O A, Gzara L, Bassyouni M, Abdel-Aziz M H, Soliman M F and Albeirutty M 2018 Assessment of blend PVDF membranes, and the effect of polymer concentration and blend

composition *Membranes* **8**(1)
[23] Fang M, He Z, Merkel T C and Okamoto Y 2018 High-performance perfluorodioxolane copolymer membranes for gas separation with tailored selectivity enhancement. *Journal of Materials Chemistry A* **6**(2) 652–658
[24] Cui Y, Kundalwal S I and Kumar S 2016 Gas barrier performance of graphene/polymer nanocomposites *Carbon* **98** 313–333.
[25] Mah K H, Yussof H W, Abu Seman M N and M A W 2018 Polyester thin film composite nanofiltration membranes via interfacial polymerization: influence of five synthesis parameters on water permeability *Journal of Mechanical Engineering and Sciences* **12**(1) 3387–3398
[26] Back J O, Brandstätter R, Spruck M, Koch M, Penner S and Rupprich M 2019 Parameter screening of PVDF/PVP multi-channel capillary membranes *Polymers* **11**(3) 1–17

IOP Conf. Series: Materials Science and Engineering **702** (2019) 012042 doi:10.1088/1757-899X/702/1/012042

Characterization of bioactive compounds extracted from patchouli: Comparing microwave-assisted hydrodistillation and hydrodistillation methods

N H Abdurahman and S Sundarajan*

Faculty of Chemical & Natural Resources Engineering, Universiti Malaysia Pahang, 26300 Gambang, Pahang, Malaysia

*E-mail: ssgeetha18rajan@gmail.com

Abstract. Microwave-assisted hydrodistillation (MAHD) is a promising and advanced technology that uses microwave heating to extract essential oils from dry leaves. Currently available conventional methods are time consuming, thermally unsafe and require costly solvents. In the current study, MAHD and traditional hydrodistillation (HD) techniques were compared and their effectiveness in the separation of the essential oil from dry patchouli leaves have been studied. The patchouli oil that obtained from MAHD and HD extraction methods were characterized using gas chromatography-mass spectrometry (GC-MS) analysis and the impact of the two extraction methods on the plant morphology were observed using scanning electron microscopy (SEM) analysis. The results indicated that MAHD method extracted 5 % of essential oil at solid to solvent ratio of 1:12 g/ml from patchouli leaves as compared to 4 % obtained using HD method at 45 minutes of extraction time. In addition, MAHD method can retain 16 compounds while only 11 chemical compounds were extracted using HD. Thus, MAHD showed a greater potential for better production of the essential oil from patchouli leaves as compared to HD method.

1. Introduction

Essential oils are fragrant, slick and unstable concentrates from different parts of plants, for example, leaves, seeds, barks, and roots. Over 150 sorts of essential oils currently being exchanged in international markets [1]. Among these essential oils, patchouli oil has higher demand due to its various applications. Patchouli (*Pogostemon cablin*) oil can be extracted from the leaves of patchouli plants which are commonly cultivated in tropical areas especially Indonesia, Philippines [2] and Malaysia [3]. Patchouli oil has many applications in various industries. The oil plays an important role in Chinese traditional medicine [4]. It acts as antidepressant, antiseptic, antiphlogistic, and astringent [5-7]. These pharmacological properties might be due to several chemical compositions of the oil [8]. Moreover, patchouli oil can be used in the cosmetic [2] and fragrance industry to provide the base and lasting character to the fragrance because of its fixative properties [9]. Furthermore, the oil can be used as a fragrance in soaps, detergents and deodorants [10]. Currently the demand for the patchouli oil is about 587 tons per annum [4]. Since there is no synthetic replacement for the patchouli oil and new applications of the oil continuously being introduced, the interest for patchouli oil will be increasing annually and the price will remain high in future. Hence, to meet international demands and contribute to country's economic growth, an efficient extraction technology that is rapid, energy-efficient, thermally safe, and capable of recovering higher yield of oil is inevitable for the separation of patchouli essential oils.

Steam distillation, hydrodistillation and Soxhlet extraction are the conventional methods that are currently used in the separation of patchouli oil. In hydrodistillation, the plant sample to be extracted is soaked in solvent and then boiled using a heating mantle [11] while in steam distillation, steam passes through the bed of plant material to be extracted [9]. For Soxhlet extraction, the plant sample is set in a thimble while condensed solvent topped off the thimble and siphoned down into the container of solvent continuously until the oil has been extracted from solid in the thimble into the extracting solvent [12]. Yield of patchouli essential oil obtained from hydrodistillation was 2.61 % for 6 hours [4] and 3.4% for 4 hours of extraction [2] while 1.5 % of patchouli oil was obtained when using steam distillation for 2 hours [9]. When using Soxhlet extraction for one hour, 0.84 % of patchouli oil was obtained [13]. However, the traditional methods possess several drawbacks such as long extraction time, cause thermal degradation, consumption of high energy, and give a low yield of patchouli oil [12, 14]. Most of the conventional techniques employed in the extraction of oil are being conducted at high temperatures, hence the chemical compounds might have been degraded which can reduce the quality of oil. Since conventional techniques require longer time of extraction, the amount of heat absorbed by the solvent will be higher which can lead to high temperature [15]. Thus, an advanced extraction technique that can extract higher amount of essential oil at shorter time need to be studied.

In this study, MAHD was studied for its effectiveness in extracting oil from patchouli leaves. Fast heating is one of the advantages of microwave extraction because it can heat up the plant materials simultaneously at higher rate [11]. MAHD had been reported to recover higher yield of essential oils from Mentha longifolia L., rosemary and damask rose [16-18]. According to [4], the obtained yield of patchouli essential oil was 2.72 % when using microwave extraction for two hours while 2.61% of patchouli oil was obtained after six hours of hydrodistillation. This shows that higher amount of patchouli oil can be obtained at shorter time when using microwave extraction. The results may vary from this project because patchouli leaves were obtained from Sabah, Malaysia for this current experiment. Cultivation region can change the qualitative and quantitative properties of patchouli leaves [9]. In addition, [19] reported that it took 15 minutes for the full recovery of essential oil from orange peels when using MAHD method while [20] obtained 5% of essential oil from Tongkat Ali when using 20 min of MAHD. Therefore, in this research, 60 minutes was used for MAHD to study the effect of longer extraction time on the yield of patchouli oil. Although methanol [21-22], CaO catalyst [22] and ethanol [21] can accelerate the extraction process, only distilled water was used as a solvent for this study to promote an environmentally-friendly extraction method. Thus, this study investigated the potential of MAHD as the alternative extraction method compared with HD method in the separation of oil from patchouli leaves. The effects of MAHD and HD parameters (extraction time and solvent to feed ratio) on the recovery of oil were studied. Moreover, the chemical compositions of the extracted oil at the highest extraction parameters for both methods were examined using GC-MS and the influence of the extraction techniques on the morphological nature of patchouli leaves before and after extraction were as well investigated using SEM analysis.

2. Materials and Methods

2.1. Raw Material

Dried patchouli leaves were obtained from Gaya Naturals Company, Tawau, Sabah. Before starting the experiment, the leaves isolated from stems and dried using oven for four days. Then, the leaves were blended using a grinder into fine powder.

2.2. Moisture Content of Patchouli Leaves

The moisture content was determined by weighing 5g of dry patchouli leaves, the plant sample was further dried for 4 days at room temperature until stable weight was attained. After 4 days of drying, the dry patchouli leaves were weighed again and the moisture content was calculated using equation (1).

$$Moisture\ content\ (\%) = \frac{Weight\ before\ drying - Weight\ after\ drying}{Weight\ before\ drying} \times 100 \quad (1)$$

2.3. Operating Procedure of Essential Oil Extraction Using MAHD and HD

A modified domestic microwave extractor comprising of power and time controller was used for this experiment. Figure 1 (a) shows the set up for the MAHD study. 30 g of patchouli powder was added to 180 ml of distilled water in a flat-bottomed glass flask. According to a previous study by [20], 250W was used for MAHD and the highest yield of essential oil obtained was 5% after 20 minutes of extraction. Thus, to study the effect of higher microwave power on the yield, the microwave power and time of extraction were kept constant at 400W and 60 minutes respectively. After one hour, the hydrosol (water + essential oil) was gathered from the Clevenger. The hydrosol was placed in a separating funnel to separate essential oil from water. The oil was filled in a vial and stored at -4 °C for further analysis. Essential oils need to be stored at below -4 °C to prevent oxidation. The experiment was repeated for 240, 300, 360, and 420 ml of distilled water.

The experiment for HD [figure 1 (b)] was carried out in the same manner as MAHD, however instead of modified microwave, heating mantle was used. 30 g of patchouli powder was added to 180 ml of distilled water in a flat-bottomed glass flask. The experiment was conducted for three hours and was repeated using 240, 300, 360, and 420 ml of distilled water. The extracted oil was filled in a vial and stored at -4 ° C until further analysis. According to [27] research, highest yield (1.3%) was obtained at 150 minutes using HD, thus 3 hours of extraction was used for this project to study the effect of longer extraction time on the yield of patchouli oil.

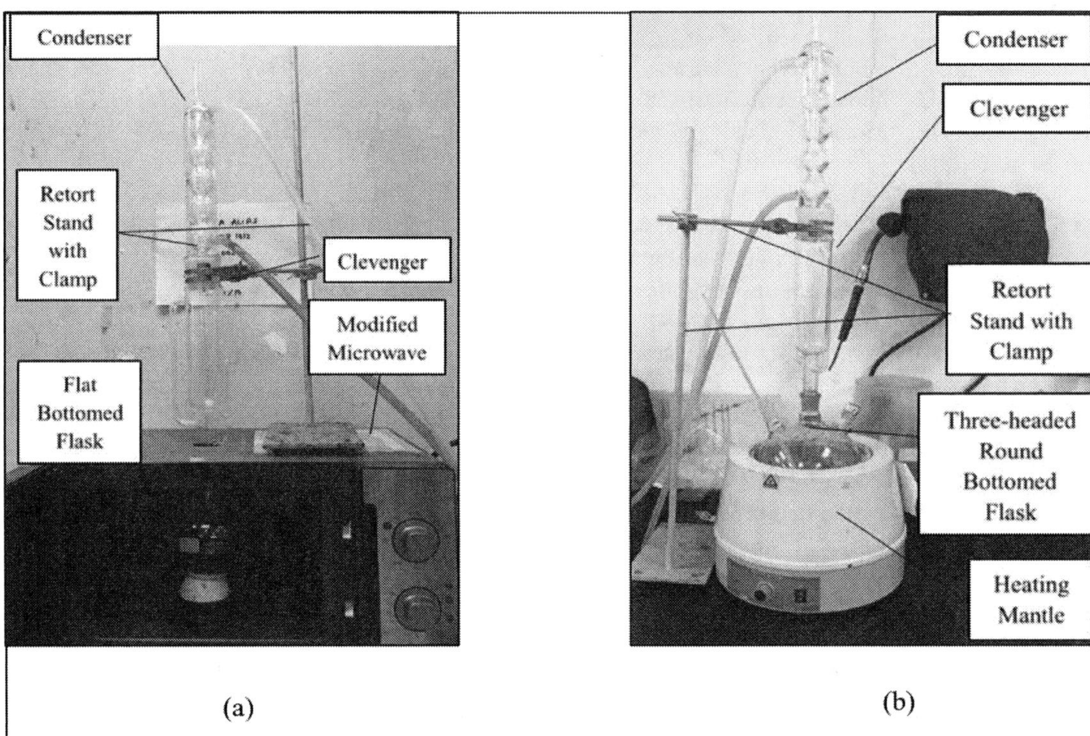

Figure 1. (a) MAHD set up and (b) HD set up.

2.4. Yield of Patchouli Essential Oil

The yield of patchouli essential oil was calculated as shown below:

$$\textit{Mass of patchouli essential oil (g)} = \textit{Density } (\rho) \times \textit{volume (ml)} \quad (2)$$
$$\textit{Density of patchouli oil} = 0.963\ g/ml$$
$$\textit{Yield} = (\textit{Weight of essential oil (g)}) / (\textit{Weight of plant sample (g)}) \times 100\ \% \quad (3)$$

2.5. GC-MS Analysis
GC-MS analysis of the extracted oil was carried out using Hewlett–Packard 6890 gas chromatograph coupled to a 5973A mass spectrometer, comprised of two fused-silica-capillary columns with different stationary phases. HP5MS™ was used as a non-polar column (30 m length, 0.25 mm diameter and 0.25μm film thickness) and Stabilwax ™ as polar column consisting of Carbowax™-PEG (60 m length, 0.25 mm diameter and 0.25μm film thickness). The following conditions were used to obtained GC-MS spectra: Helium as the carrier gas; flow rate of 1.0 ml/min; split ratio of 1:50; injection volume of 1.0μL; injection temperature of 300 °C; oven temperature progress from 100 to 250 °C at 10 °C/min; the ionization mode used was electronic impacted at 70 eV. The identification of chemical compounds in the patchouli oil was performed by comparing the mass spectral fragmentation patterns with those stored in MS database (National Institute of Standards and Technology and Wiley libraries).

2.6. SEM Analysis
SEM analysis of raw dried patchouli leaves, residue of patchouli sample where highest yield was obtained for MAHD and HD method. The samples were fixed on the specimen holder and then sputtered with gold in a sputter coater (BAL-TEC SCD 005, Balzers, Switzerland). All the specimens were examined with a Philips XL 30 scanning electron microscope under high vacuum condition and at an accelerating voltage of 20.0 keV and at a working distance of 8–9 mm.

2.7. Statistical Analysis
Each experimental run was performed in triplicate, and average values were recorded as mean ± standard deviation.

3. Results and Discussion

3.1. Effect of Solvent Volume on the Patchouli Oil Yield
Figure 2 shows the results obtained for the influence of varying solvent volume (180-420 ml) on patchouli oil yield. The microwave power and extraction time were fixed at 400 W and one hour for MAHD, respectively. However, the extraction time was fixed at three hours for HD method. From Figure 2, the highest yield of patchouli oil was acquired when 360 ml of solvent (distilled water) was used. At ratio of 12:1 ml/g, the highest yield of oil was extracted for both MAHD and HD methods. However, when using MAHD, the yield of patchouli oil obtained was higher (5 %) compared to HD (4 %). Higher volume of solvent caused more heat absorption that transferred to plant samples; hence, higher yield can be extracted from the oil gland [23]. In addition, solvent act as a pumping agent that can extract the essential oil from oil glands of the plant sample. [24] had reported 1.9437 % of the yield of essential oil obtained at the solid to solvent ratio of 1:10 g/ml using microwave extraction. The oil yield obtained in this current study (5 % at ratio of 12:1 ml/g) was higher compared to the previous study. The yield of patchouli oil increased when solvent volume increased. Although the volume of extracted essential oil from both extraction methods were different, the composition percentage of oils were similar (Table 1). The number of compounds identified when using both extraction methods depend on the ability of the extraction technique to extract the bioactive compounds from the oil glands. However, beyond 360 ml of solvent, the yield of patchouli oil declined because once the solvent hit the maximum volume use, each extraction technique could reach maximum oil yield [25].

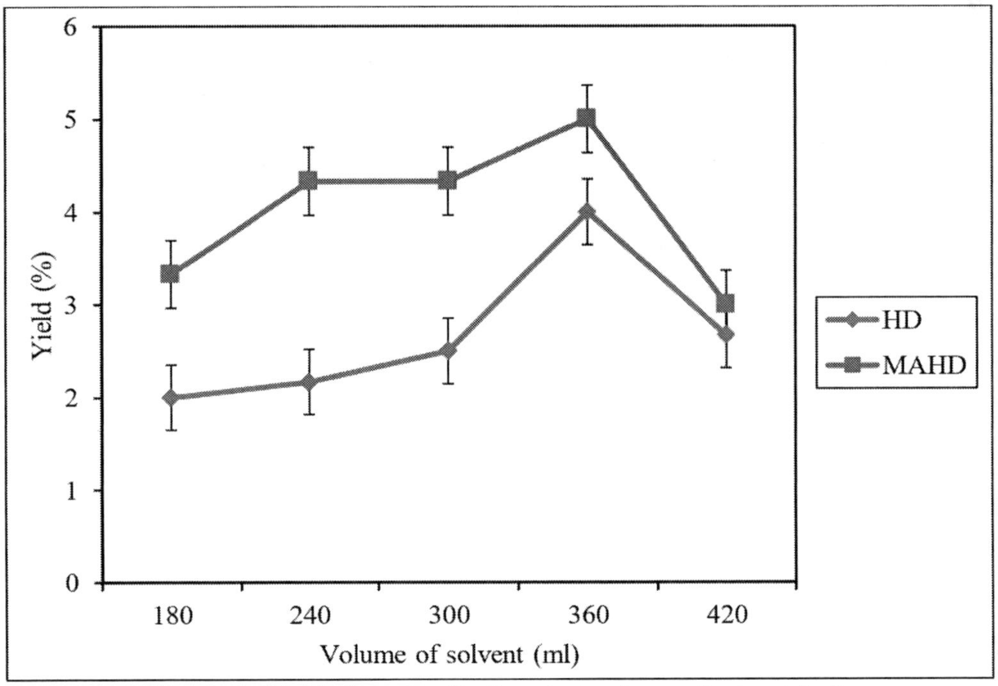

Figure 2. Yield of patchouli oil against volume of solvent (ml).

3.2. Effect of Extraction Time on the Patchouli Oil Yield

Figure 3 shows the effect of extraction time between 15-60 min on the yield of patchouli leaves at fixed microwave power of 400 W and solvent to solid ratio of 12:1 ml/g for MAHD and extraction time of three hours at fixed solvent to sample ratio of 12:1 ml/g for HD. According to [11], 4 hours and 2 hours of extraction time used for HD and MAHD methods respectively to extract essential oil from *Thymus vulgaris* L. Although 2 hours of extraction time was used for MAHD by [11], in this study, 1 hour of extraction time was used because all the water in the flask was evaporated and the extraction was stopped after 60 minutes of extraction. On the other hand, after 3 hours of extraction of patchouli oil using HD method, there is no essential oil extracted as shown in Figure 3. These are due to the difference of dielectric constants of the solvent and the sample [15]. The highest yield of patchouli oil was obtained at 45 minutes for both MAHD and HD. It can be obviously observed that the yield acquired utilizing MAHD was higher (2.33 %) compared to HD (0.9 %). This might be because of the different heat and mass gradients in the extraction techniques [15]. In MAHD, the heat and mass gradients work in the similar direction, whereas in HD, the gradients are working in opposite directions. The same direction of heat and mass gradients can heat up the samples faster and simultaneously, hence the amount of patchouli oil extracted by MAHD is greater than conventional hydrodistillation [15]. Yield of patchouli oil decreases after 45 minutes where the yield of oil from MAHD decreases from 2.33% to 1% and oil from HD goes from 0.9% to 0.43% at 60 minutes of extraction. Longer extraction time can degrade heat sensitive bioactive compounds, hence decreases the yield obtained. Exposure to high temperature for a longer time also can vaporize the solvent and eventually lower the yield of patchouli oil [26].

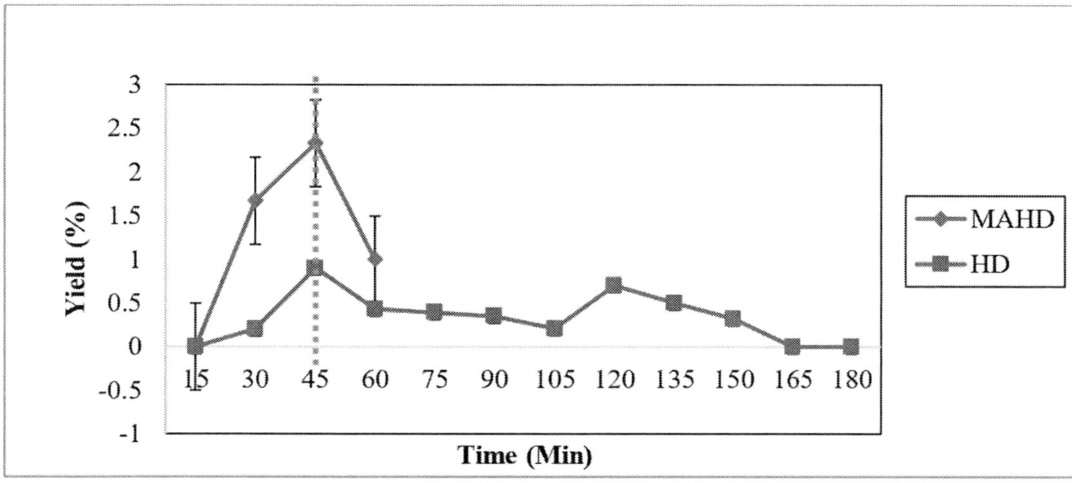

Figure 3. Yield of patchouli oil versus Time (min).

3.3. Composition of Oil from Patchouli Leaves

Patchouli essential oil at the optimal conditions was analyzed using GC-MS to identify the bioactive compounds in the oil. Table 1 illustrates the results obtained for both extraction methods. Sixteen chemical compounds were identified using MAHD while eleven chemical compounds were identified using HD. Oil obtained through HD contained higher contents of sesquiterpenes while oil obtained through MAHD packed with oxygenated compounds. This is because the heat transfer in MAHD is from the oil gland to the surrounding solvent while the heat transfer in HD is vice versa [15]. Hence, maximum quantity of essential oil can be extracted from patchouli oil gland using MAHD. Since the higher amount of essential oil can be extracted utilizing MAHD technique, higher number of oxygenated compounds were recognized in the oil. Patchoulol is the main oxygenated compound extracted from patchouli leaves. Moreover, the higher the number of oxygenated compounds, the higher the quality of essential oil. MAHD method shows a greater probability in the production of natural fragrance of the patchouli oil. Thus, MAHD method could be a better alternative extraction technique for the extraction of essential oil.

Table 1. Chemical compositions of patchouli oil extracted from patchouli leaves through HD and MAHD using GC-MS analysis

No.	Compounds	Molecular Formula	Area (%)	
			HD	MAHD
Sesquiterpenes				
1	δ-Elemene	$C_{15}H_{24}$	nd	0.2
2	β-caryophellene	$C_{15}H_{24}$	4.94	4.64
3	α-Gurjunene	$C_{15}H_{24}$	1.45	11.01
4	Aromandendrene	$C_{15}H_{24}$	nd	0.25
5	Valencene	$C_{15}H_{24}$	nd	1.91
6	β-Patchoulene	$C_{15}H_{24}$	2.4	2.6
7	β-Elemene	$C_{15}H_{24}$	1.17	nd
8	α-Patchoulene	$C_{15}H_{24}$	5.4	nd
9	Selina-3,7(11)-diene	$C_{15}H_{24}$	nd	0.7
10	α-Guaiene	$C_{15}H_{24}$	15.94	12.2
11	β-Maaliene	$C_{15}H_{24}$	nd	0.74
12	δ-Guaiene	$C_{15}H_{24}$	25.06	17.5
13	Seychellene	$C_{15}H_{24}$	7.1	8.3
14	β-Chamigrene	$C_{15}H_{24}$	3.72	nd

IOP Conf. Series: Materials Science and Engineering **702** (2019) 012042 doi:10.1088/1757-899X/702/1/012042

No.	Compounds	Molecular Formula	Area (%)	
			HD	MAHD
Oxygenated terpenes				
15	Iso-. alpha. -cedren-15-al	$C_{15}H_{22}O$	nd	0.4
16	Patchoulol	$C_{15}H_{26}O$	27.1	26.8
17	Hexahydrothunbergol	$C_{20}H_{40}O$	nd	0.05
18	Aromandendrene oxide	$C_{15}H_{24}O$	nd	2.78
19	2,6,6-Trimethyl-cyclohex-2-en-1-yl	$C_{14}H_{24}O$	0.3	nd
20	Viridiflorol	$C_{15}H_{26}O$	nd	6.1
Total non-oxygenated compounds (%)			67.18	60.05
Total oxygenated compounds (%)			27.4	36.13
Total identified (%)			94.58	96.18

3.4. Structural Changes Before and After Extraction Processes

The morphological images of the powdered patchouli leaf before and after extraction using SEM analysis are shown in Figure 4. In Figure 4, the oil glands were identified and circled. The vegetal cell of patchouli leaf was closely packed before the extraction as shown in Figure 4(a). The morphological variations in the cell walls after MAHD and HD are illustrated in Figures 4(b) and 4(c), respectively. It can be observed that the oil gland had undergone complete rupture at 60 min of MAHD, whereas, a partial rupture was observed using HD at 180 min of extraction time. These reflected that higher quantity of essential oil can be extracted in shorter time using MAHD method. During the MAHD extraction, the microwave energy is transferred into oil gland and converted into heat energy. The heat energy then transferred to plant samples to heat up the plant matrix to expel oil from the gland. Furthermore, heat is being transferred through three ways: conduction, convection and radiation. Hence, the oil glands get heated faster and essential oil can be extracted completely [4]. However, the heat energy will first reach the surface of the solvent before being transferred to plant samples in a conventional HD. In HD, the heat is transferred through conduction and convection. So, it will take longer time for the oil gland to be heated and quantity of essential oil extracted will be less as compared to MAHD method.

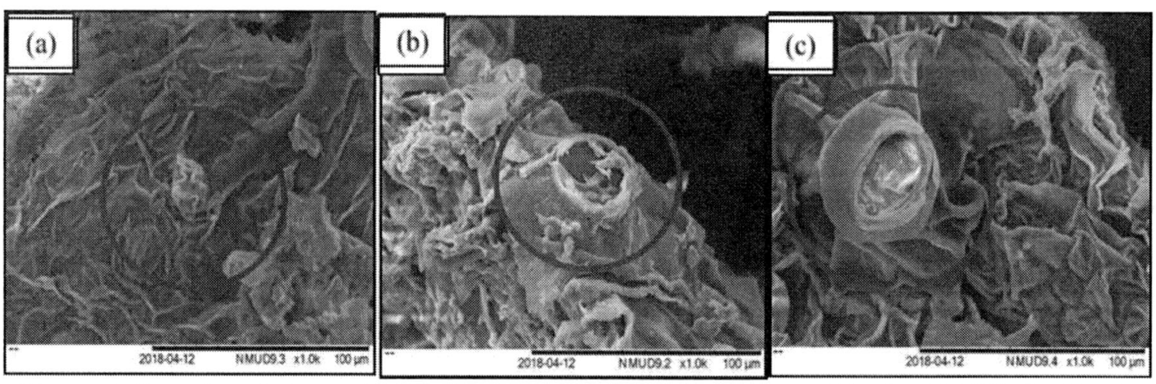

Figure 4. Scanning electron micrographs of patchouli samples; (a) raw, (b) after MAHD for 60 min and (c) after HD for 180 min.

4. Conclusion

This study investigated the potential of MAHD as compared to conventional HD method in extracting essential oil from patchouli leaves. The results showed that higher oil yield of 5 % was obtained from patchouli leaves in 60 min of extraction time as compared to HD (4 % of oil) in 180 min using a solid to solvent ratio of 1:12 (Figure 2). Moreover, higher number of chemical compounds were observed in the oil through MAHD as compared to HD method. 16 bioactive compounds were extracted using MAHD and 11 compounds were obtained using HD, still the percentage of identified compounds were not too distinct between both methods. The yield of patchoulol when using MAHD and HD extraction techniques were 26.8% and 27.1% respectively. In addition, SEM results confirmed that higher recovery of oil through MAHD as compared to HD method was due to complete rupture of oil glands in the plant matrix. Thus, MAHD showed a greater potential for better production of the essential oil from patchouli leaves as compared to HD method. In addition, MAHD can be proposed industrially as an alternative technique to separate essential oil in a larger scale.

Acknowledgement
The authors are thankful to Universiti Malaysia Pahang who provide financial support (PGRS180352).

References
[1] Kusuma H S and Mahfud M 2016 Chemical composition of essential oil of Indonesia sandalwood extracted by microwave-assisted hydrodistillation *AIP Conference Proceedings* 1755
[2] Karimi A 2014 Characterization and Antimicrobial Activity of Patchouli Essential Oil Extracted from Pogostemon cablin [Blanco] Benth **8**(May) 2301–2309
[3] Battaglia S 2003 The Complete Guide to Aromatherapy *International Centre of Holistic Aromatherapy* 1–602
[4] Kusuma H S and Mahfud M 2017a Microwave-assisted hydrodistillation for extraction of essential oil from patchouli (Pogostemon cablin) leaves *Periodica Polytechnica Chemical Engineering* **61**(2) 82–92
[5] Benth B, Sundaresan V, Singh S P and Mishra A N 2009 Composition and Comparison of Essential Oils of Pogostemon travancoricus Bedd. var. travancoricus **21**(June) 2008–10
[6] Bunrathep S, Lockwood G B, Songsak T and Ruangrungsi N 2006 Chemical constituents from leaves and cell cultures of pogostemon cablin and use of precursor feeding to improve patchouli alcohol level *Science Asia* **32**(3) 293–96
[7] Ramya H G, Palanimuthu V and Rachna S 2013 An introduction to patchouli (Pogostemon cablin Benth.) - A medicinal and aromatic plant: It's importance to mankind *Agricultural Engineering International: CIGR Journal* **15**(2) 243–50
[8] Hamidi N 2016 Extraction of essential oils from patchouli plant using advanced techniques of microwave-assisted hydrodistillation *ARPN Journal of Engineering and Applied Sciences* **11**(2) 796–99
[9] Donelian A, Carlson L H C, Lopes T J and Machado R A F 2009 Comparison of extraction of patchouli (Pogostemon cablin) essential oil with supercritical CO2 and by steam distillation *Journal of Supercritical Fluids* **48**(1) 15–20
[10] Buré C M and Sellier N M 2004 Analysis of the essential oil of Indonesian patchouli (pogostemon cabin benth.) using GC/MS (EI/CI) *Journal of Essential Oil Research* **16**(1) 17–19
[11] Golmakani M T and Rezaei K 2008 Comparison of microwave-assisted hydrodistillation with the traditional hydrodistillation method in the extraction of essential oils from Thymus vulgaris L *Food Chemistry* **109**(4) 925–30
[12] Guan W, Li S, Yan R, Tang S and Quan, C2007 Comparison of essential oils of clove buds extracted with supercritical carbon dioxide and other three traditional extraction methods *Food Chemistry* **101**(4) 1558–64
[13] Fan L, Jin R, Liu Y, An M and Chen S 2011 Enhanced extraction of patchouli alcohol from Pogostemon cablin by microwave radiation-accelerated ionic liquid pretreatment *Journal of Chromatography B: Analytical Technologies in the Biomedical and Life Sciences* **879**(30) 3653–57
[14] Da Porto C, Porretto E and Decorti D 2013 Comparison of ultrasound-assisted extraction with conventional extraction methods of oil and polyphenols from grape (Vitis vinifera L.) seeds. *Ultrasonics Sonochemistry*, **20**(4) 1076–80
[15] Nitthiyah J, Nour A H, Kantasamy R and Akindoyo J O 2017 Microwave assisted hydrodistillation – an overview of mechanism and heating properties *Australian Journal of Basic and Applied Sciences* **11**(3) 22–29

[16] Fadl A, M S Abdel-Hameed, E S Elkhateeb, A A El-Awady M and S Salman M 2018 Chemical Composition of Hydrodistillation and Solvent Free Microwave Extraction of Essential Oils from Mentha Piperita L Growing in Taif, Kingdom of Saudi Arabia, and Their Anticancer and Antimicrobial Activity *Oriental Journal of Chemistry* **34**(1) 222–233
[17] Manouchehri R, Saharkhiz M J, Karami A and Niakousari M 2018 Extraction of essential oils from damask rose using green and conventional techniques: Microwave and ohmic assisted hydrodistillation versus hydrodistillation *Sustainable Chemistry and Pharmacy* **8**(Nov) 76–81
[18] Moradi S, Fazlali A and Hamedi H 2018 Microwave-assisted hydro-distillation of essential oil from rosemary: Comparison with traditional distillation *Avicenna Journal of Medical Biotechnology* **10**(1) 22–28
[19] Golmakani M and Moayyedi M 2015 Comparison of heat and mass transfer of different assisted extraction methods of essential oil from Citrus limon (Lisbon variety) peel.
[20] Najmuldeen G F, Narayanan K, Kirubananthan D A P, Faisal G G and Narayan K 2018 Comparison of Tongkat Ali Root Chemical Composition Extracted by Soxhlet, Conventional Steam and Microwave Assisted Extraction Techniques **10**(5) 916–20
[21] Hemalatha S, Mandal V and Mohan Y 2007 Microwave assisted extraction - An innovative and promising extraction tool for medicinal plant research *Pharmacognosy Reviews* **1**(1) 7–18
[22] Ismanto A W, Kusuma H S and Mahfud M2018 Solvent-free microwave extraction of essential oil from Melaleuca leucadendra L *03007* 1–4
[23] Sinaga M S, Rico G, Nababan A N and Manullang T A 2018 Effect of solvent volume ratio and time extraction of glycerol purification *IOP Conference Series: Materials Science and Engineering* **309**(1).
[24] Kusuma H S and Mahfud M 2017b The extraction of essential oils from patchouli leaves (Pogostemon cablin benth) using a microwave air-hydrodistillation method as a new green technique *RSC Advances* **7**(3) 1336–47
[25] Hakimi M I, Ilham Z and Abdul Kohar R A 2018 Enhancement of Agro-Industrial Copra Residue Oil Yield Using Microwave-Assisted Extraction *Waste and Biomass Valorization* 1–8
[26] Durling N E, Catchpole O J, Grey J B, Webby R F, Mitchell K A, Foo L Y and Perry N B 2007 Food Chemistry Extraction of phenolics and essential oil from dried sage (Salvia officinalis) using ethanol – water mixtures **101** 1417–1424
[27] Shah K A, Bhatt D R, Desai M A, Jadeja G C and Parikh J K 2017 Extraction of essential oil from Patchouli leaves using hydrodistillation: Parametric studies and optimization **24**(July) 405–410

Appendix
Microwave-assisted hydrodistillation

a) Effect of volume of solvent

Volume of solvent (ml)	Yield of Patchouli oil (ml)	Yield of Patchouli Oil (g)	Yield of Patchouli oil (%)
180	1.04	1.0	3.33
240	1.35	1.3	4.33
300	1.35	1.3	4.33
360	1.56	1.5	5
420	0.93	0.9	3

IOP Conf. Series: Materials Science and Engineering **702** (2019) 012042 doi:10.1088/1757-899X/702/1/012042

b) Effect of time of extraction

Time of extraction (min)	Yield of Patchouli oil (ml)	Yield of Patchouli Oil (g)	Yield of Patchouli oil (%)
15	0	0	0
30	0.52	0.5	1.67
45	0.73	0.7	2.33
60	0.31	0.3	1

Hydrodistillation

a) Effect of volume of solvent

Volume of solvent (ml)	Yield of Patchouli oil (ml)	Yield of Patchouli Oil (g)	Yield of Patchouli oil (%)
180	0.62	0.6	2
240	0.68	0.65	2.167
300	0.78	0.75	2.5
360	1.25	1.2	4
420	0.83	0.8	2.67

b) Effect of time of extraction

Time of extraction (min)	Yield of Patchouli oil (ml)	Yield of Patchouli Oil (g)	Yield of Patchouli oil (%)
15	0	0	0
30	0.062	0.06	0.2
45	0.28	0.27	0.9
60	0.135	0.13	0.43
75	0.121	0.12	0.39
90	0.11	0.11	0.35
105	0.065	0.063	0.21
120	0.22	0.21	0.7
135	0.156	0.15	0.5
150	0.1	0.096	0.32
165	0	0	0
180	0	0	0

IOP Conf. Series: Materials Science and Engineering **702** (2019) 012043 doi:10.1088/1757-899X/702/1/012043

Performance profiling of UV-grafted forward osmosis polyethersulfone membrane using multivariate classical scaling technique

A F H A Rahman[1], M Y Mohd Yunus[1,2], T S L M T A Nasaruddin[1,2], N Harun[1] and M N Abu Seman[1*]

[1]Faculty of Chemical and Process Engineering Technology, College of Engineering Technology, Universiti Malaysia Pahang, Lebuhraya Tun Razak, 26300 Gambang, Kuantan, Pahang, Malaysia.
[2]Earth Resource & Sustainability Center, Universiti Malaysia Pahang, Malaysia.

*E-mail: mazrul@ump.edu.my

Abstract. In this work, Classical Scaling (CMDS) method was used to develop a set of membrane profile performance for various types of UV-grafted polyethersulfone (PES) membrane. Previously, there is no such profiling has been considered. The main motivation of the profiling is to determine what is the most preference impact factor which significantly influence the membrane strength during application. Initially, nine (9) different samples of modified PES membranes were prepared mainly by adopting numerous characterization parameters of acrylic acid monomer concentration and grafting time. Three (3) performance indicators namely water permeability (A), solute permeability (B) and structural parameter (S) were used as representing the foundation of membrane performance profile. In particular, the Euclidean and City-block scales were utilized to build up the multivariate profile based on two dimension configurations. The key finding suggests that the modified membranes were easily clustered based on its grafting mechanism. From the general observation on the compressed CMDS dimensional space, samples that fall above the x-axis in Euclidean scale configuration and those scattered samples in the City-block scale have relatively larger pores. Thus, the CMDS profiling tends to favor the pore size as the dominant impact factor in characterizing the membrane performance based on the three specified parameters that investigated.

1. Introduction

A membrane is a semi-permeable barrier that shows different selectivity between species. Membrane separation process is a process utilized as a part of separation between different phases coexists in one system. The main target in this process is to selectively allow a species to permeate through the membrane freely while preventing the permeation of other components. However, some limitations such as membrane fouling, scaling, and concentration polarization have make the process of choosing and cleaning of membrane to take longer time and somewhat need to spend some cost for regular maintenance.

An approach to make an initial guess on the membrane to be chosen for future works is proposed which is by creating a membrane profile. Based on the literature survey made, the membrane profiling is only limited to the plasma membrane which have been studied by Peirce et al. [1], Weekes et al. [2], and Hsu et al. [3]. However, no attempt to generate a membrane profiling for forward osmosis membrane has been done. In the current work, the multidimensional scaling (MDS) method is used to create membrane profiling on pore size effect prediction of the polyethersulfone (PES) membrane.

Profiling has become a way of seeing an outline of something in a clear, simple, and more precise way. It is in general defined as a depiction of something that gives us a picture of the overall system of it. A profile can be treated as an initial guess of generic trending on the respective subject that investigated. In the field of membrane technology, however, limited studies of profiling performance are observed.

In this study, a membrane profile that derived from the intrinsic properties of the NF-PES membrane is proposed. This profile is mainly developed by means of multivariate scores as it involves the investigation of various samples based on multiple sets of contributory factors. In this respect, Classical Scaling (CMDS) is one of the multivariate techniques that typically utilized to extract the essence of pattern based on a specific data set [4]. It is usually employed for envisioning the level of similarity or dissimilarity among individual cases, which display the latent information in the form of distance configuration matrix. Thus, the main goal of this study is to create a membrane profiling for pore size effect prediction of the FO membrane based on their intrinsic properties namely water permeability coefficient (A), solute permeability coefficient (B), and structural parameter (S).

2. Profiling Frameworks based on Classical Scaling Approach

The general procedures of developing the multivariate scores of Classical Scaling (CMDS) in this work was adopted from Mohd Yunus [5]. Firstly, the squared dissimilarity matrix, $\mathbf{\Delta_2}$ is determined based on $\mathbf{X}_{nxm}$, (where *n:* samples, *m:* variables) using a particular distance measure (scale) through the application of major product moment of matrix operation. In principle, the size of matrix $\mathbf{\Delta_2}$ is directly determined by the magnitude of samples, *'n'* as shown in the equation (1).

$$\mathbf{\Delta}_{n\times n}=\mathbf{\Delta_2}=\begin{bmatrix} \delta_{1,1} & \delta_{1,2} & \cdots & \delta_{1,n} \\ \delta_{2,1} & \delta_{2,2} & \cdots & \delta_{2,n} \\ . & . & \cdots & . \\ . & . & \cdots & . \\ . & . & \cdots & . \\ \delta_{n,1} & \delta_{n,2} & \cdots & \delta_{n,n} \end{bmatrix} \tag{1}$$

In this study, the basis scale of developing the multivariate score configurations CMDS are Euclidean and City-block distances as shown in equations (2) and (3).

$$\text{Euclidean distance:} \quad \delta_{ij} = \{\sum_a (x_{ia} - x_{ja})^2\}^{\frac{1}{2}} \tag{2}$$

$$\text{City-block distance:} \quad \delta_{ij} = \sum_a | x_{ia} - x_{ja} | \tag{3}$$

where δ_{ij} is the dissimilarity measure between objects i and j, x represents the original Cartesian coordinate of those objects (samples), and a is the number of axes in the original space. The motive of applying two types of different scales is for profiling verification. Next the double-centered matrix, $\mathbf{B}_\Delta$, is computed on $\mathbf{\Delta_2}$ as in equation (4).

$$\mathbf{B}_\Delta=-\frac{1}{2}\mathbf{J}_n\Delta_2\,\mathbf{J}_n \quad \text{where} \quad \mathbf{J}_n=(\mathbf{I}_n-\mathbf{1}_n\mathbf{1}_n'/n) \tag{4}$$

where $\mathbf{I}_n$ is an identity matrix of size $n \times n$, while $\mathbf{1}_n$ is a vector of size *'n'* with all elements equal to 1. Additional remarks on $\mathbf{J}_n$ are provided in Mohd Yunus [5].

$\mathbf{B_A}$ is then decomposed into Eigen basic structure through the employment of singular value decomposition (SVD) technique particularly in producing eigenvectors and eigenvalues of $\mathbf{\Delta_2}$ that depicted by equation (5). Lastly, the multivariate scores based on Cartesian coordinates, $\mathbf{X_E}$, is developed based on equation (6).

$$\mathbf{B= UVU^T} \tag{5}$$

where U and V are the Eigen vectors and Eigen values respectively.

$$\mathbf{X_E=U_+V_+^{0.5}} \tag{6}$$

where $\mathbf{V_+^{0.5}}$ is a diagonal matrix with all positive elements of $v^{0.5}$ and $\mathbf{U_+}$ is the corresponding sets of eigenvectors.

3. Case Study: FO Using NF-PES Membrane

In the experiment, the nanofiltration polyether sulfone (NFPES) membrane was purchased from Amfor Inc (China). The membrane was modified using two parameters, namely acrylic acid (AA) monomer concentration and UV grafting time. For the profiling purpose, the data used were water permeability (A), solute permeability (B) and structural parameter for membrane (S). The membrane S-value is a measure of the resistance of the membrane's support layer towards solute diffusion. These all data of A, B, and S were taken from Rahman [6]. Sets of data obtained are shown in table 1. These data are then inserted into the workspace section in MATLAB. Euclidean distance was employed to measure the dissimilarity between all samples.

Table 1. Data used in membrane profiling [6].

Membrane type	Sample number	A (L/ m^2.h.bar)	B (m/h)	S (mm)
Unmodified NF-PES	1	9	0.01540	28.02
5AA-1	2	7.37	0.02930	8.11
15AA-1	3	10.67	0.02326	12.54
30AA-1	4	6.91	0.01915	8.2
50AA-1	5	7.91	0.07898	8.1
5AA-5	6	6.84	0.02808	7.87
15AA-5	7	8.39	0.02462	8.26
30AA-5	8	7.39	0.11725	5.89
50AA-5	9	7.89	0.02516	12.51

*5AA-1 represents (concentration in g/L)AA-(time for UV grafting in minute)

4. Results and Discussion

In this work, all data were first converted to a dissimilarity matrix according to the procedure mention in section 3. Figure 1 shows the dissimilarity matrix generated by MATLAB. The points are then being mapped down in two-dimensional space such that the 'straight line' (Euclidean) distances as shown in figure 2. It is observed that S8 has the largest distance from S1 (4.6173) in relative to the other samples. The reproduction of scores for figure 1 is illustrated in figure 3 in which location of S8 is seen diverts far away from S1 in contrast to other samples. The dissimilarities are basically due to the modifications made on the membranes.

Mat_E =

0	3.2786	2.6960	3.4207	3.6026	3.5101	2.9916	4.6173	2.4984
3.2786	0	2.8252	0.4825	1.5084	0.4432	0.8586	2.5708	0.7928
2.6960	2.8252	0	3.1921	2.8815	3.2600	1.9987	3.9783	2.3105
3.4207	0.4825	3.1921	0	1.9232	0.2698	1.2399	2.8920	1.0505
3.6026	1.5084	2.8815	1.9232	0	1.7230	1.6257	1.2350	1.6923
3.5101	0.4432	3.2600	0.2698	1.7230	0	1.2930	2.6414	1.1152
2.9916	0.8586	1.9987	1.2399	1.6257	1.2930	0	2.8327	0.7561
4.6173	2.5708	3.9783	2.8920	1.2350	2.6414	2.8327	0	2.8752
2.4984	0.7928	2.3105	1.0505	1.6923	1.1152	0.7561	2.8752	0

Figure 1. Dissimilarity matrix gained from MATLAB for Euclidean Distance.

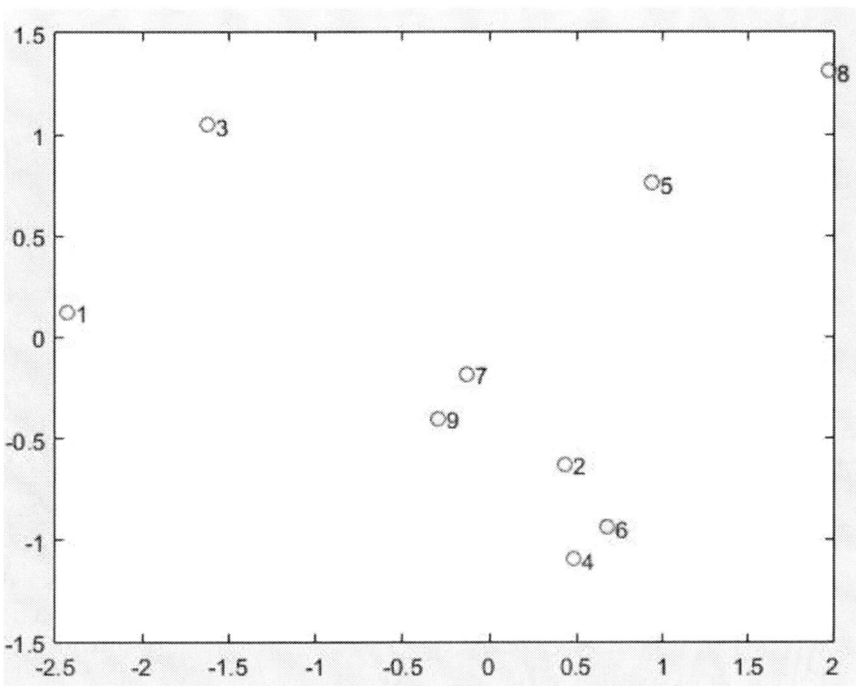

Figure 2. Membrane Profile using Euclidean Distance.

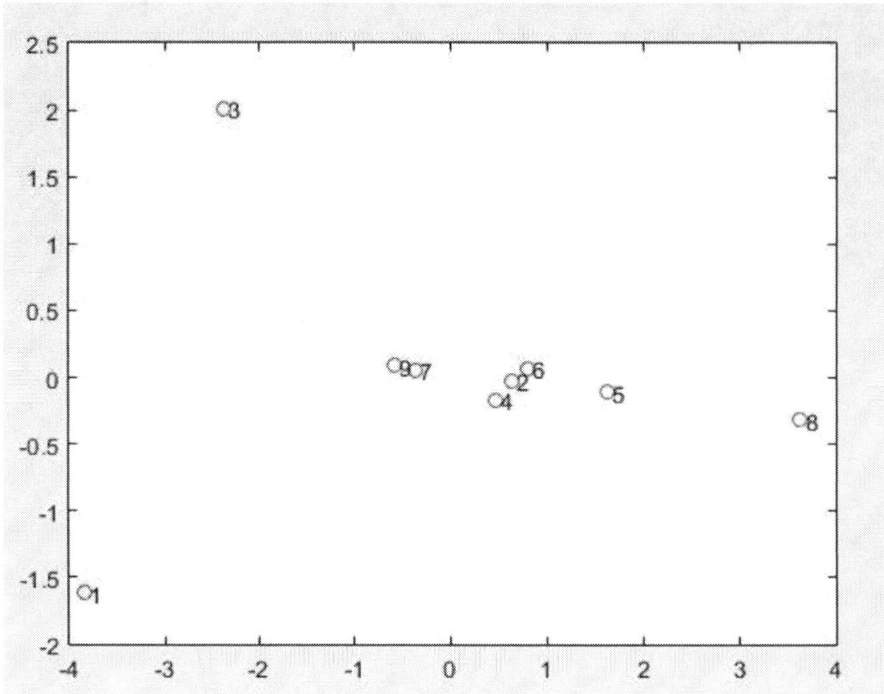

Figure 3. MDS Scores for Forward Osmosis Membranes Profile using City Block Distance.

The x-axis in figures 2 and 3 represents dimension 1, and y-axis represents dimension 2. From figure 2, it is observed that there are two division exist. The first one is those that falls below the x-axis which composed of samples 2, 4, 6, 7, and 9, and the second one is those that falls above the x-axis which composed of samples 1, 3, 5, and 8. As from figure 3 (City-block distance), there is a group cluster exists and four singletons namely sample 1, 3, 5, and 8. Comparing both figures, it is observed that the circled samples in figure 3 represents the samples that are located below the x-axis in Euclidean distance. Through result analysis for both distance measured, the overall trend suggests that the performance of City-block can be considered comparable to the Euclidean scale.

As for justifying the results shown in figures 2 and 3, for the sample 1 (unmodified membrane), Abu Seman et al. [7] found that the unmodified membrane exhibited higher flux compared to the modified membrane as the unmodified membrane has a relatively larger pores than modified membrane. The other samples in the same group are sample 5 (50AA-1min) and sample 8 (30AA-5min). According to Rahman [6], a detrimental effect was recorded for both sample in which it both membrane performances got affected. This membrane performance reduction is due to the increase of pore size as the PES membrane is dependent on the pH value [8]. Pore enlargement occurs as the pH decrease. Sample 6 (5AA-5min) has the lowest monomer concentration tested with highest UV-grafting time.

According to Rahman [6], during the immersion process when the monomer concentration is low, some molecules would penetrate the membrane and then undergo polymerization process. These molecules will lead to the reduction of membrane pores as it grafted in both membrane surface and deep within the pores. More compact membrane matrix could be developed which maximized the amount of effective grafting and its penetration inside the pores [7]. As the effective grafting time increased, the pore size will be reduced, and lead to the reduction of water permeability. Sample 4 (30AA-1min) somehow turned out the same even though the monomer concentration is huge different. Although grafted at high concentration of monomer, the effective grafting exceeds the chain scission. This is due to the low grafting time. It can be concluded that, as the effective grafting exceeds, the pore could probably reduce due to the more compact membrane was developed. This can be correlate with sample

6. For sample 3 (15AA-1min), according to Rahman [6], the water flux for the sample increased after UV-grafting which is caused by the pore enlargement mechanism. The membrane faced a severe pore enlargement due to the extensive chain scission [9].

As what has been observed and studied, the pore size effect has seemed to suit the profile so well. The pore size could either have enlarged or reduced depending on which quarter inside the profile the sample falls into, and by interpreting those data in terms of multidimensional scaling, a clearer info regarding the data could be revealed in a much simpler way. Recall back figures 2 and 3. The samples that falls above the x-axis (1, 3, 5, 8) in figure 2 shows that they all experience pore enlargement by some factors during the process. In comparing to figure 3, those scattered sample (1, 3, 5, 8) are also the group that occur pore enlargement.

5. Conclusions & Recommendations

A profiling on a set of data has been applied using MDS method. Three parameters have been considered in creating the profile namely water permeability, solute permeability, and structural parameter. Euclidean distance was used as the base of calculation. From the result, it can be concluded that, those samples that fall above the x-axis in Euclidean scale and those scattered samples in City-block scale have relatively larger pores than those below the x-axis. This happened due to some the effect of grafting time and monomer concentration. It is shows that certain limitations are seen to make an important factor to be considered when making a membrane profile.

Those factors include the sensitivity analysis and the distinctive properties of the membrane. For future works, it is suggested that those factors are being considered for a better accuracy of the result. Another method that is widely used as a multivariate tool is the Principal Component Analysis (PCA). With insignificant extra exertion PCA gives a guide to how to decrease a complex informational index to a lower dimension to uncover the occasionally covered up, improved progression that regularly underlie it [10].

Acknowledgments

The authors wish to express numerous thanks to Universiti Malaysia Pahang and the Ministry of Higher Education through the financial aid from Postgraduate Research Grant Scheme PGRS:160336, Fundamental Research Grant Scheme FRGS/1/2016/TK02/UMP/02/8 (RDU160127).

References

[1] Peirce M J, Wait R, Begum S, Saklatvala J and Cope A P 2004 Expression profiling of lymphocyte plasma membrane proteins *Mol. Cell. Proteomics* **3**(1) 56-65

[2] Weekes M P, Antrobus R, Talbot S, Hör S, Simecek N, Smith D L and Lehner P J 2012 Proteomic plasma membrane profiling reveals an essential role for gp96 in the cell surface expression of LDLR family members, including the LDL receptor and LRP6. *J. Proteome Res.* **11**(3) 1475-84

[3] Hsu J L, van den Boomen D J, Tomasec P, Weekes M P, Antrobus R, Stanton R J and Davison A J 2015 Plasma membrane profiling defines an expanded class of cell surface proteins selectively targeted for degradation by HCMV US2 in cooperation with UL141 *PLoS pathogens* **11**(4) e1004811

[4] Cox T F 2005 *An Introduction to Multivariate Data Analysis* (Hodder Education. London Great Britain)

[5] Mohd Yunus M Y 2012 *Multivariate statistical process monitoring using classical multidimensional scaling.* (Newcastle University)

[6] Rahman A F H A and Abu Seman M N 2017 Surface Modification of Polyethersulfone (PES) membrane via UV-grafting for Forward Osmosis Technology *Malaysian J. Analytic. Sci.*

[7] Abu Seman M N, Khayet M, and Hilal N 2012 Comparison of two different UV-grafted nanofiltration membranes prepared for reduction of humic acid fouling using acrylic acid and N-vinylpyrrolidone *Desalination* **287** 19-29

[8] Deng B, Yang X X, Xie L D, Li J Y, Hou Z C, Yao S, Huang Q. 2009 Microfiltration membranes with pH dependent property prepared from poly(methacrylic acid) grafted polyethersulfone powder *J. Membrane Sci.* **330**(1-2) 363-68

IOP Conf. Series: Materials Science and Engineering **702** (2019) 012043 doi:10.1088/1757-899X/702/1/012043

[9] Pieracci J, Crivello J V and Belfort G 2002 UV-assisted graft polymerization of N-vinyl-2-pyrrolidinone onto poly(ether sulfone) ultrafiltration membranes using selective UV wavelengths *Chem. Mat.* **14**(1) 256-65
[10] Shlens J 2003 A tutorial on principal component analysis: derivation, discussion and singular value decomposition *Mar.* **25**(1) 16

IOP Conf. Series: Materials Science and Engineering **702** (2019) 012044 doi:10.1088/1757-899X/702/1/012044

Comparison of optimal solvent extraction stages between P204 and [A336][P204] for the separation of europium and gadolinium

N A Ismail[1], M Y Mohd Yunus[1,2,*], M A Abdul Aziz[2] and M A Abidin[3]

[1] Earth Resources and Sustainability Center, Universiti Malaysia Pahang, Lebuhraya Tun Razak 26300 Kuantan Pahang, Malaysia
[2] Faculty of Chemical and Process Engineering Technology, Universiti Malaysia Pahang, Lebuhraya Tun Razak 26300 Kuantan Pahang, Malaysia
[3] Aras Kuasa Sdn. Bhd. Level 8, Menara Zenith, Jalan Putra Square 6, Putra Square 25200 Kuantan Pahang, Malaysia

*E-mail: yusri@ump.edu.my

Abstract. This study was designed to compare between the number of extraction stages required by 2-ethylhexyl phosphoric acid mono 2-ethylhexyl ester P204 and ionic liquid-based system of [A336][P204] in a cascade counter-current extraction process for the separation of europium (Eu) and gadolinium (Gd). Eu and Gd were chosen for this study as these metals are among the strenuous due to low availability in nature and small separation factor. The value of the separation factor (β) of both extractants, was obtained from the experimental work, while other process parameters were calculated using the counter-current model developed by Xu Guangxian. Extraction of Eu using [A336][P204] was found optimum at 3.5 M of nitric acid, with organic to aqueous (O/A) ratio of 9:1, while the optimum condition for P204 was determined at 3.0 M of hydrochloric acid and 9:1 of O/A ratio. The optimized β value for P204 was identified lower (β:2.38) compared to the condition for [A336][P204] (β:3.44). In order to obtain the optimal number of separation stages, the target purity and recovery rates were set to 99.9% and 90% respectively for both extractants. From the Xu model, the results show that [A336][P204] extractant requires at a minimum of four and two extraction and scrubbing stages respectively for maintaining the desired outputs. Meanwhile, P204 extractant needs to develop at least eight extraction stages as well as five scrubbing stages to produce the same production capacity. This finding generally suggests that [A336][P204] is favorable over P204 for the separation of europium and gadolinium particularly from the viewpoint of economic interest.

1. Introduction

Since the 1940s, the application of solvent extraction methods to separate rare earth (RE) has been studied in a laboratory and plant scale, primarily as a consequence of work created for the manufacturing of nuclear fuel. The extraction of liquid-liquid is a method of separation based on the distribution of a solution between two stages; aqueous and organic, which practically immiscible. The organic stage usually includes a diluent and an extractant that capable of complexing with the RE element and then forming more soluble compounds, i.e. showing higher affinity with the organic stage. Organophosphorus acids belong to a compound formation class and di-2-ethylhexylphosphoric acid (P204) is one of the examples. This extractant has been widely used since Peppard et al. [1] pioneering work for the discovery of the dimer for of RE-P204 metal complex. The use of P204 in RE separation was discussed by Tianchi et al. [2], Qing et al. [3] and others. This extractant has been selected due to great separation factor among other extractant and environment-friendly properties as the extraction process is saponification-free.

A fresh category of advanced extractants based on ionic liquid, such as [methyltrioctylammonium][di-(2-ethylhexyl) phosphinate] ([A336][P227]) [4], [tricaprylmethylammonium][dihexyl diglycolamate], [A336][DHDGA] [5] and [trioctylmethylammonium][2,4,4-trimethylpentyl phosphinate] [A336/Cy272] [6] has been developed in the past few years. Most of these extractants are superior in term of extraction capabilities compared to common extractant in the market. Our earlier publication revealed that [A336][P204] provide synergistic impact towards La(III), Sm(III), and Y(III) compared to the P204 due to the presence of the dual-functional group [7]. This impact could considerably improve the extraction capabilities to extract RE while offering a greener separation path by eliminating the saponification process.

To recover a high purity RE from a solvent extraction technology, a multi-stage contact for the organic and aqueous phases using a mixer settler is often used. Also, to avoid coagulation, the aqueous liquid phase and the organic phase are injected in the opposite direction, known as the countercurrent process. The extraction process starts when the RE in the acidic medium (aqueous phase) is mixed with the extractant (organic phase) in a container until the mixture reaches an equilibrium point.
In the settling stage, the mixture is then divided back into aqueous and organic phases, where some of the RE from the aqueous phase is now trapped in the organic phase, producing a complex of metal-extractants. The process of mixing, settling, and separating the organic and aqueous stages are repeated several times (1 to n times) until an adequate RE extraction purity and recovery value are achieved (Figure 1). Each cycle is counted as one extraction stage.

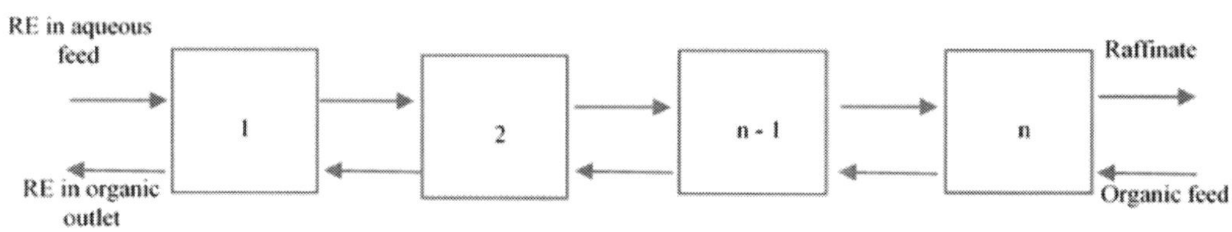

Figure 1. Schematic diagram of a counter-current extraction

This study evaluates the impact of two different extractants on the number of extraction and scrubbing stages needed for complete RE separation. An optimization study of P204 and [A336] [P204] in different parameters, such as type and concentration of acid, as well as organic to the aqueous ratio of the solution, was explored. In this study, the theory proposed by Xu Guangxian was used in calculating the theoretical stages required to separate Eu and Gd from their mixed solution. The least number of extraction and scrubbing stages is preferred as it reflects the cost and the efficiency of the overall extraction process.

2. Theoretical background

Figure 2 shows the simplified design procedure to calculate the number of extraction stage based on the counter-current theory developed by Guangxian Xu in 1978 [8]. Each stage of the design procedure is discussed in the following sections. This theory was established based on five assumptions;

1. The extractability sequence of the RE is represented by A, B, and C, in which A is the element or group with the highest extractability.
2. Since counter-current extraction is comprised of a multistage extraction process, the average separation factor is used due to the different distribution ratios in different stages.
3. The extraction factor is kept constant.
4. RE composition in the aqueous feed and the aqueous phase is assumed the same.
5. All extraction stages have a constant flow ratio.

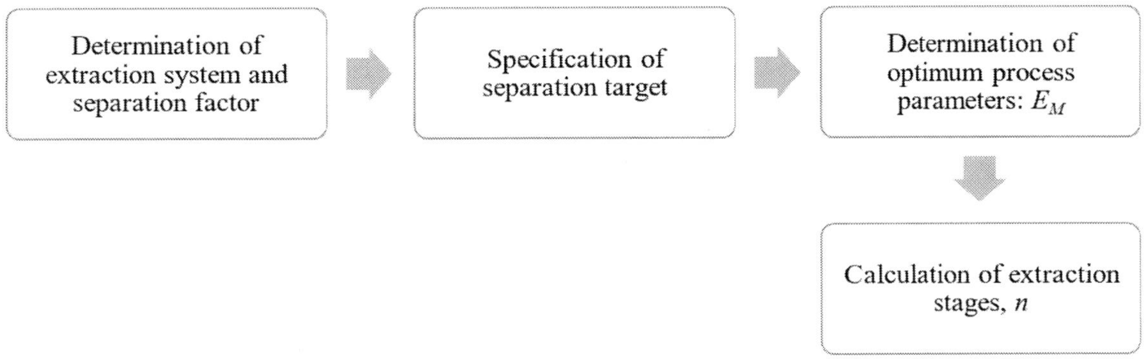

Figure 2. Simplified design procedure to calculate the number of extraction stage

2.1. Determination of Extraction System and Separation Factor

It is important to experimentally optimize an SX system by determining its parameters, such as the type and concentration of acid, the organic to the aqueous ratio (O/A), and feed concentration. Once a suitable extraction system is determined, the distribution factor, D and separation factor, β are measured, as shown by equation (1) and equation (2). D can be calculated when the RE distribution between the organic and aqueous phases has achieved an equilibrium state,

$$D_{\mathrm{A}}= \frac{[A_{[\mathrm{org}]}]}{[A_{(\mathrm{aq})}]} \tag{1}$$

From the D, the separation factor, β can be measured. β is defined as the ratio of distribution between two adjacent rare earth metals (A and B) in the separation system,

$$\beta_{\mathrm{A}/\mathrm{B}} = \frac{\mathrm{D_A}}{\mathrm{D_B}} \tag{2}$$

2.2. Separation Target Specification

To calculate the number of extraction stages in one particular SX system, the mole fraction of metal A in the feed (f_{A}), the target purity of product A in the organic outlet ($P_{\mathrm{A(O)}}$), and the target percentage of the product recovery of A (Y_{A}) must be determine beforehand. Based on the assumptions mentioned in Section 2, the REs are categorized based on their extractability in the SX system; A is the element/group of REs that are easily extracted, while B is the relatively difficult element/group to be extracted from the system. The f_A and f_B can be calculated using the following equation (3):

$$f_{\mathrm{A}}=\frac{\text{Total mole of A}}{\text{Total mole of B}} , f_{\mathrm{A}}+f_{\mathrm{B}}=1 \tag{3}$$

If A is the major product, then, the concentrating factor of A, a can be calculated using equation (4):

$$a = \frac{P_{A_{n+m}(o)} / (1-P_{A_{n+m}(o)})}{f_A / f_B} \tag{4}$$

And the concentrating factor for B can be calculated using equation (5):

$$b = \frac{a- Y_{\mathrm{a}}}{a\,(1-Y_{\mathrm{a}})} \tag{5}$$

However, if the major product is B, then, the concentrating factor of B can be calculated using equation (6):

$$b = \frac{P_{B_{1(a)}}/(1-P_{B_{1(a)}})}{f_B/f_A} \tag{6}$$

And the concentrating factor for A for the major product, B is calculated using equation (7):

$$a = \frac{b - Y_B}{b\,(1-Y_B)} \tag{7}$$

The solute fraction in the organic and aqueous outlets can be calculated using equation (8), as follows

$$f'_A = \frac{f_A Y_A}{P_{A_{n+m(o)}}}, f'_B = \frac{f_B Y_B}{P_{B_{1(a)}}}, f'_A + f'_B = 1 \tag{8}$$

2.3. Determination of Optimum Process Parameters
Next, the average extraction factors, E_M and E'_M, were measured from the optimum conditions based on either the SX system is the extraction or scrubbing controlled, as shown in table 1.

Table 1. Determining the extraction factor from the optimum conditions

If $f'_B > \frac{\sqrt{\beta}}{\sqrt{\beta}+1}$, the SX system is extraction controlled		If $f'_B < \frac{\sqrt{\beta}}{\sqrt{\beta}+1}$, the SX system is scrubbing controlled	
$E_M = \frac{1}{\sqrt{\beta}}$	(9)	$E'_M = \sqrt{\beta}$	(11)
$E'_M = \frac{E_M f'_B}{E_M - f'_A}$	(10)	$E_M = \frac{E'_M f'_A}{E'_M - f'_B}$	(12)

2.4. Determination of the Number of Stages

If B is the major component in the feed and targeted as a high purity product at the aqueous outlet, the number of extraction stage, n can be calculated using the following equation (13):

$$n = \log b / \log \beta E_M \tag{13}$$

However, if the targeted end product is a high purity A, and A is the major component in the aqueous feed, then, equation (14) can be used:

$$n = \log b / \log \beta E_M + 2.303 \log \frac{P^*_{A(a)} - P_{A_{1(a)}}}{P^*_{A(a)} - P_{A_{n(a)}}} \tag{14}$$

3. Experimental Method

3.1. Chemicals and Reagents
Aliquat 336, (<99% purity, [A336]Cl) and di-(2-ethylhexyl) phosphate (P204) were purchased from Sigma Aldrich (USA). Nitric acid (HNO_3), sulphuric acid (H_2SO_4), hydrochloric acid (HCl), sodium hydroxide (NaOH), sodium chloride (NaCl), and sodium nitrate ($NaNO_3$) were purchased from Fischer

IOP Conf. Series: Materials Science and Engineering **702** (2019) 012044 doi:10.1088/1757-899X/702/1/012044

Scientific. A stock solution consisting of Eu(III), and Gd(III) was prepared by dissolving their oxides (99.9%) in concentrated acid. The oxides are purchased from Sigma Aldrich (USA) All other chemicals used were of analytical grade.

3.2. Extraction Procedure
The [A336][P204] was prepared using the acid-base neutralization method, as described by Sun et al. (9). The extractant was dissolved in n-heptane for extraction. 5 mL of the aqueous phase containing RE(III) and 5 mL of the organic phase containing the extractant were mixed and shaken for 1 h at 200 rpm using a mechanical shaker, which was sufficient for equilibrium. The pH of the aqueous mixture was maintained at pH 1–2 using sodium salt (either NaCl or $NaNO_3$). The mixtures were then centrifuged for 5 min at 8,000 rpm to enhance the separation between the aqueous and organic layers. The separation factor was calculated for each extraction process.

3.3. Calculating the Extraction Stages
Based on the value of the separation factor, the number of extraction stages was calculated in Microsoft Office Professional Plus Excel 2013 as described in the counter-current theory. For this study, the recovery was set to 90%, whereas the purity was fixed at 99.99%.

4. Results and Discussion

4.1. System Optimization and Determining the Separation Factor
To determine the optimum condition based on the value of the separation factor, Eu and Gd were separated from their mixed solution. The results are shown in figure 3 and figure 4. The outcome of the Eu and Gd separation on different acid medium, HNO_3 and HCl are shown in figure 3 and expressed as *D* versus concentration. For both acidic media, the separation is higher using [A336][P204] as the extractant compared to the P204. In the separation using the HCl, the effect of the acid was evident in lower acid concentration. Once the acid concentration exceeds 3.0 M, the distribution for both acids was quite similar. The order of the distribution coefficient are as follows; [A336][P204] in HNO_3 > P204 in HNO_3 > [A336][P204] in HCl > P204 in HCl. Guo et al. reported a similar result showing a high separation of RE(III) in the nitrate medium compared to the chloride medium when [A336][P204] was used as an extractant (10).

Sun et al. [9] suggested that neutral dissociation is the primary extraction mechanism for [A336][P204] in the HNO_3 medium [11]. They reported that the distribution percentage of Eu(III) was 11 times greater in HNO3 than in HCl ($DHNO_3$ = 21.7, $DHCl$ = 1.97). However, in this research, when the same extractant was used, the largest distribution ratio for HNO_3 compared with HCl was only 0.7 times greater. The changes in the extraction capabilities in a different medium were due to the property of the anion in the system (NO_3^-, Cl^-, and SO_4^{2-}), which regulate the stability and selectivity of the extractant in the system [10]. While it can be verified that HNO_3 is the best extraction medium for this study, the detailed mechanism is not fully understood.

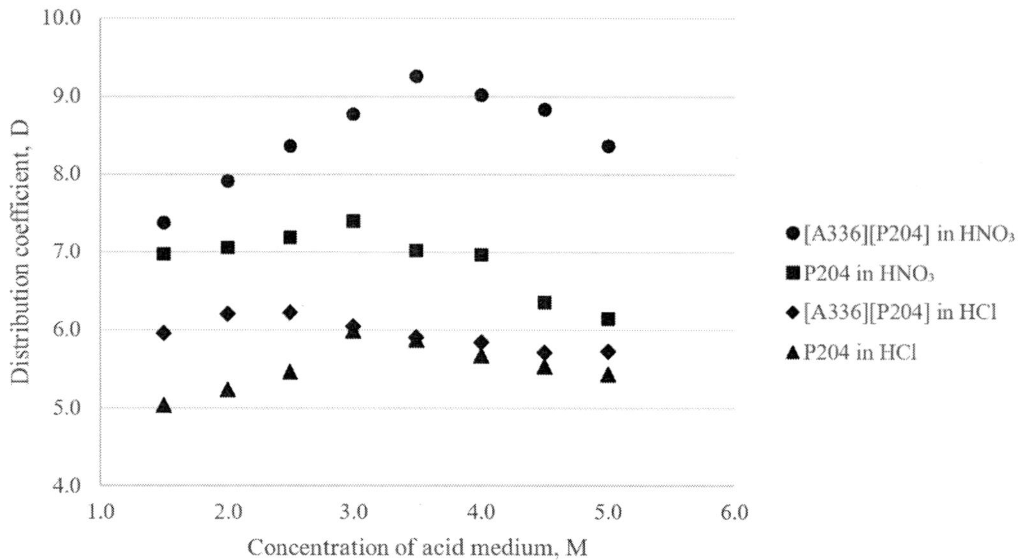

Figure 3. Effect of acid concentration on the distribution coefficient of Eu and Gd

Figure 4 explores on the effect of different O/A phase ratios on the separation of Eu and Gd. As displayed on the previous graph, since both [A336][P204] and P204 showed greater separation capabilities in the nitrate medium, the effect of O/A phase ratio was studied in their preferred medium. The trend showed increasing distribution coefficient with higher organic content for both extractants. The larger volume of organic compared to aqueous, increase the possibility of the metal to move and complex with the extractant in the organic medium. A similar finding was also discovered by other researchers [12] and [3]. Based on figure 3 and figure 4, the best condition for Eu and Gd separation using [A336][P204] (β: 3.44) would be in HNO_3, with O/A ratio of 9:1. Meanwhile, HCl and O/A ratio of 9:1 would be the best condition for Eu separation using P204 (β: 2.38).

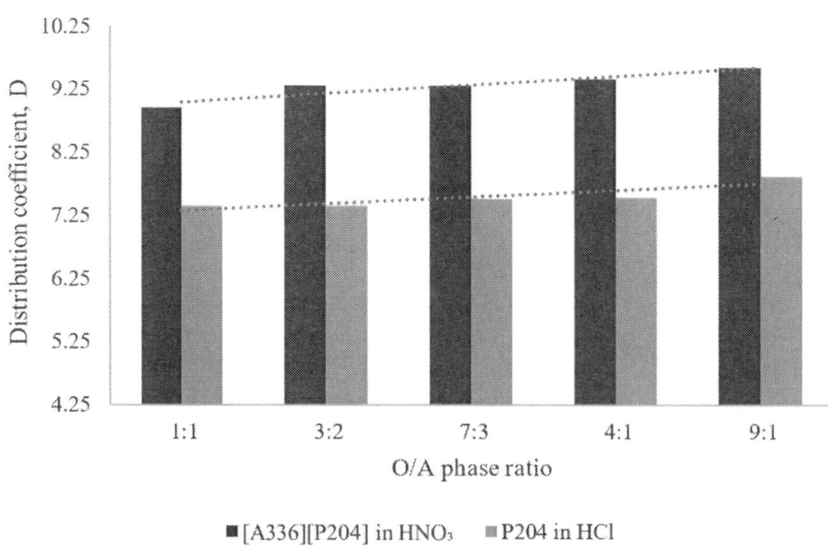

Figure 4. Effect of O/A phase ratio on the separation of Eu and Gd mixed solution

4.2. Extraction Stages of P204 and [A33][P204]

Two extractants, a common extractant, P204, and a synthesized bifunctional ionic liquid, [A336][P204], were chosen to compare and evaluate the effectiveness for Eu and Gd separation. The optimum condition for both extractants was explored and the value of β was determined. The amount of the theoretical extraction and scrubbing stages were calculated based on the experimental information as shown in table 2. [A336][P204] needed four complete extraction stages and two scrubbing stages, to separate Eu with 99.99% purity and 90% recovery, while P204 needed eight extraction stages and five scrubbing stages. Both processes were extraction controlled and the schematic diagrams are as in figure 5.

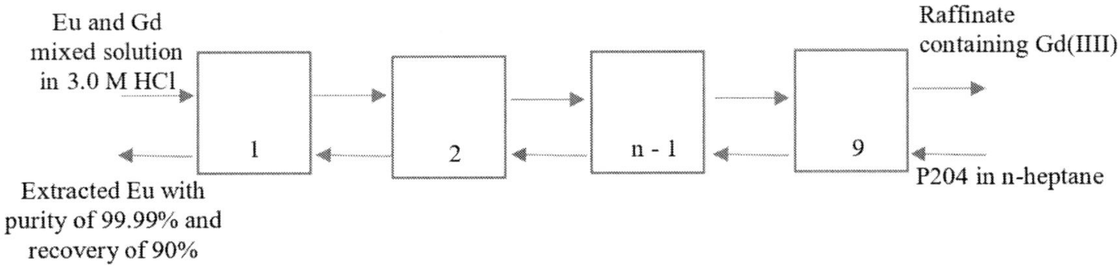

(a) P204

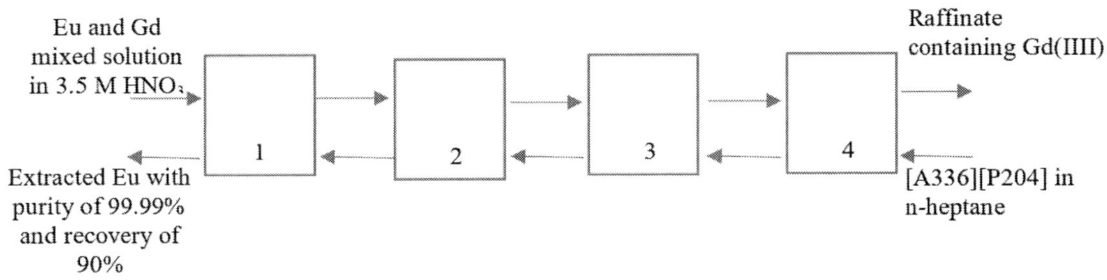

b) [A3336][P204]

Figure 5. Countercurrent process of Eu and Gd separation by (a) P204 (b) [A336][P204]

5. Conclusion

This study focuses on the model development of the countercurrent process for Eu and Gd by P204 and [A336][P204]. The modelling was based on the Counter-current Theory and the least number of extraction and scrubbing stages were preferred. Under the optimized condition of 3.0 M of HCl and 9:1 of O/A, the β for the separation of Eu and Gd using P204 was 2.38. For [A336][P204], the β was 3.44, under the optimized condition of 3.5 M of HNO_3 and 9:1 of O/A. The [A336][P204] was proved to have higher separation capabilities based on the value of the β. The total number of extraction and scrubbing stages was calculated based on the information from the experimental work. P204 required a total of thirteen stages (eight stages for extraction and five stages for scrubbing) to extract Eu from Gd with 99.99% purity and 90% recovery, while [A336][P204] required only six stages (four for extraction and two for scrubbing). To conclude, the [A336][P204] offered better separation capacities, thus reducing the number of extraction stages required.

Table 2. Simplified theoretical output for Eu and Gd separation using P204 and [A336][P204] as the extractants

STEP 1	*System condition*	*Extractant*	P204	[A336][P204]
		Medium	3.0 M HCl	3.5 M HNO_3
		O/A phase ratio	9:1	9:1
		Group A	Eu	Eu
		Group B	Gd	Gd
	Separation Factor	$\beta_{Eu/Gd}$	2.38	3.44
STEP 2	*Mole fraction in feed solution*	f_A	0.36	0.37
		f'_A	0.32	0.32
		f_B	0.64	0.63
		f'_B	0.68	0.68
	Target purity of the product	P_A	0.9999	0.9999
		P_B	0.0001	0.0001
	Percentage of the product recovery	Y_A	0.90	0.90
		Y_B	0.1	0.1
	Concentrating factor	a	17,776	17,025
		b	9.999	9.999
STEP 3	*Average extraction factor*	E_M	0.65 (extraction controlled)	0.54 (extraction controlled)
STEP 4	*Number of extraction stage*	n	≈ 8 stages	≈ 4 stages
	Number of scrubbing stage	m	≈ 5 stages	≈ 2 stages

Acknowledgement
The research was supported by the Universiti Malaysia Pahang under the Internal Grant Scheme RDU 160339 and RDU 190160.

References

[1] Peppard DF, Mason GW, Maier JL, Driscoll WJ. Fractional extraction of the lanthanides as their di-alkyl orthophosphates.1957 *J Inorg Nucl Chem.* **4**(5–6) 334–43.

[2] Liu T, Chen J, Li H, Li K, Li D. Further improvement for separation of heavy rare earths by mixtures of acidic organophosphorus extractants. 2019 *Hydrometallurgy* **188** 73–80.

[3] Ye Q, Li G, Deng B, Luo J, Rao M, Peng Z, et al. Solvent extraction behavior of metal ions and selective separation Sc^{3+} in phosphoric acid medium using P204. 2019 *Sep Purif Technol* **209** 175–81.

[4] Chen L, Chen J, Li H, Yang M, Zou D, Deng Y, et al. Applying basic research on a dialkylphosphoric acid based task-specific ionic liquid for the solvent extraction and membrane separation of yttrium. 2018 *Sep Purif Technol.* **207** 179–86.

[5] Khodakarami M, Alagha L. Separation and recovery of rare earth elements using novel

ammonium-based task-specific ionic liquids with bidentate and tridentate O-donor functional Groups. 2019 *Sep Purif Technol* **232** 1-12
[6] Devi N, Sukla LB. studies on liquid-liquid extraction of yttrium and separation from other rare earth elements using bifunctional ionic liquids.2019 *Miner Process Extr Metall Rev.* **40**(1):46–55.
[7] Ismail NA, Hisyam A, Shariff S, Aziz BA. Interaction, mechanism and synergism of rare earth in [A336][DEHPA] extraction system. 2016 *Aust J Basic Appl Sci* **10**(17):48–54.
[8] Xu G. Theory of countercurrent extraction: equations of optimization and their application. 1978 *Rare Earth Niobium I.* 67–75.
[9] Sun X, Yang J, Yu L, Ji C, Deqian L. An engineering-purpose preparation strategy for ammonium type ionic liquid with high purity. 2010 *Am Inst Chem Eng.* **56**(4) 989–996.
[10] Guo L, Chen J, Shen L, Zhang J, Zhang D, Deng Y. Highly selective extraction and separation of rare earths (III) using bifunctional ionic liquid extractant. 2014 *ACS Sustain Chem Eng.* **2**(8) 1968–1975.
[11] Sun X, Ji Y, Hu F, He B, Chen J, Li D. The inner synergistic effect of bifunctional ionic liquid extractant for solvent extraction. 2010 *Talanta.* **81**(4–5):1877–1883.
[12] Ye S, Jing Y, Wang Y, Fei W. Recovery of rare earths from spent FCC catalysts by solvent extraction using saponified 2-ethylhexyl phosphoric acid-2-ethylhexyl ester (EHEHPA). 2017 *J Rare Earths*. 35(7) 716–722.

Diffusion coefficients of trivalent rare earth/actinide ions in acid and alkaline aqueous solutions

M S Liyananadirah, M R Deraman, W H Ibrahim* and N Harun

Faculty of Chemical and Process Engineering Technology, College of Engineering Technology, Universiti Malaysia Pahang, Lebuhraya Tun Razak, 26300 Gambang, Kuantan, Pahang, Malaysia.

*E-mail: anisah@ump.edu.my

Abstract. Leaching is one of the key process of rare earth (RE) element recovery. This paper was set out to validate the effectiveness of acid/alkali in RE leaching by measuring the limiting ionic conductivity (λ_o), self-diffusion coefficient (D_i) and overall diffusion coefficient (D_{AB}) to obtain a clear insight on the transport properties of ions and salts in aqueous solution. The calculation diffusivity of D_i which is diffuse individual ions in pure water and D_{AB} diffuse in chemical reagents were determined using Nernst-Haskell equation, applicable for 25°C. The results of D_i were found to be consistent with experimental data from the literature where most of calculation and experiment data are less than 5% error. The final outcome had managed to confirm that NaOH is best reagent for most RE element leaching, which suits the common practice of the industry.

1. Introduction

Technically, Rare Earth (RE) elements are shown as a series of 15 lanthanide elements, as well as scandium and yttrium, in the periodic table [1]. In the new era of globalization, the industry has a renewed interest in benefiting RE elements as main components in electronics due to the unique physicochemical properties that the elements possess [2]. However, along with its popular demands, the extraction of RE elements from mineral ores remains a major issue. Considering the recent trends in technology, science experts have been urged to explore ways that would be effective to obtain RE elements and overcome the concerns for supply [3-4]. Studying the mechanism of RE extraction will provide important insights to manage the resources in the most cost-effective approach.

Generally, the overall RE processing consists of six important steps including deposit mining, beneficiation, chemical treatment, separation, refining and purification [5]. The pure RE elements exist in complex rocks, known as mineral ores and recovered in the beneficiation process [5-6]. Commercial sources of minerals and the respective elements contain bastnesite (La, Ce) FCO_3, monazite, (Ce, La, Y, Th) PO_4 and xenotime YPO_4 [2, 6]. Central to the subject of extracting pure RE elements is the chemical treatment step, in which RE elements are isolated from the mineral ore. The chemical treatment (or RE extraction) consists of two crucial sequential steps; first is solid liquid extraction (leaching) and liquid-liquid extraction (solvent extraction process). Leaching is the key process to maximize RE extraction and its recovery from ores. As highlighted by several authors, the chemical reagent used for the leaching process is the most essential factor to ensure effective RE recovery [7-8]. Chemical reagents that are strong acid and strong alkali are much favoured [9]. There are several common types of acid/alkali treatment including sulphuric acid (H_2SO_4), nitric acid (HNO_3), hydrochloric acid (HCl), sodium hydroxide (NaOH), and sodium carbonate (Na_2CO_3) [10]. Sulfuric acid (H_2SO_4) and caustic soda (NaOH) are mostly used by the industry. Chemical treatment in leaching process may take over to leach the RE elements concentration.

In order to apply hydrometallurgy for RE leaching, it is essential to acquire the thermodynamic aspects and quantities such as enthalpies, entropies, standard partial molar volume, resistivity, thermoelectric power, and thermal conductivity; in a case concerning RE elements, the data for trivalent RE and actinide ions have been studied by several authors [11-12]. What remains scarce is the information on limiting transport properties like limiting ionic conductivity (λ_o), and the self-diffusion coefficient (D_i) [12]. The study by Mauerhofer et al. [12] calculated the λ_o based on Stokes-Einstein law and D_i using Nernst-Einstein relation for trivalent RE and actinide ions dissolved in pure water at 25 °C and found that the calculation results match the experimental data. It is important to unravel the mechanism of the process by measuring the diffusivity of electrolytes by considering the ionic dissociation of acid/alkali in the medium, relative for the individual lanthanide and actinide ions. To date, the impact of acid/alkali in RE extraction and diffusivity has long become the subject of interest. Current literature successfully tested the effectiveness of reagents in RE diffusivity, many of which using the value of D_i as a method of measure. Surprisingly, most scholars have yet to provide a systematic evaluation of leaching RE using acid/alkali in terms of the mechanism of the process. In other words, most have neglected to extend their scope to include the behaviour of the ionic liquid which could be assessed using the D_{AB} value. This study was designed specifically to understand the transport of RE elements in the leaching medium, as well as substantiate the diffusivity of cations and anions of acid/alkali reagents by determining the overall diffusion coefficient (D_{AB}) as a measure of efficacy for RE leaching. The objective of this paper is to describe RE leaching process with involvement of acid/alkali reagents in terms of the self-diffusion D_i and overall diffusion coefficient D_{AB} by using Nernst-Haskell equation.

2. Methodology

Figure 1 shows proposed method for acid/alkaline reagents selectivity of RE leaching proses. The relative effect of acid and alkali as transport reagents to assist diffusion of RE salt was evaluated in this study. The self-diffusion coefficient (D_i) resembles the diffusivity of the lanthanide and actinide ions, while overall diffusion coefficient (D_{AB}) measures the ability of an acid/alkali in contributing to diffusion of the RE ions in the ionic medium. First, the values of D_i were calculated using λ_o data from Spedding et al. [13]. The D_i for individual ion, in a diluted single salt solution at 25°C can be calculated from Nernst-Haskell equation as shown in equation (1) [14-15].

$$D_i = 2.66\ x\ 10^{-7}\ \frac{\lambda_o}{n_o} \tag{1}$$

where λ_o is the limiting ionic conductance and n_o is the charge of valence of the ion species.

The limiting ionic conductivity (λ_o) represents the magnitude of conductivity caused by per mole ions in a diluted solution at 25 °C [16]. A tabulated data on the value of λ_o for various ions are easily available from Astle and Beyer [17]. Meanwhile, the experimental value of λ_o for RE and actinide ions have long been reported by Spedding et al. [18]. Meanwhile, Mauerhofer et al. [12] studied λ_o and D_i of lanthanide ions in pure water at 25°C. The calculated D_i were then compared with the experimental data from literature data.

After that, the values of D_i were used to calculate the values of D_{AB} which refers to the diffusivity of cation and anion components in an acid/alkali used for leaching. The equation of molecular diffusivity in a diluted solution (D_{AB}) as shown in equation (2) includes the valence of the cation and anion, as well as ionic conductance respective of each cation and anion species. In a case where two or more chemical species are present at different concentrations, inter-diffusion (counter diffusion) must be included to satisfy electro-neutrality [13]:

$$D_{AB} = 8.928 \, X \, 10^{-10} T \frac{\frac{1}{n_+} + \frac{1}{n_-}}{\frac{1}{\lambda_+} + \frac{1}{\lambda_-}} \quad (2)$$

where D_{AB} is the molecular diffusivity in a diluted solution in cm^2/s while n_+ and n_- are the valence of cation and anion. λ_+ and λ_- are the limiting ionic conductance for cation and anion, respectively.

The purpose to calculate D_i is to validate λ_0 value, which is obtained from Spedding et al [13]. After validation of the results, λ_0 is employed in D_{AB} calculation where the results and findings are discussed and presented in Section 3.

3. Results & Discussion

3.1. Transport Properties of Self Diffusion Coefficient (D_i)

The experimental values of D_i for trivalent RE and actinide ions were obtained from literatures that had performed an empirical study on RE extraction employing the open-end-capillary tube method [13, 19-20]. The experiment was performed under standard condition at 25°C and 1 atm. The calculated values for D_i was determined by using equation 1 and the results are demonstrated in figures 1 and 2, while detailed numbers are tabulated in table 1.

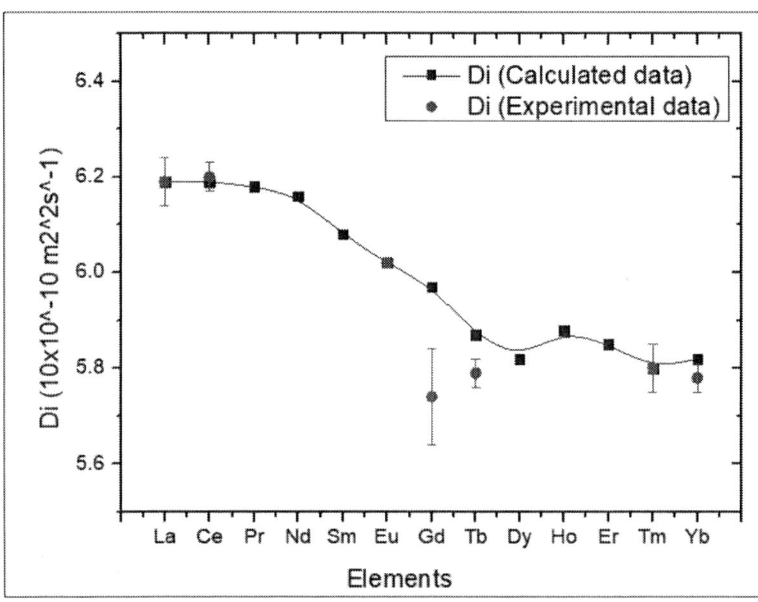

Figure 1. Self-diffusion coefficient (D_i) for lanthanide elements. Mean value and standard deviation error of experiment data (circle).

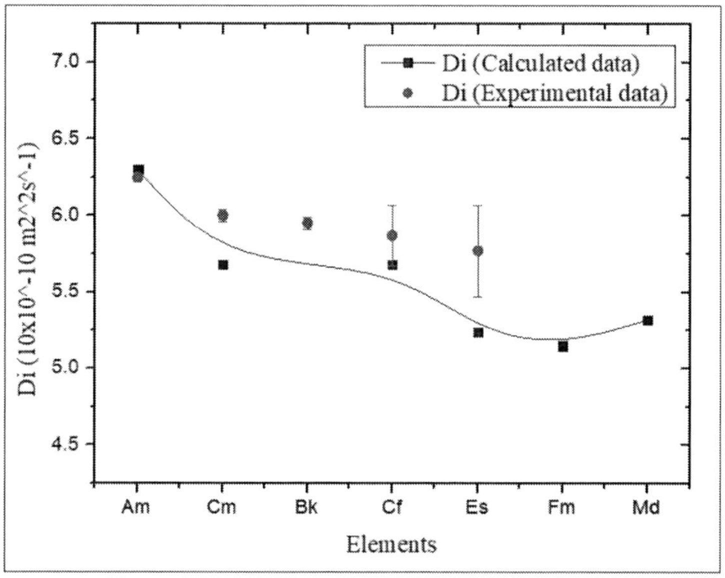

Figure 2. Self-diffusion coefficient (D_i) for actinides elements. Mean value and standard deviation error of experiment data (circle).

The plotted graphs in figures 1 and 2 clearly indicate that D_i is an important predictor of the efficacy of a leaching system. It appears that the results obtained by calculation are positively good agreement with the experimental data except for Gd^{3+} and Tb^{3+}. This high error is due to some experimental error to find λ_0 for both elements. However, this error can be tolerated because only two elements are deviate compare to the other five elements which is positively agreement with experiment data.

On the other hand, the predicted values for C_m, E_a and F_m elements from actinide series were underestimated with minimum errors of less than 10%. Despite the deviation, the trend of plots of the calculated results is consistent with the experimental data. It is rather motivating to find that the projection exhibited in both figures is in fact similar with finding by Mauerhofer et al. [12]. Looking from another perspective, it is somewhat noticeable that the decreasing trend of D_i for both lanthanide and actinide cations may have a correlation with the atomic weight of each individual ions. Literature studies revealed that smaller elements possess lower atomic weight, which eventually affect the value of its D_i.

3.2. Overall Diffusion Coefficient (D_{AB}) in Acid/Alkaline Aqueous Solution

The value of D_{AB} is calculated to define the characteristic of H_2SO_4, HCl, HNO_3, NaOH, or Na_2CO_3 in leaching each of the RE and actinide ions. Nernst-Haskell equation was used to determine the value of overall diffusivity coefficient (D_{AB}) with λ_o obtained from Spedding et al. [16], calculated for standard condition at 25°C and 1 atm. Currently, the experimental values of D_{AB} for RE and actinide ions are not available in literature except for D_{AB} value $EuCl_3$ as shown in table 1. A satisfactory agreement between the predicted result and the experimental data were observed for element $EuCl_3$ at 25°C. The D_{AB} value of $EuCl_3$ calculated that the model is in good prediction which is the result similar to experimental data by Ribeiro et al. [21]. However, the others experimental values of D_{AB} are not available. Considering the important of knowledge contribution in RE extraction process, the prediction of D_{AB} values could be useful in preliminary study and as a reference in industrial application.

Table 1. The experimental data and the calculated results of D_i and D_{AB} for trivalent RE and actinide series ions at standard condition (T=25°C and P =1 atm).

RE Elements	λ_o [20]	D_i (Experimental Data)	D_i (This study)	$\Delta[D_{i(exp)} - D_{i(cal)}]$	D_{AB} (This Study) (m^2/s)				
(charges)	10^{-4} m^2S eq^{-1}	10^{-10} m^2s^{-1}	10^{-10} m^2s^{-1}	%	H_2SO_4	HCl	HNO_3	NaOH	Na_2CO_3
Lanthanide									
La (3+)	69.7	6.19 c	6.19	0	8.26E-10	1.29E-09	1.25E-09	1.74E-09	7.47E-10
Ce (3+)	69.8	6.20 a	6.19	0.2	8.27E-10	1.29E-09	1.25E-09	1.74E-09	7.45E-10
Pr (3+)	69.6	-	6.18	-	8.26E-10	1.29E-09	1.25E-09	1.73E-09	7.42E-10
Nd (3+)	69.4	-	6.16	-	8.24E-10	1.29E-09	1.25E-09	1.72E-09	7.40E-10
Pm (3+)	NA	-	NA	-	NA	NA	NA	1.71E-09	7.38E-10
Sm (3+)	68.5	-	6.08	-	8.19E-10	1.28E-09	1.24E-09	1.71E-09	7.38E-10
Eu (3+)	67.8	6.02 a	6.02	0	8.14E-10	1.27E-09	1.23E-09	1.72E-09	7.38E-10
Gd (3+)	67.3	6.04 b	5.97	1.2	8.11E-10	1.27E-09	1.23E-09	1.70E-09	7.34E-10
		5.74 a		4	7.76E-10	1.22E-09	1.18E-09	1.70E-09	7.34E-10
Tb (3+)	66.1	5.79 a	5.87	1.4	8.00E-10	1.25E-09	1.21E-09	1.70E-09	7.33E-10
Dy (3+)	65.6	-	5.82	-	8.04E-10	1.26E-09	1.22E-09	1.70E-09	7.35E-10
Ho (3+)	66.3	-	5.88	-	8.02E-10	1.26E-09	1.22E-09	1.70E-09	7.35E-10
Er (3+)	65.9	-	5.85	-	7.98E-10	1.25E-09	1.21E-09	1.70E-09	7.34E-10
Tm (3+)	65.4	5.80 a	5.8	0	8.00E-10	1.25E-09	1.21E-09	1.70E-09	7.35E-10
Yb (3+)	65.6	5.96 b	5.82	2.7	NA	NA	NA	1.71E-09	7.36E-10
		5.78 a		0.7	7.94E-10	1.24E-09	1.21E-09	1.74E-09	7.00E-10
Lu (3+)	NA	-	NA	-					
Sc (3+)	64.7	-	5.74	-					
Actinide									
Ac (3+)	-	-	-		NA	NA	NA	NA	NA
Th (3+)	-	-	-		NA	NA	NA	NA	NA
Pa (3+)	-	-	-		NA	NA	NA	NA	NA
U (3+)	-	-	-		NA	NA	NA	NA	NA
Np (3+)	-	-	-		NA	NA	NA	NA	NA
Pu (+3)	-	-	-		NA	NA	NA	NA	NA

RE Elements	λ_o [20]	D_i (Experimental Data)	D_i (This study)	$\Delta[D_{i(exp)} - D_{i(cal)}]$	D_{AB} (This Study) (m^2/s)				
(charges)	10^{-4} m^2S eq^{-1}	10^{-10} m^2s^{-1}	10^{-10} m^2s^{-1}	%	H_2SO_4	HCl	HNO_3	NaOH	Na_2CO_3
Am (+3)	71	6.25	6.3	0.8	1.09E-09	1.31E-09	1.26E-09	1.85E-09	7.78E-10
Cm (+3)	64	6	5.68	5.3	1.01E-09	1.24E-09	1.20E-09	1.72E-09	7.38E-10
Bk (+3)	-	5.95	-	-	NA	NA	NA	NA	NA
Cf (+3)	64	5.87	5.68	3.2	1.01E-09	1.24E-09	1.20E-09	1.72E-09	7.38E-10
Es (3+)	59	5.77	5.24	9.2	9.56E-10	1.18E-09	1.15E-09	1.61E-09	7.07E-10
Fm (3+)	58	-	5.15	-	9.44E-10	1.17E-09	1.14E-09	1.59E-09	7.01E-10
Md (+3)	60	-	5.32	-	9.68E-10	1.19E-09	1.16E-09	1.63E-09	7.13E-10
No (+3)	-	-	-	-	NA	NA	NA	NA	NA
Lr (+3)	-	-	-	-	NA	NA	NA	NA	NA

Furthermore, the results from this calculation are exhibited in figures 3 and 4. These results are obtained from table 1 and demonstrated in graph for further understanding. It appears that NaOH (pink bar) exhibits the highest value of D_{AB} for each lanthanide and actinide components compared to the others reagent (refer figure 3). The high quantity of D_{AB} shows the high potential acid/alkaline reagent extract RE elements. These results manage to provide a clear quantitative evidence that NaOH is the best reagent, as per claimed by a great deal of research [22]. NaOH had managed to obtain 90% REE, leached within 5 hours with mass ratio of 1:2–1:4 NaOH/RE [10]. The contribution of NaOH concentration towards the extraction kinetics and alkali consumption for 5 hours of extraction time, increasing the concentration of NaOH by 50% of the initial concentration had rapidly improve time consumption to become 1 hour.

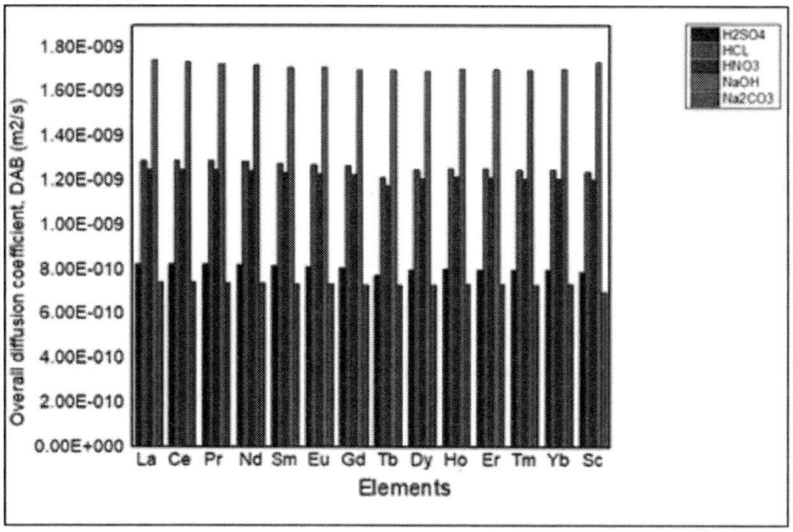

Figure 3. The values of D_{AB} versus RE elements for different types of leaching reagent.

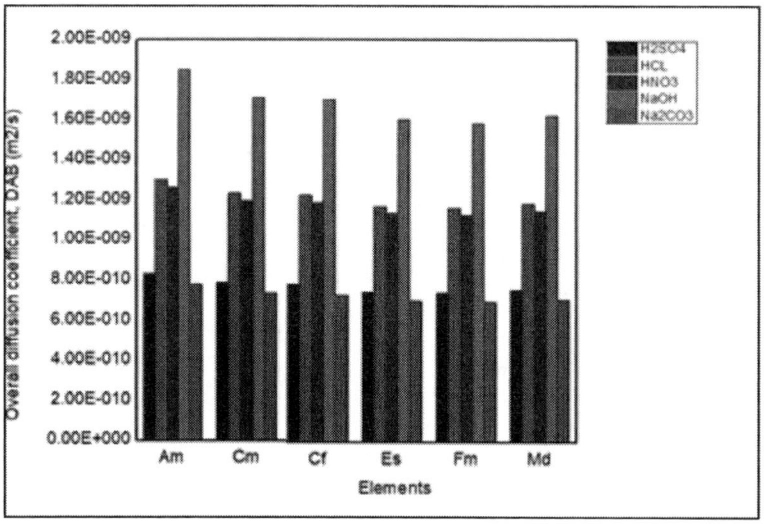

Figure 4. The values of D_{AB} versus actinide series for different types of leaching reagent.

Nonetheless, the industry extracting RE from monazite and bastnaesite are indeed using NaOH, among others [23]. Another significant analysis by Peelman et al. [23] is also in accord with our finding that NaOH is undeniably practical, as they recovered RE elements from bastnaesite with 90% efficiency. It is demonstrated that the bastnaesite was milled with NaOH powder, washed with water to remove the Na compounds, and finally leached with HCl.

In this study, the pattern in figure 3 is identical with figure 4. From these figures show that both RE and actinide series appeared to be affected by H_2SO_4 and HCl in the same manner. A study by researchers had compared these acids while leaching apatite-base ore [24]. It is simulating to discover that our calculation of D_{AB} for HCl is higher than H_2SO_4, which is parallel with the observation of Kim and others [24]. As clearly shown in figures 3 and 4, each ion of the lanthanide and actinide series are undeniably more influenced by HCl than H_2SO_4. Based on the calculated data, we concluded that RE elements are favour to combined with NaOH or HCl. These findings related to the limiting anionic conductivity (λ_-), which is NaOH has high λ_- which is 197.6 [14], thus, increasing the mobility of RE element from its ore. To make the comparison, the limiting anionic conductivity (λ_-) of Cl^-, NO_3^-, SO_4^- and CO_3^- are 76.3, 71.4, 80 and 69.3, respectively [14, 16]. Literature findings say that limiting ionic conductivity (λ_o) and D_{AB} are interrelated, by way that ionic conductivity acts as a driving force which stimulates higher diffusivity rate of RE elements [25].

According to these data, it shows that the Nernst-Haskell equation can be used to validate that D_{AB} is directly influenced by value of λ_-. Thus, it could conceivably be hypothesized that higher D_{AB} suggests better leaching performance as a result of the counter ion effect [25]. This phenomenon can be clearly explained by the ionic diffusion theory [26]. For so many years, diffusion is often considered as a neutral process involving only neutral atoms or molecules. Recent practice in RE leaching has included charged particles to improve the kinetics of the process as the result of electric field induced within the system or even externally imposed [26]. In a case of RE extraction, presence of anions from the acid/alkali attracts cation particles in the mineral ore, thus, explains the higher diffusivity rate.

Consider a salt from RE, for example $LaOH_3$ (a=1, b=3). As the molecule dissolves in water, it separates into 1 La^{3+}, and 3 OH^- in a process known as ionization. The chemical equation as shown in equation (3) simplifies the ionization of particles in water.

$$Aa + Bb \rightarrow aA^{c+} + bB^{c-} \quad (3)$$

Diffusion of Aa and Bb (denotes A and B ions) in the medium may be regarded as molecular or a coupled diffusion of AaBb, wherein 'a' number of cations moving in a homogeneous medium with 'b' number of anions.

5. Conclusions

The objective of this paper is to describe RE leaching with involvement of acid/alkali reagents in terms of the self-diffusion (Di) and overall diffusion coefficient (D_{AB}) by using Nernst-Haskell equation. We can conclude that, along with this calculation and the findings of this research have strengthened the idea that presence of ionic particles from the acid/alkali reagents could increase the leaching rate of RE elements from its ore. The calculated value of D_{AB} using Nernst-Haskell equation established here are comparative with experimental findings from various sources of literature, valid for most of the trivalent RE and actinide ions. However, contrary to expectation, a further work is required to confirm and assess the viability of limiting transport properties for the actinides.

Other than that, these findings provide some support knowledge that a strong alkali such as NaOH has a high D_{AB}, and thus, a better reagent compared to the other reagents. Recent study has successfully applied a mathematical approach to determine D_i and D_{AB} data. An implication of these findings is the possibility to study the thermodynamic properties for unique elements such as the RE elements. Nevertheless, a preliminary work to obtain data for D_{AB} through experiments is emphasized in the future to fill the research gap.

Acknowledgements
The first author is grateful for support by the Fundamental Research Grant Scheme under the Ministry of Education, Malaysian Government (RDU160123 and GRS1403175). The authors would like to thank Dr. Sureena Abdullah and Mohd Ikhwan bin Suhaimi for helpful discussions. The authors also wish to acknowledge Mybrain15 and Faculty of Chemical and Process Engineering Technology, Universiti Malaysia Pahang.

References
[1] Liao C, Sheng W U, Cheng F, Songling W A N G, Yan L I U, Zhang B and Chunhua Y A N 2013 *J. Rare Earths* **31** 331–36
[2] Xie F, An T, Dreisinger D and F Doyle 2014 *Miner. Eng.* **56** 10–28
[3] Hugo Royen I and Uwe Fortkamp 2016 *Rare Earth Elements - Purification, Separation and Recycling*
[4] Kumari A, Panda R, Jha M K, Kumar J R, and Lee J Y 2015 *Miner. Eng.* **79** 102–115
[5] Suli L M, Ibrahim W H W, Aziz B A, Deraman M R and Ismail N A 2017 *Chem. Eng. Res. Bull.* 19 20
[6] Jordens A, Cheng Y P and Waters K E 2013 *Miner. Eng.* **41** 97–114
[7] Tong S, Zhao X, Song N, Jia Q, Zhou W and Liao W 2009 *Hydrometallurgy* **100** 15–19
[8] Habashi F 1985 *The Recovery of the Lanthanides from Phosphate Rock* **4** 5–14
[9] Lucas J, Lucas P, Rochelle L and Davenport W 2014 *Rare Earths Rare Earths Production and Use*
[10] Zhang J and Edwards C 2013 *Mineral decomposition and leaching processes for treating rare earth ore concentrates* **52** 243–48
[11] Amato A 1987 *J. Low Temp. Phys.* **68** 371–97
[12] Mauerhofer B E, Zhernosekov K and Rösch F 2003 *Limiting transport properties of lanthanide and actinide ions in pure water* **477** 473–77
[13] Fourest B, Duplessis J and David F 1984 *Radiochim. Acta* **36(4)** 191-96
[14] Geankoplis C J 2003 *Transport processes and separation process principles* (Prentice Hall Professional Technical)
[15] Reid R C, Prausnitz J M and Sherwood T K 1997 *The properties of gases and liquids* 3rd ed (New York: McGraw-Hill)
[16] Lou C 1999 *Curr. Sep.* **18** 91–96
[17] Astle M J and Beyer W H 1988 *CRC Handbook of Chemistry and Physics 69th Edition (*CRC)
[18] Spedding F H, Porter P E and Wright J M 1952 *J. Am. Chem. Soc.* **74** 2055–58
[19] Ouerfelli N, Ammar M and Latrous H 1994 J. de chimie physique **91** 1786-95
[20] Weingaertner H, Braun B M and Schmoll J M 1987 J. Physic. Chem. **91(4)** 979-85
[21] Ribeiro A, Valente A and Lobo V 2011 *Chemistry diffusion behaviour of trivalent metal ions* **5** 2
[22] Gupta C K and Krishnamurthy N 2016 *Extr. Metall. rare earths* **37** 5
[23] Peelman S, Sun Z H I, Sietsma J and Yang 2015 *Leaching of Rare Earth Elements: Review of Past and Present Technologies*. Elsevier Inc.
[24] Kim R, Cho H, Han K N, Kim K and Mun M 2016 *Optimization of Acid Leaching of Rare-Earth Elements from Mongolian Apatite-Based Ore*
[25] Kashyap H K, Pradhan T and Biswas R 2006 *J. Chem. Phys.* **125** 17
[26] Onsager L 1945 *Diifusion Electrolytes Macromol. Solut.* **46** 241–265

IOP Conf. Series: Materials Science and Engineering **702** (2019) 012046 doi:10.1088/1757-899X/702/1/012046

2^4 fractional factorial design for PVDF/Pebax film composite synthesis on gas selectivity study

M S Abdul Wahab[1], A R Sunarti[1*], N H Ramli[1] and A L Ahmad[2]

[1]Faculty of Chemical and Process Engineering Technology, College of Engineering Technology, Universiti Malaysia Pahang, 26300 Gambang, Pahang, Malaysia
[2]School of Chemical Engineering, Engineering Campus, Universiti Sains Malaysia, 14300 Nibong Tebal, Pulau Pinang, Malaysia

*E-mail: sunarti@ump.edu.my

Abstract. One of the key indicator for membrane film excellencies is by having a good ideal selectivity, and it is defined as the permeability ratio of two pure gases which be separated. This study focused on factors screening and interaction between involved factors in membrane film development on gas selectivity by employing 2^4 fractional factorial design (FFD). A total of four factors; Pebax concentration, bath temperature, evaporation times and quenching times were chosen to run simultaneously and ideal selectivity was chosen as a single respond. Order of contribution was found to be as follows; Pebax concentration > quenching times > bath temperature > evaporation times, while the most impacting factors towards gas selectivity was Pebax concentration (main effect), evaporation times - Pebax concentration (interaction effect), quenching times (main effect). The composite film produced via this design of experiment managed to obtain higher barrer of CO_2 permeability of 393 and selectivity of 64.75. The outcome of the study indicates a very strong judgement in utilizing FFD to minimize and eliminate factors by considering the interaction among the factors involves in membrane film development for an ideal gas selectivity.

1. Introduction

Many research has been done on how to improve gas selectivity and performance in film composite development. Material concentration contributes to a better membrane synthesis. Some said that porosity and void can be minimized with an increasing polymer concentration [1]. But that is in the case of porous media. Selective material works a bit different way as they will be coated on top of the porous substrate. The selective material was suggested to be very dilute as a concentrated solution could thicken the layer, increase the mass transfer resistant and lead to loss in gas selectivity.

Defect in composite membranes is either incomplete coverage of the porous support or solution penetration by the thin selective solution. These two factors related to each other due to relationship between solution concentration and morphological changes. If the concentration is too low, the tendency of penetration is high thus incomplete coverage of the film structure could occur, but, if it is too high, the TFC will have a thicker structure and the selectivity will drop drastically. Perfect range of Pebax suggested from recent study is around 3% to 5% in its respective binary solution [2-5]. In this range, researcher believed that even though there is a possibility of pore penetration, yet it still allows the gas transport excellently.

Solvent and non-solvent interaction in membrane occur in water coagulation bath where phase inversion is taking place. The condition of precipitation bath could affect the morphological changes of the film form. Every structural change might be due to the molecular entanglement which restrict the change mobility during film formation. The process of de-mixing of the solvent and non-solvent rapidly occurs in the coagulation environment where the verification takes place and the structure are completely frozen. Oprea and Ciobanu [6] in their work of the effect of bath temperature towards membranes morphology found out that at low bath temperature there will be a cellular porous structure enclosed with continuous polymer matrix phase and as the bath temperature increase, they start to observe a very symmetric 'sponge structure' across the film.

The aggregation of molecules at low bath temperature decrease the formation of uniform porous structure and possibly form a stiff dense film. At elevated temperature, molecules have a better chance and great mobility to arrange themselves faster and form a symmetrical film [6]. The morphological in dry-wet phase inversion techniques are strongly depends on the demixing rate either it is an instantaneous or delay. Macrovoid (finger-like cavities) region mostly associated with the instantaneous demixing of the solvent and non-solvent while the formation of spongy porous structure appointed to the delay demixing. The study of effect of Coagulant bath temperature (CBT) done by Amirilargani et al. [7] jotted a significant finding when they said increase of CBT in range of 0 – 25 °C have improved the formation of surface porosity from a dense region to a fine pores structure across the film top to bottom.

Phase inversion during film development also found to influenced the film physical morphology thus influenced the gas selectivity during separation process. The first period where the polymer precipitate is left to vaporize in an open air for a specific period on the casting support (glass) and then move to the second period which is the quenching step in a non-solvent bath. The longer evaporation time will cause the film interface concentration and area with high polymer concentration form a thicker and dense skin. Tsay and McHugh [8] observed an increasing anisotropic spongy sublayer with an increasing of evaporation time. This pores formation behaviour is related to the nucleation and growth mechanism of the polymer. The quenching step can take up to 48 hours as the procedure lead to membrane precipitation. The period of quenching depends on the material used to prepare the dope solution and the dynamic movement of the solvent and non-solvent.

The FFD is a screening method where all variables are varied together rather than the traditional 'one-at-a-time' techniques to stimulate the interaction between N design variables [9]. A range of minimum and maximum values of each N variables must be identified. The experiment design with two-level upper bound and lower bound of the N variables is called 2^{N-1} factorial. It is the first order-model that have less number of runs. In the past, application of FFD in membranes development has shown a great potential for this purposed [10-11]. Design Expert was used to initiate the experimental design and study the interaction effect among the factors.

2. Materials and methods

2.1. Materials

Polyvinylidene fluoride (PVDF) pallets and Polyether block amide under trade name Pebax 1657 was supplied by Sigma Aldrich and Arkema France respectively. Two analytical grade solvents, ethanol and N-Methyl-2-pyrrolidone (NMP) was purchased from Fisher Scientific. Two tanks of 99.9% purify CO_2 and CH_4 gas was used for gas permeation experiment. Both polymer pellets were oven dried at 60 °C for 24 hours before the dope preparation.

2.2. PVDF Substrate Preparation

15 wt% of PVDF pellets was uniformly stirred in 85 wt% NMP at 90 °C for 7 hours. After ensuring the complete homogeneity of the dope solution, it was then left for 24 hours at room temperature for degassing purpose to release any air bubble build up in the solution. The procedure was proceeded with dope casting with film applicator setting set to 0.03 mm knife gap, room temperature and 100 rpm of casting speed. Utilizing the dry/wet phase inversion technique, the casting plate was left to vaporize for 10 s to 240 s before it was completely immersed in a non-solvent medium (water bath) for 6 hours to 24

hours to let the liquid-liquid demixing take place. The wet PVDF film was then hanging to dryness at room temperature for another 24 hours. The complete experimental design with all parameter range was tabulated in Table 1.

2.3. PVDF/Pebax film composite development

Pebax 1657 coating solution was prepared first by dissolving 3 – 5 wt% of the pellets in 70:30 ethanol:water mixture at 80 °C until a clear homogenous solution was obtained. To ensure the consistency of the layer made, 5 mL of the Pebax 1657 coating solution was prepared in a flat dish. PVDF film was dip coated into the dish for 3 times, where there is a 15 minutes times interval in between the dip procedure. For a complete solvent vaporization, the TFC was further dried at 60 °C for 12 h.

2.4. Gas Permeability Setup

The selectivity data was obtained by measuring the permeation rate of gas species through the develop membrane film. Single gas permeation setup was prepared as in figure 1, CO_2 and CH_4 were tested individually by flowing each species at 2 bars toward the membrane cell, and the flow rate of the permeate was measured by the bubble flow meter. The effective cell area was 19.63 cm^2 with 5 cm film diameter.

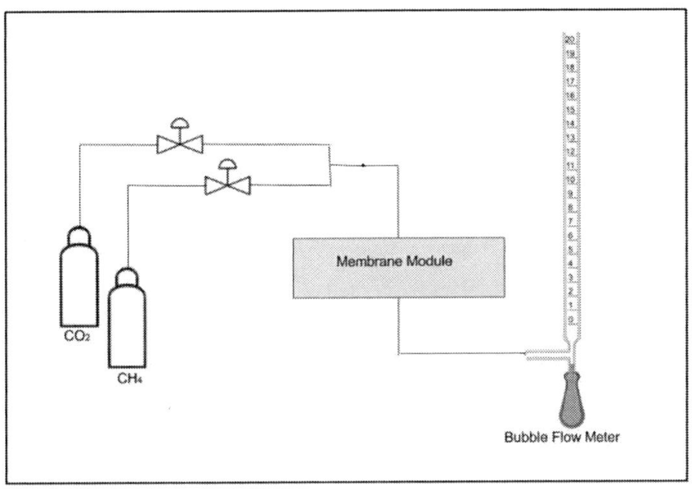

Figure. 1. Gas permeability experimental set-up

The permeability calculation was done by Equation 1 and it was expressed Barrer, where 1 Barrer = 1 x 10^{-10} cm^3 (STP) cm/(cm^2scmHg). Taking the conversion factor into account, Equation 2 can be directly used to find the gas permeability in Barrer.

$$P = \frac{lV}{At\Delta p} \ (\mathrm{cm^3 cm(stp))/(cm^2 scmHg)}] \quad (1)$$

$$P(Barrer) = \frac{1X10^{10} lV}{At\Delta p} \quad (2)$$

where P represent the film permeability in barrer, l is the film thickness in cm, A is the effective membrane area in cm^2, V is the volume in cm^3 displaced in time t(s) and Δp is the transmembrane pressure expressed in cmHg. The membrane selectivity is given in Equation 3, it is the membrane ability to separate two gases (A and B). It is the permeability ratio of species A and B respectively.

$$\alpha_{AB} = \frac{P_A}{P_B} \quad (3)$$

2.5 Factorial Design Methodology

For this purpose of study, the factors screening was aided by Design Expert 7.6.1 to analyze the factors interaction for the gas selectivity output. The end result of first order polynomial was expected from this study to fit the following model;

$$y = b_o + \sum_{i=1}^{n} b_i x_i \qquad (4)$$

Where y represents the predicted response, b_o is the constant coefficient, n is the number of factors involved, b_i is the linear parameter coefficient and x_i denoted the two-way interaction parameter coefficient. The FFD experimental factors and running were performed as shown in Table 1.

Table 1. FFD experimental design

Runs	Factor 1	Factor 2	Factor 3	Factor 4	Response
	A: Evaporation time (s)	B: Quenching time (h)	C: Bath temperature (°C)	D: Pebax concentration (%)	Selectivity
1	10	6	30	3	53.57
2	240	6	30	3	31.03
3	10	24	30	3	55.88
4	240	24	30	3	33.33
5	10	6	55	3	59.21
6	240	6	55	3	27.58
7	10	24	55	3	64.75
8	240	24	55	3	35.32
9	10	6	30	5	63.80
10	240	6	30	5	56.66
11	10	24	30	5	62.63
12	240	24	30	5	59.60
13	10	6	55	5	59.92
14	240	6	55	5	56.66
15	10	24	55	5	63.42
16	240	24	55	5	58.90

3. Result and discussion

3.1. Statistical Analysis of Gas Selectivity

Gas selectivity obtained from this study was tabulated in Table 1 with the lowest and highest data jotted at 27.58 and 64.75 respectively. The highest selectivity came from the film samples cast at the following condition; 10 s evaporation time, 24 h quenching time, 55 °C bath temperature and 3% wt Pebax concentration.

The input response was analysed for the significance of the regression model in the Design Expert Software and the analysis of variance (ANOVA) was given in Table 2. For a chemical process design, R^2 for an accepted model must be in range of 0.9 to 1. From the ANOVA analysis in Table 2, this design experiment has generated 0.99 for R^2, 0.9263 for "Pred R-Squared" and 0.9698 for "Adj R-Squared". The "Pred R-Squared" is in sensible assertion with the "Adj R-Squared", this two must be within 0.2 from each other to ensure the reliability of the model. If both R are widely differed the model might be unstable for future used [12]. The indication of model satisfactory gave by the "Adeq Precision" and it does measures the signal to noise ratio. It is the judgement factor whether the model is perfect or not to navigate the design space and able to predict the response. A ratio greater than 4 is desirable [13]. From the analysis, the "Adeq Precision" of 20.850 indicates an adequate signal.

The Model F-value of 61.15 implies the model is significant. It is a test for comparing model variance with the residual or error variance. It is calculated by dividing model mean square by residual mean square. There is only a 0.01 % chance that a "Model F-Value" this large could occur due to noise. Values of "Prob > F" less than 0.05 indicate model terms are significant. In other word, if the Prob > F value is less than 0.05, then the term in the model have a significant effect towards the response. In this case A, B, D, and AD are significant model terms for this design.

Coefficient of the variation, C.V. % expressed the standard deviation as a percentage of the mean. It is used to measure the spread of data to describe amount of variability related to the mean and a good fit should have C.V. less than 10 %. PRESS is a Predicted Residual Error Sum of Squares where it measures the ability of the model of this design experiment to perform and predict the response of a new experiment. A small figure is desirable for this part. The final mathematical models developed for this experimental design as given by design expert are as shown in equation 5.

$$\alpha = 52.65 - 7.76A + 1.59B + 0.58C + 7.56D - 0.85AC + 5.51AD + 0.79BC - 1.05CD \quad (5)$$

where A is the evaporation time, B is the quenching time, C is the bath temperature and D is the Pebax concentration while AC, AD, BC and CD is the interaction between the factors.

Table 2. ANOVA table of the regression model

Source	Sum of Squares	Mean Square	F Value	p-value Prob > F	
Model	2447.74	305.97	61.15	< 0.0001	significant
A-Evaporation time	962.78	962.78	192.43	< 0.0001	
B-Quenching time	40.27	40.27	8.05	0.0252	
C-Bath Temperature	5.38	5.38	1.07	0.3343	
D-Pebax Concentration	913.98	913.98	182.68	< 0.0001	
AC	11.55	11.55	2.31	0.1724	
AD	486.07	486.07	97.15	< 0.0001	
BC	9.99	9.99	2.00	0.2005	
CD	17.73	17.73	3.54	0.1018	
Residual	35.02	5.00			
Cor Total	2482.76				
Std. Dev.	2.24		R-Squared	0.9859	
Mean	52.65		Adj R-Squared	0.9698	
C.V. %	4.25		Pred R-Squared	0.9263	
PRESS	182.98		Adeq Precision	20.850	

3.2. Effect of Process Parameters and Interaction on Selectivity

The selectivity is straightforwardly identified with all the process variables investigated, either as a main or as a part of an interaction effect, as expressed prior. The explanation behind anticipating the selectivity is to build up a model, to help in the choice of a proper range for the TFC synthesis optimization (at a later stage). The pareto chart for the selectivity, as outlined in figure 2, highlights the order of the main and interaction parameter effects, which at last influence the selectivity model.

Figure 3 shows a perturbation plot highlighting the effect of all parameters on the selectivity. The perturbation plot allows the effect of all factors at a particular point in the design space to be compared. This type of display does not show the effect of interactions. Instead, the lines represent the behaviours of each factor while holding the others in a constant ratio (centre point by default). In the case of more than one factors this type of display could be used to find those factors that most affect the response.

The order of the level of significance of the positive effects of the TFC synthesis parameters on the selectivity follows the order: (D) > (AD) > (B) > (BC) > (C) while the order of the level of significance of the negative effects on the selectivity is as follows: (A) > (CD) > (AC). Henceforth, if large amounts of selectivity are to be gotten, the main three most impacting impacts for control are as per the following:

- Pebax Concentration (Main Effect)
- Evaporation Time × Pebax Concentration (Two-way Interaction Effect)
- Quenching Time (Main Effect)

The primary factor most affecting the selectivity appears to be the main effect of the Pebax Concentration. The coating solution was manipulated 3 % (level -1) and 5% (level +1), which later dip coated with PVDF film for 3 times. CO_2 permeability increase with increase in Pebax 1657 concentration mainly cause by the reducing of pore blockage. Because of large voids and free volume in PVDF substrate, the lower Pebax 1657 concentration make the solution very dilute and penetrate deep into membranes matrix. Beside the pore blockage of the porous support, having a different concentration could make the polymer itself become compact and decrease in gas transport ability. From a general view, the skin layer solution concentration highly dependent on the porous support structure in a way that the larger the void form in support film, the higher Pebax 1657 skin solution concentration needed to achieve a perfect film composite but still few factors regarding Pebax solution preparation needed to be taking care such as solution gelation and mixture stability.

The second factor most affecting the selectivity is the two-way interaction effect between the Pebax concentration and the Evaporation Steps. The relationship between the two factors is as illustrated in figure 4. The understanding of perfect film composite is a simple analogy, of which the perfect composite made by the combination of excellent well shaped porous support layer and a beautifully coated dense skin layer with a wide adsorption side. From the interaction plot, the highest response came from the shorter evaporation time (10 s) with high Pebax 1657 concentration (5%) while the lowest response from the combination of highest evaporation time (240 s) and lowest Pebax 1657 concentration (3%). The trend of the effectiveness of the factors towards the response mostly influenced by the evaporation time, as selectivity decrease with increase of evaporation time. A conclusion can be made saying that 10 s is enough for the solution to vaporize and create a perfect top thin dense skin layer for the PVDF film before it can be transfer to the quenching bath.

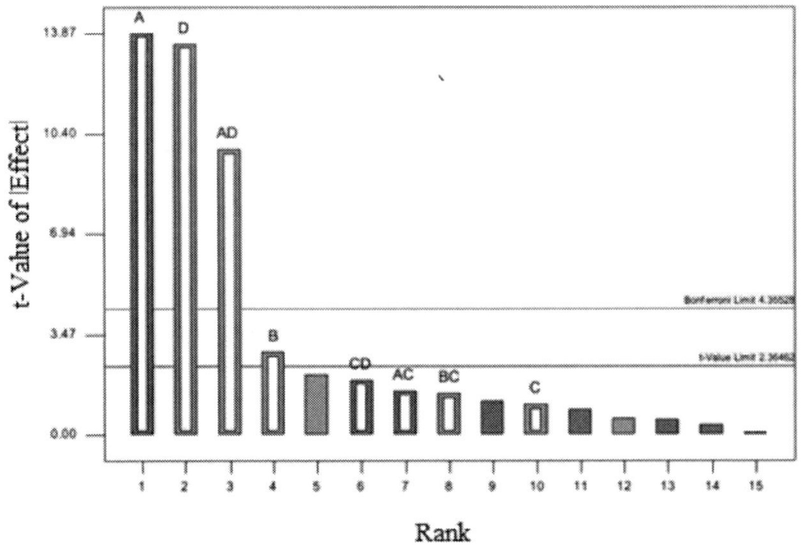

Figure 2. Pareto chart of main and interaction effect of TFC synthesis parameter (orange: positive effect and blue: negative effect)

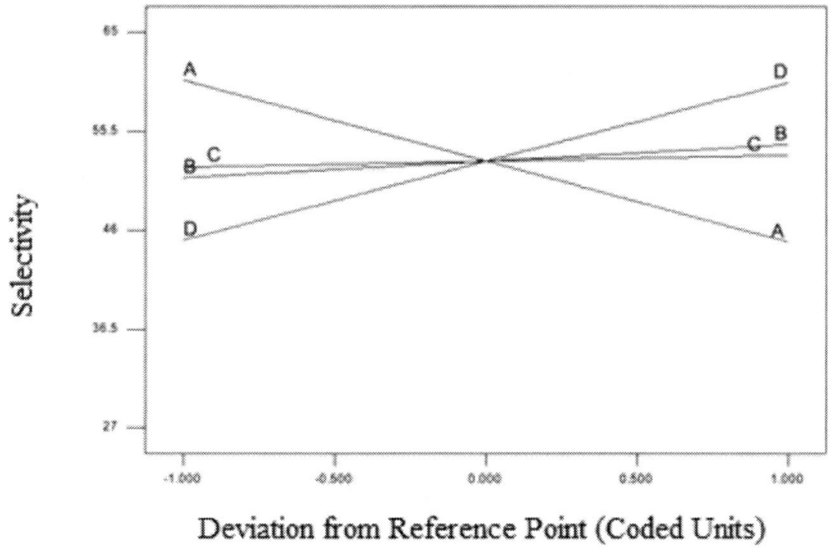

Figure 3. Perturbation plot showing the effect of all parameters on selectivity

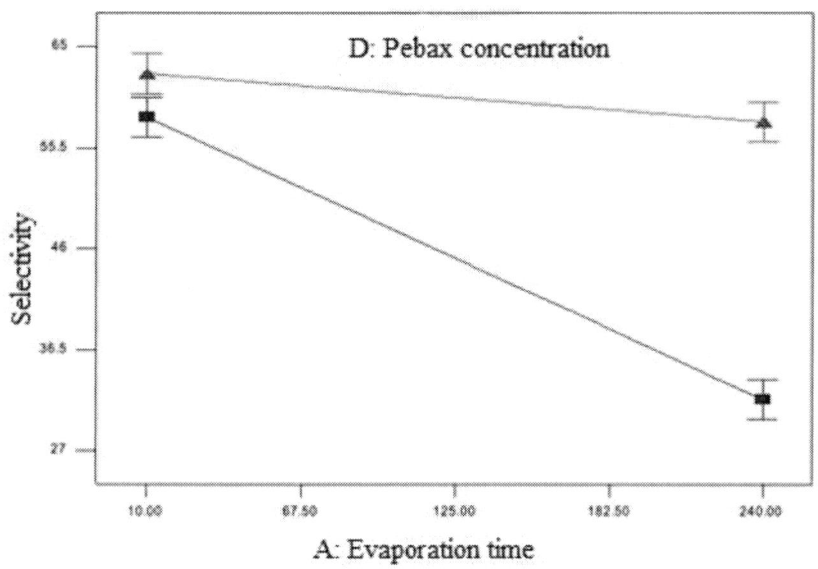

Figure 4. Interaction plot of the most significant interaction effect towards selectivity

3.3. PVDF/Pebax Film Composite for CO_2/CH_4 Separation

This study illustrated the potential of PVDF/Pebax film composite for biogas separation or in methane purification industry. Table 3 summarized the permeability-selectivity result from this work compared to the latest research regarding the use of Pebax 1657 for CO_2/CH_4 separation. This work is at the top of the list with CO_2 permeability of 393 barrer and selectivity up to 64.75. Pebax is a good CO_2 adsorbent due to the polar characteristic of this coating material that can interact with the permeating species during the separation process. Besides, the higher condensability rate of CO_2 compared to CH_4 is due to its high critical temperature that help in channeling the gas species pass through the membrane film. It is normal for rubbery and glassy polymeric membranes to scale the gas solubility (or condensability) with gas critical temperature. Besides the two facts, the gas-polymer interaction (condensability and solubility) and penetrant dependent, the interaction of dipole-quadrupole between CO_2 and EO in Pebax 1657 also a driven factor [14]. It is an intermolecular interaction between the penetrant and the medium where the gas solubility increase with the strong ion attraction.

Table 3. CO_2 permeability and CO_2/CH_4 selectivity data for various work on Pebax 1657

Film Materials	CO_2 Permeability-CO_2/CH_4 Selectivity	References
PVDF/Pebax	393 Barrer - 64.75	This work
TiO2/Pebax 1657/PVC	167 Barrer - 51	[15]
Pebax-1657/4A zeolite	155.7 Barrer - 41.3	[16]
ZIF-8/Pebax 1657/PES	758 Barrer - 16.1	[17]
PVC/Pebax	259.26 Barrer - 34.74	[18]
PEBAX/PEG/POSS	200 Barrer - 15	[19]
Pebax/PEG/MWCNT	70 Barrer - 30	[20]
Pebax/amino	361 Barrer - 16	[22]
PEG/Pebax	80 Barrer - 15.9	[3]
Pebax/SAPO-34	338 Barrer - 20	[22]

4. Conclusion

Membranes gas selectivity in response to film development factors was successfully studied by using 2^4 fractional factorial design. The statistical analysis managed to screen down and eliminate few factors which not so effective in film development for a good gas selectivity. Among the factors, Pebax concentration, quenching time and bath temperature showed positive effect towards the response. The interaction of Pebax concentration and evaporation times was discovered to be the only interaction that contribute to the good gas selectivity in PVDF/Pebax film composite. The composite film produced via this design of experiment managed to scored higher compared to recent work, which is 393 barrer of CO_2 permeability and selectivity of 64.75.

Acknowledgements
The authors wish to thank Universiti Malaysia Pahang for the grant No. RDU 150311 (UMP Grant), RDU150901 (LRGS), Faculty of Chemical and Natural Resources Engineering for the Gas Engineering lab facilities and Design Expert 7.6.1 license.

References

[1] Bakeri G, Ismail A, Shariaty-Niassar M and Matsuura T 2010 *J. Membrane Sci.* **363** 103
[2] Kim K, Ingole P, Kim J and Lee H 2013 *Chem. Eng. J.* **233** 242
[3] Car A, Stropnik C, Yave W and Peinemann K 2008 *Sep. Purif. Technol.* **62** 110
[4] Ahmadpour E, Shamsabadi A, Behbahani R, Aghajani M and Kargari A 2014 *J. Nat. Gas Sci. Eng.* **21** 518
[5] Akhtar F, Kumar M and Peinemann K 2017 *J. Membrane Sci.* **525** 187
[6] Oprea S and Ciobanu C 2007 *High Perform. Polym* 1
[7] Amirilargani M, Saljoughi E, Mohammadi T and Moghbeli M 2009 *Polym. Eng. Sci.* **50** 885
[8] Tsay C and McHugh A 1991 *J. Polym. Sci.* B. **29** 1261
[9] Azizi Namaghi H, Haghighi Asl A and Pourafshari Chenar M 2015 *J. Ind. Eng. Chem.* **31** 61
[10] Yu C, Hu C, Bai A and Yang Y 2007 *Surf. Coat Technol.* **201** 7259

[11] Chen Y, Chen H, Lee H, Chan H and Hu Y 2006 *Biomaterials.* **27** 2222
[12] El-Gendy N, Madian H and Amr S 2013 *Int. J. Microbiol.* 1
[13] Kaur Walia N, J. Pet 2014 *Environ. Biotechnol.* **05**
[14] Habibiannejad S, Aroujalian A and Raisi A 2016 **6** 79563
[15] Ahmadpour E, Sarfaraz M, Behbahani R, Shamsabadi A and Aghajani M 2016 *J Nat. Gas Sci. Eng.* **35** 33
[16] Surya Murali R, Ismail A, Rahman M and Sridhar S 2014 *Sep. Purif. Technol.* **129** 1
[17] Jomekian A, Behbahani R, Mohammadi T and Kargari A 2016 *J. Nat. Gas Sci. Eng.* **31** 562
[18] Ahmadpour E, Shamsabadi A, Behbahani R, Aghajani M and Kargari A 2014 *J. Nat. Gas Sci.* **21** 518
[19] Mushfequr Rahman M, Filiz V, Shishatskiy S, Abetz C, Neumann S, Bolmer S, Munir Khan M and Abetz V 2013 *J. Membrane Sci.* **43** 286
[20] Wang S, Liu Y, Huang S, Wu H, Li Y, Tian Z and Jiang Z 2014 *J. Membrane Sci.* **460** 62
[21] Zhao D, Ren J, Li H, Li X and Deng M 2014 *J. Membrane Sci.* **467** 41
[22] Zhao D, Ren J, Li H, Hua K and Deng M 2014 *J. Energy Chem.* **23** 227

IOP Conf. Series: Materials Science and Engineering **702** (2019) 012047 doi:10.1088/1757-899X/702/1/012047

Supported liquid membrane for acetic acid extraction: Screening of membrane support preparation factors

S A A Malid, S M Saufi* and N Harruddin

Faculty of Chemical and Process Engineering Technology, Universiti Malaysia Pahang, 26300 Gambang, Pahang, Malaysia.

*E-mail: smsaufi@ump.edu.my

Abstract. It is essential to remove acetic acid from the biomass hydrolysate during the biorefinery process. The product yield can be effected by the acetic acid due to it inhibition to the microorganism used during the fermentation. Supported liquid membrane (SLM) can be used to remove the acetic acid. The membrane support plays important role in the SLM process. However, due to the use of the commercial membrane support, lack of studied was conducted on the fabrication of custom made membrane support for the SLM process. In the current study, two level full factorial design was employed to screen three fabrication factors during preparation of polyethersulfone-graphene membrane support via vapor induced phase separation technique. The factors screened were temperature of water bath (A) (30-60 °C), exposure time (B) (10-60 s) and air humidity (C) (70-90 RH%). The response was evaluated based on the extraction percentage of the acetic acid from 10 g/L aqueous acetic acid feed solution. All three main factors were significant to the SLM performance. Air humidity factor (C) gave the highest contribution of 28.96% among the main factors. In term of the interaction between factors, water bath temperature (A) and exposure time (B) give the most significant effect with 45.01% percentage of contribution. The highest extraction percentage of the acetic acid using SLM system was 75.95% using the membrane prepare at 30 °C water bath temperature, 10 s exposure time and 70% air humidity.

1. Introduction

Energy consumption and demand in the world constantly increased every year due to the continuous development of many countries worldwide and increased growth of the human population. Majority of energy currently used globally is derived from non-renewable sources [1] and almost 80% is provided from the fossil fuel. The main sector that used energy is transportation industry which accounts more than 58% of the total amount of energy produced [2]. However, the supplies of the fossil fuels are limited and will be depleted one day. In addition, some hazard gases were release with the use of fossil fuels, thus give harmful impact to the environment [3]. These concerns have led to increased global attention in supporting the development of alternative energy based on renewable sources.

Recently, production of biofuels especially bioethanol using lignocellulose biomass has gained a lot of attention to overcome the problems of the nonrenewable energy resources. The main component of the lignocellulose biomass which are cellulose, hemicelluloses and lignin can be broke down into sugar component and later can be converted into bioethanol. Hydrolysis is used to release these sugars from the biomass matrix. However, some inhibitors byproduct also formed along with the sugars during the hydrolysis process. These compounds are considered to be toxic to the fermenting organism used during the bioethanol fermentation process. These inhibitor compounds are mainly grouped into three classes which are furan, phenolic derivatives and weak acids such as formic and acetic acid [4]. Among them, acetic acid was found in significant amount in the biomass hydrolysate [5]. Therefore, it is critical to reduce the level of acetic acid concentration in the biomass hydrolysate to the acceptable level which is less than 5 g/L [6].

Different methods were used to remove acetic acid from biomass hydrolysates in the previous study. Acetic acid can be removed by ion exchange resins, but the resins have limitations in which the pore diffusion is often slow that leads to long processing time [7]. Membrane process such as nanofiltration and reverse osmosis have been investigated to separate acetic acid from hydrolysate under severe operating condition. Although the retentions of glucose obtained is high, but negative retention were observed with acetic acid, thus making it as inefficient method [8,9]. Liquid membrane is a promising technology for the recovery of various carboxylic acids [10]. Supported liquid membrane (SLM) represents one of the feasible types of liquid membrane. It uses a porous membrane support impregnated with complexing carriers to interact specifically with the targeted solute. It use of relatively small volume of organic liquid membrane phase but lead to high separation factor and low energy process [11].

The liquid membrane formulation and the properties of the membrane support are the important aspect of the SLM process. The composition of the liquid membrane phase should be formulated based on the specific interaction between the carrier and the targeted solute. Meanwhile the membrane support should have adequate porosity and hydrophobic to retain the organic liquid membrane within the membrane pore. In addition, the support should be chemically and thermally stable on exposure to the feed, stripping and organic liquid membrane phases. Most of the previous SLM processes are using commercial membrane as their support. Therefore, limited study has been conducted on the fabrication of the custom made membrane support for the SLM process. In this study, vapour induced phase separation (VIPS) method was used to fabricate polyethersulfone (PES) flat sheet membrane support for the SLM process. Small amount of graphene was added into the PES matrix to improve the hydrophobicity, porosity and mechanical properties of the membrane. The morphology and structure of the membrane is heavily dependent on the fabrication process parameters. During VIPS process, the dope polymer solution is transformed from the liquid phase to the solid membrane phase by exchanging the solvent from the control environment to the dope solution. The rate of this phase inversion process will determined the structure of the membrane formed. Therefore, three important VIPS fabrication parameters which are exposure time to non-solvent water vapors, relative humidity and coagulation bath temperature [12] were screen using two-level full factorial design to determine the significant factor toward producing good membrane support for SLM process for acetic acid removal.

2. Material and methods

2.1. Chemicals

The membrane dope solution was made from the mixture of PES (Radel A300, Amoco Chemicals), polyethylene glycol 200 (PEG 200, Sigma Aldrich, St. Louis, MO), dimethyl acetamide (DMAc, Sigma Aldrich), and graphene (Low Dimensional Material Research Centre, Universiti Malaya). The chemical used in liquid membrane formulation were tri-n-octylamine (TOA, Sigma Aldrich) and 2-ethyl-1-hexanol (Sigma Aldrich). Acetic acid (Sigma-Aldrich) and sodium hydroxide (Merck, Darmstadt, Germany) were used as feed and stripping phase, respectively.

2.2. Preparation of dope polymer solution

The base polymer solution of 15 wt.% PES, 42.5 % DMAc, 42.5 % PEG 200 was used. The amount of graphene added to the base polymer solution was 0.1 wt.% relative to the PES content in the dope solution. All the components were mixed continuously at room temperature using motorized stirrer (IKA C-MAG

HS 7) until homogenous dope solution was formed. The air bubbles trapped in the dope solution was removed by degassing the dope in ultrasonic machine for 24 hours.

2.3. Flat sheet membrane casting and factorial screening

Semi-automatic casting machine was used to produce the flat sheet PESgraphene membrane through VIPS process. The casting machine was placed in the closed box equipped with Deerma F430 air humidifier to control the air relative humidity. The dope solution was casted on the glass plate with a casting gap of 380 μm. The casted film was exposed to the humid air at room temperature at specified exposure time and air relative humidity. Then, the cast film was immersed into water coagulation bath at specific temperature to induce solidification process. During solidification process, the casted film changed their color from transparent to white immediately after the immersion into the coagulation bath and detached out from the glass plate after some time. The solidified film was transferred to another water coagulation bath for 24 hours to remove residual solvent and was dried at room temperature for 48 hours [13].

The screening of parameters involved during membrane support preparation was performed using 2^3 Full Factorial Design in Design-Expert software (Version 7.1.6, 2008; Stat-Ease, Minneapolis, MN, USA) was used to screen the VIPS fabrication parameters. There parameters was selected as the factor which are temperature of water coagulation bath, exposure time and air humidity. The extraction percentage of acetic acid was set as a response. The screening was based on 2^3 Full Factorial Design as shown in table 1. The number of experiment generated for the screening experiment was 14.

Table 1. Experimental ranges and levels of the factors used in the factorial design

Independent variables	Coded symbol	Range and level		
		−1	0	+1
Temperature of water coagulation bath (°C)	A	30	45	60
Air exposure time (sec)	B	10	35	60
Air humidity (%)	C	70	80	90

2.4. Supported liquid membrane system

The membrane support was impregnated with organic liquid membrane phase for 24 hours. The organic liquid membrane phase was formulated from 0.5 M tri-n-octylamine (TOA) carrier dissolved in 2-ethyl-1-hexanol solvents. SLM system consist of membrane cell made up of two Teflon blocks, two channel peristaltic pump Masterflex L/S, feed channel and strip channel as shown in figure 1. Supported membrane impregnated with the liquid membrane was placed in between these two blocks membrane cell. The feed and stripping phase used in the SLM were 10 g/L of acetic acid and 0.5 M sodium hydroxide, respectively. The flow rate of the feed and stripping phase was set at 75 ml/min and 50 ml/min, respectively at countercurrent flow direction within the system. SLM was operated for 8 h and the concentration of the acetic acid in the feed phase was analyzed using High Performance Liquid Chromatography (HPLC) column.

2.5. High performance liquid chromatography

The concentration of acetic acid in the feed phase samples after the SLM experiments was analyzed to calculate the acetic acid extraction percentage as the response in the screening experiment. Synergy 4μ Hydro-RP 80 (Phenomenex) HPLC column was connected to the Water Acquity UPLC System. The mobile phase and UV detection wavelength were 0.02M potassium dihydrogen phosphate and 221nm, respectively. The percentage of acetic acid extraction was calculated using equation (1).

IOP Conf. Series: Materials Science and Engineering **702** (2019) 012047 doi:10.1088/1757-899X/702/1/012047

$$\text{Acetic Acid Extraction (\%)} = \frac{c_0 - c_f}{c_0} \times 100\% \quad (1)$$

Where c_0 and c_f are the initial and final concentration of acetic acid in the feed phase, respectively.

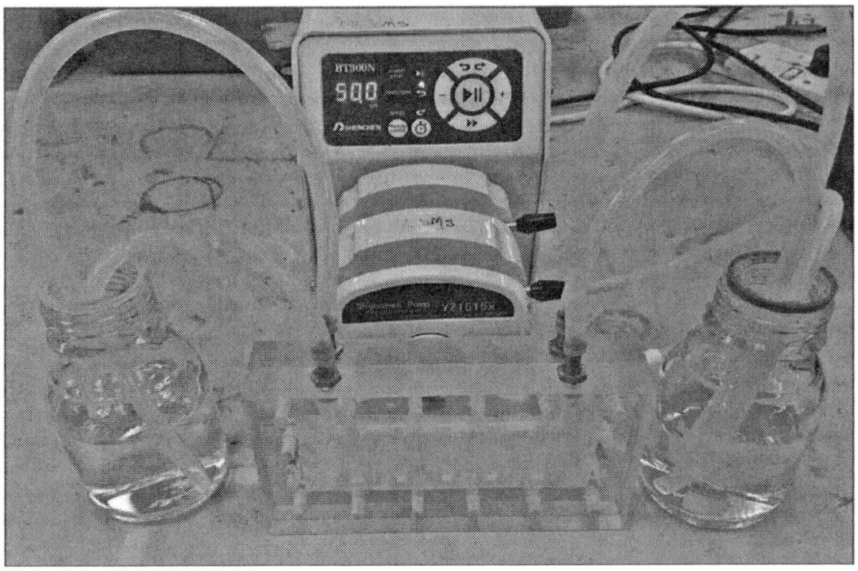

Figure 1. SLM system

3. Result and discussion

3.1. Design of experiment (DOE) and data analysis

Table 2 shows the response obtained from the SLM experiment. Based on these result, the following mathematical model was derived by the software:

$$Y = 128.12 - 0.256x_A - 0.862x_B - 0.5348x_C + 2.35 \times 10^{-3}x_Ax_B - 1.9533 \times 10^{-3}x_Ax_C + 3.98 \times 10^{-3}x_Bx_C + 1.0367 \times 10^{-4}x_Ax_Bx_C \quad (2)$$

Where Y is the extraction percentage of acetic acid and the letters A, B, C represent individual factors of temperature of water coagulation bath, air exposure time, and air humidity, respectively.

The relationships between variables were elicited using Pareto chart as shown in figure 2. The significance of the factors and their interactions was obtained by employing "Bonferroni limit" and "t-value limit" test. The effects above the "Bonferroni limit" were almost certainly significant, while effects above the "t-value limit" were possibly significant [14]. The main factors (A, B and C) and interaction of factor AB and BC showed t-value of effect beyond the "Bonferroni limit". They are the most significant factor and have a positive effect. Interaction of factor ABC sit on the "Bonferroni limit" are considered probably significant. Meanwhile, the factor of AC is considered insignificant as it locates slightly above the "t-value limit". This factor could be ignored as its contribution effect including small "t-value" will not affect the result of the experiment.

Table 2. Experimental design of removal of acetic acid using 2^3 full factorial designs

Run	Values of independent variables			Y (Extraction % of acetic acid)
	A (°C)	B (s)	C (%RH)	
1	60	60	90	65.90
2	30	60	70	61.19
3	30	10	90	65.50
4	60	60	70	66.70
5	45	35	80	62.68
6	45	35	80	62.61
7	30	10	70	75.95
8	60	10	90	56.05
9	45	35	80	63.20
10	45	35	80	62.45
11	60	10	70	67.05
12	30	60	90	57.83
13	45	35	80	62.76
14	45	35	80	62.91

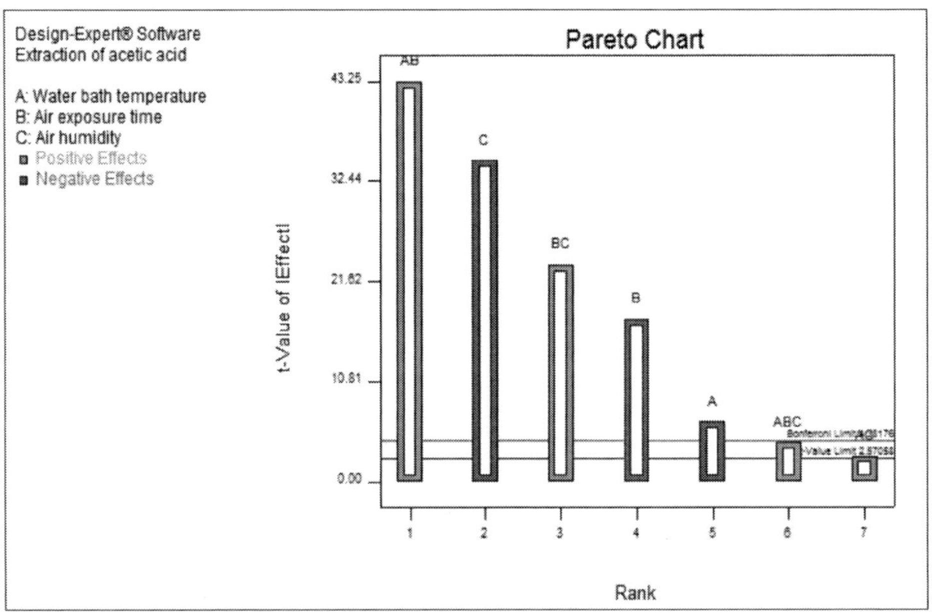

Figure 2. Pareto chart

Analysis of variance (ANOVA) was conducted to examine the reliability of the model. Based on table 3, the regression model had a high coefficient of determination ($R^2 = 0.9988$), implying that 99.88% of the variations in acetic acid extraction can be explained by the model and the model does not explain only 0.12% of the variations. The model F-value which is 570.80 implies the model is significant and there is only 0.01% chance that model F-value this large could occur due to noise. Besides, values of p-value less than 0.05 indicate model terms are significant. In this case, all of the factors are significant in the model term. The adequate precision for the ANOVA was obtained to be 95.084, which is favorable. The preferred value for adequate precision is >4 [15]. The lower coefficient of variation (CV % = 0.41) which is less than 10% clearly demonstrated that the deviations between actual and predicted values were low and affirmed the precision and reliability of conducted experiments.

Table 3. Analysis of variance model

Factor	Degrees of freedom	Sum of squares	Mean of squares	F value	p value
Model	7	272.25	38.89	570.80	< 0.0001
A	1	2.84	2.84	41.74	0.0013
B	1	20.90	20.90	306.71	< 0.0001
C	1	81.98	81.98	1203.23	< 0.0001
AB	1	127.44	127.44	1870.37	< 0.0001
AC	1	0.51	0.51	7.41	0.0417
BC	1	37.37	37.37	548.43	< 0.0001
ABC	1	1.21	1.21	17.74	0.0084

3.2. Main effect analysis

Table 4 shows the percentage contribution of main factor and their interaction with other factors during preparation of membrane support for the extraction of acetic acid via SLM process. As shown in Table 4, air humidity factor (C) gave the highest contribution of 28.96% among the main factor. In VIPS, the relative humidity is a very important parameter that influence the structure of the membrane formed. The vapor humidity greatly influenced the hydrophobicity of the membrane surface in the VIPS process, thus affecting the SLM process [16]. Low relative humidity will form dense skin on the membrane surface resulting in low pure water permeation flux. In contrast, high relative humidity will result in cellular structure that has high pure water permeation flux [16].

Table 4. Percentage contribution of each factor and their interaction

Factor	Contribution (%)
A Water bath temperature	1.00
B Air exposure time	7.38
C Air humidity	28.96
AB	45.01
AC	0.18
BC	13.20

*AB: Water bath temperature and air exposure time
*AC: Water bath temperature and air humidity

The second highest contribution for the main factor was air exposure time with 7.38% contribution. Dense film will be formed at long exposure time [17]. Low exposure time will form cellular opening structure, hence membrane with low pure water permeation flux and rejection capabilities will be produced. The last factor with 1.00% contribution was water bath temperature. Although the percentage of contribution is low, Xu et al. [18] reported that the increase in the coagulation water bath temperature resulted in great formation of macrovoids. This produces more porous structure in accommodate more organic liquid phase, thus can increase the extraction percentage in the SLM process.

3.3. Interaction between factors

The interaction between factors of water bath temperature (A) and air exposure time (B) contributed to the 45.01% of the proposed model. The interaction graph between factors AB was illustrated in Figure 3. At high water bath temperature of 60 °C, the extraction of acetic acid increased from 62% to 66% when the exposure time increased from 10 s to 60 s. Factor A and B were interact at water bath temperature of 50 °C. High water bath temperature results in greater formation of the macrovoids and more porous structure was formed [18]. In addition, complete vapor-induced phase separation occurred at high air exposure time that increased the pore formation [17]. High porosity membrane support can accommodate more liquid membrane organic phase, thus increases the percentage removal of acetic acid.

The second highest interaction effect with 13.20% contribution was the interaction between air exposure time (B) and air humidity (C). The interaction between these factors was illustrated in Figure 4. High percentage of extraction of acetic acid (72%) was obtained from the membrane prepared at 10 s exposure time and 70% relative humidity. As the relative humidity was decreased, membrane with dense skin was formed and hydrophobicity of the membrane was improved. This will lower the pure water permeation flux, thus increase the extraction of acetic acid [16].

As shown in Pareto chart previously, the interaction effect of factor water coagulation bath (A) and air humidity (C) is insignificant. The interaction of these factors to the acetic acid extraction was shown in Figure 5. The line for both factors are parallel and were not intercept with each other even at wide range of factor. An increment for both factors will result in low extraction of acetic acid, thus proving that both factors are insignificant.

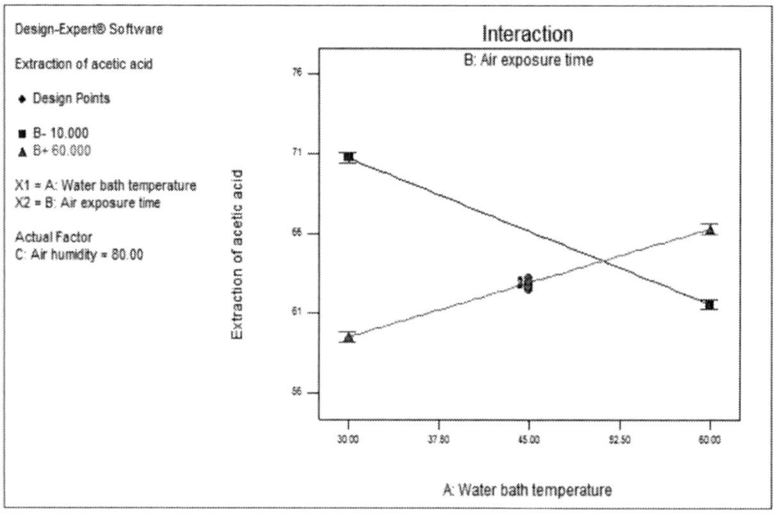

Figure 3. Interaction graph between water bath temperature (A) and exposure time (B).

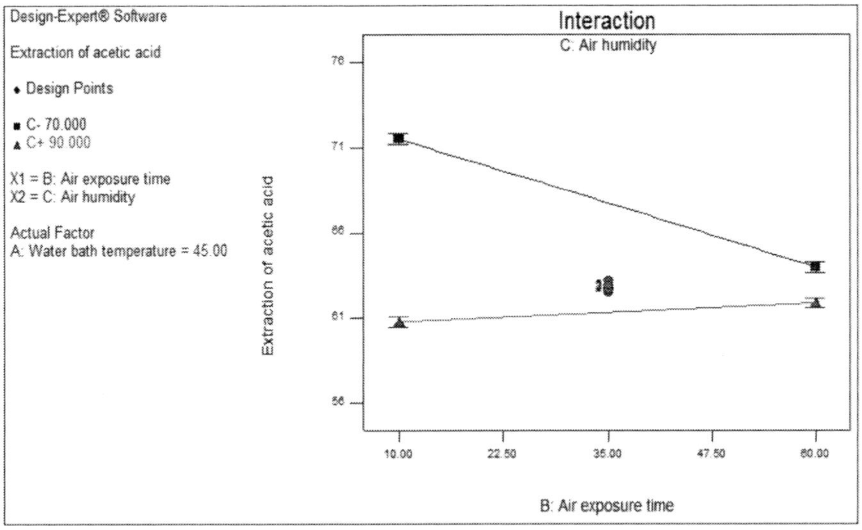

Figure 4. Interaction graph between air exposure time and air humidity

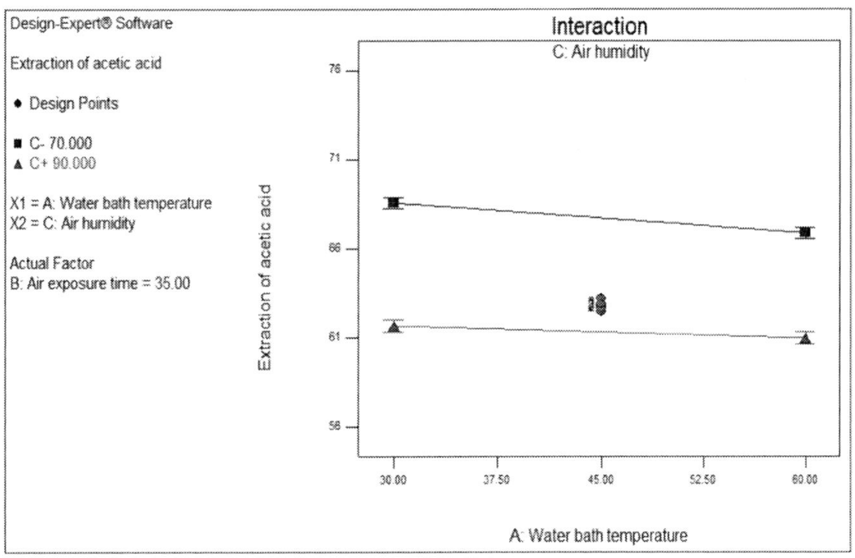

Figure 5. Interaction graph between water bath temperature and air humidity

4. Conclusion

FFD was successfully applied for screening of the important factors involved during VIPS process for the preparation of PES graphene membrane support in SLM system. All the main factors played important role on the membrane formation. The air humidity factor (C) gave the highest contribution of 28.96% among the main factor. The interaction effects of the water bath temperature with exposure time (AB) were the most significant with 45.01% contribution. Meanwhile, the interaction effect of water bath temperature (A) and air humidity (C) is insignificant as extraction of acetic acid decrease as both factors were increased. The highest extraction percentage for the acetic acid using SLM system was 75.95% which is using the membrane prepare at 30 °C water bath temperature, 10 s exposure time and 70 % air humidity.

Acknowledgement
This work was financially supported by the Fundamental Research Grant Scheme (FRGS) (RDU170117-FRGS/1/2017/TK02/UMP/02/8) awarded by Ministry of Education, Malaysia.

References
[1] Irandoust M 2016 *Ecol. Indic.* **69** 118
[2] Voloshin R A, Rodionova M V, Zharmukhamedov S K, Nejat Veziroglu T and Allakhverdiev S I 2016 *Int. J. Hydrogen Energy* **41** 17257
[3] Hossain A, Jewaratnam J and Ganesan P 2016 *Int. J. Hydrogen Energy* **41** 16637
[4] Grzenia D L, Schell D J and Wickramasinghe S R 2008 *J. Memb. Sci.* **322** 189
[5] Lee S C 2013 *Sep. Purif. Technol.* **118** 540
[6] Delgenes J P, Moletta R and Navarro J M 1996 *Enzyme Microb. Technol.* **19** 220
[7] Ghosh R 2002 *J. Chromatogr. A* **952** 13
[8] Teella A, Huber G W and Ford D M 2011 *J. Memb. Sci.* **378** 495
[9] Zhou F, Wang C and Wei J 2013 *J. Memb. Sci.* **429** 243
[10] Hassoune H, Rhlalou T and Verchère J F 2008 *J. Memb. Sci.* **315** 180
[11] Kocherginsky N M, Yang Q and Seelam L 2007 *Sep. Purif. Technol.* **53** 171
[12] Venault A, Chang Y, Wu J and Wang D 2014 *J. Taiwan Inst. Chem. Eng.* **45** 1087
[13] Harruddin N, Saufi S M, Faizal C K M and Mohammad A W 2018 *RSC Adv.* **8** 25396
[14] Jadhav S B, Kumar C K, Bandichhor R and Bhosale P N 2016 *J. Pharm. Biomed. Anal.* **118** 370
[15] Regti A, Rachid M, Stiriba S and El M 2017 *J.* **130** 129
[16] Peng Y, Fan H, Ge J, Wang S, Chen P and Jiang Q 2012 *Appl. Surf. Sci.* **263** 737
[17] Peng Y, Dong Y, Fan H, Chen P, Li Z and Jiang Q 2013 *Desalination* **316** 53
[18] Xu J, Tang Y, Wang Y, Shan B, Yu L and Gao C 2014 *J. Memb. Sci.* **455** 121

IOP Conf. Series: Materials Science and Engineering **702** (2019) 012048 doi:10.1088/1757-899X/702/1/012048

Modeling of vanillin adsorption from aqueous solution using resin H103 by artificial neural network

W S Chan[1], R A Samah[1,*], N Zainol[1], A S Fakharudin[2], S A Aziz[3] and L Y Phang[3]

[1]Faculty of Chemical and Process Engineering Technology, College of Engineering Technology, Universiti Malaysia Pahang, Lebuhraya Tun Razak, 26300 Gambang, Kuantan, Pahang, Malaysia.
[2]Faculty of Computing, Universiti Malaysia Pahang, Lebuhraya Tun Razak, 26300 Gambang, Kuantan, Pahang, Malaysia.
[3]Department of Bioprocess Technology, Faculty of Biotechnology and Biomolecular Sciences, Universiti Putra Malaysia, 43400 UPM Serdang, Selangor, Malaysia.

*E-mail: rozaimi@ump.edu.my

Abstract. Vanillin adsorption onto resin H103 was modelled using artificial neural network (ANN) approach and the best ANN algorithm was determined in this work. The first step of ANN modeling was ANN set up, followed by the optimization of ANN. The parameters for the input layers are contact time, initial vanillin concentration, resin dosage, pH, and temperature while the response is residual vanillin concentration. The neural network was trained using backpropagation (BP) algorithm. The result shows that the Levenberg–Marquardt algorithm was best suited the training function and the optimized ANN involved seven neurons at the hidden layer. This model can produce a correlation of determination value of 0.9999 with the mean square error (MSE) value of 0.0277. The best adsorption efficiencies for each factor were 98.11%, 96.03%, 98.14%, 98.2%, and 98.10% at 2.0 g of adsorbent dosage, 30 min of contact time, 100 mg/L of initial vanillin concentration, pH 5, and 25 °C, respectively. The outcomes of this work proved that ANN is excellent in predicting experimental data of vanillin adsorption by resin H103.

1. Introduction

Vanillin is one of the most important flavoring materials which has been widely used in the food industry, perfumes, cosmetics, and pharmaceuticals [1]. Vanillin can be recovered using several techniques such as distillation, extraction using gas-expanded liquid, ultrafiltration, and adsorption. The adsorption process is preferable to separate vanillin from lignin due to its nature of environmentally friendly and more economical compared to the other separation technologies [2].

The world annual production of vanillin is more than 12,000 tons. However, less than 1% of the material is natural vanillin originated from botanical sources such as cured vanilla pods and *Java citronella* [3]. The price of synthetic vanillin is less than USD15/kg, in contrast to the price of natural vanilla between USD1,200.00/kg and USD4,000.00/kg [4]. Because of the limited supply and high price of natural vanillin combined with the increasing demand for natural flavors, it becomes a motivation for the researchers to find an alternative method to produce vanillin.

Modeling is done before the implementation of a new method to predict its feasibility and explain the possibilities of response in particular circumstances. In the past few decades, it is challenging to do modeling and optimization of adsorption because they have to carry out many experiments to obtain the desired and optimum value. In the 50s, response surface methodology (RSM) was proposed by Box and Behnken [5] for the purpose modeling and optimization, and this helps the researcher save a lot of time. However, modeling and optimization of the heterogeneous photon-Fenton process using RSM showed that the degradation efficiency of the experiment was within the range of 70% to 90%, not high enough to describe the behavior of the experiment [6]. The problem faced by the same data can be solved by using ANN due to its ability to achieve a degradation efficiency of 98.11% [7]. The improvement occurred because ANN consists of three layers and a training algorithm to adjust the error and repeat until the desired output is obtained [8].

ANN is a computing system which imitates the biological neural network of living organisms. It is based on a collection of connected units called artificial neurons. These artificial neurons are typically organized in three groups – input, hidden, and output layers. Activities of the hidden layers depend on the activities of the input layers, while the hidden layers will affect the output layer. The results are then processed by an input–output function (transfer function) [9]. The number of hidden layers is directly proportional to the accuracy of the result. Other than predicting the output variables, it is also used to eliminate irrelevant input, called pruning method. When the significant variable is determined, other irrelevant variables are eliminated from the model which can help to reduce the size of the network and minimize the redundancy in the training data [10].

This work aimed to model the vanillin adsorption from aqueous solution using resin H103 by ANN, and the results describe that a good agreement between the experimental data and predicted output was achieved. Resin H103 was selected due to its high adsorption capacity and low cost [11]. The effects of adsorbent dosage, contact time, vanillin initial concentration, pH and temperature were also studied [11]. The designation of a neural network, selection of the best training method, and identification of the suitability of ANN for modeling of vanillin adsorption played an essential role in completing this research.

2. Methodology

2.1. Data Preparation

The experimental data of vanillin adsorption from aqueous solution using resin H103 were taken from a previous study [11]. The experimental data show the effect of adsorbent dosage, contact time, initial vanillin concentration, pH, and temperature on the adsorption performance of resin H103. The selection of resin H103 was based on its highest capacity and low cost. For this modeling, the data of model variables are shown in table 1.

Table 1. Data used for ANN modeling [11].

Variables	Range
Input layer	
Adsorbent dosage (g)	0.5–5.0
Contact time (min)	3–300
Initial vanillin concentration (mg/L)	6.25–800
pH	3.00–7.00
Temperature (°C)	25–55
Output layer	
Vanillin residual concentration (mg/L)	0.41–44.92

2.2. Artificial neural network model

2.2.1. Backpropagation algorithm. Backpropagation (BP) was used as the training algorithm for the modeling of vanillin adsorption from aqueous solution using resin H103 [12]. It was used to change the weights and biases in a network to improve the overall output of the network. The working principle of the backpropagation algorithm is iteration. The iteration will only stop when the error between the predicted output and the experimental data is acceptable [13]. There are 13 backpropagation algorithms exist in the ANN, and all were performed to determine the best suited BP training algorithm for this model. These training algorithms were performed with an initial ANN design of three layers with tangent sigmoid transfer function (TANSIG) at the hidden layer and linear transfer function (PURELIN) at the output layer. All the data were normalized to be between 0 and 1 to suit the application of sigmoid function in the ANN [14]. A linear activation function (PURELIN) is normally used for the output layer when the ANN is trained for regression before optimization. Ten neurons were used at this stage to determine the best-suited training algorithm. During the training, the experimental data were randomly divided into training, validation, and test subsets at a ratio of 2:1:1 [15]. The validation set was used to determine when to stop the iteration of the training set to prevent overfitting [16]. The iteration of the BP algorithm was stopped when the MSE of the validation set reached the minimum value and it started to bounce back.

In the next stage, the MSE for each training algorithm was compared. The performance of the BP training algorithm was measured using the MSE value. A smaller MSE value indicates that the predicted output is very close to the experimental data. Therefore, the BP training algorithm with the minimum value of MSE was selected as the best-suited training algorithm and to be used in the subsequent modeling.

2.2.2. Optimization of ANN. The best-suited training algorithm determined from the previous part was used in the three layers ANN with tangent sigmoid transfer function at the hidden layer and linear transfer function at the output layer to perform the optimization of ANN. The optimization of ANN was performed as the function of the number of neurons in the hidden layer and was evaluated based on the MSE and linear regression [10]. Two neurons were used as an initial guess to evaluate the performance of the ANN then it was repeated by increasing the number of neurons with an interval of one. The performance of ANN is improved with the increase in the number of neurons, but when the number of neurons exceeds the threshold number, overfitting of the model will occur and causes the model to fail to predict the experimental data [17]. Therefore, the number of neurons is increased until the MSE of the model started to increase after the minimum point was achieved.

Then, the MSE of the model for each number of neurons was compared. The optimum number of neurons was determined with the minimum value of MSE and the highest linear regression of the model. Finally, the three-layer ANN with tangent sigmoid transfer function at the hidden layer and linear transfer function at the output layer was evaluated at the best-suited training algorithm and the optimum number of neurons.

2.2.3. Validation of effect of variables on adsorption efficiency. The effects of adsorbent dosage, contact time, initial vanillin concentration, pH, and temperature on the experimental output were studied and validated by the predicted output from the ANN model. The output from the experimental data is residual vanillin concentration, which is correlated to the adsorption efficiency for the ease of comparison. The adsorption efficiency was calculated using equation (1):

$$Adsorption\ efficiency\ (\%) = \frac{100(C_1 - C_2)}{C_1} \qquad (1)$$

where C_1 and C_2 are the initial and final vanillin concentrations in the solution, respectively. The adsorption efficiency from the experimental data was compared with the predicted output from the ANN. The predicted output which is deviated from the experimental data is justified. Finally, the suitability of ANN for the modeling of vanillin adsorption from aqueous solution using resin H103 was evaluated based on the regression of the model.

IOP Conf. Series: Materials Science and Engineering **702** (2019) 012048 doi:10.1088/1757-899X/702/1/012048

3. Results and Discussion

3.1. Selection of backpropagation (BP) algorithm

Thirteen BP algorithms were compared to determine the best training function to be used in the modeling. A three-layer of ANN with a tangent sigmoid transfer function (TANSIG) at the hidden layer and a linear transfer function (PURELIN) at the output layer was used to perform all the training algorithms. The hidden layer consisted of 10 number of neurons. Gradient descent with momentum learning function (LEARNGDM) was used for this network and mean squared error (MSE) was used to measure the performance of the network. The result shows that the Levenberg–Marquardt backpropagation algorithm (LMA) with a minimum MSE was the best training function for this modeling.

The comparison of the different algorithms was based on the deviation of the predicted output from the experimental data. A large deviation of some algorithms from the experimental data was observed, while some of the algorithms were able to converge to the experimental data. The results show that LMA was able to produce the smallest deviation compared to the other backpropagation algorithms. In contrast, the predicted output by gradient descent algorithm (GDA) has the largest deviation. As shown in table 2, a minimum MSE with a value of 0.024 was produced by TRAINLM, which is followed by TRAINBR with the MSE value of 0.114. The MSE values produced by other training algorithms such as TRAINBFG, TRAINCGB, TRAINCGF, TRAINCGD, TRAINGDM, TRAINGDA, TRAINGDX, TRAINOSS, TRAINRD, and TRAINSCG are greater than one. The largest MSE was produced by TRAINGD with the MSE value of 1500. The deviation produced by the algorithm can be attributed to the nonlinear structure and combinatorial nature of the experimental data. Therefore, LMA is the best training function for this modeling because it can solve nonlinear least-squares problems and produce a least-squares curve fitting [18].

Table 2. Comparison of 13 backpropagation algorithms.

Backpropagation (BP) algorithms	Function	Mean square error (MSE)	Iteration number (IN)	R^2	Best linear equation (BLE)
BFGS quasi-Newton	TRAINBFG	9.910	23	0.919	$y = 0.92x - 0.055$
Bayesian regularization	TRAINBR	0.114	1000	1.000	$y = x + 0.0075$
Powell–Beale conjugate gradient	TRAINCGB	20.700	12	0.896	$y = 0.73x + 1.9$
Fletcher–Reeves conjugate	TRAINCGF	29.000	18	0.855	$y = 0.76x + 1.8$
Polak–Ribière conjugate gradient	TRAINCGP	28.600	10	0.818	$y = 0.59x + 2.7$
Gradient descent	TRAINGD	1500.000	6	0.405	$y = 0.47x + 40$
Gradient descent with momentum	TRAINGDM	706.000	6	0.130	$y = 0.22x + 20$
Gradient descent with adaptive learning rate	TRAINGDA	22.000	260	0.918	$y = 0.82x + 2.3$
Gradient descent with momentum and adaptive learning rate	TRAINGDX	105.000	29	0.327	$y = 0.1x + 6.6$
Levenberg–Marquardt	TRAINLM	0.024	6	1.000	$y = 0.99x + 0.013$
One-step secant	TRAINOSS	15.700	22	0.911	$y = 0.8x + 2.7$
Resilient (Rprop)	TRAINRP	44.800	21	0.821	$y = 0.67x + 1.2$
Scaled conjugate gradient	TRAINSCG	39.000	26	0.813	$y = 0.68x + 2.5$

IOP Conf. Series: Materials Science and Engineering **702** (2019) 012048 doi:10.1088/1757-899X/702/1/012048

3.2. Optimization of ANN Structure

The optimization of the ANN structure was determined based on the MSE value and the regression of the model. In the optimization, two neurons were used as the initial guess in the hidden layer for this network. When the number of neurons was increased, the MSE produced by the network became smaller. At a certain number of neurons, the MSE of the network started to increase from the minimum value of MSE. Figure 1 shows that the MSE of the network was the highest at 2 neurons with a value of 5.98. When the number of neurons was increased from 3 to 6, MSE decreased significantly from 5.35 to 0.188. The network achieved the minimum value of MSE of 0.0277 with 7 neurons at the hidden layer. When the number of neurons was increased from 7 to 14, the MSE of the network gradually increased from 0.0277 to 0.879. A further increase in the number of neurons from 14 to 15 resulted in a sharp increase in the MSE.

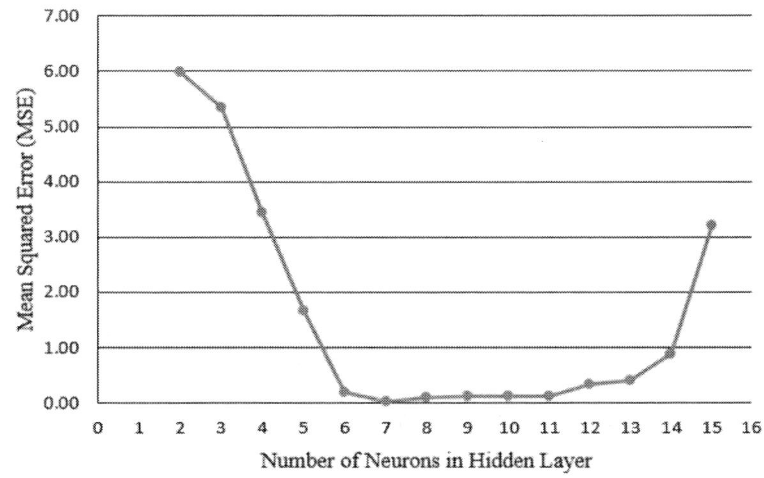

Figure 1. Relationship between MSE and number of neurons in the hidden layer.

At the same time, when the number of neurons is increased, the regression of the predicted outputs will increase. Figure 2 illustrates the dependence of regression and the number of neurons in the hidden layers. When the number of neurons was increased from 2 to 6, the regression of the predicted outputs increased from 0.9706 to 0.9992, but there is an exception when there are 3 neurons at the hidden layer where the regression decreased from 0.9706 to 0.9597 due to the combinatorial nature of the experimental data. The combinatorial experimental data sent to training was scattered which reduced the regression of the predicted output. The highest regression of predicted output of 0.9999 was achieved with 7 neurons at the hidden layer. A further increase in the number of neurons from 7 to 15 reduced the regression of the predicted output. The performance of the network is better when the number of neurons is increased. However, overfitting will occur when there is too much number of neurons. Overfitting will lead to the predicted outputs to be exact with or close to the experimental data and therefore may fail to fit additional data [17].

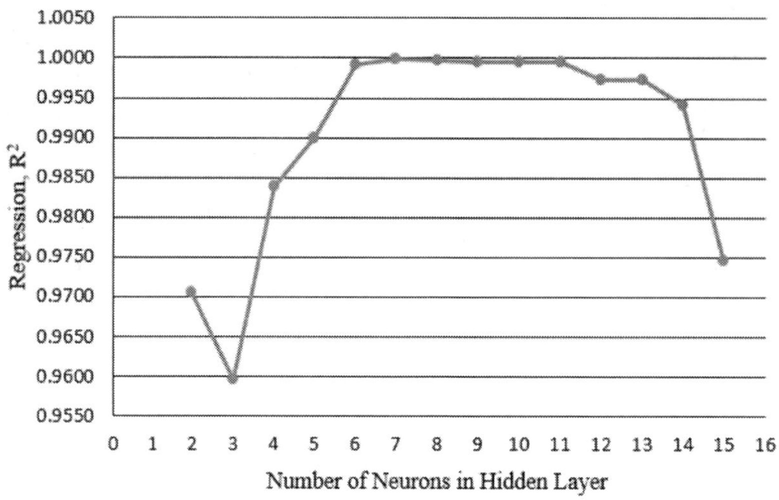

Figure 2. Relationship between regression and number of neurons in hidden layer.

The number of iteration determines the training of the network. When the predicted output deviates from the experimental data, the weight of the deviation will be adjusted, and the input of the network will be reinitialized and the network is trained again with the new input. Training will stop when the predicted output is close to the experimental data (the error was set to 1%). In this modeling, the training was stopped after 43 iterations out of 1000 (Epochs 43/1000) for LMA because further iterations show that the MSE started to increase. Figure 3 shows the MSE of training, validation, and test at different epochs for TRAINLM. The validation set was used to determine when to stop the iteration of the training set to prevent overfitting. The best epoch was at 37 because the MSE of validation started to increase after 37 epochs. At 37 epochs, the optimum MSE of the training was 0.0277.

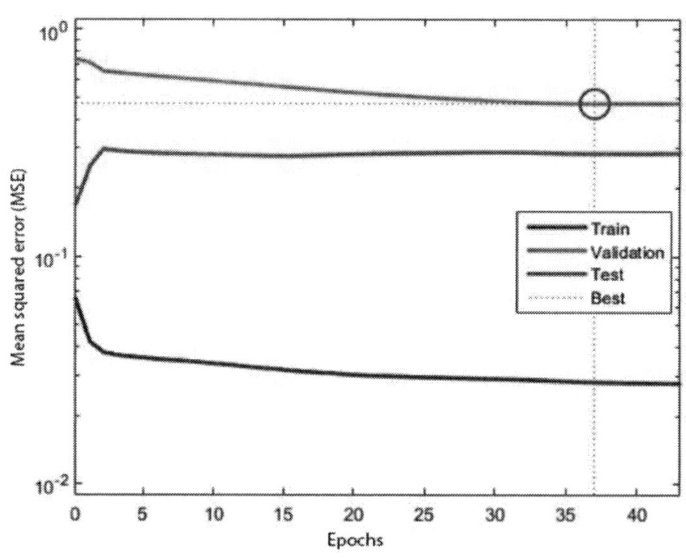

Figure 3. MSE of training, validation, and test against epochs.

Finally, the optimum artificial neural network with the BP algorithm was designed for this modeling. The network consists of three layers, which are input layer, hidden layer, and output layer. There are seven neurons at the hidden layer with tangent sigmoid transfer function (TANSIG) and an output layer with linear transfer function (PURELIN). Figure 4 shows the optimum structure of ANN, together with the BP algorithm flow chart for the prediction of residual vanillin concentration.

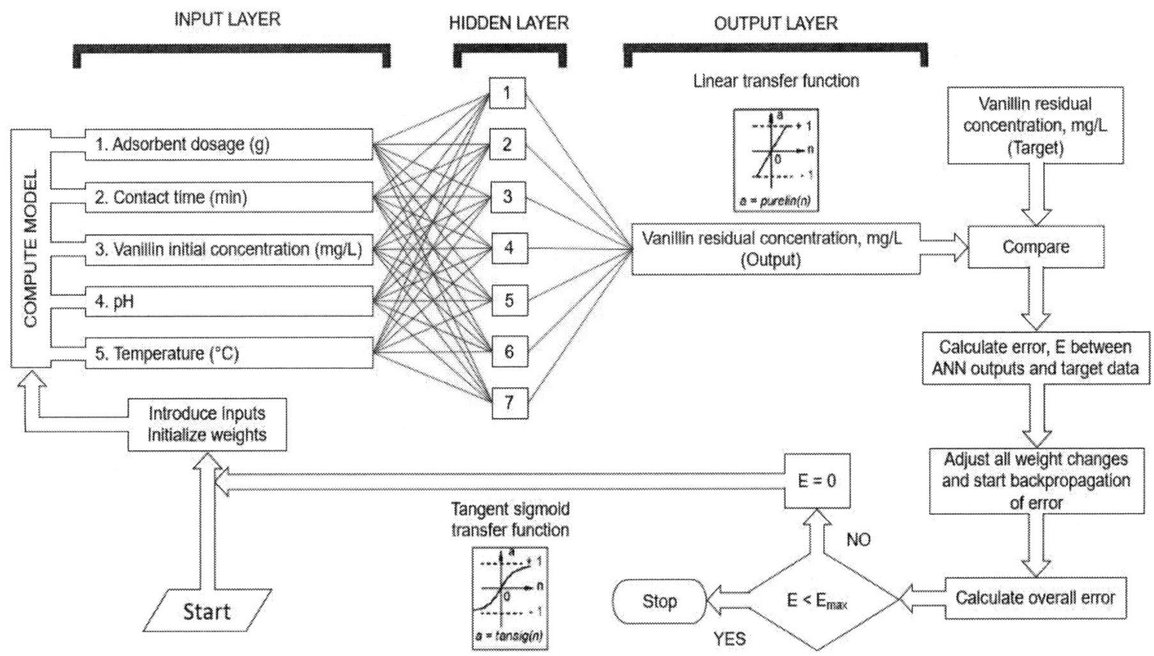

Figure 4. Optimum ANN structure with BP algorithm flow chart for prediction of residual vanillin concentration

3.3. Modeling of Vanillin Adsorption

In this modeling, the optimum ANN structure was used to test the performance of five variables which are adsorbent dosage, contact time, initial vanillin concentration, pH, and temperature. Their performance was evaluated based on the MSE of the training and regression of the model. A good model is a model with low MSE and high regression. For this modeling, the training network provided an MSE of 0.024 and the correlation of determination for the model is 0.9999, which proved that ANN is suitable to model the vanillin adsorption using resin H103. ANN is excellent in solving nonlinear model because it can produce a precise and effective response in a short time. The precision of the model was achieved by the BP algorithm. It calculated the error of the predicted output from the experimental data and propagated the error backwards, so that the input could be reinitialized and to be trained again. This process was repeated until the predicted output was close to the experimental data. On the other hands, LM algorithm which was used as a training transfer function can solve the nonlinear problem of the experimental data. It can minimize the problems of least-square curve fitting. Figure 5 shows the predicted residual vanillin concentration against the experimental residual vanillin concentration. The predicted vanillin concentration is very close and almost equal to the experimental residual concentration which means that the modeling of vanillin adsorption by ANN is successful [19].

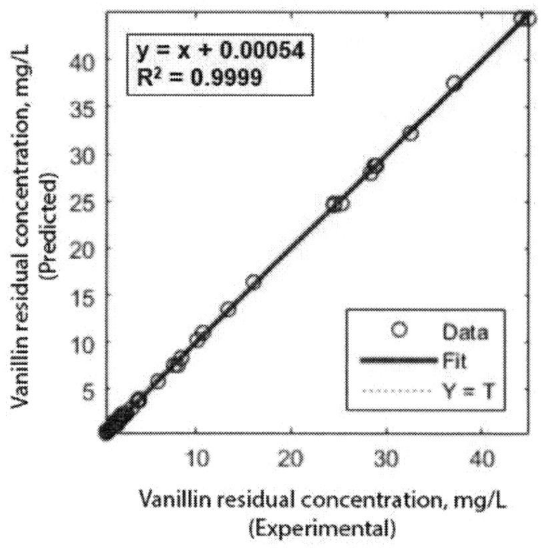

Figure 5. Predicted vanillin residual concentration versus experimental residual concentration.

3.4. Effect of Adsorbent Dosage on Adsorption Efficiency

The effect of adsorbent dosage on adsorption efficiency was studied by increasing the adsorbent dosage from 0.5 to 5.0 g while the contact time, initial vanillin concentration, pH, and temperature were maintained at 90 min, 50 mg/L, 5.94, and 25 °C, respectively. Figure 6 shows the efficiency between the experimental data and predicted output as a function of adsorbent dosage. The experimental data show that the adsorption efficiency increased from 92.37% to 98.11% when the adsorbent dosage was increased from 0.5 to 2.0 g. However, a further increase of adsorbent dosage to 5.0 g reduced the adsorption efficiency from 98.11% to 96.69%. According to Garg et al. [20], this can be attributed to the increase of surface area and the availability of adsorption sites for the removal of vanillin when the weight of the adsorbent is increased. However, a further increase of adsorbent dosage may cause aggregation and leads to overlapping of adsorption sites. Consequently, the available adsorption sites may decrease as well due to the adsorption density. The maximum adsorption efficiency was found to be 98.11% at 2.0 g of adsorbent dosage.

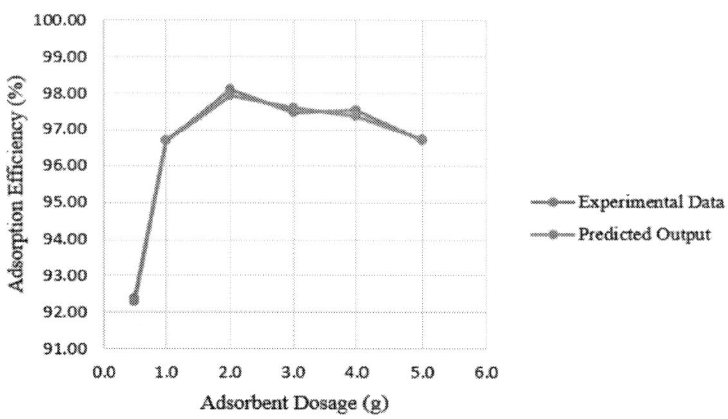

Figure 6. Adsorption efficiency between experimental data and predicted output as a function of adsorbent dosage

3.5. Effect of Contact Time on Adsorption Efficiency

The effect of contact time on adsorption efficiency was studied by increasing the contact time up to 300 min while the adsorbent dosage, initial vanillin concentration, pH, and temperature were maintained at 0.5 g, 50 mg/L, 5.94, and 25 °C, respectively. The experiment was repeated by replacing the adsorbent dosage to 2.0 g to compare the effect of adsorbent dosage and contact time on the adsorption efficiency. The experimental data show that a contact time of 60 and 30 min at 0.5 and 2.0 g of adsorbent dosage, respectively, are sufficient for the adsorption efficiency to achieve equilibrium (figure 7). A further increase of contact time did not yield a significant increase in the adsorption efficiency. Therefore, a further increase in contact time is unnecessary by considering the cost of energy consumed in agitation. The adsorption efficiency at 60 min with 0.5 g of adsorbent dosage was 88.17% while the adsorption efficiency at 30 min with 2.0 g of adsorbent dosage was 96.03%. The adsorbent dosage of 2.0 g was chosen because it could achieve higher adsorption efficiency with lesser energy consumed in agitation due to decrease in contact time.

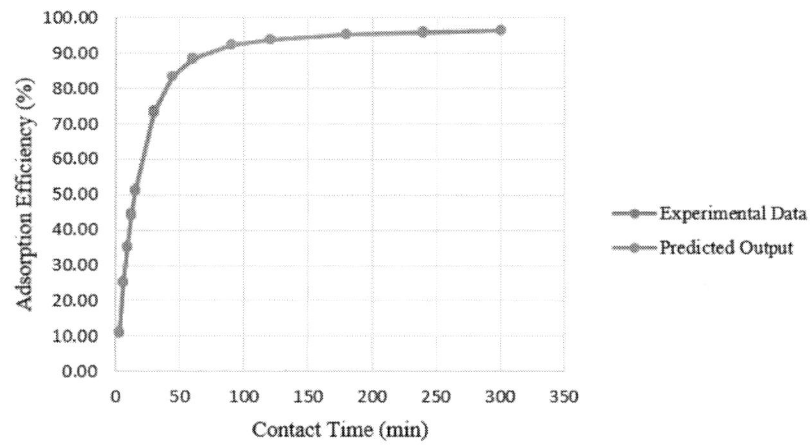

(a)

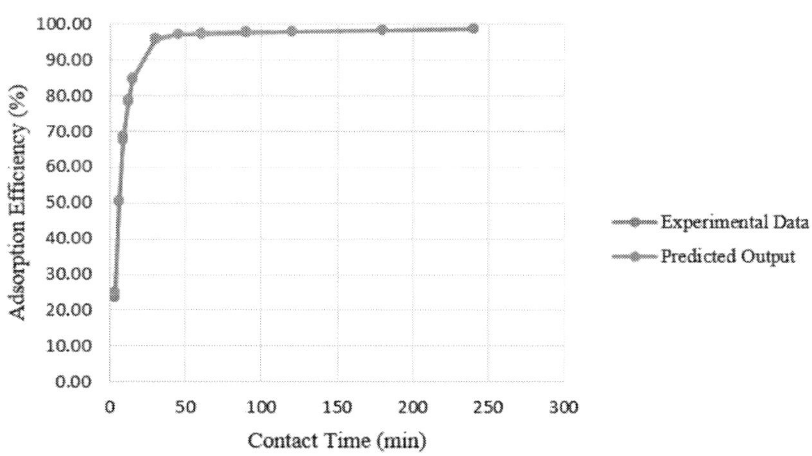

(b)

Figure 7. Adsorption efficiency between experimental data and predicted output as a function of contact time for (a) 0.5 and (b) 2.0 g of adsorbent dosage

3.6. Effect of Initial Vanillin Concentration on Adsorption Efficiency

The effect of initial vanillin concentration on adsorption efficiency was studied by increasing the initial vanillin concentration from 6.25 to 800 mg/L while the adsorbent dosage, contact time, pH, and temperature were maintained at 0.5 g, 90 min, 5.94, and 25 °C, respectively. The experimental data in figure 8 show that the adsorption efficiency increased gradually from 93.14% to 98.14% when the initial vanillin concentration was increased from 6.25 to 100 mg/L. The adsorption efficiency decreased from 98.14% to 96.40% as a result of further increase of initial vanillin concentration because beyond 100 mg/L of initial vanillin concentration, the adsorbent was already saturated where all of the adsorption sites were fully occupied. Besides, Michael and Ayebaemi [21] reported that the adsorbate takes a shorter time to adsorb themselves to the adsorption sites at low concentration of adsorbate. In contrast, it takes a longer time when the concentration of adsorbate is increased because the adsorbate needs to diffuse to the adsorbent site through the intraparticle diffusion mechanism. The optimum initial vanillin concentration was found at 100 mg/L with 98.14% adsorption efficiency.

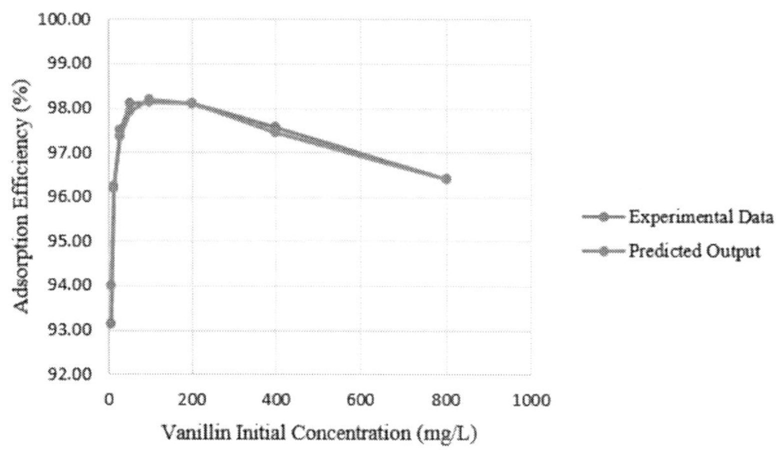

Figure 8. Adsorption efficiency between experimental data and predicted output as a function of initial vanillin concentration

3.7. Effect of pH on Adsorption Efficiency

The effect of pH on adsorption efficiency was studied by increasing the pH from 3.00 to 7.00 while the adsorbent dosage, contact time, initial vanillin concentration, and temperature were maintained at 0.5 g, 90 min, 50 mg/L, and 25 °C, respectively. The experimental data show that when pH was increased from 3.00 to 5.00, the adsorption efficiency slightly increased from 97.93% to 98.17% because at low pH, there is a high concentration of H^+ ions to compete with vanillin to bind themselves with the actives sites. Therefore, when pH was increased, H^+ ions decreased, resulting in the more available active site for vanillin, thus increasing the efficiency of vanillin removal. However, no substantial difference in the adsorption efficiency within the tested pH range, which might be due to the neutral state of vanillin in the pH range of 2.0–6.0 [22]. The optimum pH was found to be 4.00 for this modeling. Figure 9 shows the comparison of efficiency between the experimental data and predicted output as a function of pH. Different from the other factors, the ANN model shows some deviation from the experimental data which might be caused by the insufficient experimental data to be trained in the ANN model which made the ANN model to fail to predict the experimental data accurately [23].

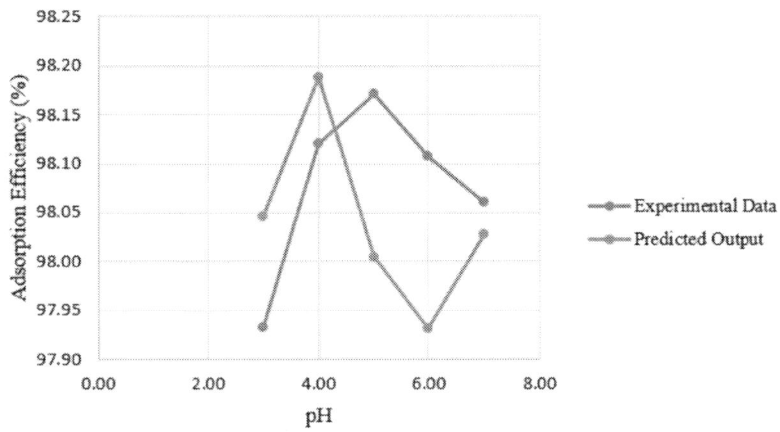

Figure 9. Adsorption efficiency between experimental data and predicted output as a function of pH.

3.8. Effect of Temperature on Adsorption Efficiency

The effect of temperature on adsorption efficiency was studied by increasing the temperature from 25 to 55 °C while the adsorbent dosage, contact time, initial vanillin concentration, and pH were maintained at 0.5, 90 min, 50 mg/L, and 5.94, respectively. The experimental data (figure 10) show that the adsorption efficiency kept decreasing from 98.10% to 96.03% when the temperature was increased from 25 to 55 °C. Because adsorption is a physisorption process [11], when temperature increases, vanillin molecules gain enough kinetic energy to overcome the electrostatic force of attraction between the adsorbent and themselves and escapes from the surface of the adsorbent which leads to a decrease in the adsorption efficiency. The optimum temperature for the adsorption process for this model was found to be 25 °C.

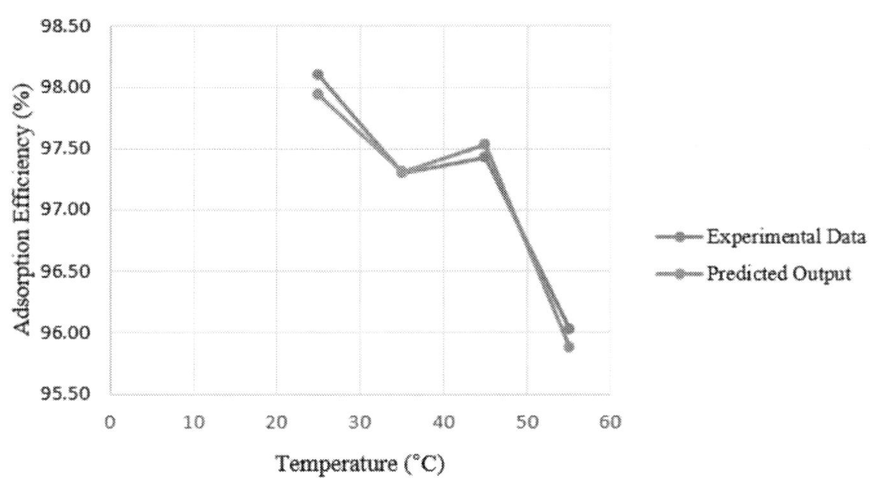

Figure 10. Adsorption efficiency between experimental data and predicted output as a function of temperature.

4. Conclusions

The optimum backpropagation ANN for this modeling is three layers with tangent sigmoid transfer function (TANSIG) at the hidden layer and linear transfer function (PURELIN) at the output layer. A minimum error of predicted data from experimental data was obtained by using the Levenberg–Marquardt algorithm with the MSE of only 0.024. The optimum ANN structure is with 7 neurons at the hidden layer with a coefficient of determination of 0.9999 which means that the regression of the model is very high and the MSE value of 0.0277. The best adsorption efficiencies for each factor were 98.11%, 96.03%, 98.14%, 98.17%, and 98.10% at 2.0 g of adsorbent dosage, 30 min (2.0 g of adsorbent dosage), 100 mg/L of initial vanillin concentration, pH 5, and 25 °C, respectively. From the result, ANN shows an excellent performance in predicting the experimental data and suitable to be used in the modeling of vanillin adsorption.

Acknowledgment
We would like to thank you Universiti Malaysia Pahang for providing MATLAB for the modeling throughout the research.

References
[1] Zhang Q F, Jiang Z T, Gao H J and Li R 2008 *Eur. Food Res. Technol.* **226**(3) 377–83
[2] Mota M I F, Pinto P C R, Loureiro J M and Rodrigues A E 2016 *Sep. Purif. Rev.* **45**(3) 227–59
[3] Chambers C M 1997 *Prooc. of The Internal Structure of Black Holes and Spacetime Singularities* 490–96
[4] Lomascolo A, Stentelaire C, Asther M and Lesage-Meessen L 1999 *Trends Biotechnol.* **17**(7) 282–89
[5] Box G E P and Behnken D W 1960 *Technometrics* **2**(4) 455–75
[6] Ba D and Boyaci I H 2007 *J. Food Eng.* **78**(3) 836–45
[7] Elemen S, Akçakoca Kumbasar E P and Yapar S 2012 *Dye. Pigment.* **95**(1) 102–11
[8] Kasiri M B, Aleboyeh H and Aleboyeh A *Environ. Sci. Technol.* **42**(21) 7970–75
[9] Singh V, Khan M, Khan S and Tripathi C K M 2009 *Appl. Microbiol. Biotechnol.* **82**(2) 379–85
[10] Gevrey M, Dimopoulos I and Lek S 2003 *Ecol. Modell.* **160**(3) 249–64
[11] Samah R A 2016 *Optimization of Batch Adsorption and Fixed-Bed Adsorption of Vanillin onto Resin H103.* Universiti Putra Malaysia, Selangor, Malaysia.
[12] Ortega-Zamorano F, Jerez J M, Munoz D U, Luque-Baena R M and Franco L 2016 *IEEE Trans. Neural Networks Learn. Syst.* **27**(9) 1840–50
[13] Karnin E D 1990 *IEEE Trans. Neural Networks* **1**(2) 239–42
[14] Gueguim Kana E B, Oloke J K, Lateef A and Adesiyan M O 2012 *Renew. Energy* **46** 276–81
[15] Basheer I A and Hajmeer M 2000 *J. Microbiol. Methods* **43**(1) 3–31
[16] Mele B and Altarelli G 1993 *Phys. Lett. B* **299**(3–4) 345–50
[17] Kumar K V and Porkodi K 2009 *Chem. Eng. J.* **148**(1) 20–25
[18] Yetilmezsoy K and Demirel S 2008 *J. Hazard. Mater.* **153**(3) 1288–300
[19] Satoh H and Namerikawa T 2007 *Proc. IEEE Int. Conf. Control Appl.* 539–44
[20] Garg V K, Gupta R, Yadav A B and Kumar R 2003 *Bioresour. Technol.* **89**(2) 121–24
[21] Michael H J and Ayebaemi I S 2005 *Electron. J. Biotechnol.* **8**(2) 162–69
[22] Li R, Jiang Z, Mao L and Shen H 1998 *Talanta* **47**(5) 1121–27
[23] Demiral Y F and Öztürk N 2013 *Proc. of ICOEST* 246–55

IOP Conf. Series: Materials Science and Engineering **702** (2019) 012049 doi:10.1088/1757-899X/702/1/012049

2^3 fractional factorial design for polymer based thin film composite (TFC) membrane synthesis for CO_2/ CH_4 separation

A A Ghazali, R K Roshan, S Abd Rahman*, M S Abdul Wahab, N S Hazleen and N Mandayar

Faculty of Chemical and Process Engineering Technology, Universiti Malaysia Pahang, 26300 Kuantan, Pahang, Malaysia.

*E-mail: sunarti@ump.edu.my

Abstract. Membrane technology is dominating the industry as an attractive approach for biogas purification due to its outstanding performance. Recently, there have been intensive efforts in the development of better separation efficiency of membrane which include altering the materials and modifying the methods in preparing the membrane. A well-formed membrane is when they achieved both high permeability and excellent separation ability. Therefore, this study is focusing on identifying the best processing factors in PVC/Pebax thin film composite (TFC) membrane development towards CO2/CH4 separation by employing 23 fractional factorial design (FFD). A total of three factors; immersion times (5 & 15 min), Pebax concentration (1 & 5 wt%) and number of coating layers (1 & 4) were chosen to run simultaneously with CO2 permeability and ideal selectivity as the responses for this study. The results show that the most influential factors that affect the permeability are immersion time and Pebax concentration, while for selectivity are Pebax concentration and number of coating layer. The best condition was known to maximize the permeability and selectivity. The identified conditions were immersion time for 15 min, Pebax concentration at 5wt% and 4 layers of coating which gave CO2 permeability and gas selectivity of 19612 Barrer and 7.15, respectively. The outcome of this study indicates that FFD was suitable to minimize and eliminate factors by considering the interaction among the factors involves in membrane film synthesis for excellent gas separation performance. Besides, the existence of Pebax layer was verified by visual interpretation using Scanning electron microscopy (SEM) and Fourier Transform Infrared (FTIR).

1. Introduction

Biogas is a form of renewable energy sources that can be found during anaerobic digestion of organic substrates in industry waste, agriculture residues, sewage sludge as well as household [1]. The composition of biogas depends on the type of feedstock being digested. Primarily, it consists of methane (CH_4) which is around (35%–70%) and carbon dioxide (CO_2) (30%–65%) [2]. Due to richness of CH_4, biogas can act as vehicle fuel, electricity and power production as well as for injection into grid as an alternative of natural gas [3]. Nevertheless, the presence of high concentration of the acidic gas CO_2 may cause equipment and pipeline corrosion as well as reduction effect on the heating value of the gas, which restricts its practical application. Therefore, it is essential to separate CO_2 from the gas stream to meet the pipeline specification and the standard requirements for biogas utilization, which requires CO concentration below 2 mol% for pipeline system and high purity of CH_4 (> 95 mol%) for use as vehicle fuel [4-5]. There is a wide range of gas separation according to different circumstances used for the CO_2 removal from gaseous compounds, such as absorption, adsorption, membranes and cryogenic distillation [6]. Among these technologies, membrane–based technology is promising which offers several advantages over others competing separation such as low operating cost, ease of scale-up and maintenance operation, lack of chemical disposal [7].

Different types of membrane materials are commercially available in the market and polymeric membranes have attracted maximum attention by the researchers due to its good mechanical properties, easier process, simplicity, and thermal stability [8]. Despite having excellent properties, they also have their own inherent limitation. An obvious issue is the trade-off between permeability and selectivity as shown in Robeson's upper bound curves [9]. Over the last few decades, many researchers have been working on developing a new kind of membrane called thin film composite (TFC) membrane. Commonly, TFC membrane comprises of two main layers; a microporous polymeric substrate and a thin selective layer on top of a thick substrate's surface as shown in Figure 1. The combination of stable support layer and high permeable selective layer was believed to enhance the selectivity–permeability trade-off [10]. However, few characteristics of the membrane such as pore diameter, thickness and surface roughness influenced the performance of the TFC membrane. Therefore, selecting the best conditions in preparing the TFC membrane is important in order to fabricate an excellent separation efficiency of TFC membrane.

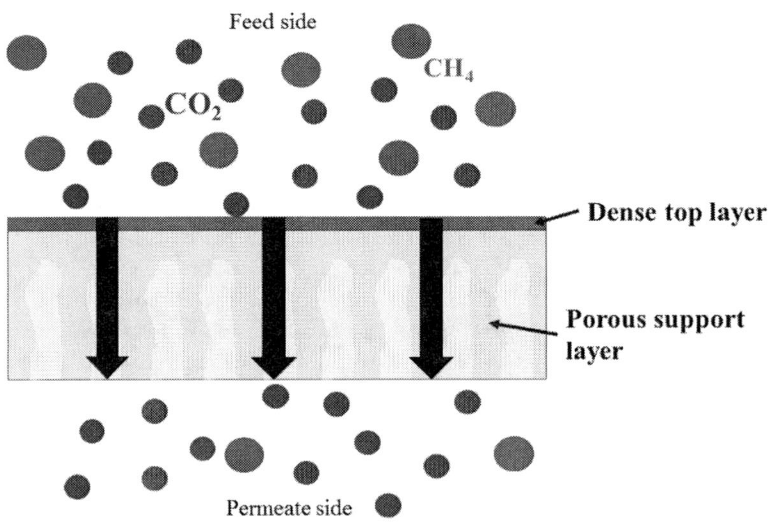

Figure 1. Schematics of thin film composite membrane.

Membranes can be synthesized in a different manner via modifying the key factors which influence the morphology of the membrane. TFC development consists of several key factors starting from material selection, casting technique and coating procedure. Commonly, the porous support layer was prepared via dry/wet phase inversion method. The process includes casting polymer solution on a glass plate and subsequently immersing in a non-solvent bath, where mass transfer takes place involving the interchange of solvent and non-solvent, by diffusion and convection [11]. Explained by Strathmann and Kock [12], in terms of thermodynamic and kinetic, the duration of the immersion, either fast or delayed demixing processes can lead to various types of membrane morphological structure. Hence, providing an optimum length of immersion time is required in order to produce defect free TFC membrane.

In many real structures of membrane, some non-ideal morphology such as polymer chain rigidification, particle pore blockages and interface void formation often occurred that reduce the performance of the membrane. According to Zarshenas et al. [13], higher polymer concentration produced dense membrane which led to the reduction of separation performances. Similar to the preparation of top selective layer in TFC membrane, having a higher concentration of coating solution will increase the viscosity which led to a thicker layer, hence, increasing the mass transfer resistant throughout the membrane causing lower gas permeability. On top of that, a dilute solution is suggested for the top selective top to be coated on top of the support layer. However, the presence of large voids and free volume in porous support layer might lead to solution penetration by the selective solution that causes pore blockage and reduce the gas permeability drastically [14]. Therefore, choosing a perfect range of polymer loading for selective layer is important to prepare an ideal morphology of TFC membrane. In recent study, at lower concentration of Pebax (< 5 wt% in its binary solution), a membrane defects free can be produced.

All common composite membrane preparation method includes separate casting, direct coating [15], gas phase deposition [16] and interfacial polymerization [17]. At present, dip–coating is one of the most widely used techniques to prepare composite membranes due to its simplicity [18]. Different number of coating layers produced different thickness. The thickness of the selective layer is also found to influence the gas separation performance [19]. In Abdul Wahab et al. [20] recent work, they identified that too thin selective layer could not make any different to the gas selectivity. However, too thick selective layer may cause an increase in mass transfer resistance throughout the membrane, causing membrane's performance drop significantly. Generally, selecting the best thickness of selective layer is vital to form a defect–free TFC membrane with perfect separation efficiency.

Consequently, the aim of this study is to identify the independent factors (input variables) that have the highest contribution on the responses (output variables). The independent factors that have been selected in this study are immersion time of polyvinyl chloride (PVC) cast solution in non-solvent bath (water), concentration of polyether block amide (Pebax) and number of coating layer while CO_2 permeability and ideal selectivity were chosen as the responses. Currently, factorial designs are one of the promising methods used to obtain empirical models relating process response to process factors, and have proved their effectiveness. Therefore, two level factorial analysis (TLFA) factorial design was employed in this study to find the best factor that contributes the most in fabricating a perfect PVC/Pebax TFC membrane.

2. Methods

2.1 Materials

Polyvinyl Chloride (PVC) in powder form and dimethylformamide (DMF) solvent in liquid form were purchased from Sigma Aldrich. Polyether block amide under trade name Pebax 1657 and ethanol solvent were supplied by Arkema France and Fisher Scientific, respectively. Two tanks of CO_2 and CH_4 gases used for gas permeation experiments with purity of 99.9% were acquired from Air Products.

2.2 PVC substrate preparation

13 wt% of PVC powder was magnetically stirred in 87 wt% DMF at 90 °C for 7 h till complete dissolution, it was then left for 24 h at room temperature for degassing purpose to release any air bubbles formed in the solution. Then, the dope solution was casted on a clean glass plate with film applicator setting set to 0.03 mm knife gap, room temperature and 100 rpm of casting speed. Employing the dry/wet phase inversion technique, the casting plate was immediately immersed in a non-solvent medium (water bath) at 25 °C for 5 to 15 min to let the liquid–liquid demixing take place. Water is selected as the bath medium due to its high miscibility with DMF which can cause a fast solvent–non-solvent exchange. Then, the PVC substrate film was dried at room temperature for another 24 h to achieve complete drying.

2.3 PVC/Pebax Film Composite Membrane formation

Pebax 1657 coating solution was prepared by dissolving 1–5 wt% of the pellets in 70:30 ethanol:water mixture at 80 °C until a clear, dilute and homogenous solution was formed. Then, for the preparation of TFC membrane in laboratory scale, dip–coating method was used. 5 mL of the Pebax 1657 coating solution was poured on a flat dish. Then, the PVC substrate film was dip coated into the dish for 1 to 4 times, with 30 min time interval in between the dip procedure. Then, the prepared TFC membrane was dried in an oven at 60 °C for 24 h for complete solvent vaporization.

2.4 Single Gas Permeation Test Setup

Permeability test were run individually for each CO_2 and CH_4 through single gas permeation test by allowing them to flow at 2 bar and room temperature (25 ± 5 °C) toward the membrane cell and bubble flow meter was used to measure the flow rate of permeate. The single gas permeation setup was prepared as illustrated in Figure 2. Test module membranes with effective cell area and diameter, 19.63 cm^2 and 5 cm, approximately were used. The selectivity data was obtained by measuring the permeability of each gas species through the newly synthesized membrane. To measure the permeability of single gas, Equation (1) was expressed in GPU [1 Barrer= $1\times10^{-10}cm^3(STP)cm/(cm^2 scmHg)$] [21]. After considering the conversion factor, equation (2) was obtained which can be directly used to calculate the gas permeability in Barrer.

$$P\,(\text{GPU}) = \frac{lV\ (\text{cm}^3\text{cm(stp)})}{At\Delta p\ (\text{cm}^2\text{scmHg})} \quad (1)$$

$$P\,(\text{Barrer}) = \frac{1\times10^{10} lV}{At\Delta p} \quad (2)$$

where P is the permeability, l is the thickness of the membrane in cm, V is the volume in cm^3, t is the time taken (s), A is the effective membrane area in cm^2, and Δp is the transmembrane pressure expressed in cmHg. The ideal selectivity (α) of a membrane between two gases, A and B, can be expressed through permeability ratio of fast gas (A) to slow gas (B) as shown in equation (3) [22].

$$\alpha_{AB} = \frac{P_A}{P_B} \quad (3)$$

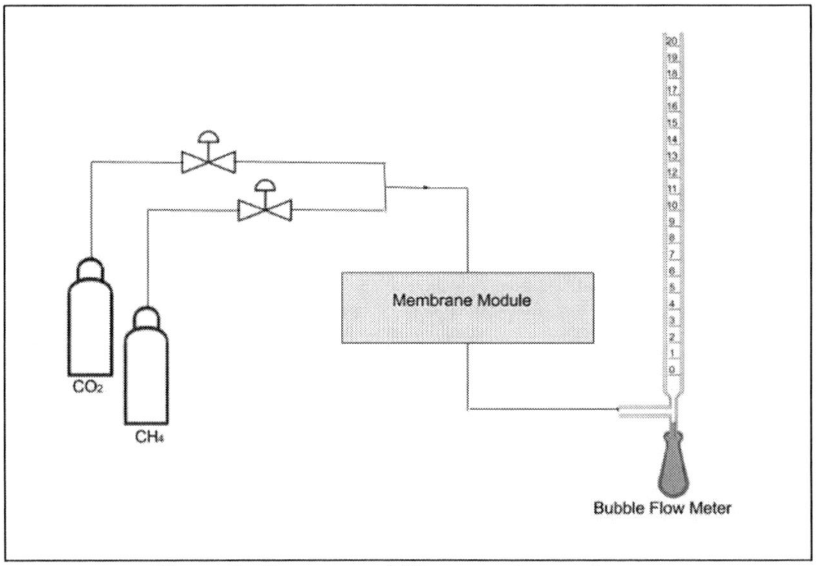

Figure 2. Gas permeability experimental setup

2.5 Factorial Design Methodology

The experimental matrix design and analysis of data were performed using Design Expert software (Version 7.1.6). The effects of several independent factors to process responses were explored by using a 2^3 fractional factorial design. Three factors that were chosen include immersion time, concentration of Pebax and number of coating layer and the responses were based on the CO_2 permeability and gas selectivity. A total of 8 runs were performed in this study. The summary of factors and responses utilized in the design are shown in table 1. Multiple linear regression analysis (MLRA) and analysis of variance (ANOVA) were used to develop mathematical models containing the notable terms.

IOP Conf. Series: Materials Science and Engineering **702** (2019) 012049 doi:10.1088/1757-899X/702/1/012049

Table 1. Summary of design, factors and responses.

Study type	Initial design		Design model		Runs
Factorial	2 Level Factorial		3FI		8
Factors	Symbols	Unit	Type	Low actual	High actual
Immersion time	A	min	Numeric	5	15
Pebax concentration	B	wt%	Numeric	1	5
Number of coating layer	C	layer	Numeric	1	4
Responses	Symbols	Unit	Analysis		
CO_2 permeability	P	Barrer	Factorial		
CO_2/CH_4 selectivity	S		Factorial		

2.6 Membrane Characterization

The cross-section morphology of the membrane was analyzed using scanning electron microscope (SEM, Hitachi/TM3030 Plus, Japan). To obtain a clear image and smooth cutting surface of the TFC membrane, it was freeze and fractured in liquid nitrogen without any part of the membrane affecting the cutting surface. Next, fourier transform infrared spectroscopy (FTIR, Nicolet AVATAR 370 DTGS with measurement wavelength range of 4000–500 cm^{-1}) was used to study the chemical information and examine the existence of each functional group of molecules in the TFC membrane. It allowed the membrane to absorb different kinds of wavelength produced from a beam and the output data were computerized to get the desired data in the form of wavenumbers, cm^{-1}. The high wavenumber region is best for investigating the thin selective layers that may present on top of the PVC layer, while the lower wavenumber region can be used to study the chemistry of both the Pebax selective layer as well as the PVC support layer.

3. Results and Discussions

3.1 Screening of Factors Affecting on the Gas Selectivity

The screening of factors that contributed on gas permeability and selectivity were carried out using 2^3 fractional factorial designs as shown in Table 2. The single gas permeation measurement is used to obtain the selectivity of CO_2/CH_4. To increase accuracy, three samples of membranes for each run were prepared and tested. Then, the average of gas permeation rate and selectivity results were reported and utilized in this work. The CO_2 permeability ranged from 7994.09 to19612.4 Barrer and the selectivity ranged from 2.55 to 7.15.

Table 2. 2^3 Fractional Factor Design analysis using Design Expert software with its respective responses.

Run	Factor 1: Immersion time (min)	Factor 2: Pebax concentration (wt%)	Factor 3: Coating layer	Response 1: CO_2 permeability (Barrer)	Response 2: CO_2/CH_4 Selectivity
1	5	1	1	15329.1	2.55
2	15	1	1	13336.5	3.48
3	5	5	1	7994.09	7.03
4	15	5	1	13319.8	7.05
5	5	1	4	10817.3	4.03
6	15	1	4	9614.5	4.83
7	5	5	4	14420.2	7.15
8	15	5	4	19612.4	7.15

3.2 Analysis of Variance of (ANOVA)

The input response was analyzed for the significance of the regression model in the Design Expert Software. The significance of the model was confirmed by the analysis of variance (ANOVA) as shown in Table 3 and 4. The significance of a regression equation was checked using F-values while the *p*-values were used to check the significance of each coefficient towards the response. The model terms are significant if the *p*-value is less than 0.05.

Based on the ANOVA results, the F-values obtained were high, 230.62 for CO_2 permeability and 2176.06 for CO_2/CH_4 selectivity while the *p*-values identified were very low, 0.0043 for CO_2 permeability and 0.0005 for CO_2/CH_4 selectivity, indicating that good models are implied. The model terms that have significant effects towards permeability and selectivity were A (immersion time), B (Pebax concentration), C (coating layer), AB (immersion time and Pebax concentration), and BC (Pebax concentration and coating layer).

R-squared (R^2) from the ANOVA was used to indicate how close the data to the fitted regression line. For a chemical process design, a good fitting model should have R^2 value in range of 0.9 to 1. The design experiment for both CO_2 permeability and selectivity have generated $R^2 > 0.9983$, predicted $R^2 > 0.9939$ and adjusted $R^2 > 0.9723$. The predicted R^2 is in reasonable agreement with the adjusted R^2 since the difference between this two are within 0.2. In other words, the design model of the experiment is in accurate description of experimental data which indicated the relationship between the variables and the responses is reliable. The indication of model satisfactory gave by the adequate precision where it measures the signal to noise ratio. A ratio greater than 4 is desirable. From the analysis, the adequate precision for permeability and selectivity were 47.37 and 108.98, respectively, indicating an adequate signal. To express the permeability and selectivity as a function of independent variables, the final empirical model in terms of actual factors are shown in Equation (4) and (5).

$$P = +\,20808.95 - 331.19A - 3506.15B - 2245.32C + 171.42\,AB + 873.02BC \qquad (4)$$

$$S = +\,0.36 + 0.11A + 1.33B + 0.58C - 0.02\,AB - 0.11\,BC \qquad (5)$$

where P and S are referred as the responses, CO_2 permeability and CO_2/CH_4 selectivity, respectively. A is immersion time, B is Pebax concentration and C is number of coating layer.

Table 3. ANOVA table for the regression model (CO_2 permeability).

Source	Sum of Squares	Mean Square	F- Value	*p*-value Prob> F	
Model	9.248E+007	1.850E+007	230.62	0.0043	significant
A- Immersion time	6.702E+006	6.702E+006	83.57	0.0118	
B- Pebax Concentration	4.881E+006	4.881E+006	60.86	0.0160	
C- Coating Layer	2.514E+006	2.514E+006	31.35	0.0304	
AB	2.351E+007	2.351E+007	293.10	0.0034	
BC	5.488E+007	5.488E+007	684.23	0.0015	
Residual	1.604E+005	80201.12			
Cor Total	9.264E+007				
R-Squared	0.9983				
Adj R-Squared	0.9939				
Pred R-Squared	0.9723				
Adeq Precision	47.37				

Table 4. ANOVA table for the regression model (CO_2/CH_4 selectivity).

Source	Sum of Squares	Mean Square	F- Value	*p*-value Prob> F	
Model	25.50	5.10	2176.06	0.0005	significant
A- Immersion time	0.38	0.38	161.44	0.0061	
B- Pebax Concentration	22.75	22.75	9707.71	0.0001	
C- Coating Layer	1.16	1.16	494.43	0.0020	
AB	0.36	0.36	154.83	0.0064	
BC	0.85	0.85	361.89	0.0028	
Residual	4.688E-003	2.344E-003			
Cor Total	25.51				
R-Squared	0.9998				
Adj R-Squared	0.9994				
Pred R-Squared	0.9971				
Adeq Precision	108.98				

3.3 Pareto Chart

Pareto chart highlights the order of the main and interaction parameter that affects the permeability and selectivity models. The pareto chart in Figure 3(a) illustrates the main and interaction effects of the factors for CO_2 permeability. The immersion time (A) shows the highest contribution towards permeability followed by Pebax concentration (B) and number of coating layer (C). There were two positive interaction effects between BC and AB. Positive effect referred to increase of permeability as the value of factor increased and vice versa. Figure 3(b) demonstrates the main and interaction effects of the factors for CO_2/CH_4 selectivity. The order of the TFC membrane synthesis parameters that give positive effects towards selectivity is B > C > A, while the order of the parameters that bring negative effects on the selectivity is BC > AB.

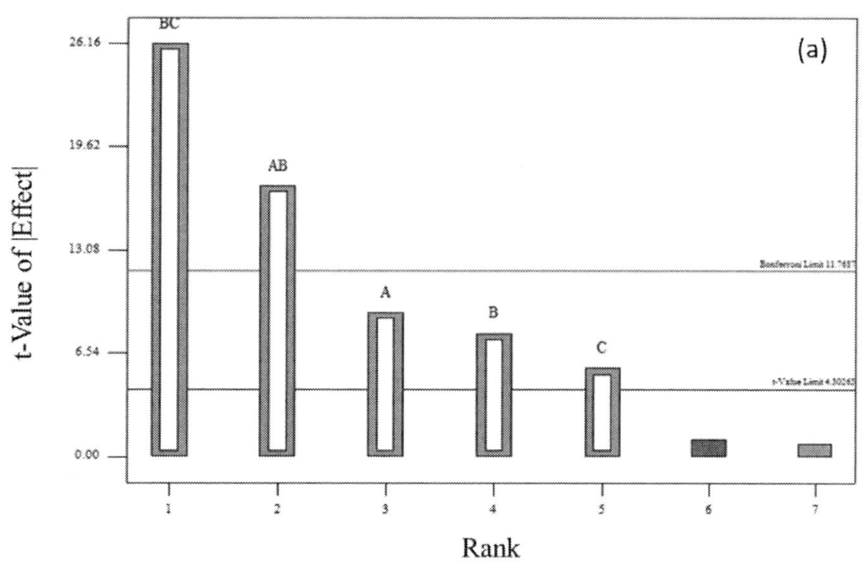

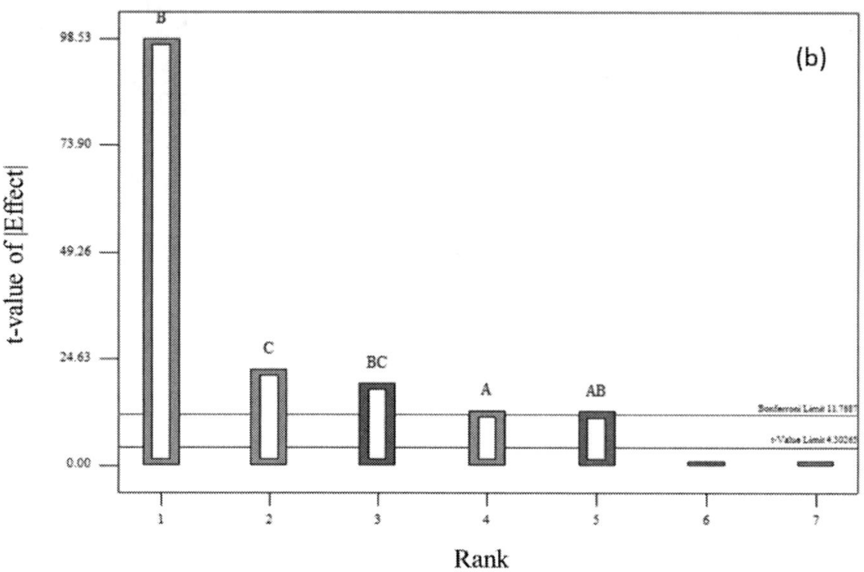

Figure 3. Pareto chart of main and interaction effect between TFC membrane synthesis parameters for (a) CO_2 permeability and (b) CO_2/CH_4 selectivity (orange: positive effect and blue: negative effect).

3.4 Effect of Process Parameters on CO_2 Permeability

The primary factor that provides a significant change to the membrane permeability was the length of the immersion time, which is the duration for the polymer precipitation process to occur. The large finger-like pores structure presence in the substrate is formed by the rapid rate of precipitation of the polymer from the cast solution during its immersion in the bath of non-solvent. The morphological changes are directly related to the immersion time where the longer it takes to immerse, the well-dispersed structure can be obtained. Therefore, a longer immersion time is required to allow the rest of the solvent slowly come out from the polymer. In this study, the immersion time was varied from 5 to 15 min for each sample. Based on Figure 4(a), the permeability increased with increased in immersion time which was due to complete dissolved solvent, forming a defect free porous membrane that creates an easy path for CO_2 gas to pass through the membrane. These results are in good agreements with the results in SEM as shown in Figure 9.

Figure 4(b) shows the effect of Pebax concentration to permeability. Similar trend was observed, where increased in Pebax concentration improved the permeability. The high increment of CO_2 permeability was ascribed to the good quadrupolar interactions existed between the CO_2 molecules and the polar groups in Pebax. Even though this TFC is a combination of porous and dense film, the separation achieved successfully due to the existence of Pebax 1657 which is polar material and good CO_2 adsorbent.

IOP Conf. Series: Materials Science and Engineering **702** (2019) 012049 doi:10.1088/1757-899X/702/1/012049

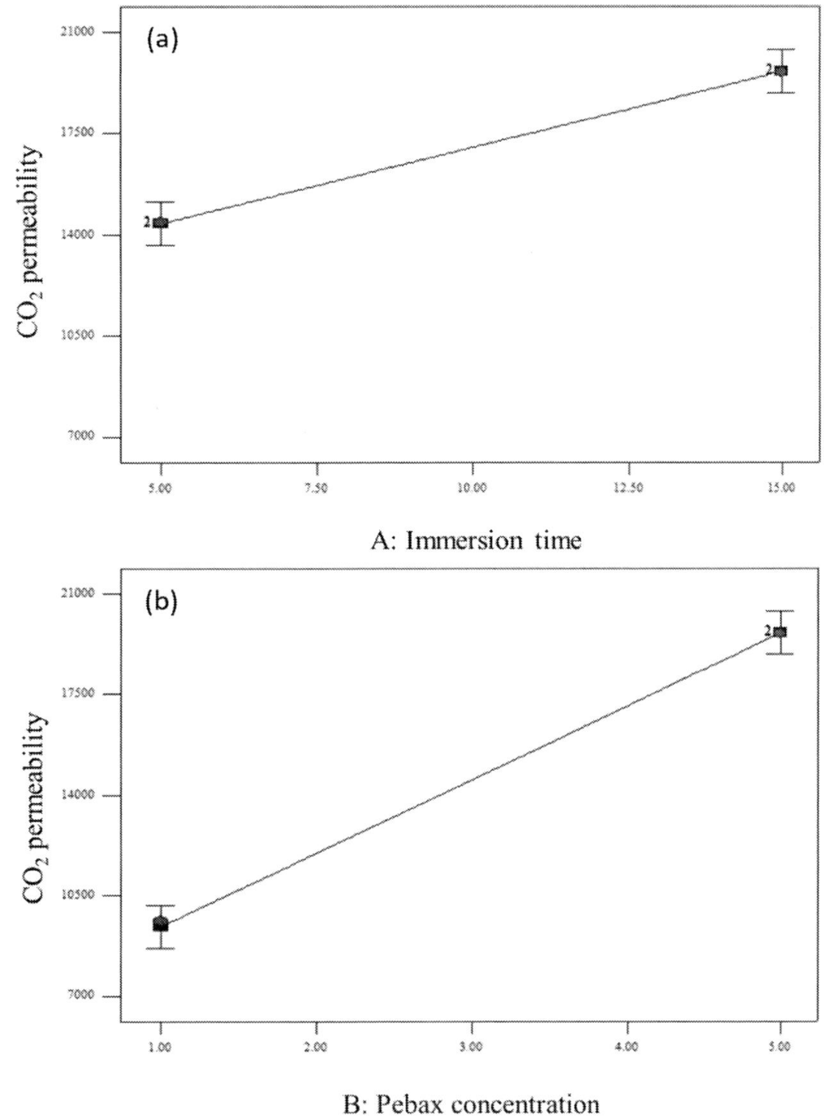

Figure 4. Most effective independent factors in CO_2 permeability; (a) immersion time (A) and Pebax concentration (B).

3.5 Interaction Effects between factors on CO_2 Permeability

Figure 5 shows the interaction effect between Pebax concentration and number of coating layer on CO_2 permeability. It was observed that greater permeability was obtained for 4 coating layers as compared to 1 layer at high Pebax concentration (5 wt%). This observation is due to the presence of more polar groups of Pebax at higher coating layer of Pebax causing more CO_2 molecules being absorbed, resulting in increment of CO_2 permeability. At lower Pebax concentration (1 wt%), permeability of a layer of Pebax was higher than 4 coating layers. This is due to the low Pebax concentration and Pebax coating layer producing thinner selective layer which allows more CO_2 gas passing through the membrane. However, variation of values of Pebax concentration at 1 coating layer of Pebax does not give impact to permeability since the thickness of selective layer is too thin, which does not give any resistant to the gas molecules to diffuse through the membrane layers.

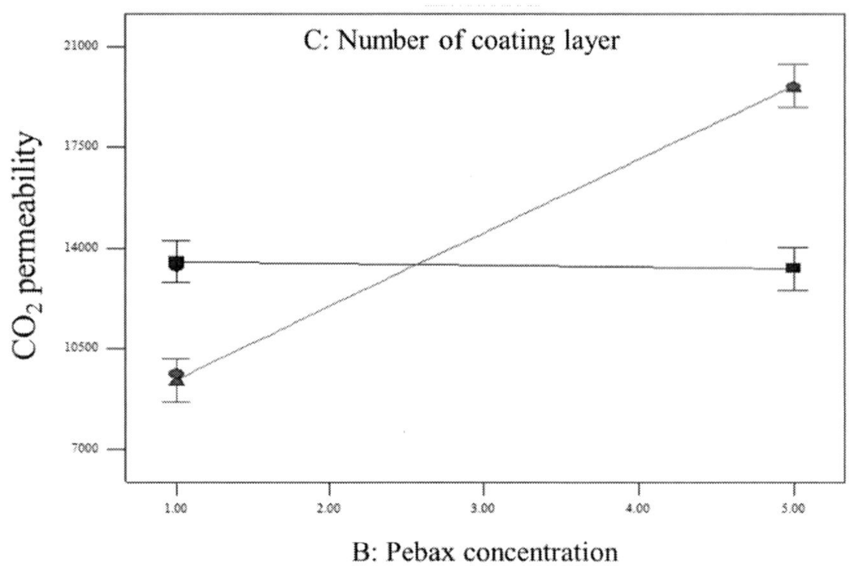

Figure 5. Most effective interaction effect between BC in CO_2 permeability.

3.6 Effect of Process Parameters on CO_2/CH_4 Selectivity

The primary factor most affecting the selectivity is the Pebax concentration. This is attributed to changes of viscosity at different Pebax concentration, where higher polymer concentration increased the viscosity of the solution. The Pebax coating solution was varied from 1 to 5 wt% of Pebax which was later dip-coated on top of PVC substrate in order to further study the role of Pebax 1657 on TFC membrane. According to Zarshenas et al. [13] a lower polymer concentration is preferred in preparing a polymeric membrane because concentrated polymer solution results in a denser membrane which is followed by lower gas permeability. However, different case is applied for the preparation of selective layer where lower concentration of Pebax produced more dilute solution that caused the solution to penetrate deep into large porous structure of the PVC substrate [23]. Hence, causing pore blockage that increased the mass transfer resistant of the gas, reducing the gas transport ability. As can be seen in Figure 6(a), at low Pebax concentration (1 wt%), the selectivity obtained is at the lowest which is 3.72. By increasing the Pebax concentration up to 5 wt%, the selectivity increased up to 7.10. Generally, the selective layer concentration highly depends on the porous substrate pore structure in a way that larger the pore structure in support film requires higher Pebax 1657 concentration solution to achieve a perfect TFC membrane.

The second factor that contributes the most in the separation properties (or selectivity) is the number of coating of Pebax. Different number of coating layers produced different thickness of selective layer. Mentioned by An et al. [24], the thickness of the selective layer affects the mass transfer resistance that influenced the entire membrane performance. In homogenous membrane, the researchers agreed that thicker membrane produced greater permeation resistance and lower the flux. The relationship between the selective layer thickness and flux is inversely proportional to each other [25]. As seen in Figure 6(b), it shows that the selectivity increased with an increase in number of coating layers indicating that thicker selective layer enhanced the separation ability of the membrane. This is due to too thin selective layer does not provide any significant resistance to the gas transport for separation to occur leading to high permeation of gases, accompanied by an undesired loss in selectivity. Therefore, the optimum layer was found at 4 coating layers with thickness 13.49 μm, approximately and has given the best selectivity of 5.79. Similar observations are reported in literature during the preparation of Pebax 1657 selective layer on top of PVDF substrate in CO_2/CH_4 separation [20].

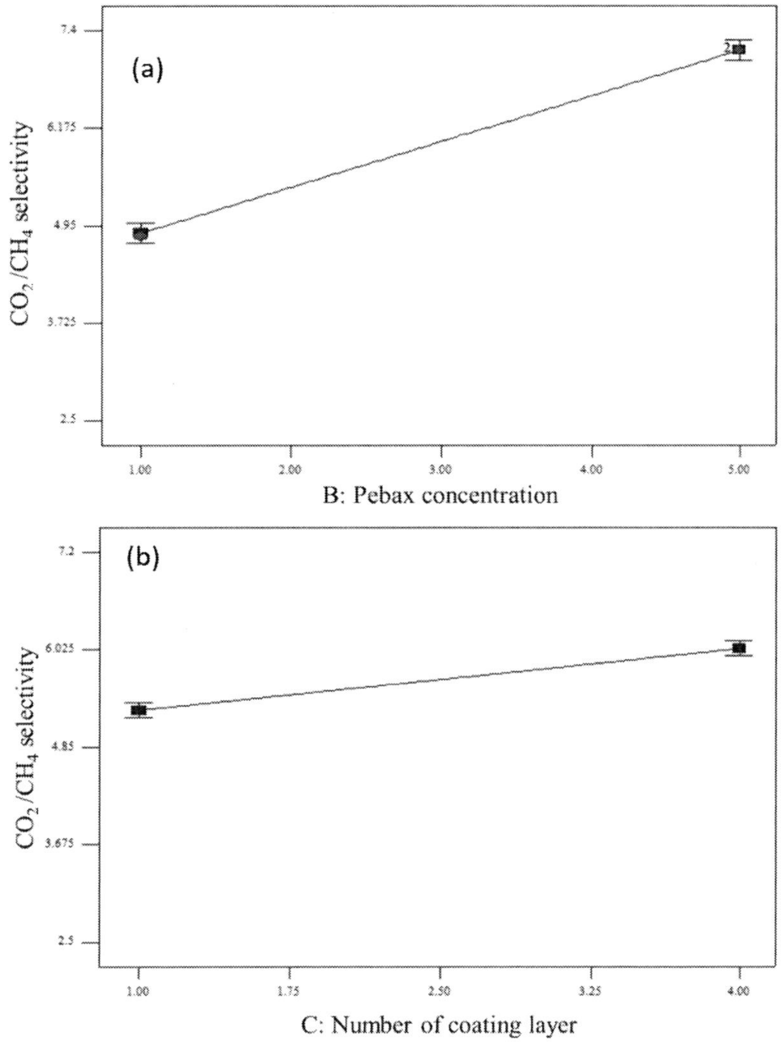

Figure 6. Most effective independent factors in CO_2/CH_4 selectivity; (a) Pebax concentration (B) and number of coating layer (C).

3.7 Interaction Effects between factors on CO_2/CH_4 Selectivity

The interaction effect between Pebax concentration and number of coating layer on CO_2/CH_4 selectivity is illustrated in Figure 7. From the interaction plot, the selectivity was observed to increase with Pebax concentration for both number of coating layers. This phenomenon occurred due to increase in Pebax concentration formed denser selective layer of TFC membrane which restricts the diffusion of gas molecules through the membrane especially CH_4 gas because of its big molecular size [26]. Fortunately, the quadrupolar nature of CO_2 molecules combined with the rubbery nature of Pebax makes it more permeable than CH_4, resulting in increased CO_2/CH_4 selectivity. The highest selectivity was obtained at 5 wt% Pebax concentration with 4 layers of Pebax coating, while the lowest selectivity was from the combination of the lowest Pebax concentration (1 wt%) and number of coating layer (1 layer). At low Pebax concentration, higher number of coating layers shows better separation ability. This observation is due to every layer of Pebax coated on the PVC substrate created a barrier which acts as a resistant and selective mechanism for gas transport, resulting in reduction of CH_4 gas permeability. However, these effects did not significantly affect the permeability of CO_2 due to its smaller kinetic diameter as compared to CH_4 (3.3 Å for CO_2 < 3.8 Å for CH_4), causing an increased in CO_2/CH_4 selectivity [27].

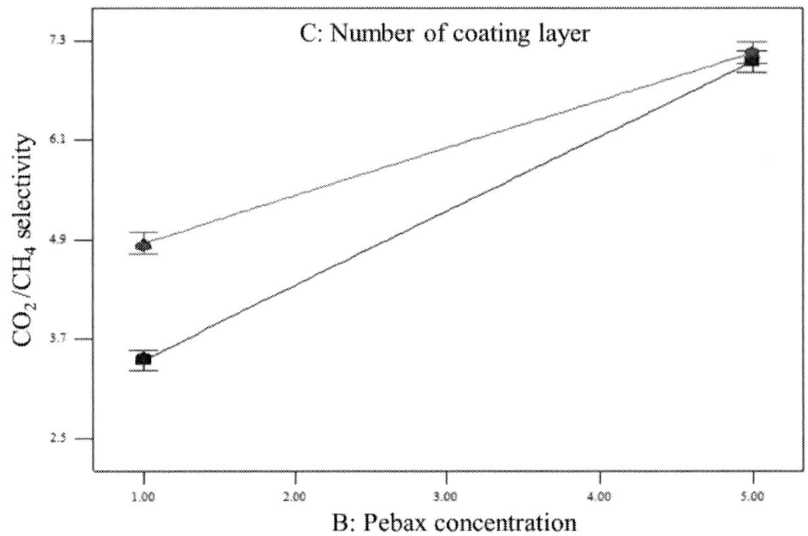

Figure 7. Most effective interaction effect between BC in CO_2/CH_4 selectivity.

One of the motivations of current membrane research is to develop membranes that surpass the Robeson upper bound, which satisfies both high separation ability (selectivity) as well as high separation speed (permeability). Regarding the CO_2/CH_4 separation, it is shown in Figure 8 that the introduction of Pebax layer on the PVC substrate resulted in a significant enhancement of membrane performance as compared to the uncoated PVC membrane. Besides, two TFC membranes were able to fall beyond the upper bound, and the highest is the TFC membrane prepared at the optimum conditions; 15 min immersion time, 5 wt% Pebax concentration and 4 Pebax coating layers.

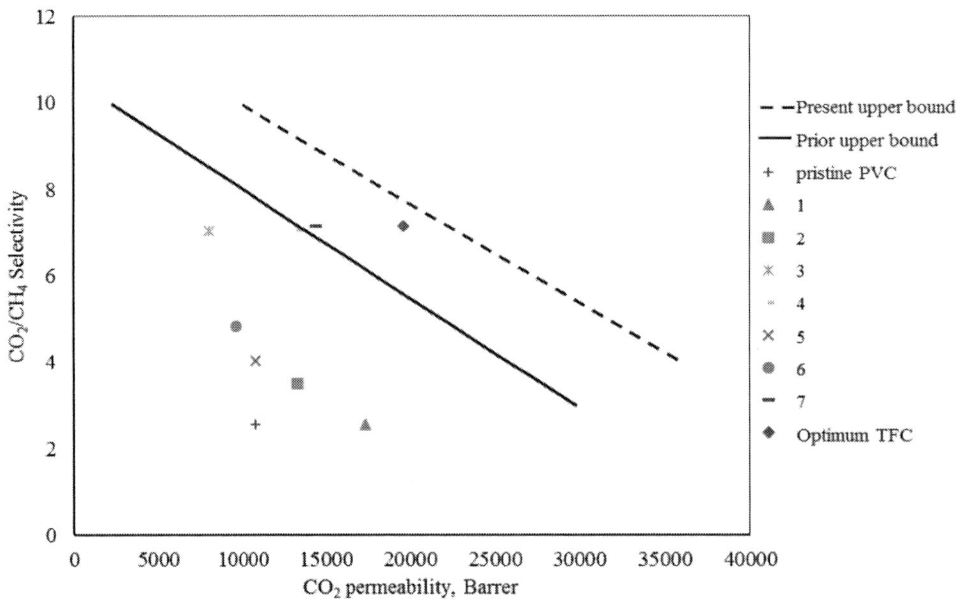

Figure 8. Robeson Upper Bound plot of the pristine PVC and TFC membranes produced in this work.

3.5 Membrane Characterization

3.5.1 Membranes Morphology.
In order to investigate the morphology of the uncoated PVC and PVC/Pebax TFC membranes, SEM images were taken from the cross-section of the membranes. The SEM images revealed that the prepared PVC substrate structures varied with different immersion time. The PVC membrane prepared with 5 min immersion time comprises of spongy like top layer and large void at the bottom (Figure 9(a)). The formation of large void was due to the development of bubble during the short immersion time. Next, the PVC membrane synthesized with 15 min immersion time shows a typical asymmetrical structure, with combination of spongy-like layer and finger-like porous structure (Figure 9(b)). Spongy porous layer is like an open–structure with minimum resistance against the transport of gas molecules while the formation of top finger-like voids in the membrane is believed due to the rapid movement of solvent DMF coming out from the film during phase inversion in water bath. These results indicate that immersion time significantly influenced the physical morphology of the membrane. Hence, 15 min was considered to be the optimal membrane immersion time for the fabrication of a well-defined membrane structure.

As seen from Figure 9(b), a clearly defined morphology with all layers were visible and well adhered to each other. It is obviously observed that the TFC membrane composed of two different layers due to the different physical nature of the polymers; i) a very thin and dense layer on top and ii) a thick and porous membrane, which mainly serves as the support. Besides, there is no adjustment in the PVC inner pore and structure both before (not shown in this paper) and after coating. This is due to the modification was performed by surface–coating, consequently just the surface of the PVC membrane is directly involved with Pebax. The thickness of the selective layer was found to be 13.49 μm.

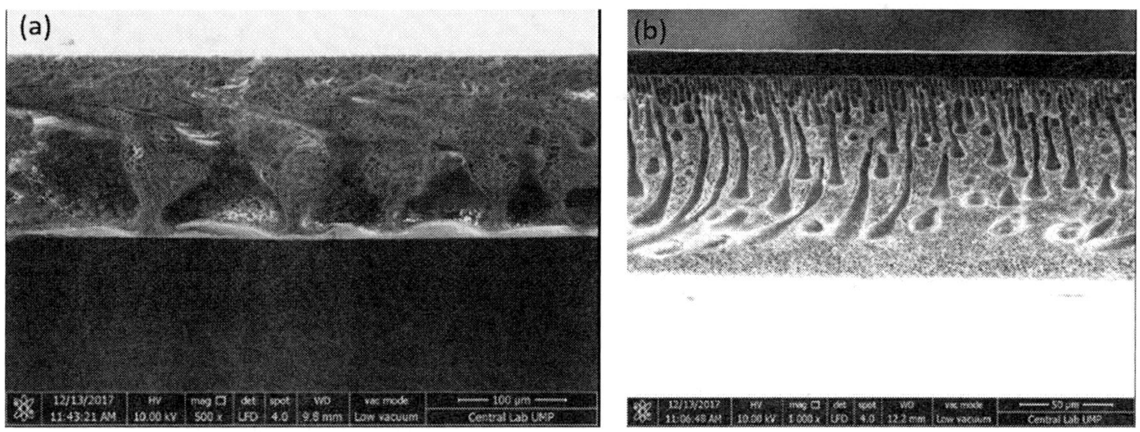

Figure 9. SEM images representing cross-sectional morphologies of (a) pristine PVC membrane and (b) optimum PVC/Pebax TFC membrane.

3.5.2 Fourier Transform Infrared Spectroscopy (FTIR) Analysis.
Figure 10 represents the chemical formulas of both PVC and Pebax 1657 which were referred to study and identify the bonding structures in the membrane. The FTIR spectra of pristine PVC and optimum PVC/Pebax TFC membranes over wavenumbers of 4000−500 cm^{-1} are presented in Figure 11. A peak observed at 1425.64 cm^{-1} belongs to aliphatic CH_3-CH_2 deformation vibration and peaks found at 686.70 cm^{-1} represents the C–Cl stretching, revealing the existence of PVC [28]. New peaks were appeared in TFC membrane spectrum which can be attributed to the addition of Pebax 1657 selective layer on PVC substrate. The presence of Pebax 1657 was observed through several molecule bonding; ether stretching (C–O), amide segment (C=O) and N–H at 1253.86 cm^{-1}, 1640 cm^{-1} and 3404.06 cm^{-1}, respectively. Other peaks in the spectrum of the TFC membrane between 2869 and 2939 cm^{-1} represents

the asymmetric and symmetric stretching of the C–H bond. Table 5 summarized the characteristics absorption of the membranes.

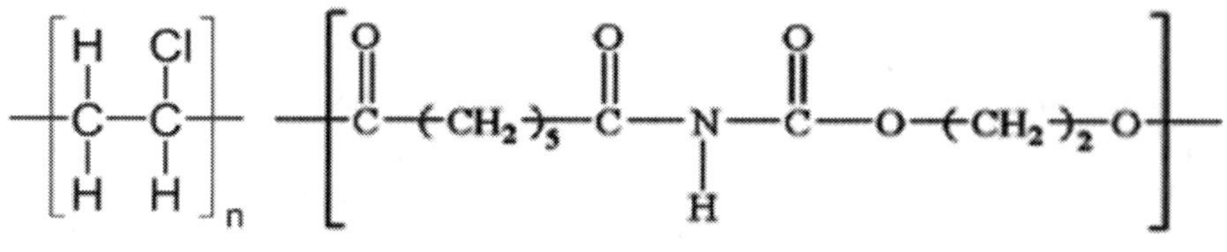

(a) (b)

Figure 10. The chemical structure of (a) PVC and (b) Pebax 1657.

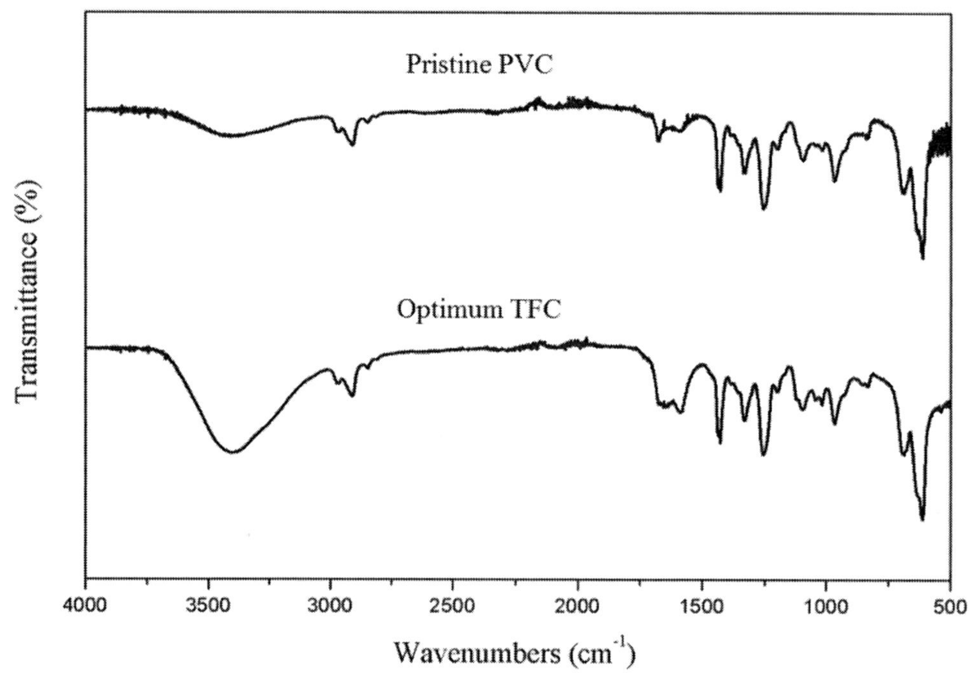

Figure 11. FTIR spectra of pristine PVC and optimum TFC membranes (4000–500 cm^{-1}).

Table 5. Vibrational frequencies of functional groups in PVC and TFC membrane.

Functional Group	Wavenumber (cm^{-1})
PVC	
CH_3–CH_2 vibration	1425.64
C–H stretch	2867–2817
C–Cl stretching	686.70
C–H bend	1350–1470
Pebax	
N–H stretch (amide)	3404.06
C–O stretch (ether)	1253.86
C=O stretch (amide)	1640
C–H stretch	2860–2925
C–H bend	1350–1470

4. Conclusion

Thin film composite (TFC) membranes comprised of PVC and Pebax 1657, have been successfully developed by dry/wet phase inversion method for PVC substrate and dip–coating technique for the TFC. SEM characterization revealed that different selective layer thicknesses of Pebax 1657 were formed based on different number of coatings made on the PVC film. By having the selective layer, the CO_2 permeability and gas selectivity of the original uncoated PVC film have also been improved. The FTIR analysis confirmed the presence of PVC and Pebax 1657 structures in the TFC. The optimum number of layers was 4 with thickness of 13.49 μm which gave CO_2 permeability and gas selectivity of 19612 Barrer and 7.15, respectively. With the increase in Pebax 1657 concentration, the fabricated TFC surpassed the Robeson upper boundary. To have a good TFC, the selective layer must not be too thin and the coating solution must also not be too diluted as they would cause defects in the TFC.

Acknowledgement

The authors wish to express their sincere gratitude to Universiti Malaysia Pahang (UMP) for the grant of (RDU 1703203) as well as Faculty of Chemical & Process Engineering Technology for the Gas Engineering lab facilities.

References

[1] Spitoni M, Pierantozzi M, Comodi G, Polonara F and Arteconi A 2015 *J Nat Gas Sci Eng* **62** 132
[2] Durán I, Rubiera F and Pevida C 2018 *Chem Eng J* **353** 197
[3] Ferella F, Puca A, Taglieri G, Rossi L and Gallucci K 2017 *J Clean Prod* **164** 1205
[4] Zhang Y, Sunarso J, Liu S and Wang R 2013 *Int J Greenh Gas Con* **12** 84
[5] Meng Y, Jiang J, Gao Y, Aihemaiti A, Ju T and Xu Y 2019 *Chem Eng J* **361** 294
[6] Freeman B, Hao P, Baker R, Kniep J, Chen E, Ding J and Rochelle GT 2014 *Energy Procedia* **63** 605
[7] Michal Ž, Bendová H, Friess K, Bara J E and Izák P 2018 *Sep Purif Technol* **203** 36
[8] Kang G and Cao Y 2014 *J Membr Sci* **463** 145
[9] Mahmoudi A, Asghari M and Zargar V 2015 *J Ind Eng Chem* **23** 238
[10] Wu Y, Zhao D, Ren J, Qiu Y and Deng M 2019 *Sep Purif Technol* **215** 480
[11] Idris A, Man Z, Maulud A S and Khan M S 2017 *Membranes* **7** 2017
[12] Strathmann H and Kock K 1977 *Desalination* **21** 241
[13] Zarshenas K, Raisi A and Aroujalian A 2016 *J Membrane Sci* **510** 270
[14] Abdul Wahab M S, Abdul Rahman S and Ahmad A L 2017 *J Phys Sci* **28** 39
[15] Danglad J, Eickelmann S and Riegler H 2018 *Chem Eng Sci* **179** 257
[16] Filpo G De, Pantuso E, Armentano K, Id P F, Profio G Di I and Nicoletta F P 2018 *Membranes* **8** 35
[17] Baroña G N B, Lim J, Choi M and Jung B 2013 *Desalination* **325** 138
[18] Yave W, Car A and Wind J 2010 *Nanotechnology* **21** 395301
[19] Shi M, Dong C, Wang Z, Tian X, Zhao S and Wang J 2018 *Chin J Chem Eng* **20** 18
[20] Abdul Wahab M. S and Abd Rahman S 2018 *Chiang Mai J Sci* **45** 484
[21] Fam W, Mansouri J, Li H and Chen V 2017 *J Membr Sci* **537** 54
[22] Sutrisna PD, Hou J, Li H, Zhang Y and Chen V 2017 *J Membrane Sci* **524** 266
[23] Shin H, Seok W, Bae S, Hak J and Kim J 2017 *J Ind Eng Chem* **53** 127
[24] An X, Xu G, Xie B and Hu Y 2019 *J Mater Chem A* **11** 311
[25] Perera D H N, Nataraj S K, Thomson N M, Sepe A, Hüttner S, Steiner U and Sivaniah E 2014 *J Membr Sci* **453** 212
[26] Alvi M A U R, Khalid M W, Ahmad N M, Niazi M B K, Anwar M N, Batool M, Cheema W and Rafiq S 2019 *Adv Polym Tech* **2019** 1
[27] Khalilinejad I, Sanaeepur H and Kargari A 2015 *J Membr Sci Res* **1** 124
[28] Tooma M A, Najim T S, Alsalhy Q F, Marino T, Criscuoli A, Giorno L and Figoli A 2015 *Desalination* **373** 58

IOP Conf. Series: Materials Science and Engineering **702** (2019) 012050 doi:10.1088/1757-899X/702/1/012050

Assessment on the severity of low density polyethylene (LDPE)/Ethylene hybrid explosion

H Badli, M N Arifin, S Z Sulaiman*, J lyas Susaidevas and N A Mahamad Khariol Nizar

Faculty of Chemical and Process Engineering Technology, Universiti Malaysia Pahang, 26300 Kuantan, Pahang, Malaysia

*E-mail: szubaidah@ump.edu.my

Abstract. The potential explosion for low-density polyethylene (LDPE) dust is greatly increased due to the presence of ethylene gas. Thus, this study was aimed to investigate the explosion severity of polyethylene/ethylene hybrid mixture in a 20-L spherical vessel. The explosion setup is equipped with two chemical ignitors and Kistler piezoelectric pressure sensors to ignite the hybrid mixture and to quantify the propagation of pressure wave during the explosion process respectively. The following hybrid explosion characteristics including the maximum explosion overpressure (Pmax), the maximum rate of pressure rise (dP/dT) and deflagration index (KST) were evaluated. This study concluded that the explosion severity of polyethylene/ethylene hybrid mixture steeply increases as the gas concentration and particle size increase to 15 g/m3 and 125 μm, beyond which the Pmax and dP/dT steadily decrease. The highest Pmax and dP/dt were recorded at 5.5 bar and, 50.5 bar/s with the KST value is 50 bar.m/s indicating the LDPE/ethylene hybrid dust explosion possess a weak explosion.

1. Introduction

Hybrid mixtures explosions can result in catastrophic personal injury and devastating damage. Hybrid mixture explosion occurs when dust, gas and vapor concentrations are respectively lower than their Minimum Explosion Concentration, (MEC) or Lower Flammability Limit, (LFL). According to Sanchirico et al. [1-2], the severity of hybrid mixture explosions was stronger than dust or gas explosions. For instance, the maximum explosion pressure and rate of pressure rise hybrid explosions (methane/coal) increased because of the presence of methane gas. They also reported that the explosion severity also increased with the increase of the flammable gas concentration. However, hybrid explosions property would be suppressed when the flammable gas was beyond the flammability range [3-4]. This phenomenon can be attributed to the limit of oxygen content.

Polyethylene is combustible dust. The presence of ethylene gas in polyethylene combustion may lead to the series of major fire and explosion. For instance, in 2015, Petronas Chemical LDPE (Low-density polyethylene) experienced a major fire incident. The main factor is due to ethylene gas ignite by the inherent static discharge from the LDPE. Other than that, [5-7] observed that the effect of polyethylene dust size and concentration also contributed to the severity of the hybrid explosion. However, their studies on the hybrid PE/Ethylene explosion were restricted in the flame propagation behavior. No attempt was made to study the explosion characteristic such as explosion pressure, rate of pressure and deflagration index. These characteristics are important to scale the severity of the explosion for prevention and mitigation purpose[8]. Thus, this research is carried to investigate the LDPE/ethylene hybrid dust explosion characteristics including maximum explosion overpressure, rate of pressure rise and dust deflagration index at various concentrations of ethylene gas on different polyethylene dust size and concentration.

2. Methodology

The flammability and severity data were obtained by employing 20-L spherical vessel as shown in figure 1. The range of LDPE concentrations tested in this vessel were 5, 15, 20, 30, 40 and 50 g/m^3. The explosion experiments were performed by using two 5 kJ chemical igniters as the standard ignition source that were connected to the ignition leads. The ignition delay time, t_v was fixed at 60 ms. The pressure inside the spherical vessel was measured by two "Kistler" piezoelectric pressure sensors. The LDPE dust was loaded directly to the storage container and were dispersed with the rebound nozzle connected to an outlet valve located at the bottom of the vessel by using compressed air pressurized at 20 barg. The vessel was interfaced with a computer, which controls the dispersion or firing sequence and data collection by using the control system named KSEP. As part of the experimental program, three repeated tests were performed to ensure a good data reproducibility.

Figure 1. The standard 20 L spherical vessel

3. Result and Discussion

3.1 Effect of LDPE concentration on the maximum overpressure, rate of pressure rise and the deflagration index, K_{ST}.

Figure 2 shows the pressure evolution when LDPE concentration varies function of maximum overpressure. The displayed maximum trend indicates that the maximum overpressure is LDPE concentration dependent. The highest value obtained for the maximum overpressure, P_{max} is 5.5 bar. From the lowest point at 3.7 ± 1.1 bar, the maximum overpressure steeply increases with an increase of LDPE concentration and steadily decreases once passing the maxima at 15 g/m^3 LDPE concentration. This maximum P_{max} is called optimum dust concentration. We speculated the plausible explanation for the decreasing trend after the concentration at 15g/m^3 is; amount of LDPE dispersed into the vessel overwhelmed the oxidant environment and instead of reacting, the excessive dusts limited the mass transfer of dust-oxidant contact. This result is in sync with the finding by other researcher who also found the inversion trend once the dust concentration is higher than the minimum explosible limit (MEC) [9].

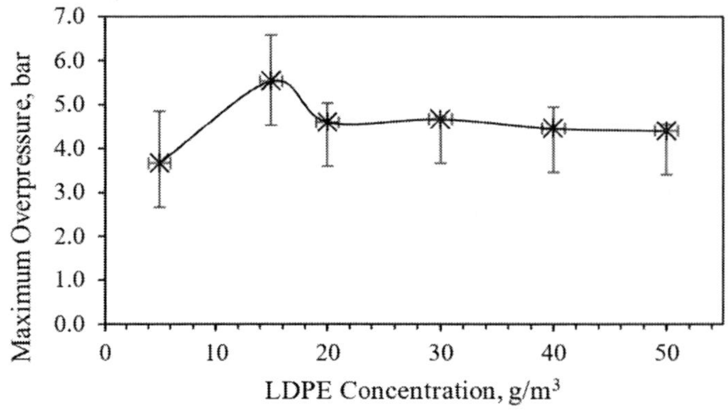

Figure 2. Effect of LDPE dust concentration on maximum overpressure. Results are expressed in bar ± standard deviations.

The severity of explosion was investigated by the determination of pressure rise rate, dP/dt and deflagration index, K_{ST} as a function of LDPE concentration. (refer Figure 3). The maximum value of $(dP/dt)_{max}$ and K_{ST} obtained are 50.5 bar and 13.71 *m*bar respectively, corresponding to 15 g/m^3 LDPE concentration. In general, the rate of pressure rise which reflects the explosion violence increases with an increase of LDPE concentration. The rate of pressure rise is also a function of particle size [4]. The K_{ST} criterium was calculated by the following equation from the cubic root law relationship[10]:

$$K_{ST} = V^{1/3}\left(\frac{dP}{dt}\right)_{max} \tag{1}$$

where V and P are the volume (m^3) and pressure (bar) in the spherical vessel respectively. From the maximum K_{st} value obtained from this experiment, the level of severity of LDPE and ethylene explosion can be categorized as moderate, class I (within the tested parameters). However, the K_{st} of value greater than zero is still a threat and pose the explosion potential. It can be seen that the rate of pressure rise (dP/dT) follows the trend of maximum overpressure, but with more declining slope after the peak point.

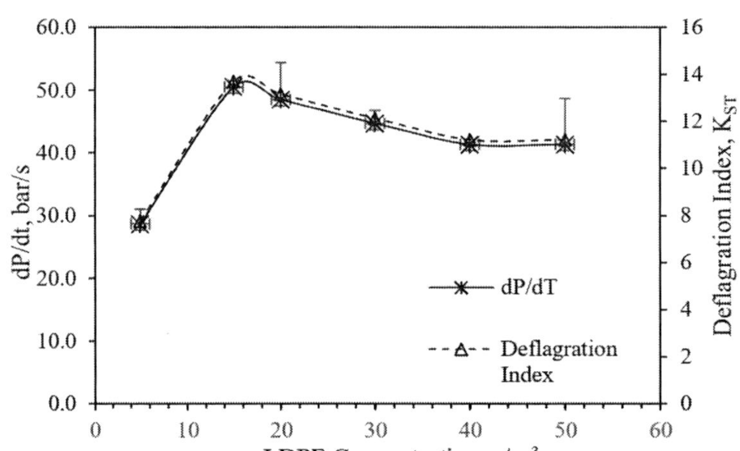

Figure 3. Effect of LDPE dust concentration on rate of pressure rise (dP/dt) and deflagration index. Results are expressed in bar ±standard deviation

3.2 Effect of LDPE particle size on the maximum overpressure (P_{max}) and rate of pressure rise, dP/dt

Figure 4 illustrates the effect of LDPE dust particles on the maximum overpressure, with the particles size ranging from 63 to 250 μm. The results of this particular test indicate the particle size affecting P_{max}

IOP Conf. Series: Materials Science and Engineering **702** (2019) 012050 doi:10.1088/1757-899X/702/1/012050

in the same pattern as the LDPE concentration vs P_{max}. From this figure, the P_{max} is at the highest when the particles size of 125 μm was tested. The smallest particle gives the lowest P_{max} at 2.78 bar, almost 50% lower than the maximum point. In this experiment, the collected data show the effect of gravity sedimentation and dispersibility of LDPE dust particles is negligible to the P_{max}. This could be attributed by the drag of air exerted on the dust particles counteracting the gravity effect on the particles.

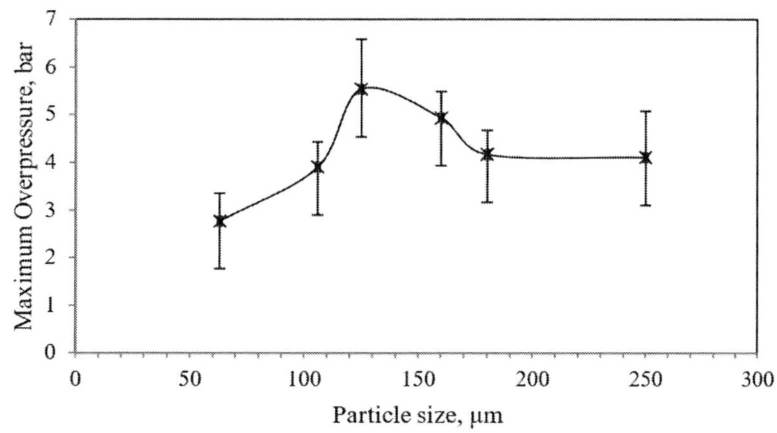

Figure 4. Effect of LDPE dust size on the maximum overpressure. Results are expressed in bar ±standard deviations.

The results of the particles size effect on the dP/dT and K_{st} are depicted in Figure 5. It can be seen that the rate of pressure rise and K_{st} depend on the LDPE particle size. The optimal particle size contributed to the $(dP/dT)_{max}$ is 125 μm. As the particles size increase, the dP/dt also increases except after passing the maxima point where the dP/dt and Kst start to decrease. This finding could be explained by having the most effective suspension in the vessel when the size of the particles is 125 μm. At this size, it is speculated that there were less particles settle at the bottom of the vessel as compared to the larger size dusts. As for the smaller particles, the mass and volume of polyethylene involved in combustion and explosion is less effective leading to the less dP/dT and K_{st}.

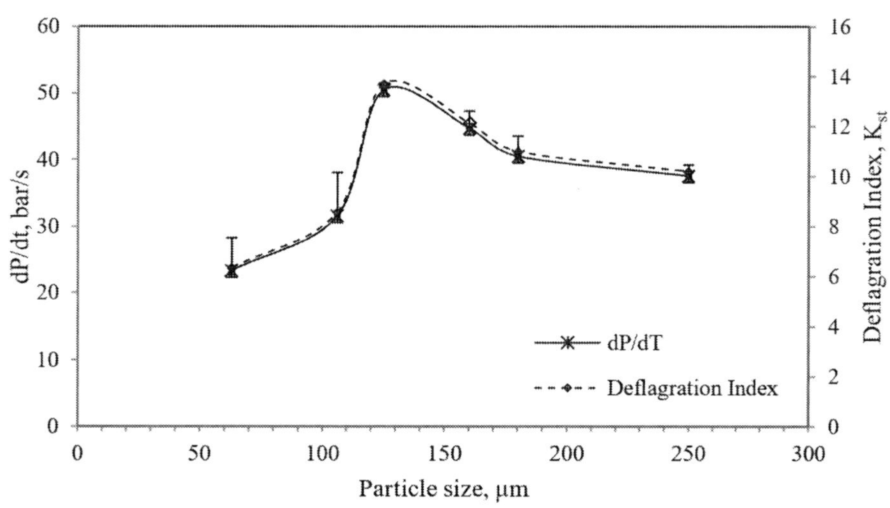

Figure 5. Effect of LDPE dust size on the rate of pressure rise (dP/dt) and deflagration index. Results are expressed in bar ±standard deviations.

IOP Conf. Series: Materials Science and Engineering **702** (2019) 012050 doi:10.1088/1757-899X/702/1/012050

3.3Effect of ethylene concentration on the maximum overpressure, rate of pressure rise and the deflagration index, K_{ST}.

The effect of ethylene concentration on the maximum overpressure is uptrend from 0 to 3 vol. % (refer Figure 6). The range of ethylene concentration was selected based on the flammability limit to gauge the hybrid effect of the LDPE. The maximum overpressure, P_{max} reaches the highest point when 3 vol.% of C_2H_4 injected into the system for hybrid explosion with LDPE. The particles size tested in this experiment is 125 μm. In the absence of ethylene, the maximum overpressure shows the lowest, 5.5 bar, but the P_{max} starts to rise with an introduction of ethylene owing to the synergistic effect[6]. The reduction of P_{max} when the ethylene was tested at 4 and 5 vol.% is in a good agreement with other researchers[11, 12] who also found the quite similar pattern of P_{max} with gas concentration. Another way of looking at the increasing trend of P_{max} can be related to the flammability limit where the P_{max} only increases when the explosion occurred at the lower flammability limit (LFL), below approximately 3 vol. % at which the explosion possible occurred at the gas-driven region. The equation to represent the overpressure as a function of ethylene concentration from 2 to 4 vol% is as follows:

$$P_{max} = -5.5105C^2 + 34.948C - 37.398 \tag{2}$$

where P_{max} and C is the maximum overpressure and concentration of the ethylene respectively.

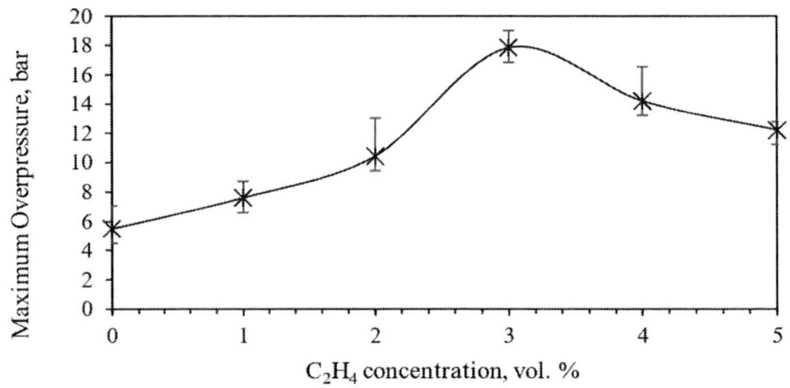

Figure 6. Effect of ethylene gas on the maximum overpressure. Results are expressed in bar ± standard deviations.

Experimental results as shown in Figure 7 demonstrated that dP/dt and K_{st} increase as C_2H_4 concentration increases. The optimal point for both parameters are 137.3 bar/s and 45.41 bar. m/s. As compared to the dust concentration and size particle effects on dP/dt, the rate of pressure rise for ethylene concentration is 137.3 bar/s, greater than double of the other two effects. As a consequence, the severity of explosion, K_{st} also increases to 45.41, more than three times higher than K_{st} values when particle size and dust concentrations were measured. These results indicate higher potential violence and severity if the ethylene concentration is present at the high concentration in the confined space.

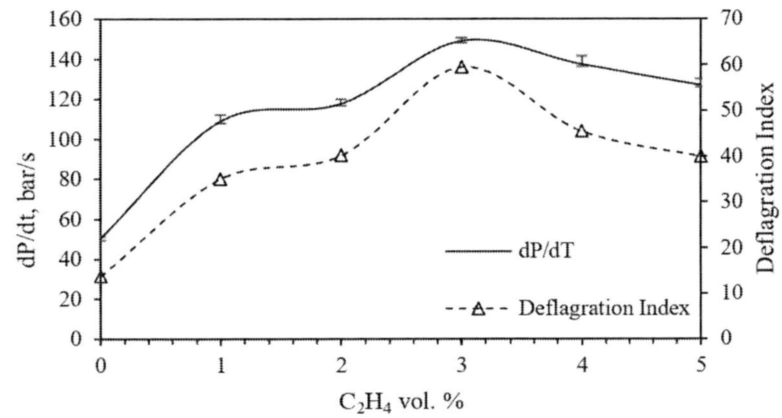

Figure 7. Effect of ethylene gas on the the rate of pressure rise (dP/dT) and deflagration index, K_{st}. Results are expressed as bar ± standard deviations.

4. Conclusion

The highest P_{max} recorded over the whole range of concentrations (5-50 g/m^3) and particle sizes (63-250 μm) is 5.5 bar at 15 g/m^3 and 125 μm respectively. The highest rate of pressure rise (dP/dT) is calculated as 50.5 bar/s, corresponding to the optimal concentration and particle size. This study concluded that the explosion severity of polyethylene/ethylene hybrid mixture steeply increases as the dust and gas concentration and particle size increase to 15 g/m^3, 3 % vol/vol and 125 μm, beyond which the P_{max} and (dP/dT) steadily decrease.

Acknowledgments

The authors would like to thank Faculty of Chemical and Process Engineering Technology, Universiti Malaysia Pahang for supporting this study.

References

[1] Sanchirico R, Di Benedetto A, Garcia-Agreda A and Russo P, 2011 *J Loss Prev Process Ind.* **24** 648-655
[2] Sanchirico R, Russo P, Di Sarli V and Di Benedetto A, 2015 *Process Saf Environ.* **94** 410-419
[3] Li Q, Lin B, Dai H and Zhao S, 2012 *Powder Technol.* **229** 222-228
[4] Addai E K, Gabel D and Krause U, 2015 *Process Saf Environ.*, **98** 72-81
[5] Gan B, Gao W, Jiang H, Li Y, Zhang Q and Bi M, 2018 *Powder Technol.* **328** 345-357
[6] Gan B, Li B, Jiang H, Zhang D, Bi M and Gao W, 2018 *J Loss Prev Process Ind.* **54** 93-102
[7] Gan B, Li B, Jiang H, Zhang D, Bi M and Gao W, 2018 *J Loss Prev Process Ind.* **55** 134-143
[8] Amyotte P R, Pegg M J and Khan F I, 2009 *Process Saf Environ.* **87** 35-39
[9] Garcia-Agre *Powder Technol. Powder Technology*, **205** 81-86
[10] Bartknecht W, 1971 *Bundesinstitut fur Arbeitsschutz (Bifa), Koblenz,*
[11] Bao Q, Fang Q, Zhang Y, Chen L, Yang S and Li Z, 2016 *Fuel,* **175** 40-48
[12] Dong C, Bi M and Zhou Y, 2012 *Safety Sci.* **50** 1786-1791

IOP Conf. Series: Materials Science and Engineering **702** (2019) 012051 doi:10.1088/1757-899X/702/1/012051

A road map to generative safety culture: An integrated conceptual model

A Shahid, M S Zaidi* and R Azizan

Faculty of Chemical and Process Engineering Technology, College of Engineering Technology, Universiti Malaysia Pahang, 26300 Gambang, Pahang, Malaysia.

*E-mail: shaiful@ump.edu.my

Abstract. Evidence has shown that maturity models are a popular means of assessing safety culture in organizations. Maturity models involve defining maturity stages or levels from less to more advanced safety cultures. A maturity model is a descriptive model in the sense that it describes essential, or key, attributes that would be expected to characterize an organization at a particular level. Existing culture maturity models still lacks a concise roadmap of strategies that can guide an organization to progress through the maturity stages from less to advanced. This work proposes a concept to integrate the present general and workplace models and techniques that provides a roadmap to generative safety culture. In doing so, a summary of the current general and workplace literature will be provided, method to integrate them to develop an integrated conceptual model for a generative safety culture, as well as the final integrated model is discussed. This conceptual model can be the basis for further research in order to provide a comprehensive picture of the safety culture improvement and maturity process. Lastly, implications for specific interventions to develop targeted safety culture improvement practices and work towards achieving generative safety culture will be discussed.

1. Introduction

The cultural aspects of safety have been one of the main focus in process industries to improve safety in the organizational settings. This is due in part to the findings of investigations into some major disasters in process industries (e.g. Texas City), other industries such as nuclear power (e.g. Chernobyl) and space travel (Challenger & Columbia), reported safety culture as one of the major causes. The term 'Safety Culture' officially emerged from the initial analysis report of the 1986 Chernobyl nuclear power plant accident [1]. This analysis first time introduced safety culture as a concept, to manage safety in a more advanced way and to look beyond the direct engineering and technical failures by dwelling more deeply into the underlying causes of accidents. Most explicitly defined by HSE (2005) as: "Organisations with positive safety culture are characterised by communications founded on mutual trust, by shared perceptions of the importance of safety, and by the confidence in the efficacy of preventive measures" [2].

A lot of research is available what safety culture consists of and on application of performance measurement frameworks, [3-7], their influence on safety outcomes i.e., frequency of accidents or lost time injuries and as well as on the use of safety culture maturity models for assessment and multi-level characterization [8-9]. Mostly, these measurement methods and maturity models of safety culture lack motivation for continuous improvement and theoretical framework [8, 10-11]. These signs clearly direct us to move from the contemporary approaches and explore new strategies for continuously improving safety culture.

In order to understand the nature of a continuously improving safety culture it is inevitable to know what are the most representative factors that make up a safety culture and how the factors can be identified, what steps should be done before embarking into a safety culture assessment, what strategies should be adopted to have an always monitoring and improving safety culture?
The questions stated above calls for a simplified integrated process for development of a conceptual integrated framework that can help to identify steps towards a generative safety culture. Therefore, this paper focuses on the integration of general and workplace models relevant to selection process of most representative factors, methods of assessment and improvement strategies that can pave a way towards a continuously improvement safety culture.

1.1. Selection of Factors for Safety Culture

Identifying and selecting the most representative factors for safety culture is challenge, particularly those factors that help in proactively improving the safety culture. Since, these factors are the one that will be assessed, evaluated and applied across the organization settings, it requires organizations to develop deep knowledge and understanding of how and what methods can be used for the identification and selection process.

There has been a voluminous work done on identifying and selecting factors of safety culture. Selection process of factors depends on various types of statistical and analytical tools, methods such as linear or arithmetic mean, aggregate mean. Furthermore, the selection process also depends on type of industry and country. Depending on organizational setup some factors might be of importance in one while others factors in a different organization [12].

A meta-analytic study done by Flin et al. [13] , analysed 18 studies and came out with the five most common themes, that are, work pressure, management/supervision, risk, competence and safety system. Similarly, Seo et al. [14], suggested that there are five main factors of safety culture, co-worker safety support, management commitment to safety, supervisor's safety support, competence level of employees with regard to safety and finally employee participation in safety-related decision making and activities. While in meta-analytic study by Clarke [15], analysed 16 studies and performed factor analysis and extracted the dominant themes common across the studies and he ended up with five main categories, i.e. Work task/work environment, Management attitude, Management actions, individual responsibility and involvement, and lastly safety management system. A literature review of 10 studies conducted by Wiegmann et al. [16], identified five indicators of safety culture namely, reporting system, management involvement, rewarded system, employee empowerment, and organizational commitment. Farrington et al. [17] after reviewing 15 studies, identified reporting system, management commitment, immediate supervisors and supervisor sub-ordinate relationships, involvement, competency, training, attitude, behaviour rules, procedures and communication were the common factors. A report prepared for the Health and Safety Executive, 2005 (HSE) identified two-way communication, leadership, involvement of staff, existence of learning culture and existence of just culture as five core dimensions [15]. The summary shows that no strict consistency in the factorial structure and no definitive set of factors has been reported.

1.2. Methods of Safety Culture Assessment

Safety culture can be assessed using different Safety Culture assessment tools. The assessment tool gives response on how well some organizational factors related to safety are working and what factors need to be improved and the effect of attitudes and behaviours on frequency of accidents or other safety outcomes. The most common assessment method being used are listed as:

a) Universal Assessment Instrument (UAI)
b) Safety Element Method (SEM)
c) Self-Diagnostic OHS Tool
d) Tripod Delta

The essential features of these methods are tabulated as follow in table.1.

Table 1. Safety performance related methods

Method	Author	Objective	No. of Elements	Safety Outcome
SEM	B.Alteren, 1998 [11]	Evaluation and improvement of OSH management performance	Six key elements: management, feedback systems and learning, safety culture, documentation, results, goals and ambitions	lost time injuries frequency, and a severity rate of injuries
UAI	Redinger and Levine, 1998 [18]	Evaluate the performance of OSH Management System	Consists of 27 sections, 118 OSH MS principles, and 486 measurement criteria	Nor reported
Self-Diagnostic OHS Tool	Cadieux et al., 2006, [19]	Subjective self-evaluation of OSH MS performance	Nine subject areas namely: organizational systems, prevention-oriented activities, management commitment, organizational structures, employee responsibility, norms and behaviours, communication, continuous improvement, and workplace compliance	The need for alteration to attain a satisfactory validation level
Tripod Delta	Cambon et al. 2006 [20]	Accident are prevented by identifying weak areas in working environment	Questionnaire-based tool, with 1500 validated questions	Corrective actions, the improvement of communication between the management staff and employees

All these methods are being validated, provide basic theoretical framework and provide relation between technical, organizational and human. Furthermore, these assessment methods can be used in different work settings and have the ability of motivation for safety improvements. [10].

1.3. Safety Culture Maturity Model

Maturity Models consists of maturity stages which assess the completeness of a process in an organization by using a different sets of criteria [8-21]. Westrum [22-23], proposed a concept that there exist a different level of cultures within an organization. He proposed three different organisational cultures, which he called pathological, bureaucratic and generative. The most immature stage of organisation has a pathological culture, which is one in which information is hidden, failure is covered up, new ideas are actively crushed and discouraged, sharing and learning from each other is actively discouraged. In such a culture, organization do not spend time risk management and safety issues.

A next higher-level in organisational culture is the bureaucratic one that handles the flow of information in a systematic way but may probably then be ignored, new ideas to improve the safety are seen to create problems, and learning and sharing might be listened but not encouraged. The generative organisation characterizes the most progressive stage of cultural maturity. Here information is actively pursued, and there are trained safety staff members collect it. New ideas are being appreciated and welcomed, and failures are promptly investigated rather than cover-up or blame. Westrum's tripartite typology was later modified to a five-level model and adapted by Parker and Hudson [24] specifically by adding up a reactive, and proactive level, and replacing the label bureaucratic with calculative with respect to safety culture (figure 1).

Reports from many accidents investigations reveals that the approach to safety within the industries in general is mostly 'reactive'. Organizations with reactive approach to safety adjust their safety policies after an event has occurred. At calculative level, organizations have systems in place to manage all possible risks, and learn to use the information they possess to manage safety professionally [24]. Proactive organizations look actively for safety gaps in the system and places safety as a priority before injury or incidents occur. At this level, organizations comply with safety performance standards, have clear safety objectives and there is active safety participation at all levels of management [25].

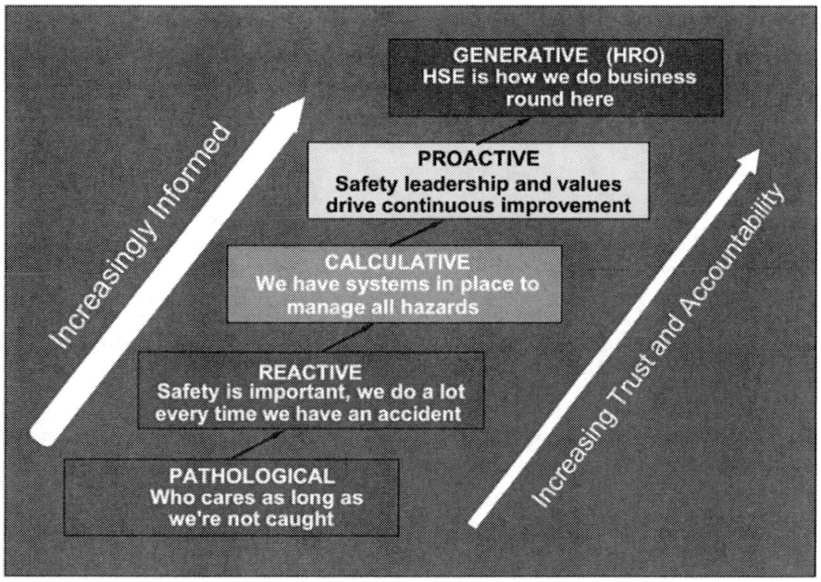

Figure 1. HSE Safety Culture Maturity Model [24].

1.4. PDSA Cycle for Safety Improvement

The PDSA method focuses on the translation of ideas and intentions into learned and informed actions. The PDSA (Plan-Do-Study-Act) is a tool that has been used widely by healthcare for quality improvement projects and bring positive changes in the processes effecting favourable outcomes [26-30].

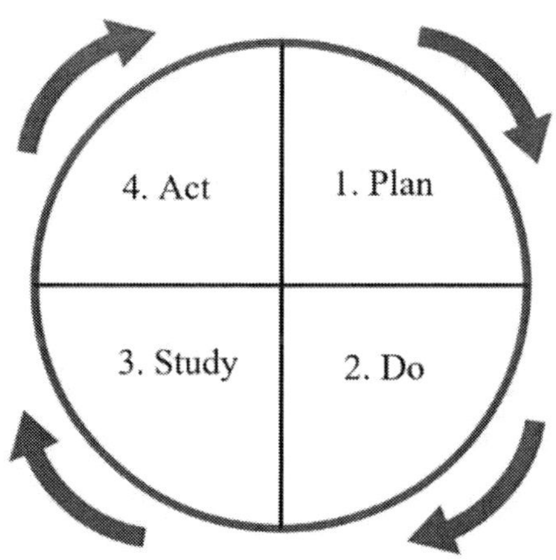

Figure 2. Demings' PDSA Cycle [31].

The PDSA cycle, shown in figure 2, is a model for learning and improvement of a product or a process. The brief description of the key components of the model are:

a) The very first step is 'Plan' a change, clear objectives, or test, aimed at improvement.
b) The second step is 'Do' that is, carry out the change or test.

IOP Conf. Series: Materials Science and Engineering **702** (2019) 012051 doi:10.1088/1757-899X/702/1/012051

c) The third step is 'Study' the results of the second step. The core objective of this step is to study or analyse the data and the process itself. Some critical questions in this step are:
   - Did it work out as planned?
   - What were the lessons learned?

d) The last step is 'Act'. Act is a vital element that ensures what measures and procedures are available to adopt the change or abandon it. Also enquiring on the state of readiness to make another change, which leads on to generating a plan for the next PDSA cycle.

The PDSA model is flexible, holistic in nature and easy to learn and use. The model supports continuous improvement efforts in a full range from the very informal to the most complex. It can be used to improve a new product line and for services improvements in a major organization setting.

2. Methodology

Based on the gaps and similarities of both the general and workplace models discussed above will help to build a new conceptual framework. The data collection for developing framework was done in five stages.

i. **First Stage:** It is the primary stage which describes the process and phenomenon of identifying and selecting different models from the existing literature.
ii. **Second Stage:** The data is structured together in new ways. Identifying a fit model from general or workplace models by looking deeply into identifying relationships between selected models. The goal is to make distinct connections between different dimensions, known as the "paradigm model". This stage helps in understanding and explaining the relationships between different categories of the models and helps to understand the phenomenon or dimensions to which they relate.
iii. **Third Stage:** This stage involves the integration of all the processes related to each model based on the similarities and overlapping features.
iv. **Fourth Stage:** This stage is systematically finalizing the overall integrated framework and the processes for the new model.
v. **Fifth Stage:** The final stage is validation and justification of the framework based on literature-based evidence provided.

The criterion to develop the integrated conceptual model is general or work place models shall be from different disciplines, including Occupational Health and safety management system, Safety culture assessment methods, Safety Culture Maturity Models and process improvement studies. The criterion not only offers new perspectives but increases the validity of the conceptual integrative framework by bringing together evidence from different disciplines. The search was to find simple methods or models and found a total of three general and workplace models which met the above criteria. Different stages of the process are described as below in table 2.

Table 2. General and Workplace Models

No.	Andersen [32]	OGP [25]	Brett & Jeffrey [33]
1.	Plan	Establish implementation team	Safety Culture Assessment
2.	Do	Assess performance	Analysis of Assessment & Survey Results
3.	Study	Confirm critical process and barriers	Recommended Practice Selection
4.	Act	Monitor weaknesses	Implementation Plan
5.		Set improvement actions	Safety Culture Assessment
6.		Regular monitoring	Analysis of Assessment & Survey Results

IOP Conf. Series: Materials Science and Engineering **702** (2019) 012051 doi:10.1088/1757-899X/702/1/012051

3. Discussion

3.1. Conceptualizing the integrated framework for generative safety culture

Based on findings from literature, a conceptual framework for generative improvement of safety culture is developed by utilizing various stages discussed in section 2. This conceptual Model is not proposed to be a final rigid structure for application purposes but is rather a flexible mean to provide guidance for discussion and further research. Further, the different stages in framework model are being discussed as follow.

- *Plan for Safety Culture Objectives.* The goal towards a generative safety culture is the continuous monitoring and improvements on every aspect of safety culture. Some main steps are discussed as below:

- Before embarking on an intervention for safety culture, the very first critical step is to ensure what being measured is specific, representative and relevant to the system. A better solution is to select vast set of factors based on validated studies. A Delphi method can be used to get experts opinion in the field for building up consensus on given set of factors. Analytical Hierarchy process (AHP) can be used to prioritize, or choose the most relevant and representative set of factors based on their importance by a definitive criterion while ANP (Analytical Network Process) can be used to find the structural influence.

- Identification of causal and influential effect in between the factors serves as a basis to track path for accidents before it happens by detecting deficiencies, errors and gaps in the safety and at the same time serves to design strategies to fill up the safety gaps and proactively improve the safety. Methods such Fuzzy Decision Making Trial and Evaluation Laboratory (F-DEMATEL) [34] and Analytical Network Process (ANP) [35], can be used to identify the influential and improvement paths. The benefit of using such methods that it can convert the relations between cause and effect of factors into a visual structural model, but also can be used as a way to handle the inner dependences within a set of factors.

- Review in detail the factors that are to be selected and compared across an organization. Systematically understand how these factors can be used to conduct a safety culture assessment in the company.

- Find partners in the same field, acknowledge them and obtain their acceptance for embarking in the progressive journey or identify inter-organizational departments for improvements and learning process.

3.1.2. Do the safety Culture assessment process. Establish a team to carry out the safety culture assessment process throughout the whole organization from top management to bottom personnel. Conduct the inter-organizational or intra-organizational safety culture assessment for benchmarking purposes, which serves as improvement strategies. The process of benchmarking is more than just a means of gathering data on how well a company performs against others. Benchmarking is the process by which companies look at the "best" in the industry and try to imitate their styles and processes. This helps companies to determine what they could be doing better. Since, organizations progress sequentially through the maturity stages, by building on the strengths and re- moving the weaknesses of the previous levels the assessment process shall be designed in a way to carry out the safety culture maturity status.

3.1.3. Study the Results. The effectiveness of safety culture data as a tool for safety improvement requires processes for developing a shared organizational understanding of the underlying meanings and causes, and it also requires processes for identifying the range of potential actions relevant to those interpretations. Rather than viewing the assessment results as an end point, the information should be

IOP Conf. Series: Materials Science and Engineering **702** (2019) 012051 doi:10.1088/1757-899X/702/1/012051

considered the starting point from which action and organizational safety changes emerge. Analyze the results from the assessment process helps identify strategies, set safety targets and key success areas for safety improvement. The safety areas identified as problematic can provide material for further analysis of underlying "root causes" and for generating improvement ideas from staff directly involved in the issues. Benchmark the process for inter-organizational and intra-organizational improvements. Data from safety culture assessments can be used to compare units within one organization or to examine differences across different organizations or systems.

3.1.4. Act on the improvement plan. Improvements strategies developed and acted upon based on the findings from the observation and analysis of the safety culture assessment and benchmarking partners. Regularly monitor the action plan, as it can be a basis for learning and speeding up positive developments and can be considered a promising step in improving safety culture. The outcomes at this stage can be used for the next assessment study with the purpose of continuous improvements and to learn from others as a basis for developing measures and programs which are aimed at increasing their own performance.

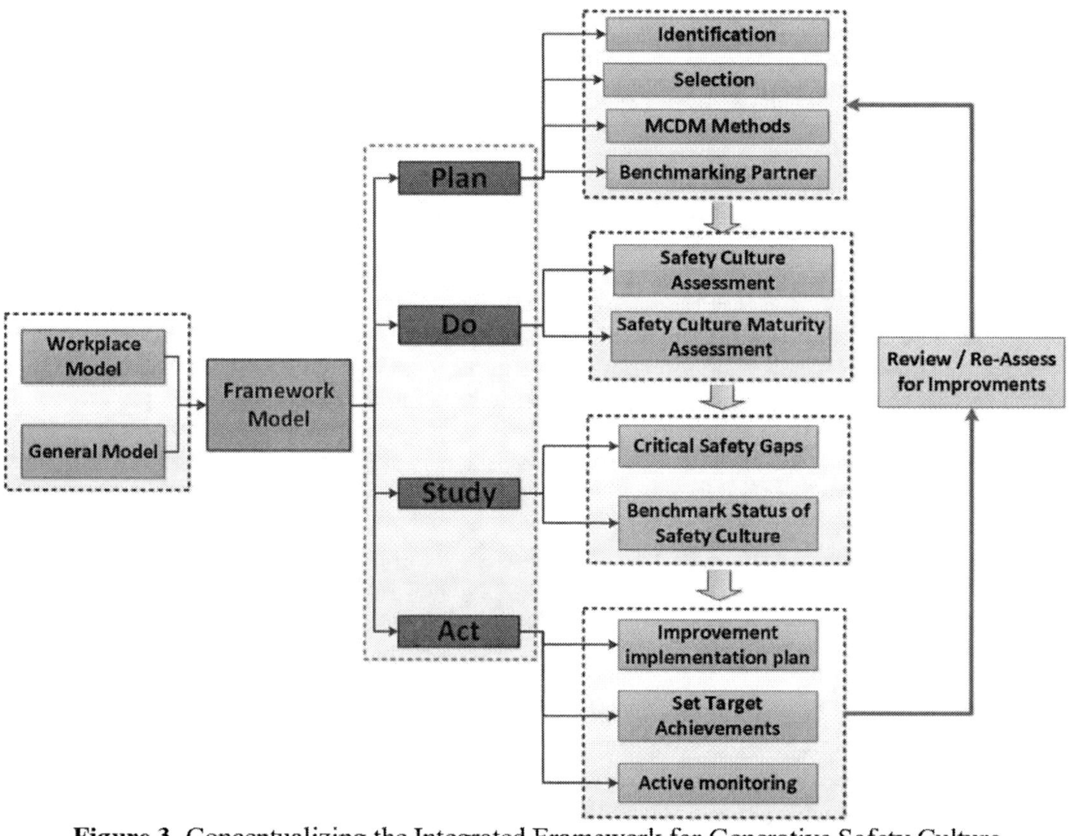

Figure 3. Conceptualizing the Integrated Framework for Generative Safety Culture

4. Conclusion

Our review of general and workplace models offers a new perspective into developing an integrated conceptual framework model for generative safety culture. PSDA model, OGP's KPI selection model, Safety culture improvement map fulfilled the criteria to be included in the study. As shown in figure 3, we propose an integrated conceptual model for generative safety culture. The integration process utilizes and explains the relationships and the overlapping features of PSDA model, OGP's KPI model and Safety Culture improvement map. By integrating work from different sources, we have systematically structured different concepts and related them to each other. The paper therefore has a theoretical contribution by combing different areas of research into a more comprehensive model.

IOP Conf. Series: Materials Science and Engineering **702** (2019) 012051 doi:10.1088/1757-899X/702/1/012051

The paper also discusses the critical importance of various steps and strategies that can guide an organization to progress through maturity stages from less to advanced.
The principal contribution of this paper is to provide a comprehensive integrated conceptual model for a generative safety culture. This is a flexible framework model and the various steps involved can be modified and accordingly validated using expert's opinion for further research or practical study.

References

[1] International Atomic Energy Agency 1992 *IAEA Report INSAG-7 Chernobyl Accident: Updating of INSAG-1* (Vienna)
[2] Health and Safety Executive 2010 *Safety Climate Measurement: User Guide and Toolkit* (Leicestershire)
[3] Frazier C B, Ludwig T D, Whitaker B and Roberts D S 2013 A hierarchical factor analysis of a safety culture survey *J. Safety Res.* **45** 15–28
[4] Di Gravio G, Mancini M, Patriarca R and Costantino F 2015 Overall safety performance of the air traffic management system: Indicators and analysis *J. Air Transp. Manag.* **44–45** 65–9
[5] Goodheart B J and Smith M O 2014 Measurable Outcomes of Safety Culture in Aviation - A Meta-Analytic Review *Int. J. Aviat. Aeronaut. Aerosp.* **1** 1–30
[6] Amrina E and Vilsi A L 2015 Key Performance Indicators for Sustainable Manufacturing Evaluation in Cement Industry *Procedia CIRP* vol 26 (Johor Bahru: Elsevier B.V.) pp 19–23
[7] Mannan M S, Mentzer R A and Zhang J 2013 Framework for creating a Best-in-Class safety culture *J. Loss Prev. Process Ind.* **26** 1423–32
[8] Goncalves Filho A P and Waterson P 2018 Maturity models and safety culture: A critical review *Saf. Sci.* **105** 192–211
[9] Filho A P G, Andrade J C S and Marinho M M D O 2010 A safety culture maturity model for petrochemical companies in Brazil *Saf. Sci.* **48** 615–24
[10] Sgourou E, Katsakiori P, Goutsos S and Manatakis E 2010 Assessment of selected safety performance evaluation methods in regards to their conceptual , methodological and practical characteristics *Saf. Sci.* **48** 1019–25
[11] Alteren B 1999 Implementation and evaluation of the Safety Element Method at four mining sites *Saf. Sci.* **31** 231–64
[12] Nardo M, Saisana M, Saltelli A and Tarantola S 2008 *Handbook of Contructing Compsoite Indicators: Methodology and user guide* (Paris: OECD Publishing)
[13] Flin R, Mearns K, Connor P O and Bryden R 2000 Measuring safety climate: identifying the common features *Saf. Sci.* **34** 177–92
[14] Dong-Chul Seo, Mohammad R. Torabi, Earl H. Blair N T E and Department 2004 A cross-validation of safety climate scale using confirmatory factor analytic approach. *J. Safety Res.* **35** 427–45
[15] Clarke S 2006 The relationship between safety climate and safety performance: A meta-analytic review *J. Occup. Health Psychol.* **11** 315–27
[16] Wiegmann D, Zhang H and Von T L 2004 Safety Culture : An Integrative Review *Int. J. Aviat. Psychol.* **8414** 37–41
[17] Farrington-Darby T, Pickup L and Wilson J R 2005 Safety culture in railway maintenance *Saf. Sci.* **43** 39–60
[18] Redinger C F and Levine S P 1998 Development and evaluation of the Michigan Occupational Health and Safety Management System Assessment Instrument: A universal OHSMS performance measurement tool *Am. Ind. Hyg. Assoc. J.* **59** 572–81
[19] Cadieux J, Roy M and Desmarais L 2006 A preliminary validation of a new measure of occupational health and safety *J. Safety Res.* **37** 413–9
[20] Cambon J, Guarnieri F and Groeneweg J 2006 Towards a new tool for measuring Safety Management Systems performance *2nd Symposium on Resilience Engineering* (Juan-les-Pins: Mines ParisTech) p 34
[21] Becker J, Knackstedt R and Pöppelbuß J 2009 Developing Maturity Models for IT Management – A Procedure Model and its Application *Bus. Inf. Syst. Eng.* **1** 213–22

IOP Conf. Series: Materials Science and Engineering **702** (2019) 012051 doi:10.1088/1757-899X/702/1/012051

[22] Westrum R 1993 Cultures with Requisite Imagination *Verification and Validation of Complex Systems: Human Factors Issues* ed J A Wise et al (Berlin: Springer Berlin Heidelberg) pp 401–16

[23] Westrum R 2004 A typology of organisational cultures *Qual. Saf. Heal. Care* **13** 22–7

[24] Hudson P 2007 Implementing a safety culture in a major multi-national *Saf. Sci.* **45** 697–722

[25] IOGP 2011 *Process Safety – Recommended Practice on Key Performance Indicators* (London)

[26] Docimo A B, Pronovost P J, Davis R O, Concordia E B, Gabrish C M, Adessa M S and Bessman E 2000 Using the online and offline change model to improve efficiency for fast-track patients in an emergency department *Jt. Comm. J. Qual. Improv.* **26** 503–14

[27] Dodds S, Chamberlain C and Williamson G R 2006 Modernising chronic obstructive pulmonary disease admissions to improve patient care: Local outcomes from implementing the Ideal Design of Emergency Access project *Accid. Emerg. Nurs.* **14** 141–7

[28] Dunbar A E, Cupit M, Vath R J, Pontiff K, Evans N, Roy M and Bolton M 2017 An improvement approach to integrate teaching teams in the reporting of safety events *Pediatrics* **139** e1–8

[29] Coury J, Schneider J L, Rivelli J S, Petrik A F, Seibel E, D'Agostini B, Taplin S H, Green B B and Coronado G D 2017 Applying the Plan-Do-Study-Act (PDSA) approach to a large pragmatic study involving safety net clinics *BMC Health Serv. Res.* **17** 1–10

[30] Donnelly P and Kirk P 2015 Use the PDSA model for effective change management *Educ. Prim. Care* **26** 279–81

[31] Moen R and Norman C 2009 Evolution of the PDCA Cycle *Proceedings of the Asian Network for Quality Congress* (Tokyo) pp 1–11

[32] Wretstrand A, Holmberg B and Berntman M 2014 Safety as a key performance indicator: Creating a safety culture for enhanced passenger safety, comfort, and accessibility *Res. Transp. Econ.* **48** 109–15

[33] Doherty B A and Price J J 2014 From a Multi-National to a Generative Safety Culture - A Road Map *SPE International Conference on Health, Safety, and Environment* (Long Beach: Society of Petroleum Engineers)

[34] Chi-Jen L and Wei-Wen W 2004 A Fuzzy Extension of the DEMATEL Method for Group Decision-Making *Japanis J.* **1** 10

[35] Wu W-W and Lee Y-T 2007 Selecting knowledge management strategies by using the analytic network process *Expert Syst. Appl.* **32** 841–7

IOP Conf. Series: Materials Science and Engineering **702** (2019) 012052 doi:10.1088/1757-899X/702/1/012052

Integrated chemical, technology & equipment process knowledge management system based on risk based process safety

N I P Anuar[1], H A Aziz[1,*] and R Ahmad[2]

[1]Faculty of Industrial Sciences and Technology, Universiti Malaysia Pahang, 26300 Kuantan, Pahang, Malaysia.
[2]RP Chemicals (Malaysia) Sdn Bhd, Lot 116 Gebeng Industrial Estate, Jalan Gebeng 1/11, 26080 Kuantan, Pahang, Malaysia.

*E-mail: hanidaaziz@ump.edu.my

Abstract. Major accidents in the industry such as fire, gas explosion, and unexpected toxic gas release are continuously reported worldwide. Lacking Process Knowledge Management (PKM) and failure to update and access the information was identified as one of the contributing factors. One of the established elements that intended to minimize this issue is PKM. However, lacking a systematic system to implement the PKM program has hindered the improvement of process safety level at workplace. The study focuses on development of PKMS database based on Risk Based Process Safety (RBPS) guideline. PKMS was developed by using Plan-Do-Check-Act (PDCA) concept. Validation of system was conducted through inserting real plant data from two chemical plants and feedback from process safety expert for further improvement. The developed PKMS allows the end user to store, review, modify and update the data regarding hazardous chemicals, technology and equipment information that involve with the operation. PKMS emphasis hazards related with the process in systematic ways thus minimizing potential incomplete control of hazard. PKMS interfaces allows management to do a self-check audit and helps the management to track the information. As leading indicator, PKMS allows the user to review overall status of the company compliance to the safety standard.

1. Introduction

1.1. Process Safety Incidents in Chemical Plants

Major accidents in the industry such as fire, gas explosion, unexpected toxic gas release continuously occurs due to lack of process knowledge management and failure to properly apply the available process knowledge directly during the event [1]. On October 21, 2016, a chemical release occurred at the MGPI Processing plant in Atchison, Kansas due to the Chemicals, was inadvertently connected to a tank containing incompatible material [8]. Final report investigation found that there are several issues closely related with the lack of chemical hazard information and lack of process equipment information in the organization. In Malaysia, a fire and explosion occurred at an Acetylene Gas Processing Factory on 22 May 2008 causing damages to the factory building structures and more than 360 cylinders have exploded. One of the gas refilling workers suffered burns on 30% of his body [2]. Lack of chemical hazard information is one of underlying factor of this accident. This problem is in need to fix to improve the knowledge management in chemical process plant.

IOP Conf. Series: Materials Science and Engineering **702** (2019) 012052 doi:10.1088/1757-899X/702/1/012052

1.2. Process Safety Management (PSM)

From the past major accidents, some new governmental regulations had been developed to manage the root cause of these accidents such as human errors, technical factor and management factor [3]. In 1992, the Occupational Safety and Health Association (OSHA) US had introduced Process Safety Management (PSM) of Highly Hazardous Chemicals standard (29 CFR 1910.119). This regulation aimed to prevent accidental releases of highly hazardous chemicals in the process industry, and to protect employees and the nearby community from exposure to those hazards [4]. The Process Safety Information (PSI) is the first element in the Process Safety Management (PSM). PSI focuses on process chemicals 29 CFR 1910.119(d) (1), technology 29 CFR 1910.119(d)(2) and equipment 29 CFR 1910.119(d)(3) management [5]. Reliable and accurate PSI data provide important resource to the other user that involved in Process Hazard Analysis (PHA) and Operating Procedure (OP) teams. Previous research study already developed PSI4MS that could fulfil all the PSI requirements based on OSHA PSM standard.

1.3. Risk Based Process Safety (RBPS)

RBPS framework is an extension effort for increasing the process safety management performance [7]. Some RBPS elements seem similar with PSM elements [3]. The purpose of RBPS Guideline is to help organizations design and implement more effective PSM systems. It provides methods and ideas on how to design a process safety management to correct deficiency and provide safety management practices. Understand hazards and risk is one from 20 pillars in the RBPS guidelines. There two elements in this pillar which are Process Knowledge Management (PKM) and Hazard Identification and Risk Analysis (HIRARC).

1.4. Process Knowledge Management (PKM)

PKM is an extended version of Process Safety Information (PSI) from PSM regulations. PKM element was introduced to provide details guidelines to comply with PSI standards. Process Knowledge Management (PKM) is one from the two elements in Understanding Hazard and Risk pillar. PKM focuses on information that can easily be recorded in documents and supports any effort to apply the risk-based methods. Prassl, Peden, & Wong in year 2005 reported that the developing of Knowledge Management System (P-KMS) as it permits the combination of engineering simulations with expert knowledge under the inclusion of uncertainty, as well as providing basic knowledge management capabilities [6]. Updating process knowledge in the manufacturing plant is quite common or can be categorized as daily or weekly routine, especially when changes involving chemicals substances, raw materials, facility issues and changing of technology are made. Too often, key data such as design bases, specification, manufacturer's drawing or data report and other process knowledge are thrown away because the document are not well organized [1]. Therefore, there is a need to develop the Process Knowledge Management System (PKMS) that strives for continuous improvement and prevent safety accidents.

2. Research Methodology

2.1. Research Process Flow

Figure 1 shows the process flow in constructing Process Knowledge Management (PKM) framework and system. First, the elements of PKM need to be analysed. After that, construct the framework for the PKMS by using Microsoft Word application. Next, develop the Process Knowledge Management System (PKMS) by using Microsoft Access and Microsoft Excel. Then, the PKMS should be validated by using data from case studies. Lastly, it is a must to determine whether the developed framework and PKMS have fulfilled with the RBPS guidelines. If it is 'Yes 'then the process will end meanwhile if it is 'No', then these steps need to be repeated starting from analysing the PKM elements.

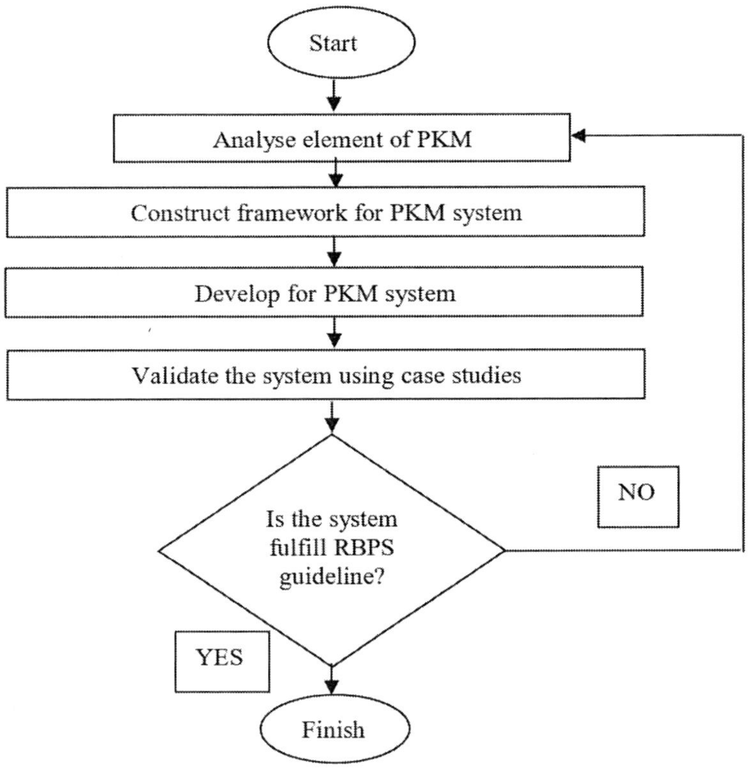

Figure 1. PKMS flowchart

2.2. Plan, Do, Check, Act (PDCA) Cycle Concept
In this study the concept of Plan Do Check Act (PDCA) management system was adopted as shown in figure 2.

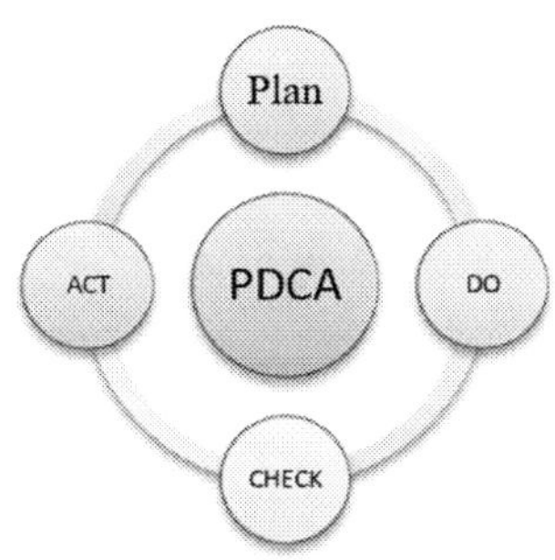

Figure 2. PDCA Cycle

- Plan: Plan for the new Process Knowledge Management System (PKMS) by analysing the requirements of PKM element from the RBPS guidelines.

- Do: Develop PKM Framework and PKMS by using Microsoft Word, Microsoft Access and Microsoft Excel.

- Check: Validate the developed PKM system by inserting raw that obtained from two chemical plants. The raw data including hazardous chemical information, technology information

IOP Conf. Series: Materials Science and Engineering **702** (2019) 012052 doi:10.1088/1757-899X/702/1/012052

(according to unit or system) and equipment information (according to tagging) into the PKMS using.

- Act: Do improvement to the system based on feedback got from validation process with safety expert.

3. Result and Discussion

3.1. Comparison of Requirement between PSI Element of PSM and PKM Element of RBPS

Both PSI & PKM element was classified under explicit knowledge [3]. Since PKM is an extended version of PSI, so there is must be slightly difference in term of requirements for both elements. Table 1 tabulates the comparison of requirements on information pertaining hazards of highly hazardous chemicals in the process between PSI and PKM. From this table, it can be explained that the extended data requirements of PKM compared to the PSI are thermodynamic data, calorimetric data, shock sensitivity, pyrophoric properties, chemical stabilizer or other chemical species and maximum deflagration/detonation pressure and flame speed.

Next, table 2 tabulates the information pertaining hazards to technology of the process for both PSI and PKM element. From the analysis, there are three requirements of PKM that did not stated in PSI which are adiabatic reaction temperature and the corresponding system pressure, separation equipment design information & design basis and the description of control system logic.

Lastly, table 3 tabulates information pertaining hazards to the equipment in the process plant for both PSI and PKM element. Based on this table, there are several requirements of PKM that PSI did not address which are mechanical data/design basis sheets for process equipment shop, fabrication drawings, piping specifications, isometric drawings, control system logic diagram, loop sheets & interlock tables, instrument data, electrical data, facility data, design basis & analysis for fixed or dedicated hoists, location of safety showers/eye wash station, fire extinguishers and other safety equipment and portable multi-unit equipment information. It can be concluded that, if industries comply with all the PKM requirements, the industry is indirectly complying with the PSI standard. It can be concluded that, PSI4M which was developed by previous researcher only cover PSI of PSM requirements meanwhile PKMS cover both of PSI & PKM requirements.

Table 1. Information pertaining hazards of highly hazardous chemicals in the process

Requirement		PSI	PKM
Information pertaining hazards of highly hazardous chemicals in the process	Toxicity information	✓	✓
	Permissible exposure limit	✓	✓
	Reactivity data	✓	✓
	Physical data	✓	✓
	Industrial hygiene data		✓
	Thermal and chemical stability data	✓	✓
	Hazardous effect of inadvertent mixing of typical contaminants with different materials contained in process streams & utility systems.	✓	✓
	Thermodynamic data		✓
	Calorimetric data		✓
	Shock sensitivity		✓
	Pyrophoric properties		✓
	Corrosivity data	✓	✓
	Chemical stabilizer or other chemical species.		✓
	Maximum deflagration/detonation pressure and flame speed		✓

Table 2. Information pertaining hazards of technology in the process

	Requirement	PSI	PKM
Information pertaining hazards to technology of the process	Simplified process flow diagram or block flow diagram for very simple processes	✓	✓
	Process chemistry	✓	✓
	Hazard related to credible undesired chemical reactions		✓
	Material and energy balances	✓	✓
	Maximum intended inventory		✓
	Safe upper & lower limit for process parameters	✓	✓
	Adiabatic reaction temperature and the corresponding system pressure		✓
	Separation equipment design information & design basis		✓
	Description of control system logic		✓
	Maps/tables showing zone/distance of concern for over-pressurization or toxic exposure hazard based on consequences analysis	✓	✓

Table 3: Information pertaining hazards to the equipment in the process

	Requirement	PSI	PKM
Information pertaining hazards to the equipment in the process	Materials of construction	✓	✓
	Piping & instrumentation diagrams(P&IDs)	✓	✓
	Electrical classification diagrams	✓	✓
	Relief system design basis & calculations	✓	✓
	Ventilation system design basis & calculations	✓	✓
	Listing of design codes & standards applicable to the process	✓	✓
	Safety systems (interlocks, detection or suppression systems)	✓	✓
	Material and energy balances	✓	
	Mechanical data/design basis sheets for process equipment		✓
	Shop fabrication drawings		✓
	Piping specifications		✓
	Isometric drawings		✓
	Control system logic diagram, loop sheets & interlock tables		✓
	Instrument data		✓
	Electrical data		✓
	Facility data		✓
	Design basis & analysis for fixed or dedicated hoists		✓
	Location of safety showers/eye wash station, fire extinguishers and other safety equipment		✓
	Portable multi-unit equipment		✓

3.2. Process Knowledge Management System (PKMS)

PKMS was validated at three chemical plants. PKMS was presented to a group of safety experts from the chemical plants to inform what kind of information that PKMS tend to capture and validate the system.

3.2.1. Login Form and Main Menu Interface.

Figure 3 shows that the login form which will automatically be popping out when user open the PKMS. The function of this interface is to ensure only authorized plant personnel can access the system. When user enter the right username and password, the main menu interface (figure 4) will automatically generate that helps user to navigate which type of process knowledge or information that they want to review.

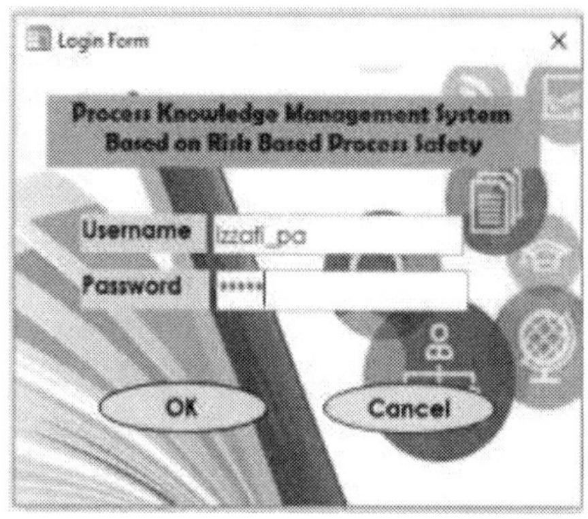

Figure 3. Login Form Interface

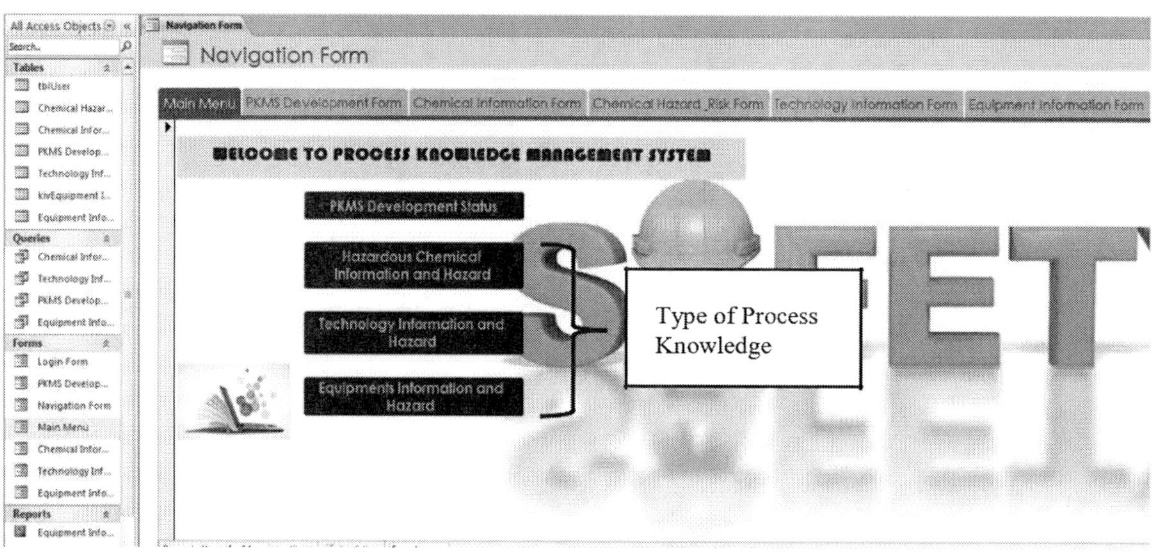

Figure 4. Main Menu Interface

3.2.2. PKMS Development Interface.

Figure 5 shows the main interface in the development of PKMS database. This interface contains 5 columns which are 'Process Knowledge', 'Status', 'Remark', 'Date of Review' and 'PIC'. In summary, this interface gives the user an overview regarding company compliances to the RBPS guidelines and PSM standard. Figure 6 shows the table interface of PKMS development. This interface allows the management to do a self-check audit and eases the management to track the information regarding to the PKM.

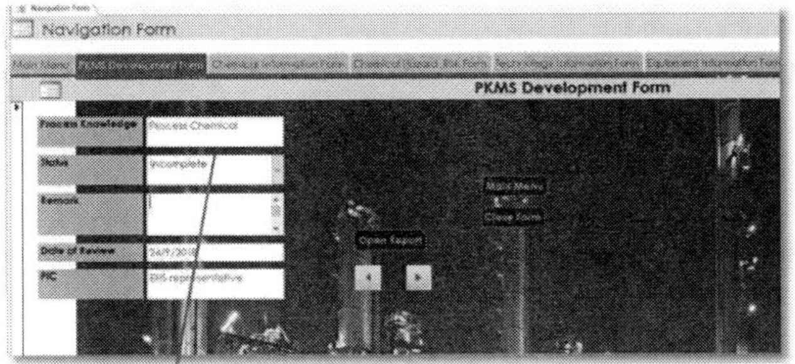

Figure 5. PKMS Development Interface (Form)

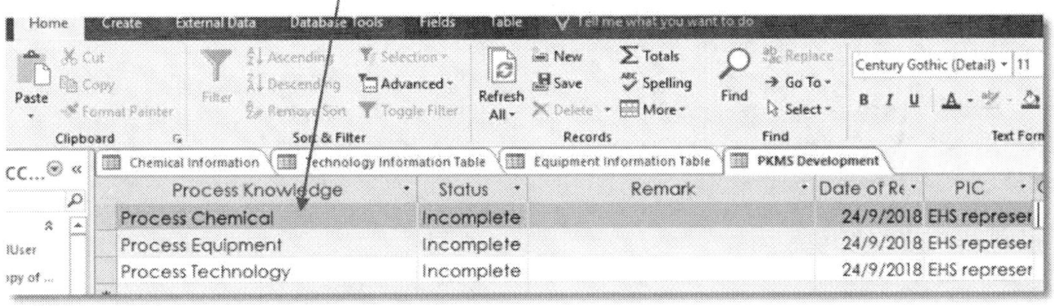

Figure 6. PKMS Development Interface (Table)

3.2.3. Chemical Information Interface.

Figure 7 shows the interface of PKMS for chemical information. This interface contains 136 hazardous chemicals as mentioned in 29 CFR 1910.119(d). The user is guided by a checklist to determine of which process chemical information that needed to be compiled. Evidence which can be store at the 'Document' column such as Safety Data Sheet (SDS). The personnel who familiars with the information involved in the plant can be either a plant coordinator, safety representatives and operation manager acted as the authorized personnel who can modify or update any information related to chemical in this PKMS [7].

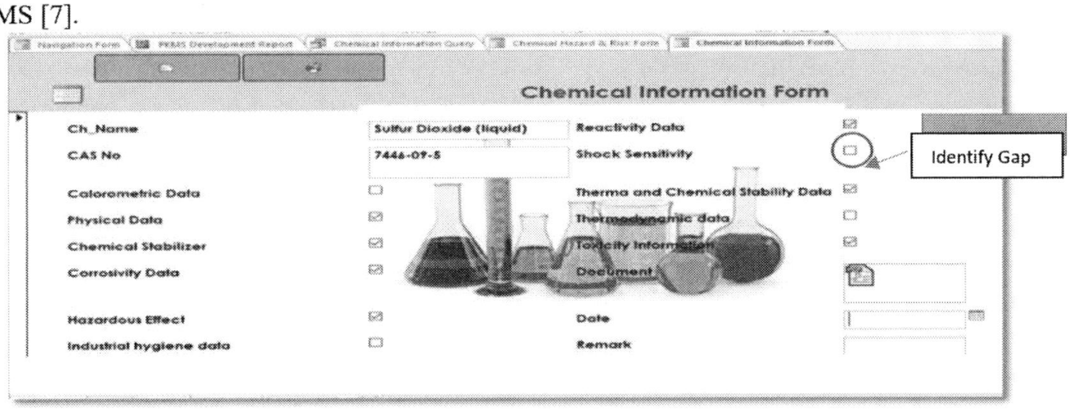

Figure 7. Chemical Information Interface (Form)

3.2.4. Chemical Hazard & Risk Interface.

Figure 8 illustrates the chemical hazard and risk interface. The previous PSI4MS tend to provide complete information concerning management of process chemical in the pilot plant operation meanwhile PKMS tend to cover information concerning the management of process chemical with emphasizing more on the hazard and risk [5]. In addition, the user should key in likelihood and severity of the consequences of the chemical hazard based on the plant standard, reported cases and risk matrix. By doing this, it will increase workers' awareness about the chemical hazard, control measure that need to be followed and who is person in charge in case of accident happen. Some of data from Plant A such as type of control measure, PIC and date of review cannot be share due to confidentiality. PKMS also contains print button which eases the user to print overall report related to chemical hazard and risk for report documentation purposes.

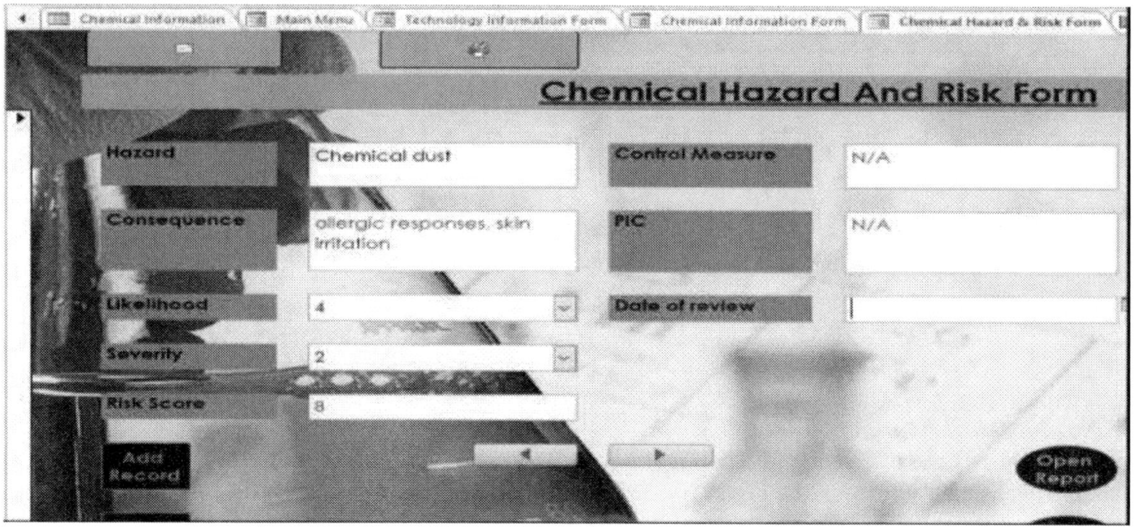

Figure 8. Chemical Hazard and Risk Interface (Form)

3.2.5. Technology Information Interface.

Technology Information was divided into work unit. This interface tends to capture information pertaining to process/ block flow diagram, material and energy balance and process chemistry and others as showing in figure 9. Finally, the 'Hazard Sheet' column was attached with a simple Hazard Identification, Risk Assessment and Risk Control (HIRARC) table also embedded in this system (refer figure 10).

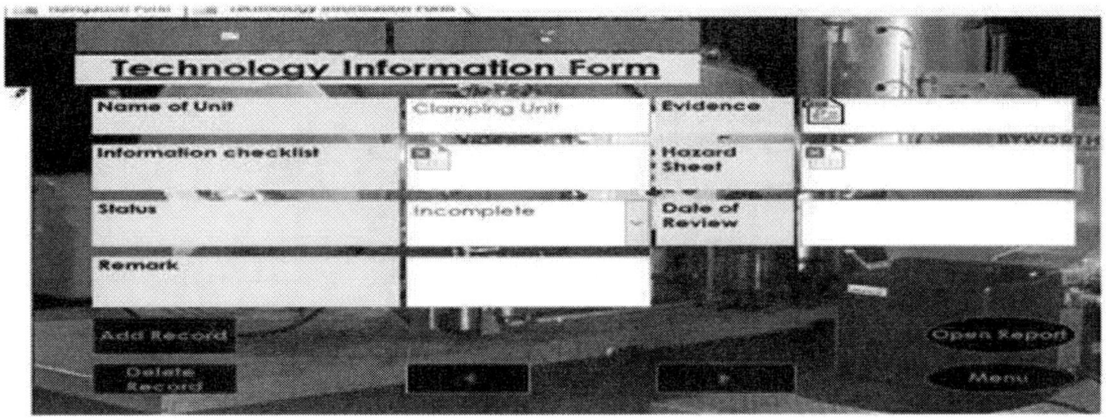

Figure 9. Technology Information Interface (Table)

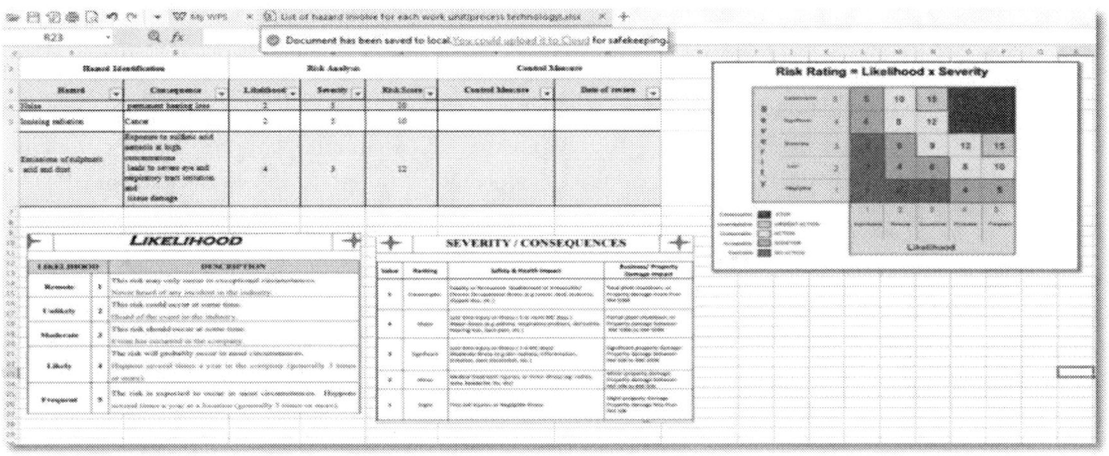

Figure 10. Hazard Identification, Risk Assessment and Risk Control table (attachment)

3.2.6. Process Equipment Interface.

To ensure the worker could operate and maintain the system appropriately, it is important to provide and maintain process equipment information so that worker can get an access to a specific information when needed [5]. Based on Figure 11, the equipment information was captured according to equipment tagging. The user can click 'Add Record' button if there is a new equipment purchased by the company with specific record and press 'Delete Record' button to delete the record. Besides that, this interface contains print and email button as a well communication medium to ensure the hazard information related to equipment is well communicated within the workplace. This interface captures information about equipment's' materials of construction, Piping & instrumentation diagrams (P&IDs, Electrical classification diagram, etc. Next, the user is required to do a process equipment checklist to determine what kind of equipment's information that contain in the organization. Then all the document of available information should attach at 'Evidence' column so the user can review the information easier and effectively. The information obtains from this PKMS also can assist other implementation of PSM elements in Plant such as Management of Change (MOC) and Process Hazard Analysis (PHA) element.

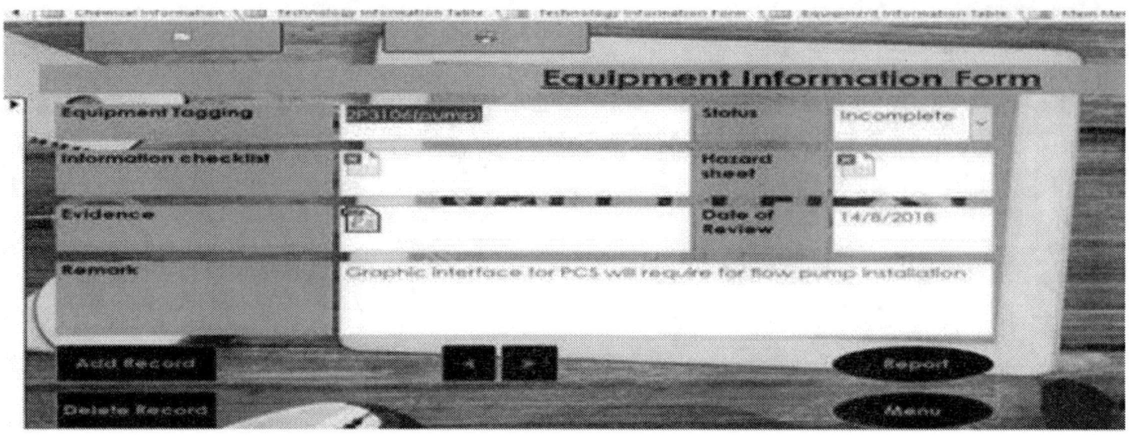

Figure 11. Equipment Information Interface (Form)

4. Conclusion

PKMS allows the end user to store, review, modify and update the data regarding to the PSI and PKM elements. PKMS also can guide the company to comply with the RBPS guideline. Most importantly, PKMS emphasis the hazard related to the process chemical, technology and equipment within company in systematic way and vital to control the process hazard. PKMS implementation helps company to identify the gap in systematic manner. In addition, it provides complete information about PSI that assists other PSM element such as Management of Change (MOC).

Acknowledgement

The author would like to acknowledge Faculty of Engineering Technology, Universiti Malaysia Pahang for the financial support and resources (RDU1703204) provided also, the research information from companies and practitioners to make this study feasible.

References

[1] Center for Chemical Process Safety 2007 *Guidelines for Risk Based Process Safety* (American John Wiley's & Sons) 169-208

[2] Department of Occupational Safety & Health Malaysia 2016 *Fire and Explosion at Acetylene Gas Processing Factory.*

[3] Chen M 2016. Process Safety Knowledge Management in the Chemical Process Industry. *American Journal of Chemical Engineering,* **4**(5), 131

[4] U.S. Occupational Safety and Health Administration (OSHA) 2000 Process safety management guidelines for compliance OSHA 3132 Washington, DC: United States. U.S. OSHA

[5] Aziz H A, Shariff A M, Rusli R, and Yew K H 2014. Managing process chemicals, technology and equipment information for pilot plant based on Process Safety Management standard. *Process Safety and Environmental Protection*, **92**(5), 423–29

[6] Prassl W F, Peden J M and Wong K W 2005 A process-knowledge management approach for assessment and mitigation of drilling risks. *Journal of Petroleum Science and Engineering*, **49**(3–4), 142–61

[7] Aziz A and Shariff A M 2017 A Journey of Process Safety Management Program for Process Industry. *International Journal ff Engineering Technology and Sciences (IJETS)* **8**(1), 1–9

[8] Chemical Safety Board 2017, April 12. *MGPI Processing, Inc. Chemical Release.*

IOP Conf. Series: Materials Science and Engineering **702** (2019) 012053 doi:10.1088/1757-899X/702/1/012053

Risk analysis of offshore pipeline due to third party damages

E Amri, N S Sulaiman* and F A Alaw

Faculty of Chemical and Process Engineering Technology, College of Engineering Technology, Universiti Malaysia Pahang, Lebuhraya Tun Razak, 26300 Gambang, Kuantan, Pahang, Malaysia

*E-mail: saaadah@ump.edu.my

Abstract. The third party damage also known as external factor that contributed to pipeline failure was an unexpected event that considered as threats to human safety. In this work, two risk assessment approaches namely qualitative and quantitative were adopted to predict and measure the causes of offshore pipeline failure due to the third party damages. Qualitative analysis involving causal and consequences of an event was developed using Bow-tie (BT) model to determine the contributing parameter of an event. The parameters obtained were then converted to Bayesian Network (BN) for a quantitative approach. Based on the qualitative analysis, the major threats identified in third party damage were categorized into "anchor impact" and "impact" meanwhile the consequences were human safety, marine life ecosystem and economical loss. Statistical data from PARLOC 2012 were utilized for BN model and transformed into conditional probability table (CPTs). the results generated indicate that the major contribution to pipeline damage was "trawling". The utilization of Bow-tie and BN analysis may complement the risk analysis of offshore pipeline due to third party damage for more informed pipeline maintenance decisions.

1. Introduction

The Pipeline is an essential carrier in oil and gas industries. Since it is carrying hazardous material for the global use, safety precaution is crucial to avoid any failure that could lead to fatalities. There are lot of causes lead to pipeline failure which will gives impact to industrial oil and gases around the world. Oil spillage happened due to pipeline leakage or ruptured will arouse world's concern. Many party will involve directly and indirectly. Meanwhile, accidents involving fire or explosion and carbon dioxide emission gives number of severe consequences, including loss of lives, injuries, environmental damage and various economic costs [1-2]. For the company itself, they will have bad reputation among society. In fact, the company also will experience with great losses in term of costing considering of product lost, public, private, and operator property damage along with clean-up and recovery costs. Therefore, maintaining pipeline integrity become the main priority in offshore pipeline because the consequences may result in catastrophic event.

From the literature, with 38% occurrence, third party damage giving the most contributing factor to pipeline failure due to anchoring activities and impact comprises with dropping object and fishing activities [3]. An extensive length of offshore pipeline become the major obstruction to detect or inspecting any problem regarding to the pipeline. In fact, the maintenance planning of oil and gas pipelines is a challenging task because of the complexity and high cost of maintenance operations

IOP Conf. Series: Materials Science and Engineering **702** (2019) 012053 doi:10.1088/1757-899X/702/1/012053

[4]. An extensive length of offshore pipeline become the major obstruction to detect or inspecting any problem regarding to the pipeline. Other than that, limited data availability become the main concern for the risk assessment.

Risk assessment is a tool designed to establish a proactive safety strategy by investigating potential risks [5]. Risk assessment was developed to overcome common problem faced in offshore pipeline. Risk assessment of pipeline failure can be a challenging task whereby scenarios leading to failure, the probability of a failure and the consequences of a failure have to be assessed in a risk assessment process. A number of qualitative and quantitative techniques have been implemented for risk assessment of oil and gas pipeline [6–10]. Fault tree analysis (FTA) and Event tree analysis (ETA) are two well established and most commonly used techniques for risk and reliability studies. Bow-tie, on the other hand, is a common platform which couples ETA and FTA by considering the common top-event named as critical event and reflects a corresponding relationship between reasons and consequences. The concept of a fault tree or event tree is following the Boolean logic, which in some circumstances may not reflect the reality. Fault tree is as well difficult to be used for quantitatively assessing system reliability because of its limited expressive power. In addition, fault tree and event tree allow the one-way inference that limits their applicability in the field of systematic risk mitigation and management.

Bow-tie (BT) is a graphical method that is broadly utilized in process industry accident modelling. Bow-tie diagram comprises of an integration of a FTA and ETA on a same platform. This method is able to portray a complete accident occurrence scenario from the causes, hazards and the consequence [11]. This bow-tie model however is limited to model mutual cause failures and conditional dependencies due to the static characteristic [12].

Another method so called the Bayesian Network is different from bow-tie method. It is an inference probabilistic method, that can overcome the static limitation of bow-tie method due to its updating mechanism. Furthermore, this method can also carry out a forward and backward prediction as well as diagnosis analysis [13]. The adaptation of Bayesian networks approaches in safety and risk assessment is due to several features offered by Bayesian network. In addition, communication of the model is strongly facilitated by means of transparent graphical representation of the undesired events and their causal relationships. Based on the above criteria, the Bayesian network is often implemented it can take full advantage of its flexible structure to fit a wide variety of accidents [14].

The Bayesian theory also is coupled with bow-tie model in order to use new evidence precursor data in updating the prior failure probability of primary events and safety barriers [15]. Qualitative analysis involving causal and consequences of an event was developed using Bow-tie model to determine the contributing parameter of an event. The parameters obtained were then proceed to quantitative analysis using by mapping bow tie model into Bayesian Network [16]. This work implements the last approach in developing a risk assessment used to predict and measure the root of offshore pipeline failure, qualitative and quantitatively.

2. Methodology

2.1. Bow-Tie method

A Bow-tie model was developed under few steps using BowtieXP software. The first step was to identify the top event. Top event is basically known as an incident that occurs when a hazard is released. In this work, the top event is the pipeline failure due to the third party damages. The next step was the assessment of the threats and consequences where the threats were positioned on the left hand side in the Bow-tie diagram. Meanwhile, the consequences were positioned on the right hand side of the diagram as the result from the top event. Barriers were then introduced in order to control the threats. For each threat, barriers were assessed to control the escalation factor from resulting to the top event.

2.2. Bayesian Network development

Bayesian Network is mapped from the proposed Bow Tie model. Each node established represents a variable while each arc representing a causal relation among two variables [17]. Conditional probability tables were assigned to the nodes denote conditional dependencies. Each CPT holds conditional probabilities of a child node being in a specific state of its parent's nodes. The conditional probability table (CPT's) can be elicited from the domain expert [18], by conversion of Boolean operations [3-19] or through the framework, literature, or observational experience [20]. In this study, the best available source to carry out the CPT elicitation is from the Pipeline and Riser Loss of Containment PARLOC 2001 [21]. Once the information (evidence) available obtained, a software programme HUGIN provide the parameter learning algorithms was utilized [3], [18]. Lastly, the essential steps in performing the probability assessment is to prove the model to help assess whether the model selected is reliable or not.

3. Results and discussions

3.1. Qualitative analysis

Third party damage is known as damages due directly to the acts of man and includes all activities not directly related to the pipeline of study. As reported in [21], the scenarios that lead to the third party damages are categorized into two main factors namely impact and anchor smash. Impact on the pipeline may be caused by the trawling activities and dropped object. Meanwhile anchor smash impact are due to ship passing and construction activities. Due to the third party activities, three consequences that are human loss, oil spill and economical loss can be envisaged for the accident scenario depending on the success or failure of the safety barriers. Based on the collected data, a BT was developed as shown in figure 1.

3.2. Quantitative analysis using Bayesian Network

3.2.1. Risk analysis. The BT diagram was converted and mapped into a BN as shown in figure 2. Once the network model was developed, Conditional Probability Tables (CPTs) are elicited for further analysis. In this work, CPTs were assigned based on the data gathered from various existing reports and literatures. Table 1 shows example of data collected from [21]. Other than that, the consequences of failure data are collected from various existing literature [22-23]. The proposed BN model is then analyzed using HUGIN.

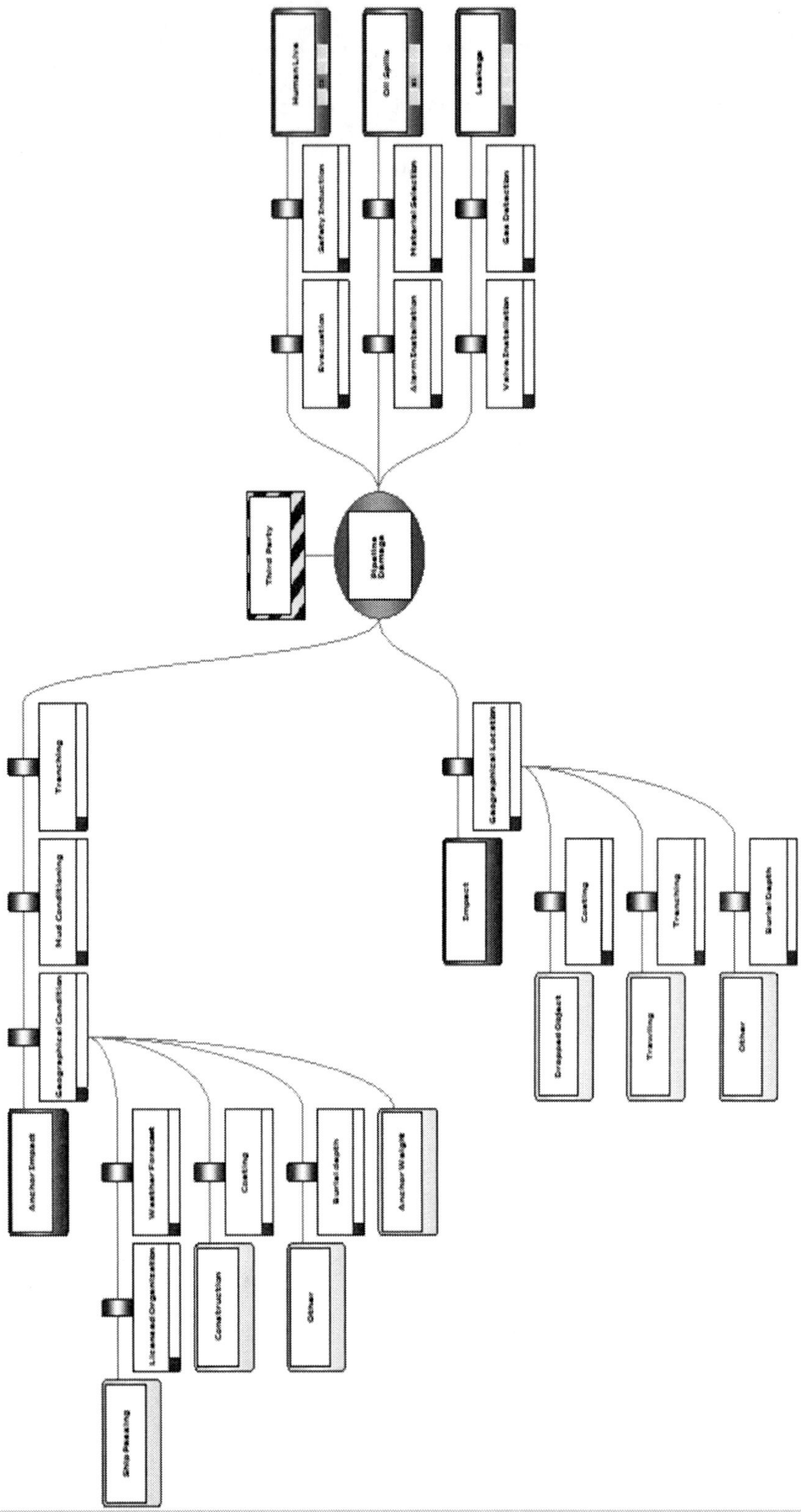

Figure 1. Qualitative Analysis by Bow-Tie Model

Table 1. Data of incidents happen [21].

Parameter	Sources	Number of incidents
Anchor	Ship/Supply Boat	6
	Rig or Construction	3
	Other/Unknown	10
	Total	19
Impact	Ship on Riser	-
	Trawl	23
	Dropped Object	-
	Wreck	1
	Construction	-
	Other/Unknown	9
	Total	33

3.2.2. Probability analysis

The Bayesian Network model is used to revise the prior probabilities by taking into account the evidence which is a task not supported by BT. Entering the probabilities, the BN is now available to perform different types of analysis. In this study, the most significant purpose is in updating probabilities in the light of actual observations of events. In BN modelling, these are called evidences for the third party damages BN. By running the proposed model for this third party damages scenario, as can be seen in figure 2, the conditional probability that "death" happened from pipeline failure is 0.92% given 95% evacuation success. Meanwhile the "oil spill" is 2.33% given alarm is 95% work and probability of "economical loss" is 21.32%.
Entering pieces of evidence and using them to revise the probabilities in this way is called propagation. figure 3 shows the results with "*evidence*" node for anchor impact represented by an *evidence bar* with 100% occurrence. As would be expected, the probability of occurrence for pipeline damaged increases from 26.65% to 35.34% when anchor impact has been observed.

IOP Conf. Series: Materials Science and Engineering **702** (2019) 012053 doi:10.1088/1757-899X/702/1/012053

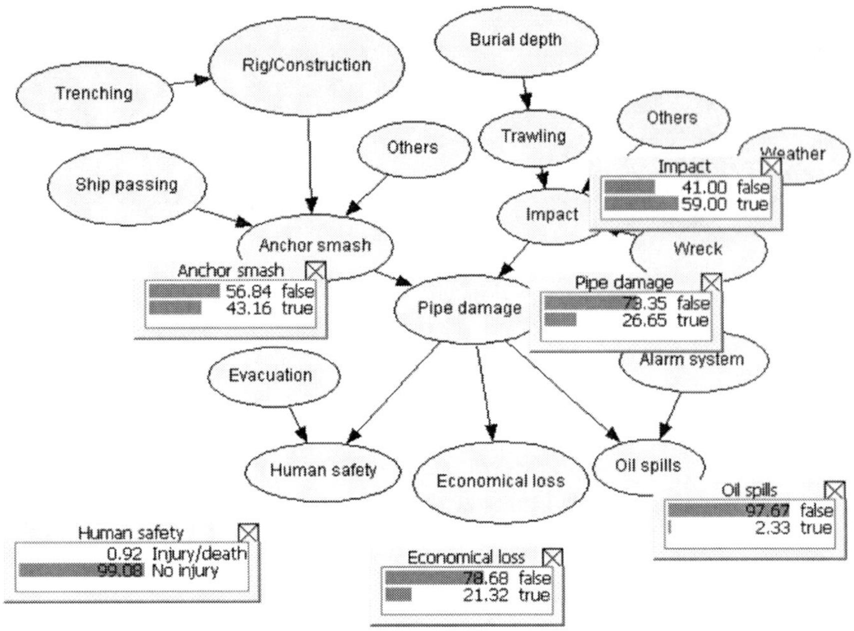

Figure 2. BN showing results for Pipeline Damage

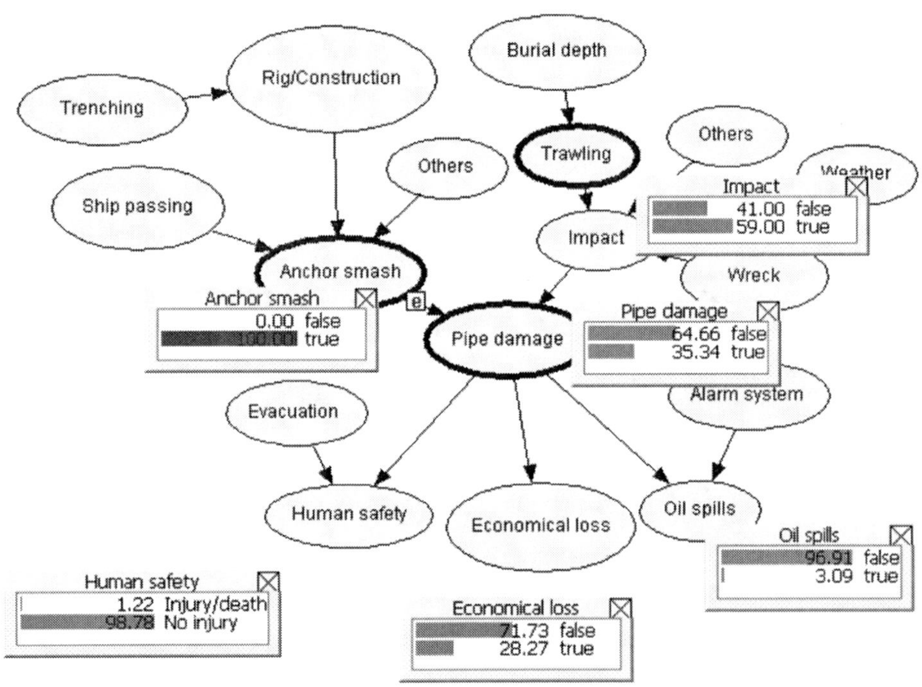

Figure 3. BN showing propagated result when anchor impact is 100% occurred.

From figure 4, it shows that propagation of 100% pipeline damage happened will affect the results of its dependence variables. Based on *"human safety"* node, the significant effect is "*injury"* where it increased to 96.55% with 95.0% evacuation success. While for "*oil spills"* and "*economical loss"* are 3.09% with and 28.27% loss probability, respectively. Considering the after-event consequences, pipeline failure gives significant effect towards human safety compared to economy and oil spillage. Furthermore, from the result obtained, "*impact*" node gives the higher probability risk factor compared to *"anchor smash"* and the major factor that causing the higher probability of impact is trawling activities. Thus the authorities may monitor the trawling activities to be in a safe area and away from the buried pipeline.

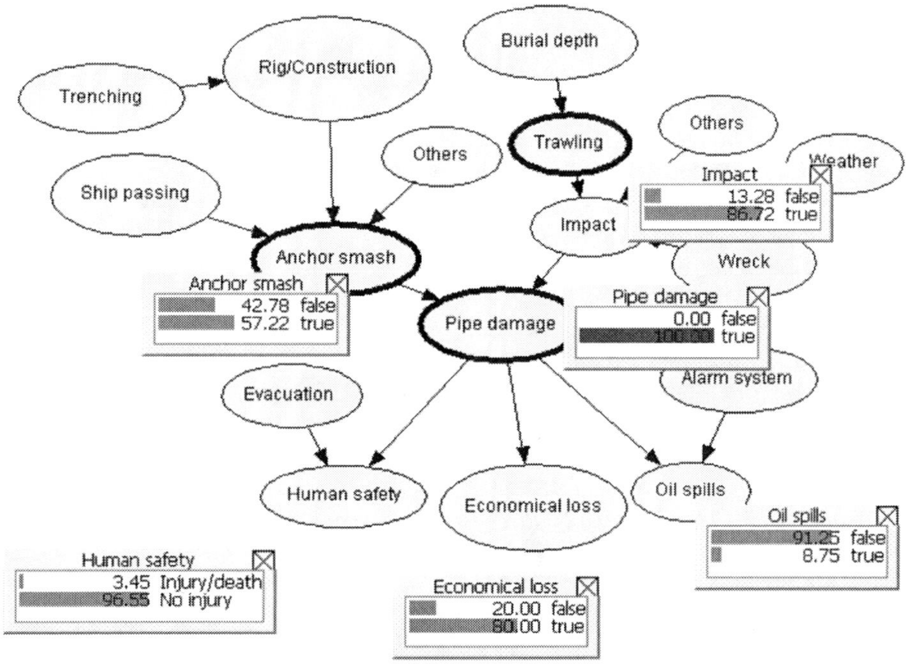

Figure 4. BN showing propagated result when pipeline damages is 100% occurred

4. Conclusions

The Bow-tie model applied in figure 1 may help to determine the activities that could prevent and control pipeline damages from happened. The proposed model shown in figure 2 explicitly represent dependencies of events, updating probabilities and representation of uncertain knowledge. There is major difference between BT and BN analysis but both of the method complementing each other in risk assessment. In addition to generating a numerical value that can be employed in estimation of pipeline damage due to the third party activities, a BN model also provides the means to access how the underlying factors may affect the pipeline condition probability. The scenario analysis was able to assess the impact of evidence or knowledge for probability updating. Diagnosis analysis was conducted to determine the critical risk factor that may lead to catastrophic accidence occurrence. It was found that the most contributing factor lead to pipeline damage due to third party was *"trawling"* activities. These analyses are appropriate to be used during model construction to gain detailed insight of the accidental occurrence. It is envisaged that this risk analysis approach may assist the operator by prioritizing the major contributor of pipeline failure, thus making the monitoring and inspection process of offshore pipeline at the right place and right time.

Acknowledgement
Authors would like to thank Universiti Malaysia Pahang for the financial assistance through research grants RDU1703169.

References

[1] Chen L, Arzaghi E, Abaei M M, Garaniya V and Abbassi R 2018 *J. Loss Prevent Proc.* **51** 178-185

[2] Wu J, Zhou R, Xu S and Wu Z 2017 *J. Loss Prevent Proc.* **46** 126-136
[3] Sulaiman N S and Tan H 2014 *J. Ener Chal Mech.* **1** 13-23
[4] Tang Y, Jing J, Zhang Z and Yang Y 2018 *Energies* **11(1)** 14-32
[5] Singh M and Hetlevik S 2017 *Int J Syst Assur Eng Manag.* **8(2)** 1588–1595.
[6] Yang Y, Khan F, Thodi P and Abbassi R 2017 *Reliab Eng and Syst Safe.* **159** 214–222.
[7] Seo J K, Cui Y, Mohd M H, Ha Y C, Kim B J and Paik J K 2015 *Ocean Eng.* **109** 539–552.
[8] Badaruddin M F, Zhang F and Bai Y 2009 *Proceedings of the Int. Conf. on Offshore Mech. and Arctic Eng. (OMAE)* vol 3 p 139
[9] Liu Y, Hu H and Zhang D 2013 *Transp. Res. Rec. J. Transp. Res. Board* **2326 (1)** 24–31.
[10] Li X, Chen G, Khan F and Xu C 2019 *Ocean Eng.* **178** 156-169
[11] Khakzad N, Khan F and Amyotte P 2012 *Reliab Eng Syst Safe.* **104** 36-44
[12] Li X, Chen G and Zhu H 2016 *Process Saf Environ.* **103** 163-173
[13] Bhandari J, Abbassi R, Garaniya V and Khan F 2015 *J. Loss Prevent Proc.* **38** 11–23
[14] Khakzad N, Khan F and Amyotte P 2013 *Process Saf Environ.* **91** 46–53.
[15] Kalantarnia M, Khan F and Hawboldt K 2009 *J. Loss Prevent Proc.* **22** 600–606.
[16] Khakzad N, Khan F and Amyotte P 2011 *Reliab Eng Syst Safe.* **96** 925–932.
[17] Alaw F A and Sulaiman N S 2018 *J. Chem Eng Ind Biotech.* **4** 1-9.
[18] Lee D and Pan R 2018 *Reliab Eng Syst Safe.* **171** 57–66.
[19] Sulaiman N S 2017 *Pipeline integrity management system (PIMS) using Bayesian Networks for lifetime extension* University of Aberdeen UK.
[20] Zarei E, Azadeh A, Khakzad N, Aliabadi M M and Mohammadfan I 2017 *J. Hazard Mater.* **321** 830-840.
[21] HSE 2003 *PARLOC 2001: The Update of Loss of Containment Data for Offshore Pipelines* Health Safety Executive UK.
[22] Mihailidou E K, Antoniadis K D and Assael M J 2012 *Int. Rev. of Chem. Eng. (I.RE.CH.E)* vol 4 p 1
[23] Eleye-Datubo A G, Wall A, Saajedi A and Wang J 2006 *Risk Anal.* **26 (3)** 695–721.

IOP Conf. Series: Materials Science and Engineering **702** (2019) 012054 doi:10.1088/1757-899X/702/1/012054

Development of compliance checklist system for landed housing construction activities

S N H S M Annuar, E H Sukadarin*, J Zakaria and H A Aziz

Occupational Safety and Health Program, Faculty of Industrial Sciences and Technology, Universiti Malaysia Pahang, 26300 Gambang, Pahang, Malaysia

*E-mail: ezrin@ump.edu.my

Abstract. In the construction industry (landed housing), most of the accidents and near-misses in occur due to unsafe behavior of the employers and employees. Lacking compliance with regulations is also one of the causes of accidents and near-miss occurrence in the construction industry including for landed housing development). This paper aims to identify Building Operations and Works of Engineering Construction (BOWEC) requirements under Factories and Machinery Act with Regulations (Act 139), Law of Malaysia for landed housing construction activities. Behavior-Based Safety (BBS) approach is used to develop a compliance checklist based on the BOWEC requirements. The developed checklists were validated in terms of its validity by using case study in the construction industry. After the validation processes, the final checklist is ready to be used. The final checklists are divided into two parts; (1) the checklist for the employer and (2) the checklists for employees. Workplace assessment was conducted using the developed checklist. The finding reveals that more than half of the construction activities are comply with the legal requirements.

1. Introduction

In developing countries, construction industry is one of the essential national backbones. There are various types of activities in construction industry and indirectly have a highly complex and hazardous environment that may cause more injuries and fatalities [1]. The construction industry is third in the ranking of occupational sectors with regards to the accident rate [2]. The construction industry is known as the place that possesses high level of hazard and risk to the employees [3]. It is because the employees in the construction industry work in high dealing with machinery equipment and external activities.

In the construction industry, there are three levels of construction which are: building construction, civil/heavy construction and industrial construction. However, in Malaysia, there are many construction sites which are involved with the building construction of landed housing. Many accidents are associated with building construction of landed housing such as falling from height, drowning, struck by support formwork and machine-related [4].

To maintain safety and health in construction industry, a Behavior-Based Safety (BBS) is the proposed tool to manage the employer and employees work activities, which is from unsafe behavior to safe behavior that can prevent accident and injury from occurring. BBS checklist is used to evaluate behavior of employer and employees at the workplace [5]. For the BBS implementation and procedures, it is based on "DO IT" application process [6]. In addition, the BBS checklist that is developed in a system is more easy for the observer to make the observations, and makes the finding process convenient because the data is saved timely.

The regulation used in the Malaysia construction industry is Factories and Machinery Act 1967 under the scope of (Building Operations and Works of Engineering Construction, BOWEC) (Safety) Regulations 1986 [2, 7]. So, to ensure safety and health of employer and employees in construction sites, it must comply with BOWEC regulations and other supporting regulation in Malaysia 's legislation [6, 8]. Therefore, to support the above requirement, this paper is aimed to highlight the BOWEC 1986 requirement, and then apply it into the BBS approach for landed housing construction activities In addition, Occupational Safety and Health Act (OSHA) 1994 [9], also reviewed to further understand the needs of legal requirements at Malaysia construction industry.

2. Methodology

2.1. Analyze the elements and requirements of BOWEC Regulation 1986
In this study, the Deming Cycle (figure 1) is adopted to develop the BBS checklist. The cycle starts from Plan, Do, Check and Act. This cycle provides a structure for iterative testing of changes to improve quality of systems. The method is widely accepted and being applied in numerous occupational safety and health related publications such as in Azmi and Aziz [10] and Hang and Sukadarin [7]. The elements of BOWEC Regulations 1986 (FMA 1967) as specified by DOSH Malaysia were analyzed for construction landed housing. The elements of requirements were served as the basis for the information that is used to develop the BBS checklist.

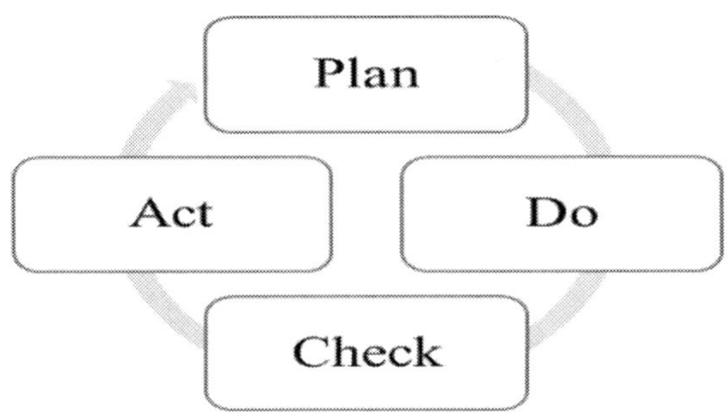

Figure 1. *The Deming Cycle.*

2.2. Development of BBS checklist
After the requirements of BOWEC Regulations, 1986 are extracted for the construction industry (landed housing), The BBS checklist was developed by using Microsoft Words 2016. The development of BBS checklist was divided into two parts which are BBS checklist for employer and BBS checklist for employees.

2.3. Development of system BBS checklist
After the development of BBS checklists for employer and employees were validated by subject matter expert in occupational safety and health field, the system of BBS checklists is developed by using Microsoft Access 2016 based on the BOWEC requirements in the construction industry (landed housing) that serves as guide in complying the requirements. The development process took about two months to be completed.

2.4. Validation of system BBS checklist
After the BBS system is completed and tested so that it can be run smoothly with the subject matter expert, the system was sent to be validated and used at a construction industry (landed housing) which is located at Kuantan, Pahang. The adapted techniques including a case study supporting with series of observation and interviews with employer and employees at the chosen construction site.

2.5. The calculation of the compliance percentage by BOWEC Regulation 1986 requirements
The compliance percentage of the construction industry (landed housing) with BOWEC requirements was then conducted using equation (1) below.

$$\frac{C}{N}\times 100 \qquad (1)$$

where C and N are the number of compiled and total number of questions by each part, respectively.

3. Results and Discussion

3.1. Analysis of the elements and requirements of BOWEC Regulation 1986
Table 1 shows the elements in BBS checklist, which is for construction industry (landed housing) based on BOWEC Regulation 1986 requirements. As can be seen, it contains 13 parts from BOWEC Regulations 1986 that focusing for construction industry (landed housing).

Table 1. BBS element for construction industry (landed housing) based on BOWEC Regulation 1986.

Part	BBS Elements
1	General provisions
2	Concrete work in construction site
3	Assembly of structural steel and precast concrete in construction site
4	Good and safe condition of roof, gutters, windows, louvres and ventilators in construction site
5	Use of safety belts in construction site
6	Use of ladders and step-ladders in construction site
7	Use of scaffolds in construction site
8	Demolition work in construction site
9	Excavation work in construction site
10	Handling and storage, use and disposal of material in construction site
11	Use of piling in construction site
12	Blasting work and use of explosives in construction site
13	Use hand and power tools in construction site

3.2. Development of BBS checklist
The developed of BBS checklists were validate. BBS checklists were validated in two phases. The first phase of validation is named as face validity. This face validity is conducted by subject matter expert (SME). The second phase of the validation process is named as content validity. The content validation is conducted with on-duty Safety and Health Officer in the construction industry (landed housing). For the content validity, the validation process is repeated for two times which are at the off-site and on-site.

3.2.1. Face validity. Face validity has been conducted with subject matter expert (SME). One of the recommendations by SME during the validation process is to make an improvement of BBS checklist such as to change of parts and elements in the BOWEC regulations requirements with a simple and more understandable statement that may be more convenient to be used by low educated workers. Figure 2 shows the example of an improvement that had been done in the BBS checklist as recommended by SME. It is shown an example of BBS checklist before and after face validation process by SME.

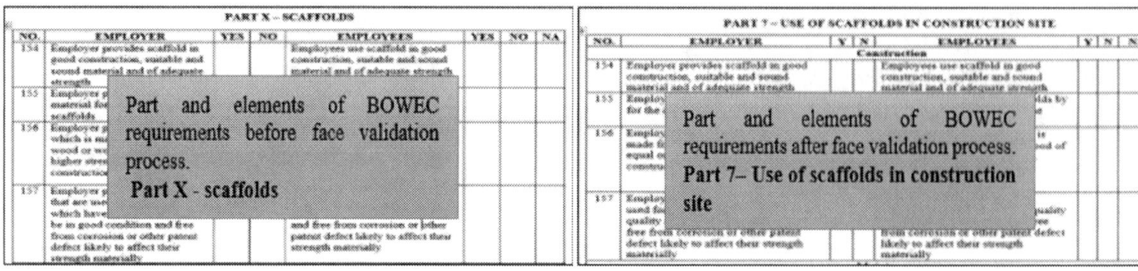

(a) **(b)**

Figure 2. (a) BBS checklist before the validation process (b) BBS checklist after the face validation process.

3.2.2. Content validity. The second phase of validation is content validity has been done by on-duty Safety and Health Officer from the construction industry (landed housing) background. For the second phase of validation, the validation process was conducted two times which is at off-site and on-site construction project. These two times of validation are conducted by the different on-duty Safety and Health Officer.

Off-site content validity has been done with an on-duty Safety and Health Officer who comes from the construction industry (landed housing) background. The recommendations by on-duty Safety and Health Officer were taken into consideration and improvements were made. For example, one of the suggestions is to add a column for a "remarks" section. The purpose of this "remarks" section is to make notes if some information needs to be added when the process of data collection is ongoing. The involved Safety and Health Officer also agreed all the elements in the BBS checklist that have been summarized based BOWEC Regulation 1986 requirements are valid to be used in construction sites. figure 3 shows the example of improvement of BBS checklist that has been made based on recommendation by the Safety and Health Officer.

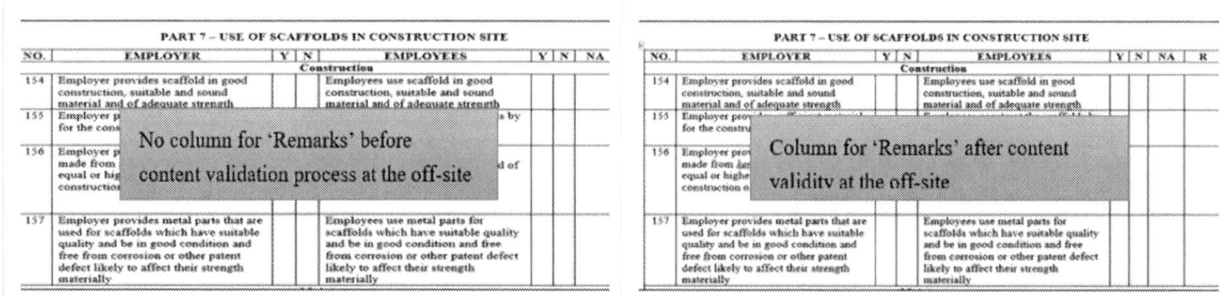

(a) **(b)**

Figure 3. (a) BBS checklist before content validation process at the off-site and (b) BBS checklist after content validation process at the off-site.

On-site content validation has been done with an on -duty Safety and Health Officer who also from construction industry (landed housing) background. The recommendation has been made based on the suggestions by the Safety and Health Officer. Among the suggestions, it would be beneficial if the BBS checklist can be prepared as short as possible but can be understood by all level of organization in construction site. The on-duty Safety and Health Officer also suggested improving the scope of elements in BBS checklist to becoming smaller and specific by only focusing on one type of construction.

At the beginning of extraction process of BOWEC Regulation 1986 requirements, the elements were covers from part 2 until part 16 in the BOWEC Regulation 1986 document. So, the checklist became too general to be used at the construction site that focuses on the work of developing landed housing. Based on the needs of landed housing development, there are only a few elements in BOWEC Regulation 1986 that are applicable. So, at the early stage, the drafted BBS checklist, was consists of 310 checklists. So, an improvement has been done by reducing it to 45 checklists of employer and 43 checklists of employees. Figure 4 shows the examples of improvements of the BBS checklist has been made based on recommendations by Safety and Health Officer.

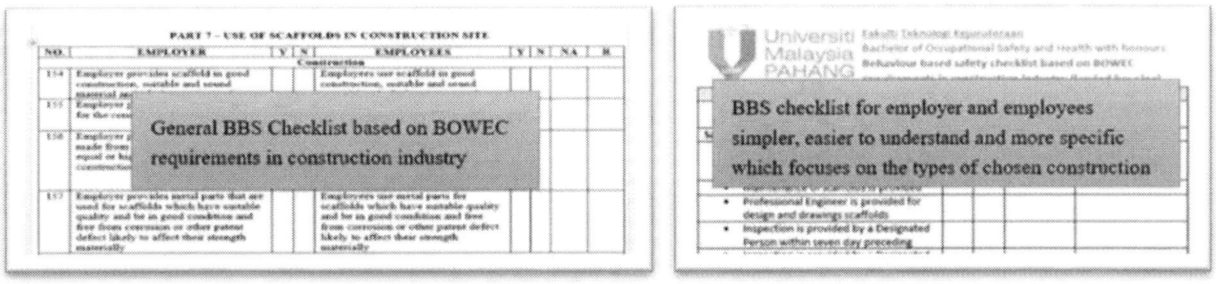

(a) (b)

Figure 4. (a) BBS checklist before content validation process at the on-site, (b) BBS checklist after content validation process at the on-site

3.3. System developed for Behaviour-Based Safety Checklist

Computerized BBS checklist has been developed by using Microsoft Access 2016. The Microsoft Access 2016 was used to create a system and store information for reporting, referencing and analysis of the compliance audit based on BOWEC Regulation 1986 requirements for landed housing construction work. Figure 5 shows the example of BBS checklist interface in a system.

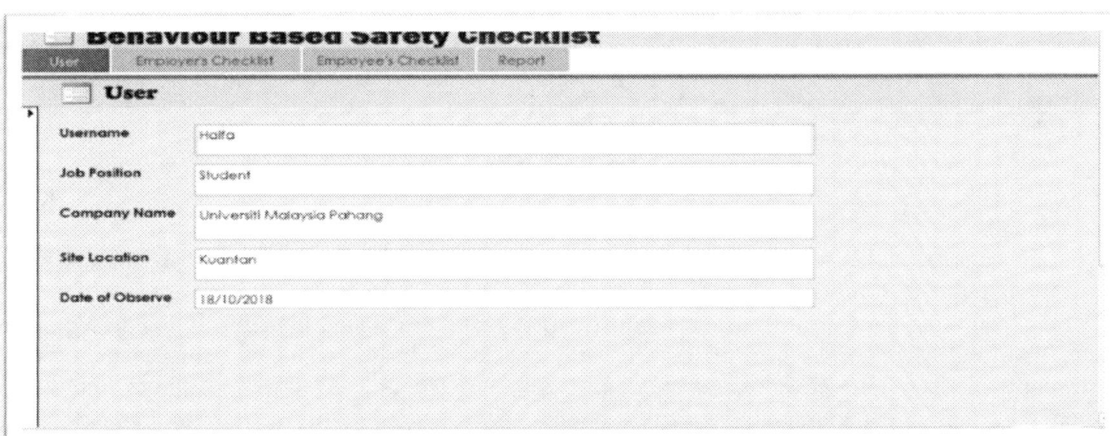

Figure 5. The example of BBS checklist interface.

3.4. Compliance of landed housing construction site with the BOWEC Regulation 1986.

Simple percentage calculation used to measure the compliance of landed housing construction site with the BOWEC Regulation 1986. As can be seen in the table 2, there are 61 questions has been developed for employer's checklist, and 60 questions has been developed for employee's checklist.

IOP Conf. Series: Materials Science and Engineering **702** (2019) 012054 doi:10.1088/1757-899X/702/1/012054

However, during the site visit, only 45 questions that applicable to the chosen construction site for employer's checklist. Meanwhile, for employee's checklist only 43 questions that found applicable.

There are three parts which were not applied in collection of data's site. The parts are part 8, part 11 and part 12. Based on BOWEC Regulations 1986, part 8 is demolition work. Demolition work is not applied in the involved site. For part 11, is piling. Piling was not applied because this site only uses a rough foundation as to reduce cost. The last part is part 12, whereby the requirement that needs to be followed for blasting work and the usage of explosives material. This is because of the study construction site was not built at the hilly area.

There are three parts which complied 100% with the BOWEC Regulation 1986 requirements for employer's and employees' checklist. The parts are part 2, part 3 and part 4. Overall percentage of employer and employees complied with the BOWEC Regulation 1986 requirements are 82% and 77%, respectively.

Table 2. Details number of Parts in the BBS checklist system and finding on the legal requirement.

Part	Employer's Checklist			Employee's Checklist		
	Total no. of the question (N)	Comply = YES (C)	C/N (%)	Total no. of the question (N)	Comply = YES (C)	C/N (%)
1	11	10	91	10	8	80
2	1	1	100	1	1	100
3	1	1	100	1	1	100
4	3	3	100	3	3	100
5	3	2	67	2	1	50
6	1	0	0	1	0	0
7	6	6	100	4	3	75
8	5	-	-	5	-	-
9	7	4	57	7	4	57
10	7	6	86	8	7	88
11	6	-	-	7	-	-
12	5	-	-	5	-	-
13	5	4	80	6	5	88
Total of questions		61			60	
Item's measured	45	37	82	43	33	77

4. Conclusions

This study concludes the result of the objectives. The BOWEC Regulation 1986 requirements in construction industry (landed housing) was analysed. The BBS checklist was developed based on BOWEC Regulation 1986 requirements in construction industry (landed housing) by using Microsoft Words 2016 and Microsoft Access 2016. The BBS checklist was validated by using case studies in construction industry (landed housing) located at Kuantan, Pahang, Malaysia. Authors believe the developed system will encourage the construction industry to comply with the current legislations and at the same time may encourage more research works in order to help construction industry to manage their site and keep them safe.

Acknowledgments

The authors would like to thank Universiti Malaysia Pahang and Ministry of Education Malaysia for funding the project through a grant (RDU181001 – Knowledge Transfer Program in Construction Industry) and making this study feasible.

IOP Conf. Series: Materials Science and Engineering **702** (2019) 012054 doi:10.1088/1757-899X/702/1/012054

References

[1] Skeepers N C and Mbohwa C 2015 *Procedia Manuf.* **4** 10–16
[2] Department of Occupational Safety and Health (DOSH) Malaysia 2018 *Occupational Accidents Statistics by Sector Until October 2018*
[3] Hoła B and Szósta M 2014 *Procedia Engin.* **91** 429–34
[4] Samuel O, Adul Hamid R and Saidin Misnan M 2017 *J. Multidiscip. Eng. Sci. Technol.* **4(8)** 2458–9403
[5] Oostakhan M, Mofidi A and A D Talab 2012 *Iran. Rehabilitation J.* **10** 21–25
[6] Chen D and Tian H 2012 *Procedia Engin.* **43** 528–34
[7] Law of Malaysia. FMA 1967 *Factories and Machinery Act with Regulations* (MDC Publishers Sdn Bhd)
[8] Hang L P and Sukadarin E H 2020 *Advances in Intelligent Systems and Computing*, vol 969 (Springer, Cham)
[9] Law of Malaysia. OSHA 1994 *Occupational Safety and Health Act and Regulations* (MDC Publishers Sdn Bhd)
[10] Azmi N and Aziz H A 2018 *Ser. Mater. Sci. Eng.* **458** 012049

IOP Conf. Series: Materials Science and Engineering **702** (2019) 012055 doi:10.1088/1757-899X/702/1/012055

Psychological impact of noise exposure among machine and non-machine operators in construction industry

H Y Kee , N S Fauzan* and M Widia

Occupational Safety and Health Program, Faculty of Industrial Sciences and Technology, Universiti Malaysia Pahang, 26300 Kuantan, Pahang, Malaysia

*E-mail: syafiqah@ump.edu.my

Abstract. Noise hazard is reported as a serious issue among the construction industry in Malaysia. The aim of the study is to determine the personal noise exposure level and the prevalence of psychological health effect symptoms among the machine and non-machine operators on construction site in Malaysia. Sixty-one (61) construction workers were participated as respondents and categorized into machine and non-machine operators group. All respondents were monitored for 8 hours working time using personal noise dosimeter for personal noise monitoring. Questionnaires and interviewed sessions were used to determine the prevalence of psychological health effect symptoms. The results obtained reveal that the construction workers have high personal noise exposure level where machine operators are significantly higher compared to non-machine operators. The prevalence of psychological health effect symptoms among machine operators is 89% higher than the non-machine operators. Lastly, there is a significant positive relationship between the personal noise exposure level and the prevalence of psychological health effect symptoms among the construction workers.

1. Introduction

Noise is one of the primary pollutants around the world that affects human working and living environment [1]. Noise is described as a common occupational hazard in most of the workplaces, especially in the construction industry [2]. The construction activity categorized as one of the sources of noise pollution which could lead to the occurrence of psychological and physical health effect such as stress towards the people work or live near the construction site [3].

The construction activities have led to several hazards, especially the noise hazard where it has become a common and serious source of environmental noise that harmed human's health [4]. According to the Social Security Organization (SOCSO) reports in 2016, there are 7338 accident cases occur in the Malaysian construction industry. It is the third-highest number of cases reported industry among all other sectors [5].

The Department of Occupational Safety and Health (DOSH) has shown a statistical report whereby the occupational noise-related hearing disorders are the highest occurrence occupational disease among the occupational diseases in Malaysia. The statistic of occupational disease among the occupational disease and poisoning by type of disease statistics for 1995–2009 from the DOSH, it shows that noise-induced hearing loss (NIHL) was the most commonly notified occupational disease. In year 2016, the analysis statistic shows the occupational noise-related hearing disorder consists of 2876 cases [6]. Ali [7] found that nearly 81% of the construction workers exposed to high noise exposure level that could contribute to the hearing impairments and the psychological health effect symptoms. The psychological health effect symptoms indirectly contribute to the increase in the rate of accidents owing to interference with the sound signals and other non-hearing effects caused by noise.

Therefore, this study is to investigate the impact of noise exposure on the psychological health of the machine and non-machine operators in the construction industry.

2. Methodology

A walk-through observation conducted and guided by the supervisor. A noise exposure observation checklist from the Noise Hazard Identification Form by University of Melbourne [8] was used as a preliminary assessment to obtain deeper understanding regarding the safety practice and the exposure of noise among the construction workers. About 61 construction workers were selected randomly to participate in the study as respondents. The construction workers categorized into two (2) groups; machine operators group (n=31) and non-machine operators group (n=30).

Personal noise monitoring was conducted to obtain the personal noise exposure level among the respondents in the construction site. The Spark 706RC-ATEX personal noise dosimeter was used to measure the personnel noise exposure level among construction workers [9]. The data of 8 hours' time-weighted average (TWA) for every respondent obtained from the personal noise dosimeter for further analysis.

OSHA Recordable Questionnaire was used to measure the prevalence of psychological health effect symptoms among the construction workers [10]. The questionnaires distributed among the respondents during the interview sessions.

The data obtained were analyzed using Statistical Package of Social Science (SPSS) Version 20.0. Independent t-test was carried out to compare the personal noise exposure level between machine operators and non-machine operators. Independent Chi-square test was performed to compare the prevalence of psychological health effect symptoms on noise exposure between machine operators and non-machine operators. Binary Logistic Regression test performed for identifying the relationship between the personal noise exposure level and the prevalence of psychological health effect symptoms among the construction workers in the selected construction site.

3. Result and Discussion

Figure 1 and Figure 2 show the 8 hours time-weighted average (TWA) personal noise exposure monitoring for machine and non-machine operators. The personal noise exposure level among the construction workers was analyzed and compared with the exposure limit stated in regulations especially the Factories and Machinery (Noise Exposure) Regulations 1989 in Factories and Machinery Act (FMA) 1967 [11] and NIOSH [12]. The result shows that about 19.7% of construction workers identified expose to above the action levels as stated in Factories and Machinery (Noise Exposure) Regulations 1989 which is 85dBA. The construction industry generates noise from different types of heavy machines used in the construction site on a range between 80dBA until 120dBA where it puts the construction workers in the risks of over noise exposure [13].

IOP Conf. Series: Materials Science and Engineering **702** (2019) 012055 doi:10.1088/1757-899X/702/1/012055

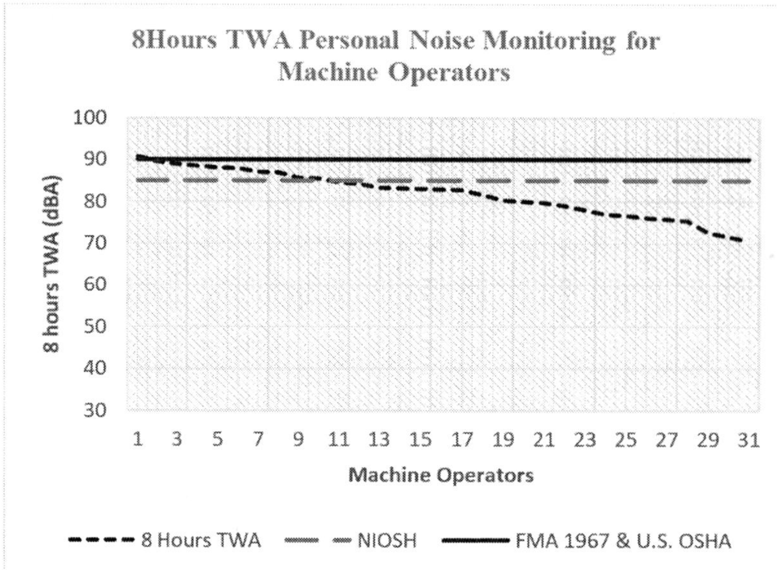

Figure 1. 8 Hours TWA Personal Noise Monitoring for Machine Operators

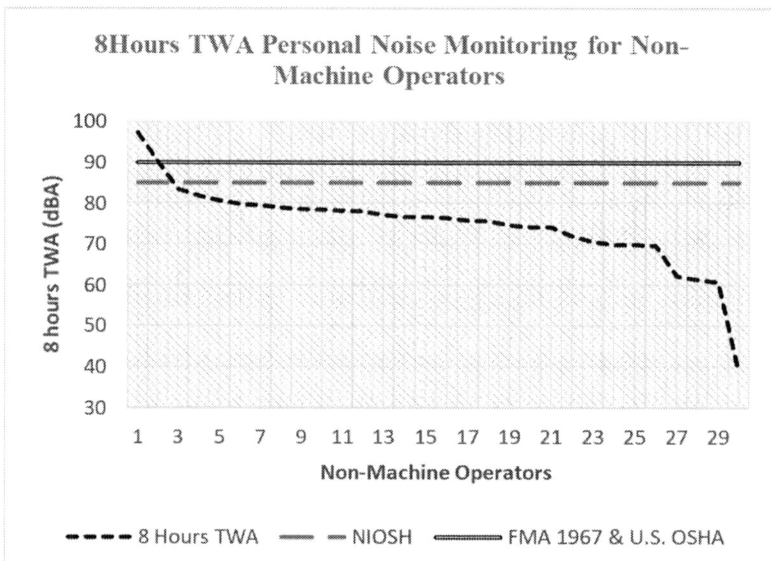

Figure 2. 8 Hours TWA Personal Noise Monitoring for Non-Machine Operators

Table 1 shows the minimum, maximum, means, and standard division of results for the personal noise monitoring among the machine operators and non-machine operators on the 8 hours time-weighted average (TWA). Based on the result, the mean value of TWA 8 hours (dBA) among machine operator recorded was 81.81 dBA while the non-machine recorded was 74.71 dBA. Besides, the results reveal that the maximum value of personal noise level of 90.7 dBA, which is above the permissible exposure limit (PEL). Main sources of noise at the construction site is the construction machines where most of the machines produce impacts such as devices for breaking of concrete, steel bar machines, electronic grinders, electronic mosaic cutters, pile drivers, and earth moving machines that produce unwanted sound [1]. These mechanisms considered as point or linear sources of noise based on the level of movement at the construction site in the studies of noise.

IOP Conf. Series: Materials Science and Engineering **702** (2019) 012055 doi:10.1088/1757-899X/702/1/012055

Table 1. Minimum, Maximum, Means and Standard Deviation of Personal Noise Monitoring

Category	Time-Weighted Average TWA (8) (dBA)			
	Minimum	Maximum	Mean	Standard Deviation
Machine Operators	70.6	90.7	81.81	5.54
Non-Machine Operators	38.7	97.3	74.71	10.13

N=61

Table 2 shows the independent t-test result that found significant differences (t=3.411, df=59, p=0.001) of personal noise exposure level between machine and non-machine operators group. The mean of personal noise exposure level for the machine operators (81.81dBA) was significantly (p=0.001, <0.05) higher compared to the mean of personal noise exposure level for non-machine operators (74.71dBA). The machine operators were exposed to loud noise most of the working hours compared to the non-machine operators at the construction site especially while handing the bar tying machine

Table 2. Comparison of Personal Noise Exposure Level between Machine Operators and Non-Machine Operators

Variable	T-test Statistics^2 (df)	*P*-value
Personal Noise Exposure Level (dBA)	3.411 (59)	***0.001

N=61 *Independent T-Test ** Significant at p<0.05 *** Significant at p<0.001

Table 3 shows the prevalence of psychological health effect symptoms among the machine operators and non-machine operators. Tension in a noisy work environment identified as the highest prevalence of health effect symptom among machine operators (90.3%) and non-machine operators (43.3%). The tension issue could arise when they are unable to withstand the demand of work environment [14]. For the least prevalence of psychological health effect symptom among both machine operators (12.9%) and non-machine operators (3.3%) is the experience of any sleep disturbance.

Table 3. Prevalence of Psychological Health Effect Symptoms of Construction Workers

Symptoms of Health Effect	Yes N (%)	
	Machine Operators	Non-Machine Operators
Ear pain	17 (54.8)	6 (20.0)
Sudden hearing loss	5 (16.1)	2 (6.7)
Tinnitus	24 (77.4)	5 (16.7)
Noticeable change in hearing	13 (41.9)	2 (6.7)
Anger and aggressiveness	23 (74.2)	10 (33.3)
Dizziness	11 (35.5)	6 (20.0)
Experience of any sleep disturbance	4 (12.9)	1 (3.3)
Stressful in noisy work environment	28 (90.3)	12 (40.0)
Tension in noisy work environment	29 (93.5)	13 (43.3)
Difficulty in concentrating or decision making	16 (51.6)	8 (26.7)

N=61

The result also shows that there were significantly ($p<0.05$) differences between the machine and non-machine operators. The related prevalence of psychological health effect symptoms was ear pain, tinnitus, noticeable change in hearing, anger and aggressiveness, stressful in the noisy work environment, the tension in a noisy work environment and the difficulty in concentrating or decision making. While Table 4 shows that machine operators group has a significantly higher prevalence of all these health effect symptoms compare to non-machine operators group. Noise exposure is associated with the psychological health effect and could vary with the source of environmental noise exposure [15].

Table 4. Prevalence of Psychological Health Effect Symptoms Between Machine Operators and Non-Machine Operators

Symptoms of Health Effects	Machine Operators	Non-Machine Operators	Statistics	
	Yes N (%)	Yes N (%)	Value[a] (x^2) (df)	*p*-value
Ear pain	17 (73.9)	6 (26.1)	7.878 (1)	0.005*
Sudden hearing loss	5 (71.4)	2 (28.6)	1.344 (1)	0.246
Tinnitus	24 (82.8)	5 (17.2)	22.563 (1)	0.000*
Noticeable change in hearing	13 (86.7)	2 (13.3)	10.227 (1)	0.001*
Anger and aggressiveness	23 (69.7)	10 (30.3)	10.250 (1)	0.001*
Dizziness	11 (64.7)	6 (35.3)	1.818 (1)	0.178
Experience of any sleep disturbance	4 (80.0)	1 (20.0)	1.856 (1)	0.173
Stressful in noisy work environment	28 (70.0)	12 (30.0)	17.102 (1)	0.000*
Tension in noisy work environment	29 (69.0)	13 (31.0)	17.926 (1)	0.000*
Difficulty in concentrating or decision making	16 (66.7)	8 (33.3)	3.976 (1)	0.046*

N=61 *Independent chi-square test * Significant at $p<0.05$

From the data obtained through the Binary Logistic Regression test shows that there is a significant relationship between the personal noise exposure level – 8 hours time-weighted average (TWA) and the prevalence of psychological health effect symptoms among the construction workers. The most strength and significant positive relationship among machine operators is the noticeable change in hearing (p-value=0.007). While the most strength and significant positive relationship among non-machine operators is anger and aggressiveness (p-value=0.027). The prevalence of anger and aggressiveness could increase as the people exposed to noise pollution and noise act as a stressor that causes unwanted expression of human such as anger [16-17].

Table 5. Relationship between Personal Noise Exposure Level and Prevalence of Psychological Health Effect Symptoms

Variable	Personal noise exposure level (8 Hours TWA) (*p*-value)	
	M.O.	N.M.O.
Ear pain	0.011*	0.095
Sudden hearing loss	0.072	0.047*
Tinnitus	0.809	0.055
Noticeable change in hearing	0.007*	0.057
Anger and aggressiveness	0.062	0.027*
Dizziness	0.016*	0.050*
Experience of any sleep disturbance	0.058	0.993
Stressful in noisy work environment	0.436	0.031*
Tension in noisy work environment	0.767	0.028*
Difficulty in concentrating or decision making	0.481	0.071

N = 61 * Binary Logistic Regression Test * Significant at $p<0.05$

4. Conclusion

The study had found the personal noise exposure level among machine and non-machine operators in construction industry. The machine operators group has significantly higher prevalence of psychological health effect symptoms compare to non-machine operators group where there were significant differences ($p<0.005$) between machine operators and non-machine operators for the prevalence of tinnitus, noticeable change in hearing, anger and aggressiveness, stressful, and tension in noisy work environment. The significant relationship between the personal noise exposure level and the prevalence of psychological health effect symptoms among the construction workers was identified in the study. The presented findings have important implications in industrial practice where the contractor should provide further safety training, required hearing protection devices and even implement the safety culture at the construction site to increase the safety awareness among the construction workers. The result presented here may facilitate improvement in determining the impact of noise exposure towards worker's health in construction industries.

Acknowledgement

The research study funded by Universiti Malaysia Pahang (UMP). We are very grateful due to the support of UMP. We want to give our utmost appreciation to the construction employer and employees for their cooperation throughout the study.

References

[1] Kantová R 2017 Construction Machines as a Source of Construction noise *Procedia. Eng.* **190** 92–99
[2] Gerges S N Y, Sehrndt G A and Parthey W 1992 *Noise Sources (Pfeiffer)* 103–124.
[3] Baba I, Md Ali Z, Bakar A, Ashraf N and Ramly M Z 2011 Determination of noise exposure level at construction area *ISASM* **1**(3) 1–3
[4] Liu Y, Xia B, Cui C and Skitmore M 2017 Community response to construction noise in three central cities of Zhejiang province, China *Environ Pollut* **230** 1009–1017
[5] Social Security Organization 2016 *SOCSO Annual Report 2016* SOCSO.
[6] Jaafar N I, Md Daud M K, Mohammad I and Abd Rahman N 2017 Noise-induced hearing loss in grass-trimming workers *Egypt. J Ear Nose Throat Allied Sci.* **18**(3) 227–29
[7] Ali S A 2011 A case study of construction noise exposure for preserving worker's hearing in Egypt *Acoust Sci Tech* **32**(5) 211–15

IOP Conf. Series: Materials Science and Engineering **702** (2019) 012055 doi:10.1088/1757-899X/702/1/012055

[8] University of Melbourne 2015 *Noise Hazard Identification Form* Melbourne Retrieved from https://safety.unimelb.edu.au/__data/assets/word_doc/0009/1716750/noise-hazard-identification-form.docx

[9] Spark ® ATEX Noise Dosimeters and Blaze ® Software Technical Reference Manual 2016

[10] AdventistHealth. *OSHA Recordable Questionnaire* Retrieved from https://www.adventisthealth.org/documents/portland/osha_recordable_questionnaire.pdf

[11] Government of Malaysia 2014 *Factories & Machinery (Noise Exposure) Regulation 1989 Factories and Machinery Act* 1967 (pp. 451–470) Damansara Perdana City: Member of the Malaysian Book Publisher Association.

[12] NIOSH 1998 *Occupational Noise Exposure Revised Criteria 1998* Education and Information Division National Institute for Occupational Safety and Health, (Criteria For A Recommended Standard)

[13] Said K M, Haron Z A, Saim A A, Abidin M Z, Yahya K and Han L M 2014 Occupational Noise Exposure Among Road Construction Workers *J. Teknol* **70**(7)

[14] Oliveira R C, Santos J N, Rabelo A T and Magalhães Mde C 2015 The impact of noise exposure on workers in Mobile Support Units *CoDAS* **27**(3) 215–22

[15] Hammersen F, Niemann H and Hoebel J 2016 Environmental noise annoyance and mental health in adults: Findings from the cross-sectional German health update (GEDA) study 2012 *Int. J. Environ. Res. Public Health* 1**3**(10) 1–12

[16] Jones D M, Chapman A J and Auburn T C 1981 Noise in the environment: A social perspective. *J. Environ. Psychol* **1**(1) 43–59

[17] Ramirez J M Alvarado and Santisteban C 2004 *Individual differences in anger reaction to noise Individual* Differences Research 2(2) 125–136

IOP Conf. Series: Materials Science and Engineering **702** (2019) 012056 doi:10.1088/1757-899X/702/1/012056

Influence of natural smoke vent opening in stairway of multi-storey building

S S Nurud[1,*], M Saari[2], I Abdullah[1], A NorMariah[2] and B Rafee[2]

[1]Faculty of Engineering Technology, Universiti Malaysia Pahang, Lebuhraya Tun Razak, 26300 Gambang, Pahang Darul Makmur.
[2]Faculty of Engineering, Universiti Putra Malaysia, 43400 UPM Serdang, Selangor Darul Ehsan.

*E-mail: nurud@ump.edu.my

Abstract. Stairway used as an escape as well as firefighter's route during evacuation. In addition, stairway connecting different floors of a building and becomes a path for the smoke spread in fire event. In a building, every escape route should be installed with smoke control system to ensure the prevention of dangerous smoke accumulation at those areas. The fire perimeter in terms of heat output and smoke generation is highly depends on building occupancy and the efficacy of smoke confinement may have a great challenge. In this paper, numerical simulations were conducted to study the efficacy of natural smoke vent to confine fire-induced smoke transportation in the stairways of multi-storeys buildings. The simulation used Fire Dynamic Simulator (FDS) was conducted on a full-scale building where the influences of smoke vent opening at different fire size were discussed. When the value of heat release rate (HRR) were kept constant, the different vent's size opening had a different influence on the efficacy of smoke vent and an appropriate opening size was obtained and proposed for further action. The finding of this study can assist the fire engineer to ensure that the smoke vent installation play a good role in confinement of smoke diffusion.

1. Introduction

In a recent year, urban population and urbanization level increases rapidly in Malaysia. Due to large number of peoples and complex building environment, fires could induce many casualties and property loss. Fire statistics in Malaysia increase from 21, 524 fire cases in 2008 to 54,540 fire cases in 2014 with total cost damage nearly RM1 billion and RM3 billion, respectively [1]. Nevertheless, until June 2015, there is 25, 978 fire cases with total cost damage more than RM3 billion. This statistic shows the huge increment of property loss to be incurred in 2015 although the fire statistic is lower from year 2013 due to high cost of construction and building furnishing nowadays. Beside property losses, a number of injuries and fatality cases in fires also increase with 88 fatalities and 79 injuries in 2008 to 98 fatalities and 152 injuries in 2012. During fire in a multi-story building, despite being one of the main routes for evacuation and rescue operation, stairways also become the means of smoke spread between floors. Flame can cause to injury and property damage however majority of fire deaths resulted from smoke inhalation during evacuation or trapped in the building [2-5]. A case had been reported recently where a couple had died in stairway of level 31st while escaping from level 38th when their high-rise apartment occurred to fire [6]. This case is similar with fire happened previously in 1998 at Upper West Side apartment tower, New York city which killed four peoples due to smoke inhalation in stairway during escaping [7]. Smoke generated from combustion depends on type of building occupancy and material used for construction. As a product innovation for fire risk reduction, a future alternative of construction material from natural waste product which has low flame and smoke has been studied and reported by researchers [8-10]. Hence for a building design, it should be constructed based on fire safety design as stated in building code and standards [11].

As stairway is the only measure for egress path during fire, it is worthwhile to study the fire and smoke behaviour in this area for multi-story building. Quantitative tools to analyze the smoke condition in a stairway could assist fire-safety engineer to explore different methods of risk reduction. Most of researchers are generally aware that full-scale fire test are the best way to obtain valuable information about chemistry , engineering, management, human behaviour and psychology which primarily related to fire origin and location, fire spread and growth, smoke propagation, material properties and structural integrity [12]. However, they also learn that full-experiment is expensive, time consuming [13-14] and poor repeatability due to unstable nature of fire. An alternative to study a certain fire phenomena with traditional full- and small-scale experiment is by using a numerical model as reported by Rahman et al., [15].

Fire safety in building is a set of measures designed to reduce the fire risk from its origin and risk of injury towards building occupants as well as fire fighters. Beside fire code regulation and standard, fire risk management is one of the measures for fire safety in building. In general, if a building design is constructed based on corresponding prescriptive codes, the level of fire risk is considered in acceptable risk range (tolerable) [16-17]. According to the current prescriptive code practiced locally, an openable vent outlet at the top of protected stairway without mechanically pressurized may be used as a smoke control system. However, the codes only specified the vent opening size without declared the effectiveness of the opening size in reducing the impacts and venting the smoke outside especially for a different type of building occupancy. As a result, in real fire event, some stairway with natural smoke vent opening cannot effectively vent the smoke to the outside. Currently, there are many researches focused on the smoke venting in road tunnels [18-22] and subway stations [5, 23-25], but there are only a few study reported on the smoke venting in protected stairway [26-28]. Thus, the main objective for this study is to simulate the smoke control in protected stairway with openable vent at the top of stairway where the size and location of smoke vent will be varied. It will provide evidence for the proper design of smoke vent as a smoke control system for protected stairway.

2. Methodology

2.1 Fire Dynamic Simulator (FDS)

The Fire Dynamics Simulator (FDS) is type of fire simulation software develop by the Building and Fire Research Laboratory (BFRL) under the U.S. National Institute of Standards and Technology (NIST). The software resolves numerical equations for low speed fluid flow concentrating on the smoke and radiation from the fire. It is most widely used to model smoke movement and fire scenarios in complicated buildings with many obstructions and complex geometry. The FDS can be used to simulate 3D fire scenarios, and is capable to estimate physical data; such as temperature, pressure, smoke layer height and smoke movement at a fire site [29]. Smoke view is a software that designed for FDS to produce animations and images of the output results [30]. The decision to use FDS was made based on the availability of the software and ease of use [31]. The input files for FDS6 were created in notepad with randomly size of shaft compartments and opening. The model was a protected stairway of multi-story building consisted of two parts: room with dimension of 2 m (W) x 4 m (L) x 3 m (H) and stairway with dimension 2.5 m (W) x 5.5 m (L) x 27 m (H) as shown in Figure 1.

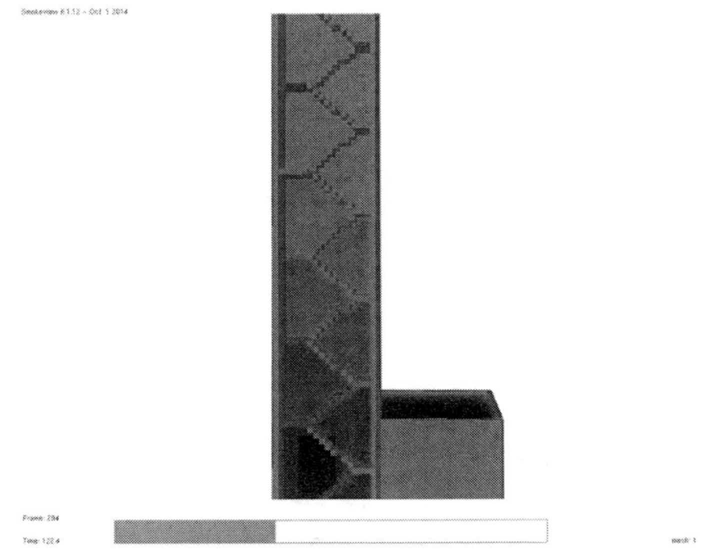

Figure 1. Room and stairway model

For this study, three parameters were varied in the input files in FDS and smoke temperature as the output quantity was considered as dependent variables. The fire source was located in a center of fire room and HRR value was selected ranging from 300 to 1000 kW. Beside opening size, the efficacy of vent was study on different value of HRR as it produced different amount of smoke according to Shi et al. [32]. The radiation fraction was kept constant at 0.35 in all simulations. Different vent opening values were selected based on the total area of shaft compartment as stated in standards which is not less than 5% of the cross sectional of stairway area and the values for those parameters were given in table 1. The mesh size for the simulation was set according to grid resolution's result and ran for 600s. The computer used in the analysis was: Intel® Core™ i5-2400 CPU @ 3.10GHz, RAM 4.00 GB. Data from simulation were taken when a stable smoke layer had formed in the stairway shaft.

Table 1. Parameters that were varied in all simulations

Variables	Values
Fire size (kiloWatt)	300
	500
	900
	1000
Size of opening, meter (vent)	1/3.5

For the design of fire protection system, the characteristics of design fire such as the value of HRR, CO and CO_2 production rates are important [33]. Thus it is important to conduct experiments to measure these three parameters. Nonetheless, full-scale experiments are expensive and resource demanding. Alternatively, computer simulations which also referred as numerical experiment could be used as a research tools to study fire phenomenon in compartment fire and have been applied in different studies in fire science during the last decades [34-39]. Additionally, this type of experimental method had been reviewed by Johansson [40] as a complementary to traditional experiment and is a promising method in fire science research. In his scope to explore it as a research method, the advantages and challenges of numerical experiments were compared to traditional experiments using several examples of previously performed numerical experiments of compartment fires. Less expensive, reduce time; able to control the experiment and fire measurement not affected by the instrument are the main advantages discussed using numerical experiments.

3. Result and Discussion

3.1 Varied Heat Release Rate (HRR)

In order to study the effect of the heat release rate (HRR), four sets of fire test using heptane as fuel with steady HRR of 300kW, 500 kW, 900 kW and 1000 kW were simulated in a protected stairway with vent opened. According to the experiment, the fire is sharply increased to a stable value of maximum heat release rate (HRR) and was then kept at the maximum value to the end as shown in figure 2. The duration of the four tests was 600 s (10 minutes). In general, when the HRR increases, the increase of the air drawn into the stairway accelerates the spread of the hot smoke. Thus the driving force for the flow in close stairway is the energy input from a fire originated in the lower compartment. figure 2 shows temperature curves in the top storey of stairway at various HRR.

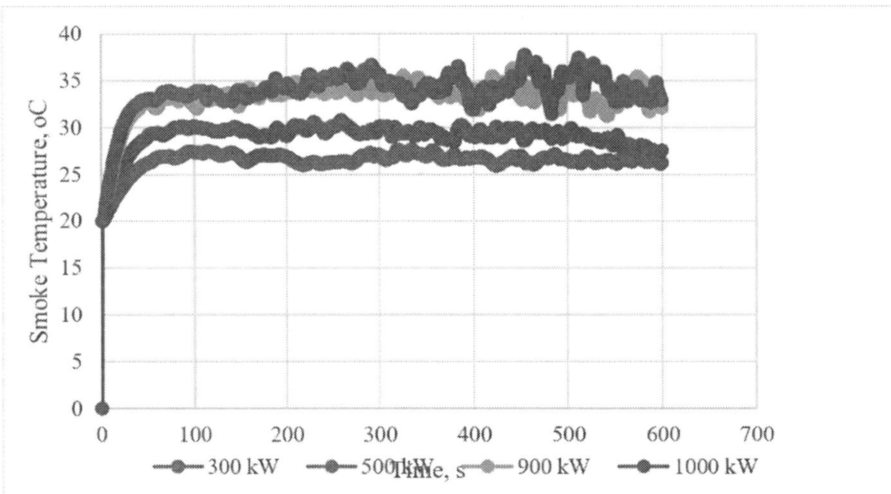

Figure 2. Temperature curves at the top storey of stairway at various HRR; 300 kW, 500 kW, 900 kW and 1000 kW.

The heat release rate has huge effect on distributions of smoke temperature. The maximal smoke temperature at top storey of stairway shaft as shown in Figure 2 ranged from 28 °C to 38 °C under HRR from 300kW to 1000kW, respectively. However, the smoke temperature curve for 900 kW is nearly similar with 1000 kW due to slightly different of fire size. As these temperatures were measured at the top of stairway with vent opened, the temperature is still remained close to ambient temperature as the smoke was well mixed with air even after 10 minutes of fire ignition.

3.2 Smoke movement in the stairway with and without vent opening

Smoke movement in a stairway with and without vent opening were tested in this study to study the smoke movement from lower level to the top storey. Figure 3 and 4 shows smoke movement in smokeview resulted from numerical simulation using FDS in a stairway produced by fire size, HRR = 500 kW at 10s, 100s, 200s, 300s and 500s after fire ignited.

IOP Conf. Series: Materials Science and Engineering **702** (2019) 012056 doi:10.1088/1757-899X/702/1/012056

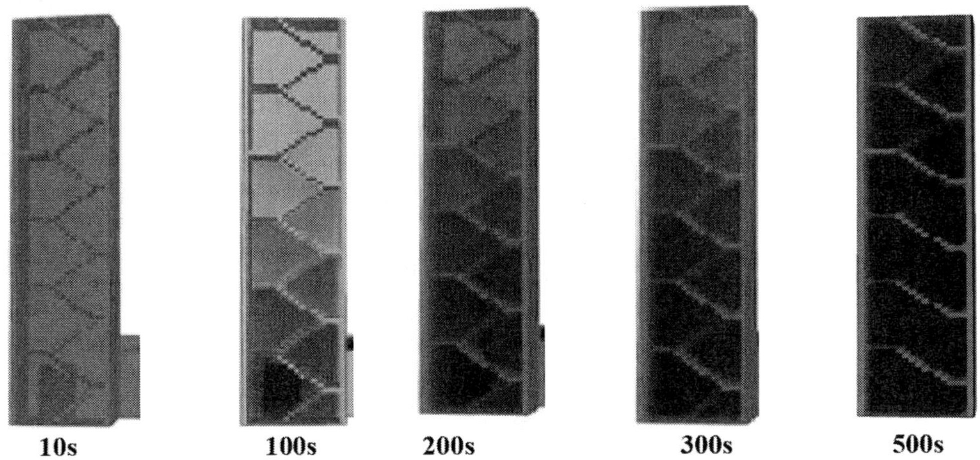

Figure 3. Smoke movement in the stairway without vent opening

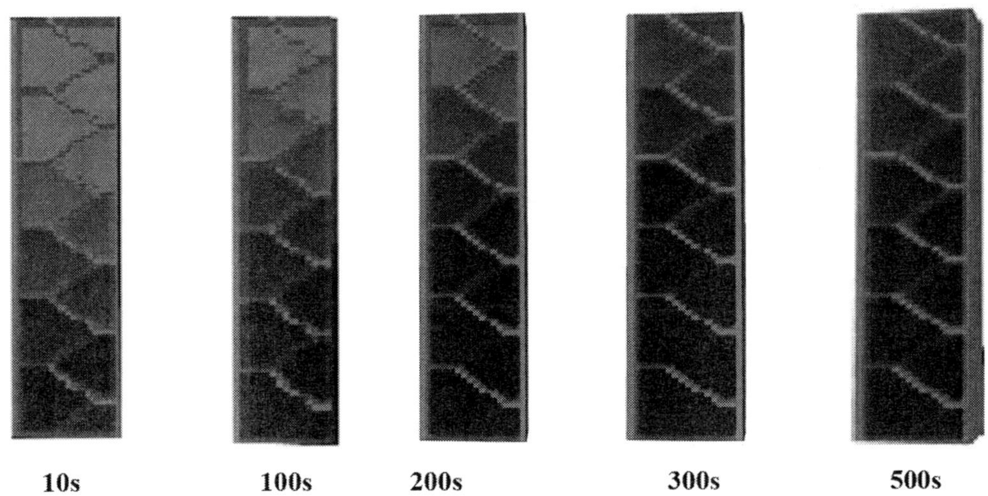

Figure 4. Smoke movement in the stairway with vent opening

As shown in Figure 3, it took nearly 100 s for the smoke to enter the first floor of stairway without vent opening. At the same time, for the stairway with vent opening the smoke begin to filled and diffuse in the second and third floor (Figure 4). It shows that for stairway with no vent opening, it takes longer time for the smoke to enter and spread through the stairway. Thus, the lower floor is filled with smoke and after a certain duration the smoke becomes dense as it will reduce visibility during evacuation. After 500 s since the fire break out, the smoke diffused slowly due to turbulent diffusion into the 5th and 6th floor and the whole stairway were filled with smoke after the smoke arrived at the top and flew downward at lower storey of stairway. As a result, the smoke in the stairway accumulated more and more and diffused to the lower space of stairway. The purpose of installing natural vent is to provide fresh air to the stairway [41], however during fire event the air pressure inside the stairway was higher due to smoke diffusing cause by stack effect and as the smoke from fire room entered the stairway accordingly, there was still part of the smoke flowing outside the opened vent. In the situation of stairway with vent opening, the smoke diffused faster into the higher floor at 10 s after fire ignited and spread to the 5th and 6th floor at 300 s which finally flowed to the outside through vent opening figure 4). Consequences, the smoke distribution in each level of stairway could be uniform even though after a few minutes.

3.3 Varied vent opening

Figure 5 shows the influence of vent opening area on the gas temperature rise at different fire size; 300kw, 700 kW and 900 kW. From previous observation, the smoke temperature measured for fire size 1000 kW is nearly similar with 900 kW, thus for this part of study fire size of 1000 kW is excluded. The fire size has a strong influence on the average rate of smoke temperature rise in the stairway as increasing fire size will increase smoke temperature rise as discovered by previous research [25][42]. Furthermore, compare to all fire sizes; 300 kW, 700 kW and 900 kW in Figure 5 (a), (b), and (c) respectively, smoke temperature decrease when vent opening area increase from 1m x 1m to 2.2m x 2.7m. However, with further increase in vent opening area to 3.5m x 3.5m the smoke temperature reaches a value which same as vent opening area; 2.2m x 2.7m. It indicates that increase vent opening area is helpful to slow down the development of fire as well as reduce smoke entrainment thus resulted in increasing visibility. Therefore, when a fire occurs in a room and smoke diffuse to the adjacent stairway, vent opening at the top storey of stairway is conducive to provide more time for evacuation, firefighting and rescue work.

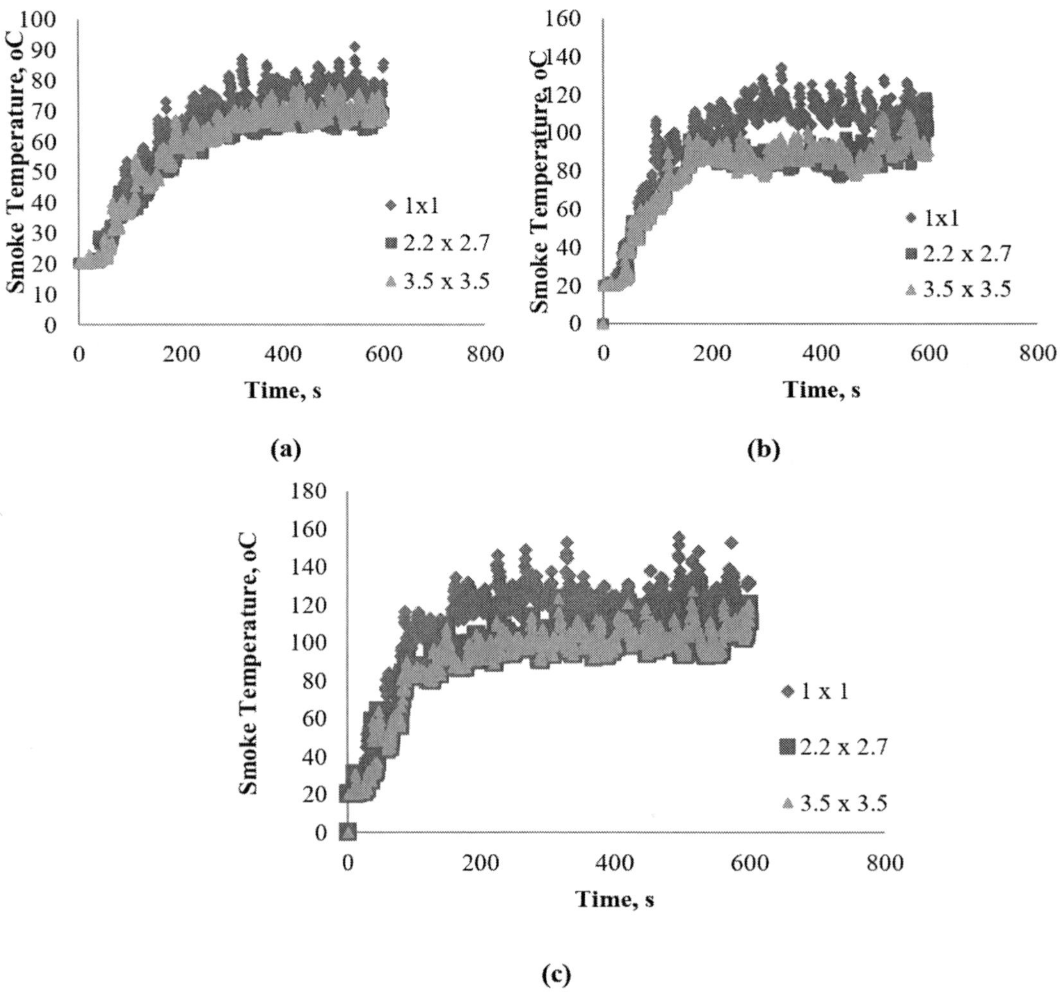

Figure 5. Temperature curve in first storey of stairway at various HRR a) HRR = 300 kW b) HRR -= 700 kW c) HRR = 900 kW

IOP Conf. Series: Materials Science and Engineering **702** (2019) 012056 doi:10.1088/1757-899X/702/1/012056

4. Conclusion

Vent opening area at the top storey on the smoke temperature rise in stairway was studied in this paper. Three different vent sizes were adopted in FDS simulation with four different fire sizes and the effects of these parameters on smoke temperature rise in stairway were discussed. The fire size has a strong influence on the average rate of smoke temperature rise in the stairway. According to simulation result, natural vent really played a good role in preventing smoke entrainment and different opening size had different influence on the efficacy of smoke venting from stair shaft. The bigger the vent opening, the more smoke could escape from the stairway to the outdoor. It was also revealed that no matter what size of fire, the smoke temperature in the stairway decrease significantly once the vent opening area is increased. In order to support this study using numerical study, physical experiment with full-scale fire experiment could be conducted in future for multi-storey building. The finding of this study could assist the fire engineer to ensure that the smoke vent installation could play a good role in confinement of smoke diffusion.

Acknowledgement

Author would like to express her gratitude to all co-researchers for their guidance and support in terms of document preparation as well as data collection. No financial implication involved in this study.

References

[1] FRDM 2015 *Fire Statistic in Malaysia 2008-2014.*

[2] Bennetts I D and Poh K W 2006 Design of Sprinklered Shopping Centre Buildings for Fire Safety OneSteel-Market Mills (Newcastle, Australia) p 42

[3] Tharmarajan P 2007 *The Essential Aspects of Fire Safety Management in High-Rise Buildings.* (Universiti Teknology Malaysia)

[4] Cebela A 2012 *Fire safety education for staff members - case study* (Lund University Sweden).

[5] Luo N, Li A, Gao R, Tian Z and Hu Z 2014 Smoke confinement utilizing the USME ventilation mode for subway station fire *Saf. Sci.* **70** 202–10

[6] New York Daily News 2014

[7] The New York Times 1998

[8] Mohammed A A, Bachtiar D, Siregar J P and Rejab M R 2016 Effect of sodium hydroxide on the tensile properties of sugar palm fibre reinforced thermoplastic polyurethane composites *J. Mech. Eng. Sci.* **10**(1) 1765–77

[9] Aniza N, Hassan S, and Inayat M 2016 Thermogravimetric kinetic analysis of Malaysian poultry processing waste material under inert and oxidative atmospheres *J. Mech. Eng. Sci.* **10**(2) 1943–55

[10] Fajrin J, Zhuge Y, Bullen F and Wang H Flexural behaviour of hybrid sandwich panel with natural fiber composites as the intermediate layer *J. Mech. Eng. Sci.* **10**(2) 1968–83

[11] UBBL, Uniform Building By-Laws 1984

[12] Buchanan A H and Abu A K 2017 *Structural Design for Fire Safety*, 2nd Ed. (United Kingdom: John Wiley & Sons).

[13] Li Y Z and Ingason H 2015 A New Methodology of Design Fires for Train Carriages Based on Exponential Curve Method *Fire Technol.* p 16

[14] Yang P, Tan X and Xin W 2011 Experimental study and numerical simulation for a storehouse fire accident *Build. Environ.* **46**(7) 1445–59

[15] Rahman M M, Alim M A, Mamun M A H, Chowdhury M K and Islam A K M S 2007 Numerical Study Of Opposing Mixed Convection **2**(2) 25–36

[16] Chu G and Sun J 2008 Decision analysis on fire safety design based on evaluating building fire risk to life *Saf. Sci.* **46**(7) 1125–36

[17] Ramachandran G and Charters D 2011 *Quantitative Risk Assessment in Fire Safety* (London, United Kingdom: Taylor & Francis LTD)

[18] Ji J, Gao Z H, Fan C G and Sun J H 2013 Large Eddy Simulation of stack effect on natural smoke exhausting effect in urban road tunnel fires *Int. J. Heat Mass Transf.* **66** 531–42

[19] H. Bjelland 2013 Engineering Safety with application to fire safety design of buildigns and road

tunnels (University of Stavanger)
[20] Horváth I, van Beeck J and Merci B 2013 Full-scale and reduced-scale tests on smoke movement in case of car park fire *Fire Saf. J.* **57** 35–43
[21] Chen J, Fang Z and Yuan J 2013 Numerical Simulation about Smoke Vent Arrangement Influence on Smoke Control in Double-decked Tunnel *Procedia Eng.* **52** 48–55
[22] Meng N. *et al.* 2014 Numerical study on the optimization of smoke ventilation mode at the conjunction area between tunnel track and platform in emergency of a train fire at subway station *Tunn. Undergr. Sp. Technol.* **40** 151–59
[23] Bartlett N 2012 Optimization of Smoke Control Systems in Underground Subway Stations (Ghent University)
[24] Luo N, Li A, Gao R, Song T, Zhang W and Hu Z Performance of smoke elimination and confinement with modified hybrid ventilation for subway station *Tunn. Undergr. Sp. Technol.* **43** 140–47
[25] Qu L and Chow W K 2013 Common practices in fire hazard assessment for underground transport stations *Tunn. Undergr. Sp. Technol.* **38** 377–84.
[26] Lambert K and Merci B 2014 Experimental Study on the Use of Positive Pressure Ventilation for Fire Service Interventions in Buildings with Staircases *Fire Technol.* **50**(6) 1517–34.
[27] Mokhtarzadeh-Dehghan M R 2011 Numerical simulation and comparison with experiment of natural convection between two floors of a building model via a stairwell *Int. J. Heat Mass Transf.* **54**(1–3) 19–33
[28] Ji J, Wan H, Li Y, Li K and Sun J 2015 Influence of relative location of two openings on fire and smoke behaviors in stairwell with a compartment *Int. J. Therm. Sci.* **89** 23–33
[29] Khairul Zaimy M, Anika Zafiah M, Najibah A and Nurulsaidatulsyida A 2013 Mechanical And Thermal Properties Of Waste Bio-Polymer Compound By Hot Compression Molding Technique *J. Mech. Eng. Sci.* **5** 582–591
[30] Mcgrattan K, Peacock R and Overholt K 2016 Validation of Fire Models Applied to Nuclear Power Plant Safety *Fire Technol.* **52** 5–24
[31] Audouin L, Rigollet L, Prétrel H, Le Saux W and Röwekamp M 2013 OECD PRISME project: Fires in confined and ventilated nuclear-type multi-compartments - Overview and main experimental results *Fire Saf. J.* **62** 80–101
[32] Shi C L, Li Y Z, Huo R, Yao B, Chow W K and Fong N K 2005Mechanical smoke exhaust for small retail shop fires *Int. J. Therm. Sci.* 44(5) 477–90
[33] Fauzi F A, Ghazalli Z, and Siregar J P 2016Effect of various kenaf fiber content on the mechanical properties of composites *J. Mech. Eng. Sci.* **10**(3) 2226–33
[34] Phillips W G B 1994 Simulation models for fire risk assessment *Fire Saf. J.* **23**(2) 159–69
[35] Novozhilov V, Harvie D J E, Kent J H, Apte V B and Pearson D 1997 A computational fluid dynamics study of wood fire extinguishment by water sprinkler *Fire Saf. J.* **29**(4) 259–82
[36] Prasad K, Patnaik G and Kailasanath K 2002 A numerical study of water-mist suppression of large scale compartment fires *Fire Saf. J.* **37**(6) 569–89
[37] Bittern A 2004 Analysis of FDS Predicted Sprinkler Activation Times with Experiments (University of Canterbury)
[38] Huang H, Ooka R, Chen H and Kato S 2009 Optimum design for smoke-control system in buildings considering robustness using CFD and Genetic Algorithms *Build. Environ.* **44**(11) 2218–27
[39] Johansson N 2015 Fire Dynamics of Multi-Room Compartment Fires (Lund University)
[40] Johansson N 2014 Numerical experiments and compartment fires *Fire Sci. Rev.* **3**(2) 1–12
[41] Zhang J, Lu S, Li C and Yuen R K K 2014 Vent Location Impacts on Building Compartment Fire under Natural Roof Ventilation *APCBEE Procedia.* **9** 360–64,.
[42] Harish R and Venkatasubbaiah K 2014 Effects of buoyancy induced roof ventilation systems for smoke removal in tunnel fires *Tunn. Undergr. Sp. Technol.* 42 195–205

IOP Conf. Series: Materials Science and Engineering **702** (2019) 012057 doi:10.1088/1757-899X/702/1/012057

The implementation of lean manufacturing and ergonomics in Small Medium Enterprise – Case study

A N M Rose*, M F F Ab Rashid, N M Z Nik Mohamed, M A S Jubri

Faculty of Mechanical Engineering, Universiti Malaysia Pahang, 26600 Pekan, Pahang, Malaysia

*E-mail: nasser@ump.edu.my

Abstract. Productivity can be increased by eliminating of all wastes. The objective of lean manufacturing is to eliminate wastes. Commonly, there are seven wastes known as transport, inventory, motion, waiting, over processing, overproduction and defects. However, ergonomics is a process to improve the working posture by designing or arranging a workplace, system and material handling. The aim of this study is to improve the productivity through the integration of basic lean manufacturing and ergonomics in the food company. The walkthrough and distribution of questionnaire have been conducted in this company. Each activity was observed to identify non-value-added activity and work related muscular skeletal disorder (WMSD) that affect the productivity by using questionnaire, motion study and Rapid Upper Limb Assessment (RULA). The questionnaire result has shown that the most of wasted motion is in weighing and measuring activity. In addition, all workers in this process were affected by the musculoskeletal disorder at their neck and shoulder. All movement were simulated through Delmia software. The RULA score shows that the worker had experienced pain at the arm and wrist. The score also suggested that the process should be changed immediately. The changes might reduce the worker's pain and increase the productivity.

1. Introduction

As a developing country, Malaysia is rapidly growing in various industrial sectors. One of the sectors involved in driving the Malaysian economy is the food industry. The food industry in Malaysia is predominantly owned by Malaysian small and medium enterprise (SME) [1]. It is expected the food industry will grow very fast especially on halal food. Therefore, Malaysian SMEs are encouraged to prepare and ready with the challenges of halal food. SME is categorised based on a company which has 200 permanents workers or sales turn over less than RM50 million [2]. Most of the SMEs are lack of capital, skills, expertise and resources [3]. They are concentrating and focusing more on production activity, did not aware much on ergonomics factors [4]. As a result, the workers might become disable due to prolong affected of musculoskeletal disorders [5]. It is considered the most common work-related problem in SME due to working posture and design of workstations. Most of the work done in SME are stills carried out manually due to the lack of technology capability [6]. Therefore, issues of work related musculoskeletal disorders (WMSD) and injuries of the workers are great concern. The risk of the workers in this industry exposed to WMSD is very high [7]. WMSD definitely will affect worker productivity and efficiency. Besides, SMEs also do not apply the good manufacturing practice [8], thus both of the problems were affecting the productivity of the company. Productivity is the most valued aspect of all companies including SME. It is a measurement to determine the company performance based on production volume compared to the provided inputs. The integration of basic lean manufacturing and ergonomics could enhance worker productivity. Both of the systems are concerned on wastes which due to availability of non-value added activities. Non value added activity is anything not add value to the product. The common non value added activities are motion, inventory, defects, over processing, over production, waiting and transportation [9]. It is including of poor handling due to body movement or posture [10]. The effectiveness of lean manufacturing implementation is complemented by the ergonomic features such as the employee interface with tools, parts, and environmental factors in the workplace [11]. This study aims to propose the integration of basic LM and ergonomics in the case study company.

Lean manufacturing (LM) and ergonomics concerned are working towards free of wastes. The objective of LM is to eliminate all possible wastes whereas ergonomics is to enhances workplace health, safety, and work design issues. The integration of these two systems will give better performance to the practitioner compared to a single practice. The workers will feel comfortable with ergonomic working zone and safe to work in a long duration. Hence, the organization could improve labour productivity through the efficient system and proper handling. Among the LM tools and practices could be implemented in SME are 5S, continual improvement, reduce set up time, visual display and value stream mapping [12]. These tools and practices are considered easy to be implemented and less investment. In addition, 5S is a basic tool and must be implemented before other tools can be implemented. Lack of human factor in the organisation will decrease productivity due to pained and fatigued. Indirectly it could increase the absenteeism among the workers. Therefore, it could affect the efficiency and productivity. In other words, the demands of LM and ergonomics must be balanced to achieve high productivity. 5S is the most used tool in the LM discipline. This tool also can be implemented universally through any field of work due to its simplicity and easy to be implemented. 5S is generally focusing on the tidiness of the workplace that helps increasing the agility of worker but it is not included one aspect of safety.

Ergonomics comes from the old Greek word of 'ergo' and 'nomos' that means work and natural law respectively. Ergonomics is a discipline of the usage of scientific principles, methods, and data drawn from a variety of discipline for the development of systems in which humans play a significant role. The field also extends from a single person of a simple tool to multi-person with complex machine and organisation [13]. Musculoskeletal disorder (MSD) is affecting the human musculoskeletal system due to exposing to the ergonomics risk factor. Upper limbs (the hand, wrist, elbow and shoulder), the neck and lower back are particularly vulnerable to MSDs. Lower limbs and the upper back may also be affected [14]. Work-Related Musculoskeletal disorder (WMSD) tends to decrease the productivity of the worker due to the MSD disease. The employee absenteeism will be increased if the MSD is not controlled effectively [15].

Rapid Upper Limb Assessment (RULA) is one of ergonomics assessment tools available for evaluating the risk level of WMSD. The tool referred to body posture diagram that provides evaluation criteria for exposure to the risk factor. The assessment was divided into three stages which focusing on the desired placed that want to be assessed. As an example, RULA is conducted to investigate the risk based on the results of the questionnaire that lead to the upper limb body part. DELMIA software is used to evaluate the RULA score of body movement through a simulation.

6S is the extension of 5S, with the addition of another 'S' which known as Safety. The rest of 'S' still the same i.e. sorting, set in order, shining, standardise and sustain. It is aiming for creating a safe workplace for the employees from injuries due to repetitive work. MSD is a disease that developed by time due to exposing of ergonomics risk factor and could affect worker health. Thus, in the long term, this symptom will affect overall productivity. Therefore, through the integration of lean manufacturing and ergonomics, an organisation could enhance productivity through the elimination of waste which due to non-value added activities including poor posture movement. 6S is in the new perspective is to create and maintain organisation in clean, safe and efficient that enables the highest level of value-added performance. The awareness of worker safety from MSD disease also part of 6S.

The 6S methodology is a progressive expansion of traditional 5S which design to reduce damages, accidents and optimize productivity in the workplace. The safety in 6S is a way of ensuring the worker and workplace are safe which free from accident, injuries, fatigue and pain [16].

IOP Conf. Series: Materials Science and Engineering **702** (2019) 012057 doi:10.1088/1757-899X/702/1/012057

2. Methodology

The methodology of this study comprised of three phases. Firstly, the questionnaire is distributed to the workers. The questionnaire used in this study is a Nordic Questionnaire. The main purpose of this questionnaire is to evaluate the musculoskeletal disorder in the perspective of the ergonomics context and for occupational healthcare service [17]. The workers were asked to rate their musculoskeletal symptoms based on the provided questionnaire. They are encouraged to ask on any unclear information for clarification. Only three workers were evaluated with the range of age between 19 to 53 years old. Similarly, Fazi et al. also conducted an ergonomic study in food company with three workers [7]. Total working postures evaluated comprised of 11 processes. The researcher also observed and identified an awkward posture. RULA assessment was selected for this analysis because most of the workers were discovered upper extremity MSD. Then, Delmia software was used to generate an assessment score. All movements were simulated according to the real workers' movement in their working area. The simulation was followed exactly based on recorded movement which was taken by the video recorder. Before recording, the workers were informed to work as usual without interruption. Total score will determine whether the job needs an immediate change to reduce MSD risk. Table 1 shows the reference of RULA assessment for level of MSD risk.

Table 1. RULA Score [13]

Score	Level of MSD Risk
1-2	Negligible risk, no action required
3-4	Low risk, change may be needed
5-6	Medium risk, further investigation change soon
7+	Very high risk, implement change now.

However, the walkthrough was carried out to observe any wastes incurred in this company. Any abnormal movement, process or activity were recorded for improvement.

3. Result

The main objective of lean manufacturing is to eliminate waste. The observation shows that all of the workers were felt pain to their body due to uncomfortable while performing the job. As a result, they have to stop frequently for a rest about one to two minutes. Table 2 shows the questionnaire result of pain gained by the workers. The effect of the awkward posture has caused productivity low. In average they were operated 77 minutes for 50 packs per cycle which in total 150 packs per day. Three workers were working here, two female and one male. All of them were having the same pain at their neck and shoulders since the last 12 months. However, only one worker had pain at her wrist since last 12 months, especially when performing the heavy lifting (HL).

Table 2. Questionnaire Result

Sex	Height (cm)	Weight (kg)	Neck	Shoulder	Wrist
M	167	67	Yes	Yes	No
F	158	57	Yes	Yes	Yes (HL)
F	161	60	Yes	Yes	No

*Note: M=Male; F=Female

The Delmia simulation analysis shows the details of the score as in table 3.

Table 3. The RULA Scores by using DELMIA Software

Process	Arm and Wrist Score	Neck, Trunk and Legs Score	Final Score	Risk level
Weighing and Measuring	5	7	7	High
First Rolling	7	2	5	Medium
Second Rolling	5	3	4	Low
Boiling	5	2	4	Low
Cooling	6	3	5	Medium
Packaging	7	2	5	Medium

Table 3 shows RULA assessment and risk level for every process. The highest score is 7 for weighing and measuring. This score means high risk which needs an immediate change of current working posture. It was due to the worker need to bend their body at 90 degrees while weighing and measuring as shown in Figure 1(a). Figure 1(b) is based on Delmia simulation. The worker continues to bend for a long period with frequently has a short period of break. It was observed that this working posture has influenced the neck and trunk due to bending and twisting while filling for the weighing process. This working position is classified as an awkward posture. The increasing level of fatigue might effects productivity. The worker agreed that they were discovered mild pain at their neck and back.

a) 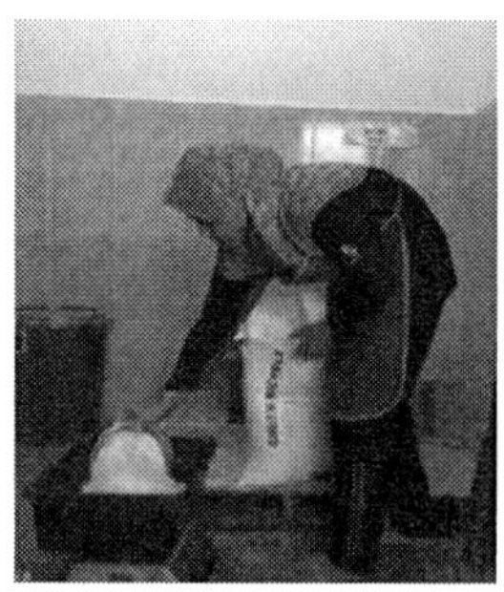

b) 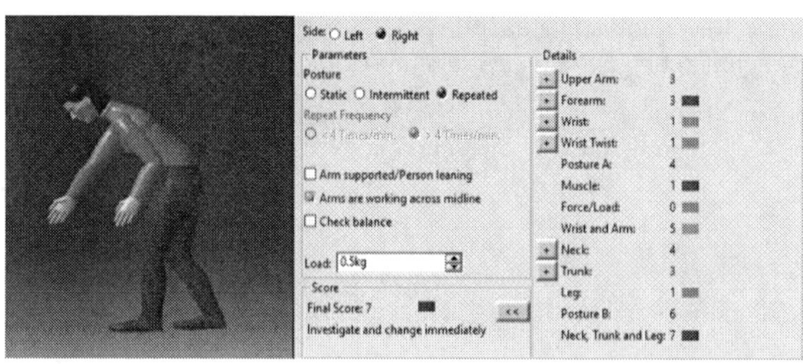

Figure 1. (a) Real working condition for weighing and measuring (b) Delmia Simulation of Weighing and Measuring processes

Whereas three processes, first rolling, cooling and packaging were classified as medium risk. These processes need to be carried out further investigation as an effort to increase productivity. Other two processes second rolling and boiling were categorised as low risk. Although this level is considered safe for the worker, it still has a tendency to increase the risk level if the worker keeps using the same method, especially on boiling. As for safety and health precautions, the bending posture could be avoided through the development of jig or use table with a recommended height of 0.8m to 0.9m for weighing and measuring [18]. Figure 2(a) shows the sketch of a suggested working table.

The suggested working table height was simulated in the DELMIA software as in figure 2(b) and the result shows positive with the final score of 3. The height of the table should be around 0.8 m to 0.9 m below to the average of worker elbow height or suitable with the worker working height according to guidelines that may vary [19].

(a)

(b)

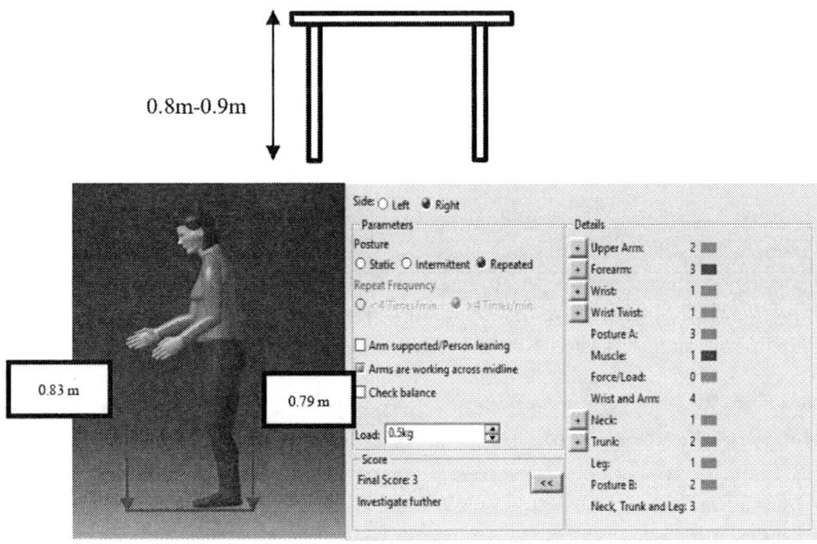

Figure 2. (a) Suggested working table height (b) DELMIA analysis with a suggestion of table height

The study also identified non-value-added activity which known as scooping up the flour and weighing the noodle. These processes consume too much time to get the right weight. It took about 20 minutes for every batch. It was due to manually filled in without using any standard container. Therefore, this study has suggested to the company to use the right scoop with almost the right weight as needed compared to the current process which requires repeating process before getting the right weight. Due to time constraint, this study is limited to the suggestion for improvement only.

4. Conclusion

The integration of basic lean manufacturing and ergonomics could enhance company performance. Some companies especially SME are lack of expertise in lean manufacturing and ergonomics. Therefore, in this case study, the research has concentrated on how basic lean manufacturing system and ergonomics could be implemented in SME-food's company. The research has discovered the problems on ergonomic and lean manufacturing which are awkward working posture and non-value added activities. After the investigation carried out thoroughly, the safety and working condition has been proposed to a better working posture. The RULA evaluation has proved that the current working position for weighing and measuring need to change immediately. As in this study, the working posture can be improvised by using a table with a proper height between 0.8 m to 0.9 m. The simulation has validated the immediate change of working height that could reduce RULA score from 7 to 3. It is expected that the improvement of working posture could enhance company performance by reducing the lead time to less than 77 minutes for 50 packs. It could be achieved once the non-value activity on bending too long is eliminated. As overall production, this company is predicted to produce more than 150 packs per day.

Embedding ergonomic features in lean manufacturing implementation will lessen mistakes, improve productivity and reduce WMSD risk factors.

Acknowledgment
The authors would like to acknowledge Universiti Malaysia Pahang for supporting this research.
Ref: FRGS/1/2018/TK03/UMP/03/3)-(RDU190193)

References

[1] MITI. 2016. Food industry in Malaysia. Retrieved from Ministry of International Trade and Industry: www.miti.gov.my/miti/resources/11._Food_Industry_.pdf
[2] http://www.smecorp.gov.my accessed by 20th June 2019
[3] Rose A N M, Deros, B M.and Rahman,M N A. 2013. Lean manufacturing practices implementation in Malaysian's SME automotive component industry. *Applied Mechanics and Materials* **315** 686-90
[4] Siong V Y et al 2018 Ergonomic Assessment in Small and Medium Enterprises (SMEs). *Journal of Physics: Conference Series*
[5] Kelli D. Allen 2017 Musculoskeletal Health: Addressing the Leading Causes of Disability N C *Medical Journal* 78 **5** 306-9
[6] Radzi N M, Shamsuddin A and Wahab E 2017 Enhancing the Competitiveness of Malaysian SMES Through Technological Capability: A Perspective; *The Social Sciences***12**(4) 719-24
[7] Fazi H M, Nik Mohamaed N, Ab Rashid M F and Rose A N M 2016 Ergonomics study for workers at food. *MATEC Web of Conferences 90*
[8] Anthony J and Kumar M 2005 Six sigma in small and medium sized UK manufacturing enterprises *International Journal of Quality and Reliability Management* **22**(8) 860-74
[9] Ohno T 1988 The toyota production system. English translation *Productivity Press*
[10] Suryoputro M R, Sari A D, Burhanudin R, Sugarindra M 2017 Lean production design using value stream mapping and ergonomics approach for waste elimination on buffing panel upright process *IOP Conference Series Material Science Engineering*
[11] Kim I J 2017 The Function of Ergonomics in Lean Manufacturing Design and Control *Journal of ergonomic*, **7**(5) 1-3
[12] Rose A N M, Ab Rashid M F F, Nik Mohamed N M Z and Ahmad H 2017 Similarities of lean manufacturing approaches implementation in SMEs towards the success: Case study in the automotive component industry *MATEC Web of Conferences 87*
[13] Elbert K H A 2018 Ergonomics*: How to Design for Ease and Efficiency* Third Ed. (Academic Press)
[14] Yusuff R M and Abdullah N S 2016 Ergonomics as a lean manufacturing tool for improvements in a manufacturing company Proceedings *of the International Conference on Industrial Engineering and Operations Management* 581-88
[15] Deros B M, Daruis D D I and Basir I M 2015 A Study on Ergonomic Awareness among Workers Performing Manual Material Handling Activities. *Procedia - Social and Behavioral Sciences* **195** 1666 –73
[16] Gautam V S, Shah A R, Parmar A N, and Kedariya V D 2014 Study of 6S Concept and its Effect in Industry *International Journal of Emerging Technology and Advanced Engineering* **4**(10) 272-77
[17] Kuronika I, Jonsson B, Kilbom A, Vibterberg H, Biering-Sorensen F, Andersson G, and Jorgensen K 1987 Standardise Nordic questionnaire for the analysis of musculoskeletal symptom *Applied Ergonomics* 233-37
[18] https://ergo-plus.com/rula-assessment-tool-guide/ accessed by 20th June 2019.
[19] Department of Occupational Safety and Health Malaysia 2002 *Guideline of Occupational Safety and Health for Standing at Work* Department of Occupational Safety and Health Malaysia

IOP Conf. Series: Materials Science and Engineering **702** (2019) 012058 doi:10.1088/1757-899X/702/1/012058

A review on possible approaches of anaerobic biological processes for palm oil mill effluent: Process, quality, advantages, and limitations

S K Al-Amshawee[1] **and M Y M Yunus**[1,2*]

[1] Faculty of Chemical and Process Engineering Technology, Universiti Malaysia Pahang, Gambang, Pahang 26300, Malaysia
[2] Earth Resource and Sustainability Centre (ERAS), Universiti Malaysia Pahang, Gambang, Pahang 26300, Malaysia

*E-mail: yusri@ump.edu.my

Abstract. Palm oil mills play an essential role in the economic development of many countries. Though, it is the primary source of environmental pollution and degradation. Water scarcity and the high cost of new water supply technologies are the two major factors responsible for the increasing recognition of the importance to conserve water resources by wastewater treatment and reuse. Sustainability of sanitation systems should be related to low requirements in cost, energy consumption, and maintenance. Anaerobic biotreatment is the preferred industrial choice for mediating high strength wastewater. Anaerobic biotreatments for wastewater are increasingly being researched as cost-effective alternatives to deliver low sludge accumulation, efficient biodegradation and mineralisation, microbes' reduction, and solids-free effluents. In the last decade, many studies investigated various types of anaerobic reactors in combination with membranes. This review shows the potential of anaerobic bio mediations for palm oil mill effluent. Also, the paper discusses the impact of various factors on both biological and filtration performances and identifying strengths and limitations.

1. Introduction

Human activities directly pollute water courses and the environment by toxic compounds, nutrients, microorganisms, and pathogenic substances. Luckily, global industries, such as heavy industry, livestock farming, mining, and plantation are turning into the economic mode. Despite that, the pollution amount is still increasing per time due to unsuccessful willing. Pollutants involve liquid and solid wastes that are harmful, and danger on the healthy life, and the lifestyle. Hence, a significant call for the need of marvellous treatment to decrease pollution amount, but we have not accomplished that yet because of numerous difficulties. For that, pollution awareness is an excellent alternative caution against water contamination [1–6].

Wastewater requires robust developed technology to handle the massive contamination loading with no significant issues. Thus, operating, and the capital cost will increase. Malaysian palm oil industry provides thousands employment with USD 20 billion revenue [6]. Palm oil mill effluent (POME) is a discharged wastewater from palm oil industry. POME is observed as a thick brownish liquid containing a high concentration of biochemical oxygen demand (BOD), chemical oxygen demand (COD), oil &

grease, and solids, resulting in severe worldwide issues like global warming. The generated POME amount in 2011 was around 60 million ton, while it was around 30 million ton in year 2004 and 44 million ton in year 2008 [7–9]. Usually, aerobic and anaerobic ponds are utilized for treating POME within 100 days. Although, the effluent has not met the enforced rules and regulations by the environment department of Malaysia DOE for industrial discharges [10–13].

Wastewater could be processed under physical (mechanical forces), chemical (reactions), or/and biological (microorganisms) [14]. Physical and chemical treatments have not stood well for processing high strength wastewater i.e. high concentration of ammonia, suspended solids, heavy metals, high COD and BOD ranges, and shock loadings. Therefore, tertiary treatment is needed for discharging high-quality effluent. Numerous researches have been conducted on the potential of biological wastewater treatment, such as bacteria, yeast, and microalgae [15–17]. Membrane, anaerobic, and aerobic are the current treatments for POME [9]. Table 1 shows the advantages and disadvantages of using aerobic, anaerobic, or membrane processing for wastewater.

Table 1. Benefits and drawbacks of using membrane, aerobic, and anaerobic systems

Type	Benefits	Drawback	References
Membrane	1) Produce consistent and good water quality after treatment, 2) smaller space required, 3) can disinfect wastewater effluent	1) Short lifetime, 2) fouling, 3) expensive	[18]
Aerobic	1) Short retention time, 2) efficient in handling toxic wastes	1) Unsuitable for land application, 2) low pathogen inactivation, 3) high energy consumption	-
Anaerobic	1) Low energy consumption, 2) no aeration, 3) biogas production, 4) low sludge generation, 5) high COD removal	1) Large area required, 2) slow startup, 3) long retention time	[19–22]
Evaporation	1) The remaining solids can be applied as fertiliser	1) High energy consumption	-

It has been proposed to employ anaerobic treatment for POME due high COD removal, low sludge production (5-20%), capital cost, biogas recovery, and energy requirement. Additionally, methane gas is generated during POME anaerobic treatment, which is a valuable revenue if its captured. Although, anaerobic techniques can barely handle the strength of slaughterhouse wastewater [21]. Hence, anaerobically treated water requires extra processing to achieve complete removing of pathogenic organisms, total phosphorus (TP), and total nitrogen (TN) [23–25]. Finally, it was found that using a combination system (e.g. UV/H_2O_2, hybrid anaerobic baffled bioreactor) can discharge a satisfying effluent quality [21,26].

2. Anaerobic Biological Treatment

Microorganisms break down different kinds of contaminates in POME under anaerobic conditions, this process is named as anaerobic biological treatment. Organic matters and nutrients are considered as significant food for microbes in POME to increase their population and size. Anaerobic biological treatment is an essential process for POME to achieve high removal amount of pollutants with low costs. It is considered friendly and economical because it produces biomass which could be used as fertiliser, and biogas source. Although, not all anaerobic treatments capture and store the discharged biogas. The below section reviews possible POME anaerobic mediations showing their mechanisms and pointing to their drawbacks that require more investigations.

2.1 Anaerobic Contact Digestion (ACD)

ACD is considered kind of anaerobic digester. ACD comprises of sedimentation tank and digester where digester sludge is discharged to the sedimentation tank to settle, while the effluent is pumped to the digester. It has been occupied for POME, fermented olive mill wastewater, alcohol distillery wastewater, and ice-cream wastewater [27–29]. If the digester and the sedimentation tank are not combined, it is named as conventional method, but if they are combined, it is termed as non-conventional system. ACD treatment is considered an appropriate method for highly contaminated wastewater remediation. ACD is stable, and quick start treatment compared to anaerobic filter method because it delivers sufficient mixing [29,30]. During using pilot plant ACD for POME treatment, a scum layer was produced [27]. Although, it delivered about 80% COD reduction. Despite that, it was found requiring long solid retention time (SRT) and can process only half of the influent.

2.2 Microbial Fuel Cell (MFC)

Microorganisms oxidize organic solids within oxygen existence in the anodic chamber which produce electrons and protons that travel to the cathode side [31]. In another definition, MFC is a biochemical device that utilizes microorganisms as biological tool to transform organic solids' chemical energy (e.g. glucose) into electricity [32,33]. Usually, anode and cathode sections are separated by membrane layer (e.g. proton exchange membrane) or salt bridge which permits proton (H^+) transfer from the anode to the cathode side to form water by combining with oxygen [34]. Studies had reported on electricity generation, microbial communities, electrodes materials, degradation process, and membrane material in MFC treatment [31,35]. Its worth mentioning that cathode, anode, and membrane materials are still expensive. MFC has been required to have good scalability, low cost membrane, high quality electrodes, and high electricity generation to suit human demand of energy. Despite that, most of the anode materials are built up from high resistivity carbon which largely contributes in wasting energy [36–38]. Luckily, using Si, Ni, TiO_2, and Cu as edged metals for producing low resistivity composite based carbon as MFC anode electrode may overcome the negative factors. Currently, carbon felt, carbon paper, carbon cloth, graphite fiber brush, and graphite rod are the common electrodes for MFC, but carbon felt is the most preferred.

Many sorts of microorganisms were investigated inside MFC treatment such as *Shewanella Putrefaciens*, *Pseudomonas Aeruginosa*, and *Geobacter Metallireducens*. Electrogens such as *Shewanella Putrefaciens*, and *Geobacter Metallireducens* which are pure cultures able to transport electrons for current generation might be less or more than mixed culture. Hence, organic converting process into electricity is made in MFC by electrogens (pure cultures) and mixed bacteria cultures [32,39–42]. Electrical conductive pili, electron mediators, and direct outer membrane c-type cytochrome transfer are the common ways of electron transfer in MFC by extracellular electron transfer mechanism [43,44].

Attachment process of microbes with MFC electrode is important to achieve stable electricity generation. Communications are performed among microorganisms during immobilization stage at the biofilm or extracellular polymeric substance (EPS). High energy conversion, efficient operation during low and ambient temperatures, and no aeration is required are the major advantages of using MFC [45,46]. It has been utilized for textile wastewater containing azo dyes, mustard tuber wastewater, chocolate industry wastewater, starch processing wastewater, food processing wastewater, domestic wastewater, and paper recycling wastewater [47–54]. POME is an appropriate substrate for MFC due high COD, and BOD concentrations. A study achieved 45 Mw/m^2 power generation and 45% COD reduction during 15 days of employing POME as influent for double chamber-MFC [55]. Another research achieved 3004 Mw/m^2 power generation from using synthetic wastewater containing acetate in double chamber-MFC, while it produced only 622 Mw/m^2 from utilizing POME [56].

MFC has been integrated with other technologies to avoid system weaknesses like Up flow membrane less microbial fuel cell (UML-MFC) combination. Successfully, UML-MFC delivered higher microorganisms removal than using anaerobic digestion system [57]. In addition, ammoniacal nitrogen, and COD removal were 93.6, and 96.5%, respectively. The main drawbacks of scaling up MFC are limited power output, membrane fouling, high operation cost, and high internal resistance which make MFC being long-term Laboratory scale [58].

2.3 Anaerobic Ponds (AnPs)

AnPs are neither mixed nor aerated ponds, they are structured to process high organic loading rates with lack of dissolved oxygen (DO). It is a traditional way employed for wastewater treatment sludge digestion, especially in the tropical countries [59]. AnPs are not installed in low or moderate temperature countries because it requires a long time and ends up in delivering unsuccessful digestion and low gas generation. Strong wastewater containing high pollutants amount is fed to AnPs for solids settlement and BOD reduction. Anaerobically, solids are digested and settled at the pond bottom, while the partially clarified supernatant liquor is pumped to another treatment stage. Methanogenic (methane production) and acid-forming bacteria (acid formation) are related phases working groups in AnP to deliver biological degradation. Biodegradation process leads to CH_4 and CO_2 emission by anaerobic microbes. Complex organic solids are being transformed at the acid phase by acid forming bacteria to short-chain volatile organic acids. Then, the bacteria utilise short chain organic acids to produce CO_2, hydrogen gas, and acetate. Finally, methanogenic bacteria transform the final components to methane, as shown in equation 1, and 2.

$$CH_3COOH \rightarrow CH_4 + CO_2 \tag{1}$$

$$CO_2 + 4H_2 \rightarrow CH_4 + 2H_2O \tag{2}$$

Population and activity balance is important to accomplish high performance of wastewater pollutants degradation as dynamic equilibrium. Also, pH over 6, and temperature higher than 15 °C are the main circumstances impacting the biological activities [23]. Therefore, the optimal pH and temperature ranges were 6.6-7.6, and 25-40 °C, respectively [60]. Furthermore, it was reported that sludge amount is minimal due to high-performance bioactivity. Organic molecules reduction, sludge digestion, and solids settlement are achievable through using longer hydraulic retention time (HRT). Usually, AnPs operation should be within 5-50 days retention time, 2-4 meters depth, and 2 m/d hydraulic surface loading [24,61]. It is expected to deliver 80-90% reduction of BOD5, require none or low energy, generate energy, be high flexible, accumulate low sludge, handle high organic loading rate, and be long lifetime operation [60]. Despite that, AnPs produce hydrogen sulphide, odorous compounds, and require maintenance.

2.4 Up Flow Anaerobic Sludge Blanket Reactor (UASB)

Anaerobic treatments became primary treatment among other methods for domestic, mixed, high strength i.e. contains great amount of organic compounds, oil and grease, heavy metals, ammonia, and fats (e.g. industrial, agro-industrial, sugar beet, coffee, pharmaceutical, POME, ice cream, slaughterhouse, and potato wastewater) [62–70]. Frequently, UASBs have been installed to vastly reduce suspended solids (SS), and organic loading ranges. In year 1970, UASB reactors were developed to become high rate anaerobic digesters for tropical and subtropical countries such as India, Colombia, and Brazil [71]. It is worth mentioning that UASB reactors are still not recommended for non-tropical regions due to limited digestion, slow biodegradation, sludge deterioration, and sludge flotation unless there are suspended solids pretreatment and two stages of anaerobic treatment [72-73].

UASB treats wastewater within 6-12 h of retention time by distributing wastewater influent across the reactor base, then flowing through the sludge layer to the upper level to assure high contact performance between the anaerobic bacteria and the wastewater pollutants (see Figure 1). UASB reactor is separated into upper (settling) and lower (digestion) zone. Biogas bubbles are produced due to the biodegradation process, where UASB's deflectors prevent delaying in sedimentation process by stopping gas bubbles entering the settling layer. Also, phase separators in UASB reactor are employed for gas collecting. About 0.2 kg sludge is generated for every 1.0 kg reduced of BOD which is higher than anaerobic ponds, and less than conventional activated sludge treatment. In some cases, UASB reactor having low HRT (2-5 h) can be used to prevent methanogenesis reaction, its named as hydrolytic up-flow sludge blankets (HUSB). Despite that, hydrolysis process requires over 15 days to deliver high performance in UASB reactors.

At 25 °C, AnP can remove 70% BOD in one-day retention time, while UASB reactor produces 70% BOD removal in 6 h. Despite that, UASB reactors are considered smaller and higher cost than AnPs, but AnPs generate odours. Additionally, UASB reactors suffer from sludge granulation, long startup time, methane production, reactor clog and biomass aggregation [65,68,74,75]. Although, it has been recommended to implement toxic and inhibition detection, organic loading rate determination, and survey sludge development to prevent sludge accumulation.

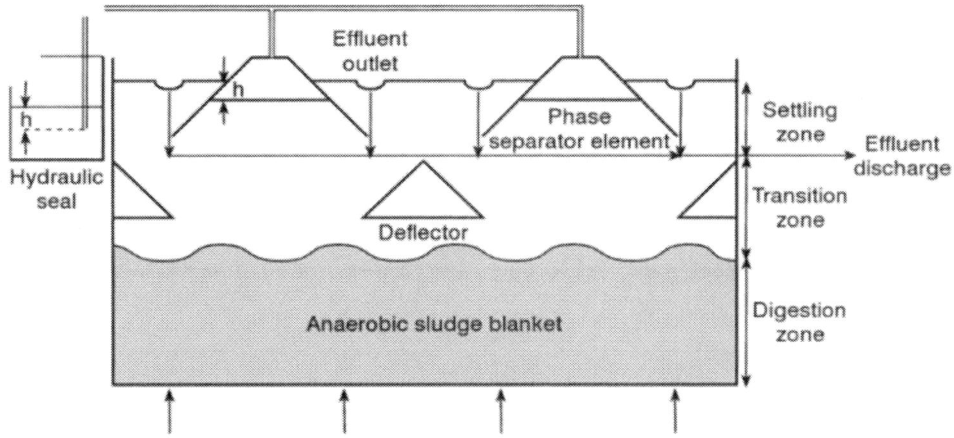

Figure 1. Schematic of UASB reactor [4]

A study proved that UASB reactors deliver more performance in removing suspended solids, COD, and organic solids than settling tanks. Also, two stages UASB system delivered 71% COD removal at 15 °C. About 98.4% COD removal was achieved by treating POME through UASB reactor with 10.63 kg COD/m^3 day organic loading rate (OLR) [64]. Although, treating wastewater containing high volatile fatty acid overloads and destabilizes the UASB reactor after 15 days of processing. Hence, UASB structure is endorsed to be fitted with high organic loading rates, and capable of processing POME effluent. For POME processing, a researcher investigated granule formation from high OLR by two-staged UASB reactor [76]. Hence, the study showed an importance to employ a pair of UASB reactors for segregating acidogenesis, and methanogenesis stages. Also, it was reported that high methane conversion with 90% COD removal is possible for 30 kg COD/m^3 day of OLR. Also, another study proved that reactors seeding process with granulated sludge deliver short startup time with high reduction performance, and require less time to process gradual rise in influent OLR [74,77].

2.5 Up Flow Anaerobic Sludge Fixed Bed Reactor (UASFF)

UASFF is a combined system between the anaerobic filter and UASB reactor. It is designed to achieve high performance, eliminate drawbacks, and combine benefits. Many advantages are reported by using UASFF such as clog elimination, prevent biomass washout, handling high OLR, stability, and excellent biomass retention [30]. A study proved that using UASB or anaerobic filter provide lower performance than UASFF for wood fibre wastewater [78]. Also, UASFF was examined for POME, brewery, coffee, virgin olive oil, slaughterhouse, dairy, and sugar wastewater [79–86]. It was reported that methane emission from UASFF is acceptable, and the removal efficiency is at least 70% of COD concentration except wood fibre wastewater because it is considered difficult to be composed. The stability of UASFF for POME processing is decided by the internal packing and effluent recycle ratio [79].

2.6 Anaerobic Baffled Bioreactor (ABR)

The development of ABR has been initiated in the 1980s. It is applied widely for wastewater due to having an anaerobic filter, low bacterial washout, up-flow anaerobic sludge blanket, no fixed media required, low up flow liquid velocity, high contact time, high stability towards toxic, no requirement for particular sludge or gas separator, and low energy required [87,88]. Also, ABR is presented as series of Up flow anaerobic sludge blanket (UASB) reactors [89]. ABRs behave partially as trickling submerged fixed, activated sludge reactor, UASB reactor, septic tanks, as well as fluidised bed reactor. ABRs comprise from series of baffles and compartments where the influent flows under and over.

The significant attention towards ABRs is to produce capable bioreactors of high capacity of solids retention. Therefore, vertical baffled to a plug flow was illustrated for treating high solids slurry [90]. The design of ABR provides sufficient time for influents to be decomposed and discharge acceptable effluents. Well-structured and designed ABR brings a considerable contact between microorganisms and substrates in short time and deliver high removal rate without occupying massive volumes. A series of vertical baffles force the influent to move over and under which helps microorganisms to thrive (see Figure 2).

Vertical ABRs presented good ability in retaining biomass and methane gas generation. Upflow chamber is the Methanogenesis zone, while the downflow chamber is acidification zone. Compared to continuous stirred tank reactors (CSTRs), less ABR's volume can achieve higher reaction rate per unit reactor volume by strongly retaining the biomass separately in the reactor from the inlet wastewater [91]. The physical design of ABR makes it capable of treating contaminated water in one unit with using circulation pattern and requires low capital costs. Figure 2 shows ABR structure and procedure for wastewater treatment. Unlike UASB, ABR does not require granulation technique to increase the bioreactor efficacy.

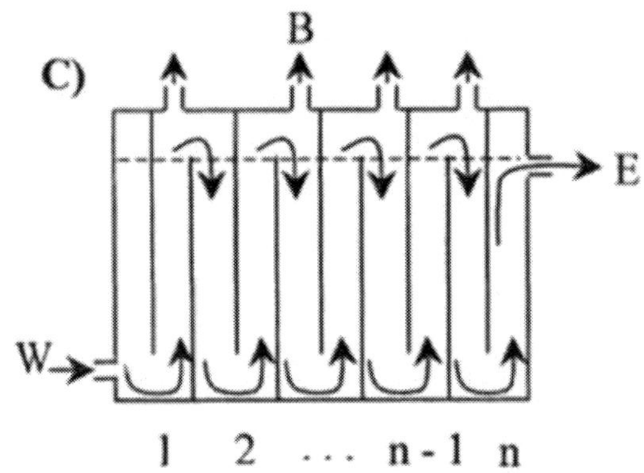

Figure 2. Vertical Baffled Reactor (W= wastewater, B= biogas, E= effluent) [92]

In year 1987, hybrid ABRs (HABRs) were introduced. HABRs involve suspended and attached growth initiated by boosting flocculent and granular biomass growth by using 0.46 m/h of liquid up-flow velocities, 0.97 kg COD/kg VSS d of low initial loading rate, and 4.01 g VSS/l [93]. A stable formation of 0.5 mm granules presented in all HABRs after the one-month duration. Despite that, the flocs were found weak and less than 1.5 mm, but it reached 3.5 mm after three months and it is determined by the substrate kind. Acetoclastic methanogens (Methanosarcina cluster) were the granules content and it raises to the reactor surface because it is full of gas cavities and being low-density.

ABR got a further modification from Boopathy and Sievers [94] for treating swine wastewater containing a high range of small particulate materials. Another study occurred on ABR modification by using two compartments, where the first section size was 10 L, and the second section was half the first compartment size. It was recorded that the three-chamber ABR collected half the solid amount (10.45 g/L) of the two compartments ABR. Although, three-chamber ABR showed higher treatment efficiency and solid washout than two compartment ABR [94]. COD and BOD removal were 90% at the up-flow compartments [95–97]. Laboratory scale ABR was examined for slaughterhouse wastewater (SWW) characterised by 63.38 mg/l of TN, and 183.35 mg/l of total organic carbon (TOC) [98]. The highest reduction percentage was 51.52% of TN and 88.88% of TOC. It was found that operating costs increase with TOC range, according to Bustillo-Lecompte et al. [26]. Another research reported about ABR and UV/H_2O_2 combination for treating SWW containing 973 mg/l of TOC in laboratory scale [99]. It is concluded that using hybrid system delivers higher removal efficiency than employing an individual method. After three days of treatment, the combined system delivered 95% TOC reduction. Table 2 presents ABR treatment for various sorts of wastewater.

Table 2. Anaerobic baffled bioreactor treatment for different sorts of wastewater [92]

Wastewater	HRT (h)	Influent COD (mg/L)	COD Removal (%)	OLR (kg/m^3.d)	Gas produced (L/day)
Dilute wastewater	80	50	80	-	
Brewery wastewater	15	-	92	5.6	24.01
Soybean wastewater	39.5	2000	97	1.2	-
Heavy oil produced water	60	50	65	-	-
Domestic wastewater	48	305.18	74	-	-
Municipal wastewater	6	350	86	2.62	0.34
Palm oil mill wastewater	3	16000	77.31	1.60	27.4
Undiluted brewery wastewater	15	5500	31.82	2.75	55.7
Synthetic substrate wastewater	24	3000	82	3.00	-
Pulp and paper mill black liquor	48	4020	68	5.00	2.95
Penicillin reduction wastewater	64	8.00	65	2.64	-
Domestic wastewater	22	716	72	-	-
High sulphate wastewater	240	6.6	82.71	0.66	0.29
Low strength wastewater	12	550	89	1.69	-
Soybean protein processing wastewater	39.5	10000	97	6	-
Complexed wastewater	8	500	88	2	0.31
Synthetic wastewater	10	501	90.7	1.2	0.36

2.7 *Anaerobic Fluidised Bed Reactor (AFBR)*

AFBR is a reactor that can be occupied to handle different sorts of multiphase chemical reactions. The operation of AFBR has minimal issues in gas hold up, plugging, and channelling matters [8,100,101]. Moreover, it possesses large surface area and requires low HRT to handle high OLR, and high polluted wastewater [100–103]. Anaerobic expanded bed development got initiated to convert diluted organic wastes to methane at high hydraulic and organic loading rates, and low temperatures. During year 1988, anaerobic expanded bed followed by post-treatment was scaled up into pilot scale 10,000 gal/day and studied. It was reported that the assessment occurred for two years on two AFBR, where sand was the carrier in the first reactor, and the second reactor used granular activated carbon (GAC) [104]. It was shown that using sand as carrier delivers less attached biomass with slow biofilm growth and higher BOD removal.

Contact time between the inlet and the bed is deciding by influent up-flow velocity. Therefore, attachment and growth of biomass will occur on the reactor support media [30]. AFBR treatment efficiency can be manipulated by bed material sort that can impact the whole process [102,103]. The performance of AFBR has been examined to treat POME, pharmaceutical effluent, slaughterhouse wastewater, ice-cream wastewater, brewery wastewater, wine and distillery wastewater, real textile wastewater, and cutting-oil wastewater [63,100,102,103,105–110]. In comparison to up flow configuration, a study showed that inverse flow AFBR is capable of high OLRs, with high stability even at overload situations [107]. Even AFBRs are capable of high OLRs, it is preferred to occupy less AFBR size for lower OLRs [30].

A study combined ovoid saponite with AFBR. It resulted 94.4% COD removal, while using granular activated carbon produced 60% COD reduction [8]. AFBR has been determined as a suitable method for POME mediation process than anaerobic filters because of the good strength to treat high OLRs. Also, shorter HRT 6 h is required for AFBR to mediate POME than anaerobic filters HRT which is 1.5 to 4.5 days [30]. AFBR treatment for different wastewater kinds with COD removal and OLR ranges are tabulated in below. From the organised data in the table, it declared that 65-95% is the COD removal range by AFBR.

IOP Conf. Series: Materials Science and Engineering **702** (2019) 012058 doi:10.1088/1757-899X/702/1/012058

2.8 Continuous Stirred Tank Reactor (CSTR)

CSTR is a closed tank acts as a digester with a mixer. It has been occupied to treat coke wastewater, jam wastewater, and dilute diary wastewater, but in some cases, it was operated under aerobic circumstances [111–113]. CSTR operation is a continuous flow of outlets, and inlets. Gas generation through the biological activity increases due to agitation process which delivers high contact in short time.

CSTR capability was investigated for removing pollutants from POME [114]. It produced effluent with range 93.6 to 97.7% of COD reduction. In Masai, Malaysia, Keck Seng Berhad utilised CSTR as a mill for POME mediation [115]. It delivered 83% COD removal while occupying CSTR for dairy wastewater resulted 60% COD reduction. Keck Seng Berhad got different results from [114].

Operating conditions like temperature have high impact on wastewater treatment performance. Another factor known as insufficient mixing properties led Keck Seng Berhad to receive low COD removal. Another research worked on combining CSTR with biofilm support system (BSS) [116]. BSS was implemented by using low-density nylon mesh as supporting material for biomass growth. The hybrid system proved high ability towards discharging effluent with lower COD ranges.

2.9 Constructed Wetlands

Wetlands are usually land areas which are kept wet during part or all of the times. They are shallow earthen tanks involve a soil layer for plants growth which performs attached growth biological wastewater mediation. Often, wetlands change among deeply flooded or uplands and continuously systems [117]. It can treat industrial wastewater, mine wastewater, animal wastewater, domestic wastewater, and municipal wastewater.

Historically, natural wetlands have received wastewater discharge as a convenient method for waste treatment in southeastern United States coastal plain areas, and poorly drained fens and marshes of the north. For instance, Florida cypress dones, or Michigan Houghton Lake fen were extensively examined and recognised as a treatment method for wastewater [118]. During the 1950s, the first practices of wetland vegetation were performed to reduce wastewater contamination, while the first full-scale free water surface wetland was structured in between 1967 to 1969 in the Netherlands for camping site wastewater treatment. An advantageous development occurred on wetlands performance by occupying coarse materials (washed gravel) instead of wetlands soil in the late 1980s [119]. Also, during the 1980s, a thorough understanding had been made for improving numerous wetland benefits and weaknesses. Hence, it got strong global attention as a robust technology for wastewater mediation [118]. Then, vertical flow constructed wetlands have been employed in the 1990s as a significant process for ammonia and nitrogen reduction since it provides massive oxygenation for the nitrification process. After few years, it showed the importance of using a hybrid system involving constructed wetlands to accomplish nitrification and denitrification process [119].

In the northern and the central part of Europe, constructed wetlands were found popularly employed for thirty years in wastewater treatment plants for small populated cities (see Figure 3). Developing countries consider constructed wetlands more attractive method in southern and northern countries than conventional wastewater techniques. In addition, constructed wetlands are considered flexible, low operation duty, less susceptible to OLR variation, low sludge formation, low maintenance and construction cost, and require low experienced workers than conventional method [20,21,120,121]. Despite that, employing massive land area, and low performance at low temperatures are considered significant drawbacks, but it is still a concrete way for tropical regions [121,122].

Constructed wetlands are considered unique and ecofriendly among other technologies [117]. It was reported that full scale constructed subsurface flow wetland able to achieve 30% organic matter reduction [95,123]. Also, the enforced regulations and laws are not satisfied with the discharged effluent quality by constructed wetlands where TSS, COD, and BOD5 reduction percentage are about 85, 89, and 91%, respectively. Constructed wetland efficacy for mediating slaughterhouse wastewater (SWW) was investigated by Soroko[124] via utilising three basins, where the first one was horizontal flow (HFCW), and the other two were vertical flow (VFCW). They produced 78.20, 99.90, and 97.40% of TN, BOD, and COD removal with using gravel and sand beds, where SWW influent contained 500, 2500, and 3188 mg/l of TN, BOD, and COD. Typha latifolia was combined with constructed wetland which had 89% active volume and 111 days HRT for SWW treatment [125]. The hybrid system accomplished 87, 88, 72, and 95% removal performances of TN, TP, TSS, and BOD, respectively. Differently constructed wetlands were examined by [126] for SWW treatment. SWW influent was characterised by 56-64, 79-87, and 293-314 mg/l of TN, BOD, and COD, respectively. It delivered 5.20-25.40, 9.27-71.40, and 28.28-75.03% removal ranges of TN, BOD, and COD, respectively [126].

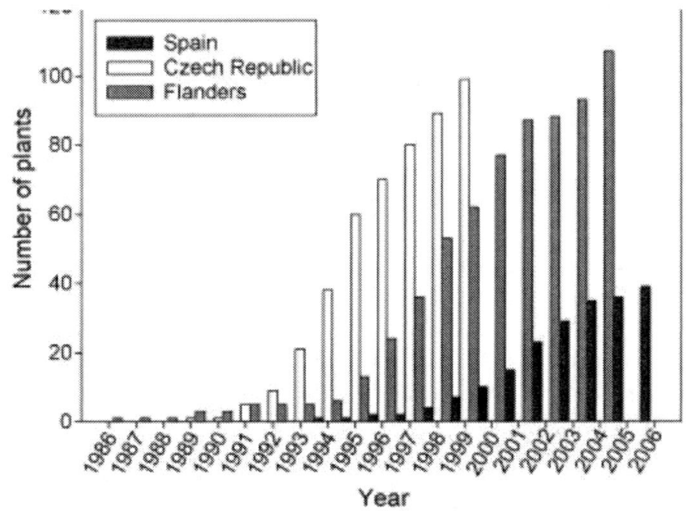

Figure 3. Cumulative number of WW treatment plants based in CWs over the last years in various European regions [127]

2.10 Anaerobic Filters

Anaerobic digestion has been utilised for wastewater treatment since over 100 years ago. The development started from airtight vessel and septic tank to full mixed digester equipped with temperature control and ended as high rate biological reactor containing active biomass. The advancement of anaerobic wastewater digestion led to variety of changes such as growth technique, fluidised bed (attachment on mobilised carriers), and anaerobic filter (attachment on static carriers) [104]. To date, all these advancements are still in developing stage.

The anaerobic filters were employed to treat various types of wastewater such as ice-cream manufacture wastewater, beet sugar water, distillery wastewater, drug wastewater, slaughterhouse wastewater, brewery wastewater, municipal wastewater, landfill leachate, wine vinases, and soybean processing wastewater [104]. It is preferred in some cases because it delivers high pollutants reduction, handles high loadings, produces high contact rate between liquid medium and biomass without impacting the process performance, needs inexpensive construction, requires short hydraulic retention time, and demands small reactor volume [8]. Although, clogging is a significant drawback for continuous run anaerobic filters [128–130]. During processing POME with 20 g COD/l/day of OLR, and processing SWW with 6 g COD/l/day of organic loading rate OLR, the anaerobic filter got clogging. Anaerobic filters utilise the packing surface for biomass attachment, development, and growth. Influent enters the process from the bioreactor bottom, while the generated biogas and the effluent discharge from the top.

POME was processed by the anaerobic filter, according to Borja and Banks [110,131]. Anaerobic filters were found able to discharge effluents with at least 70% COD removal [30]. Empirical examinations on improving anaerobic filters efficacy got occurred. For instance, it was reported that optimal recycle ratio differs based on OLR range which will produce enhanced COD removal and biomass high retention time. Thus, increasing the optimal recycle ratio delivers higher methane amount [132]. Open-pored support media which involve high porosity were suggested by Bala et al. [8] for biomass retention optimisation by trapping it. Moreover, continuous fed anaerobic filters were found able to accomplish great biological decomposition, and stability [133]. The highest record of POME

COD removal was 94% with 63% of methane at 4.5 kg COD/m^3/day of OLR, while 90% was overall COD removal with 60% of methane gas composition [131].

2.11 Membrane Bioreactor (MBR)

Between the late 1960's and early 1970's, membrane bioreactor (MBR) was introduced for processing wastewater. Then, it faced massive evolution and extensive usage since the early 1990s. In Mansfield, Ohio, United States, the first large MBR was structured. Also, in the 1990s, the first submerged MBR was built, and it was found requiring anaerobic or aerobic circumstances, and lower operational cost to deliver higher active biomass settling, higher mixed liquor, higher suspended solids concentration and efficient longer processing in comparison to other MBR types [134]. Later, the first large-scale internal MBR system was built during 1998 in North America for industrial food wastewater treatment [135]. However, high capital cost and difficult maintenance reduced the overall research on MBR.

MBR process comprises from biological treatment (aerobic or anaerobic suspended growth) and membrane separation (like ultrafiltration, and microfiltration) [136]. The membrane layer is a thin film porous structure employed as a filter which rejects anything larger than the pores open. The pressure difference between membrane layer (transmembrane pressure) generates potential energy allowing soluble components and treated wastewater to pass, while particulate matters (organic compounds, graphite, glass, metals, ceramics) are retained based on membrane pore size, and membrane structure. Activated sludge process is part of MBR treatment where microporous semipermeable pressure-driven rejection membranes are retaining biomass on their surfaces.

There are two types of MBR, the first type includes immersed MBR in Bioreactor, and they co-occur, while the second type recirculates part of the effluent for the filtration system after the biological treatment, while the rest is pumped back to the bioreactor for extra treatment. Thus, it is a high-cost operation due to the enormous need of efficient pumping system [137].

During membrane process, multiple mechanisms are happening in removing antibiotics. At the initial filtration stages, adsorption can occur by membranes to the hydrophobic antibiotics (who are possessing strong hydrogen bonding). In the other side, stable rejection can deliver removal process due to uncharged solutes steric effects or charged solutes combined electrostatic and steric effects. These mechanisms are varying based on the membrane features (pore size, surface morphology, material), the solution (ionic strength, pH), and the compound (hydrophilicity/hydrophobicity, pKa, molecular weight cut-off (MWCO)) [138]. During wastewater treatment, reverse osmosis (Pore size > 0.001 mm), and nanofiltration (Pore size less or equal 0.0001 mm) are substantial membranes in removing low molecular weight (pharmaceutical, antibiotics), while ultrafiltration and microfiltration are large enough to be micropollutants passages (see table 3). Additionally, many examinations presented antibiotics removal involving trimethoprim, tetracyclines, sulfonamides, and quinolones. Hence, there is a demand to combine reverse osmosis (RO) with nanofiltration (NF) membranes to achieve overall high efficacy [139,140]. Fouling (By inorganic matter, SMP, EPS) can lead to rejection improvement of micropollutants due to the transforming to negative surface charge which contributes the ionic species electrostatic rejection [141,142]. Hence, non-ionic solutes adsorptive capacity has increased eventually.

MBR biological treatment is similar to the conventional wastewater treatments. For instance, cell growth, metabolism, nitrification, denitrification, and phosphorus removal are occurring inside MBR system. However, MBR system permits accumulation of better nitrification rates, very high sludge volume index, lower sludge production, high amounts of predators like metazoan, and protozoa, high biomass concentration, and slow-growing microorganisms [134,143]. It was proposed that metazoan, and protozoa gazing on the nitrifying bacteria can decrease MBR nitrifying capacities.

MBR nitrification and denitrification rates are affected by sludge age. Previously, it was shown that the removal efficiency of total nitrogen increased from 49 to 73% by increasing sludge age from 20 to 60 days during treating black water [144]. Also, the treatment occurred under low COD/nitrogen ratio, high nitrogen amount, and low dissolved oxygen (0.1-0.2 mg/l). The process delivered 100% reduction during denitrification process, and 40% reduction within nitrification process. In contrast, another researcher found the different outcome in decreasing sludge age from 29 to 16 days which resulted higher total nitrogen removal from 89 to 91%. The study had occurred by three stages anaerobic anoxic and aerobic system of the pilot plant [145].

IOP Conf. Series: Materials Science and Engineering **702** (2019) 012058 doi:10.1088/1757-899X/702/1/012058

Pre-denitrification is universal wastewater system for biological nitrogen removal. It includes anaerobic and anoxic tank followed by aeration tank for nitrification process. The aeration process generates nitrate which is pumped back to the anoxic tank for denitrification. Then, the provided electron donors are employed with the nitrate to achieve denitrification. Sometimes, different recirculation rates are used to improve MBR performance.

Traditionally, MBRs are used in conventional activated sludge process (CASP) as a replacement for the final sedimentation process or the secondary clarifier [146]. For that, micro or ultrafiltration MBR maintain higher concentration of mixed liquor suspended solids (MLSS) compared to conventional activated sludge process (ASPs). It resulted 99% solids removal and almost complete clarification. Additionally, semipermeable MBR is capable of being a good barrier against bacterial cells and colloids. It was reported 2-5 log removals of human enteric viruses, and 2-7 log removals of coliform bacteria for different range membranes and MBR [147,148]. In conclusion, operating conditions (gel layer formation), pore size, membrane type, and membrane material are essential factors in MBR performance against colloidal, viruses, and bacteria retention.

The produced water can be used as an inlet for heat integration, landscape purposes, and industrial sanitary. Hence, A researcher had examined ultrafiltration MBR efficacy against organics and nutrients removal from SWW [149]. SWW was characterised by having 102, 16, and 571 mg/l of TN, TP, and COD, respectively. The process resulted 97, 96, 65, and 44% removal of COD, TOC, TP, and TN, respectively. Also, UF MBR produced a successful organics reduction, but high nitrate amount passed with the effluent. For that, the produced effluent requires a complete denitrification.

Table 3. Comparison of different membrane dimensions and pore size exclusion used in SWW treatment [95].

Membrane type	Pore size (mm)	TOC removal (%)	COD removal (%)	BOD removal (%)	TN removal (%)
Microfiltration (MF)	0.080-0.550	44.81	90.63	-	45.22
Ultrafiltration (UF)	0.030	75.00-96.00	83.00-97.00	-	27e44
Ultrafiltration (UF)	0.010-0.100	-	94.52-94.74	97.80-97.89	-
Reverse Osmosis (RO)	0.001-0.005	-	85.80	50.00	90.00

2.12 Membrane Bioelectrochemical Reactor (MBERs)

MBERs are combined system involving Microbial fuel cell (MFC) and Membrane filtration (MF or UF). The membrane system can be installed within either the cathodic or anodic compartment. MBERs Hybrid treatment could be a remarkable option for wastewater treatment evolution and competitive among conventional aerobic MBRs (AeMBRs) or anaerobic MBRs with successful nutrients removal, low dissolved methane, no aeration, minimal maintenance requirement, and low energy requirement [150]. Additionally, scaling up (over 1 m^3), construction cost (like ion exchange membrane, and catalyst), discharging high suspended solid concentration, no effective treatment against nutrients (like nitrogen, and phosphorus), and extended treatment period (11 h are required to achieve 65-70% degradation) are considered weaknesses for Bioelectrochemical systems (BES). In processing wastewater, MFCs approach is an advantageous method. Also, MFC-MBR integration is an approach for sustainable wastewater treatment. Cathodic and anionic compartments are semi-separated by involving anion exchange membrane (AEM). Its named as microbial desalination cell (MDC), if used as desalination solution.

The bioelectrochemical concept can be defined as the transformation of organic chemical energy to hydrogen gas or bioelectrical energy by microbial electrolysis cells (MECs) or MFCs. MFC does not require aeration system, for that sludge production, is less than CAS process [45]. Biodegradation is occurring for organic matters in the anodic MFC section. Then, electrons are generated and transferred to a solid electrode through NADH/NAD+, then to the serial cytochrome agents at the outer cell membrane. Finally, ferricyanide, nitrate, or oxygen in the cathodic compartment are accepting the transferred electrons through an external circuit.

If the theoretical open circuit voltage (OCV) is about 1.1 V, the carbon source is sodium acetate in the anodic compartment, and oxygen is employed as an electron acceptor in the cathodic chamber, then 0.8 V was observed as the low potential due ohmic loss, electrolyte diffusion resistance, and overpotential.

It was found a significant reduction in fouling matter via using granular activated carbon (fluidized bed support material), and membrane in the cathodic section. Also, AEM film can be employed for nutrients removal and make MBER versatile process.

MFC ability to direct transfer of chemical energy into bioelectrical energy with flexibility and low carbon footprint has shortened long way of using treatments for biodiesel production (see figure 4). Also, applying an external voltage inside MFC leads to gaseous hydrogen evolution or hydrogen peroxide in the cathodic compartment, known as MEC process.

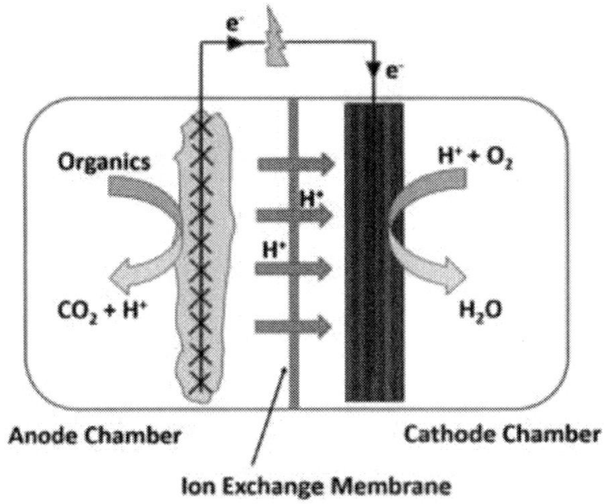

Figure 4. schematic of two chambers MFC

2.13 *Fungi*

Varying circumstances are not huge matter for fungal organisms because it has significant capability for augmenting their metabolism based on the environmental conditions. Their presence is vigorously reliant on metabolism climatisation feature. Also, fungal metabolic activity is supported by extra and intracellular enzymes.

Textile wastewater treatment starts with dyes decomposition to another chemically forms by fungal enzymes. Laccase, manganese peroxidase (MnP), and lignin peroxidase (LiP) are utilised enzymes by fungal organisms for biodegradation process [151]. Also, the used fungal species for azo dyes decomposition is the white rot fungal cultures. Hence, wastewater COD concentration faced a removal process with dyes decomposition by white rot fungus *Pleurotus eryngii, Penicillium simplicissimum,* and White rot fungi *Coriolopsis sp* [152–154]. Though, long holding time, unstable treatment, large reactor, unreliable enzyme generation, nitrogen necessity, and long growth are a significant drawback of wastewater fungal treatment [155]. Also, after 20 to 30 days of fungal wastewater treatment, bacteria dominate the medium by their fast growth while fungi predominate wastewater medium with idle dyes degradation [156]. Despite that, it was found that *azoreductase* enzyme can provoke Green macroalgae *Cladophora* species to degrade wastewater azo dyes [157-158].

2.14 *Anaerobic Digestion*

Its degradation process of wastewater organic compounds under anoxic or anaerobic circumstances. Also, it can be defined as the engineered methanogenic anaerobic decomposition of organic matter. In 1859, the first anaerobic digester was built in Bombay, India by leper colony. Then, sludge processing and sedimentation occurred with a dual basin in 1904 [159]. Initially, anaerobic digestion process was used in treating sludge, industrial wastewater, and municipal by organic matter decomposition approach. Time is the matter with high or standard rate anaerobic digestion process because the responsible bacteria consortia and other microorganisms require an undetermined period for the bioaugmentation, adaption, and organic matter biodegradation [8,104]. Anaerobic digestion requires long retention time, stable operation, and large reactor size to accomplish full digestion process (see table 4). Hence, it was proposed to develop high rate anaerobic bioreactors via biogas capturing, delivering shorter retention time, and employing less reactor volume [104]. Standard rate anaerobic digestion is processing wastewater with no requirement for mixing or heat, while high rate digester requires heating and mixing. As a result of that, high rate digestion process needs 15 days or less of hydraulic retention time (HRT), while standard process requires 30 to 60 days [160].

Hydrolysis, acidogenesis (including acetogenesis), and methanogenesis are theoretical biological digestion phases of POME, which discharge treated water, carbon dioxide, and methane [161]. First of all, proteins, lipids, carbohydrates, nucleic acids and other complex molecules (high molecular mass) are processed into substrates such as amino acid, fatty acids, simple carbohydrates, purines and pyrimidines, and sugar by Hydrolysis process. Then, lower molecular mass like organic acids, lactate, succinate, methylamine, acetate, carbon dioxide, hydrogen gas, methanol, and other fermentation products are produced through utilizing acidogenic bacteria via breaking amino acids, fatty acids, sugar, and other compounds where carbon dioxide, hydrogen, and acetic acid (from acetogenesis) are main components of the generated organic acids. Finally, acetolactic methanogens utilise carbon dioxide and acetic acid for methane production as the final product, while carbon dioxide and hydrogen are metabolised by hydrogenotrophic [95,161]. In comparison with alternative technologies, anaerobic digestion process claims to be cost-effective, significant waste stabilisation, low energy requirement, high COD removal, and low sludge generation for POME treatment, also it does not require aeration system. Additionally, methane gas is generated during the anaerobic digestion process, which can be captured, and utilised for additional worthy revenue. In the seventeenth century, biogas from organic matter biodegradation got scientific attention.

Table 4. Comparison between anaerobic and aerobic digestion [21].

Characteristic	**Aerobic**	**Anaerobic**
Organic loading rate	Moderate	High
Organic removal efficiency	High	High
Nutrient requirement	High	Low
Energy requirement	High	High
Alkalinity requirement	Low	Low to moderate
Sludge production	High	Low
Temperature sensitivity	low	High
Odor production	Less opportunity for odours	Potential odour problems
Bioenergy and nutrient	No	Yes
Startup time	2-4 weeks	2-4 months

It was reported that POME and high strength wastewater (high organic carbon concentration) could be processed under anaerobic digestion method [76,99].

2.15 Biological Activated Carbon (BAC) Filtration

BAC filtration involves biodegradation process and sorption for removing CECs (Recognized as potential hazardous) [162,163]. Microorganisms development and growth are supported by a granular activated carbon (GAC) fixed bed. Eventually, bacteria initiate their evolution on the bed surface.

Previously, BAC presented a bioremediation process by biodegradation and sorption course on secondary treated wastewater where nitrogen and dissolved organic carbon (DOC) concentrations were decreased [162]. Treatment procedure starts with sorption process by activated carbon at the beginning, then biodegradation process dominants the medium by the attached bacteria on the GAC bed. According to Neptune project, biologically activated coke was utilised as sorbent material with the fixed bed system in BAC filtration succeeding to conventional bioactivated sludge treatment. It produced 70 to 90% removal rates for numerous pharmaceutical components including the hardly degradable compounds, such as diclofenac, and carbamazepine (unpublished data, http://www.aqua-biocarbon.de/aktuelles.html). After ozonation treatment, BAC was employed instead of UV treatment for removing NDMA, thus its cost-effective process [164]. Though, it is recommended to check transformation products (TPs) rate to stop their passing through activated carbon (AC) filter [165]. In comparing to many wastewater technologies, AC produces no TPs because of its dependent on sorption process, but its limited capability. Hence, AC is relevant for exchanging or regenerating during determined durations. Also, pollutants removal can be varied with the AC sort, amount, sorption mechanism, and quality. The biological performance and AC efficacy are highly dependent on the medium pH level. Hence, sulfamethoxazole, sulfamethazine, and acetaminophen compound sorption and biodegradation can be significantly varied [166]. It was reported that neutral materials like carbamazepine were processed in constant removal rate, while negatively charged compounds like diclofenac, naproxen, and ketoprofen had advanced reduction [167]. Another research showed a race on different spots of AC surface to accomplish DOC sorption [164]. High polar compounds like anionic organic compounds and X-ray contrast media are used to indicate AC loading during wastewater treatment to avoid unrequired consequences. Despite that, removing high polarity compounds, and examination are hard challenges. In comparison to liquid chromatography (LC) analysis, ion exchange chromatography (IC) technique offers more efficient separation capacity for cationic and anionic compounds. Hydrophilic interaction liquid chromatography (HILIC) is used for high polar CECs like illicit drugs, pesticides, and pharmaceuticals [168]. AC filter operation can be applied before the oxidation for removing potentially toxic compounds generated by oxidative treatment, like chlorine, or ozone [169]. Though, AC has a contribution in N-nitrosamines generation from secondary amines. Also, hydrophilic natural organic matter (NOM) fractions might pass through AC filter. However, GAC performance is highly based on the matrix efficiency and GAC particle size if NOM is presented [170].

3. Conclusion

This review study showed the weaknesses of anaerobic treatments towards treating high strength POME, like clogging, specific species domination, instability, high retention time, expensive, and employing large surface area. On the other hand, they are promising methods, since they can capture biogas, and generate low sludge amount. Majority of wastewater industries use ponding systems due to their low cost. Despite that, it produces greenhouse gases and causing soil clogging. The advanced concept of anaerobic treatments is the combined system like MBR. It is a promising way to pass numerous drawbacks and discharging reusable water. Finally, applying the sustainable approach, and zero waste energy can develop, and produce eco-friendly treatment.

4. Acknowledgment

This research work is financially supported by the Fundamental Research Grant Scheme (FRGS/UMP.05/25.12/04/01/1) with the RDU number RDU190160 which is awarded by the Ministry of Higher Education Malaysia (MOHE) via Research and Innovation Department, Universiti Malaysia Pahang (UMP) Malaysia.

5. References

[1] Rawat I, Ranjith Kumar R, Mutanda T and Bux F 2011 Dual role of microalgae: Phycoremediation of domestic wastewater and biomass production for sustainable biofuels production *Appl. Energy* **88** 3411–24
[2] Kim T H, Lee Y, Han S H and Hwang S J 2013 The effects of wavelength and wavelength mixing ratios on microalgae growth and nitrogen, phosphorus removal using Scenedesmus sp. for wastewater treatment *Bioresour. Technol.* **130** 75–80
[3] Markou G and Georgakakis D 2011 Cultivation of filamentous cyanobacteria (blue-green algae) in agro-industrial wastes and wastewaters: A review *Appl. Energy* **88** 3389–401

[4] Park J, Jin H F, Lim B R, Park K Y and Lee K 2010 Ammonia removal from anaerobic digestion effluent of livestock waste using green alga Scenedesmus sp. *Bioresour. Technol.* **101** 8649–57

[5] Kumar Y P, King P and Prasad V S R K 2006 Removal of copper from aqueous solution using Ulva fasciata sp.-A marine green algae *J. Hazard. Mater.* **137** 367–73

[6] Kamarudin K F, Tao D G, Yaakob Z, Takriff M S, Rahaman M S A and Salihon J 2015 A review on wastewater treatment and microalgal by-product production with a prospect of palm oil mill effluent (POME) utilization for algae *Der Pharma Chem.* **7** 73–89

[7] Awalludin M F, Sulaiman O, Hashim R and Nadhari W N A W 2015 An overview of the oil palm industry in Malaysia and its waste utilization through thermochemical conversion, specifically via liquefaction *Renew. Sustain. Energy Rev.* **50** 1469–84

[8] Bala J D, Lalung J and Ismail N 2014 Palm Oil Mill Effluent (POME) Treatment "'Microbial Communities in an Anaerobic Digester'": A Review *Int. J. Sci. Res. Publ.* **4** 1–24

[9] Wu T Y, Mohammad A W, Jahim J M and Anuar N 2010 Pollution control technologies for the treatment of palm oil mill effluent (POME) through end-of-pipe processes *J. Environ. Manage.* **91** 1467–90

[10] Miao X and Wu Q 2006 Biodiesel production from heterotrophic microalgal oil *Bioresour. Technol.* **97** 841–6

[11] Ahmad A L, Chong M F, Bhatia S and Ismail S 2006 Drinking water reclamation from palm oil mill effluent (POME) using membrane technology *Desalination* **191** 35–44

[12] Sathish A and Sims R C 2012 Biodiesel from mixed culture algae via a wet lipid extraction procedure *Bioresour. Technol.* **118** 643–7

[13] Song D, Fu J and Shi D 2008 Exploitation of Oil-bearing Microalgae for Biodiesel *Chin. J. Biotechnol.* **24** 341–8

[14] Bishop DF, O'farrell TP and Stamberg JB 1972 Physical- chemical treatment of municipal wastewater *J. Water Pollut. Control Fed.* **44** 361–71

[15] Ruiz-Marin A, Mendoza-Espinosa L G and Stephenson T 2010 Growth and nutrient removal in free and immobilized green algae in batch and semi-continuous cultures treating real wastewater *Bioresour. Technol.* **101** 58–64

[16] Jacinto M L J A J, David C P C, Perez T R and De Jesus B R 2009 Comparative efficiency of algal biofilters in the removal of chromium and copper from wastewater *Ecol. Eng.* **35** 856–60

[17] Zainal A, Yaakob Z, Takriff M S, Rajkumar R and Ghani J A 2012 Phycoremediation in anaerobically digested Palm Oil Mill Effluent using cyanobacterium, Spirulina platensis *J. Biobased Mater. Bioenergy* **6** 704–9

[18] Sajjad A-A, Yunus M Y B M, Azoddein A A M, Hassell D G, Dakhil I H and Hasan H A 2019 Electrodialysis Desalination for Water and Wastewater: A Review *Chem. Eng. J.* **380** 122231

[19] McHugh S, O'Reilly C, Mahony T, Colleran E and O'Flaherty V 2003 Anaerobic granular sludge bioreactor technology *Rev. Environ. Sci. Biotechnol.* **2** 225–45

[20] Bustillo-Lecompte C F, Mehrvar M and Quiñones-Bolaños E 2014 Cost-effectiveness analysis of TOC removal from slaughterhouse wastewater using combined anaerobic-aerobic and UV/H2O2 processes *J. Environ. Manage.* **134** 145–52

[21] Chan Y J, Chong M F, Law C L and Hassell D G 2009 A review on anaerobic-aerobic treatment of industrial and municipal wastewater *Chem. Eng. J.* **155** 1–18

[22] Mittal G S 2006 Treatment of wastewater from abattoirs before land application - A review *Bioresour. Technol.* **97** 1119–35

[23] Gomec C Y 2010 High-rate anaerobic treatment of domestic wastewater at ambient operating temperatures: A review on benefits and drawbacks *J. Environ. Sci. Heal. - Part A Toxic/Hazardous Subst. Environ. Eng.* **45** 1169–84

[24] Oliveira S C and Von Sperling M 2009 Performance evaluation of UASB reactor systems with and without post-treatment *Water Sci. Technol.* **59** 1299–306

[25] Chernicharo C A L 2006 Post-treatment options for the anaerobic treatment of domestic wastewater *Rev. Environ. Sci. Biotechnol.* **5** 73–92

[26] Bustillo-Lecompte C F, Mehrvar M and Quiñones-Bolaños E 2014 Cost-effectiveness analysis of TOC removal from slaughterhouse wastewater using combined anaerobic-aerobic and

UV/H2O2 processes *J. Environ. Manage.* **134** 145–52
[27] Ibrahim A, Yeoh B G and Cheah S C 1985 Thermophilic anaerobic contact digestion of palm oil mill effluent *Water Sci. Technol.* **17** 155–66
[28] Vlissidis A and Zouboulis A I 1993 Thermophilic anaerobic digestion of alcohol distillery wastewaters *Bioresour. Technol.* **43** 131–40
[29] Hamdi M and Garcia J L 1991 Comparison between anaerobic filter and anaerobic contact process for fermented olive mill wastewaters *Bioresour. Technol.* **38** 23–9
[30] Poh P E and Chong M F 2009 Development of anaerobic digestion methods for palm oil mill effluent (POME) treatment *Bioresour. Technol.* **100** 1–9
[31] Pant D, Van Bogaert G, Diels L and Vanbroekhoven K 2010 A review of the substrates used in microbial fuel cells (MFCs) for sustainable energy production *Bioresour. Technol.* **101** 1533–43
[32] Zhang T, Cui C, Chen S, Yang H and Shen P 2008 The direct electrocatalysis of Escherichia coli through electroactivated excretion in microbial fuel cell *Electrochem. commun.* **10** 293–7
[33] Kim B H, Chang I S and Gadd G M 2007 Challenges in microbial fuel cell development and operation *Appl. Microbiol. Biotechnol.* **76** 485–94
[34] Li W W, Yu H Q and He Z 2014 Towards sustainable wastewater treatment by using microbial fuel cells-centered technologies *Energy Environ. Sci.* **7** 911–24
[35] Solanki K, Subramanian S and Basu S 2013 Microbial fuel cells for azo dye treatment with electricity generation: A review *Bioresour. Technol.* **131** 564–71
[36] Mehdinia A, Ziaei E and Jabbari A 2014 Multi-walled carbon nanotube/SnO2 nanocomposite: A novel anode material for microbial fuel cells *Electrochim. Acta* **130** 512–8
[37] Karra U, Manickam S S, McCutcheon J R, Patel N and Li B 2013 Power generation and organics removal from wastewater using activated carbon nanofiber (ACNF) microbial fuel cells (MFCs) *Int. J. Hydrogen Energy* **38** 1588–97
[38] Mink J E, Rojas J P, Logan B E and Hussain M M 2012 Vertically grown multiwalled carbon nanotube anode and nickel silicide integrated high performance microsized (1.25 μl) microbial fuel cell *Nano Lett.* **12** 791–5
[39] Rezaei F, Xing D, Wagner R, Regan J M, Richard T L and Logan B E 2009 Simultaneous cellulose degradation and electricity production by Enterobacter cloacae in a microbial fuel cell *Appl. Environ. Microbiol.* **75** 3673–8
[40] Ren Z, Steinberg L M and Regan J M 2008 Electricity production and microbial biofilm characterization in cellulose-fed microbial fuel cells *Water Sci. Technol.* **58** 617–22
[41] Liu M, Yuan Y, Zhang L X, Zhuang L, Zhou S G and Ni J R 2010 Bioelectricity generation by a Gram-positive Corynebacterium sp. strain MFC03 under alkaline condition in microbial fuel cells *Bioresour. Technol.* **101** 1807–11
[42] Hassan S H A, Kim Y S and Oh S E 2012 Power generation from cellulose using mixed and pure cultures of cellulose-degrading bacteria in a microbial fuel cell *Enzyme Microb. Technol.* **51** 269–73
[43] Singh R, Paul D and Jain R K 2006 Biofilms: implications in bioremediation *Trends Microbiol.* **14** 389–97
[44] O'Toole G, Kaplan H B and Kolter R 2000 Biofilm Formation as Microbial Development *Annu. Rev. Microbiol.* **54** 49–79
[45] Rabaey K and Verstraete W 2005 Microbial fuel cells: Novel biotechnology for energy generation *Trends Biotechnol.* **23** 291–8
[46] Zuo J, Cui L, Fan M and Song W 2007 Production of electricity from artificial wastewater using a single chamber microbial fuel cell *Taiyangneng Xuebao/Acta Energiae Solaris Sin.* **28** 320–3
[47] Li Z, Zhang X, Lin J, Han S and Lei L 2010 Azo dye treatment with simultaneous electricity production in an anaerobic-aerobic sequential reactor and microbial fuel cell coupled system *Bioresour. Technol.* **101** 4440–5
[48] Sun J, Hu Y, Bi Z and Cao Y 2009 Improved performance of air-cathode single-chamber microbial fuel cell for wastewater treatment using microfiltration membranes and multiple sludge inoculation *J. Power Sources* **187** 471–9
[49] Sun J, Bi Z, Hou B, Cao Y qing and Hu Y you 2011 Further treatment of decolorization liquid of

azo dye coupled with increased power production using microbial fuel cell equipped with an aerobic biocathode *Water Res.* **45** 283–91
[50] Guo F, Fu G, Zhang Z and Zhang C 2013 Mustard tuber wastewater treatment and simultaneous electricity generation using microbial fuel cells *Bioresour. Technol.* **136** 425–30
[51] Patil S A, Surakasi V P, Koul S, Ijmulwar S, Vivek A, Shouche Y S and Kapadnis B P 2009 Electricity generation using chocolate industry wastewater and its treatment in activated sludge based microbial fuel cell and analysis of developed microbial community in the anode chamber *Bioresour. Technol.* **100** 5132–9
[52] Oh S E and Logan B E 2005 Hydrogen and electricity production from a food processing wastewater using fermentation and microbial fuel cell technologies *Water Res.* **39** 4673–82
[53] Min B and Logan B E 2004 Continuous electricity generation from domestic wastewater and organic substrates in a flat plate microbial fuel cell *Environ. Sci. Technol.* **38** 5809–14
[54] Huang L and Logan B E 2008 Electricity generation and treatment of paper recycling wastewater using a microbial fuel cell *Appl. Microbiol. Biotechnol.* **80** 349–55
[55] Baranitharan E, Khan M R, Prasad D M R and Salihon J Bin 2013 Bioelectricity generation from palm oil mill effluent in microbial fuel cell using polacrylonitrile carbon felt as electrode *Water. Air. Soil Pollut.* **224** 1533
[56] Jong B C, Liew P W Y, Juri M L, Kim B H, Mohd. Dzomir A Z, Leo K W and Awang M R 2011 Performance and microbial diversity of palm oil mill effluent microbial fuel cell *Lett. Appl. Microbiol.* **53** 660–7
[57] Cheng J, Zhu X, Ni J and Borthwick A 2010 Palm oil mill effluent treatment using a two-stage microbial fuel cells system integrated with immobilized biological aerated filters *Bioresour. Technol.* **101** 2729–34
[58] You S, Zhao Q, Zhang J, Jiang J and Zhao S 2006 A microbial fuel cell using permanganate as the cathodic electron acceptor *J. Power Sources* **162** 1409–15
[59] Del Mundo Dacera D, Babel S and Parkpian P 2009 Potential for land application of contaminated sewage sludge treated with fermented liquid from pineapple wastes *J. Hazard. Mater.* **167** 866–72
[60] Cairl McCarty P M and G F P 2000 *Chemistry for environmental engineering and Science 5th Ed.* (McGraw-Hill)
[61] Mara D 1997 *Design Manual for Waste Stabilization Ponds in India* (Leeds : Lagoon Technology International)
[62] Sayed S, de Zeeuw W and Lettinga G 1984 Anaerobic treatment of slaughterhouse waste using a flocculant sludge UASB reactor *Agric. Wastes* **11** 197–226
[63] Hawkes F R, Donnelly T and Anderson G K 1995 Comparative performance of anaerobic digesters operating on ice-cream wastewater *Water Res.* **29** 525–33
[64] Borja R and Banks C J 1994 Anaerobic digestion of palm oil mill effluent using an up-flow anaerobic sludge blanket reactor *Biomass and Bioenergy* **6** 381–9
[65] Stronach S M, Rudd T and Lester J N 1987 Start-up of anaerobic bioreactors on high strength industrial wastes *Biomass* **13** 173–97
[66] Dinsdale R M, Hawkes F R and Hawkes D L 1997 Comparison of mesophilic and thermophilic upflow anaerobic sludge blanket reactors treating instant coffee production wastewater *Water Res.* **31** 163–9
[67] Lettinga G, van Velsen A F M, Hobma S W, de Zeeuw W and Klapwijk A 1980 Use of the upflow sludge blanket (USB) reactor concept for biological wastewater treatment, especially for anaerobic treatment *Biotechnol. Bioeng.* **22** 699–734
[68] Kalyuzhnyi S, Estrada De Los Santos L and Martinez J R 1998 Anaerobic treatment of raw and preclarified potato-maize wastewaters in a UASB reactor *Bioresour. Technol.* **66** 195–9
[69] Barbosa R A and Sant'Anna G L 1989 Treatment of raw domestic sewage in an UASB reactor *Water Res.* **23** 1483–90
[70] Behling E, Diaz A, Colina G, Herrera M, Gutierrez E, Chacin E, Fernandez N and Forster C F 1997 Domestic wastewater treatment using a UASB reactor *Bioresour. Technol.* **61** 239–45
[71] Lettinga G 1995 Anaerobic digestion and wastewater treatment systems *Antonie Van*

Leeuwenhoek **67** 3–28
[72] Sayed S K I and Fergala M A A 1995 Two-stage UASB concept for treatment of domestic sewage including sludge stabilization process *Water Sci. Technol.* **32** 55–63
[73] Haandel A C Van and Lettinga G 1995 Anaerobic Sewage Treatment: A Practical Guide for Regions with a Hot Climate (9780471951216): Adrianus C. Van Haandel, Gatze Lettinga: Books *Anaerob. Sew. Treat. a Pract. Guid. Reg. with a hot Clim.* 236
[74] Kalyuzhnyi S V., Sklyar V I, Davlyatshina M A, Parshina S N, Simankova M V., Kostrikina N A and Nozhevnikova A N 1996 Organic removal and microbiological features of UASB-reactor under various organic loading rates *Bioresour. Technol.* **55** 47–54
[75] Fang H H P and Chui H K 1994 Comparison of startup performance of four anaerobic reactors for the treatment of high-strength wastewater *Resour. Conserv. Recycl.* **11** 123–38
[76] Borja R, Banks C J and Sánchez E 1996 Anaerobic treatment of palm oil mill effluent in a two-stage up-flow anaerobic sludge blanket (UASB) system *J. Biotechnol.* **45** 125–35
[77] Goodwin J A S, Wase D A J and Forster C F 1992 Pre-granulated seeds for UASB reactors: How necessary are they? *Bioresour. Technol.* **41** 71–9
[78] Ganjidoust H and Ayati B 2007 Comparing the efficiency of hybrid reactor in treating wood fiber , cellulose and lignin wastewater *Most* **3** 23–7
[79] Najafpour G D, Zinatizadeh A A L, Mohamed A R, Hasnain Isa M and Nasrollahzadeh H 2006 High-rate anaerobic digestion of palm oil mill effluent in an upflow anaerobic sludge-fixed film bioreactor *Process Biochem.* **41** 370–9
[80] Yu H and Gu G 1996 Biomethanation of brewery wastewater using an anaerobic upflow blanket filter *J. Clean. Prod.* **4** 219–23
[81] Bello-Mendoza R and Castillo-Rivera M F 1998 Start-up of an anaerobic hybrid (UASB/filter) reactor treating wastewater from a coffee processing plant *Anaerobe* **4** 219–25
[82] Borja R, Alba J and Banks C J 1996 Anaerobic digestion of wash waters derived from the purification of virgin olive oil using a hybrid reactor combining a filter and a sludge blanket *Process Biochem.* **31** 219–24
[83] Lo K V., Liao P H and Gao Y C 1994 Anaerobic treatment of swine wastewater using hybrid UASB reactors *Bioresour. Technol.* **47** 153–7
[84] Borja R, Banks C J and Wang Z 1995 Performance of a hybrid anaerobic reactor, combining a sludge blanket and a filter, treating slaughterhouse wastewater *Appl. Microbiol. Biotechnol.* **43** 351–7
[85] Córdoba P R, Francese A P and Siñeriz F 1995 Improved performance of a hybrid design over an anaerobic filter for the treatment of dairy industry wastewater at laboratory scale *J. Ferment. Bioeng.* **79** 270–2
[86] Guiot S R and van den Berg L 1985 Performance of an upflow anaerobic reactor combining a sludge blanket and a filter treating sugar waste *Biotechnol. Bioeng.* **27** 800–6
[87] Fia R, Pereira E L, Fia F R L, Emboaba D G and Gomes E M 2015 Start-up of anaerobic reactors for slaughterhouse wastewater treatment *Eng. Agric.* **35** 331–9
[88] Wang J, Huang Y and Zhao X 2004 Performance and characteristics of an anaerobic baffled reactor *Bioresour. Technol.* **93** 205–8
[89] McCarty P L 1982 One hundred years of anaerobic treatment. *Anaerob. Dig. 1981. Proc. Symp. Travemunde* 3–22
[90] Fannin K F, Srivastra V J, Conrad J R and Chynoweth D P 1981 *Marine biomass program: anaerobic digester system development* (Annual Report for General Electric Company)
[91] Iza J, Colleran E, Paris J M and Wu W M 1991 International workshop on anaerobic treatment technology for municipal and industrial wastewaters: Summary paper *Water Sci. Technol.* **24** 1–16
[92] Hassan S R and Dahlan I 2013 Anaerobic wastewater treatment using anaerobic baffled bioreactor: A review *Cent. Eur. J. Eng.* **3** 389–99
[93] Boopathy R and Tilche A 1991 Anaerobic digestion of high strength molasses wastewater using hybrid anaerobic baffled reactor *Water Res.* **25** 785–90
[94] Boopathy R and Sievers D M 1991 Performance of a modified anaerobic baffled reactor to treat

swine waste *Trans. Am. Soc. Agric. Eng.* **34** 2573–8
[95] Bustillo-Lecompte C F and Mehrvar M 2015 Slaughterhouse wastewater characteristics, treatment, and management in the meat processing industry: A review on trends and advances *J. Environ. Manage.* **161** 287–302
[96] Kuşçu Ö S and Sponza D T 2005 Performance of anaerobic baffled reactor (ABR) treating synthetic wastewater containing p-nitrophenol *Enzyme Microb. Technol.* **36** 888–95
[97] Barber W P and Stuckey D C 1999 The use of the anaerobic baffled reactor (ABR) for wastewater treatment: A review *Water Res.* **33** 1559–78
[98] Bustillo-Lecompte C F, Mehrvar M and Quiñones-Bolaños E 2013 Combined anaerobic-aerobic and UV/H2O2 processes for the treatment of synthetic slaughterhouse wastewater *J. Environ. Sci. Heal. - Part A Toxic/Hazardous Subst. Environ. Eng.* **48** 1122–35
[99] Cao W and Mehrvar M 2011 Slaughterhouse wastewater treatment by combined anaerobic baffled reactor and UV/H2O2 processes *Chem. Eng. Res. Des.* **89** 1136–43
[100] Toldrá F, Flors A, Lequerica J L and Vallés S 1987 Fluidized bed anaerobic biodegradation of food industry wastewaters *Biol. Wastes* **21** 55–61
[101] Borja R, González E, Raposo F, Millán F and Martín A 2001 Performance evaluation of a mesophilic anaerobic fluidized-bed reactor treating wastewater derived from the production of proteins from extracted sunflower flour *Bioresour. Technol.* **76** 45–52
[102] Garcia-Calderon D, Buffiere P, Moletta R and Elmaleh S 1998 Anaerobic digestion of wine distillery wastewater in down-flow fluidized bed *Water Res.* **32** 3593–600
[103] Sowmeyan R and Swaminathan G 2008 Performance of inverse anaerobic fluidized bed reactor for treating high strength organic wastewater during start-up phase *Bioresour. Technol.* **99** 6280–4
[104] Abdurahman N H, Rosli Y M and Azhari N H 2013 The Performance Evaluation of Anaerobic Methods for Palm Oil Mill Effluent (POME) Treatment: A Review *Int. Perspect. Water Qual. Manag. Pollut. Control*
[105] Perez M, Rodriguez-Cano R, Romero L I and Sales D 2007 Performance of anaerobic thermophilic fluidized bed in the treatment of cutting-oil wastewater *Bioresour. Technol.* **98** 3456–63
[106] Şen S and Demirer G N 2003 Anaerobic treatment of real textile wastewater with a fluidized bed reactor *Water Res.* **37** 1868–78
[107] Alvarado-Lassman A, Rustrián E, García-Alvarado M A, Rodríguez-Jiménez G C and Houbron E 2008 Brewery wastewater treatment using anaerobic inverse fluidized bed reactors *Bioresour. Technol.* **99** 3009–15
[108] Borja R and Banks C J 1995 Response of an anaerobic fluidized bed reactor treating ice-cream wastewater to organic, hydraulic, temperature and pH shocks *J. Biotechnol.* **39** 251–9
[109] Saravanane R, Murthy D V S and Krishnaiah K 2001 Treatment of anti-osmotic drug based pharmaceutical effluent in an upflow anaerobic fluidized bed system *Waste Manag.* **21** 563–8
[110] Borja R and Banks C J 1995 Comparison of an Anaerobic Filter and an Anaerobic Fluidized Bed Reactor Treating Palm Oil Mill Effluent *Process Biochem.* **30** 511–21
[111] Chen T H and Shyu W H 1996 Performance of four types of anaerobic reactors in treating very dilute dairy wastewater *Biomass and Bioenergy* **11** 431–40
[112] Vázquez I, Rodríguez J, Marañón E, Castrillón L and Fernández Y 2006 Simultaneous removal of phenol, ammonium and thiocyanate from coke wastewater by aerobic biodegradation *J. Hazard. Mater.* **137** 1773–80
[113] Mohan S and Sunny N 2008 Study on biomethonization of waste water from jam industries *Bioresour. Technol.* **99** 210–3
[114] Ugoji E O 1997 Anaerobic digestion of palm oil mill effluent and its utilization as fertilizer for environmental protection *Renew. Energy* **10** 291–4
[115] Tong S L and Jaafar a. B 2006 *POME Biogas capture, upgrading and utilization* (Palm Oil Engineering Bulletin)
[116] Ramasamy E V. and Abbasi S A 2000 Energy recovery from dairy waste-waters: Impacts of biofilm support systems on anaerobic CST reactors *Appl. Energy* **65** 91–8

[117] Kadlec, R.H. & Wallace S D 2009 *Treatment Wetlands* (CRC Press)

[118] Qasim S and Kamal M 2006 Natural Systems for Wastewater Treatment *Encyclopedia of Environmental Science and Engineering, Fifth Edition, Volumes One and Two* (agris.fao.org) pp 737–45

[119] Water and Sanitation Program 2008 Constructed Wetlands : A promising wastewater treatment system for small localities *Water* **2** 105–11

[120] Oller I, Malato S and Sánchez-Pérez J A 2011 Combination of Advanced Oxidation Processes and biological treatments for wastewater decontamination-A review *Sci. Total Environ.* **409** 4141–66

[121] Kadlec R H and Hey D L 1994 *Constructed wetlands for river water quality improvement* vol **29** (Lewis Publishers)

[122] Rousseau D P L 2005 Performance of Constructed Treatment Wetlands : Model-Based Evaluation and Impact of Operation and Maintenance Werking Van Aangelegde Zuiveringsmoerassen : Modelgebaseerde Evaluatie En Impact Van *Environ. Technol.* **300**

[123] Gutiérrez-Sarabia A, Fernández-Villagómez G, Martínez-Pereda P, Rinderknecht-Seijas N and Poggi-Varaldo H M 2004 Slaughterhouse Wastewater Treatment In a Full-scale System With Constructed Wetlands *Water Environ. Res.* **76** 334–43

[124] Soroko M 2007 Treatment of wastewater from small slaughterhouse in hybrid constructed wetlands systems *Ecohydrol. Hydrobiol.* **7** 339–43

[125] R. Carreau, S. VanAcker, A. C. VanderZaag, A. Madani, A. Drizo, R. Jamieson and R. J. Gordon 2012 Evaluation of a Surface Flow Constructed Wetland Treating Abattoir Wastewater *Appl. Eng. Agric.* **28** 757–66

[126] Odong R, Kansiime F, Omara J and Kyambadde J 2013 The potential of four tropical wetland plants for the treatment of abattoir effluent *Int. J. Environ. Technol. Manag.* **16** 203–22

[127] Puigagut J, Villaseñor J, Salas J J, Bécares E and García J 2007 Subsurface-flow constructed wetlands in Spain for the sanitation of small communities: A comparative study *Ecol. Eng.* **30** 312–9

[128] Bodkhe S 2008 Development of an improved anaerobic filter for municipal wastewater treatment *Bioresour. Technol.* **99** 222–6

[129] Jawed M and Tare V 2000 Post-mortem examination and analysis of anaerobic filters *Bioresour. Technol.* **72** 75–84

[130] Parawira W, Murto M, Zvauya R and Mattiasson B 2006 Comparative performance of a UASB reactor and an anaerobic packed-bed reactor when treating potato waste leachate *Renew. Energy* **31** 893–903

[131] Borja R and Banks C J 1994 Treatment of palm oil mill effluent by upflow anaerobic filtration *J. Chem. Technol. Biotechnol.* **61** 103–9

[132] Yu H Q, Hu Z H, Hong T Q and Gu G W 2002 Performance of an anaerobic filter treating soybean processing wastewater with and without effluent recycle *Process Biochem.* **38** 507–13

[133] Nebot E, Romero L I, Quiroga J M and Sales D 1995 Effect of the feed frequency on the performance of anaerobic filters *Anaerobe* **1** 113–20

[134] Mutamim N S A, Noor Z Z, Hassan M A A, Yuniarto A and Olsson G 2013 Membrane bioreactor: Applications and limitations in treating high strength industrial wastewater *Chem. Eng. J.* **225** 109–19

[135] Sutton P M 2012 Membrane Bioreactors for Industrial Wastewater Treatment: the State-of-the-Art Based on Full Scale Commercial Applications *Proc. Water Environ. Fed.* **2003** 23–32

[136] Widjaja T, Soeprijanto and Altway A 2010 Effect of Powdered Activated Carbon Addition on a Submerged Membrane Adsorption Hybrid Bioreactor with Shock Loading of a Toxic Compound *J. Math. Technol.* 139–46

[137] Judd S and Judd C 2008 *The MBR book: Principles and Applications of Membrane Bioreactors in Water and Wastewater Treatment* (Butterworth-Heinemann)

[138] Le-Minh N, Khan S J, Drewes J E and Stuetz R M 2010 Fate of antibiotics during municipal water recycling treatment processes *Water Res.* **44** 4295–323

[139] Dolar D, Gros M, Rodriguez-Mozaz S, Moreno J, Comas J, Rodriguez-Roda I and Barceló D

2012 Removal of emerging contaminants from municipal wastewater with an integrated membrane system, MBR-RO *J. Hazard. Mater.* **239–240** 64–9
[140] Alturki A A, Tadkaew N, McDonald J A, Khan S J, Price W E and Nghiem L D 2010 Combining MBR and NF/RO membrane filtration for the removal of trace organics in indirect potable water reuse applications *J. Memb. Sci.* **365** 206–15
[141] EPA 2011 Principles of Design and Operations o f Wastewater Treatment Pond Systems f or Plant Operators , Engineers , and Managers Principles of Design and Operations of Wastewater Treatment Pond Systems for Plant Operators , Engineers , *Natl. Risk Manag. Res. Lab.*
[142] Selmane D, Christophe V and Gholamreza D 2008 Extraction of proteins from slaughterhouse by-products: Influence of operating conditions on functional properties *Meat Sci.* **79** 640–7
[143] Rachmani. A. 2013 Cost and Performance Comparison of a Membrane Bioreactor (MBR) Plant and a Bardenpho Plant for Wastewater Treatment *Res. Commons Univ. Waikato* **1994**
[144] Hocaoglu S M, Insel G, Ubay Cokgor E and Orhon D 2011 Effect of sludge age on simultaneous nitrification and denitrification in membrane bioreactor *Bioresour. Technol.* **102** 6665–72
[145] Galil N I and Jacob L 2009 Comparative characterization of biosolids from a membrane bioreactor and from a sequencing batch reactor *Environ. Eng. Sci.* **26** 1001–8
[146] Mohhammad S, Keong L W, Kandiah S and Top A G M 2007 Membrane Bioreactor Technology for Tertiary Treatment of Palm Oil Mill Effluent (POME) *MPOB Inf. Ser.* **366** 1–4
[147] Simmons F J, Kuo D H W and Xagoraraki I 2011 Removal of human enteric viruses by a full-scale membrane bioreactor during municipal wastewater processing *Water Res.* **45** 2739–50
[148] Hirani Z M, DeCarolis J F, Adham S S and Jacangelo J G 2010 Peak flux performance and microbial removal by selected membrane bioreactor systems *Water Res.* **44** 2431–40
[149] Gürel L and Büyükgüngör H 2011 Treatment of slaughterhouse plant wastewater by using a membrane bioreactor *Water Sci. Technol.* **64** 214–9
[150] Dottorato A C, Universit I I and Ciclo C X Degradation of Refractory Organic Compounds in Aqueous Wastes employing a combination of biological and chemical treatments *Univ. degli Stud. di Cagliari*
[151] Holkar C R, Jadhav A J, Pinjari D V., Mahamuni N M and Pandit A B 2016 A critical review on textile wastewater treatments: Possible approaches *J. Environ. Manage.* **182** 351–66
[152] Chen S H and Yien Ting A S 2015 Biodecolorization and biodegradation potential of recalcitrant triphenylmethane dyes by Coriolopsis sp. isolated from compost *J. Environ. Manage.* **150** 274–80
[153] Chen S H and Yien Ting A S 2015 Biosorption and biodegradation potential of triphenylmethane dyes by newly discovered Penicillium simplicissimum isolated from indoor wastewater sample *Int. Biodeterior. Biodegrad.* **103** 1–7
[154] Hadibarata T, Teh Z C, Rubiyatno, Zubir M M F A, Khudhair A B, Yusoff A R M, Salim M R and Hidayat T 2013 Identification of naphthalene metabolism by white rot fungus Pleurotus eryngii *Bioprocess Biosyst. Eng.* **36** 1455–61
[155] Anastasi A, Parato B, Spina F, Tigini V, Prigione V and Varese G C 2011 Decolourisation and detoxification in the fungal treatment of textile wastewaters from dyeing processes *N. Biotechnol.* **29** 38–45
[156] Jonstrup M, Kumar N, Guieysse B, Murto M and Mattiasson B 2013 Decolorization of textile dyes by Bjerkandera sp. BOL 13 using waste biomass as carbon source *J. Chem. Technol. Biotechnol.* **88** 388–94
[157] Meng X, Liu G, Zhou J and Fu Q S 2014 Effects of redox mediators on azo dye decolorization by Shewanella algae under saline conditions *Bioresour. Technol.* **151** 63–8
[158] Khataee A R, Dehghan G, Zarei M, Ebadi E and Pourhassan M 2011 Neural network modeling of biotreatment of triphenylmethane dye solution by a green macroalgae *Chem. Eng. Res. Des.* **89** 172–8
[159] Zeb B S, Mahmood Q and Pervez A 2013 Characteristics and performance of anaerobic wastewater treatment (A review) *J. Chem. Soc. Pakistan* **35** 217–32
[160] MOORE B A 2012 *Investigation Into The Technical Feasibility Of Biological Treatment Of Precious Metal Refining Wastewater* (Rhodes University)

[161] Gerardi M H 2003 *The Microbiology of Anaerobic Digesters* (Wiley-Interscience)
[162] Reungoat J, Escher B I, Macova M and Keller J 2011 Biofiltration of wastewater treatment plant effluent: Effective removal of pharmaceuticals and personal care products and reduction of toxicity *Water Res.* **45** 2751–62
[163] Gerrity D, Gamage S, Holady J C, Mawhinney D B, Quiñones O, Trenholm R A and Snyder S A 2011 Pilot-scale evaluation of ozone and biological activated carbon for trace organic contaminant mitigation and disinfection *Water Res.* **45** 2155–65
[164] Prasse C, Stalter D, Schulte-Oehlmann U, Oehlmann J and Ternes T A 2015 Spoilt for choice: A critical review on the chemical and biological assessment of current wastewater treatment technologies *Water Res.* **87** 237–70
[165] Prasse C, Wagner M, Schulz R and Ternes T A 2012 Oxidation of the antiviral drug acyclovir and its biodegradation product carboxy-acyclovir with ozone: Kinetics and identification of oxidation products *Environ. Sci. Technol.* **46** 2169–78
[166] Nam S W, Choi D J, Kim S K, Her N and Zoh K D 2014 Adsorption characteristics of selected hydrophilic and hydrophobic micropollutants in water using activated carbon *J. Hazard. Mater.* **270** 144–52
[167] Nguyen L N, Hai F I, Kang J, Price W E and Nghiem L D 2013 Coupling granular activated carbon adsorption with membrane bioreactor treatment for trace organic contaminant removal: Breakthrough behaviour of persistent and hydrophilic compounds *J. Environ. Manage.* **119** 173–81
[168] van Nuijs A L N, Tarcomnicu I and Covaci A 2011 Application of hydrophilic interaction chromatography for the analysis of polar contaminants in food and environmental samples *J. Chromatogr. A* **1218** 5964–74
[169] Hanigan D, Zhang J, Herckes P, Krasner S W, Chen C and Westerhoff P 2012 Adsorption of N-nitrosodimethylamine precursors by powdered and granular activated carbon *Environ. Sci. Technol.* **46** 12630–9
[170] Corwin C J and Summers R S 2010 Scaling trace organic contaminant adsorption capacity by granular activated carbon *Environ. Sci. Technol.* **44** 5403–8

IOP Conf. Series: Materials Science and Engineering **702** (2019) 012059 doi:10.1088/1757-899X/702/1/012059

Compression power requirements for Oxy-fuel CO_2 streams in CCS

N K Daud[1*], S Martynov[2], H Mahgerefteh[2] and N Nasuha[4]

[1]Faculty of Chemical and Process Engineering Technology, Universiti Malaysia Pahang, Lebuhraya Tun Razak, 26300 Gambang, Pahang, Malaysia.
[2]Department of Chemical Engineering, University College London, London WC1E7JE.
[3]Faculty of Chemical Engineering, Universiti Teknologi MARA, Cawangan Pulau Pinang, Kampus Permatang Pauh, 13500 Permatang Pauh, Pulau Pinang.

*E-mail: khonisah@ump.edu.my

Abstract. CO_2 compression systems are commonly designed assuming negligible amount of impurities in CO_2 fluid, it is of practical interest to evaluate the impact of impurities in oxy-fuel streams on the compression power requirements. Compared to more traditional post-combustion and pre-combustion capture methods, oxy-fuel technology produces a CO_2 stream with relatively high concentration of impurities that may require partial or a high level of removal and whose presence can be expected to increase the costs of CO_2 compression. Four types of compression technologies employed include four-stage compressor with 4 intercoolers, single-stage supersonic shockwave compressor, three-stage compressor combined with subcritical liquefaction and pumping and three-stage compressor combined with supercritical liquefaction and pumping. The study depicts that decrement of the impurities content from 15 to 0.7%v/v in the CO_2 streams reduced the total compression power in the compression system. The study also concludes that three-stage compressor combined with subcritical liquefaction and pumping can potentially offer higher efficiency than four-stage compressor with 4 intercoolers for almost pure CO_2 streams. In the case of raw oxy-fuel mixture, that carries relatively large amount of impurities, subcritical liquefaction proved to be less feasible, while supercritical liquefaction efficiency is only marginally lower than that in the four-stage compressor with 4 intercoolers.

1. Introduction

In future Carbon Capture and Sequestration (CCS) projects, CO_2 captured from large industrial emission sources will be purified and compressed for long-term sequestration in geological formations. Pipelines offer the most economically viable option for moderate and long distance transportation of large quantities of CO_2. In order to ensure single-phase transportation of CO_2 via pipelines, the pipeline pressures need to be maintained in the range from 85 to 150 bar [1], i.e. above the critical pressure of the fluid, which in case of pure CO_2 is 73 bar. At such pressures the fluid can be either in dense-phase state, at temperatures below the pseudo-critical temperatures (the temperature at which the heat capacity reaches its maximum) or in supercritical phase at temperatures above the pseudo-critical temperatures. With the increase in the fluid pressure and decrease in the temperature, the fluid compressibility and specific volume decrease, making the CO_2 pipeline transportation more economically efficient for dense-phase liquid CO_2. In order to compress CO_2 gas to the dense-phase state, several options have been previously analysed for pure CO_2 [2]. In particular, multistage compression combined with liquefaction and subsequent pumping of liquid phase CO_2 was found to be more efficient than conventional gas phase compression. This becomes practically possible in case of pure CO_2 due to its relatively high boiling point (ca. 20 °C at 60 bar pressure), which allows using utility streams for liquefaction; also, pumping of a liquid is less energy demanding than gas-phase compression [3]. In case of industrial grade CO_2 stream, the boiling point of the liquid can either be increased or decreased depending on the nature and amount of impurities present in the CO_2 fluid. However, the impact of stream impurities on the CO_2 compression is not clear.

This paper evaluates the impact of impurities from oxy-fuel capture technology on power consumptions for compression options previously recommended by [3] for pure CO_2.

In the present study, the content and amount of impurities in CO_2 streams captured using oxy-fuel technology have been identified by Porter et al. [4]. Impure components present in CO_2 stream determine the thermodynamic and phase equilibrium properties of the fluid, and hence may significantly affect the power requirements for compression and pipeline transportation. In particular, density and compressibility are two properties which determine the compression work and both depend on the amount and concentration of impurities in CO_2 fluid [5]. Also, the power requirements for liquefaction of CO_2 stream largely depend on the vapour-liquid equilibrium of CO_2 mixtures.

Our preliminary analysis showed that impurities listed in Porter et al. [4] which amount does not exceed 0.1% v/v have no significant impact on the thermo-physical properties relevant for compression/transportation of CO_2 fluid. Therefore, for the purpose of the present study, the average of concentration of each component is calculated from Porter et al. [4] within the given range. To investigate the highest possibility of the effect of impure components, the composition of impure CO_2 from raw/dehumidified, double flashing and distillation purification processes from oxy-fuel are adopted in this study as shown in table 1.

Table 1. Composition of CO_2 streams captured from oxy-fuel combustion technology [4].

Oxy-fuel	Component (% v/v)								
combustion	CO_2	O_2	N_2	Ar	NO_2	SO_2	SO_3	H_2O	CO
Raw/dehumidified	85	4.70	5.80	4.47	0.0100	0.0050	0.002	0.010	0.0050
Double flashing	96.70	1.20	1.60	0.40	0.0150	0.0036	-	-	-
Distillation	99.30	0.40	0.20	0.10	0.0033	0.0037	-	-	-

In this study, in order to predict physical properties of CO_2 mixtures involved in equations (2) and (4), the Peng-Robinson Equation of State (PR EoS) with standard mixing rules in REFPROP package is applied as one of the simplest and accurate equations developed for CO_2 mixtures [6]. As discussed earlier, the presence of impure components in CO_2 stream can change the thermodynamic properties, affecting the compression work and cooling power requirements for impure CO_2.

2. Methodology

The thermodynamic analysis of the N-stage compression with intercooling involves calculating the total power consumed by compression:

$$W_{Comp} = \sum_{i=1}^{N} \frac{G}{\eta_{comp,i}} \int_{P_i^{in}}^{P_i^{out}} \left(\frac{dp}{\rho}\right)_s \tag{1}$$

$$W_{comp} = G \sum_{i=1}^{N} \frac{1}{\eta_{comp,i}} \left(h_i^{out} - h_i^{in}\right) \tag{2}$$

where W_{comp}, G, η_{comp}, h^i_{in} and h^i_{out} are the compression work, mass flow rate, isentropic compression efficiency, enthalpy at compressors' inlet and discharge, respectively.

And the total power required for removing heat from the flow at the intercooling stages:

$$Q_{cool} = G\sum_{i=2}^{N}\left(h_{i-1}^{out} - h_i^{in}\right) \tag{3}$$

$$W_{cool} = \frac{Q_{cool}}{\eta_{cool}}\left(\frac{T_{cond} - T_{ev}}{T_{ev}}\right) \tag{4}$$

where Q_{cool} is the total cooling duty associated with removing heat from the compressor and W_{cool} is the power demand for cooling/liquefaction while η_{cool}, T_{cond} and T_{ev} are respectively the heat transfer efficiency of the isobaric cooling process, condensation and evaporation temperatures.

Using equations (2) and (3) implies knowledge of the fluid thermodynamic properties, namely density and enthalpy of CO_2 mixtures. These properties are calculated using REFPROP package [7]. The integral in equation (1) defines the compression work done on the fluid which is valid irrespective of the fluid phase state, and hence can be applied to evaluate compression work for the gas and pumping for the liquid.

3. Results and discussion

Results of calculations of the power consumed in the multistage compression of pure and impure CO_2 streams are presented in this section. The study is performed for impure CO_2 streams with the compositions described in table 1 and available compression technology options. To enable cross-comparison of compression power for the various compression technologies and various CO_2 mixtures, the stream conditions at the inlet and discharge of compressor are kept identical in all the cases studied. In particular, at the inlet of compressor, the pressure and temperature are set to 15 bar and 38 °C, while the compressor discharge pressure is set to 151 bar with the discharge temperature is depending on the compression option applied. The 15 bar inlet pressure is chosen as typical value corresponding to pressure of CO_2 stream coming from oxy-fuel capture unit [8]. Following the study by Witkowski et al. [9], the compressors and inter-stage coolers' efficiencies are set to $\eta_{c,i}=\eta_{h,i}=075\text{-}0.85$ with the streams' mass flow rate of 156.4 kg/s.

3.1. Adaptation of industrial compressor options of impure CO_2.

Figures 1-4 show the compression pathways plotted in pressure-enthalpy phase diagrams of pure CO_2 and impure CO_2 mixtures, for the four-stage compressor with 4 intercoolers (option A), single-stage supersonic shockwave compressor (option B), three-stage compressor combined with subcritical liquefaction/refrigeration and pumping (option B) and three-stage compressor combined with supercritical liquefaction and pumping (option D), respectively.

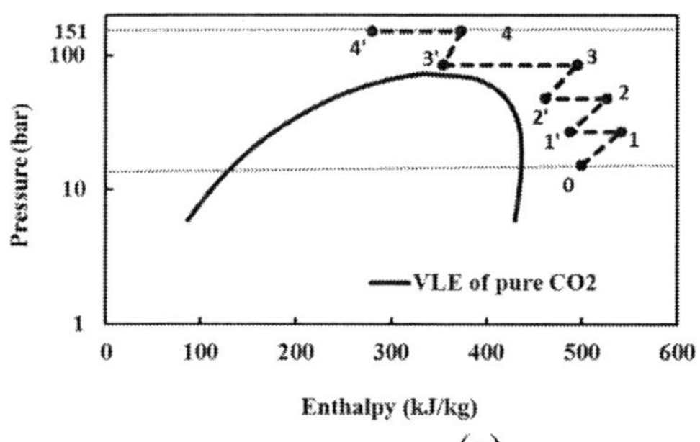

(a)

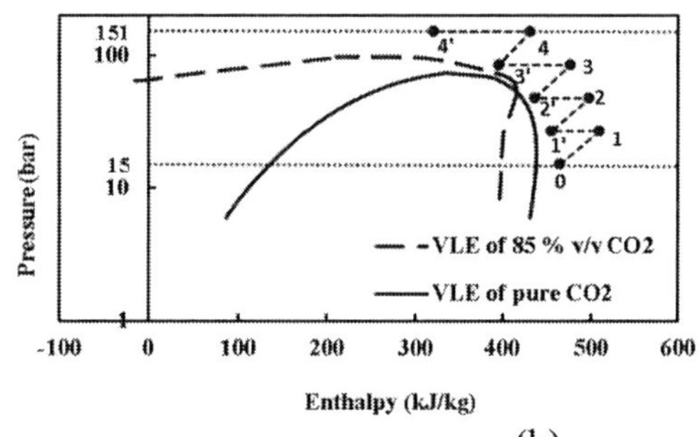

(b)

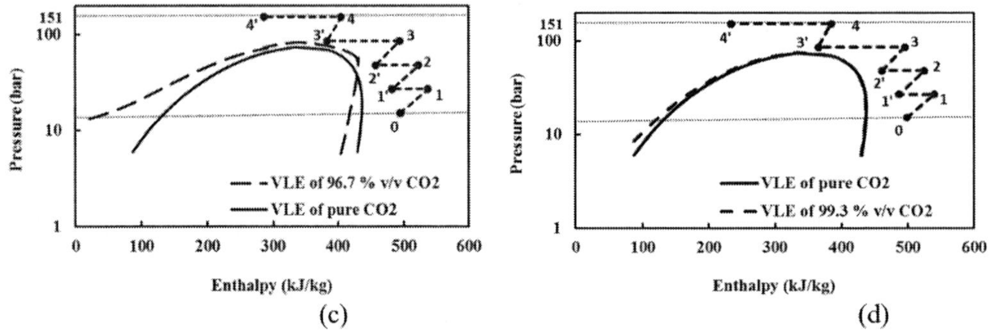

Figure 1. Pressure-enthalpy diagrams for compression of pure CO_2 (a), raw/dehumidified (b), double flashing (c) and distillation (d) from oxy-fuel captures using compression option A (dashed lines).

Figure 1 presents the application of the conventional multistage compression to the pure CO_2 and oxy-fuel mixtures. In this case, the compression ratio for each stage is set to 1.78 with the CO_2 streams enter the compressor at 15 bar that corresponds to typical pressure level at the exit of oxy-fuel capture unit [8].

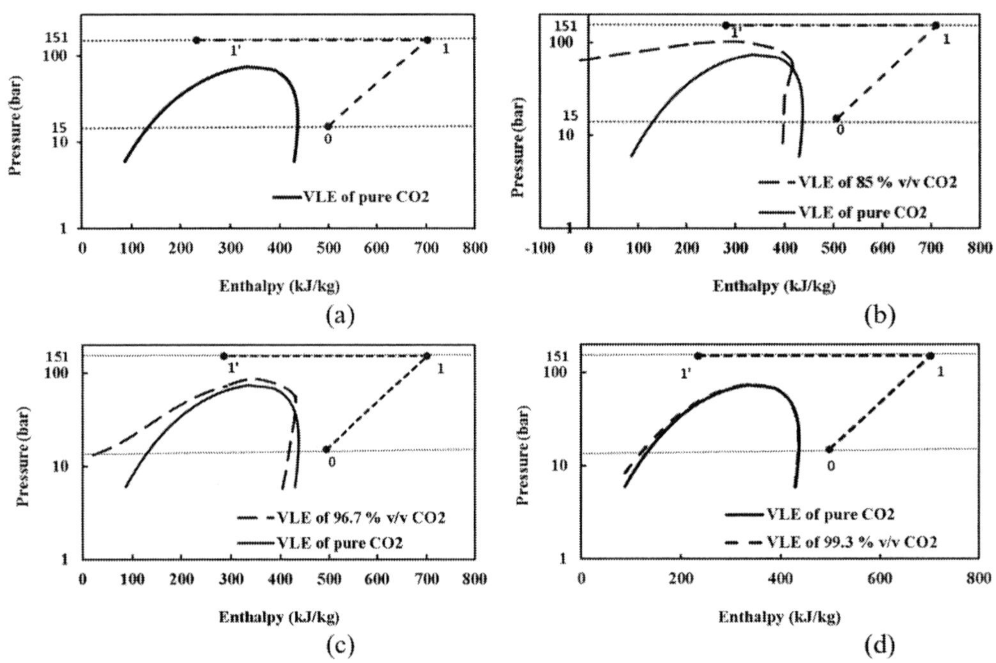

Figure 2. Pressure-enthalpy diagrams for compression of pure CO_2 (a), raw/dehumidified (b), double flashing (c) and distillation (d) from oxy-fuel captures using compression option B (dashed lines).

Figure 2 shows the compression pathways in pressure-enthalpy diagrams the option B achieved using the supersonic shockwave compressor with the compression ratio of 10 per stage. In this case, only single-stage compression is applied to compress the pressure from 15 bar to 151 bar discharge pressure. The constant pressure intercooling is used to reduce the temperature after the compression from ca. 267-295 °C back to the inlet temperature of 38 °C, respectively.

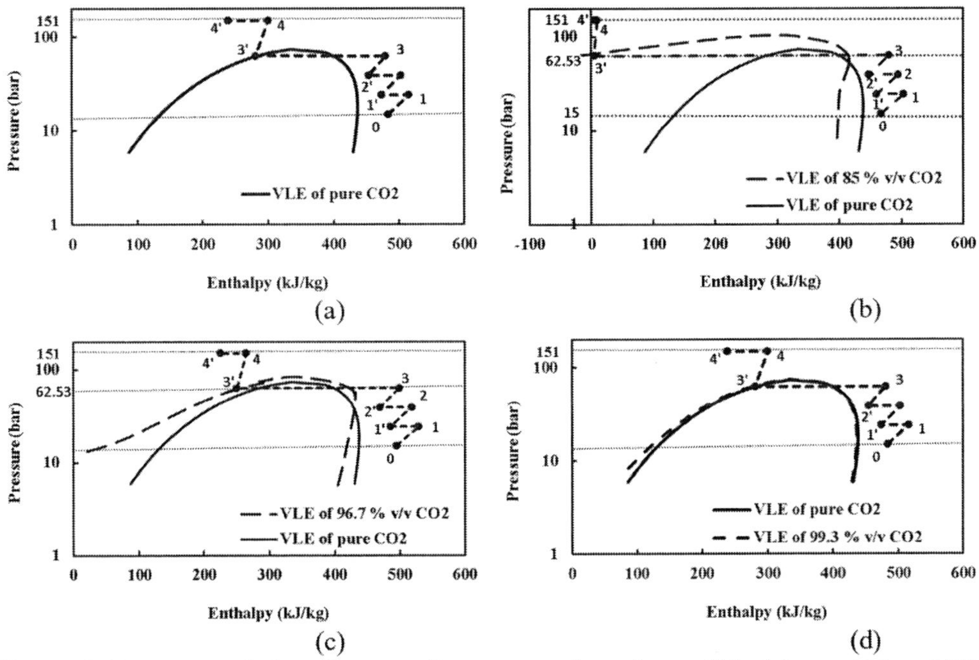

Figure 3. Pressure-enthalpy diagrams for compression of pure CO_2 (a), raw/dehumidified (b), double flashing (c) and distillation (d) from oxy-fuel captures using compression option C (dashed lines).

In figure 3, the compression pathways are shown for the compression option C, which integrates the three-stage compressor with pumping following liquefaction of CO_2 stream at subcritical temperature. In this study, the liquefaction pressure is set to 62 bar for CO_2 streams, while the liquefaction temperature is determined by the bubble point temperature of the fluid at this pressure:

Pure CO_2: $T_{liq} = 20$ °C
Oxy-fuel mixture:
Raw/dehumidified with 85 % v/v: $T_{liq} = -45$ °C
Double flashing with 96.7 % v/v: $T_{liq} = 14$ °C
Distillation with 99.9 % v/v: $T_{liq} = 19$ °C

In case of raw/dehumidified stream with 85 % v/v of CO_2 purity, liquefaction would require using cryogenic coolants, increasing significantly the cost of the whole process.

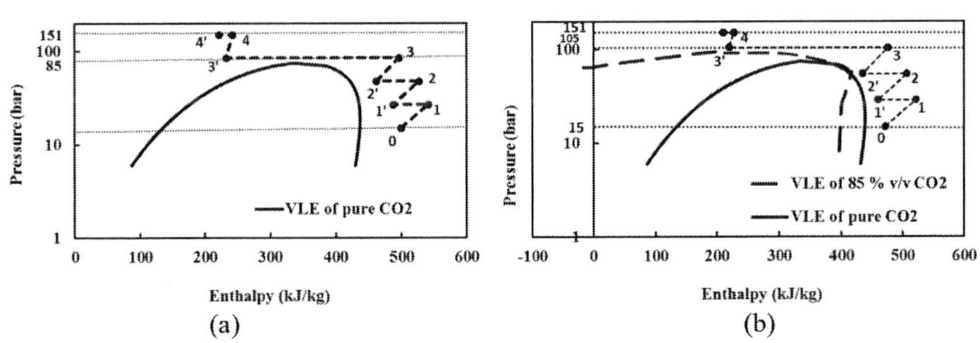

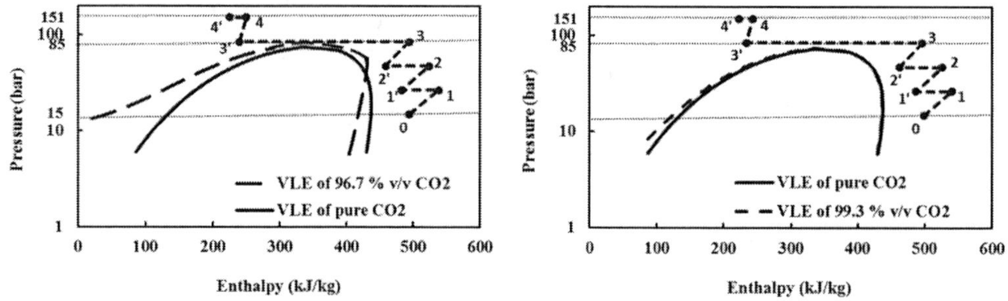

(c) (d)

Figure 4. Pressure-enthalpy diagrams for compression of pure CO_2 (a), raw/dehumidified (b), double flashing (c) and distillation (d) from oxy-fuel captures using compression option D (dashed lines).

Figure 4 shows the compression pathways in case of compression option D, where the three-stage compressor is combined with supercritical liquefaction and pumping. In this option, the liquefaction cooling is applied at 85 bar, 15 °C for pure CO_2 and impure from double flashing and distillation oxy-fuel captures and 110 bar, 5 °C for raw/dehumidified CO_2 stream.

3.2. CO_2 compression power requirements

Table 2 shows the results of total power consumption in terms of compression and intercooling power as required for the multistage compression options A, B, C and D.

Table 2. Power consumption for the compression technology options employed.

	CO_2 (% v/v)	Compression technology options			
		A	B	C	D
		Compression power (kW)			
Pure CO_2	100	29674	35273	20508	24340
Oxy-fuel:					
Raw/dehumidified	85	34512	37867	23128	29012
Double flashing	96.7	31092	35825	21661	24829
Distillation	99.3	30029	35383	20860	24438
		Intercooling work (kW)			
Pure CO_2	100	10750	51649	12869	16208
Oxy-fuel:					
Raw/dehumidified	85	11163	54883	43646	21395
Double flashing	96.7	11055	54753	13357	16574
Distillation	99.3	10825	54232	13120	16504
		Total power = compression + intercooling work (kW)			
Pure CO_2	100	40424	86922	33377	40548
Oxy-fuel:					
Raw/dehumidified	85	45675	92750	66774	50407
Double flashing	96.7	42147	90578	35018	41403
Distillation	99.3	40854	89615	33980	40942

As shown in table 2 for pure CO_2, compression work is largest in case of single-stage supersonic shock wave compression (option B). This can be explained by higher compression ratios (10:1) applied compared to the other options. Data from table 2 also show that the cooling duty is also largest in case of option B, which is due to the significant increase in the temperature at the discharge of the compressor in this case of 285 °C compared to the relatively low discharge temperatures in the other compression options (90 °C). Applying liquefaction as can be expected reduces the compression work, but increases the energy spend on intercooling (compare options C and D with A).

As can be seen from table 2, the presence of impure components in CO_2 stream affects both the compression power and cooling requirements for multistage compression. In particular, the compression work and intercooling work rise as concentration of impurities increase. The rising demand of compression work is related to the changes of bubble point curves for that CO_2 streams. For options C and D, the intercooling work increases as liquefaction temperature declines from 20 to -45 °C. This is because the coefficient of performance (COP) of refrigerator decreases and the required energy level of the heating steam increase as the liquefaction temperature drop [10]. In case of option B, increasing the impurities in the stream requires enormous amount of power to cool down the system to acceptable level due to high compression ratio applied. Importantly, the subcritical compression technology (option C) is more economically efficient than other options, resulting with lower total power consumption for the mixtures with low concentration of impure components (< 3.3 % v/v). The total power consumption required for oxy-fuel distillation with 99.3 % v/v CO_2 is about the same compared with pure CO_2 for all compression strategies applied.

4. Conclusions

The present study describes the result analysis of the impact of impurities from oxy-fuel capture technology on the compression power requirements for CO_2 pipeline transportation. The study compares various compression technologies including the four-stage compressor with 4 intercoolers, single-stage supersonic shockwave compressor and three-stage compressor combined with liquefaction and pumping. The results of this study indicate the presence of impurities show significant impacts on compression and intercooling work. In particular, the CO_2 stream from raw/dehumidified presents considerably different total power consumption, in comparison to double flashing and distillation processes. It was found that the compression power is increased by 7-19 % more than that for the compression of pure CO_2, while for intercooling power, the demand can be up to 32 % for streams with impurities less than 3.3 % v/v. In case of raw dehumidified stream with 15 % v/v of impurities, the intercooling power was found nearly 2 times higher than intercooling power of pure CO_2. From this study, the integration of multistage compression with liquefaction and pumping can greatly decrease the compression power as compared to all compression options. However, further declines of liquefaction temperature will reflect the intercooling work with increase the energy consumption. In case of single-stage supersonic shockwave compressor, the highest total power consumption is observed due to high compression ratio applied.

Acknowledgement

Financial support from Universiti Malaysia Pahang (UMP) through research university grant under agreement number RDU1803177 is gratefully acknowledged.

References

[1] Lemontzoglou A, Pantoleontos G, Asimakopoulou A G, Tsongidis N I and Konstandopoulos A G 2017 *Int. J. Greenh. Gas Con.* **66** 10-24
[2] Moore J J, Lerche A, Delgado H, Allison T and Pacheco J 2011 *The 40th Turbomachinery Symposium* 107–20
[3] Witkowski A and Majkut M 2012 *Arch. Mech. Eng.* **3** 1–18
[4] Porter R T J, Fairweather M, Pourkashanian M and Woolley R M 2015 *Int. J. Greenh. Gas Con.* **36** 161–74
[5] Chima O and Dmitriy K 2018 *Appl. Energy* **230** 816–35
[6] Mahmoud N, Chapoy A, Burgass R and Tohidi B 2017 *J. Chem. Thermodyn,* **111** 157-72
[7] Goos E, Riedel U, Zhao L and Blum L 2011 *Energy Procedia* 4 3778–85.
[8] Gusca J and Blumberga D 2011 *Energy Procedia* **4** 2526–32
[9] Witkowski A, Rusin A, Majkut M, Rulik S and Stolecka K 2013 *Energy Convers. Manag.* **76** 665–73
[10] Duan L, Chen X and Yang Y 2013 *Int. J. Energy Res.* **37** 1453–64
[11] Martynov S B, Daud N K, Mahgerefteh H, Brown S and Porter R T J 2016 *Int. J. Greenh. Gas Con.* **54** 652-61

IOP Conf. Series: Materials Science and Engineering **702** (2019) 012060 doi:10.1088/1757-899X/702/1/012060

Oil emulsions and the different recent demulsification techniques in the petroleum industry - A review

S M Abed*, N H Abdurahman, R M Yunus, H A Abdulbari, S Akbari

Faculty of Chemical and Process Engineering Technology, Universiti Malaysia Pahang, 26300 Gambang, Kuantan, Pahang, Malaysia.

*E-mail: saadkirkuk83@gmail.com

Abstract. The emulsions are formed during the production of crude oil, which is often accompanied by water. These emulsions are undesirable and should treat to remove the dispersed water and accompanying inorganic salts .so, many researchers have over the years focused to probe of the appropriate demulsification techniques in the petroleum sector in order to meet production and transportation requirements, and to reduce corrosion and catalyst poisoning and to maximize the overall profitability of the crude oil production. Therefore, this study presents an overview of the emulsions, formation, classification, Stability, and properties. practical demulsification techniques in the petroleum sector, including chemical, microwave irradiation, biological, thermal, membrane, electrical, and ultrasonic techniques for both oilfield and synthetic emulsions. Amongst these techniques, chemical demulsification has been the most widely applied and reported in the literature. Moreover, a more effective demulsification process could be attained by leveraging synergistic effects by combining one or more of these techniques

1. Introduction

Oil production and transportation via pipelines results in the formation of water-oil emulsions due to the presence of shear forces and pressure [1-2]. They are formed as water droplets or brine which are continuously dispersed in crude oil and are referred to as water-in-oil (w/o) emulsions. There are also emulsions which consist of crude oil droplets in brine or water (o/w) although they are rarely encountered [3-4]. When crude oil is extracted from a reservoir, it contains several impurities like sediments and water [5]. It may also contain hydrocarbons like asphaltenes, waxes, resins, solids from crude, and carboxylic acids which serve as natural emulsifiers [6]. There may be an accumulation of some of these components at the water-oil interface, resulting in the formation of a stable film that surrounds the droplets. This film promotes the formation of stable w/o emulsion by hindering the water droplets from coalescence [7].

Several problems may arise from the co-production of water and crude oil. This may include the cost associated with the pumping or transportation of the water by tanker or via the pipeline, corrosion of production facilities (like pipes, pumps, and downstream distillation columns), poisoning of the downstream by industrial catalysts, the need to install the extra equipment for the production of crude oil of export quality, as well as the issues related to improved crude oil viscosity as a result of the presence of small water droplets in the crude oil [8]. From the economic and operational perspectives, there is a need to ensure complete separation of water from crude oil prior to its transportation and subsequent refining [9]. The available crude oil demulsification techniques can be grouped into mechanical, chemical, thermal, and electrical techniques [10]. Chemical demulsification involves the use of chemicals to induce the destabilization of the shielding interfacial films, resulting in an improved rate of emulsion breakage. As an economical and effective method, chemical additives are widely used in dewatering of crude oils emulsions [11]. Several parameters are involved in oil demulsification and the alteration of each parameter can have a significant influence on the stability of the emulsion. Such parameters include salinity, temperature, conc. of the emulsifier/demulsifier, agitation rate and its duration, pH, oil/water content, size and distribution of droplet [12, 13].

This review is a comprehensive survey of the related literature (previous and current studies) on petroleum emulsions, types, formation, stability, properties and braking of emulsions. The review also covered surfactant, their structure, type, and properties. Demulsifiers, their mechanism, influencing parameters, and water separation techniques from crude oils were also reviewed. The importance of separating water from petroleum was also reviewed from both technical and environmental perspectives.

2. Crude Oil Emulsions

2.1. Definition of Emulsions

An emulsion is defined as a phase that contains two immiscible liquids such as oil and water. In an emulsion system, one of the stages is dispersed in the second as globules; the phase that exists as globules is called the sparse phase, while the continuous phase is the phase that produced the dispersion model wherein the droplets are suspended. The diameter of the emulsion droplets usually falls within 0.5 μm or more [14]. The widespread occurrence of emulsion in our daily lives (such as in food, pulp, cosmetics, and pharmaceutical industry) made them a thing of great practical interest [15]. Although petroleum emulsions may not be well known, they are widespread and of significant importance in the petroleum industry. They are encountered in almost all aspects of the petroleum recovery and production processes, ranging from crude oil drilling, production, transportation, and processing [16]. Being that they are undesirable in the petroleum sector, they are often associated with the high pumping costs, reduced throughput, and corrosion of pipeline and other equipment [17].

2.2. Classification of Emulsions

Kokal et al. [18] inferred that oil is produced in a combined form with water and this mixture referred to as oil emulsion or oil field emulsion. A crude oil emulsion is often produced when the oil and water phases are vigorously stirred; nevertheless, once the stirring is stopped, there is an immediate breakdown in the emulsion [19]. There are three wide groups of emulsion - oil-in-water (o/w), water-in-oil (w/o), and multiple emulsions [like water-in-oil-in-water (w/o/w) and oil-in-water-in-oil (o/w/o)]. For instance, o/w/o implies the dispersion of oil droplets in water droplets which are further dispersed in a continuous oil phase, as depicted in figure. 1 [20]. Salomon (2006) suggested that there are other complex structures of emulsion. For instance, the disperse phase in double emulsions consists of another stage which may differ from the continuous phase in composition. Water-in-oil (w/o) emulsions normally contain less than or equal to 50 % of water while o/w emulsions have more than 80 % of water (o/w emulsion can also exist with less than 20 % of water) [21].

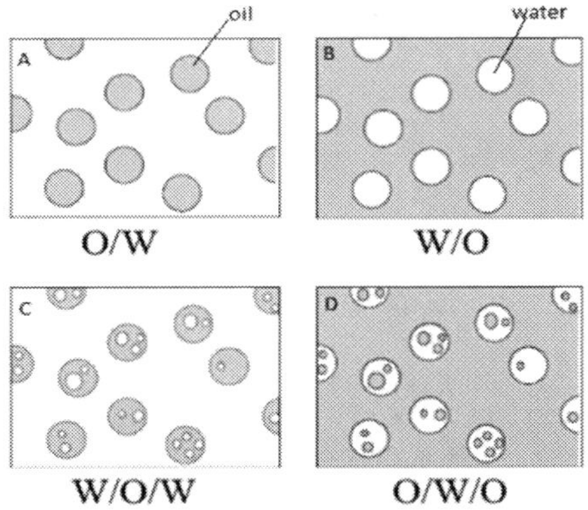

Figure 1. Types of emulsion (a) oil-in-water (b) water-in-oil (c) & (d) multiple emulsion.

IOP Conf. Series: Materials Science and Engineering **702** (2019) 012060 doi:10.1088/1757-899X/702/1/012060

2.3. The Formation of Crude Oil Emulsions

The formation of crude oil emulsions is usually observed in all stages of production, starting from the oil reservoirs through transportation. The formation of crude oil emulsion is initiated when there is contact between two immiscible liquids in the presence of an emulsifying agent provided there is adequate blending or an agitator that facilitates the dispersal of one liquid as droplets [22]. Several studies have shown that the formation of crude oil emulsions is strongly dependent on the composition of the crude oil which often consist of natural emulsifying agents such as asphaltenes, acidic compounds (naphthenic acids, fatty acid acids, aromatic acid), and resins [34]. These natural emulsifiers are abundant in heavy and extra heavy crude oils.

Authors have established that the emulsion formation and behavior can vary from one crude oil to another. A study by Delgado-Linares et al. [35] revealed that the natural surfactants in crude oil facilitate the reduction of interfacial tension of oil-water which in turn leads to the formation of the interfacial film. This interfacial film is characterized by mechanical strength and serves an important function in the formation and stability of the emulsion [36]. Hence, the variation in crude oil compositions might influence the significant changes in emulsification behavior. During the emulsification process, each of the natural surfactants plays a unique role. Asphaltene, which consists of a myriad of non-hydrocarbon compounds, helps to improve emulsion stability as a result of its good interfacial activity [37]. Analogous to asphaltene is resin, which consists of a myriad of macromolecular non-hydrocarbon compounds (relatively low molecular weight and polarity) and plays a significant role in the formation of crude oil emulsion [43]. Ashrafizadeh et al. [23-25] suggested that there are many ways of producing oil droplets to form emulsions, such as the utilization of devices like colloid mills, homogenizers (high pressure), dispersing machines, application of high shearing stresses, etc. Alwadani [26] also mentioned the possibility of automatic emulsification via chemical processes or via temperature-mediated nucleation processes in one of the phases. Table 1 shows some of the important parameters used by researchers to prepare emulsions.

Table 1. Some important parameters researchers used by previous researchers to prepare emulsions.

Type of emulsion	Water content (v/v)	Mixing time (min)	Mixing speed (rpm)	Type of emulsifier	Emulsifiers concentration	Ref
W/O	70/30	5-15	1800	Low sulfur waxy residue and Triton X-100.		[3]
O/W	50/50	0.5-1	1000	Ethoxylated nonylphenol family (RENEX) with an ethoxylated amine (Ultramina, 5 EO groups)		[27]
O/W	69, 72, and 66 vol.%	5-15	1000-2000	Triton X-100	0.3-2.5wt.%	[28]
W/O	70/30	30	800	Tallowalkylamine acetates or ARMACT	0.036 m3	[29]
O/W	Oil ratio 50-70% by total volume	5-15	10000-15000	SDDS, Triton X-100, Span 83, Tween 80, and Coco-Amide	0.5-1%	[30]
W/O	4/1	5	10000	n-heptane (analytical grade) and Toluene		[31]
O/W	90/10, 70/30 and 50/50	30 min.	9000			[32]
W/O	70/30	5	1300	Span 80	3%	[33]

2.4. The Stability of the Crude Oil Emulsions

Crude oil is made up of several components but mainly composed of carbon, hydrogen, and small amounts of oxygen, nitrogen, sulfur, and other common metals like nickel, iron, and vanadium in trace amounts. The role of these elements is to encourage hydrogen bond formation and polar interactions between the asphaltene components. This interaction gives rise to the formation of the interfacial film and accounts for its strength and elasticity [34]. Except for microemulsions, all other emulsions are thermodynamically unstable. Meanwhile, emulsion destabilization requires a lot of time, and it may be impossible for a stable emulsion to resolve itself within a defined period without undergoing some chemical or mechanical treatments [35]. After the formation of some emulsions, they can be easily decomposed into oil and water phases; however, some other may persist for prolonged periods. Stability results from the small-sized droplet, the ratio of oil to water, the presence of some factors such as temperature, water pH, salt concentration, surfactants, etc. According to Mat et al. [36], stability refers to the "persistence of an emulsion in the environment" and has been considered an important property of w/o emulsions. The formation of a stable w/o emulsion requires the presence of an emulsifying agent like clay particles, chemicals, asphaltenes, naphthenic acid, waxes, and resins [37] which will help in the rigid layer formation around the dispersed droplets to prevent the occurrence of coalescence.

3. Emulsion Rheology and Shear Viscosity

The flow property of emulsions is their most important property, i.e., their rheology which is generally defined as the study of the flow and deformation of materials that are subjected to the influence of shear stress. The study of the rheological characteristics of emulsions is of great interest for both its scientific understanding and industrial application [38, 39]. The rheology characteristics of any material is an important physical attribute. According to Becker [50], the rheological characteristics of emulsions can be affected by six factors, including the surface phase viscosity, the division of the droplet volume in the continuous part, the indoor stage viscosity, the size and concentration of the dispersed phase, the type of emulsifying agent used and the created interfacial film at the interface, as well as the shear rate. As per Alwadani [40], the rheology behavior of an emulsion, based on its structure, can either be Newtonian or non-Newtonian. Emulsions may generally exhibit Newtonian behavior when the concentration of the disperse is low to medium but may behave as shear-thinning fluids when the dispersed amount is high. The interaction between droplets accounts for the increase in the viscosity as the amount of the dispersed phase increases. For the shear-thinning liquids, emulsions can be described by apparent shear viscosity η, which is similar to that of pure fluids. From Newton's law, $\eta = \tau ij / \gamma ij$. η is the proportionality coefficient (or ratio) between shear stress (τij) and rate of strain (or shear rate, γij). Factors affecting the shear viscosity of an emulsion are the viscosity of the constant phase, the dispersed part content, the viscosity of the dispersed phase, shear rate, temperature and the average size, and size distribution of droplets. The viscosity increases with the dispersed phase content due to interactions among droplets. If the pressure of a given emulsion sample is stable, the major influencing factor on the viscosity becomes the temperature and the dispersed phase concentration [38].

4. Phase Inversion Temperature (PIT)

The phase inversion phenomenon is described as an operation in which the continuous and dispersed phases of an emulsion abruptly exchanges phases from o/w to w/o emulsion or vice versa. Shinoda & Arai [41] devised an approach which depends on the determination of the phase inversion temperature (PIT) of a surfactant-oil-water mixture stirred under heat. In the presence of a surfactant, the dispersed phase is dependent on either the surfactants' relative affinity to the competent phases or on the arrangement of the phase additions that make up the emulsion. For instance, an o/w emulsion will be formed by a surfactant that has a high affinity for water at low temperatures, but when there is a change in the status of the emulsion, there may be phase obverse which implies a situation where the dispersed phase becomes the continuous phase, and vice versa. Similarly, any change in the temperature will cause phase inversion due to the altering of the surfactants' affinity to each phase. As per Friberg, the PIT is the temperature where the preferential solubility of an emulsifier or a surfactant shifts from water to oil or oil to water [42]. Being that this approach considers several variables that influence the PIT, it is a more trusted approach compared to the earlier methods. One thing is that the PIT method is only applicable to nonionic surfactants due to the low sensitivity of ionic surfactants to temperature [43, 44]. Other factors which may be considered to affect phase inversion are the type and concentration of the emulsifiers, as well as the influence of physical parameters like temperature and mechanical shear. Phase inversion may also be said to depend on the characteristics of the emulsification protocol which portrays the way of making or modifying the emulsion, or the changes in the composition of the emulsion with time or space [45].

5. Demulsification techniques

Demulsification refers to the breaking down of emulsion into it's component incompatible phases (i.e. water and oil). It is an important process in the petroleum sector where emulsions are always encountered as deliberate or natural occurrences. The separation of water from crude oil is an important process in the petroleum refineries as it must be performed prior to oil refining. The current method of breaking w/o emulsion via the use of chemical additives called emulsion breakers but technically, the resistance of w/o emulsions to coalescence, as well as their response to demulsification processes is a function of the physical and chemical properties of the oil from which such emulsions are formed. They also depend on aging and emulsification conditions. This suggests variability in the methods for the optimization of w/o emulsion demulsification from one setting to the other [46]. Various methods of emulsion separation have been suggested by the previous researchers; however, there are two major demulsification approaches in current use – physical and chemical approaches. With the chemical method, a suitable demulsifier is added into the emulsion while the physical technique involves heating, mechanical or electrical methods [47]. The thermochemical method is the commonly used method in the industries; it involves heating with the addition of a demulsifier [48]. Emulsions can also be demulsified using methods like filtration, membrane separation, and pH adjustment [49]. The subsequent sections will focus on the survey of the existing and future methods of demulsification.

5.1. Chemical demulsification

This involves the flocculation and/or alteration of the characteristics of the interfacial films using chemicals to cause coalescence. It is the common method for the treatment of w/o and o/w emulsion where the breaking process is facilitated by the chemical additives [3]. Several factors like the type of oil and its viscosity, the size distribution of the water phase, and the presence of solids can affect the performance of a demulsifier [50]. A study by Ben Mya et al. [51] reported the use of REB09305 OS demulsifier to treat Algerian crude oil; the study focused on the effects of the contact time, temperature, and centrifugation speed on the w/o separation efficiency. The results of the study showed a 100% demulsification ratio to be achieved at the process condition of demulsifier dose 20 ppm, ambient temperature, centrifugation speed 1200 rpm, and contact time of 45 min. Furthermore, Yau et al. [52] evaluated the performance of different demulsifiers (including 37 different types of anionic, cationic and nonionic emulsifiers) on the demulsification of Marpol oil waste. From the results, the best o/w separation performance was achieved by the anionic and nonionic surfactants after 30 h at a temperature of 70°C.

Another study by Liu et al. [53] used magnetic graphene oxide (M-GO) to separate o/w emulsions. The results showed the capability of M-GO to facilitate the separation of the o/w emulsions within a short period. The M-GO was recycled 6–7 times and still achieved 99.98% demulsification efficiency at the optimum concentration. Alginite was used as a demulsifying agent by Hippmann et al. [6] to demulsify Hungarian crude oil shale. During the study, the w/o emulsion used was made up of equal concentrations of crude oil and brine. From the results, the addition of alginite to the emulsion caused phase separation within a few minutes. Li et al. [54] demulsified water-in-aging crude oil (WACO) emulsions using a novel type of tannic acid phenol-amine (TAPA) polyether with numerous branches and aromatic rings. From the results, the laboratory tests showed a demulsification efficiency of 91.7% in 90 min while the offshore test showed 97.9% within 45 min toward w/o emulsions. Hazrati et al. [55] demulsified w/o emulsion samples using different ionic liquids ILs. The results showed the maximum water separation performance of 86–95% and interfacial tension (IFT) of 0.7–6.26 mN/m to be achieved within 24 h.

Table 2. Summary of the recent studies on chemical demulsifiers.

Type of emulsion	Type of demulsification	Emulsifier/ demulsifier concentration	Ratio & effect of demulsifier on separation	Year of study	Ref.
W/O	REB09305 OS demulsifier	5-40 ppm	100% demulsification ratio achieved at the process condition of demulsifier dose 20 ppm, ambient temperature, centrifugation speed 1200 rpm, and contact time of 45 min	2017	[51]
W/O	Anionic, cationic and nonionic.		Anionic and nonionic surfactants performed best after 30 h at 70°C.	2017	[52]
O/W	Magnetic graphene oxide	0.04-0.25 wt%,	99.98% demulsification efficiency achieved at an optimal dosage.	2017	[53]
W/O	Alginite	0.5 wt.%	The water removal capacity was >98%.	2018	[6]
WACO	Polyether	50-150 mg/L	laboratory test showed 91.7% demulsification efficiency within 90 min and 97.9% within 45 min on an offshore platform.	2018	[54]
W/O	Ionic liquids	500-3500 ppm	86–95% water separation efficiency achieved within 24 h.	2018	[55]

5.2. Biological demulsification

The early 1980s witnessed the start of researches on biological demulsification [56] when it was suggested as potential substitutes to chemical demulsifiers in the petroleum sector. Microbial demulsification is based on the concept that the existing hydrocarbons in the emulsion can serve as the source of nutrients to microbes. A study by Wen et al. [57] reported the use of S-XJ-1 as a biological demulsifier. The investigation, with the aid of a Turbiscan stability analyzer, showed a similar performance to chemical demulsification as they S-XJ-1 cells participated in the breakdown of the emulsion, thereby achieving 81.3% demulsification of w/o emulsion within 24 h at the cell conc. of 500 mg/L. Xuwei et al. [58] reported the use of novel biodemulsifier, rhamnolipid to demulsify waste crude oil. The results showed that waste crude oil treatment with the novel rhamnolipid achieved a dewatering efficiency of over 90%. Another study by Roostaie et al. [59] had investigated the use of cellulose-based compounds as demulsifier. The study focused on the performance of different forms of cellulose at different viscosity levels. It also investigated the breakdown of crude oil emulsion using a mixture of ethylcellulose and ethoxylated coco amine. From the results, all the cellulose variants exhibited poor dehydration efficiency when used alone. However, a dehydration efficiency of 96.8% was achieved with the mixture of 500 ppm ethoxylated coco amine and 1000 ppm ethylcellulose 46 (EC46) within 12 h.

Saxena et al. [60] studied the use of a biodegradable anionic as a chemical to enhance the process of oil recovery. The surfactant exhibited good wetting properties by changing the surface of the quartz and carbonate rocks from oil-wet to water-wet. The surfactant slugs were used at different concentrations to perform flooding experiments and an additional 25%–27% oil recovery was reported after the conventional flooding. Jiang et al. [61] studied α-amylase and its effect as a bio demulsifier on the de-emulsification rate and viscosity of amphiphilic polymer solution. The conc. of α-amylase, salinity, and temperature were altered during the experiments and the results show α- amylase to be a good bio demulsifier as it can improve the rate of de-emulsification of such emulsions. Akbari & Biria [62] investigated *Acinetobacter calcoaceticus* for the production of bio-demulsifiers and the ability of such demulsifiers to breakdown w/o emulsions. From the results, the microbe, at the optimal conditions, produced an extracellular biodemulsifier which caused a reduction of the surface tension to 38.6 mN/m and facilitated the breakdown of 95% of a surfactant-reinforced emulsion.

Table 3. Summary of the recent studies on bio-demulsifiers.

Type of emulsion	Name of biodemulsifier	Demulsifier concentration	Ratio & effect of demulsifier on separation	Year of study	Ref
W/O	*Candida sphaerica* UCP 0995		40% demulsification ration achieved.	2015	[63]
W/O	Cellulose-based compounds	100-1500ppm	Ethylcellulose is efficient in breaking the emulsion but its major drawback is the low dehydration rate.	2017	[59]
Enhanced oil recovery process	Alpha sulfonated ethyl ester (α-SEE)		The surfactant exhibited good wetting characteristics	2017	[60]
O/W	α-amylase	0-6000 mg/L	α- amylase exhibited good bio demulsification performance.	2018	[61]
W/O	Biodemulsifier		95% of the emulsion was demulsified.	2018	[62]

5.3. Mechanical demulsification
According to Auflem & Jiang [64, 65], the mechanical demulsification process involves the use of mechanical forces to breakdown the physical barrier/difference in the water and oil phase densities to achieve separation. Among the tools that can be used to achieve mechanical separation during crude oil, demulsification include cyclones, centrifugal separators, gravity settling tanks, etc. The centrifuge has not been as widely used as the other methods for treating emulsions due to its high capital cost and low capacity [66]. A study by Hao et al. [67] focused on the removal of oil from water using centrifugal contactors. The aim of the study is to address the over-standard oil content problem in electric desalting wastewater. The parameters considered during the study include the effects of the rotor speed, the water temperature, the inlet flow rate, the crude oil density, and the oil content. From the results, there were increases in oil removal efficiency as the inlet flow rate was decreasing and increasing oil conc. with decreasing oil density. Under gravity, the emulsions were allowed to sediment under the influence of normal gravitational force. This sedimentation promoted coalescence by bringing the dispersed phase droplets closer. Centrifugation is utilized in some processes to achieve closer contact with droplets in a rapid manner [68].

5.4. Thermal demulsification
Thermal demulsification involves the use of heat to promote the breakdown of emulsion in an oil field or refinery. In the lab scale, the commonly used method is the conventional hot plate to generate the required temperature. The rigidity of the interface which promotes the coalescence of droplets when they collide may be affected high temperature. Furthermore, a higher temperature can encourage the rate of collision between the droplets and cause a reduction in the stability of the emulsion. One effect of heating is that light compounds that help increase the density of crude oil could be lost, thereby affecting gravity settling [69, 70]. Heat treatment is often used in conjunction with chemicals in most cases to enhance the efficiency of the process since the viscosity of emulsions can be tremendously reduced by temperature [71]. The effectiveness of thermal demulsification has been reported by various researchers; for instance, Tov et al. [72] studied the effect of a thermo-chemical demulsification process on w/o emulsion resolution and reported as follows: the thermo-chemical method is more efficient in breaking down w/o emulsions of paraffinic petroleum compared to asphaltene oil; the application of

heat could facilitate a slow breakdown of the emulsion but the addition of the chemicals will progress the separation process rapidly. Abdurahman et al. [73] compared hot-plate w/o emulsions breakdown with microwave-assisted demulsification and reported the microwave-assisted process to have demonstrated a great separation power compared to the hotplate heating possibly due to the microwave heating pattern where materials are volumetrically heated based on their dielectric characteristics. Binks and Rocher [74] investigated the effect of temperature on the breakdown of wax-stabilized w/o emulsion. For the stability test, the amount of water or oil resolved from the emulsion was tested over a given period. The results showed a temperature-mediated increase in the rate of coalescence because of the fusing, melting, and varnishing of the wax particles at the interface at an elevated temperature, thereby inducing the droplets to come together easily and merge (meaning separation) [74]. Taylor et al. [75] investigated the temperature effect on the breakdown of bitumen emulsions and reported a gradual breakage of the emulsion as the temperature of the system approaches the cloud point. Furthermore, there was a decrease in the bitumen-water interfacial tension as the temperature was increasing; this decreases almost varnished at the cloud point temperature. The effect of temperature on demulsification processes was investigated by Al-Sabagh et al. [76] using TE18P. From the results, the demulsification efficiency of asphaltenic crude oil improved by 8.7% asphaltene content as the temperature was increased from 50°C to 70°C.

5.5. Microwave demulsification

The first microwave irradiation-assisted demulsification process was reported by Wolf in 1986 [77]. Microwave heating has been reported to be superior to the traditional heating schemes in the breakdown of emulsions as a result of its less energy consumption and faster separation rate [78]. The microwave heating technique can be applied as a standalone demulsification process in the petroleum sector as it can minimize the use of chemical demulsifiers [79]. For instance, Martínez-Palou et al. [48] comparatively studied the performance of microwave and oil bath heating in terms of the water separation time of o/w emulsion. They also studied the effect of salt content and a chemical demulsifier on the demulsification of o/w emulsion. From the results, a greater degree of water separation was achieved by the microwave heating technique in less time compared to the traditional oil bath heating; increase in the microwave power and salt content also increased the separation of o/w emulsions under the co-influence of chemical demulsifier [48]. Binner et al. [80] studied the use of microwave heating and natural gravity settling techniques to separate w/o emulsions. The result of the study showed microwave heating to improve the coalescence of the water phase droplets in w/o emulsions, while the separation time was reported to reduce significantly compared to the untreated emulsions. da Silva et al. [81] applied microwave technology to the breakage of water-in-heavy crude oil emulsions. The study focused on the effects of the heating mode, pH, temperature, and process time on the acidic species migration from crude oil to water during heavy crude oil demulsification. From the results, the temperature was found to be the most influential variable as it favors the separation of a larger number of acidic species. The use of a microwave to breakdown emulsion was reported by Santos et al. [82]. The study focused on the influence of the modes of microwave application (single and multiple modes) on the efficiency of a synthetic w/o emulsion separation. The results showed the microwave efficiency to be influenced by the heating temperature and water content of the emulsion. Regarding the modes, the single-mode reached highest demulsification efficiency using lesser energy compared to the multimode; however, the reproducibility of the multimode was better than that of the single-mode.

5.6. Electrical demulsification

Electrical demulsification involves the deformation of droplets and generation of an attraction force between drops (thus leading to coalescence) by applying an electric field. The application of the electric field makes it easier for small water droplets to quickly fuse into larger ones. The droplets deform in the electric field as they migrate to each other. Coalescence occurs more quickly as the droplets are elongated and deformed [83]. Electrical demulsification is considered a better option to chemical and thermal demulsification; however, the way to adapt this method to the different emulsion properties is yet to be understood properly [84]. So, in this review, some of the previous studies that used the electrical method as a demulsifier for emulsions will be reviewed. Mousavi et al. [85] studied the effect of pulsatile electric fields (PEF) on the formation of the secondary droplets. From the results, the production of secondary droplets proceeded at a low-frequency PEF; the study also approved that increases in the PEF frequency will suppress the DC electric field. Another study by Zhang et al. [86] focused on the effect of surfactants on the breakdown of water droplets suspended in a dielectric oil (immiscible) under an electric field. The experiments proceeded on a single drop in a microscopic cell under the influence of an externally applied electric field. Different concentrations of Tween 20 and SDS were also investigated as surfactants. From the results, the extent of deformation of the big drops was faster under the influence of electric field compared to the smaller drops.

Wang et al. [87] investigated two charged droplets for electro-coalescence behaviors under different fields (constant and pulsed direct current) using molecular simulations. The outcome of the simulations showed the exertion of an attractive force on the two droplets upon the application of the constant DC electric field. This electric field was reported to accelerate the approach velocity between the two droplets. However, the approach velocities of the two droplets under the pulsed DC electric field accelerated within the pulse duration but was almost stable during the pulse interval. Qiao et al. [88] investigated the demulsification of steam-assisted gravity drainage (SAGD) by applying the demulsifier FC20 under an electric field. The results indicated the process of the water content separation from the emulsion to be reduced by 1.0% upon the addition of 30% diesel oil and 300 mg/L FC20. This percentage reduction occurred under an electric field of 750 V/cm for 20 min and 4 h settling time at 90°C. Li et al. [89] investigated the effects of power voltage, duty ratio, electric frequency, operating time and operating temperature on the use of the DTS-4C airtight oil dehydration apparatus to dehydrate ultra-heavy oil. From the results, the best performance was achieved at the following condition: 5.5 kV voltage magnitude; 0.5 duty ratio 0.5; 120°C operating temperature, 4.0 kHz electric frequency; and 3 h operating time. This condition achieved a dehydration efficiency of 97.81% and post-electric dehydration water content of 0.37 wt%; these are acceptable results for oil refineries.

5.7. Ultrasonic demulsification

Much attention has been given to ultrasonic demulsification processes in the last couple of years due to its associated simplicity and efficiency [90]. Several studies have been reported on the use of the ultrasonic method as a demulsifier of emulsions. For instance, Check and Mowla [91] investigated the effect of irradiation input power, irradiation time, injected water volume, and temperature on the efficiency of salt and water removal from crude oil via ultrasonic irradiation. The results showed an input power of 57.7 W, the temperature of 100°C, and irradiation time of 6.2 min as the optimal condition for the ultrasonic irradiation of the crude oil. The desalting/dehydration efficiencies under these optimum conditions were recorded as 84% and 99.8%, respectively. Antes et al. [92] studied the possibility of crude oil demulsification via an indirect application of low-frequency ultrasound without any chemical demulsifier. Under the reported experimental condition, a demulsification efficiency of about 65% was achieved on an emulsion with a water content of 50% and DSD of 10 lm. The study reported higher demulsification efficiency at an ultrasound temperature of 45°C and power of 160 W within 15 min. Xie et al. [93] reported the use of a pulsed ultrasonic irradiation technique to develop a method for oil phase separation from waste refinery oil emulsion. The results of the experiments using the method showed the developed method to be better in coalescing the water droplets, thereby dehydrating the waste refinery oil. An 8% decrease in the waste oil water content was achieved which complied with the refinery requirements on the acceptable water content of waste oil.

Khajehesamedini et al. [94] worked on the reduction of the utilization of chemical demulsifiers in crude oil processing by using low-frequency ultrasonic waves as an alternative method. The study focused on the effects of the process parameters (ultrasonic field intensity, initial water content, and irradiation time) on the separation efficiency. From the results, it was concluded that chemical emulsifiers must be involved in demulsification processes, but the rate of its utilization can be reduced by 50% when working on a w/o emulsion by ensuring the use of a suitable ultrasonic waves intensity and irradiation time. Wang et al. [95] reported the use of ultrasonic irradiation to demulsify super-heavy crude oil using static experiments at high temperatures. The experiment investigated the influence of both ultrasonic

IOP Conf. Series: Materials Science and Engineering **702** (2019) 012060 doi:10.1088/1757-899X/702/1/012060

(frequency, power, time, and sound intensity) and external factors (temperature, water ratio, and sedimentation time) on the demulsification process. The results showed an increase in the rate of dehydration as the ultrasonic sound intensity increases but up to a certain level. The increase in sedimentation time was also reported to increase the dehydration rate at a given level. It was suggested that higher temperatures improved the ultrasonic dehydration and demulsification processes, while the effect of water ratio on the dehydration rate was reported as heavy and complex.

5.8. Membrane demulsification

Membrane separation has been considered as a treatment method for oily wastewater since 1973. The major advantages of this process include the homogeneity in the quality of the permeate irrespective of influent differences, non-involvement of any chemical, efficiency in the separation of small droplets, high separation efficiency, cost-effectiveness, and compact setting. However, its application is restricted by the high capital costs required for large volumes of effluent, polymeric membranes degradation, and fouling during operation, fairly low permeate flux, as well as sensitivity to chlorinated and polar solutions [96]. Several studied have been dedicated to the use of membranes as demulsifiers. For instance, Zhang et al. [97] focused on the feasibility of treating different emulsion wastewaters using an integrated ultrafiltration process (UF) that consists of a combined chemical and vibratory shear breakage of emulsions in an anaerobic/aerobic biofilm reactor. It was observed from the results that better oil removal was achieved with calcium chloride compared to the other inorganic salts. It was also suggested that chemical pre-treatment prior to demulsification could improve membrane filtration and oil removal in wastewater demulsification via a vibratory shear-enhanced technique.

Lin et al. [98] reported the removal of 4-methoxyphenyl (4-MP) from wastewater using a w/o/w Pickering emulsion liquid membrane. The study reported that the demulsification of the water content of oil can be achieved via magnetic force or centrifugation without significantly affecting the chemistry of the oil phase. From the results, the extraction efficiency of >86% was achieved when functionalized Fe_3O_4 nanoparticle was used as the emulsifier. Skale et al. [99] worked on the feasibility of using ultrafiltration to separate w/o Pickering emulsions. The study reported an unexpected flux behavior of the Pickering emulsions, while the permeability was increasing with higher pressure levels. Pressures between 1 and 4 bars recorded fluxes ranging from 3 to 40 L/ (m^2 h). It was also reported that the emulsions can be concentrated to about 80% water phase fraction. A study by Othman et al. [33] reported the use of palm oil-based organic phase as diluents in an emulsion liquid membrane process (ELM) for the treatment of artificial wastewater that contains 300×10^{-6} of phenol. From the results, the palm oil-based ELM showed efficient phenol removal from the wastewater at an optimum condition that consists of 500 rpm agitation rate, 3% span 80 conc., external phase pH of 8, the contact time of 5 min, NaOH conc. of 0.1 mol/L as stripping agent, and treat ratio of 1:10. This condition achieved an extraction performance of about 83%, representing 11 folds' enhancement.

6. Conclusions

Water mix with petroleum oil inside the well or during the process of oil pumping from the reservoir. It also be encountered during desalting processes in the form of different forms of emulsion, such as w/o, o/w, w/o/w, and o/w/o. Such emulsions are sources of economic and operational concerns to the industry; hence, several solutions have been proposed for addressing these concerns. Finding the solution to emulsions demands detailed knowledge and serious studies on both the possible consequence of its formation and the factors that facilitate its formation; factors that contribute to its stabilization must also be considered. Several studies have been performed from both the Industrial and academic perspectives on the role of several factors in the formation of stable petroleum emulsions. This paper strived to overview emulsions, their formation, kinds, constancy, and properties. Also reviewed are effective demulsification parameters and methods, including chemical, biological, thermal, microwave irradiation, electrical, ultrasonic, and membrane methods which are used for both oilfield and synthetic emulsions. Amongst these techniques, chemical demulsification has been the most widely applied and reported in the literature It is worth noting that the efficiency of each of these techniques is dependent on the operating parameters and their interplay. Moreover, a more effective demulsification process could be attained by leveraging synergistic effects by combining one or more of these techniques, as discussed in this review.

Acknowledgment
This study was sponsored and funded by Universiti Malaysia Pahang under the grant number PGRS190362.

References

[1] Nour A H, Hassan M A A and Yunus R M 2007 *J. Appl. Sci.* **7** 1437-41
[2] Al-Sabagh A M, Kandile N G and Noor El-Din M R 2011 *Sep. Sci. Technol.* **46** 1144-63
[3] Abdurahman H N, Yunus R M and Jemaat Z 2007 *J. Appl. Sci.* **7** 196-201
[4] Saad M A, Kamil M, Abdurahman N, Yunus R M and Awad O I 2019 *Proc.* **7** 470
[5] Mohayeji M, Farsi M, Rahimpour M R and Shariati A 2016 *J. Taiwan Inst. Chem. Eng.* **60** 76-82
[6] Hippmann S, Ahmed S S, Fröhlich P and Bertau M 2018 *Colloid. Surf. A Physicochem. Eng. Asp.* **553** 71-79
[7] Biniaz P, Farsi M and Rahimpour M R 2016 *Fuel* **184** 325-33
[8] Thompson D, Taylor A and Graham D 1985 *Colloids Surf.* **15** 175-89
[9] Nour A H, Yunus R M and Anwaruddin H 2007 *J. Appl. Sci.* **7** 3512-17
[10] Sun T, Zhang L, Wang Y, Zhao S, Peng B, Li M and Yu J 2002 *J. colloid interface sci.* **255** 241-47
[11] Liu J, Li X, Jia W, Li Z, Zhao Y and Ren S 2015 *Energ. Fuel.* **29** 4644-53
[12] Fortuny M, Oliveira C B, Melo R L, Nele M, Coutinho R C and Santos A F 2007 *Energ. Fuel.* **21** 1358-64
[13] Sheng J, 2010 *Modern chemical enhanced oil recovery: theory and practice*. Gulf Professional Publishing
[14] Pal R 1994 *Colloid. Surf. A: Physicochem. Eng. Asp.* **84** 141-93
[15] Verzaro F, Bourrel M, Garnier O, Zhou H and Argillier J 2002 *SPE International Thermal Operations and Heavy Oil Symposium and International Horizontal Well Technology Conference*: Society of Petroleum Engineers.
[16] Jamaluddin A and Nazarko T 1995 *J. Can. Petrol. Technol.* **34**
[17] Aske N 2002 *Characterisation of crude oil components, asphaltene aggregation and emulsion stability by means of near infrared spectroscopy and multivariate analysis* (Norwegian University of Science and Technology Trondheim)
[18] Kokal S, Dawood N, Fontanilla J, Al-Ghamdi A, Nasr-El-Din H and Al-Rufaie Y 2002 *Annual Technical Conference and Exhibition*: Society of Petroleum Engineers
[19] Langevin D, Poteau S, Hénaut I and Argillier J 2004 *Oil Gas Sci. Technol.* **59** 511-21
[20] Schramm L L 2005 *Emulsions, Foams, and Suspensions: Fundamentals and Applications* **1** 201-21
[21] Salomon D R 2006 *Asphalt emulsion technology* (TRC E-102, Washington, DC) Transportation Research Board
[22] Zeidani K 2007 *Heavy Oil-in-water Emulsion Flow and Blocking Mechanism in Porous Medi* ProQuest
[23] Hasan S W, Ghannam M T and Esmail N 2010 *Fuel* **89** 1095-100
[24] Ashrafizadeh S and Kamran M 2010 *J. Petrol. Sci. Eng.* **71** 205-11
[25] Lin C Y and Chen L W 2006 *Fuel Process. Technol.* **87** 309-17
[26] Alwadani M S 2009 *Characterization and rheology of water-in-oil emulsion from deepwater fields* Rice University
[27] Dos Santos R G, Bannwart A C, Briceño M I and Loh W 2011 *Chem. Eng. Res. Des.* **89** 957-67
[28] Abdurahman N, Azhari N and Yunus Y 2013 *Int. J. Eng. Sci. Innov. Technol.* **2** 170-79
[29] Al-Yaari M, Hussein I A and Al-Sarkhi A 2014 *Appl.Clay Sci.* **95** 303-9
[30] Liyana M, Nour A, Rizauddin D and Jolius G 2014 *Int. J. Res. Eng. Technol.* **3** 489-96
[31] Hajivand P and Vaziri A 2015 *Brazilian J. Chem. Eng.* **32** 107-18
[32] Atta A M, Al-Lohedan H A, Abdullah M M S and ElSaeed S M 2016 *J. Ind. Eng. Chem.* **33** 122-

30
[33] Othman N, Noah N F M, Shu L Y, Ooi Z Y, Jusoh N, Idroas M and Goto M 2017 *Chinese J. Chem. Eng.* **25** 45-52
[34] Zaki N N, Carbonell R G and Kilpatrick P K 2003 *Indus. Eng. Chem. Res.* **42** 6661-72
[35] Schramm L L 1992 *Adv. Chem. Ser.* **321** 1-51
[36] Mat H, Samsuri A, Rahman W and Rani S 2006 *Research Paper, Department of Chemical Engineering, University Teknologi Malaysia.*
[37] Mehta S D 2006 *Making and breaking of water in crude oil emulsions* (Texas A&M University)
[38] Tadros T F 1994 *Colloid. Surf. A: Physicochem. Eng. Asp.* **91** 39-55
[39] Barnes H A 1994 *Colloid. Surf. A: Physicochem. Eng. Asp.* **91** 89-95
[40] Alwadani M S 2010 Characterization and rheology of water-in-oil emulsion from deepwater fields Rice University
[41] Shinoda K and Arai H 1964 *J. Physic. Chem.* **68** 3485-90
[42] Adams F, Walstra P, Brooks B, Richmond H, Zerfa M, Bibette J, Hibberd D, Robins M, Weers J, and Kabalnov A 2007 *Modern aspects of emulsion science* (Royal Society of Chemistry)
[43] Davis H 1994 *Colloid. Surf. A: Physicochem. Eng. Asp.* **91** 9-24
[44] Nguyen D, Sadeghi N and Houston C 2012 *Energ. Fuels* **26** 2742-50
[45] Salager J 2006 *Surf. Sci. Ser.* **132** 185
[46] Strom-Kristiansen T, Lewis A, Daling P S and Nordvik A B 1995 *Spill Sci. Technol. Bull.* **2** 133-41
[47] Kukizaki M and Goto M 2008 *J. Membrane Sci.* **322** 196-203
[48] Martínez-Palou R, Cerón-Camacho R, Chávez B, Vallejo A A, Villanueva-Negrete D, Castellanos J, Karamath J, Reyes J and Aburto J 2013 *Fuel* **113** 407-14
[49] Gafonova O V 2001 *Role of asphaltenes and resins in the stabilization of water-in-hydrocarbon emulsions* Calgary
[50] Mikula R and Munoz V 2000 *Characterization of demulsifiers* (Cambridge University Press: Cambridge)
[51] Ben Mya O, Houga S, Chihouba F and Asla B 2017 *Egypt. J. Petrol.*
[52] Yau Y H, Rudolph V, Ho K C, Lo C C and Wu K 2017 *J. Water Process Eng.* **17** 40-49
[53] Liu J, Wang H, Li X, Jia, W, Zhao Y and Ren S 2017 *Fuel* **189** 79-87
[54] Li Z, Geng H, Wang X, Jing B, Liu Y and Tan Y 2018 *Chem. Eng. J.* **354** 1110-19
[55] Hazrati N, Miran Beigi A A and Abdouss M 2018 *Fuel* **229** 126-34
[56] Stewart A, Gray N, Cairns W and Kosaric N 1983 *Biotechnol. lett.* **5** 725-30
[57] Wen Y, Cheng H, Lu L J, Liu J, Feng Y, Guan W, Zhou Q and Huang X F 2010 *Bioresour. Technol.* **101** 8315-22
[58] Long X, Zhang G, Shen C, Sun G, Wang R, Yin L and Meng Q 2013 *Bioresour. Technol.* **131** 1-5
[59] Roostaie T, Farsi M, Rahimpour M R and Biniaz P 2017 *Sep. Purif. Technol.* **179** 291-96
[60] Saxena N, Pal N, Dey S and Mandal A 2017 *J. Taiwan Inst. Chem. Eng.* **81** 343-55
[61] Jiang J, Wu H, Lu Y, Ma T, Li Z, Xu D, Kang W and Bai B 2018 *Bioresour. Technol.* **259** 349-56
[62] Akbari N and Biria D 2018 *J. Environ. Chem. Eng.* **6** 4144-50
[63] Luna J M, Rufino R D, Jara A M A T, Brasileiro P P F and Sarubbo L A 2015 *Colloids Surf. A: Physicochem. Eng. Asp.* **480** 413-18
[64] Jiang T 2010 *Diluted bitumen emulsion characterization and separation* Rice University
[65] Auflem I H 2002 *Influence of asphaltene aggregation and pressure on crude oil emulsion stability Norwegian University of Science and Technology* (Doktor Ingeniør Thesis)
[66] Sherman P 1968 *Emulsion science* (Academic Press)
[67] Hao M, Bai Z, Wang H and Liu W 2013 *J. Petrol. Sci. Eng.* **111** 37-41
[68] Pal R, Yan Y, Masliyah J and Schramm L 1992 *Adv. Chem. Ser.* **231** 295-312
[69] Kokal S and Al-Juraid J 1998 *Annual Technical Conference and Exhibition*: Society of Petroleum Engineers
[70] Sztukowski D M 2005 *Asphaltene and solids-stabilized water-in-oil emulsions*
[71] Issaka S A, Nour A H, and Yunus R M 2015 *J. Petrol. Environ. Biotechnol.* **6** 1
[72] Daling P S, Moldestad M Ø, Johansen Ø, Lewis A and Rødal J 2003 *Spill Sci. Technol. Bull.* **8**

123-36
[73] Abdurahman H and Rosli M 2006 *Appl. Sci.* **6** 2307-11
[74] Binks B P and Rocher A 2009 *J. colloid interf. Sci.* **335** 94-104
[75] Taylor S E 2011 *Fuel* **90** 3028-39
[76] Al-Sabagh A M, Nasser N M, Khamis E A and Abd-El-Raouf M 2015 *Egypt. J. Petrol.* **24** 363-74
[77] Wolf N O 1986 *Use of microwave radiation in separating emulsions and dispersions of hydrocarbons and water* ed: Google Patents
[78] Evdokimov I N and Losev A P 2014 *J. Petrol. Sci. Eng.* **115** 24-30
[79] Nour A H, Yunus R M and Nour A H 2010 *World Acad. Sci. Eng. Technol.* **4**
[80] Binner E R, Robinson J P, Silvester S A, Kingman S W and Lester E H 2014 *Fuel* **116** 516-21
[81] da Silva E B, Santos D, de Brito M P, Guimarães R C L, Ferreira B M S, Freitas L S, de Campos M C V, Franceschi E, Dariva C, Santos A F and Fortuny M 2014 *Fuel* **128** 141-47
[82] Santos D, da Rocha E C L, Santos R L M, Cancelas A J, Franceschi E, Santos A F, Fortuny M and Dariva C 2017 *Sep. Purif. Technol.* **189** 347-56
[83] Less S, Hannisdal A, Bjørklund E and Sjöblom J 2008 *Fuel* **87** 2572-81
[84] Lundgaard L, Berg G, Ingebrigtsen S and Atten P 2006 *Electrocoalescence for oil-water separation: Fundamental aspects* ed: Taylor & Francis
[85] Mousavi S H, Ghadiri M and Buckley M 2014 *Chem. Eng. Sci.* **120** 130-42
[86] Zhang L, He L, Ghadiri M and Hassanpour A 2015 *J. Petrol. Sci. Eng.* **125** 38-47
[87] Wang B B, Wang X D, Wang T H, Lu G and Yan W M 2016 *Int. J. Heat Mass Transf.* **98** 10-16.
[88] Qiao Y, Yang J, Cai H, Cheng P and Xu X 2016 *J. Petrol. Sci. Eng.* **145** 148-153
[89] Li B, Fan Y, Sun Z, Wang Z and Zhu L 2017 *Powder Technol.* **316** 338-44
[90] Leong T, Johansson L, Juliano P, McArthur S L and Manasseh R 2013 *Ind. Eng. Chem. Res.* **52** 16555-76
[91] Check G R and Mowla D 2013 *Ultrason. Sonochem.* **20** 378-85
[92] Antes F G, Diehl L O, Pereira J S, Guimaraes R C, Guarnieri R A, Ferreira B M, Dressler V L and Flores E M 2015 *Ultrason. Sonochem.* **25** 70-5
[93] Xie W, Li R and Lu X 2015 *Ultrason. Sonochem.* **26** 136-41
[94] Khajehesamedini A, Sadatshojaie A, Parvasi P, Reza Rahimpour M and Mehdi Naserimojarad M 2018 *Ultrason. Sonochem.* **48** 383-95
[95] Wang Z, Gu S and Zhou L 2018 *Ultrason. Sonochem.* **40** 1014-20
[96] Hong A C, Fane A G and Burford R P 2002 *Desalination* **144** 185-91
[97] Zhang W, Xiao P and Wang D 2014 *Bioresour. Technol.* **159** 150-6
[98] Lin Z, Zhang Z, Li Y and Deng Y 2016 *Chem. Eng. J.* **288** 305-11
[99] Skale T, Hohl L, Kraume M and Drews A 2017 *J. Membrane Sci.* **535** 1-9

AUTHOR INDEX

Institute of Physics
Dirac House, Temple Back
Bristol BS1 6BE UK

ISSN: 1757-8981
ISBN 978-1-7138-1032-2
9 781713 810322